Protein Stability and Folding

Springer-Verlag Berlin Heidelberg GmbH

W. Pfeil

Protein Stability and Folding

A Collection of Thermodynamic Data

With 7 Figures

 Springer

Professor Dr. W. Pfeil

University of Potsdam
Institute of Biochemistry and Molecular Physiology
Maulbeerallee 2 a
D-14469 Potsdam

ISBN 978-3-642-63716-2

Die Deutsche Bibliothek – CIP-Einheitsaufnahme

Pfeil, Wolfgang : Protein stability and folding: a collection of thermodynamic data / W. Pfeil :
Berlin ; Heidelberg ; New York ; Barcelona ; HongKong ; London ; Milan ; Paris ; Santa Clara ;
Singapore ; Tokyo : Springer, 1998
ISBN 978-3-642-63716-2   ISBN 978-3-642-58760-3 (eBook)
DOI 10.1007/978-3-642-58760-3

Production Editor: Christiane Messerschmidt, Rheinau
Typesetting: Fotosatz-Service Köhler OHG, Würzburg
Cover design: Design & Production GmbH, Heidelberg
SPIN: 10634877   2/3020 – 5 4 3 2 1 0 – Printed on acid-free paper

Contents

Preface

Protein folding remains one of the most exclusive problems of modern biochemistry. Structure analysis has given access to the wealth of the molecular architecture of proteins. As architecture needs static calculations, protein structure is always related to thermodynamic factors that govern folding and stability of a particular folded protein over the non-organized polypeptide chain.

During the past decades a huge amount of thermodynamic data related to protein folding and stability has been accumulated. The data are certainly of importance in dechiffring the protein folding problem. At the same time, the data can guide the construction of modified and newly synthesized proteins with properties optimized for particular application.

The intention of this book is a generation of a data collection which makes the vast amount of present data accessible for multidisciplinary research where chemistry, physics, biology, and medicine are involved and also pharmaceutical and food research and technology.

It took several years to compile all the data and the author wishes to thank everyone who provided data, ideas or even unpublished results. The author is, in particular, indebted to Prof. Wadsö (Lund, Sweden) and IUPAC's Steering Committee on Biophysical Chemistry. Furthermore, support by the Deutsche Forschungsgemeinschaft (INK 16 A1-1) is acknowledged.

Berlin, 1998   W. Pfeil

Abbreviations

ACES	N-[2-acetamido]-2-aminoethanesulfonic acid
ANS	8-anilino-1-napthhalenesulfonic acid
BICINE	N,N-bis[2-hydroxyethyl]glycine
bis-ANS	4,4′-bis-(1-anilino-8-napthhalenesulfonic acid)
dansyl	2-(dimethylamino)napththalene-5-sulfonyl chloride
cal	(superscript) calorimetrically determined value
calc	(subscript) calculated value
CD	Circular Dichroism
conc.	Concentration
DSC	Differential Scanning Calorimetry
DTE	Dithioerythritol
DTT	Dithiothreitol
EDTA	Ethylenediaminetetraacetic acid
EGTA	Ethyleneglycol-bis(β-aminoethylether)N,N,N′,N′-tetra-acetic acid
EPR	Electron Paramagnetic Resonance
FTIR	Fourier Transform Infrared spectrosopy
GuHCl	Guanidine hydrochloride
GuHSCN	Guanidine isothiocyanate
HEPES	N-[2-hydroxyethyl]piperazine-N′-[2-ethanesulfonic acid]
HEW	Hen Egg White (Lysozyme)
HPLC	High Performance Liquid Chromatography
IAEDANS	N-(iodoacetyl)-N′-(5-sulfo-1-naphthyl)-ethylenediamine
IR	InfraRed Spectroscopy
ITC	Isothermal Titration Calorimetry
kDa	Kilo Dalton
MES	2-[N-Morpholino]ethanesulfonic acid
m.g.	molten globule
MOPS	3-[N-Morpholino]propanesulfonic acid
MW	Molecular Weight
n.d.	not determined
NMR	Nuclear Magnetic Resonance
ox.	oxidized
PIPES	Piperazine-N,N′-bis[2-ethanesulfonic acid]
red.	reduced
red., alkyl.	reduced and alkylated
TES	2-([2-Hydroxy-1,1-bis(hydroxymethyl)ethyl]amino)ethanesulfonic acid
Tris	Tris[hydroxymethyl]aminomethane
TWEEN	Polyoxyethylenesorbitan
v. H.	(superscript) equilibrium treatment by means of the van't Hoff equation
w. t.	wild-type
wt*	pseudo wild-type

Symbols

T	temperature
T^0	standard temperature, $T = 298.16$ K
T_{trs}	transition temperature
pH	pH value
pH^0	standard pH value, $pH = 7.0$
I	ionic strength
I^0	standard value of ionic strength, $I = 0.1$
$\Delta_{unf} G$	Gibbs energy change at protein unfolding
ΔG	abbreviation for $\Delta_{unf} G$
$\Delta_{unf} g$	specific value of Gibbs energy change at protein unfolding
Δg	abbreviation for $\Delta_{unf} g$
$\Delta (\Delta_{unf} G)$	Gibbs energy change at protein unfolding, difference value, refers to the reference (e.g., wild-type) protein if not otherwise indicated
$\Delta (\Delta G)$	abbreviation for $\Delta (\Delta_{unf} G)$
$\Delta_{unf} G_{res}$	Gibbs energy change at protein unfolding per amino acid residue
$\Delta_{unf} G^0$	Gibbs energy change at protein unfolding at standard conditions and in the absence of denaturant
$\Delta_{unf} G(T)$	temperature function of Gibbs energy change at protein unfolding
$\Delta_{trs} G$	Gibbs energy change at a structural transition which is different from protein unfolding
$\Delta_{unf} H$	enthalpy change at protein unfolding
ΔH	abbreviation for $\Delta_{unf} H$
$\Delta_{unf} h$	specific value of enthalpy change at protein unfolding
$\Delta_{unf} H_{res}$	enthalpy change at protein unfolding per amino acid residue
$\Delta_{unf} H^{cal}$	enthalpy change at protein unfolding determined by calorimetry
$\Delta_{unf} H^{v.H.}$	enthalpy change at protein unfolding determined by van't Hoff treatment
$\Delta_{unf} H^0$	enthalpy change at protein unfolding at standard conditions
$\Delta_{unf} H(T)$	temperature function of enthalpy change at protein unfolding
$\Delta_{unf} S$	entropy change at protein unfolding
ΔS	abbreviation for $\Delta_{unf} S$
$\Delta_{unf} s$	specific value of entropy change at protein unfolding
$\Delta_{unf} S^0$	entropy change at protein unfolding at standard temperature
$\Delta_{unf} S(T)$	temperature function of entropy change at protein unfolding
$\Delta_{unf} Cp$	heat capacity change at protein unfolding
$\Delta_{unf} c_p$	specific value of heat capacity change at protein unfolding
Cp	heat capacity at constant pressure
$\Delta_{unf} \nu$	difference number of protons between unfolded and native protein states
$\delta g_{tr, i}$	Gibbs energy change at transfer of the i-th side chain from water to denaturant
K_{unf}	equilibrium constant for protein unfolding
K	abbreviation for equilibrium constant
K^0_{unf}	equilibrium constant for protein unfolding at zero denaturant concentration
K_b	binding constant
k_i	rate constant
R	gas constant, $R = 8.3143$ J/K/mol
a	activity
a_+	mean ion activity
$c_{1/2}$	denaturant concentration at which the transition midpoint occurs

m	dependence of ΔG on denaturant concentration (see linear extrapolation method)
n	number of moles
n_i	number of groups of the i-th type
Δn	preferential denaturant binding parameter
α	degree of conversion in equilibrium treatment
ϵ	average degree of exposure of amino acid residues in native protein
N	native state of protein
D	denatured state of protein
U	unfolded state of protein
I or X	intermediate states at protein unfolding
A	acid form (intermediate)
*	asterisk characterizing data which were calculated on the basis of other thermodynamic quantities in the original paper

Dimensions

T	K
T (°C)	degree centigrade
$\Delta_{unf} H$	kJ/mol
$\Delta_{unf} h$	J/g
$\Delta_{unf} G$	kJ/mol
$\Delta_{unf} g$	J/g
$\Delta_{unf} S$	kJ/mol/K
$\Delta_{unf} s$	J/g/K
$\Delta_{unf} Cp$	kJ/mol/K
$\Delta_{unf} c_p$	J/g/K
Cp	kJ/mol/K
c_p	J/g/K
$c_{1/2}$	$M = mol_{(denaturant)}$
m	$kJ/mol/M = kJ/mol_{(protein)}/mol_{(denaturant)}$

Introduction

Introduction

Protein stability may be understood as the resistance of a protein against the denaturing action of physical or chemical factors (e.g., temperature, pH, addition of solutes etc.). Usually, various factors may induce quite different changes in the protein such as loss of activity, changes of spectral properties and other physical parameters or even cleavage of the polypeptide chain. To quantitate protein stability, the conformational stability is commonly used. Under physiological conditions the folded (native) and unfolded (denatured) states of a protein are in equilibrium. The Gibbs energy change (ΔG) for the equilibrium reaction taking place without cleavage of covalent bonds

$$\text{folded (N)} \rightleftharpoons \text{unfolded (D)} \qquad\qquad\qquad (1)$$

is referred to as the conformational stability of a protein. Thus, protein stability is defined on the sound basis of thermodynamic quantities. Therefore, it is possible to compare proteins irrespective of their different biological activity, molecular mass etc. Moreover, related thermodynamic quantities such as enthalpy change (ΔH), entropy change (ΔS), and heat capacity change (ΔCp) at unfolding can be determined which allows to gain a deeper insight into the forces that stabilize the unique three-dimensional structure.

Below the basic equations and a brief description of the approaches used for the determination of conformational stability are given. Additional information is included to facilitate a critical evaluation of the thermodynamic data. However, it is not the intention of a data collection to give a review of general aspects of the thermodynamics of protein folding. For this purpose, the reader will find references below.

Table 1 contains tabulated experimental data of Gibbs energy change ΔG at protein folding that represent the conformational stability. For reasons of convenience ΔG is given as positive value. Positive ΔG values characterize stable proteins, and positive $\Delta(\Delta$G$)$ values designate mutant protein being more stable than the wild-type or the reference protein.

Table 2 contains tabulated thermodynamic key values, i.e., molar enthalpy changes at protein unfolding (ΔH) and molar heat capacity changes (ΔCp) along with additional data defining the external conditions for which the data are valid. The tabulated data are condensed in such a manner that complex thermodynamic functions can be calculated (see the equations given in the first part the introduction).

Table 3 corresponds to the preceding one. However, it contains specific values, i.e., Δh, Δcp in J/g and J/g/K.

The data contained in Tables 1–3 were taken from the literature without making changes except for the conversion to Joules. No additional corrections or changes of trivial names of proteins were made. For reasons of completeness, data obtained by various approaches have been included in the data collection. Some of the approaches contain *a priori* assumptions which are briefly explained below.

Protein stability is related to a great variety of factors. For the diversity of aspects, the reader is referred to reviews and monographs.

- Previous data collections containing thermodynamic quantities: I-75P, I-81P, I-86P2.
- Thermodynamics of protein folding, general aspects: I-79P3, I-82P, I-88P1, I-88P2, I-90D, I-90P2, I-92M2, I-92O, I-92P, I-92S4, I-93M1, I-93P2, I-95M1.
- Instrumental and methodical aspects in determining thermodynamic quantities: I-86H, I-86P1, I-87B, I-87S1, I-87S2, I-89C, I-89P3, I-89W, I-90B, I-94C.
- Protein stability and mutational aspects: I-89A, I-92G, I-92S3, I-93M2, I-93M3, I-95M2, I-95S4, I-95T.

1 Evaluation of the Approaches for the Determination of Conformational Stability and Related Thermodynamic Quantities

1.1 Thermal Unfolding

Van't Hoff treatment. When an N to U transition (eq. 1) is followed by a suited signal recorded *versus* temperature, the classical melting curve is obtained. Then the degree of conversion between the states N (nativ, folded) and U (denatured, unfolded) can be obtained. Designating the degree of conversion as α, the equilibrium constant K_{unf} follows according to eq. (2b). From the temperature dependence of K_{unf} the van't Hoff heat $\Delta_{unf}H^{v.H.}$ can be obtained (eq. 2c). Since the van't Hoff heat may be temperature dependent itself there follows the heat capacity change $\Delta_{unf}Cp$ from Kirchhoff's Law (eq. 2d).

$$\Delta_{unf}G = - RT \ln (K_{unf}) \tag{2a}$$

$$K_{unf} = \alpha/(1 - \alpha) \tag{2b}$$

$$\Delta_{unf}H^{v.H.} = - R \ [\partial(\ln K_{unf})/\partial(1/T)]_{pH} \tag{2c}$$

$$\Delta_{unf}Cp = [\partial(\Delta_{unf}H^{v.H.})/\partial T]_{pH} \tag{2d}$$

Equation (2e) is a more general expression. It must be integrated a second time to obtain $\Delta_{unf}G(T)$.

$$(d \ln K_{unf}/dT)_{pH} = \frac{1}{R \ (T_{trs})^2} [\Delta_{unf}H^{v.H.}(T_{trs}) - \int_T^{T_{trs}} \Delta_{unf}Cp \ dT]. \tag{2e}$$

van't Hoff treatment according to eqs.(1–2e) is based on the *a priori* assumption of an all-or-none transition between native and denatured protein. The approach is, therefore, an indirect way for the determination of thermodynamic quantities since the two-state assumption is testable only by comparison with the calorimetrically determined enthalpy change (I-66L, I-74P). For the thermodynamic background of the van't Hoff equation it is referred to I-97H.

Guidelines for the procedure are contained in I-89P1. An improvement of the performance of thermal denaturation is possible by simultaneous registration of different signals during the temperature scan (I-83W, I-83S, I-94E, I-94R1, I-94R2, I-95S5). For error propagation, which might have considerable influence on $\Delta_{unf}Cp$, see I-79P2, I-87B. An improved procedure for the determination of $\Delta_{unf}Cp$ by combining different approaches was proposed in I-89P2. Systematic errors occurring in thermal denaturation will be discussed below.

Differential Scanning Microcalorimetry (DSC). In DSC the heat capacity of protein in solution is directly measured *versus* temperature. It yields a curve like that shown in Fig. 1.

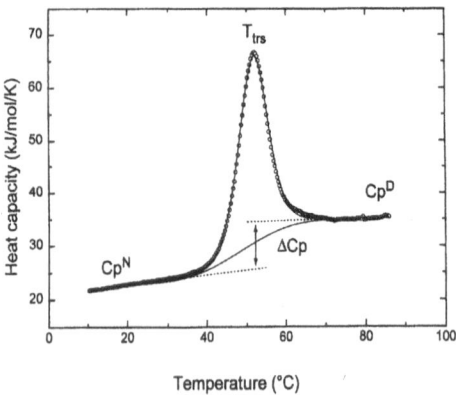

Fig. 1. Heat capacity *versus* temperature obtained by scanning microcalorimetry. Shown are the raw data along with a two-state fit. For the determination of $\Delta_{unf}H$ and $\Delta_{unf}Cp$, see introduction

From Fig. 1, the enthalpy change can be obtained from the area under the peak. This is the calorimetric enthalpy change $\Delta_{unf}H^{cal}$ which is free of any assumptions. At the same time, from peak height or peak half width the van't Hoff heat $\Delta_{unf}H^{v.H.}$ can be obtained from the same calorimetric recording. This makes DSC a valuable tool for testing the validity of the two-state assumption.

The heat capacity change $\Delta_{unf}Cp$ can be determined as follows:

- by extrapolation of pre- and postdenaturational heat capacity (Cp^N, Cp^D) into the transition region as shown in Fig. 1,
- plotting $\Delta_{unf}H^{cal}$ *versus* transition temperature T_{trs} according to Kirchhoff's law (see eq. 2d, except use of $\Delta_{unf}H^{cal}$ instead of $\Delta_{unf}H^{v.H.}$),
- by direct heat capacity measurements of native and denatured protein since $\Delta_{unf}Cp = Cp^D - Cp^N$.

Studying thermodynamics of protein unfolding, DSC is widely used as a tool that does not need any models or *a priori* assumptions. DSC enables precise determination of the thermodynamic key values T_{trs}, $\Delta_{unf}H$, and $\Delta_{unf}Cp$. Technical characteristics of modern instruments and guidelines for the performance of studies on biopolymers are contained in I-80P2, I-86P4, I-89P5, I-94C, I-95F, I-95P2, I-97P2. For mathematical analysis of melting peaks and deconvolution of superimposed transitions the reader is referred to I-78B, I-78F1, I-78F2, I-86P4, I-87K, I-88K, I-93P1, I-94F, I-94S, I-95F. Formulation of thermal transitions in proteins including effects of ligand binding, domain interaction, and oligomerization is given in I-86H, I-87S2, I-89B, I-89R, I-90S2, I-92S4, I-94F, I-94S.

In experimental studies, the key values can be determined with sufficient accuracy. Usually, the standard deviation amounts to about ± 0.1 K in T_{trs}, $\pm 2\%$ in $\Delta_{unf}H$, and $\pm 5\%$ in $\Delta_{unf}Cp$. However, more precise measurements were performed, and detailed error analyses were reported (see, e.g., I-83F, I-88S2, I-89S, I-91X, I-92H1, I-92S5, I-93T for well documented examples). Corrections of $\Delta_{unf}H$ for the heat of protonation of buffer substances need not be large when amino acids are used as buffer substances (I-74P). For precise correction at elevated temperature when using GOOD buffers, we refer to I-92Z.

Systematic errors in DSC may arise from the determination of concentration (see, e.g., the ribonuclease T1 example, I-94Y). Errors due to irreversible denaturation are considered in I-86P4. The influence of the so called residual structure in thermal unfolding is considered in I-88P1. The two sources of error may not necessarily have any significant influence on the $\Delta_{unf}H$ determined by DSC according to I-86P4.

Calculation of thermodynamic functions. Thermal unfolding usually takes place at elevated temperature whereas the conformational stability $\Delta_{unf}G$ is generally of interest at the standard temperature $T° = 298.16$ K $= 25°C$ or at the physiological temperature. Necessarily, temperature functions such as $\Delta_{unf}G(T)$ are to be determined. At the transition temperature T_{trs} (transition midpoint)

$$\Delta_{unf}G(T_{trs}) = \Delta_{unf}H(T_{trs}) - T_{trs}\Delta_{unf}S(T_{trs}) = 0. \tag{3}$$

$\Delta_{unf}H$, $\Delta_{unf}S$, and $\Delta_{unf}G$ are temperature dependent:

$$\Delta_{unf}H(T) = \Delta_{unf}H(T_{trs}) - \int_{T}^{T_{trs}} \Delta_{unf}Cp(T)\,dT \tag{4}$$

$$\Delta_{unf}S(T) = \frac{\Delta_{unf}H(T_{trs})}{T_{trs}} - \int_{T}^{T_{trs}} \Delta_{unf}Cp(T)\,d\ln T \tag{5}$$

$$\Delta_{unf}G(T) = \Delta_{unf}H(T) - T\Delta_{unf}S(T) = \Delta_{unf}H(T_{trs})\frac{T_{trs}-T}{T_{trs}} - \int_{T}^{T_{trs}} \Delta_{unf}Cp(T)\,dT$$
$$+ T\int_{T}^{T_{trs}} \frac{\Delta_{unf}Cp(T)}{T}\,dT \tag{6}$$

In case the heat capacity change $\Delta_{unf}Cp(T)$ at protein unfolding does not depend significantly on temperature (see below), the above equations can be simplified:

$$\Delta_{unf}H(T) = \Delta_{unf}H - \Delta_{unf}Cp\,(T_{trs} - T) \tag{7}$$

$$\Delta_{unf}S(T) = \frac{\Delta_{unf}H}{T} - \Delta_{unf}Cp\,\ln\frac{T_{trs}}{T} \tag{8}$$

$$\Delta_{unf}G(T) = \Delta_{unf}H\frac{T_{trs}-T}{T_{trs}} - \Delta_{unf}Cp(T_{trs}-T) + \Delta_{unf}Cp\,T\ln\frac{T_{trs}}{T}. \tag{9}$$

Equation (9) is important for transforming $\Delta_{unf}G$ from one temperature to another. At the same time, eq. (9) is the main equation in determining $\Delta_{unf}G(T)$ from calorimetrically measured transition temperature, enthalpy change and heat capacity change.

Calculating $\Delta_{unf}G(T)$, error propagation is mainly due to $\Delta_{unf}Cp$ (I-79P2, I-87B). However, $\Delta_{unf}Cp$ determined from single calorimetric recordings can be verified additionally by Kirchhoff's Law (eq. 2d) or by separate heat capacity measurements on the native and unfolded states. Usually, heat capacity changes at protein unfolding are regarded as temperature independent.

According to I-90P3, $\Delta_{unf}Cp$ itself was found to be temperature dependent (see also the examples of cytochrome c, lysozyme HEW, myoglobin, and ribonuclease A in the following tables). In experimental work, the temperature dependence of heat capacity change remains almost undetected. As shown in Fig. 2, the experimental window (with proper pre- and postdenaturational heat capacity functions) for the exact determination of $\Delta_{unf}H^{cal}$ might be too narrow to see the curvature of $\Delta_{unf}H^{cal}$ versus T_{trs}. In Fig. 2, the curvature follows from heat capacity measurements of native and unfolded proteins (inset). The error in $\Delta_{unf}G(T)$ due to the temperature dependence of $\Delta_{unf}Cp$ need not be large and can be neglected in most cases (I-89P3, I-92P).

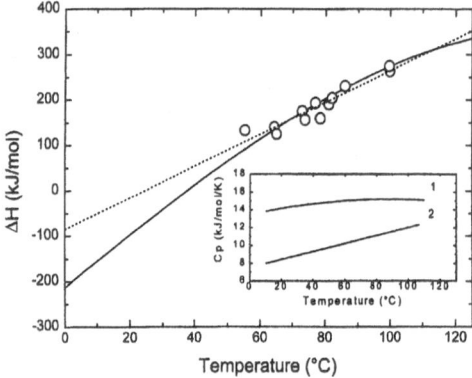

Fig. 2. Experimental $\Delta_{unf}H^{cal}$ values plotted versus T_{trs}.
Broken line: linear fit corresponding to a temperature independent $\Delta_{unf}Cp$.
Solid line: Temperature function of $\Delta_{unf}Cp$ constructed by means of the heat capacities of native unfolded protein shown in the insert (curve 1: unfolded protein, curve 2: native protein).
The data were taken from Ref. I-97P1

1.2 Denaturant-Induced Unfolding

This approach includes both the above mentioned two-state assumption and assumptions on the mechanism of protein denaturant interaction. The procedure for the determination of conformational stability, however, is rather simple and sensitive, in particular, in difference $\Delta_{unf}G$ determination. For guidelines and necessary properties of aqueous denaturant solutions see I-86P1, I-89P1, I-95S3. For multi-dimensional spectroscopic data correlation see I-95R.

The approach is based on the determination of equilibrium data for protein unfolding in denaturant containing solution (GuHCl, urea, etc.). The extrapolation to zero denaturant concentrations is of critical importance for the evaluation of $\Delta_{unf}G$ values obtained in this way.

a) Denaturant binding model:

$$K_{unf} = K^{\circ}_{unf}(1 + K_b\,a_{\pm})^{\Delta n} \tag{10}$$

In eq. (10) $a_{\pm}$ stands for the mean ion activity of GuHCl and must be replaced by activity (a) in case of unfolding by urea (for alternative expressions see I-69A).

b) Linear extrapolation:

$$\ln K^\circ_{unf} = \ln K_{unf} + m_{(denaturant)} \qquad (11a)$$

or

$$\Delta_{unf} G = \Delta_{unf} G^{H_2O} - m_{(denaturant)} \qquad (11b)$$

c) Transfer model:

$$\Delta_{unf} G = \Delta^\circ_{unf} G + \epsilon \sum n_i \, \delta g_{tr,1} \qquad (12)$$

Case (a) – Denaturant binding model: Among the equations introduced in I-69A is eq. (10) agrees best with independent findings. However, the choice of the denaturant binding constant may be a source of systematic errors (I-75P, I-79P1, I-89P1). For recent results on the validity of the denaturant binding model, see I-92M1.

Case (b) – Linear extrapolation: Equation (11) is currently the most widely applied extrapolation procedure for obtaining $\Delta_{unf} G$ from denaturant induced unfolding transitions. For the thermodynamic basis of the linear extrapolation procedure, the reader is referred to I-75P, I-78S, I-79P1, I-87B, I-87S1, I-90S1, I-92S1. Error propagation is analyzed in I-88B, I-88S1, I-90P. Note, linear extrapolation and denaturant binding model produce different results on extrapolation to zero denaturant concentration (Fig. 3).

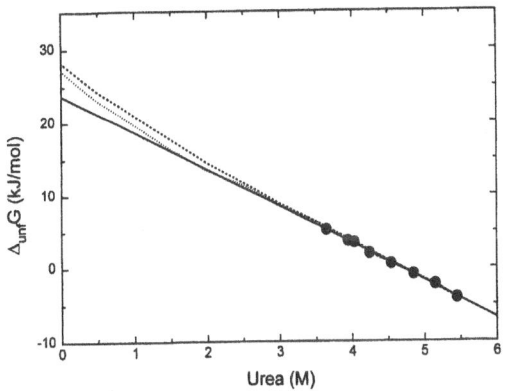

Fig. 3. The various procedures for the extrapolation to zero denaturant concentration at denaturant induced protein unfolding. Solid line: linear extrapolation, dotted line: transfer model, broken line: denaturant binding model. The data that concern ribonuclease T1 were taken from Ref. I-89P1

The linear extrapolation procedure may be improved by a method that includes the pre- and postdenaturational baselines for a non-linear regression (I-88B, I-88S1).

Recently, the validity of extrapolation to zero denaturant concentration has been reconsidered (I-94B, I-95Y, I-95S1, I-95J, I-96I, I-96G, I-96N). Deviation from linearity have been reported (I-94B, I-95J, I-96I). Differences in estimates of $\Delta_{unf} G^\circ$ obtained by urea and GuHCl denaturation may be produced by electrostatic interactions (I-94M, I-96S).

For the relation between denaturant m values and heat capacity changes ΔCp please see I-95M3, I-95S4.

Case (c) – Transfer model: For use of eq. (12), tabulated values of the Gibbs energy change at transfer of amino acid side chains from water to the aqueous denaturant solution ($\delta g_{tr,1}$) are available (I-68T, I-70T, I-75P). The weak point of the approach is the determination of ϵ, the average degree of exposure of the side chains. Usually values around 0.25 are assumed (I-75P). The determination of ϵ might be improved using experimental approaches or atomic coordinates of proteins. For a more recent consideration see I-90P.

1.3 Unfolding by Protonation

Acid or alkaline denaturation is based on the existence of two conformation states (of charge Z_1 and Z_2) with different degrees of protonation:

$$(d \ln K_{unf}/dT) = Z_2 - Z_1 = \Delta_{unf}\nu \tag{13}$$

The Gibbs energy change can be obtained as a pH dependent function:

$$\Delta_{unf}G(\ln a_{H^+}) = \Delta_{unf}G° + RT \int_{a_0}^{a_1} \Delta_{unf}\nu\,(\ln a_{H^+})\,d\ln a_{H^+} \tag{14}$$

Here, a_0 and a_1 denote $(\ln a_{H^+})°$ and $(\ln a_{H^+})$ which correspond to the initial and final pH values at transfer of the protein from one to another. Equations (13,14) are valid for a reversible two-state unfolding transition. The approach is mainly dependent on the accuracy of determination of the difference number of protons $\Delta_{unf}\nu$ of the states N and D. For the procedure, examples, and estimation of standard deviation see I-76P1, I-79P2, I-88B.

1.4 Urea Gradient Gel Electrophoresis

This approach is rather a diagnostic tool for the detection of intermediates at protein unfolding and might, in that way, substantiate the two-state assumption in urea induced protein unfolding. At the same time, $\Delta_{unf}G$ can be roughly estimated using the linear extrapolation procedure (with implications given above). For guidelines see I-86C, I-89G, I-89P1.

1.5 Isothermal Calorimetry

Isothermal calorimetry can be applied to determine $\Delta_{unf}H$ at denaturant induced unfolding. For the procedure and approximation of the preferential interaction heat see I-76P2, I-79P2, I-86P3, I-89W, I-94C.

1.6 Hydrogen Exchange

Values of $\Delta_{unf}G$ obtained by hydrogen exchange have been included in the following tables as far as global unfolding is concerned. For implications of the approach see I-93K1, I-93K2, I-94B, I-94Q, I-95P1, I-95S2. Recent developments (I-96B, I-96E, I-97C) permit the determination of ΔG at individual residues along the polypeptide chain. For further details please see examples of barnase (I-96C2), ribonuclease A (I-93M4), ribonuclease H* (I-96C1), and trypsin inhibitor (I-93K1) listed in Table 1.

2 Comments on the Tabulated Data

The thermodynamic quantities on protein stability and folding are listed in three tables: Table 1 contains Gibbs energy change, Table 2 contains related molar enthalpy and heat capacity changes, and Table 3 contains specific enthalpy and heat capacity changes.

Within the tables the proteins are arranged in alphabetical order using the same nomenclature as in the original papers. For amino acids the three-letter code is used throughout. For proteins characterized by a large amount of data, the listings are subdivided into small tables, e.g., for species, liganded and modified forms, mutants, or with respect to specific experimental details. In cases of point mutations, the proteins are arranged with increasing position numbers starting from the N-terminus. Original data, if given in calories, were transformed into Joules using 1 cal = 4.184 J. Temperature is given in degrees Celsius (°C) if not otherwise indicated. Some data, that were not explicitly given in the original papers but were represented in figures, were included into the tables along with remarks. Data that were calculated on the basis of other thermodynamic quantities in the original papers are indicated by asterisks (*). Thermodynamic quantities from former data collections (I-75P, I-81P, I-86P2) were critically checked and included, thus making the following tables as complete as possible.

Table 1 contains Gibbs energy change $\Delta_{unf}G$. Since the Gibbs energy change at protein unfolding corresponds to the (negative) stabilization energy ($\Delta G_{unfolding} = - \Delta G_{stabilization}$), ΔG is given as a positive value for reasons of convenience. A higher positive ΔG value characterizes more stable protein.

Sometimes, in particular in connection with mutant proteins, $\Delta(\Delta G)$ values are given. These data usually refer to the Gibbs energy change of a wild-type (reference) protein according to $\Delta(\Delta G) = \Delta G_{(mutant)} - \Delta G_{(wild-type)}$, and positive $\Delta(\Delta G)$ values designate mutant proteins being more stable than the wild-type. In such cases, usually the wild-type protein is the reference protein assuming $\Delta(\Delta_{unf}G) = 0$.

The accuracy in determination of difference values $\Delta(\Delta_{unf}G)$ may be enhanced when referring to the transition midpoint. If this is the case there will be a note in the tables. In case of thermal denaturation, this respective applies to the following equation:

$$\Delta(\Delta_{unf}G) = \Delta T \times \Delta S \tag{15}$$

(I-87B) or the assumption of an identical heat capacity change of wild-type and mutant protein (I-87B). For an error analysis see I-92H2. The following equation,

$$\Delta(\Delta_{unf}G) = \Delta T/T_{trs(wild-type)} \times [\Delta_{unf}H_{(wild-type)} + \Delta_{unf}Cp \times \Delta T], \tag{16}$$

was recommended by I-93L.

In case of denaturant-induced unfolding and use of the linear extrapolation model, $\Delta(\Delta_{unf}G)$ refers to

$$\Delta(\Delta_{unf}G) = \Delta_{unf}G°_{(mutant)} - \Delta_{unf}G°_{(wild-type)}. \tag{17}$$

For simplicity, the slope may be assumed to be the same for wild-type and mutant protein. The accuracy in determining $\Delta_{unf}G$ may be enhanced when the denaturant concentration at the transition midpoint ($D_{50\%}$ or $C_{1/2}$) is used:

$$\Delta_{unf}G^{D50\%} = m \times \Delta D_{50\%} \tag{18}$$

(see I-92S2, I-93C).

With respect to the implications contained in some methodical variants, it might be useful to refer to $\Delta_{unf}G$ values obtained by the same approach, when conformational stability of various proteins, mutants, liganded forms, etc. are to be compared.

Table 2 contains tabulated thermodynamic key values, i.e., molar enthalpy changes at protein unfolding (ΔH) and molar heat capacity changes (ΔCp) along with additional data defining the external conditions for which the data are valid (e.g. thermal transition temperatures T_{trs}, given in degrees Celsius, °C). In a remark on the approach it is clearly stated whether the enthalpy changes are calorimetric values or van't Hoff enthalpies. In case of calorimetric values we usually refer to experimental data without mathematical treatment. In case of deconvolution of overlapping peaks of multidomain proteins, the procedure is indicated.

If possible, the listed data refer to either the maximum thermal transition temperature or standard temperature (25°C). The tabulated data are condensed in such a manner that complex thermodynamic functions can be calculated (see eqs. 3–9). Therefore, preference was given to complete data sets consisting of T_{trs}, ΔH, and ΔCp.

In order to condense available data on heat capacity and enthalpy change, single pairs of data are usually represented in the following tables, even if more extended listings of $\Delta_{unf}H$ versus T_{trs} are available. As mentioned above, the data refer, where possible, to the highest observed transition temperature or to standard temperature. Some more extended protocols, however, are given for those proteins that may serve as a matter of reference.

Table 3 corresponds to the preceding one but it contains specific values, i.e., Δh, Δcp in J/g and J/K/g. This is due to the fact that heat capacity changes are often represented in Joules per gram whereas enthalpy changes are given as molar values.

The tables are completed by remarks on the approaches and experimental conditions to make the data self consistent. The arrangement of the data in tables, however, requires some shortening. The reader will always find more details in the original papers.

3 Why to Prefer Complex Thermodynamic Functions

The data listed in this book enable the construction of phase diagrams. The phase diagrams provide a more general description of $\Delta_{unf} G$ *versus* temperature and pH. As can be seen by the example of cytochrome b_5 (Fig. 4) and lysozyme (Fig. 5), the dependence of conformational stability on pH and temperature may considerably vary from protein to protein. The same applies to mutant proteins as shown by the example of lysozyme phage T4 in Fig. 6.

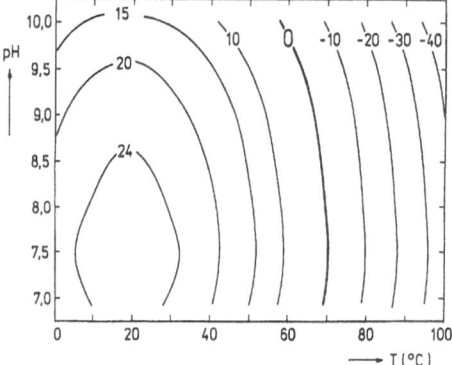

Fig. 4. Phase diagram of cytochrome b_5. The isolines represent $\Delta_{unf} G$ *versus* temperature and pH. $\Delta_{unf} G$ was calculated using data from Ref. I-80P1

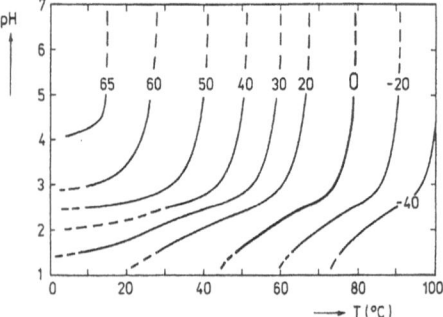

Fig. 5. Phase diagram of hen egg white lysozyme. The isolines represent $\Delta_{unf} G$ *versus* temperature and pH. $\Delta_{unf} G$ was calculated using data from Ref. I-74P and I-76P3

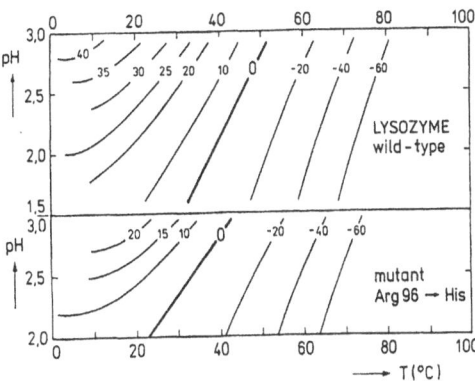

Fig. 6. Phase diagram of lysozyme phage T4, wild-type protein and mutant Arg96→His.
The isolines represent $\Delta_{unf}G$ *versus* temperature and pH. $\Delta_{unf}G$ was calculated using data from Ref. I-89K

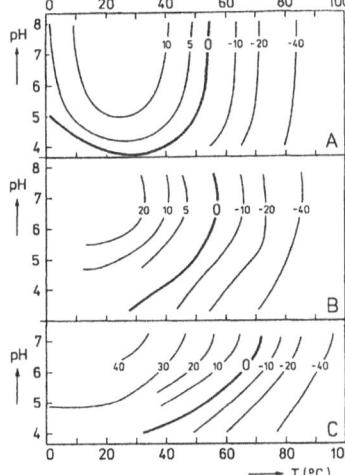

Fig. 7. Phase diagram of staphylococcal nuclease.
A - without ligand, **B** - in the presence of 20 mM calcium, **C** - in the presence of 20 mM calcium and 2 mM thymidine 3',5'-diphosphate.
The isolines represent $\Delta_{unf}G$ *versus* temperature and pH. $\Delta_{unf}G$ was calculated using data from Ref. I-85C

Comparing e.g. staphylococcal nuclease stability in ligand-free and various liganded states (Fig. 7), remarkable stability differences become obvious around neutral pH and 25°C. However, at about 50°C and pH 5, close to half conversion, all forms show nearly the same $\Delta_{unf}G$. That means, data points taken by accident in this region would never visualize the specific differences between these forms.

Seen in that light, a simple comparison of single $\Delta_{unf}G$ values taken from Table 1 might in some cases produce misleading conclusions that can be avoided when using phase diagrams.

References

References were labelled with an I- to distinguish them from the references of the Tables.

I-66L   Lumry, R., Biltonen, R., Brandts, J.F.: Biopolymers 4 (1966) 917.
I-68T   Tanford, C.: Adv. Protein Chem. 23 (1968) 121.
I-69A   Aune, K.C., Tanford, C.: Biochemistry 8 (1969) 4586.
I-70T   Tanford, C.: Adv. Protein Chem. 24 (1970) 1.
I-74P   Privalov, P.L., Khechinashvili, N.N.: J. Mol. Biol. 86 (1974) 665.
I-75P   Pace, C.N.: CRC Crit. Rev. Biochem. 3 (1975) 1.
I-76P1   Pfeil, W., Privalov, P.L.: Biophys. Chem. 4 (1976) 23.
I-76P2   Pfeil, W., Privalov, P.L.: Biophys. Chem. 4 (1976) 33.
I-76P3   Pfeil, W., Privalov, P.L.: Biophys. Chem. 4 (1976) 41.
I-78B   Biltonen, R.L., Freire, E.: CRC Crit. Rev. Biochem. 5 (1978) 85.
I-78F1   Freire, E., Biltonen, R.L.: Biopolymers 17 (1978) 463.
I-78F2   Freire, E., Biltonen, R.L.: Biopolymers 17 (1978) 481.
I-78S   Schellman, J.A.: Biopolymers 17 (1978) 1305.
I-79P1   Pace, C.N., Vanderburg, K.E.: Biochemistry 18 (1979) 288.
I-79P2   Pfeil, W.: in: Biochemical Thermodynamics, Studies in modern thermodynamics, vol. 1, ed. by M.N. Jones, Elsevier Scientific Publ. Co., Amsterdam, Oxford, New York, 1979, p. 75.
I-79P3   Privalov, P.L.: Adv. Protein Chem. 33 (1979) 167.
I-80P1   Pfeil, W., Bendzko, P.: Biochim. Biophys. Acta 626 (1980) 73.
I-80P2   Privalov, P.L.: Pure Appl. Chem. 52 (1980) 479.
I-81P   Pfeil, W.: Mol. Cell. Biochem. 40 (1981) 3.
I-82P   Privalov, P.L.: Adv. Protein Chem. 35 (1982) 1.
I-83F   Fukada, H., Sturtevant, J.M., Quiocho, F.A.: J. Biol. Chem. 258 (1983) 13193.
I-83S   Saito, Y., Wada, A.: Biopolymers 22 (1983) 2123.
I-83W   Wada, A., Saito, Y., Ohogushi, M.: Biopolymers 22 (1983) 93.
I-85C   Calderon, R.O., Stolovich, N.J., Gerlt, J.A., Sturtevant, J.M.: Biochemistry 24 (1985) 6044.
I-86C   Creighton, T.E.: Methods Enzymol. 131 (1986) 156.
I-86H   Hinz, H.-J.: Methods Enzymol. 130 (1986) 59.
I-86P1   Pace, C.N.: Methods Enzymol. 131 (1986) 266.
I-86P2   Pfeil, W.: in: Thermodynamic Data for Biochemistry and Biotechnology, ed. by H.-J. Hinz, Springer-Verlag Berlin, Heidelberg, New York, Tokyo, 1986, p. 349.
I-86P3   Pfeil, W., Bychkova, V.E., Ptitsyn, O.B.: FEBS Lett. 198 (1986) 287.
I-86P4   Privalov, P.L., Potekhin, S.A.: Methods Enzymol. 131 (1986) 4.
I-87B   Becktel, W.J., Schellman, J.A.: Biopolymers 26 (1987) 1859.
I-87K   Kidokoro, S.-I., Wada, A.: Biopolymers 26 (1987) 213.
I-87S1   Schellman, J.A.: Ann. Rev. Biophys. Biophys. Chem. 16 (1987) 115.
I-87S2   Sturtevant, J.M.: Ann. Rev. Phys. Chem. 38 (1987) 463.
I-88B   Bolen, D.W., Santoro, M.M.: Biochemistry 27 (1988) 8069.
I-88K   Kidokoro, S.-I., Wada, A.: Biopolymers 27 (1988) 271.
I-88P1   Pfeil, W.: in: Biochemical Thermodynamics, Studies in modern Thermodynamics, vol. 8, ed. by M.N. Jones, 2nd ed., Elsevier Science Publishers B.V., Amsterdam, Oxford, New York, Tokyo, 1988, p.53.
I-88P2   Privalov, P.L., Gill, S.J.: Adv. Protein Chem. 39 (1988) 191.
I-88S1   Santoro, M.M., Bolen, D.W.: Biochemistry 27 (1988) 8063.
I-88S2   Schwarz, F.P., Kirchhoff, W.H.: Thermochim. Acta 128 (1988) 267.
I-89A   Alber, T.: Ann. Rev. Biochem. 58 (1989) 765.
I-89B   Brandts, J.F., Hu, C.Q., Lin, L.-N., Mas, M.T.: Biochemistry 28 (1989) 8588.
I-89C   Creighton, T.E.: Protein Structure – a practical approach, IRL Press, Oxford, New York, Tokyo, 1989.
I-89G   Goldenberg, D.P.: in: Protein Structure – a practical approach, ed. by T.E. Creighton, IRL Press, Oxford, New York, Tokyo, 1989, p. 225.
I-89K   Kitamura, S., Sturtevant, J.M.: Biochemistry 28 (1989) 3788.
I-89P1   Pace, C.N., Shirley, B.A., Thomson, J.A.: in: Protein Structure – a practical approach, ed. by T.E. Creighton, IRL Press, Oxford, New York, Tokyo, 1989, p. 311.

I-89P2   Pace, C.N., Laurents, D.V.: Biochemistry 28 (1989) 2520.
I-89P3   Privalov, P.L.: Ann. Rev. Biophys. Biophys. Chem. 18 (1989) 47.
I-89P4   Privalov, P.L., Tiktopulo, E.I., Venyaminov, S.Yu, Griko, Yu.V., Makhatadze, G.I., Khechinashvili, N.N.: J. Mol. Biol. 205 (1989) 737.
I-89P5   Privalov, P.L., Plotnikov, V.V.: Thermochim. Acta 139 (1989) 257.
I-89R   Robert, C.H., Colosimo, A., Gill, S.J.: Biopolymers 28 (1989) 1705.
I-89S   Schwarz, F.P: Thermochim. Acta 147 (1989) 71.
I-89W   Wiseman, T., Williston, S., Brandts, J.F., Lin, L.-N.: Analyt. Biochem. 179 (1989) 131.
I-90B   Brandts, J.F., Lin, L.-N.: Biochemistry 29 (1990) 6927.
I-90D   Dill, K.A.: Biochemistry 29 (1990) 7133.
I-90P1   Pace, C.N., Laurents, D.V., Thomson, J.A.: Biochemistry 29 (1990) 2564.
I-90P2   Privalov, P.L.: Crit. Rev. Biochem. Mol. Biol. 25 (1990) 281.
I-90P3   Privalov, P.L., Makhatadze, G.I.: J. Mol. Biol. 213 (1990) 385.
I-90S1   Schellman, J.A.: Biopolymers 29 (1990) 215.
I-90S2   Shrake, A., Ross, P.D.: J. Biol. Chem. 265 (1990) 5055.
I-91X   Xie, D., Bhakuni, V., Freire, E.: Biochemistry 30 (1991) 10673.
I-92G   Goldenberg, D.P.: in Protein Folding, ed. by T.E. Creighton, W.H. Freeman and Co., New York, 1992, pp. 353
I-92H1   Hu, C.-Q., Sturtevant, J.M., Thomson, J.A., Erickson, R.E., Pace, C.N.: Biochemistry 31 (1992) 4876.
I-92H2   Hurley, J.H., Baase, W.A., Matthews, B.W.: J. Mol. Biol. 224 (1992) 1143.
I-92M1   Makhatadze, G.I., Privalov, P.L: J. Mol. Biol. 226 (1992) 491.
I-92M2   Murphy, K.P., Freire, E.: Adv. Protein Chem. 43 (1992) 313.
I-92O   Oobatake, M., Ooi, T.: Progr. Biophys. Molec. Biol. 59 (1992) 237.
I-92P   Privalov, P.L.: in Protein Folding, ed. by T.E. Creighton, W.H. Freeman and Co., New York, 1992, pp. 83.
I-92S1   Santoro, M.M., Bolen, D.W.: Biochemistry 31 (1992) 4901.
I-92S2   Serrano, L., Kellis Jr., J.T., Cann, P., Matouschek, A., Fersht, A.R.: J. Mol. Biol. 224 (1992) 783.
I-92S3   Shortle, D.: Quart. Rev. Biophys. 25 (1992) 205.
I-92S4   Shrake, A., Ross, P.D.: Biopolymers 32 (1992) 925.
I-92S5   Straume, M., Freire, E.: Analyt. Biochem. 203 (1992) 259.
I-92Z   Zolkiewski, M., Ginsburg, A.: Biochemistry 31 (1992) 11991.
I-93C   Clarke, J., Fersht, A.R.: Biochemistry 32 (1993) 4322.
I-93K1   Kim, K.-S., Fuchs, J.A., Woodward, C.: Biochemistry 32 (1993) 9600.
I-93K2   Kim, K.-S., Woodward, C.: Biochemistry 32 (1993) 9609.
I-93L   Lin, L., Kallenbach, N.R.: Protein Engng. 6 (1993) 18.
I-93M1   Makatadze, G.I., Privalov, P.L.: J. Mol. Biol. 232 (1993) 639.
I-93M2   Matthews, B.: Annu. Rev. Biochem. 62 (1993) 130.
I-93M3   Matthews, B.W.: Current Opinion Struct. Biol. 3 (1993) 589.
I-93M4   Mayo, S.L., Baldwin, R.L.: Science 262 (1993) 873.
I-93P1   Potekhin, S.A.: Biophysics 37 (1993) 41, 227.
I-93P2   Privalov, P.L., Makhatadze, G.I.: J. Mol. Biol. 232 (1993) 660.
I-93T   Tanaka, A., Flanagan, J., Sturtevant, J.M.: Protein Science 2 (1993) 567.
I-94B   Bai, Y., Milne, J.S., Mayne, L., Englander, S.W.: Proteins: Structure, Function, and Genetics 20 (1994) 4.
I-94C   Cooper, A., Johnson, C.M.: in Methods in Molecular Biology, Vol. 22 : Microscopy, Optical Spectroscopy, and Macroscopic Techniques, ed. Jones, C., Mulloy, B., Thomas A.H., Humana Press Inc., Totowa, NJ, 1994, pp. 109, 125. 137.
I-94E   Eftink, M.R.: Biophys. J. 66 (1994) 482.
I-94F   Freire, E.: Methods Enzymol. 240 (1994) 502.
I-94M   Monera, O.D., Kay, C.M., Hodges, R.S.: Protein Sci. 3 (1994) 1984.
I-94Q   Qian, H., Mayo, S.L., Morton, A.: Biochemistry 33 (1994) 8167.
I-94R1   Ramsay, G., Eftink, M.R.: Biophys. J. 66 (1994) 516.
I-94R2   Ramsay, G.D., Eftink, M.R.: Methods Enzymol. 240 (1994) 615.
I-94S   Straume, M.: Methods Enzymol. 240 (1994) 530.
I-94Y   Yu, Y., Makhatadze, G.I., Pace, C.N., Privalov, P.L.: Biochemistry 33 (1994) 3312.

I-95F	Freire, E.: in Methods in Molecular Biology, Vol. 40: Protein Stability and Folding, Theory and Practice, ed. Shirley, B.A., Humana Press Inc., Totowa, NJ, 1995, pp. 1991.
I-95J	Johnson, C.M., Fersht, A.R.: Biochemistry 34 (1995) 6795.
I-95M1	Makhatadze, G.I., Privalov, P.L.: Adv. Protein Chem. 47 (1995) 307.
I-95M2	Matthews, B.W.: Adv. Protein Chem. 47 (1995) 249.
I-95M3	Myers, J.K., Pace, C.N., Scholtz, J.M.: Protein Sci. 4 (1995) 2138.
I-95P1	Perrett, S., Clarke, J., Hounslow, A.M., Fersht, A.R.: Biochemistry 34 (1995) 9288.
I-95P2	Privalov, G., Kavina, V., Freire, E., Privalov, P.L.: Analyt. Biochem. 232 (1995) 79.
I-95R	Ramsay, G., Ionescu, R., Eftink, M.R.: Biophys. J. 69 (1995) 701.
I-95S1	Scholtz, J.M., Barrick, D., York, E.J., Stewart, J.M., Baldwin, R.L.: Poc. Natl. Acad. Sci. USA 92 (1995) 185.
I-95S2	Scholtz, J.M., Robertson, A.D.: in Methods in Molecular Biology, Vol. 40: Protein Stability and Folding, Theory and Practice, ed. Shirley, B.A., Humana Press Inc., Totowa, NJ, 1995, pp. 291.
I-95S3	Shirley, B.A.: in Methods in Molecular Biology, Vol. 40: Protein Stability and Folding, Theory and Practice, ed. Shirley, B.A., Humana Press Inc., Totowa, NJ, 1995, pp. 177.
I-95S4	Shortle, D.: Adv. Protein Chem. 47 (1995) 217.
I-95S5	Stites, W.E., Byrne, M.P., Aviv, J., Kaplan, M., Curtis, P.M.: Analyt. Biochem. 227 (1995) 112.
I-95T	Terwilliger, T.C.: Adv. Protein Chem. 47 (1995) 177.
I-95Y	Yao, M., Bolen, D.W.: Biochemistry 34 (1995) 3771.
I-96B	Bai, Y., Englander, S.W.: Proteins: Structue, Function, and Genetics 24 (1996) 145.
I-96C1	Chamberlain, A.K., Handel, T.M., Marqusee, S.: Nature Struct. Biol. 3 (1996) 782.
I-96C2	Clarke, J., Fersht, A.R.: Folding & Design 1 (1996) 243.
I-96E	Englander, S.W., Sosnick, T.R., Englander, J.J., Mayne, L.: Current Opinion Struct. Biol. 6 (1996) 18.
I-96G	Gupta, R., Yadav, S., Ahmad, F.: Biochemistry 35 (1996) 11925.
I-96I	Ibarra-Molero, B., Sanchez-Ruiz, J.M.: Biochemistry 35 (1996) 14689.
I-96N	Nicholson, E.N., Scholtz, J.M.: Biochemistry 35 (1996) 11369.
I-96S	Smith, J.S., Scholtz, J.M.: Biochemistry 35 (1996) 7292.
I-97C	Clarke, J., Itzhaki, L.S., Fersht, A.R.: Trends Biochem. Sci. 22 (1997) 284.
I-97H	Holtzer, A.: Biopolymers 42 (1997) 499.
I-97P1	Pfeil, W., Gesierich, U., Kleemann, G.R., Sterner, R.: J. Mol. Biol. 272 (1997) 591.
I-97P2	Plotnikov, V.V., Brandts, J.M., Lin, L.-N., Brandts, J.F.: Analyt. Biochem. 250 (1997) 237.

Table 1.
Gibbs Energy Change – Molar Values

Acetylcholinesterase

Acetylcholinesterase from *Torpedo californica*, mercury derivative

pH	T	ΔG	Approach	Remarks	Ref
7.3	22	22.6±2.9	GuHCl	(1,2)	94K10
7.3	22	23.8±1.7	GuHCl	(1,3)	94K10
7.3	24	24.3±2.5	GuHCl	(1,4)	94K10

Remarks:
(1) at Cys231 labeled with the mercury derivative of a stable nitrosyl radical, the transition is a molten globule→unfolded transition, data treatment by linear extrapolation
(2) transition monitored by electron paramagnetic resonance (EPR)
(3) transition monitored by CD at 222 nm
(4) transition monitored by fluorescence intensity at 320 nm

Actin

G-Actin, various transitions

Transition	pH	T	ΔG	Approach	Remarks	Ref
overall	7.5	25	27	heat, optical methods		77C3
overall	8.0	23	43.1	DSC	(1)	90B2
trans. (1)	8.0	23	15.1	DSC	(2)	90B2
trans. (2)	8.0	23	35.1	DSC	(3)	90B2
N→U	8.0	23	54.4	urea	(4)	90B2
X→U	8.0	23	40.2	urea	(4)	90B2
N→X	8.0	23	10.0	GuHCl	(4)	90B2
X→U	8.0	23	44.4	GuHCl	(4)	90B2
X→U	8.0	23	47.3	GuHCl	(4)	90B2

Remarks:
(1) formal approximation of the transition by a two-state model
(2) the first of two subsequent transitions with transition temperature of 52.3 °C
(3) the second of two subsequent transitions with transition temperature of 56.7 °C
(4) linear extrapolation

Acyl-Coenzyme A Binding Protein

Bovine acyl-coenzyme A binding protein, recombinant

pH	T	ΔG	Approach	Remark	Ref
5.3	5	25.5±3.8	GuHCl	(1)	95K12

Remark:
(1) linear extrapolation to zero denaturant concentration by a method that includes the pre- and postdenaturational baselines for a non-linear regression of the data

Acyl-coenzyme A binding protein, species and mutants

Species	pH	T	ΔG	Approach	Remarks	Ref
bovine	5	5.6	34.4±0.5	GuHCl	(1)	96K7
bovine	5	5.6	29.5±5.7	GuHCl	(2)	96K7
bovine, mutant Tyr31→Asn	5.6	5	29.0±1.0	GuHCl	(1)	96K7
	5.6	5	26.5±2.5	GuHCl	(2)	96K7
rat	5.6	5	25.5±0.6	GuHCl	(1)	96K7
rat	5.6	5	25.3±2.0	GuHCl	(2)	96K7
yeast	5.6	5	34.4±1.3	GuHCl	(1)	96K7
yeast	5.6	5	32.6±4.7	GuHCl	(2)	96K7

Remarks:
(1) from folding kinetics
(2) from equilibrium unfolding, linear extrapolation

Acylphosphatase

Acylphosphatase from horse muscle

	pH	T	ΔG	Approach	Remarks	Ref
	3.8	37	20.8±0.7	urea	(1–3)	94T1
	3.8	37	17.2±0.7	urea	(1,2,4)	94T1

Remarks:
(1) linear extrapolation
(2) measured in D_2O
(3) transition monitored by CD
(4) transition monitored by ^{1}H-NMR

Adenylate Kinase

Adenylate kinase from baker's yeast, wild-type and mutants

Mutant	pH	T	ΔG	Approach	Remarks	Ref
wild-type	7.5	5	16.6±2.1	GuHCl	(1)	95S11
wild-type	7.5	10	20.5±2.4	GuHCl	(1)	95S11
wild-type	7.5	15	22.9±2.1	GuHCl	(1)	95S11
wild-type	7.5	25	16±3	heat	(2)	95S11
wild-type	4.6	25	9±2	heat	(2)	95S11
extended	7.5	25	18±2	heat	(2,3)	95S11
Val8→Ile	7.5	25	11±2	heat	(2)	95S11
Gln48→Glu	7.7	25	12±4	heat	(2)	95S11
Thr77→His	7.5	25	7±2	heat	(2)	95S11
Thr77→His	5.0	25	12±2	heat	(2)	95S11
Thr110→His	7.6	25	5±1.5	heat	(2)	95S11
Thr110→His	4.5	25	3±1.5	heat	(2)	95S11
Asn169→Asp	7.5	25	7±2	heat	(2)	95S11
Ile213→Phe	7.5	25	8±1.5	heat	(2)	95S11

Remarks:
(1) linear extrapolation
(2) calculated using ΔCp=8.5 kJ/mol/K
(3) extended=adenylate kinase with C-terminal extension Pro-His-His

Adrenodoxin

Adrenodoxin, wild-type (recombinant), in the presence of Na_2S

pH	T	ΔG	Approach	Remarks	Ref
8.5	25	21±4	DSC	(1,2)	95B9

Remarks:
(1) buffer: 40 mM glycine, 10 mM Na_2S, 1 mM ascorbic acid, 10 mM mercaptoethanol, pH 8.5
(2) the difference Gibbs energy between holo and apo adrenodoxin amounts to $\Delta(\Delta G)$ = 14 kJ/mol at 37°C, the transition temperature of the apo protein

Adrenodoxin from bovine adrenal cortex

pH	T	ΔG	Approach	Remarks	Ref
7.0	25	18.3±1	pressure	(1,2)	96L1

Remarks:
(1) pressure denaturation monitored by fourth derivative spectroscopy, ΔG is given at 0.1 MPa
(2) ΔV amounts to –66±5 ml/mol

Adrenodoxin (ADX), wild-type and mutants, in the presence of Na_2S

Mutant	pH	T	ΔT	Δ(ΔG)	Approach	Remarks	Ref
wild-type	8.5	25	0.0	0.0	DSC	(1,2)	95B9
apo-ADX	7.4	37	–15	–14	DSC	(1–3)	95B9
(4–108)	8.5	25	3.5	3.5	DSC	(1,2,4)	96B11
(4–114)	8.5	25	–0.8	–1.7	DSC	(1,2)	96B11
Thr54→Ala	8.5	25	–4.6	–5.6	DSC	(1,2)	96B11
Thr54→Ser	8.5	25	–0.1	0.1	DSC	(1,2)	96B11
His56→Arg	8.5	25	–3.7	–8.0	DSC	(1,2)	96B11
His56→Gln	8.5	25	–6.8	–11.1	DSC	(1,2)	96B11
His56→Thr	8.5	25	–3.8	–6.1	DSC	(1,2)	96B11
Asp76→Glu	8.5	25	1.7	1.8	DSC	(1,2)	96B11
Tyr82→Leu	8.5	25	–1.0	–1.6	DSC	(1,2)	96B11
Tyr82→Phe	8.5	25	–0.2	–0.2	DSC	(1,2)	96B11
Tyr82→Ser	8.5	25	–1.1	–1.5	DSC	(1,2)	96B11
Tyr82→Trp	8.5	25	0.0	–1.0	DSC	(1,2)	96B11
Cys95→Ser	8.5	25	3.9	3.8	DSC	(1,2)	96B11

Remarks:
(1a) see also Ref. 95P1
(1b) T is the reference temperatur for $\Delta(\Delta G)$, ΔT is the difference in T_{trs} between mutant and wild-type, T_{trs} of wild-type amounts to 51.7°C, $\Delta(\Delta G)$ was calculated using ΔCp = 7.28±0.67 kJ/mol/K except for His56→Arg (ΔCp = 4.29±0.37 kJ/mol/K)
(2) buffer: 40 mM glycine, 10 mM Na_2S, 1 mM ascorbic acid, 10 mM mercaptoethanol, pH 8.5
(3) measured in 25 mM phosphate, 1 mM DTT, pH 7.4
(4) recombinant short form consisting of residues 4–128

Aldolase

Aldolase from *Staphylococcus aureus*

pH	T	ΔG	Approach	Remarks	Ref
7.0	20	8.8±2	GuHCl	(1a,2)	92R3
7.0	20	9.4±3.1	GuHCl	(1b,3)	92R3

Remarks:
(1a) linear extrapolation, slope m = 12.3±1.6 kJ/(mol×M_D)
(1b) linear extrapolation, slope m = 13.2±2.9 kJ/(mol×M_D)
(2)   transition measured by fluorescence
(3)   transition measured by CD

Alkaline Phosphatase

Alkaline phosphatase

Form	pH	T	ΔG	Approach	Remarks	Ref
apo apophosphoryl	7.5	30	83.1	DSC		77C2
	7.5	30	23.4	DSC		77C2
apomonomer	7.5	30	31.8	DSC	(1)	79C

Remark:
(1) measured in the presence of formamide

α-Helix Forming Propensity, see helix

4-Aminobutyrate Aminotransferase

4-Aminobutyrate aminotransferase

pH	T	ΔG	Approach	Remarks	Ref
7	25	52.7	urea	(1,2)	94P2
7	25	53.6	urea	(1,3)	94P2
7	25	54.4	urea	(1,4)	94P2

Remarks:
(1) the data refer to an D→2U equilibrium (dimer→2 unfolded monomers)
(2) transition monitored by CD
(3) transition monitored by Trp fluorescence
(4) transition monitored through iodacetamide-modified aminotransferase

Anthranilate Isomerase

N-(5'-Phosphoribosyl)anthranilate isomerase, wild-type

Mutant	pH	T	$\Delta(\Delta G)$	Approach	Remarks	Ref
wild-type	6.5	23	0	GuHCl	(1)	92E1
(Ser212→Cys) crosslinked with Cys27						
	6.5	23	4.2	GuHCl	(1)	92E1

Remark:
(1) linear extrapolation

N-(5'-Phosphoribosyl)anthranilate isomerase, recombinant

pH	T	ΔG	Approach	Remarks	Ref
7.4	25	22.4*	GuHCl	(1,2)	94J2

Remarks:
(1) linear extrapolation to zero denaturant concentration by a method that includes the pre- and postdenaturational baselines for a non-linear regression of the data
(2) thermal transition at $T_{trs} = 51°C$

α_1-Antitrypsin

α_1-Antitrypsin, recombinant, wild-type and mutants

Mutant	pH	T	$\Delta(\Delta G)$	Approach	Remarks	Ref
wild-type	6.5	23	0.0	urea	(1,2)	94K12
Phe51→Cys	6.5	23	12.6	urea	(1–3)	94K12
Phe51→Leu	6.5	25	8.8	urea	(4)	96L2
Thr59→Ala	6.5	25	4.2	urea	(4)	96L2
Thr68→Ala	6.5	25	4.2	urea	(4)	96L2
Ala70→Gly	6.5	25	6.7	urea	(4)	96L2
Met374→Ile	6.5	25	9.6	urea	(4)	96L2
Ser381→Ala	6.5	25	4.2	urea	(4)	96L2
Lys387→Arg	6.5	25	4.2	urea	(4)	96L2
double mutant (Phe51→Leu and Met374→Ile)						
	6.5	25	15.1	urea	(4)	96L2
triple mutant (Thr59→Ala, Thr68→Ala and Ala70→Gly)						
	6.5	25	14.2	urea	(4)	96L2
double mutant (Ser381→Ala and Lys387→Arg)						
	6.5	25	8.8	urea	(4)	96L2

Remarks:
(1) linear extrapolation
(2) buffer: 10 mM potassium phosphate, 50 mM NaCl, 1 mM EDTA, 1 mM β-mercaptoethanol
(3) variant with increased thermostability obtained by random mutagenesis
(4) $\Delta(\Delta G)$ was calculated from $\Delta C_m \times 2.6 \times 4.184$ using the wild-type m value

Apoflavodoxin, see Flavodoxin

Apolipophorin

Apolipophorin, see also Apolipoprotein

Apolipophorin of *Manduca sexta*

pH	T	ΔG	Approach	Remarks	Ref
7.0	25	5.4±0.3	GuHCl	(1,2,3)	93R2
7.0	20	8.8±0.5	GuHCl	(1,4)	93R2

Remarks:
(1) linear extrapolation
(2) T_{trs} in the absence of GuHCl amounts to 52°C
(3) measured by CD
(4) measured by fluorescence

Apolipophorin III from *Locusta migratoria*

Form	pH	T	ΔG	Approach	Remarks	Ref
native	7.0	25	22.5	GuHCl	(1)	94W2
deglycosylated	7.0	25	4.9	GuHCl	(1)	94W2

Remark:
(1) linear extrapolation

Apolipoprotein

Apolipoprotein, see also Apolipophorin

Apolipoprotein A1

pH	T	ΔG	Approach	Remarks	Ref
7.4	25	5.9–7.5	pressure	(1)	85M1
8	23	15.5	GuHCl	(2)	80E
9.2	25	11.3	urea	(3)	76T
9.2	37	10	DSC		76T
9	37	43.9	DSC	(4)	77T

Remarks:
(1) pressure-induced unfolding
(2) denaturant binding model, k = 0.8
(3) transfer model
(4) reconstituted with dimyristoyllecithin

Apolipoprotein CII

pH	T	ΔG	Approach	Remarks	Ref
7.4	25	11.7	GuHCl	(1)	80M1

Remark:
(1) using both linear extrapolation and denaturant binding model

Apolipoprotein from the chicken *Gallus domesticus*, lipid-free apoI

pH	T	ΔG	Approach	Remarks	Ref
7.4	25	7.8	GuHCl	(1,2)	93K8
7.4	20	12.4	GuHCl	(1,3)	93K8
7.4	20	12.0	GuHCl	(1,4)	93K8

Remarks:
(1) linear extrapolation
(2) monitored by CD, the ΔG value is the most accurate one according to Ref. 93K8
(3) monitored by fluorescence, excitation 282 nm
(4) monitored by fluorescence, excitation 300 nm

Apomyoglobin, see Myoglobin

Apoplastocyanin, see Plastocyanin

Arabinose-Binding Protein

L-arabinose-binding protein (*E. coli*)

pH	T	ΔG	Approach	Ref
7.4	25	38.5	DSC	83F2

Arc Repressor

Arc repressor dimer

pH	T	ΔG	Approach	Remarks	Ref
7.3	25	46.0	urea	(1,2)	89B2
7.3	25	46.4	GuHCl	(1,2)	89B2
7.3	25	45.5*	heat	(2,3)	89B2

Remarks:
(1) linear extrapolation based on the equilibrium $N_2 \rightarrow 2U$
(2) measurement in the presence of 100 mM KCl
(3) ΔG was calculated using eq.(5) of Ref. 89B2

P22 Arc repressor

Salt Concentration	pH	T	ΔG	Approach	Remarks	Ref
100 mM	7.5	20	39.8±0.9	urea	(1,2,4)	94M13
100 mM	7.5	20	40.1±1.2	urea	(1,2,5)	94M13
100 mM	7.5	20	41.1	urea	(1,2,6)	94M13
250 mM	7.5	20	44.4±1.5	urea	(1,3,4)	94M13
250 mM	7.5	20	44.3±1.8	urea	(1,3,5)	94M13
250 mM	7.5	20	43.3	urea	(1,3,6)	94M13

Remarks:
(1) linear extrapolation
(2) buffer: 50 mM Tris-HCl, 0.2 mM EDTA, 100 mM KCl
(3) buffer: 50 mM Tris-HCl, 0.2 mM EDTA, 250 mM KCl
(4) equilibrium denaturation, monitored by CD at 234 nm
(5) equilibrium denaturation, monitored by fluorescence
(6) kinetics

P22 Arc repressor, wild-type and mutants

Mutant	pH	T	ΔG	Approach	Remarks	Ref
wild-type	7.5	25	43.6	urea	(1–4)	93M13
Arc-st5	7.5	25	45.3	urea	(1–5)	93M13
Arc-st6	7.5	25	44.2±0.8	urea	(1–4,6)	93M13
Arc-st11	7.5	25	44.2±0.9	urea	(1–4,7)	93M13
(Glu28→Ala-Arc-st11)	7.5	25	42.9	urea	(1–4,7)	93M13
(Arg31→Leu-Arc-st11)	7.5	25	34.1	urea	(1–4,7)	93M13
(Ser32→Ala-Arc-st11)	7.5	25	31.7	urea	(1–4,7)	93M13

Remarks:

(1) linear extrapolation to zero denaturant concentration by a method that includes the pre- and postdenaturational baselines for a non-linear regression of the data

(2) ΔG refers to the Arc dimer at standard state concentration of 1 M

(3) buffer: 50 mM Tris-HCl, 250 mM KCl, 0.2 mM EDTA

(4) measured by fluorescence

(5) Arc-st5: Arc with the C-terminal extension Lys-Asn-Gln-His-Glu-COOH

(6) Arc-st6: Arc with the C-terminal extension His-His-His-His-His-His-COOH, ΔG is a mean of five measurements monitored by fluorescence and CD

(7) Arc-st11: Arc with the C-terminal extension His-His-His-His-His-His-Lys-Asn-Gln-His-Glu-COOH, ΔG is a mean of seven measurements monitored by fluorescence and CD

P22 Arc repressor, wild-type and mutants

Mutant	pH	T	$\Delta(\Delta G)$	Approach	Remarks	Ref
wild-type	7.5	55.6	0.0	heat	(1)	93M13
Arc-st5	7.5	55.6	−0.6	heat	(1,2)	93M13
Arc-st6	7.5	55.6	1.0	heat	(1,3)	93M13
Arc-st11	7.5	55.6	0.6	heat	(1,4)	93M13
(Glu28→Ala-Arc-st11)	7.5	56.9	−1.9	heat	(1,4)	93M13
(Arg31→Leu-Arc-st11)	7.5	56.9	−12.8	heat	(1,4)	93M13
(Ser32→Ala-Arc-st11)	7.5	56.9	−13.8	heat	(1,4)	93M13

Remarks:

(1) buffer: 50 mM Tris-HCl, 250 mM KCl, 0.2 mM EDTA

(2) Arc-st5: Arc with the C-terminal extension Lys-Asn-Gln-His-Glu-COOH

(3) Arc-st6: Arc with the C-terminal extension His-His-His-His-His-His-COOH

(4) Arc-st11: Arc with the C-terminal extension His-His-His-His-His-His-Lys-Asn-Gln-His-Glu-COOH

Arc repressor, alanine mutants

Mutant	pH	T	ΔG	Approach	Remarks	Ref
Arc-st6	7.5	25	45.6	urea	(1)	94M14
Arc-st11	7.5	25	46.0	urea	(2)	94M14
Met1→Ala-st6	7.5	25	45.2	urea	(1)	94M14
Lys2→Ala-st6	7.5	25	46.0	urea	(1)	94M14
Gly3→Ala-st6	7.5	25	46.9	urea	(1)	94M14
Met4→Ala-st6	7.5	25	46.4	urea	(1)	94M14
Ser5→Ala-st6	7.5	25	46.0	urea	(1)	94M14
Lys6→Ala-st6	7.5	25	47.3	urea	(1)	94M14

Arc repressor, alanine mutants (continued)

Mutant	pH	T	ΔG	Approach	Remarks	Ref
Met7→Ala-st6	7.5	25	43.1	urea	(1)	94M14
Pro8→Ala-st6	7.5	25	57.7	urea	(1)	94M14
Gln9→Ala-st6	7.5	25	45.2	urea	(1)	94M14
Phe10→Ala-st6	7.5	25	34.3	urea	(1)	94M14
Asn11→Ala-st6	7.5	25	47.7	urea	(1)	94M14
Leu12→Ala-st11	7.5	25	34.7	urea	(2)	94M14
Arg13→Ala-st6	7.5	25	43.1	urea	(1)	94M14
Trp14→Ala-st11	7.5	25	29.3	urea	(2)	94M14
Pro15→Ala-st11	7.5	25	38.1	urea	(2)	94M14
Arg16→Ala-st6	7.5	25	46.4	urea	(1)	94M14
Glu17→Ala-st6	7.5	25	43.5	urea	(1)	94M14
Val18→Ala-st6	7.5	25	43.5	urea	(1)	94M14
Leu19→Ala-st6	7.5	25	37.7	urea	(1)	94M14
Asp20→Ala-st6	7.5	25	42.3	urea	(1)	94M14
Leu21→Ala-st11	7.5	25	31.8	urea	(2)	94M14
Val22→Ala-st11	7.5	25	<25	urea	(2)	94M14
Arg23→Ala-st11	7.5	25	41.0	urea	(2)	94M14
Lys24→Ala-st11	7.5	25	43.5	urea	(2)	94M14
Val25→Ala-st6	7.5	25	43.9	urea	(1)	94M14
Glu27→Ala-st6	7.5	25	47.3	urea	(1)	94M14
Glu28→Ala-st11	7.5	25	43.9	urea	(2)	94M14
Asn29→Ala-st11	7.5	25	39.3	urea	(2)	94M14
Gly30→Ala-st11	7.5	25	35.6	urea	(2)	94M14
Arg31→Ala-st11	7.5	25	31.8	urea	(2)	94M14
Ser32→Ala-st11	7.5	25	30.1	urea	(2)	94M14
Val33→Ala-st11	7.5	25	37.2	urea	(2)	94M14
Asn34→Ala-st11	7.5	25	46.0	urea	(2)	94M14
Ser35→Ala-st6	7.5	25	46.4	urea	(1)	94M14
Glu36→Ala-st11	7.5	25	<25	urea	(2)	94M14
Ile37→Ala-st11	7.5	25	<25	urea	(2)	94M14
Tyr38→Ala-st11	7.5	25	30.1	urea	(2)	94M14
Gln39→Ala-st11	7.5	25	46.4	urea	(2)	94M14
Arg40→Ala-st11	7.5	25	26.8	urea	(2)	94M14
Val41→Ala-st11	7.5	25	<25	urea	(2)	94M14
Met42→Ala-st11	7.5	25	31.0	urea	(2)	94M14
Glu43→Ala-st6	7.5	25	44.4	urea	(1)	94M14
Ser44→Ala-st11	7.5	25	39.3	urea	(2)	94M14
Phe45→Ala-st11	7.5	25	<25	urea	(2)	94M14
Lys46→Ala-st11	7.5	25	46.0	urea	(2)	94M14
Lys47→Ala-st11	7.5	25	38.5	urea	(2)	94M14
Glu48→Ala-st11	7.5	25	36.0	urea	(2)	94M14
Gly49→Ala-st11	7.5	25	37.7	urea	(2)	94M14
Arg50→Ala-st11	7.5	25	38.1	urea	(2)	94M14
Ile51→Ala-st11	7.5	25	38.1	urea	(2)	94M14
Gly52→Ala-st11	7.5	25	46.9	urea	(2)	94M14

Remarks:
General remark – ΔG was determined by the linear extrapolation method
Buffer used: 50 mM Tris, 250 mM KCl, 0.2 mM EDTA, pH 7.5
(1) st6 – carboxy-terminal extension $(His)_6$
(2) st11 – carboxy-terminal extension $(His)_6$-Lys-Asn-Gln-His-Glu

Arc repressor, wild-type and mutants

Mutant	pH	T	ΔG	Approach	Remarks	Ref
wild-type	7.5	25	43.1±1.3	urea	(1–5)	95W1
triple mutant (Arg31→Met, Glu36→Tyr and Arg40→Leu)						
	7.5	25	59.4±0.8	urea	(1–5)	95W1
triple mutant (Arg31→Met, Glu36→Trp and Arg40→Leu)						
	7.5	25	61.9±0.8	urea	(1–5)	95W1
triple mutant (Arg31→Val, Glu36→Tyr and Arg40→Ile)						
	7.5	25	57.7±0.4	urea	(1–5)	95W1
triple mutant (Arg31→Ile, Glu36→Tyr and Arg40→Val)						
	7.5	25	52.3±1.7	urea	(1–5)	95W1
triple mutant (Arg31→Gln, Glu36→Tyr and Arg40→Val)						
	7.5	25	51.9±0.8	urea	(1–5)	95W1
triple mutant (Arg31→Leu, Glu36→Met and Arg40→Ile)						
	7.5	25	49.8±0.8	urea	(1–5)	95W1
Ala substituted mutants (stabilizing buffer system):						
wild-type	7.5	25	51.0±0.8	urea	(1–4,6)	95W1
Arg31→Ala	7.5	25	42.7±0.4	urea	(1–4,6)	95W1
Glu36→Ala	7.5	25	25.9±0.4	urea	(1–4,6)	95W1
Arg40→Ala	7.5	25	38.5±1.3	urea	(1–4,6)	95W1
double mutant (Arg31→Ala and Glu36→Ala)						
	7.5	25	24.7±0.4	urea	(1–4,6)	95W1
double mutant (Arg31→Ala and Arg40→Ala)						
	7.5	25	31.0±0.8	urea	(1–4,6)	95W1
double mutant (Glu36→Ala and Arg40→Ala)						
	7.5	25	33.1±0.8	urea	(1–4,6)	95W1
triple mutant (Arg31→Ala, Glu36→Ala and Arg40→Ala)						
	7.5	25	35.1±1.3	urea	(1–4,6)	95W1

Remarks:
(1) wild-type contains Arg31, Glu36 and Arg40
(2) linear extrapolation to zero denaturant concentration by a method that includes the pre- and postdenaturational baselines for a non-linear regression of the data
(3) ΔG is the Gibbs energy change of unfolding and dimer dissociation at a 1 M standard concentration
(4) ΔG refers to the dimer
(5) buffer: 50 mM Tris-HCl, 200 mM KCl, pH 7.5
(6) buffer: 10 mM Tris, 50 mM KCl, 0.5 M Na_2SO_4, pH 7.5

Arc repressor, wild-type and fluorescence labeled variants

Variant	pH	T	ΔG	Approach	Remarks	Ref
wild-type	7.5	25	43.5	urea	(1,2)	96J3
RC23/AF	7.5	25	45.2	urea	(1–4)	96J3
RC23/AEDA NS	7.5	25	43.9	urea	(1–3,5)	96J3
MYL	7.5	25	54.0	urea	(1,2,6)	96J3
MYL-RC23/AF	7.5	25	54.4	urea	(1,2,4,6)	96J3
MYL-RC23/AEDANS	7.5	25	52.7	urea	(1,2,5,6)	96J3

Remarks:
(1) non-linear fit procedure, ΔG was calculated for a standard state of 1M, equilibrium unfolding compared with kinetics shows that the logarithm of the rate constant for dissociation/unfolding of the Arc dimer varies in non-linear fashion with denaturant concentration
(2) measured in 50 mM Tris, 50 mM KCl, pH 7.5
(3) recombinant Arc repressor containing a C-terminal sequence $(His)_6$-Lys-Asn-Gln-His-Glu
(4) AF = modified by [5-(iodoacetamido)fluorescein]
(5) AEDANS = modified by (5-[[2-[(iodoacetyl)amino]-ethyl]amino]naphthalene-1-sulfonic acid)
(6) MYL-RC23 variant encoding Met31, Tyr36, and Leu40

Arc repressor, MYL variant

Mutant	pH	T	ΔG	Approach	Remarks	Ref
MYL	7.5	25	59.7	GuHCl	(1–4)	96H3
Asn29→Ala	7.5	25	62.5	GuHCl	(1–5)	96H3
Ser44→Ala	7.5	25	63.8	GuHCl	(1–5)	96H3
Glu48→Ala	7.5	25	60.3	GuHCl	(1–5)	96H3
triple mutant (Asn29→Ala, Ser44→Ala and Glu48→Ala)	7.5	25	65.3	GuHCl	(1–5)	96H3

Remarks:
(1) in all mutants, called MYL, hydrophobic interactions between Met31, Tyr36, and Leu40 replace the wild-type salt-bridge interactions between Arg31, Glu36, and Arg40
(2) linear extrapolation, the ΔG values refer to the 1 M standard state
(3) buffer: 50 mM Tris-HCl and 50 mM KCl
(4) heat denaturation gives T_{trs} = 56.8, 58.4, 59.3, 64.3, and 66.0°C respectively, and ΔCp = –4.5±0.3 kJ/mol/K per mole of dimer
(5) variant MYL and the indicated amino acid replacement(s)

Arc repressor, single-chain variant at various concentrations

Concentration	pH	T	ΔG	Approach	Remarks	Ref
0.5 mM	7.5	25	26.4±1.7	urea	(1–4)	96R2
5.0 mM	7.5	25	26.8±2.1	urea	(1–4)	96R2
0.5 mM	7.5	25	25.9±1.3	urea	(1–4)	96R2

Remarks:
(1) two Arc monomers connected by a glycine-rich linker joining the C-terminus of one subunit to the N-terminus of an identical subunit
(2) linear extrapolation, two-state transition
(3) buffer: 50 mM Tris/HCl, 250 mM KCl, 0.1 mM EDTA
(4) ΔG for Arc-stl1 amounts to 43.1±2.1 kJ/mol at standard state concentration, obtained by a two-state transition between native dimer and two unfolded monomers

Asparaginase

Asparaginase II

Mutant	pH	T	ΔG	Approach	Remarks	Ref
wild-type	7.4	25	48	urea	(1,2)	92D1
Thr12→Ala	7.4	25	48	urea	(1,2)	92D1
Thr12→Ser	7.4	25	47	urea	(1,2)	92D1
Thr119→Ala	7.4	25	48	urea	(1,2)	92D1
Ser122→Ala	7.4	25	52	urea	(1,2)	92D1

Remarks:
(1) linear extrapolation
(2) the data were taken from Fig. 3 in Ref. 92D1

Aspartate Aminotransferase

Aspartate aminotransferase

Form	pH	T	ΔG	Approach	Remarks	Ref
apoenzyme	7.1	25	251±13	urea	(1)	73I
pyridoxal phosphate	7.1	25	331±8	urea	(1)	73I
pyridoxamine phosphate	7.1	25	243±16	urea	(1)	73I

Remark:
(1) transfer model

Aspartate aminotransferase, multiple subform α

Form	pH	T	ΔG	Approach	Ref
apoenzyme	8.2	30	126	DSC	81R
pyridoxal	6	30	226	DSC	81R
pyridoxal phosphate	8.2	30	238	DSC	81R
	8.2	30	222	DSC	81R
pyridoxamine	8.2	30	159	DSC	81R

Aspartate aminotransferase from *E. coli*, wild-type and mutants

Mutant	pH	T	ΔG	Δ(ΔG)	Approach	Remarks	Ref
wild-type	7.5	25	41.7±1.9	0.0	urea	(1)	92G1
Cys82→Ala	7.5	25	36.3±0.9	−5.4±2.1	urea	(1)	92G1
Cys191→Ala	7.5	25	32.4±1.0	−9.2±2.1	urea	(1)	92G1
Cys191→Ser	7.5	25	34.0±1.3	−7.5±2.1	urea	(1)	92G1
Cys192→Ala	7.5	25	34.8±0.9	−6.7±2.1	urea	(1)	92G1
Cys270→Ala	7.5	25	31.4±1.2	−10.5±2.1	urea	(1)	92G1
Cys401→Ala	7.5	25	55.2±3.1	13.4±3.8	urea	(1)	92G1
multiple	7.5	25	37.3±1.3	−4.2±2.1	urea	(1,2)	92G1

Remarks:
(1) linear extrapolation, the transition was monitored by enzyme activity
(2) multiple mutant (Cys82→Ala, Cys191→Ala, Cys192→Ala, Cys270→Ala, Cys401→Ala)

Aspartate Transcarbamoylase

Aspartate transcarbamoylase from *E. coli*
difference ΔG values specified for the catalytic (C) and catalytic-regulatory (CR) subunit at the respective transition temperature

Mutant	pH	Δ(ΔG)(C)	Δ(ΔG)(CR)	Approach	Ref
wild-type	7	0.0	0.0	DSC	90B10
Glu86→Gln	7	21.8	−15.5	DSC	90B10
Lys164→Ile	7	22.2	−6.3	DSC	90B10
Tyr165→Phe	7	−0.8	10.0	DSC	90B10
Glu239→Ala	7	0.8	−3.8	DSC	90B10
Glu239→Gln	7	−0.2	−8.4	DSC	90B10
Tyr240→Phe	7	−3.3	2.1	DSC	90B10
Asp271→Ser	7	16.3	−2.1	DSC	90B10

Aspartate transcarbamylase from *E. coli*

Transition	pH	T	ΔG	Approach	Remarks	Ref
dissociat.	8.3	7.5	29–42	urea	(1)	94B8
unfolding	8.3	7.5	8	urea	(1)	94B8

Remark:
(1) various models that account for dissociation and unfolding are considered in Ref. 94B8

30   Table 1. Gibbs Energy Change – Molar Values

Azurin

Azurin from *Pseudomonas aeruginosa*

pH	T	ΔG	Approach	Remarks	Ref
7.03	25	63	DSC	(1)	95L1

Remark:
(1) the data refer to a fit that accounts for the kinetics of irreversible denaturation

Bacteriorhodopsin

Bacteriorhodopsin

pH	T	ΔG	Approach	Remarks	Ref
7	25	150	urea	(1,2)	89M1
7	25	230	urea	(1,3)	89M1

Remarks:
(1) linear extrapolation
(2) the Gibbs energy change specifies the molecular swelling transition
(3) the Gibbs energy change specifies the unfolding transition

Barnase = ribonuclease from *Bacillus amyloliquefaciens*

The data entries are arranged as follows:

a) wild-type: various approaches, dependence on pH and ionic strength,
b) wild-type and mutants: various approaches,
c) wild-type and mutants: salt dependence,
d) single and multiple amino acid replacements,
e) difference energy for intermediate, transition state, and folding,
f) insertion mutants, and step-wise mutation of barnase to binase.

a) Wild-type: various approaches, dependence on pH and ionic strength

Barnase, use of different optical properties for the determination of ΔG

pH	T	ΔG	Approach	Remarks	Ref
6.3	25	38.9±3.3	urea	(1,2)	93V3
6.3	25	38.2±1.9	urea	(1,3)	93V3
6.3	25	38.1±1.3	urea	(1,4)	93V3
6.3	25	38.1±1.5	urea	(1,5)	93V3
6.3	25	38.8±1.7	urea	(1,6)	93V3
6.3	25	36.8±0.6	urea	(1,7)	93V3

Remarks:
(1) linear extrapolation to zero denaturant concentration by a method that includes the pre- and postdenaturational baselines for a non-linear regression of the data, buffer: 50 mM MES
(2) monitored by CD at 230 nm
(3) monitored by CD at 260 nm
(4) monitored by CD at 266 nm
(5) monitored by CD at 278 nm
(6) monitored by CD at 286 nm
(7) monitored by fluorescence

Barnase, wild-type, comparison of ΔG from thermal- and urea-mediated denaturation

pH	T	ΔG	Approach	Ref
4.4	25	29.9±0.2	urea, linear extrapol.	94M9
4.4	25	36.4±0.4	DSC	94M9
6.3	25	37.2±0.4	urea, linear extrapol.	94M9
6.3	25	43.9±0.4	DSC	94M9

Barnase, use of various approaches to check the validity of the linear extrapolation method for denaturant-induced unfolding

pH	T	ΔG	Approach	Remarks	Ref
6.3	25.0	36.8	urea	(1)	95J1
6.3	25.0	43.5	DSC		95J1
6.3	25.0	43.93±0.33	urea	(2)	95J1
6.3	25.0	43.93±0.33	urea	(3)	95J1

Remarks:
(1) linear extrapolation to zero denaturant concentration by a method that includes the pre- and postdenaturational baselines for a non-linear regression of the data
(2) denaturant binding model with $\Delta n = 53\pm6$ and $K = 0.088\pm0.012$ mol^{-1}
(3) data obtained by DSC and CD measurements in the presence of urea, treated with the following expression that includes a curvature into the extrapolation method:
$$\Delta G = (43.93\pm0.33) - [(11.09\pm0.21) \cdot c_{urea}] + [(0.33\pm0.04) \cdot c^2_{urea}]$$

Barnase, dependence of ΔG on pH

pH	T	ΔG	Approach	Remarks	Ref
2.0	25	−1.9	DSC	(1)	94M7
2.5	25	7.6	DSC	(1)	94M7
3.0	25	21.7	DSC	(1)	94M7
3.5	25	31.7	DSC	(1)	94M7
4.0	25	35.6	DSC	(1)	94M7
4.5	25	37.9	DSC	(1)	94M7
5.0	25	38.7	DSC	(1)	94M7

Remark:
(1) ΔG was calculated with temperature dependent
$\Delta C_p = 0.6+0.108\times T-0.00028\times T^2$ (T in K) and $\Delta H = 344$ kJ/mol at 298.2 K

Barnase, dependence of ΔG on pH and ionic strength

pH	T	ΔG	Approach	Remarks	Ref
7.00	25	37.4	urea	(1,2,5)	92P1
7.00	25	35.5	urea	(1,3,5)	92P1
2.50	25	7.4	urea	(1,3,9)	92P1
2.50	25	6.4	urea	(1,3,4,9)	92P1
2.76	25	18.7	urea	(1,3,8)	92P1
2.76	25	16.4	urea	(1,3,4,8)	92P1
2.82	25	15.8	urea	(1,3,7)	92P1
2.82	25	14.1	urea	(1,3,4,7)	92P1
2.87	25	16.2	urea	(1,2,6)	92P1
2.87	25	15.9	urea	(1,3,6)	92P1
2.87	25	13.6	urea	(1,3,4,6)	92P1
2.98	25	19.6	urea	(1,3,9)	92P1

Barnase, dependence of ΔG on pH and ionic strength (continued)

pH	T	ΔG	Approach	Remarks	Ref
2.98	25	18.4	urea	(1,3,4,9)	92P1
3.17	25	23.6	urea	(1,2,6)	92P1
3.17	25	23.6	urea	(1,3,6)	92P1
3.17	25	20.8	urea	(1,3,4,6)	92P1

Remarks:
(1) linear extrapolation
(2) monitored by fluorescence: 290 nm excitation, 315 nm emission
(3) monitored by fluorescence: 279 nm excitation, 320 nm emission
(4) including correction for the variation in pH across the transition region
(5) 30 mM MOPS buffer
(6) 30 mM diglycine buffer
(7) 30 mM diglycine buffer with 0.1 M NaCl
(8) 30 mM diglycine buffer with 0.5 M NaCl
(9) 100 mM diglycine buffer

Barnase

Ionic strength	pH	T	ΔG	Approach	Remarks	Ref
50 mM	6.0	25	42	heat	(1)	94O1
600 mM	6.0	25	46	heat	(1)	94O1

Remark:
1) for the pH dependence of ΔG, see Fig. 4 in Ref. 94O1

Barnase, wild-type, apparent free energies of exchange of amide protons at 0 and 1.2 M GuHCl AT p^2H 6.8

Residue	ΔG(0 M)	ΔG(1.2M)	Approach	Ref
10	26.8	20.5	H/D, NMR	96C13
11	35.6	22.6	H/D, NMR	96C13
12	30.1	20.9	H/D, NMR	96C13
13	40.2	25.9	H/D, NMR	96C13
14		21.3	H/D, NMR	96C13
15	41.0	22.2	H/D, NMR	96C13
16	30.5	25.1	H/D, NMR	96C13
17	33.1	22.6	H/D, NMR	96C13
19	50.6	28.9	H/D, NMR	96C13
25		18.8	H/D, NMR	96C13
26	33.1	22.2	H/D, NMR	96C13
30		23.4	H/D, NMR	96C13
31	35.6	24.3	H/D, NMR	96C13
33	35.6	20.1	H/D, NMR	96C13
35	34.3	22.6	H/D, NMR	96C13
36	24.3	17.6	H/D, NMR	96C13
44	25.9	20.1	H/D, NMR	96C13
45	33.1	24.3	H/D, NMR	96C13
46	43.5	23.4	H/D, NMR	96C13
49	41.8	24.3	H/D, NMR	96C13
50	46.4	26.4	H/D, NMR	96C13
51	39.7	22.2	H/D, NMR	96C13
52	46.4	23.8	H/D, NMR	96C13
53	43.1	26.4	H/D, NMR	96C13
56	40.2	20.9	H/D, NMR	96C13
71	31.0	22.2	H/D, NMR	96C13

Barnase, wild-type, apparent free energies of exchange of amide protons at 0 and 1.2 M GuHCl AT p^2H 6.8 (continued)

Residue	$\Delta G(0\ M)$	$\Delta G(1.2M)$	Approach	Ref
72		22.6	H/D, NMR	96C13
73		21.8	H/D, NMR	96C13
74		22.6	H/D, NMR	96C13
76		20.9	H/D, NMR	96C13
87	41.8	27.6	H/D, NMR	96C13
88	42.7	20.5	H/D, NMR	96C13
89		18.4	H/D, NMR	96C13
89		18.4	H/D, NMR	96C13
90		20.5	H/D, NMR	96C13
91	46.9	25.9	H/D, NMR	96C13
94	32.2	24.3	H/D, NMR	96C13
95	34.7	20.1	H/D, NMR	96C13
97		20.9	H/D, NMR	96C13
98	46.0	23.4	H/D, NMR	96C13
99	42.7	23.8	H/D, NMR	96C13
107	33.1	24.7	H/D, NMR	96C13

b) Wild-type and mutants: various approaches

Barnase, wild-type and mutants ΔG values at standard temperature obtained by various approaches

Mutant	pH	T	ΔG	Approach	Remarks	Ref
wild-type	6.3	25	39.4	urea	(1)	89K3
wild-type	6.3	25	37.2	GuHCl	(1)	89K3
wild-type	6.3	25	39.7	heat	(2)	89K3
wild-type	6.3	25.0	41.6±2.6	urea	(1,3)	90S3
Leu14→Ala	6.3	25	20	urea	(1)	89K3
Leu14→Ala	6.3	25	22.2	heat	(2)	89K3
Ile88→Ala	6.3	25	22.3	urea	(1)	89K3
Ile88→Ala	6.3	25	22.2	heat	(2)	89K3
Ile88→Val	6.3	25	33.3	urea	(1)	89K3
Ile88→Val	6.3	25	33.9	heat	(2)	89K3
Ile96→Ala	6.3	25	26.3	urea	(1)	89K3
Ile96→Ala	6.3	25	25.8	GuHCl	(1)	89K3
Ile96→Ala	6.3	25	25.9	heat	(2)	89K3
Ile96→Val	6.3	25	36.3	urea	(1)	89K3
Ile96→Val	6.3	25	34.3	GuHCl	(1)	89K3
Ile96→Val	6.3	25	36	heat	(2)	89K3

Remarks:
(1) linear extrapolation
(2) van't Hoff treatment
(3) mean value of 9 independent measurements documented in Ref. 90S3

Barnase, wild-type and mutants

Mutant	pH	T	ΔG	Approach	Remarks	Ref
wild-type	6.3	25	37.2±1.1	urea	(1,2)	95M9
His18→Ala	6.3	25	28.2±0.7	urea	(1,2)	95M9
His18→Gly	6.3	25	33.5±1.4	urea	(1,2)	95M9
Lys27→Ala	6.3	25	39.4±2.5	urea	(1,2)	95M9
Glu29→Ala	6.3	25	31.5±1.7	urea	(1,2)	95M9
Ala32→Gly	6.3	25	33.4±1.0	urea	(1,2)	95M9

Remarks:
(1) linear extrapolation to zero denaturant concentration by a method that includes the pre- and postdenaturational baselines for a non-linear regression of the data
(2) ΔG was calculated from the data given in Tab. 1, Ref. 95M9

Barnase, wild-type and mutants $\Delta(\Delta G)$ values obtained by heat denaturation. $\Delta(\Delta G)$ refers to 48°C

Mutant	pH	T_{trs}	ΔT	Δ(ΔG)	Approach	Remarks	Ref
wild-type	6.3	53.9	0.0	0.0	heat	(1)	89K3
Leu14→Ala	6.3	42	−11.9	−17.6	heat	(2)	89K3
Ile88→Ala	6.3	42.7	−11.2	−16.3	heat	(2)	89K3
Ile88→Val	6.3	51	−2.9	−4.6	heat	(2)	89K3
Ile96→Ala	6.3	44.9	−9	−13.4	heat	(2)	89K3
Ile96→Val	6.3	51.5	−2.4	−3.8	heat	(2)	89K3

Remarks:
(1) van't Hoff treatment
(2) van't Hoff treatment, $\Delta(\Delta G)$ refers to the wild-type at 48°C

Barnase, wild-type and mutants $\Delta(\Delta G)$ values obtained by urea-induced unfolding at 25°C

Mutant	pH	T	Δ(ΔG)	Approach	Remarks	Ref
wild-type	6.3	25	0.0	urea	(1)	89M4
Thr6→Gly	6.3	25	−5.6	urea	(1)	89M4
Thr6→Ala	6.3	25	−9.3	urea	(1)	89M4
Leu14→Ala	6.3	25	−20.1	urea	(1)	89M4
Thr16→Ser	6.3	25	−7.8	urea	(1)	89M4
His18→Gln	6.3	25	−6.7	urea	(1)	89M4
Thr26→Gly	6.3	25	−6.6	urea	(1)	89M4
Thr26→Ala	6.3	25	−9	urea	(1)	89M4
Tyr78→Phe	6.3	25	−6.3	urea	(1)	89M4
Ile88→Val	6.3	25	−6.2	urea	(1)	89M4
Ile88→Ala	6.3	25	−18.7	urea	(1)	89M4
Ile96→Val	6.3	25	−4.1	urea	(1)	89M4
Ile96→Ala	6.3	25	−14.7	urea	(1)	89M4

Remark:
(1) linear extrapolation

Barnase, wild-type and mutants

Mutant	pH	T	ΔG	Approach	Ref
wild-type	4.40	25	36.4±0.5	DSC	94M9
Asp8→Ala	4.40	25	33.3	DSC	94M9
Asp12→Gly	4.40	25	33.0	DSC	94M9
Asp22→Met	4.40	25	35.1	DSC	94M9
Thr16→Arg	4.40	25	37.1	DSC	94M9
Thr26→Gly	4.40	25	27.5	DSC	94M9
Ile51→Val	4.40	25	28.6	DSC	94M9
Asp54→Asn	4.40	25	25.9	DSC	94M9
Asp54→Asn	4.40	25	25.1	DSC	94M9
Arg69→Met	4.40	25	24.3	DSC	94M9
Arg83→Gln	4.40	25	29.6	DSC	94M9
Ile88→Val	4.40	25	30.2	DSC	94M9
Ser91→Ala	4.40	25	31.5	DSC	94M9
Ser92→Ala	4.40	25	24.4	DSC	94M9
Ile96→Val	4.40	25	31.4	DSC	94M9
Ile96→Val	4.40	25	31.5	DSC	94M9
double mutant (Thr16→Ser and Glu61→Gly)					
	4.40	25	30.1	DSC	94M9

Barnase, wild-type and mutants

Mutant	pH	T	ΔG	Approach	Remarks	Ref
wild-type	6.3	25.0	42.7±0.8	heat	(1)	95O1
Asp54→Ala	6.3	25.0	28.9±0.8	urea	(1–3)	95O1
Asp54→Asn	6.3	25.0	31.8±0.8	urea	(1–3)	95O1
Glu73→Ala	6.3	25.0	32.2±0.8	urea	(1,2,4)	95O1
Asp75→Asn	6.3	25.0	21.8±0.8	urea	(1,2,5)	95O1

Remarks:
(1) measured at pH 6.3, buffer conc. 50 mM
(2) ΔG was obtained subtracting $\Delta(\Delta G)$ of the mutant obtained by urea denaturation from the above value for the wild-type obtained by thermal denaturation
(3) based on $\Delta(\Delta G)$ from Ref. 92S8
(4) based on $\Delta(\Delta G)$ from Ref. 92M4
(5) based on $\Delta(\Delta G)$ from Ref. 93T5

Barnase, wild-type and mutants

Mutant	pH	T	$\Delta(\Delta G)$	Approach	Remarks	Ref
wild-type	6.3	25	0.0	urea	(1,2)	92M4
Lys27→Ala	6.3	25	1.5	urea	(1,2)	92M4
Asp54→Asn	6.3	25	−11.2	urea	(1,2)	92M4
Ser57→Ala	6.3	25	0.6	urea	(1,2)	92M4
Asn58→Asp	6.3	25	1.7	urea	(1,2)	92M4
Arg59→Ala	6.3	25	2.7	urea	(1,2)	92M4
Glu73→Ala	6.3	25	−9.6	urea	(1,2)	92M4
His102→Ala	6.3	25	−1.0	urea	(1,2)	92M4
Tyr103→Phe	6.3	25	0.0	urea	(1,2)	92M4
Gln104-Ala	6.3	25	−0.7	urea	(1,2)	92M4

Remarks:
(1) urea denaturation monitored by fluorescence, for details, see Ref. 89K3
(2) buffer: 50 mM MES, pH 6.3

Barnase, wild-type and mutants

Mutant	pH	T	ΔG	Δ(ΔG)	Approach	Remarks	Ref
wild-type	6.3	25	41*	0.0	urea	(1,2)	88S1
wild-type	7.5	25	37*	0.0	urea	(1,2)	88S1
His18→Gln	6.3	25	–6.7	pH	(3)	88S1	

Remarks:
(1) linear extrapolation
(2) the data were taken from Fig. 4a in Ref. 88S1
(3) NMR, pH titration

Barnase, wild-type and mutants

Mutant	pH	T	ΔG	Approach	Remarks	Ref
wild-type	6.3	25	43.1	heat	(1–3)	96Z
wild-type in ^2H_2O	6.9	25	44.8	heat	(1,2,4)	96Z
Ser91→Ala	6.3	25	36.0	heat	(1–3)	96Z
Ser91→Ala	6.3	25	35.6	DSC	(5)	96Z

Remarks:
(1) the data refer to experiments performed in the presence and absence of GroEL or SecB, see also other thermodynamic data
(2) van't Hoff treatment, transition monitored by CD at 280 nm
(3) measured in 50 mM MES (pH 6.3) at 10 µM barnase
(4) measured in 20 mM imidazole buffer, ^2H_2O, p^2H 6.9
(5) measured at 14 µM barnase

Barnase, pseudo wild-type and mutants (see remark 4)

Mutant	pH	T	ΔG	Approach	Remarks	Ref
wt*	6.3	37	22.2	heat	(1–4)	96K10
Gln15→Ile	6.3	37	22.6	heat	(1–3)	96K10
His18→Gln	6.3	37	19.2	heat	(1–3)	96K10
Asn58→Ala	6.3	37	10.0	heat	(1–3)	96K10
double mutant (Gln15→Ile and Lys108→Arg)	6.3	37	35.6	heat	(1–3)	96K10

Remarks:
(1) vant't Hoff treatment
(2) ΔG was calculated using ΔCp = 5.9 kJ/mol/K from Ref. 94G6
(3) buffer: 50 mM MES pH 6.3
(4) pseudo wild-type (wt*) containing Val36→Met and His102→Ala

c) Wild-type and mutants: salt dependence

Barnase, wild-type, at different salt concentrations

Salt Concentration	pH	T	ΔG	Approach	Remarks	Ref
0	6.3	25.0	41.9	urea	(1,2)	90S3
0	6.3	25.0	42.5	urea	(1,3)	90S3
50 mM NaCl	6.3	25.0	43.4	urea	(1,3)	90S3
500 mM NaCl	6.3	25.0	45.5	urea	(1,3)	90S3
900 mM NaCl	6.3	25.0	54.0	urea	(1,3)	90S3

Remarks:
(1) linear extrapolation
(2) 10 mM MES buffer
(3) 50 mM MES buffer

Barnase mutant (Asp12→Ala), at different salt concentrations

Salt Concentration	pH	T	ΔG	Approach	Remarks	Ref
0	6.3	25.0	42.3	urea	(1,2)	90S3
0	6.3	25.0	39.2	urea	(1,3)	90S3
50 mM NaCl	6.3	25.0	44.5	urea	(1,3)	90S3
500 mM NaCl	6.3	25.0	45.0	urea	(1,3)	90S3
900 mM NaCl	6.3	25.0	44.6	urea	(1,3)	90S3

Remarks:
(1) linear extrapolation
(2) 10 mM MES buffer
(3) 50 mM MES buffer

Barnase mutant (Thr16→Arg), at different salt concentrations

Salt Concentration	pH	T	ΔG	Approach	Remarks	Ref
0	6.3	25.0	38.6	urea	(1,2)	90S3
0	6.3	25.0	44.5	urea	(1,3)	90S3
50 mM NaCl	6.3	25.0	42.7	urea	(1,3)	90S3
500 mM NaCl	6.3	25.0	44.1	urea	(1,3)	90S3
900 mM NaCl	6.3	25.0	54.0	urea	(1,3)	90S3

Remarks:
(1) linear extrapolation
(2) 10 mM MES buffer
(3) 50 mM MES buffer

Barnase, double mutant (Asp12→Ala and Thr16→Arg), at different salt concentrations

Salt Concentration	pH	T	ΔG	Approach	Remarks	Ref
0	6.3	25.0	41.3	urea	(1,2)	90S3
0	6.3	25.0	41.8	urea	(1,3)	90S3
50 mM NaCl	6.3	25.0	43.5	urea	(1,3)	90S3
500 mM NaCl	6.3	25.0	46.6	urea	(1,3)	90S3
900 mM NaCl	6.3	25.0	45.6	urea	(1,3)	90S3

Remarks:
(1) linear extrapolation
(2) 10 mM MES buffer
(3) 50 mM MES buffer

Barnase, wild-type and mutants concerning (Asp8, Asp12, and Arg110), at different salt concentrations

Salt Concentration	pH	T	ΔG	Approach	Remarks	Ref
Wild-type						
0 M	6.3	25.0±0.1	41.5	urea	(1,2)	90H1
0.2 M	6.3	25.0±0.1	45.1	urea	(1,2)	90H1
0.5 M	6.3	25.0±0.1	46.8	urea	(1,2)	90H1
Mutant (Asp8→Ala; Asp12 and Arg110 unchanged):						
0 M	6.3	25.0±0.1	37.4	urea	(1,2)	90H1
0.2 M	6.3	25.0±0.1	42.4	urea	(1,2)	90H1
0.5 M	6.3	25.0±0.1	45.5	urea	(1,2)	90H1
Mutant (Asp12→Ala; Asp8 and Arg110 unchanged):						
0 M	6.3	25.0±0.1	40.1	urea	(1,2)	90H1
0.2 M	6.3	25.0±0.1	43.7	urea	(1,2)	90H1
0.5 M	6.3	25.0±0.1	46.3	urea	(1,2)	90H1
Mutant (Asp110→Ala; Asp8 and Asp12 unchanged):						
0 M	6.3	25.0±0.1	39.6	urea	(1,2)	90H1
0.2 M	6.3	25.0±0.1	44.1	urea	(1,2)	90H1
0.5 M	6.3	25.0±0.1	46.9	urea	(1,2)	90H1
Double mutant (Asp8→Ala and Asp12→Ala; Arg110 unchanged):						
0 M	6.3	25.0±0.1	37.8	urea	(1,2)	90H1
0.2 M	6.3	25.0±0.1	43.5	urea	(1,2)	90H1
0.5 M	6.3	25.0±0.1	46.6	urea	(1,2)	90H1
Double mutant (Asp8→Ala and Arg110→Ala; Asp12 unchanged):						
0 M	6.3	25.0±0.1	39.6	urea	(1,2)	90H1
0.2 M	6.3	25.0±0.1	43.6	urea	(1,2)	90H1
0.5 M	6.3	25.0±0.1	47.0	urea	(1,2)	90H1
Double mutant (Asp12→Ala and Arg110→Ala; Asp8 unchanged):						
0 M	6.3	25.0±0.1	43.4	urea	(1,2)	90H1
0.2 M	6.3	25.0±0.1	46.0	urea	(1,2)	90H1
0.5 M	6.3	25.0±0.1	49.2	urea	(1,2)	90H1
Triple mutant (Asp8→Ala, Asp12→Ala and Arg110→Ala):						
0 M	6.3	25.0±0.1	42.0	urea	(1,2)	90H1
0.2 M	6.3	25.0±0.1	46.1	urea	(1,2)	90H1
0.5 M	6.3	25.0±0.1	49.3	urea	(1,2)	90H1

Remarks:
(1) linear extrapolation to zero denaturant concentration by a method that includes the pre- and postdenaturational baselines for a non-linear regression of the data
(2) measurement in 50 mM MES buffer containing additionally the indicated NaCl concentration

Barnase, wild-type and mutants, dependence of ΔG on ionic strength

Mutant	pH	T	ΔG	Approach	Remarks	Ref
wild-type	6.3	25	35.5	urea	(1,2a)	91S1
wild-type	6.3	25	37.5	urea	(1,2b)	91S1
wild-type	6.3	25	42.9	urea	(1,2c)	91S1
wild-type	6.3	25	41.5	urea	(1,2d)	91S1
wild-type	6.3	25	42.8	urea	(1,2e)	91S1
Ser28→Glu	6.3	25	42.7	urea	(1,2a)	91S1
Ser28→Glu	6.3	25	39.5	urea	(1,2b)	91S1
Ser28→Glu	6.3	25	42.4	urea	(1,2c)	91S1
Ser28→Glu	6.3	25	42.1	urea	(1,2d)	91S1
Ser28→Glu	6.3	25	44.9	urea	(1,2e)	91S1
Ala32→Lys	6.3	25	38.6	urea	(1,2a)	91S1
Ala32→Lys	6.3	25	37.0	urea	(1,2b)	91S1
Ala32→Lys	6.3	25	44.7	urea	(1,2c)	91S1
Ala32→Lys	6.3	25	42.7	urea	(1,2d)	91S1
Ala32→Lys	6.3	25	43.6	urea	(1,2e)	91S1
double mutant (Ser28→Glu and Ala32→Lys)						
	6.3	25	42.8	urea	(1,2a)	91S1
	6.3	25	37.4	urea	(1,2b)	91S1
	6.3	25	44.5	urea	(1,2c)	91S1
	6.3	25	41.5	urea	(1,2d)	91S1
	6.3	25	44.8	urea	(1,2e)	91S1

Remarks:
(1) linear extrapolation
(2) buffer composition:
(2a) 7 mM MES
(2b) 10 mM MES
(2c) 50 mM MES
(2d) 50 mM MES and 250 mM NaCl
(2e) 50 mM MES and 556 mM NaCl

Barnase, mutants, in 30 mM buffer

Mutant	pH	T	ΔG	Approach	Remarks	Ref
wild-type	5.9	25	34.0	urea	(1)	92S1
wild-type	8.9	25	33.7	urea	(1)	92S1
Thr6→His	5.9	25	22.9	urea	(1)	92S1
Thr6→His	8.9	25	27.2	urea	(1)	92S1
Gly34→His	4.4	25	23.1	urea	(1)	92S1
Gly34→His	8.9	25	20.5	urea	(1)	92S1
Asp54→Asn	5.9	25	22.9	urea	(1)	92S1
Asp54→Asn	8.9	25	24.0	urea	(1)	92S1
Trp94→Leu	5.9	25	32.1	urea	(1)	92S1
Trp94→Leu	8.9	25	31.0	urea	(1)	92S1
double mutant (Thr26→His and Asp24→Asn)						
	5.9	25	10.0	urea	(1)	92S1
	8.9	25	15.8	urea	(1)	92S1
double mutant (Trp94→Leu and His18→Gly)						
	5.9	25	33.9	urea	(1)	92S1
	8.9	25	35.1	urea	(1)	92S1

Remark:
(1) linear extrapolation using an averaged slope of a plot of free energy versus urea conc., buffers used: pH 4.4 sodium acetate, pH 5.9 MES, pH 8.9 Tris

Barnase, mutants, at salt conc. 1 M

Mutant	pH	T	ΔG	Approach	Remarks	Ref
wild-type	4.4	25	43.5	urea	(1,2)	92S1
wild-type	8.9	25	41.8	urea	(1,2)	92S1
Thr6→His	4.4	25	33.9	urea	(1,2)	92S1
Thr6→His	8.9	25	34.5	urea	(1,2)	92S1
double muant (Thr26→His and Asp54→Asn)						
	4.4	25	26.8	urea	(1,2)	92S1
	8.9	25	27.2	urea	(1,2)	92S1
Gly34→His	4.4	25	31.1	urea	(1,2)	92S1
Gly34→His	8.9	25	27.2	urea	(1,2)	92S1
Asp54→Asn	4.4	25	37.3	urea	(1,2)	92S1
Asp54→Asn	8.9	25	35.4	urea	(1,2)	92S1
Trp94→Leu	5.9	25	41.8	urea	(1,2)	92S1
Trp94→Leu	8.9	25	38.9	urea	(1,2)	92S1
double mutant (Trp94→Leu and His18→Gly)						
	5.9	25	43.5	urea	(1,2)	92S1
	8.9	25	43.1	urea	(1,2)	92S1

Remarks:
(1) linear extrapolation using an averaged slope of a plot of free energy versus urea conc.
(2) 1 M salt corresponding to 0.97 M NaCl and 0.03 M buffer, buffers used: pH 4.4 sodium acetate,
    pH 5.9 MES, pH 8.9 Tris

Barnase, wild-type and mutants measured in the presence of 30 mM MES buffer

Mutant	pH	T	ΔG	Δ(ΔG)	Approach	Remarks	Ref
wild-type	5.8	25	39.0	0.0	urea	(1)	92L5
wild-type	9.0	25	33.6	0.0	urea	(2)	92L5
His18→Gly	5.8	25	34.9	−4.1	urea	(1)	92L5
His18→Gly	9.0	25	36.0	2.4	urea	(1)	92L5
Trp94→Leu	5.8	25	32.0	−7.0	urea	(1)	92L5
Trp94→Leu	9.0	25	31.0	−2.7	urea	(1)	92L5
Trp94→Phe	5.8	25	35.2	−3.8	urea	(1)	92L5
Trp94→Phe	9.0	25	31.9	−1.7	urea	(1)	92L5
Trp94→Tyr	5.8	25	33.8	−5.2	urea	(1)	92L5
Trp94→Tyr	9.0	25	30.2	−3.4	urea	(1)	92L5
double mutant(His18→Gly and Trp94→Leu)							
	5.8	25	33.9	−5.1	urea	(1)	92L5
	9.0	25	35.1	1.4	urea	(1)	92L5

Remarks:
(1) ΔG was obtained by linear extrapolation, Δ(ΔG) is based on the urea concentration at the transition mid-
    point multiplied with an average value of the slope
(2) reference value for all measurements at pH 9.0

Barnase, wild-type and mutants measured in the presence of 1 M NaCl

Mutant	pH	T	ΔG	Δ(ΔG)	Approach	Remarks	Ref
wild-type	5.8	25	46.0	0.0	urea	(1)	92L5
wild-type	9.0	25	39.4	0.0	urea	(2)	92L5
His18→Gly	5.8	25	41.9	−4.1	urea	(1)	92L5
His18→Gly	9.0	25	41.5	2.1	urea	(1)	92L5
Trp94→Leu	5.8	25	39.4	−6.6	urea	(1)	92L5
Trp94→Leu	9.0	25	36.6	−2.8	urea	(1)	92L5
Trp94→Phe	5.8	25	41.6	−4.4	urea	(1)	92L5

Barnase, wild-type and mutants measured in the presence of 1 M NaCl (continued)

Mutant	pH	T	ΔG	$\Delta(\Delta G)$	Approach	Remarks	Ref
Trp94→Phe	9.0	25	36.9	−2.5	urea	(1)	92L5
Trp94→Tyr	5.8	25	40.7	−5.3	urea	(1)	92L5
Trp94→Tyr	9.0	25	35.2	−4.1	urea	(1)	92L5
double mutant (His18→Gly and Trp94→Leu)							
	5.8	25	41.1	−4.9	urea	(1)	92L5
	9.0	25	40.7	1.3	urea	(1)	92L5

Remarks:
(1) ΔG was obtained by linear extrapolation, $\Delta(\Delta G)$ is based on the urea concentration at the transition mid-point multiplied with an average value of the slope
(2) reference value for all measurements at pH 9.0

d) Single and multiple amino acid replacements

Barnase, wild-type and mutants

Mutant	pH	T	ΔG	Approach	Remarks	Ref
wild-type	6.3	25.0	42.5	urea	(1)	91S5
Tyr13→Ala	6.3	25.0	27.0	urea	(1,2)	91S5
Tyr17→Ala	6.3	25.0	33.2	urea	(1,3)	91S5
double mutant (Tyr13→Ala and Tyr17→Ala)						
	6.3	25.0	22.6	urea	(1)	91S5
Tyr13→Phe	6.3	25.0	39.8	urea	(1,2)	91S5
Tyr17→Phe	6.3	25.0	40.0	urea	(1,3)	91S5
double mutant (Tyr13→Phe and Tyr17→Phe)						
	6.3	25.0	37.3	urea	(1)	91S5

Remarks:
(1) linear extrapolation
(2) Tyr17 unchanged
(3) Tyr13 unchanged

Barnase, wild-type and mutants, $\Delta(\Delta G)$ values

Mutant	pH	T	$\Delta(\Delta G)$	Approach	Remarks	Ref
wild-type	6.3	25.0	0.0	urea	(1)	91S5
Tyr13→Ala	6.3	25.0	−15.2	urea	(1,2)	91S5
Tyr17→Ala	6.3	25.0	−9.2	urea	(1,3)	91S5
double mutant (Tyr13→Ala and Tyr17→Ala)						
	6.3	25.0	−19.3	urea	(1)	91S5
Tyr13→Phe	6.3	25.0	−1.5	urea	(1,2)	91S5
Tyr17→Phe	6.3	25.0	−1.3	urea	(1,3)	91S5
double mutant (Tyr13→Phe and Tyr17→Phe)						
	6.3	25.0	−2.5	urea	(1)	91S5
double mutant (Thr16→Ser and Tyr17→Ala)						
			−10	urea	(1)	91H3

Remarks:
(1) the procedure for calculating $\Delta(\Delta G)$ is described in Ref. 91S5
(2) Tyr17 unchanged
(3) Tyr13 unchanged

Barnase, wild-type and mutants

Mutant	pH	T	Δ(ΔG)	Approach	Remarks	Ref
wild-type			0.0	urea	(1)	89S7
Thr6→Ala			−10.6	urea	(1)	89S7
Thr6→Asn			−5.3	urea	(1)	89S7
Thr6→Asp			0.5	urea	(1)	89S7
Thr6→Gln			−7.8	urea	(1)	89S7
Thr6→Glu			−1.1	urea	(1)	89S7
Thr6→Gly			−5.6	urea	(1)	89S7
Thr6→Ser			−0.9	urea	(1)	89S7
Thr26→Ala			−8.8	urea	(1)	89S7
Thr26→Asn			−5.4	urea	(1)	89S7
Thr26→Gln			−7.2	urea	(1)	89S7
Thr26→Glu			−0.2	urea	(1)	89S7
Thr26→Gly			−6.6	urea	(1)	89S7
Thr26→Ser			−2.3	urea	(1)	89S7
Thr26→Val			−9.7	urea	(1)	89S7

Remark:
(1) linear extrapolation, assuming identical slope for the mutant proteins

Barnase, wild-type and mutants

Mutant	pH	T	Δ(ΔG)	Approach	Remarks	Ref
wild-type	6.3	25	0.0	urea	(1)	92H5
Asp8→Ala	6.3	25	−4.1±0.1	urea	(1,2)	92H5
Asp12→Ala	6.3	25	−1.4±0.1	urea	(1,2)	92H5
Arg110→Ala	6.3	25	−1.9±0.1	urea	(1,2)	92H5
double mutant (Asp8→Ala and Asp12→Ala)						
	6.3	25	−3.7±0.1	urea	(1,2)	92H5
double mutant (Asp8→Ala and Arg110→Ala)						
	6.3	25	−2.0±0.1	urea	(1,2)	92H5
double mutant (Asp12→Ala and Arg110→Ala)						
	6.3	25	1.9±0.1	urea	(1,2)	92H5
triple mutant (Asp8→Ala, Asp12→Ala and Arg110→Ala)						
	6.3	25	0.5±0.1	urea	(1,2)	92H5

Remarks:
(1) measured in 50 mM MES buffer
(2) Δ(ΔG) is based on the difference in urea midpoint concentration of mutant and wild-type

Barnase, wild-type and mutants measured in 50 mM MES buffer

Mutant	pH	T	ΔG	Δ(ΔG)	Approach	Remarks	Ref
wild-type	6.3	25	36.9	0.0	urea	(1)	92S8
Ile4→Ala	6.3	25	33.8	−5.6	urea	(2)	92S8
Ile4→Val	6.3	25	33.5	−2.5	urea	(2)	92S8
Asn5→Ala	6.3	25	30.3	−7.7	urea	(2)	92S8
Thr6→Ala	6.3	25	28.5	−9.0	urea	(2)	92S8
Thr6→Gly	6.3	25	34.4	−5.2	urea	(2)	92S8
Asp8→Ala	6.3	25	33.3	−3.7	urea	(2)	92S8
double mutant (Asp8→Ala and Asp12→Ala)							
	6.3	25	31.0	−3.3	urea	(2)	92S8
double mutant (Asp8→Ala and Arg110→Ala)							
	6.3	25	32.2	−1.8	urea	(2)	92S8

Barnase, wild-type and mutants measured in 50 mM MES buffer (continued)

Mutant	pH	T	ΔG	Δ(ΔG)	Approach	Remarks	Ref
triple mutant (Asp8→Ala, Asp12→Ala and Arg110→Ala)							
	6.3	25	37.7	0.4	urea	(2)	92S8
Val10→Ala	6.3	25	22.4	−14.2	urea	(2)	92S8
Val10→Thr	6.3	25	28.6	−10.4	urea	(2)	92S8
Asp12→Ala	6.3	25	35.3	−1.3	urea	(2)	92S8
double mutant (Asp12→Ala and Arg110→Ala)							
	6.3	25	38.5	1.7	urea	(2)	92S8
Tyr13→Ala	6.3	25	24.4	−14.0	urea	(2)	92S8
Leu14→Ala	6.3	25	17.9	−18.1	urea	(2)	92S8
Thr16→Arg	6.3	25	40.0	2.3	urea	(2)	92S8
Thr16→Ser	6.3	25	31.2	−7.0	urea	(2)	92S8
Tyr17→Ala	6.3	25	29.9	−8.5	urea	(2)	92S8
double mutant (Tyr13→Ala and Tyr17→Ala)							
	6.3	25	20.3	−17.5	urea	(2)	92S8
	6.3	25		−20.9	urea	(1)	92S9
double mutant (Thr16→Ala and Tyr17→Ala)							
	6.3	25	28.9	−9.0	urea	(2)	92S8
double mutant (Thr16→Ser and Tyr17→Ala)							
	6.3	25		−10.7	urea	(1)	92S9
His18→Gln	6.3	25		−6.0	urea	(1)	92S9
Asn23→Ala	6.3	25	27.7	−9.4	urea	(2)	92S8
Tyr24→Phe	6.3	25	37.3	−0.1	urea	(2)	92S8
Ile25→Ala	6.3	25	21.5	−14.7	urea	(2)	92S8
Ile25→Val	6.3	25	33.4	−4.7	urea	(2)	92S8
Thr26→Ala	6.3	25	29.9	−8.1	urea	(2)	92S8
Thr26→Gly	6.3	25	32.7	−6.1	urea	(2)	92S8
Lys27→Gly	6.3	25	35.3	−1.8	urea	(2)	92S8
Glu29→Gly	6.3	25	28.3	−7.4	urea	(2)	92S8
Ser31→Ala	6.3	25	36.3	0.6	urea	(3)	92S8
Gln31→Ser	6.3	25	37.2	1.0	urea	(3)	92S8
Leu33→Gln	6.3	25	31.0	−5.5	urea	(2)	92S8
Val36→Ala	6.3	25	29.9	−5.4	urea	(2)	92S8
Val36→Thr	6.3	25	32.5	−4.8	urea	(2)	92S8
Asn41→Asp	6.3	25	25.9	−10.5	urea	(2)	92S8
Val45→Ala	6.3	25	32.0	−7.3	urea	(2)	92S8
Val45→Thr	6.3	25	27.6	−10.2	urea	(2)	92S8
Ile51→Ala	6.3	25	17.2	−19.7	urea	(2)	92S8
Ile51→Val	6.3	25	32.2	−7.5	urea	(2)	92S8
Asp54→Ala	6.3	25	24.9	−12.4	urea	(2)	92S8
Asp54→Asn	6.3	25	29.5	−10.1	urea	(2)	92S8
Ile55→Ala	6.3	25	30.7	−4.8	urea	(2)	92S8
Ile55→Val	6.3	25	34.3	−1.1	urea	(2)	92S8
Ile55→Thr	6.3	25		−3.8	urea	(1)	92S9
Val55→Thr	6.3	25	31.7	−2.5	urea	(2)	92S8
Asn58→Ala	6.3	25	28.5	−11.3	urea	(2)	92S8
Asn58→Asp	6.3	25	38.2	2.0	urea	(2)	92S8
Lys62→Arg	6.3	25	35.6	−1.8	urea	(2)	92S8
Ile76→Ala	6.3	25	30.0	−7.9	urea	(2)	92S8
Ile76→Val	6.3	25	32.8	−3.4	urea	(2)	92S8
Asn77→Ala	6.3	25	30.8	−6.9	urea	(2)	92S8
Tyr78→Phe	6.3	25	32.1	−5.6	urea	(2)	92S8
Asn84→Ala	6.3	25	30.3	−8.5	urea	(2)	92S8
Ile88→Ala	6.3	25	20.1	−16.8	urea	(2)	92S8
Ile88→Val	6.3	25	30.0	−5.6	urea	(2)	92S8
Leu89→Thr	6.3	25		−11.9	urea	(1)	92S9

Barnase, wild-type and mutants measured in 50 mM MES buffer (continued)

Mutant	pH	T	ΔG	Δ(ΔG)	Approach	Remarks	Ref
Leu89→Val	6.3	25	34.9	−1.3	urea	(2)	92S8
Val89→Thr	6.3	25	23.2	−10.7	urea	(2)	92S8
Ser91→Ala	6.3	25	26.8	−8.1	urea	(2)	92S8
Ser92→Ala	6.3	25	23.4	−11.7	urea	(2)	92S8
Ile96→Ala	6.3	25	23.7	−13.3	urea	(2)	92S8
Ile96→Val	6.3	25	32.6	−3.7	urea	(2)	92S8
Thr99→Val	6.3	25	24.7	−11.2	urea	(2)	92S8
Tyr103→Phe	6.3	25	36.1	0	urea	(2)	92S8
Thr105→Val	6.3	25	27.9	−9.4	urea	(2)	92S8
Ile109→Ala	6.3	25	31.6	−8.7	urea	(2)	92S8
Ile109→Val	6.3	25	33.5	−3.2	urea	(2)	92S8
Arg110→Ala	6.3	25	37.2	−1.7	urea	(2)	92S8

Remarks:
(1) linear extrapolation
(2) ΔG was obtained by linear extrapolation, Δ(ΔG) is based on the urea concentration at the transition mid-point multiplied with an average value of the slope
(3) see additional remarks in Ref. 92S8 concerning position 31

Barnase, mutants

Mutant	pH	T	ΔG	Δ(ΔG)	Approach	Remarks	Ref
Thr6→Pro	6.3	25	23.0	−12.9	urea	(1)	92S10
Asp8→Gly	6.3	25	31.4	−4.9	urea	(1)	92S10
Asp8→Ser	6.3	25	33.9	−4.1	urea	(1)	92S10
Asp12→Gly	6.3	25	36.0	−5.4	urea	(1)	92S10
Asp12→Ser	6.3	25	33.1	−2.7	urea	(1)	92S10
Thr16→Ala	6.3	25	34.7	−1.1	urea	(1)	92S10
Thr16→Gly	6.3	25	31.4	−6.9	urea	(1)	92S10
Tyr17→Gly	6.3	25	20.1	−16.9	urea	(1)	92S10
Tyr17→Ser	6.3	25	28.5	−10.7	urea	(1)	92S10
His18→Ala	6.3	25	28.9	−7.7	urea	(1)	92S10
His18→Arg	6.3	25	32.6	−4.9	urea	(1)	92S10
His18→Asn	6.3	25	29.3	−7.0	urea	(1)	92S10
His18→Asp	6.3	25	27.6	−10.7	urea	(1)	92S10
His18→Gly	6.3	25	33.9	−2.9	urea	(1)	92S10
His18→Lys	6.3	25	33.1	−5.0	urea	(1)	92S10
His18→Ser	6.3	25	28.0	−9.5	urea	(1)	92S10
Thr26→Asp	6.3	25	36.8	−0.3	urea	(1)	92S10
Lys27→Gly	6.3	25	37.7	−2.3	urea	(1)	92S10
Ser28→Ala	6.3	25	39.3	1.7	urea	(1)	92S10
Ser28→Gly	6.3	25	34.7	−1.9	urea	(1)	92S10
Glu29→Ala	6.3	25	31.8	−5.2	urea	(1)	92S10
Glu29→Gly	6.3	25	27.6	−7.6	urea	(1)	92S10
Glu29→Ala	6.3	25	31.8	−5.2	urea	(1)	92S10
Glu29→Ser	6.3	25	31.8	−5.2	urea	(1)	92S10
Gln31→Ala	6.3	25	35.6	0.4	urea	(1)	92S10
Gln31→Gly	6.3	25	32.2	−4.1	urea	(1)	92S10
Gln31→Ser	6.3	25	38.1	−1.0	urea	(1)	92S10
Gly34→Ala	6.3	25	25.5	−13.1	urea	(1)	92S10
Gly34→Arg	6.3	25	25.1	−10.1	urea	(1)	92S10
Gly34→Asn	6.3	25	25.5	−11.5	urea	(1)	92S10
Gly34→Asp	6.3	25	23.4	−14.2	urea	(1)	92S10
Gly34→His	6.3	25	27.2	−11.1	urea	(1)	92S10

Barnase, mutants (continued)

Mutant	pH	T	ΔG	$\Delta(\Delta G)$	Approach	Remarks	Ref
Gly34→Lys	6.3	25	25.5	−13.1	urea	(1)	92S10
Gly34→Ser	6.3	25	24.7	−13.4	urea	(1)	92S10
Gly34→Thr	6.3	25	23.4	−14.1	urea	(1)	92S10
double mutants:							
(His18→Ala and Trp94→Leu)							
	6.3	25	27.2	−9.9	urea	(1)	92S10
(His18→Arg and Trp94→Leu)							
	6.3	25	30.1	−8.9	urea	(1)	92S10
(His18→Asn and Trp94→Leu)							
	6.3	25	29.7	−7.4	urea	(1)	92S10
(His18→Lys and Trp94→Leu							
	6.3	25	25.5	−9.8	urea	(1)	92S10
(His18→Ser and Trp94→Leu)							
	6.3	25	25.1	−10.8	urea	(1)	92S10

Remark:
(1) ΔG was determined using linear extrapolation, $\Delta(\Delta G)$ was determined using ΔG at the urea concentration at which half conversion was reached for the wild-type and mutant protein, respectively. The difference value was multiplied by the average slope m = 1.95 kJ/(mol×M$_D$), buffer: 50 mM MES

Barnase, wild-type and Ala32 mutants

Mutant	pH	T	ΔG	$\Delta(\Delta G)$	Approach	Remarks	Ref
wild-type	6.3	25.0	37.3	0.0	urea	(1)	92H6
Ala32→Asn	6.3	25.0	34.5	−2.8	urea	(1)	92H6
Ala32→Arg	6.3	25.0	36.7	−0.6	urea	(1)	92H6
Ala32→Asp	6.3	25.0	34.4	−3.0	urea	(1)	92H6
Ala32→Cys	6.3	25.0	33.1	−4.2	urea	(1)	92H6
Ala32→Gln	6.3	25.0	35.3	−2.0	urea	(1)	92H6
Ala32→Glu	6.3	25.0	35.0	−2.3	urea	(1)	92H6
Ala32→Gly	6.3	25.0	33.5	−3.8	urea	(1)	92H6
Ala32→His	6.3	25.0	34.1	−3.3	urea	(1)	92H6
Ala32→Ile	6.3	25.0	33.9	−3.4	urea	(1)	92H6
Ala32→Leu	6.3	25.0	35.8	−1.5	urea	(1)	92H6
Ala32→Lys	6.3	25.0	36.5	−0.8	urea	(1)	92H6
Ala32→Met	6.3	25.0	36.0	−1.3	urea	(1)	92H6
Ala32→Phe	6.3	25.0	34.4	−2.9	urea	(1)	92H6
Ala32→Pro	6.3	25.0	20.2	−17.1	urea	(1)	92H6
Ala32→Ser	6.3	25.0	35.6	−1.7	urea	(1)	92H6
Ala32→Thr	6.3	25.0	34.0	−3.3	urea	(1)	92H6
Ala32→Trp	6.3	25.0	33.2	−4.1	urea	(1)	92H6
Ala32→Tyr	6.3	25.0	33.8	−3.4	urea	(1)	92H6
Ala32→Val	6.3	25.0	33.6	−3.7	urea	(1)	92H6

Remark:
(1) ΔG was determined using linear extrapolation. $\Delta(\Delta G)$ is based on Δconc.$_{(urea)50\%}$. Buffer: 50 mM MES

Barnase, wild-type and mutants

Mutant	pH	T	ΔG	Δ(ΔG)	Approach	Remarks	Ref
wild-type	6.3	25	36.8±0.6	0.0	urea	(1–3)	93S2
Ile76→Thr	6.3	25	25.5±1.1	−10.6±0.6	urea	(1–4)	93S2
double mutant (Ile4→Ala and Ile76→Val)							
	6.3	25	29.0±0.2	−7.8±0.6	urea	(1–4)	93S2
double mutant (Ile4→Ala and Tyr78→Phe)							
	6.3	25	26.2±2.0	−11.0±0.6	urea	(1–4)	93S2
double mutant (Ile4→Ala and Ile51→Val)							
	6.3	25	25.8±1.5	−13.3±0.8	urea	(1–4)	93S2
multiple mutant (Ile4→Ala, Ile25→Val, Ile51→Val and Tyr78→Phe)							
	6.3	25	15.0±0.5	−21.9±0.7	urea	(1–4)	93S2

Remarks:
(1) linear extrapolation
(2) Ref. 93S2 contains further ΔG values for the I→U and I→F transition based on kinetics
(3) transition monitored spectrofluorimetrically
(4) Δ(ΔG) is based on ΔG at the transition midpoint

Barnase, wild-type and disulfide mutants

Mutant	pH	T	ΔG	Δ(ΔG)	Approach	Remarks	Ref
wild-type	6.3	25	36.8±0.6	0.0	urea	(1)	93C7
double mutant (Ala43→Cys and Ser80→Cys), disulfide bond 43–80 SS:							
	6.3	25	41.8±3.3	8.8±0.8	urea	(1)	93C7
double mutant (Ala43→Cys and Ser80→Cys), residues 43–80 (SH)$_2$:							
	6.3	25	32.2±2.5	−4.6±0.4	urea	(1)	93C7
double mutant (Ser85→Cys and His102→Cys), disulfide bond 85–102 SS:							
	6.3	25	54.0±2.5	18.0±0.8	GuHCl	(1)	93C7
double mutant (Ser85→Cys and His102→Cys), residues 85–102 (SH)$_2$:							
	6.3	25	35.1±1.7	−2.1±0.4	GuHCl	(1)	93C7

Remark:
(1) linear extrapolation to zero denaturant concentration by a method that includes the pre- and postdenaturational baselines for a non-linear regression of the data, Δ(ΔG) was obtained from a similar fit making use of a direct comparison of the half transitions in order to enhance the accuracy

Barnase, cysteine mutants and newly introduced disulfide bridges

Mutant	pH	T	$\Delta(\Delta G)$	Approach	Remarks	Ref
wild-type	6.3	25	0.0	urea	(1,2)	95C5
70–92 SS	6.3	25	−12.1±0.8	urea	(1–3)	95C5
70–92 (SH)$_2$	6.3	25	−9.6±1.3	urea	(1–3)	95C5
Thr70→Cys	6.3	25	−4.2±0.4	urea	(1,2)	95C5
Ser92→Cys	6.3	25	−5.4±0.8	urea	(1,2)	95C5
43–80 SS	6.3	25	+8.8±0.8	urea	(1–3)	95C5
43–80 SS	6.3	25	+8.4	H/D exchange		95C6
43–80 (SH)$_2$	6.3	25	−4.6±0.4	urea	(1–3)	95C5
Ala43→Cys	6.3	25	−4.6±0.4	urea	(1,2)	95C5
Ser80→Cys	6.3	25	−0.4±0.04	urea	(1,2)	95C5
85–102 SS	6.3	25	+18.0±0.8	urea	(1,2,4)	95C5
85–102 SS	6.3	25	+17.6	H/D exchange		95C6
85–102 (SH)$_2$	6.3	25	+0.8±0.1	urea	(1–3)	95C5
Ser85→Cys	6.3	25	−2.5±0.4	urea	(1,2)	95C5
His102→Cys	6.3	25	−0.1±0.1	urea	(1,2)	95C5

Remarks:
(1) linear extrapolation to zero denaturant concentration by a method that includes the pre- and postdenaturational baselines for a non-linear regression of the data
(2) $\Delta(\Delta G)$ refers to the denaturant concentration at which 50% of the wild-type protein is unfolded, 4.58 M GuHCl, all measurements were carried out in 50 mM MES, pH 6.3
(3) the position numbers refer to the above single mutants in either reduced or oxidized state
(4) as indicated in (2), however, at 2 M GuHCl

e) difference energy for intermediate, transition state, and folding

Barnase, wild-type and mutants, difference energy for intermediate, transition state, and folding, measured in 50 mM MES
$\Delta(\Delta G)(UI)$: difference energy for intermediate (U→I)
$\Delta(\Delta G)(U\ddagger)$: difference energy for transition state (U→‡)
$\Delta(\Delta G)(UF)$: difference energy for folding (U→F)

Mutant	pH	T	$\Delta(\Delta G)$ (UI)	$\Delta(\Delta G)$ (U‡)	$\Delta(\Delta G)$ (UF)	Approach	Remarks	Ref
wild-type	6.3	25	0.0	0.0	0.0	urea	(1)	92M1
Ile4→Ala	6.3	25	0	0.3	6.3	urea	(2)	92M1
Ile4→Val	6.3	25	0	0	2.8	urea	(2)	92M1
Val4→Ala	6.3	25	0	0.3	3.5	urea	(2)	92M1
Asn5→Ala	6.3	25	0.7	0.8	8.6	urea	(2)	92M1
Thr6→Ala	6.3	25	2.9	2.9	9.6	urea	(3)	92M1
Val10→Ala	6.3	25	4.2	5.3	15.2	urea	(2)	92M1
Val10→Thr	6.3	25	2.9	4.3	10.8	urea	(2)	92M1
Tyr13→Ala	6.3	25	6.4	7.9	15.5	urea	(4)	92M1
double mutant (Tyr13→Ala and Tyr17→Ala)								
	6.3	25	6.3	8.3	19.4	urea	(4)	92M1
Leu14→Ala	6.3	25	9.9	11.9	19.0	urea	(3)	92M1
Thr16→Ser	6.3	25	5.4	6.1	7.0	urea	(3)	92M1
Tyr17→Ala	6.3	25	4.3	5.4	9.5	urea	(4)	92M1
His18→Gln	6.3	25	4.9	5.1	5.9	urea	(3)	92M1
Asn23→Ala	6.3	25	−0.2	−0.4	10.5	urea	(2)	92M1
Ile25→Val	6.3	25	−1.1	−1.2	4.9	urea	(2)	92M1
Thr26→Ala	6.3	25	0.1	−0.2	8.4	urea	(3)	92M1
Glu29→Gly	6.3	25	−1.0	−1.1	7.9	urea	(2)	92M1
Val36→Ala	6.3	25	−1.4	−0.5	5.6	urea	(2)	92M1

Barnase, wild-type and mutants, difference energy for intermediate, transition state, and folding, measured in
50 mM MES (continued)
$\Delta(\Delta G)(UI)$: difference energy for intermediate (U→I)
$\Delta(\Delta G)(U\ddagger)$: difference energy for transition state (U→‡)
$\Delta(\Delta G)(UF)$: difference energy for folding (U→F)

Mutant	pH	T	$\Delta(\Delta G)$ (UI)	$\Delta(\Delta G)$ (U‡)	$\Delta(\Delta G)$ (UF)	Approach	Remarks	Ref
Val36→Thr	6.3	25	−0.3	−0.2	4.8	urea	(2)	92M1
Asn41→Asp	6.3	25	−0.2	−0.3	10.5	urea	(2)	92M1
Val45→Ala	6.3	25	−2.2	−1.3	7.7	urea	(2)	92M1
Val45→Thr	6.3	25	−0.8	−0.7	10.2	urea	(2)	92M1
Ile51→Val	6.3	25	−1.2	−1.2	7.7	urea	(2)	92M1
Asp54→Ala	6.3	25	−2.0	−2.1	13.0	urea	(2)	92M1
Asp54→Asn	6.3	25	−2.1	−2.2	10.1	urea	(2)	92M1
Ile55→Ala	6.3	25	2.9	3.2	5.4	urea	(2)	92M1
Ile55→Thr	6.3	25	1.2	1.8	4.2	urea	(2)	92M1
Asn58→Ala	6.3	25	8.1	8.5	9.1	urea	(2)	92M1
Lys62→Arg	6.3	25	1.3	1.6	1.8	urea	(2)	92M1
Ile76→Ala	6.3	25	1.9	3.8	8.5	urea	(2)	92M1
Ile76→Val	6.3	25	−0.4	0.1	3.7	urea	(2)	92M1
Val76→Ala	6.3	25	2.3	3.7	4.9	urea	(2)	92M1
Asn77→Ala	6.3	25	−0.1	−0.1	7.9	urea	(2)	92M1
Tyr78→Phe	6.3	25	0.7	0.6	5.9	urea	(4)	92M1
Asn84→Ala	6.3	25	1.6	1.3	9.4	urea	(2)	92M1
Ile88→Ala	6.3	25	10.4	16.6	17.4	urea	(4)	92M1
Ile88→Val	6.3	25	4.0	5.5	5.9	urea	(4)	92M1
Val88→Ala	6.3	25	6.6	11.3	11.6	urea	(2)	92M1
Leu89→Val	6.3	25	0.5	0.1	0.1	urea	(2)	92M1
Leu89→Thr	6.3	25	6.5	12.5	12.1	urea	(2)	92M1
Val89→Thr	6.3	25	5.8	12.2	10.7	urea	(2)	92M1
Ser91→Ala	6.3	25	4.6	7.5	8.1	urea	(2)	92M1
Ser92→Ala	6.3	25	7.3	10.9	11.5	urea	(2)	92M1
Ile96→Ala	6.3	25	9.5	12.0	13.9	urea	(4)	92M1
Ile96→Val	6.3	25	2.3	2.3	4.0	urea	(4)	92M1
Val96→Ala	6.3	25	7.1	9.7	9.9	urea	(4)	92M1
Thr105→Val	6.3	25	2.9	4.6	9.4	urea	(2)	92M1
Ile109→Ala	6.3	25	3.8	5.6	9.3	urea	(2)	92M1
Ile109→Val	6.3	25	0.6	0.5	3.4	urea	(2)	92M1
Val109→Ala	6.3	25	3.2	5.1	5.9	urea	(2)	92M1

Remarks:
(1) $\Delta G(I→U) = 13.4\pm0.5$ kJ/mol, $\Delta G(\ddagger→U) = 40.2\pm0.5$ kJ/mol, $\Delta G(F→U) = -(42.7\pm0.5)$ kJ/mol, data from
    Ref. 90M3
(2) $\Delta(\Delta G)(U→I)$ and $\Delta(\Delta G)(U→\ddagger)$ are mainly based on folding kinetics data, $\Delta(\Delta G)(U→F)$ is corrected for
    4 molar urea according to Ref. 92S8
(3) data are taken from Ref. 92S8, other details see (2)
(4) data are taken from Ref. 91H3, other details see (2)

Barnase, various transitions

Transition	pH	T	ΔG	Approach	Remarks	Ref
N→D	6.3	25	42.7	heat	(1–3)	96O3
‡→D	6.3	25	−54.4		(1,2,4)	96O3
I→D	6.3	25	11.7		(1,2,5)	96O3

Remarks:
(1) the values are relative to the thermally denatured state extrapolated to 25°C
(2) at pH 6.3, $\mu = 50$ mM
(3) data from Ref. 94O1
(4) from equilibrium unfolding and unfolding kinetics
(5) from ‡ and refolding kinetics

Barnase, mutants that affect the buried salt bridges Arg69-Asp93 and Arg83-Asp75, Transition N→U

Mutant	pH	T	$\Delta(\Delta G)$	Approach	Remarks	Ref
wild-type	6.3	25	0.0	urea	(1,2)	96T3
Arg69→Lys	6.3	25	13.1±0.4	urea	(1,2)	96T3
Arg69→Met	6.3	25	8.9±0.2	urea	(1,2)	96T3
Arg69→Ser	6.3	25	11.4±0.3	urea	(1,2)	96T3
Asp75→Asn	6.3	25	20.1±0.4	urea	(1,2)	96T3
Arg83→Gln	6.3	25	8.6±0.2	urea	(1,2)	96T3
Arg83→Lys	6.3	25	17.3±0.5	urea	(1,2)	96T3
Asp93→Asn	6.3	25	17.2±0.4	urea	(1,2)	96T3
double mutant (Arg69→Ser and Asp93→Asn)						
	6.3	25	14.6±0.3	urea	(1,2)	96T3
double mutant (Arg83→Lys and Asp75→Asn)						
	6.3	25	22.6±0.7	urea	(1,2)	96T3

Remarks:
(1) linear extrapolation to zero denaturant concentration by a method that includes the pre- and postdenaturational baselines for a non-linear regression of the data
(2) $\Delta(\Delta G)$ is based on ΔG at the transition midpoint, for details, see Refs. 93S2 and 93V3

Barnase, mutants that affect the buried salt bridges Arg69-Asp93 and Arg83-Asp75, intermediate and transition states

Mutant Transition	pH	T	$\Delta(\Delta G)$	Approach	Remarks	Ref
Arg69→Met (N→I)	6.3	25	5.3±0.4	urea	(1,2)	96T3
Arg69→Met (N→‡)	6.3	25	3.0±0.3	urea	(1,2)	96T3
Arg69→Ser (N→I)	6.3	25	6.3±0.5	urea	(1,2)	96T3
Arg69→Ser (N→‡)	6.3	25	3.4±0.5	urea	(1,2)	96T3
Asp93→Asn (N→I)	6.3	25	9.8±0.5	urea	(1,2)	96T3
Asp93→Asn (N→‡)	6.3	25	3.0±0.3	urea	(1,2)	96T3
double mutant (Arg69→Ser and Asp93→Asn)						
(N→I)	6.3	25	7.4±0.4	urea	(1,2)	96T3
(N→‡)	6.3	25	1.8±0.3	urea	(1,2)	96T3
Arg83→Lys (N→I)	6.3	25	18.5±0.4	urea	(1,2)	96T3

Barnase, mutants that affect the buried salt bridges Arg69-Asp93 and Arg83-Asp75, intermediate and transition states (continued)

Mutant Transition	pH	T	$\Delta(\Delta G)$	Approach	Remarks	Ref
Arg83→Lys (N→‡)	6.3	25	18.4±0.2	urea	(1,2)	96T3
Asp75→Asn (N→I)	6.3	25	21.2±0.8	urea	(1,2)	96T3
Asp75→Asn (N→‡)	6.3	25	21.7±0.8	urea	(1,2)	96T3
double mutant (Arg83→Lys and Asp75→Asn)						
(N→I)	6.3	25	20.5±0.8	urea	(1,2)	96T3
(N→‡)	6.3	25	20.7±0.5	urea	(1,2)	96T3

Remarks:
(1) from rate constants for unfolding extrapolated to water and to 4 M urea
(2) see also table containing data on the N→U transition

f) Insertion mutants, and step-wise mutation of barnase to binase

Barnase, insertion mutants

Mutant	pH	T	ΔG	$\Delta(\Delta G)$	Approach	Remarks	Ref
wild-type	6.3	25	36.74±0.59	0.0	urea	(1–3)	94V3
wild-type	4.4	25	35.66	0.0	DSC	(4–6)	94V3
endo-[RNAseT1-(93–99)]^{102a}-barnase							
	6.3	25	24.70±1.00	−12.04±1.13	urea	(1–3)	94V3
	4.4	25	21.28	−14.38	DSC	(4–6)	94V3
endo-[RNAseT1-(95–98)]^{104a}-barnase							
	6.3	25	10.02±0.67	−17.72±0.88	urea	(1–3)	94V3
	4.4	25	16.76	−18.89	DSC	(4–6)	94V3

Remarks:
(1) endo-[RNAseT1-(93–99)]^{102a}-barnase: the mutant protein contains RNAse T1 residues at positions 93–99 inserted between residues 102 and 103 of barnase endo-[RNAseT1-(95–98)]^{104a}-barnase: the mutant protein contains RNAse T1 residues at positions 95–98 inserted between residues 104 and 105 of barnase
(2) linear extrapolation to zero denaturant concentration by a method that includes the pre- and postdenaturational baselines for a non-linear regression of the data
(3) buffer: 50 mM MES pH 6.3
(4) $\Delta(\Delta G)$ was obtained using $\Delta(\Delta G) = \Delta T \times \Delta S_{(mutant)}$
(5) ΔCp was taken as $\Delta Cp = 6.688$ kJ/mol/K
(6) buffer: 20 mM sodium acetate

Barnase, step-wise mutation of barnase to binase

Mutant	pH	T	ΔG	$\Delta(\Delta G)$	Approach	Remarks	Ref
Gln15→Ile	6.3	25	42.7	4.0	urea	(1)	93S6
Thr16→Arg	6.3	25	40.2	2.2	urea	(1)	93S6
His18→Lys	6.3	25	33.1	−5.0	urea	(1)	93S6
Lys19→Arg	6.3	25	40.6	0.9	urea	(1)	93S6
Glu23→Gln	6.3	25	36.0	−0.04	urea	(1)	93S6
Gln31→Ser	6.3	25	38.1	−1.0	urea	(1)	93S6
Asp44→Glu	6.3	25	35.6	0.3	urea	(1)	93S6
Ile55→Val	6.3	25	35.6	−1.2	urea	(1)	93S6
Lys62→Arg	6.3	25	36.8	−2.0	urea	(1)	93S6

Barnase, step-wise mutation of barnase to binase

Mutant	pH	T	ΔG	$\Delta(\Delta G)$	Approach	Remarks	Ref
Gly65→Ser	6.3	25	41.0	2.1	urea	(1)	93S6
Lys66→Ala	6.3	25	41.0	1.0	urea	(1)	93S6
Thr79→Val	6.3	25	37.7	1.2	urea	(1)	93S6
Ser85→Ala	6.3	25	38.1	−0.5	urea	(1)	93S6
Ile88→Leu	6.3	25	36.8	−1.2	urea	(1)	93S6
Leu89→Val	6.3	25	35.6	−1.1	urea	(1)	93S6
Gln104→Ala	6.3	25	36.8	−0.9	urea	(1)	93S6
Lys108→Arg	6.3	25	40.2	3.9	urea	(1)	93S6
triple mutant (Gln15→Ile, Thr16→Arg and Lys19→Arg)							
	6.3	25	45.2	6.9	urea	(1)	93S6
multiple mutant (Gln15→Ile, Thr16→Arg, Lys19→Arg, Gly65→Ser, Lys66→Ala and Lys108→Arg)							
	6.3	25	49.4	14.4	urea	(1)	93S6
Binase	6.3	25	42.7	3.9	urea	(1,2)	93S6
Barnase	6.3	25	36.8	0	urea	(1,2)	93S6

Remarks:
(1) linear extrapolation, $\Delta(\Delta G)$ is based on the midpoint concentration and the average slope, $\Delta(\Delta G)$ is re-garded in Ref. 93S6 as the more precise value
(2) reference proteins

Barstar

Barstar, recombinant

pH	T	ΔG	Approach	Remarks	Ref
8.0	4	14.2±2.1	GuHCl	(1,2)	95A1
8.0	20	21.8±2.1	GuHCl	(1,2)	95A1
8.0	50	15.5±2.1	GuHCl	(1,2)	95A1

Remarks:
(1) linear extrapolation to zero denaturant concentration by a method that includes the pre- and postdenatura-tional baselines for a non-linear regression of the data
(2) the transition was monitored by CD at 222 nm and 275 nm

Barstar, wild-type and mutants

Mutant	pH	T	ΔG	Approach	Remarks	Ref
wild-type	7	25	21.8±2.1	GuHCl	(1,2)	95K7
wild-type	7	25	21±2	urea	(1,2)	95K7
wild-type	7	25	22.6±2.1	heat, v.H.	(3)	95K7
wild-type	8	25	22.2	GuHCl	(1,5)	95K7
wild-type	9	25	15	GuHCl	(1,5)	95K7
double mutant (Cys40→Ala and Cys82→Ala)						
	7	25	18.4	GuHCl	(1,4)	95K7
	8.0	25	4.1±0.1	DSC	(6)	95W5

Remarks:
(1) linear extrapolation according to the scheme N→U (native→unfolded) to zero denaturant concentration by a method that includes the pre- and postdenaturational baselines for a non-linear regression of the data
(2) ΔG is based on a transition monitored by CD at 220 nm and fluorescence at 322 nm
(3) ΔG is based on a thermal transition monitored by CD at 220 nm and 275 nm, and absorbance at 287 nm
(4) ΔG of the double mutant remains unchanged from pH 6 to 9
(5) data from Fig. 9 in Ref. 95K7
(6) measured in the presence of 400 mM NaCl

Barstar, pseudo wild-type (Cys40→Ala, Cys82→Ala and Pro27→Ala) and mutants

Mutant	pH	T	ΔG	Approach	Remarks	Ref
wt*	8.0	2	10.0	urea	(1,2)	95N4
wt*	8.0	10	12.6	urea	(1,2)	95N4
Ile5→Val	8.0	10	8.4	urea	(1,2)	95N4
Leu16→Val	8.0	10	7.9	urea	(1,2)	95N4
Leu34→Val	8.0	10	7.9	urea	(1,2)	95N4
Leu51→Val	8.0	10	10.0	urea	(1,2)	95N4

Remarks:
(1) linear extrapolation
(2) measured in 50 mM Tris, 100 mM KCl, pH 8

Barstar, wild-type and mutants, at varying NaCl concentrations

NaCl Concentration	pH	T	ΔG	Approach	Remarks	Ref
wild-type:						
0 mM	8	25	22.1	urea	(1,2)	94S2
300 mM	8	25	24.6	urea	(1,3)	94S2
700 mM	8	25	29.4	urea	(1,4)	94S2
Asp35→Ala:						
0 mM	8	25	23.4	urea	(1,2)	94S2
300 mM	8	25	24.8	urea	(1,3)	94S2
700 mM	8	25	29.7	urea	(1,4)	94S2
Asp39→Ala:						
0 mM	8	25	23.3	urea	(1,2)	94S2
300 mM	8	25	25.1	urea	(1,3)	94S2
700 mM	8	25	30.4	urea	(1,4)	94S2
Glu76→Ala:						
0 mM	8	25	25.5	urea	(1,2)	94S2
300 mM	8	25	26.9	urea	(1,3)	94S2
700 mM	8	25	31.6	urea	(1,4)	94S2
Glu80→Ala:						
0 mM	8	25	31.0	urea	(1,2)	94S2
300 mM	8	25	30.5	urea	(1,3)	94S2
700 mM	8	25	34.0	urea	(1,4)	94S2

Remarks:
(1) linear extrapolation using an average slope
(2) buffer: 50 mM Tris-HCl pH 8, 10 mM DTT
(3) buffer: 50 mM Tris-HCl pH 8, 10 mM DTT, 300 mM NaCl
(4) buffer: 50 mM Tris-HCl pH 8, 10 mM DTT, 700 mM NaCl

Barstar, cis- and trans-peptidylprolyl isomers, double mutant (Cys40→Ala and Cys82→Ala)

pH	T	ΔG	Approach	Remarks	Ref
8	25	20.3±0.8	urea	(1)	93S5
8	25	22.6±1.3	urea	(2)	93S5
8	25	11.9±0.4	urea	(3)	93S5
8.0	25.0	21.0±1.6	DSC		95M5

Remarks:
(1) equilibrium unfolding monitored by CD and fluorescence, data treatment by a method that includes the pre- and postdenaturational baselines for a non-linear regression of the data
(2) from refolding kinetics, ΔG refers to refolding of the unfolded protein with cis-Pro48 into the native form
(3) from refolding kinetics, ΔG refers to refolding of the unfolded protein with trans-Pro48 into the intermediate form ("misfolded protein") with trans-Pro48

Barstar, mutant (His17→Gln)

Transition	pH	T	ΔG	Approach	Remarks	Ref
N→I	7	25	10.5	GuHCl	(1–4)	95N2
I→U	7	25	2.9	GuHCl	(1–4)	95N2

Remarks:
(1) non-linear fit to an equation that corresponds to a native→intermediate→unfolded (N→I→U) transition with a linear dependence of ΔG on the denaturant concentration
(2) transition monitored by fluorescence
(3) the transition midpoint monitored by fluorescence is 1.4±0.1 M, that of near UV CD 1.6±0.1 M, and that of far UV CD is 1.8±0.1 M GuHCl
(3) the thermal transition monitored by optical absorbance at 287 nm and fluorescence is 63.0±0.5°C, that of CD at 275 nm 65.0±0.5°C, and that of CD at 220 nm 68.3±0.5°C
(4) all four optical probes yield the same midpoint for the wild-type protein

Barstar, chemically modified cysteines

Mutant/Form	pH	T	ΔG	Approach	Remarks	Ref
wild-type	25	7	19.2±0.8	GuHCl	(1)	96R1
Cys40→Ala	25	7	19.7±0.8	GuHCl	(1)	96R1
Cys82→Ala	25	7	16.7±0.8	GuHCl	(1)	96R1
DTNB-modified proteins:						
wild-type	25	7	24.7±1.3	GuHCl	(1,2)	96R1
Cys40→Ala	25	7	22.6±1.7	GuHCl	(1,2)	96R1
Cys82→Ala	25	7	19.7±0.8	GuHCl	(1,2)	96R1

Remarks:
(1) linear extrapolation
(2) DTNB = 5,5'-dithiobis(2-nitrobenzoic acid)

Binase

Ribonuclease from *Bacillus intermedius 7P* (binase)

pH	T	ΔG	Approach	Remarks	Ref
6.3	25	42.7	urea	(1,2)	93S6
7	25	43*	DSC		87P4

Remarks:
(1) linear extrapolation
(2) reference protein for step-wise mutation of barnase to binase (see pages 50 and 51)

β-Sheet Forming Propensities, see β-sheet

Calbindin

Bovine calbindin D9k, wild-type and point mutations

Mutant	pH	T	ΔG	Approach	Remarks	Ref
wild-type	8	25	20±1	urea	(1)	88W1
wild-type	8	25	37±2	urea	(2)	88W1
Tyr13→Phe	8	25	15±1	urea	(1)	88W1
Tyr13→Phe	8	25	26±2	urea	(2)	88W1
Glu17→Gln	8	25	22±1	urea	(1)	88W1
Glu17→Gln	8	25	35±2	urea	(2)	88W1
Pro20→Gly	8	25	15±1	urea	(1)	88W1
Pro20→Gly	8	25	26±2	urea	(2)	88W1
Pro20→Gly and Asn21 deleted						
	8	25	16±1	urea	(1)	88W1
	8	25	17±1	urea	(2)	88W1
Pro20 deleted						
	8	25	15±1	urea	(1)	88W1
	8	25	26±2	urea	(2)	88W1

Remarks:
(1) linear extrapolation
(2) linear extrapolation, modified procedure

Calbindin D9K, wild-type and mutants

Mutant	pH	T	ΔG	Approach	Remarks	Ref
wild-type	7.0	25.0±0.1	25.0	urea	(1)	90A2
Glu17→Gln	7.0	25.0±0.1	25.6	urea	(1)	90A2
Asp19→Asn	7.0	25.0±0.1	28.8	urea	(1)	90A2
Glu26→Gln	7.0	25.0±0.1	24.4	urea	(1)	90A2
double mutants (Glu17→Gln and Asp19→Asn)						
	7.0	25.0±0.1	31.8	urea	(1)	90A2
double mutants (Glu17→Gln and Glu26→Gln)						
	7.0	25.0±0.1	25.3	urea	(1)	90A2
double mutants (Asp19→Asn and Glu26→Gln)						
	7.0	25.0±0.1	26.0	urea	(1)	90A2
triple mutant (Glu17→Gln, Asp19→Asn and Glu26→Gln)						
	7.0	25.0±0.1	29.4	urea	(1)	90A2

Remark:
(1) linear extrapolation

Chaperonin, see also GroEl, GroEs

Carbonic Anhydrase

Bovine carbonic anhydrase

Form	pH	T	ΔG	Approach	Ref
native		45		hydrogen exchange	80C
native		48		DSC	79P5
guadinated		51		hydrogen exchange	80C

Carbonic anhydrase from erythrocyte

	pH	T	ΔG	Approach	Remarks	Ref
	7.55	25	84	heat	(1)	91L2

Remark:
(1) ΔCp was estimated from the amino acid content to 3.9 kJ/mol/K

Carbonic anhydrase, wild-type and mutant (cis-Pro202→Ala)

Form/Trans.	pH	T	ΔG	Approach	Remarks	Ref
w.t./N→I	7.5	23	35.1±1.7	GuHCl	(1–3)	93T7
w.t./N→I	7.5	23	34.7±2.1	GuHCl	(1,3,4)	93T7
w.t./I→U	7.5	23	29.7±2.5	GuHCl	(1,3,4)	93T7
mutant (cis-Pro202→Ala):						
mutant/N→I	7.5	23	13.8±0.4	GuHCl	(1–3)	93T7
mutant/N→I	7.5	23	11.7±1.3	GuHCl	(1,3,4)	93T7
mutant/I→U	7.5	23	25.1±1.7	GuHCl	(1,3,4)	93T7

Remarks:
(1) approximated by a three-state model
(2) transition monitored by enzymic activity
(3) buffer: 100 mM Tris-sulfate, 1 mM DTE
(4) transition monitored by UV absorption

Human carbonic anhydrase II, cloned wild-type and mutants

Transition	pH	T	ΔG	Approach	Remarks	Ref
cloned wild-type:						
N→I	7.5	23	31.8	GuHCl	(1,2)	93M7
I→U	7.5	23	24.3	GuHCl	(1,2)	93M7
N→I	7.5	23	33.1	GuHCl	(3)	93M7
mutant (Cys206→Ser):						
N→I	7.5	23	28.5	GuHCl	(1,2)	93M7
I→U	7.5	23	49.8	GuHCl	(1,2)	93M7
N→I	7.5	23	33.1	GuHCl	(3)	93M7
double mutant (Ser56→Cys and Cys206→Ser):						
N→I	7.5	23	30.1	GuHCl	(1,2)	93M7
I→U	7.5	23	33.1	GuHCl	(1,2)	93M7
N→I	7.5	23	28.9	GuHCl	(3)	93M7
double mutant (Val88→Cys and Cys206→Ser):						
N→I	7.5	23	54.0	GuHCl	(1,2)	93M7
I→U	7.5	23	36.8	GuHCl	(1,2)	93M7
N→I	7.5	23	39.3	GuHCl	(3)	93M7

Human carbonic anhydrase II, cloned wild-type and mutants (continued)

Transition	pH	T	ΔG	Approach	Remarks	Ref
double mutant (Trp123→Cys and Cys206→Ser):						
N→I	7.5	23	28.5	GuHCl	(1,2)	93M7
I→U	7.5	23	20.1	GuHCl	(1,2)	93M7
N→I	7.5	23	28.0	GuHCl	(3)	93M7
double mutant (Ile256→Cys and Cys206→Ser):						
N→I	7.5	23	19.2	GuHCl	(1,2)	93M7
I→U	7.5	23	36.0	GuHCl	(1,2)	93M7
N→I	7.5	23	17.6	GuHCl	(3)	93M7

Remarks:
(1) linear extrapolation
(2) the transitions were monitored by the ratio in absorption A_{292}/A_{260}
(3) the transition was monitored by enzyme activity

Carbonic anhydrase II, mutations and truncations in the N-terminus

Mutant	Transition	pH	T	ΔG	Approach	Remarks	Ref
HCAII-pwt	N→I	7.5	23	38.9	GuHCl	(1–3)	95A6
HCAII-pwt	N→I	7.5	23	41.4	GuHCl	(1,2,4)	95A6
HCAII-pwt	I→U	7.5	23	32.2	GuHCl	(1–3)	95A6
trunc5	N→I	7.5	23	15.1	GuHCl	(1–3,5)	95A6
trunc5	N→I	7.5	23	8.8	GuHCl	(1,2,4,5)	95A6
trunc5	I→U	7.5	23	32.6	GuHCl	(1–3,5)	95A6
trunc17	N→I	7.5	23	7.5	GuHCl	(1–3,5)	95A6
trunc17	N→I	7.5	23	9.2	GuHCl	(1,2,4,5)	95A6
trunc17	I→U	7.5	23	32.2	GuHCl	(1–3,5)	95A6
trunc24	N→I	7.5	23	7.1	GuHCl	(1–3,5)	95A6
trunc24	N→I	7.5	23	10.0	GuHCl	(1,2,4,5)	95A6
trunc24	I→U	7.5	23	35.6	GuHCl	(1–3,5)	95A6
Trp5→Phe	N→I	7.5	23	33.9	GuHCl	(1–3,5)	95A6
Trp5→Phe	N→I	7.5	23	27.2	GuHCl	(1,2,4,5)	95A6
Trp5→Phe	I→U	7.5	23	32.2	GuHCl	(1–3,5)	95A6
Trp16→Phe	N→I	7.5	23	16.3	GuHCl	(1–3,5)	95A6
Trp16→Phe	N→I	7.5	23	8.4	GuHCl	(1,2,4,5)	95A6
Trp16→Phe	I→U	7.5	23	33.9	GuHCl	(1–3,5)	95A6
double mutant (Trp5→Phe and Trp16→Phe)							
	N→I	7.5	23	6.7	GuHCl	(1–3,5)	95A6
	N→I	7.5	23	8.4	GuHCl	(1,2,4,5)	95A6
	I→U	7.5	23	36.0	GuHCl	(1–3,5)	95A6

Remarks:
(1) human carbonic anhydrase II, pseudo wild-type with Ser rather than Cys in position 206
(2) data analysis by a non-linear fit according to a three-state folding model that is based on linear dependence of ΔG on the denaturant concentration
(3) transition monitored by fluorescence
(4) transition monitored by enzymatic activity
(5) trunc5 to trunc24: truncated forms that lack the first 5, 17, and 24 N-terminal residues

Carboxypeptidase

Procarboxypeptidase B from porcine pancreas, fragment, globular activation domain

pH	T	ΔG	Approach	Remarks	Ref
7.5	37	26	DSC	(1)	91C6

Remark:
(1) buffer: 20 mM phosphate

Activation domain of human procarboxypeptidase A2 (ADA2h)

pH	T	ΔG	Approach	Remarks	Ref
7.0	25	17.0±1.0	urea	(1)	95V6
7.0	25	15.1±2.0	DSC		95V6

Remark:
(1) linear extrapolation to zero denaturant concentration by a method that includes the pre- and postdenaturational baselines for a non-linear regression of the data

Human procarboxypeptidase A2 and mutants designed for enhanced helix propensity

Mutant	pH	T	ΔG	Approach	Remarks	Ref
wild-type	7.0	25	17.4±0.4	urea	(1–5)	95V7
helix 1	7.0	25	20.5±0.5	urea	(1,2,4,5)	95V7
helix 2	7.0	25	21.5±0.3	urea	(1,2,4,5)	95V7

Remarks:
(1) linear extrapolation to zero denaturant concentration by a method that includes the pre- and postdenaturational baselines for a non-linear regression of the data
(2) transition monitored by fluorescence
(3) thermal transition at $T_{trs} = 77°C$
(4) with the following replacements in helix 1: Asn25→Lys, Gln32→Lys, and Glu33→Lys
(5) with the following replacements in helix 2: Gln60→Glu, Val64→Ala, Ser68→Ala, and Gln69→His

Cellular Retonoic Acid-Binding Protein, see retinoic acid

Cellular Retinol-Binding Protein II, see retinol

Cellulase

Endoglucanase III, single-domain cellulase from *Trichoderma reesei*

pH	T	ΔG	Approach	Remarks	Ref
5.5	25.0	48.6±4.1	urea	(1–3)	96A3
5.5	25.0	47.6±6.7	urea	(1,4)	96A3
5.5	25.0	52.9±10.6	urea	(1,5)	96A3
5.5	25.0	48.2*	DSC		96A3
5.5	30	49	urea	(1,6)	96A3
5.5	35	41	urea	(1,6)	96A3
5.5	40	32	urea	(1,6)	96A3
5.5	45	21	urea	(1,6)	96A3

Remarks:
(1) linear extrapolation to zero denaturant concentration by a method that includes the pre- and postdenaturational baselines for a non-linear regression of the data
(2) reduction of the disulfide linkage results in $\Delta(\Delta G) = -29$ kJ/mol
(3) transition monitored by fluorescence intensity
(4) transition monitored by CD
(5) from kinetics
(6) data from Fig. 7 in Ref. 96A3

CheY

CheY, globular protein involved in chemotaxis

pH	T	ΔG	Approach	Remarks	Ref
7	25	13.0±3.8	GuHCl	(1)	93D3
7	25	28.0±7.5	urea	(1)	93D3
7	25	30.1±10.5	urea	(1)	93D3
7	25	18.4±2.5	GuHCl	(1,2)	93D3
7	25	29.7±2.5	urea	(1,2)	93D3
7	25	30.5±2.9	urea	(1,2)	93D3

Remarks:
(1) linear extrapolation to zero denaturant concentration by a method that includes the pre- and postdenaturational baselines for a non-linear regression of the data
(2) measured in the presence of 10 mM $MgCl_2$

CheY, chemotactic protein from *E. coli*

pH	T	ΔG	Approach	Remarks	Ref
2.5	25	12.1±2	DSC	(1)	93F3
2.5	25	7.4±0.9	urea	(2)	93F3
3.7	25	20.7±1.5	urea	(2)	93F3
4.4	25	26.6±1.9	urea	(2)	93F3
5.1	25	31.8±2.3	urea	(2)	93F3
5.1	25	32.5±2.0	urea	(3)	93F3
6.3	25	23.5±1.4	urea	(2)	93F3
6.3	25	14.2±0.7	GuHCl	(2)	93F3
7.0	25	23.9±1.4	urea	(2)	93F3
8.0	25	21.9±1.4	urea	(2)	93F3
10.3	25	20.5±2	DSC	(1)	93F3

Remarks:
(1) treatment of the calorimetric recording by a model that includes an association-dissociation equilibrium between unfolded monomer and intermediate dimer
(2) linear extrapolation, transition monitored by fluorescence
(3) linear extrapolation, transition monitored by CD

CheY, chemotactic protein from *E. coli*

Salt Concentration	pH	T	ΔG	Δ(ΔG)	Approach	Remarks	Ref
0	7.0	25	32.2±1.9	0.0	urea	(1)	93F3
30 mM NaCl	7.0	25		0.4	urea	(1)	93F3
80 mM NaCl	7.0	25		1.0	urea	(1)	93F3
193 mM NaCl	7.0	25		1.6	urea	(1)	93F3
327 mM NaCl	7.0	25		2.9	urea	(1)	93F3
570 mM NaCl	7.0	25		3.7	urea	(1)	93F3
0	7.0	25	32.2±1.9	0.0	urea	(1)	93F3
2.4 mM MgCl$_2$	7.0	25		2.7	urea	(1)	93F3
10 mM MgCl$_2$	7.0	25		5.4	urea	(1)	93F3
20 mM MgCl$_2$	7.0	25		5.5	urea	(1)	93F3
30 mM MgCl$_2$	7.0	25		5.7	urea	(1)	93F3
80 mM MgCl$_2$	7.0	25		5.9	urea	(1)	93F3

Remark:
(1) linear extrapolation, transition monitored by fluorescence, measured in 5 mM PIPES buffer

CheY, chemotactic protein from E. coli, wild-type and mutant

Mutant	pH	T	ΔG	Approach	Remarks	Ref
wild-type	7.0	25	21.8±1.3	urea	(1)	94M22
Phe14→Asn	7.0	25	33.9±1.7	urea	(1)	94M22
double mutant (Phe14→Asn and Pro110→Gly)						
	7.0	25	26.4±1.7	urea	(1)	94M22

Remark:
(1) linear extrapolation

CheY, chemotactic protein from E. coli, mutants designed for enhanced helix propensity

Mutant	pH	T	ΔG	Approach	Remarks	Ref
pseudo w.t.	7.0	25	35.1±0.8	urea	(1,2)	96M10
pseudo w.t.	7.0	25	35.1±1.3	urea	(1,3)	96M10
helix 1	7.0	25	31.0±0.8	urea	(1,2,4)	96M10
helix 1	7.0	25	30.5±0.8	urea	(1,3,4)	96M10
helix 3	7.0	25	34.7±1.3	urea	(1,2,6)	96M10
helix 3	7.0	25	36.8±1.3	urea	(1,3,6)	96M10
helix 4	7.0	25	37.7±1.3	urea	(1,2,7)	96M10
helix 4	7.0	25	38.9±1.3	urea	(1,3,7)	96M10
helix 5	7.0	25	38.5±1.7	urea	(1,2,8)	96M10
helix 5	7.0	25	39.7±2.1	urea	(1,3,8)	96M10
helix 2	7.0	25	30.1±1.3	urea	(1,2,5,9)	96M10
helix 2	7.0	25	26.4±1.3	urea	(1,3,5,9)	96M10
Lys39→Ala	7.0	25	31.4±0.8	urea	(1,2,9)	96M10
Lys39→Ala	7.0	25	27.6±1.3	urea	(1,3,9)	96M10

Remarks:
(1) pseudo wild-type: mutant Phe14→Asn, template for the following mutants
(2) transition monitored by fluorescence, linear extrapolation, for the procedure, see Ref. 93F3
(3) kinetic approach, for the procedure, see Ref. 94M22
(4) with the following replacements in helix 1: Thr16→Ala, Arg19→Glu, and Asn23→Arg
(5) with the following replacements in helix 2: Gly39→Ala, Asp41→Glu, and Asn44→Arg
(6) with the following replacement in helix 3: Thr71→Arg
(7) with the following replacements in helix 4: Lys91→Asn, Asn94→Ala, and Ile96→Leu
(8) with the following replacements in helix 5: Thr115→Glu, Glu118→Lys, Asn121→Ala, and Lys122→Glu
(9) Gly39→Ala is reference protein for helix 2

CheY, chemotactic protein from E. coli, mutants derived from wild-type

Mutant	pH	T	ΔG	Approach	Remarks	Ref
wild-type	7.0	25	23.4±0.4	urea	(1)	95L10
wild-type	7.0	25	23.4±0.4	urea	(1,2)	96L4
wild-type	7.0	25	21.8±1.3	urea	(1,3)	96L4
Asp12→Ala	7.0	25	33.9±1.3	urea	(1,2)	96L4
Asp13→Ala	7.0	25	34.7±1.3	urea	(1,2)	96L4
Phe14→Ala	7.0	25	26.8±0.8	urea	(1,2)	96L4
Phe14→Asn	7.0	25	35.6±1.3	urea	(1)	95L10
Ala48→Gly	7.0	25	24.3±0.4	urea	(1)	95L10
Asp57→Ala	7.0	25	37.7±1.7	urea	(1,2)	96L4
Pro61→Gly	7.0	25	20.9±0.8	urea	(1,2)	96L4
Ala74→Gly	7.0	25	21.8±0.7	urea	(1)	95L10
Ala74→Gly	7.0	25	22.2±0.8	urea	(1,2)	96L4
Ala77→Gly	7.0	25	22.2±0.5	urea	(1)	95L10

CheY, chemotactic protein from *E. coli*, mutants derived from wild-type (continued)

Mutant	pH	T	ΔG	Approach	Remarks	Ref
Ala80→Gly	7.0	25	25.9±0.9	urea	(1)	95L10
Ala88→Gly	7.0	25	23.8±0.8	urea	(1)	95L10
Ala90→Gly	7.0	25	25.5±1.0	urea	(1)	95L10
Ala99→Gly	7.0	25	21.3±0.7	urea	(1)	95L10
Ala99→Gly	7.0	25	21.3±0.8	urea	(1,2)	96L4
Ala101→Gly	7.0	25	19.2±0.8	urea	(1)	95L10
Ala101→Gly	7.0	25	19.2±0.8	urea	(1,2)	96L4
Ala113→Gly	7.0	25	20.1±1.0	urea	(1)	95L10
Ala113→Gly	7.0	25	18.0±0.8	urea	(1,2)	96L4
Ala114→Gly	7.0	25	21.3±0.7	urea	(1)	95L10
Ala114→Gly	7.0	25	20.1±0.4	urea	(1,2)	96L4

Remarks:
(1) transition monitored by fluorescence, linear extrapolation, for the procedure, see Ref. 93F3
(2) buffer: 50 mM PIPES
(3) buffer: 5 mM sodium phosphate

CheY, chemotactic protein from *E. coli*, mutants derived from pseudo wild-type (wt* = Phe14→Asn)

Mutant	pH	T	ΔG	Approach	Remarks	Ref
wt*	7.0	25	35.6±1.3	urea	(1)	95L10
wt*	7.0	25	35.6±1.3	urea	(1,2)	96L4
wt*	7.0	25	39.3±0.8	kinetics	(2)	96L4
Val10→Thr	7.0	25	11.7±0.8	urea	(1,2)	96L4
Val11→Thr	7.0	25	22.2±0.4	urea	(1,2)	96L4
Val21→Thr	7.0	25	34.7±0.8	urea	(1,2)	96L4
Val21→Thr	7.0	25	38.1±3.3	kinetics	(2)	96L4
Asn23→Gly	7.0	25	35.6±1.3	urea	(1,2)	96L4
Asn23→Gly	7.0	25	38.1±0.4	kinetics	(2)	96L4
Val33→Thr	7.0	25	29.3±0.8	urea	(1,2)	96L4
Ala36→Gly	7.0	25	22.6±0.5	urea	(1)	95L10
Ala36→Gly	7.0	25	22.6±0.4	urea	(1,2)	96L4
Asp38→Ala	7.0	25	27.6±1.2	urea	(1)	95L10
Asp38→Ala	7.0	25	27.6±1.3	urea	(1,2)	96L4
Asp38→Gly	7.0	25	31.4±0.8	urea	(1)	95L10
Asp38→Gly	7.0	25	31.4±0.8	urea	(1,2)	96L4
Gly39→Ala	7.0	25	31.4±0.8	urea	(1,2)	96L4
Val40→Thr	7.0	25	32.6±0.8	urea	(1,2)	96L4
Ala42→Gly	7.0	25	25.9±0.8	urea	(1,2)	96L4
Val54→Thr	7.0	25	15.5±0.4	urea	(1,2)	96L4
Ile55→Val	7.0	25	29.3±0.8	urea	(1,2)	96L4
Asp64→Ala	7.0	25	30.5±0.8	urea	(1)	95L10
Asp64→Ala	7.0	25	31.4±0.8	urea	(1,2)	96L4
Asp64→Gly	7.0	25	34.3±1.1	urea	(1)	95L10
Ile72→Val	7.0	25	29.3±0.8	urea	(1,2)	96L4
Gly76→Ala	7.0	25	37.7±1.0	urea	(1)	95L10
Gly76→Ala	7.0	25	37.7±1.3	urea	(1,2)	96L4
Val83→Thr	7.0	25	20.9±0.4	urea	(1,2)	96L4
Lys91→Ala	7.0	25	37.2±1.3	urea	(1)	95L10
Lys91→Gly	7.0	25	37.7±1.1	urea	(1)	95L10
Ala97→Gly	7.0	25	29.7±1.0	urea	(1)	95L10
Ala97→Gly	7.0	25	29.7±1.3	urea	(1,2)	96L4
Ala98→Gly	7.0	25	30.1±0.8	urea	(1)	95L10
Ala98→Gly	7.0	25	30.1±0.8	urea	(1,2)	96L4

CheY, chemotactic protein from *E. coli,* mutants derived from pseudo wild-type (wt* = Phe14→Asn) (continued)

Mutant	pH	T	ΔG	Approach	Remarks	Ref
Ala103→Gly	7.0	25	28.5±0.6	urea	(1)	95L10
Ala103→Gly	7.0	25	28.5±0.8	urea	(1,2)	96L4
Val108→Thr	7.0	25	31.4±0.8	urea	(1,2)	96L4
Thr112→Ala	7.0	25	29.3±2.1	urea	(1)	95L10
Thr112→Ala	7.0	25	29.3±2.1	urea	(1,2)	96L4
Thr112→Gly	7.0	25	31.4±1.8	urea	(1)	95L10
Ile123→Val	7.0	25	32.2±1.3	urea	(1,2)	96L4

Remarks:
(1) transition monitored by fluorescence, linear extrapolation, for the procedure, see Ref. 93F3
(2) buffer: 50 mM PIPES

Cholinesterase

Cholinesterase

Form	pH	T	ΔG	Approach	Remarks	Ref
native	8.4	25	>9	urea	(1)	86M
organophosphate-inhibited	8.4	25	>9	urea	(1)	86M

Remark:
(1) urea gradient electrophoresis, possibly underestimated value

Chymotrypsin

α-Chymotrypsin

pH	T	ΔG	Approach	Remarks	Ref
3	25	30.1	heat, optical method		69B1
4	25	48.5±2.1	DSC		74P1
4	25	45	GuHCl	(1)	84S1
4	25	49	GuHCl	(2)	84S1
4.3	25	57.3	GuHCl	(3)	79P1
4.3	25	47.7	GuHCl	(4)	79P1
4.3	25	49.8	GuHCl	(5)	79P1
4.3	25	32.6	GuHCl	(1)	74G1
4.3	25	35.1	urea	(1)	74G1
4.3	25	43.5	GuHCl	(2)	74G1
4.3	25	36.8	urea	(2)	74G1
7	25	58.6	heat, optical method		69L
		51.5	GuHCl	(6)	74G2

Remarks:
(1) linear extrapolation
(2) transfer model
(3) denaturant binding model, k = 1.2
(4) denaturant binding model, k = 0.8
(5) denaturant binding model, k = 0.6
(6) denaturant binding model

α-Chymotrypsin, phenylmethansulfonyl modified (PMS-Ct)

pH	T	ΔG	Approach	Remarks	Ref
4	25	36.4±1.8	GuHCl	(1)	88S4
4	25	38.5±1.9	urea	(1)	88S4
4	25	36.9±2.3		(2)	88S4
4	25	36.7		(3)	88S4
6	25	49.8		(4)	88B2

Remarks:
(1) linear extrapolation
(2) denaturant: 1,3-dimethylurea, linear extrapolation
(3) simultaneous fit of three transition curves obtained using GuHCl, urea, and 1,3-dimethylurea by linear extrapolation
(4) simultaneous fit of two transition curves obtained using GuHCl and urea by linear extrapolation

Chymotrypsin Inhibitor

Chymotrypsin inhibitor CI2, various approaches

pH	T	ΔG	Approach	Remarks	Ref
2.2	25	8.1	DSC		91J1
2.5	25	11.7	DSC		91J1
2.8	25	15.6	DSC		91J1
3.2	25	21.8	DSC		91J1
3.5	25	27.1	DSC		91J1
4.2	50	23.0	GuHCl	(1,2)	90J
4.2	50	17.2	GuHCl	(1–3)	90J
6.3	25	29.4±0.7	GuHCl	(1)	91J1
6.3	25	30.0±1.8	heat		91J1
6.3	25	30.3±0.9	kinetics	(5)	91J1
6.3	25	31.2±1.5	kinetics	(6)	91J2
6.3	25	31.8±0.4	GuHCl	(1)	95I1
6.3	25	38.5±6.3	heat	(4)	96T1
6.3	25	32.6±0.4	kinetics	(5)	96T1

Remarks:
(1) linear extrapolation
(2) monitored by fluorescence
(3) there exists a correlation between ΔG from GuHCl denaturation and amide proton exchange
(4) equilibrium value, for details see Ref. 96T1
(5) unfolding and refolding kinetics treated by a two-state model
(6) calculated from the ratio of refolding and unfolding rate constants, taking into account the equilibrium due to proline isomerization

Chymotrypsin inhibitor 2, wild-type and mutants

Mutant	pH	T	ΔG	Approach	Remarks	Ref
wild-type	6.3	25	31.5±0.7	GuHCl	(1,2)	94J1
Thr58→Ala	6.3	25	28.7±2.1	GuHCl	(1,2)	94J1
Thr58→Asp	6.3	25	31.4±1.4	GuHCl	(1,2)	94J1
Glu60→Ala	6.3	25	27.1±2.5	GuHCl	(1,2)	94J1
double mutant (Thr58→Ala and Glu60→Ala)						
	6.3	25	28.7±1.5	GuHCl	(1,2)	94J1
double mutant (Thr58→Asp and Glu60→Ala)						
	6.3	25	30.5±1.5	GuHCl	(1,2)	94J1

Remarks:
(1) linear extrapolation to zero denaturant concentration by a method that includes the pre- and postdenaturational baselines for a non-linear regression of the data
(2) for further experimental details see Ref. 93J1

Chymotrypsin inhibitor 2, recombinant

Mutant	pH	T	ΔG	Δ(ΔG)	Approach	Remarks	Ref
wild-type	7.0	30	29.5±0.7	0	GuHCl	(1)	93M8
Ile39→Leu	7.0	30	21.3±0.5	–8.2	GuHCl	(1)	93M8
Ile39→Val	7.0	30	24.6±0.7	–4.9	GuHCl	(1)	93M8
Arg67→Ala	7.0	30	24.4±0.5	–5.1	GuHCl	(1)	93M8

Remark:
(1) linear extrapolation

Chymotrypsin inhibitor 2 from Barley (CI2)

Mutant	pH	T	ΔG	Approach	Remarks	Ref
wild-type	6.25	25	31.8±0.5	GuHCl	(1,2)	94M8
Ser31→Ala	6.25	25	27.5±1.2	GuHCl	(1,2)	94M8
Ser31→Gly	6.25	25	31.5±1.0	GuHCl	(1,2)	94M8
Glu33→Asn	6.25	25	30.5±1.5	GuHCl	(1,2)	94M8
Glu33→Asp	6.25	25	28.6±1.8	GuHCl	(1,2)	94M8
Glu33→Gln	6.25	25	29.4±1.5	GuHCl	(1,2)	94M8
Glu34→Asn	6.25	25	31.5±1.5	GuHCl	(1,2)	94M8
Glu34→Asp	6.25	25	33.4±2.4	GuHCl	(1,2)	94M8
Glu34→Gln	6.25	25	29.0±1.4	GuHCl	(1,2)	94M8
double mutant (Glu33→Ala and Glu34→Ala)						
	6.25	25	31.8±2.8	GuHCl	(1,2)	94M8
triple mutant (Ser31→Gly, Glu33→Ala and Glu34→Ala)						
	6.25	25	28.4±2.2	GuHCl	(1,2)	94M8
triple mutant (Ser31→Ala, Glu33→Ala and Glu34→Ala)						
	6.25	25	28.7±1.9	GuHCl	(1,2)	94M8

Remarks:
(1) linear extrapolation, for the procedure, see Ref. 93J1
(2) measured in 50 mM MES buffer

Chymotrypsin inhibitor 2 from Barley (CI2)

Mutant	pH	T	$\Delta(\Delta G)$	Approach	Remarks	Ref
pseudo wild-type	6.25	25	0.0	GuHCl	(1,2)	94M8
Ser31→Ala	6.25	25	3.85±0.46	GuHCl	(1,2)	94M8
Ser31→Gly	6.25	25	3.43±0.17	GuHCl	(1,2)	94M8
Glu33→Asn	6.25	25	3.01±0.25	GuHCl	(1,2)	94M8
Glu33→Asp	6.25	25	2.26±0.25	GuHCl	(1,2)	94M8
Glu33→Gln	6.25	25	1.30±0.21	GuHCl	(1,2)	94M8
Glu33→Asn	6.25	25	4.60±0.29	GuHCl	(1,2)	94M8
Glu34→Asp	6.25	25	3.18±0.38	GuHCl	(1,2)	94M8
Glu34→Gln	6.25	25	2.01±0.21	GuHCl	(1,2)	94M8
double mutant (Glu33→Ala and Glu34→Ala)						
	6.25	25	3.47±0.42	GuHCl	(1,2)	94M8
triple mutant (Ser31→Gly, Glu33→Ala and Glu34→Ala)						
	6.25	25	6.99±0.54	GuHCl	(1,2)	94M8
	6.25	25	3.51±0.67	GuHCl	(1–3)	94M8
triple mutant (Ser31→Ala, Glu33→Ala and Glu34→Ala)						
	6.25	25	7.15±0.54	GuHCl	(1,2)	94M8
	6.25	25	3.68±0.67	GuHCl	(1–3)	94M8

Remarks:
(1) $\Delta(\Delta G)$ is the difference between mutant and wild-type at half conversion, for the procedure, see Ref. 93J1
(2) measured in 50 mM MES buffer
(3) $\Delta(\Delta G)$ refers to the pseudo wild-type Glu33→Ala and Glu34→Ala

Chymotrypsin inhibitor 2 from Barley (CI2)

Mutant	pH	T	$\Delta(\Delta G)$	Approach	Remarks	Ref
pseudo wild-type	3.5	73.6	0	DSC	(1)	94M8
Ser31→Gly	3.5	73.6	3.51	DSC	(1)	94M8
wild-type	3.0	63.7	0	DSC	(1)	94M8
Ser31→Ala	3.0	63.7	4.18	DSC	(1)	94M8
double mutant (Glu33→Ala and Glu34→Ala)						
	3.0	63.7	2.68	DSC	(1)	94M8
triple mutant (Ser31→Gly, Glu33→Ala and Glu34→Ala)						
	3.0	63.7	5.36	DSC	(1)	94M8
	3.0	63.7	2.68	DSC	(1,2)	94M8
triple mutant (Ser31→Ala, Glu33→Ala and Glu34→Ala)						
	3.0	63.7	5.56	DSC	(1)	94M8
	3.0	63.7	2.89	DSC	(1,2)	94M8

Remarks:
(1) $\Delta(\Delta G)$ was calculated by $\Delta(\Delta G) = \Delta T \times \Delta S$
(2) $\Delta(\Delta G)$ refers to the pseudo wild-type Glu33→Ala and Glu34→Ala

Chymotrypsin inhibitor 2 (CI2), wild-type and mutants, β-sheet propensity

Mutant	pH	T	$\Delta(\Delta G)$	Approach	Remarks	Ref
wild-type	6.3	25	0.0	GuHCl	(1–3)	95O3
Thr22→Ala	6.3	25	−3.56±0.21	GuHCl	(1–3)	95O3
Thr22→Gly	6.3	25	−4.85±0.25	GuHCl	(1–3)	95O3
Thr22→Val	6.3	25	−1.34±0.29	GuHCl	(1–3)	95O3
Ile49→Ala	6.3	25	−8.87±0.25	GuHCl	(1–3)	95O3

Chymotrypsin inhibitor 2 (CI2), wild-type and mutants, β-sheet propensity (continued)

Mutant	pH	T	Δ(ΔG)	Approach	Remarks	Ref
Ile49→Gly	6.3	25	−14.73±0.29	GuHCl	(1–3)	95O3
Ile49→Thr	6.3	25	−5.61±0.17	GuHCl	(1–3)	95O3
Ile49→Val	6.3	25	0.33±0.33	GuHCl	(1–3)	95O3
Val53→Ala	6.3	25	−2.68±0.46	GuHCl	(1–3)	95O3
Val53→Gly	6.3	25	−10.17±0.21	GuHCl	(1–3)	95O3
Val53→Thr	6.3	25	−4.31±0.21	GuHCl	(1–3)	95O3
Ala77→Gly	6.3	25	−7.87±0.33	GuHCl	(1–3)	95O3
Val79→Ala	6.3	25	−6.32±0.25	GuHCl	(1–3)	95O3
Val79→Gly	6.3	25	−13.56±0.25	GuHCl	(1–3)	95O3
Val79→Thr	6.3	25	−1.59±0.46	GuHCl	(1–3)	95O3
Val82→Ala	6.3	25	−6.07±0.29	GuHCl	(1–3)	95O3
Val82→Gly	6.3	25	−14.64±0.29	GuHCl	(1–3)	95O3
Val82→Thr	6.3	25	−4.81±0.29	GuHCl	(1–3)	95O3

Remarks:
(1) linear extrapolation to zero denaturant concentration by a method that includes the pre- and postdenaturational baselines for a non-linear regression of the data
(2) Δ(ΔG) is the difference in stability between mutant CI2 and wild-type at a mean value of $[GuHCl]^{50\%}$
(3) buffer: 50 mM MES, pH 6.3

Barley chymotrypsin inhibitor 2 (CI-2), complexes of the fragments CI-2(20–59) and CI-2(60–83)

Mutant	pH	T_{trs}	Δ(ΔG)	Approach	Remarks	Ref
wild-type	6.3	46.4	0.0	heat	(1–3)	95R7
Thr22→Ala	6.3	43.5	−3.6±0.2	heat	(1,2)	95R7
Leu27→Ala	6.3	23.9	−11.2±0.6	heat	(1,2)	95R7
Ser31→Ala	6.3	38.0	−3.7±0.2	heat	(1,2)	95R7
Glu33→Asn	6.3	42.3	−2.9±0.2	heat	(1,2)	95R7
Lys36→Ala	6.3	33.8	−2.1±0.1	heat	(1,2)	95R7
Lys37→Ala	6.3	41.2	1.0±0.5	heat	(1,2)	95R7
Lys37→Gly	6.3	35.6	−4.1±0.3	heat	(1,2)	95R7
Val38→Ala	6.3	41.7	−2.0±0.3	heat	(1,2)	95R7
Ile39→Val	6.3	31.6	−5.4±0.2	heat	(1,2)	95R7
Lys43→Ala	6.3	40.6	−2.7±0.3	heat	(1,2)	95R7
Ile48→Val	6.3	39.2	−4.6±0.2	heat	(1,2)	95R7
Ile49→Ala	6.3		−8.9±0.3	heat	(1,2)	95R7
Ile49→Val	6.3	47.8	0.3±0.3	heat	(1,2)	95R7
Leu51→Ala	6.3	37.3	−9.9±0.2	heat	(1,2)	95R7
Leu51→Val	6.3	45.9	−2.1±0.3	heat	(1,2)	95R7
Val53→Ala	6.3	43.6	−2.6±0.5	heat	(1,2)	95R7
Val53→Thr	6.3	38.1	−4.3±0.2	heat	(1,2)	95R7
Phe69→Leu	6.3	32.3	−8.8±0.3	heat	(1,2)	95R7
Val70→Ala	6.3	29.6	−8.2±0.3	heat	(1,2)	95R7
Ile76→Val	6.3	47.5	0.8±0.4	heat	(1,2)	95R7
double mutant (Ser31→Ala and Glu34→Ala)						
	6.3	38.0	−3.2±0.3	heat	(1,2)	95R7
multiple mutant (Ser31→Ala, Glu33→Ala and Glu34→Ala)						
	6.3	32.3	−7.0±0.2	heat	(1,2)	95R7

Remarks:
(1) the fragments associate to give native-like structure
(2) thermal denaturation at 5 μM complex in 10 mM phosphate buffer pH 6.3; the data were fitted to theoretical curves for simultaneous dissociation and unfolding
(3) see also Ref. 93J1 and 94O3

Chymotrypsin inhibitor 2, truncated form, wild-type and mutants, ΔG and $\Delta(\Delta G)$ determined at zero denaturant concentration

Mutant	pH	T	ΔG	$\Delta(\Delta G)$	Remarks	Ref
wild-type	6.3	25	31.5±0.7	0.0	(1–3)	93J1
Leu27→Ala	6.3	25	23.6±1.4	−8.8±1.6	(1–3)	93J1
Val38→Ala	6.3	25	31.5±2.4	−0.0±2.5	(1–3)	93J1
Ile39→Val	6.3	25	27.7±1.7	−3.8±1.8	(1–3)	93J1
Ile48→Ala	6.3	25	17.3±1.6	−14.2±1.8	(1–3)	93J1
Ile48→Val	6.3	25	28.5±2.6	−3.1±2.7	(1–3)	93J1
Val66→Ala	6.3	25	10.8±1.5	−20.8±1.7	(1–3)	93J1
Leu68→Ala	6.3	25	16.7±1.0	−14.9±1.3	(1–3)	93J1
Val70→Ala	6.3	25	26.8±2.4	−4.7±2.5	(1–3)	93J1
Ile76→Ala	6.3	25	14.4±1.6	−17.1±1.8	(1–3)	93J1
Ile76→Val	6.3	25	31.3±2.1	−0.3±2.2	(1–3)	93J1
double mutant (Ile48→Ala and Ile76→Val)						
	6.3	25	14.5±1.1	−17.1±1.3	(1–3)	93J1

Remarks:
(1) truncated protein: first 19 residues were deleted, and Leu20 replaced by Met
(2) GuHCl denaturation
(3) linear extrapolation to zero denaturant concentration by a method that includes the pre- and postdenaturational baselines for a non-linear regression of the data

Chymotrypsin inhibitor 2, truncated form, wild-type and mutants, $\Delta(\Delta G)$ determined form the denaturant half conversion concentration

Mutant	pH	T	$\Delta(\Delta G)$	Approach	Remarks	Ref
wild-type	6.3	25	0.0	GuHCl	(1,2)	93J1
Leu27→Ala	6.3	25	−11.0±0.3	GuHCl	(1,2)	93J1
Val38→Ala	6.3	25	−1.9±0.3	GuHCl	(1,2)	93J1
Ile39→Val	6.3	25	−5.3±0.3	GuHCl	(1,2)	93J1
Ile48→Ala	6.3	25	−16.1±0.4	GuHCl	(1,2)	93J1
Ile48→Val	6.3	25	−4.6±0.3	GuHCl	(1,2)	93J1
Val66→Ala	6.3	25	−20.4±0.9	GuHCl	(1,2)	93J1
Leu68→Ala	6.3	25	−16.0±0.4	GuHCl	(1,2)	93J1
Val70→Ala	6.3	25	−8.2±0.3	GuHCl	(1,2)	93J1
Ile76→Ala	6.3	25	−17.8±0.5	GuHCl	(1,2)	93J1
Ile76→Val	6.3	25	0.9±0.4	GuHCl	(1,2)	93J1
double mutant (Ile48→Ala and Ile76→Val)						
	6.3	25	−16.9±0.4	GuHCl	(1,2)	93J1

Remarks:
(1) truncated protein: first 19 residues were deleted, and Leu20 replaced by Met
(2) $\Delta(\Delta G)$ is the difference in ΔG of wild-type and mutant protein at the denaturant concentration at which the transition midpoint occurs; the $\Delta(\Delta G)$ value is indicated as the most precise one in Ref. 93J1

Chymotrypsin inhibitor 2, truncated form, wild-type and mutants, ΔT and $\Delta(\Delta G)$ determined by calorimetry

Mutant	pH	T	ΔT	$\Delta(\Delta G)$	Approach	Remarks	Ref
wild-type	3.0	25	0.0	0.0	DSC	(1,2)	93J1
Leu27→Ala	3.0	25	−15.5	−11.1	DSC	(1)	93J1
Val38→Ala	3.0	25	−2.9	−1.9	DSC	(1)	93J1
Ile39→Val	3.0	25	−6.3	−5.0	DSC	(1)	93J1
Val70→Ala	3.0	25	−12.7	−9.1	DSC	(1)	93J1
Ile76→Val	3.0	25	1.4	0.4	DSC	(1)	93J1
Ile76→Ala	3.0	25	−25.7	−17.7	DSC	(1)	93J1

Chymotrypsin inhibitor 2, truncated form, wild-type and mutants, ΔT and $\Delta(\Delta G)$ determined by calorimetry (continued)

Mutant	pH	T	ΔT	$\Delta(\Delta G)$	Approach	Remarks	Ref
wild-type	3.5	25	0	0	DSC	(1,3)	93J1
Ile39→Val	3.5	25	−6.5	−4.9	DSC	(1)	93J1
Ile48→Val	3.5	25	−4.7	−3.8	DSC	(1)	93J1
Ile48→Ala	3.5	25	−21.3	−16.3	DSC	(1)	93J1
Val66→Ala	3.5	25	−28.5	−20.5	DSC	(1)	93J1
Leu68→Ala	3.5	25	−21.5	−15.6	DSC	(1)	93J1
Ile76→Ala	3.5	25	−23.5	−18.2	DSC	(1)	93J1
double mutant (Ile48→Ala and Ile76→Val)							
	3.5	25	−21.0	−19.6	DSC	(1)	93J1

Remarks:
(1) truncated protein: first 19 residues were deleted, and Leu20 replaced by Met
(2) $\Delta G = 22.4^*$ kJ/mol
(3) $\Delta G = 28.6^*$ kJ/mol

Chymotrypsin inhibitor, wild-type and mutants

Mutant	pH	T	ΔG	$\Delta(\Delta G)$	Approach	Remarks	Ref
wild-type	6.3	25	30.0±1.3	0.0	GuHCl	(1)	94G1
Lys43→Ala	6.3	25	27.3±1.1	−2.7±0.3	GuHCl	(1)	94G1
Pro44→Ala	6.3	25	24.1±0.2	−8.9±0.5	GuHCl	(1)	94G1
Glu45→Ala	3.75	72.2		−2.0	DSC	(2)	94G1
Glu45→Ala	6.3	25	27.6±1.2	−2.4±0.6	GuHCl	(1)	94G1
double mutant (Lys43→Ala and Glu45→Ala)							
	6.3	25	25.9±1.1	−4.0±0.3	GuHCl	(1)	94G1

Remarks:
(1) linear extrapolation, data treatment by a method that includes the pre- and postdenaturational baselines for a non-linear regression of the data, $\Delta(\Delta G)$ is based on ΔG at the transition midpoint
(2) $\Delta(\Delta G)$ was obtained using $\Delta(\Delta G) = \Delta T \times \Delta S$

Chymotrypsin inhibitor, wild-type and mutants

Mutant	pH	T	ΔT	$\Delta(\Delta G)$	Approach	Remarks	Ref
wild-type	4.15	74.3±0.1	0.0	0.0	heat	(1,2)	94G1
Lys43→Ala	4.15	74.3±0.1	−1.3±0.2	−1.1	heat	(1,2)	94G1
Pro44→Ala	4.15	74.3±0.1	−9.1±0.4	−7.6	heat	(1,2)	94G1
Glu45→Ala	4.15	74.3±0.1	−1.4±0.1	−3.4	heat	(1,2)	94G1
double mutant (Lys43→Ala and Glu45→Ala)							
	4.15	74.3±0.1	−3.6±0.1	−3.0	heat	(1,2)	94G1

Remarks:
(1) $\Delta(\Delta G)$ was obtained using $\Delta(\Delta G) = \Delta T \times \Delta S$
(2) transition monitored by CD

Chymotrypsin inhibitor 2 from Barley (CI2)

Mutant	pH	T	ΔG	Approach	Remarks	Ref
M20-CI2	6.3	25	30.5±1.0	GuHCl	(1,2)	93O2
hybrid	6.3	25	10.9±0.3	GuHCl	(1,3)	93O2

Remarks:
(1) linear extrapolation
(2) mutant M20-CI2 without 19 N-terminal residues and with point mutation Leu20→Met
(3) the hybrid contains point mutations Pro52→Glu and Phe69→Ala, between Glu52 and Ala69 the subtilisin
    Carlsberg nonapeptide insert Lys136-Ala144, i.e., Lys-Gln-Ala-Val-Asp-Asn-Ala-Tyr-Ala

Chymotrypsin inhibitor 2 (CI2), 64-residue monomeric protein, wild-type and mutants

Mutant	pH	T	ΔG	Approach	Remarks	Ref
wild-type	6.25	25	31.8±0.5	GuHCl	(1,2)	95I1
Lys2→Ala	6.25	25	27.4±0.8	GuHCl	(1,2)	95I1
Lys2→Met	6.25	25	28.7±0.8	GuHCl	(1,2)	95I1
Thr3→Ala	6.25	25	27.2±1.3	GuHCl	(1,2)	95I1
Thr3→Gly	6.25	25	27.8±1.7	GuHCl	(1,2)	95I1
Thr3→Val	6.25	25	26.7±1.3	GuHCl	(1,2)	95I1
Pro6→Ala	6.25	25	46.0±10.0	GuHCl	(1,2)	95I1
Glu7→Ala	6.25	25	27.5±1.0	GuHCl	(1,2)	95I1
Glu7→Gln	6.25	25	29.4±1.8	GuHCl	(1,2)	95I1
Leu8→Ala	6.25	25	23.6±1.4	GuHCl	(1,2)	95I1
Lys11→Ala	6.25	25	26.8±1.4	GuHCl	(1,2)	95I1
Ser12→Ala	6.25	25	27.5±1.2	GuHCl	(1,2)	95I1
Ser12→Gly	6.25	25	31.5±1.0	GuHCl	(1,2)	95I1
Glu14→Asn	6.25	25	30.6±1.5	GuHCl	(1,2)	95I1
Glu14→Asp	6.25	25	28.6±1.8	GuHCl	(1,2)	95I1
Glu14→Gln	6.25	25	30.1±1.6	GuHCl	(1,2)	95I1
Glu15→Asn	6.25	25	31.5±1.6	GuHCl	(1,2)	95I1
Glu15→Asp	6.25	25	33.3±2.4	GuHCl	(1,2)	95I1
Glu15→Gln	6.25	25	29.0±1.4	GuHCl	(1,2)	95I1
Ala16→Gly	6.25	25	25.9±1.0	GuHCl	(1,2)	95I1
Lys17→Ala	6.25	25	27.1±0.7	GuHCl	(1,2)	95I1
Lys17→Gly	6.25	25	23.4±1.1	GuHCl	(1,2)	95I1
Lys18→Ala	6.25	25	27.7±2.2	GuHCl	(1,2)	95I1
Lys18→Gly	6.25	25	25.2±2.0	GuHCl	(1,2)	95I1
Val19→Ala	6.25	25	31.5±2.4	GuHCl	(1,2)	95I1
Ile20→Val	6.25	25	27.7±1.7	GuHCl	(1,2)	95I1
Leu21→Ala	6.25	25	26.7±2.1	GuHCl	(1,2)	95I1
Leu21→Gly	6.25	25	23.1±1.0	GuHCl	(1,2)	95I1
Gln22→Ala	6.25	25	29.5±2.5	GuHCl	(1,2)	95I1
Gln22→Gly	6.25	25	27.9±1.8	GuHCl	(1,2)	95I1
Asp23→Ala	6.25	25	27.4±0.7	GuHCl	(1,2)	95I1
Lys24→Ala	6.25	25	24.1±1.2	GuHCl	(1,2)	95I1
Lys24→Gly	6.25	25	17.2±1.7	GuHCl	(1,2)	95I1
Pro25→Ala	6.25	25	26.7±1.3	GuHCl	(1,2)	95I1
Glu26→Ala	6.25	25	28.5±1.0	GuHCl	(1,2)	95I1
Ile29→Ala	6.25	25	17.2±1.6	GuHCl	(1,2)	95I1
Ile29→Val	6.25	25	28.6±1.9	GuHCl	(1,2)	95I1
Ile30→Ala	6.25	25	27.0±1.1	GuHCl	(1,2)	95I1
Ile30→Gly	6.25	25	19.5±0.7	GuHCl	(1,2)	95I1
Ile30→Thr	6.25	25	27.3±1.0	GuHCl	(1,2)	95I1
Ile30→Val	6.25	25	30.0±0.8	GuHCl	(1,2)	95I1
Leu32→Ala	6.25	25	24.5±1.1	GuHCl	(1,2)	95I1

Chymotrypsin inhibitor 2 (CI2), 64-residue monomeric protein, wild-type and mutants (continued)

Mutant	pH	T	ΔG	Approach	Remarks	Ref
Leu32→Ile	6.25	25	27.6±1.4	GuHCl	(1,2)	95I1
Leu32→Val	6.25	25	27.5±1.5	GuHCl	(1,2)	95I1
Pro33→Ala	6.25	25	29.4±1.1	GuHCl	(1,2)	95I1
Val34→Ala	6.25	25	26.1±2.4	GuHCl	(1,2)	95I1
Val34→Gly	6.25	25	23.0±0.9	GuHCl	(1,2)	95I1
Val34→Thr	6.25	25	24.0±1.1	GuHCl	(1,2)	95I1
Thr36→Ala	6.25	25	25.9±0.9	GuHCl	(1,2)	95I1
Thr36→Ser	6.25	25	27.7±0.5	GuHCl	(1,2)	95I1
Thr36→Val	6.25	25	25.2±1.8	GuHCl	(1,2)	95I1
Ile37→Ala	6.25	25	26.4±1.0	GuHCl	(1,2)	95I1
Val38→Ala	6.25	25	29.0±1.6	GuHCl	(1,2)	95I1
Thr39→Ala	6.25	25	28.7±2.1	GuHCl	(1,2)	95I1
Thr39→Asp	6.25	25	31.4±1.4	GuHCl	(1,2)	95I1
Glu41→Ala	6.25	25	27.1±2.5	GuHCl	(1,2)	95I1
Arg43→Ala	6.25	25	27.5±1.6	GuHCl	(1,2)	95I1
Asp45→Ala	6.25	25	27.5±1.2	GuHCl	(1,2)	95I1
Val47→Ala	6.25	25	10.8±1.5	GuHCl	(1,2)	95I1
Leu49→Ala	6.25	25	16.7±1.0	GuHCl	(1,2)	95I1
Phe50→Ala	6.25	25	18.4±1.1	GuHCl	(1,2)	95I1
Phe50→Leu	6.25	25	23.8±1.4	GuHCl	(1,2)	95I1
Phe50→Val	6.25	25	25.9±2.1	GuHCl	(1,2)	95I1
Val51→Ala	6.25	25	26.8±2.4	GuHCl	(1,2)	95I1
Asp52→Ala	6.25	25	18.7±2.0	GuHCl	(1,2)	95I1
Asp53→Asn	6.25	25	26.7±0.5	GuHCl	(1,2)	95I1
Asn56→Ala	6.25	25	26.7±0.6	GuHCl	(1,2)	95I1
Asn56→Asp	6.25	25	25.5±1.0	GuHCl	(1,2)	95I1
Ile57→Ala	6.25	25	14.4±1.6	GuHCl	(1,2)	95I1
Ile57→Val	6.25	25	31.2±2.1	GuHCl	(1,2)	95I1
Ala58→Gly	6.25	25	27.8±2.8	GuHCl	(1,2)	95I1
Val60→Ala	6.25	25	25.8±1.5	GuHCl	(1,2)	95I1
Val60→Gly	6.25	25	25.4±1.1	GuHCl	(1,2)	95I1
Val60→Thr	6.25	25	25.9±2.2	GuHCl	(1,2)	95I1
Pro61→Ala	6.25	25	17.2±1.3	GuHCl	(1,2)	95I1
Val63→Ala	6.25	25	26.5±1.9	GuHCl	(1,2)	95I1
Val63→Gly	6.25	25	19.3±0.6	GuHCl	(1,2)	95I1
Val63→Thr	6.25	25	25.4±1.9	GuHCl	(1,2)	95I1
(Lys2→Ala and Glu7→Ala)						
	6.25	25	26.9±1.0	GuHCl	(1,2)	95I1
(Pro6→Ala and Ala16→Gly)						
	6.25	25	26.4±1.5	GuHCl	(1,2)	95I1
(Glu14→Ala and Glu15→Ala)						
	6.25	25	31.9±2.7	GuHCl	(1,2)	95I1
(Glu28→Met and Met40→Leu)						
	6.25	25	31.0±0.9	GuHCl	(1,2)	95I1
(Ile29→Ala and Ile57→Val)						
	6.25	25	14.6±0.8	GuHCl	(1,2)	95I1
(Ile29→Ala and Ile57→Val)						
	6.25	25	14.6±0.8	GuHCl	(1,2)	95I1
(Leu32→Val and Phe50→Leu)						
	6.25	25	27.3±1.5	GuHCl	(1,2)	95I1
(Leu32→Val and Phe50→Ala)						
	6.25	25	18.2±0.9	GuHCl	(1,2)	95I1
(Leu32→Ala and Phe50→Leu)						
	6.25	25	18.2±0.8	GuHCl	(1,2)	95I1

Chymotrypsin inhibitor 2 (CI2), 64-residue monomeric protein, wild-type and mutants (continued)

Mutant	pH	T	ΔG	Approach	Remarks	Ref
(Leu32→Ala and Phe50→Ala)						
	6.25	25	10.2±1.0	GuHCl	(1,2)	95I1
(Leu32→Ala and Val38→Ala)						
	6.25	25	19.5±0.6	GuHCl	(1,2)	95I1
(Leu32→Val and Val38→Ala)						
	6.25	25	23.7±0.8	GuHCl	(1,2)	95I1
(Ile37→Ala and Val38 deleted)						
	6.25	25	20.0±0.5	GuHCl	(1,2)	95I1
(Val38→Ala and Phe50→Leu)						
	6.25	25	22.6±0.8	GuHCl	(1,2)	95I1
(Val38→Ala and Phe50→Ala)						
	6.25	25	16.0±0.9	GuHCl	(1,2)	95I1
(Thr39→Ala and Glu41→Ala)						
	6.25	25	28.7±1.5	GuHCl	(1,2)	95I1
(Thr39→Asp and Glu41→Ala)						
	6.25	25	30.5±1.5	GuHCl	(1,2)	95I1
(Arg43→Ala and Asp45→Ala)						
	6.25	25	21.5±1.5	GuHCl	(1,2)	95I1
(Ser12→Gly, Glu14→Ala and Glu15→Ala)						
	6.25	25	28.4±2.1	GuHCl	(1,2)	95I1
(Ser12→Ala, Glu14→Ala and Glu15→Ala)						
	6.25	25	28.7±1.8	GuHCl	(1,2)	95I1
(Leu32→Ala, Val38→Ala and Phe50→Leu)						
	6.25	25	17.8±0.5	GuHCl	(1,2)	95I1
(Leu32→Val, Val38→Ala and Phe50→Leu)						
	6.25	25	19.9±0.5	GuHCl	(1,2)	95I1

Remarks:
(1) linear extrapolation
(2) measured in 50 mM MES pH 6.25

Chymotrypsin inhibitor CI2, difference values derived from association/folding of the segments CI2(1–40) with CI2(41–64)

Mutant	pH	T	$\Delta(\Delta G)$	Remarks	Ref
wild-type	6.3	25	0	(1)	96N1
Thr3→Ala	6.3	25	3.6	(1)	96N1
Leu8→Ala	6.3	25	11.2	(1)	96N1
Ser12→Ala	6.3	25	3.7	(1)	96N1
Glu14→Asn	6.3	25	2.9	(1)	96N1
double mutant (Glu14→Ala and Glu15→Ala)					
	6.3	25	3.2	(1)	96N1
triple mutant (Ser12→Ala, Glu14→Ala and Glu15→Ala)					
	6.3	25	7.0	(1)	96N1
Ala16→Gly	6.3	25	4.6	(1)	96N1
Lys17→Ala	6.3	25	2.1	(1)	96N1
Lys18→Gly	6.3	25	4.1	(1)	96N1
Ala18→Gly	6.3	25	5.0	(1)	96N1
Val19→Ala	6.3	25	2.0	(1)	96N1
Ile20→Val	6.3	25	5.4	(1)	96N1
Lys24→Ala	6.3	25	2.7	(1)	96N1
Ile29→Val	6.3	25	4.6	(1)	96N1
Ile30→Ala	6.3	25	8.9	(1)	96N1
Val30→Ala	6.3	25	9.2	(1)	96N1

Chymotrypsin inhibitor CI2, difference value derived from association/folding of the segments CI2(1–40) with CI2(41–64) (continued)

Mutant	pH	T	$\Delta(\Delta G)$	Remarks	Ref
Leu32→Val	6.3	25	2.1	(1)	96N1
Leu32→Ala	6.3	25	9.9	(1)	96N1
Val34→Thr	6.3	25	4.3	(1)	96N1
Val34→Ala	6.3	25	2.6	(1)	96N1
Phe50→Leu	6.3	25	8.8	(1)	96N1
Val51→Ala	6.3	25	8.2	(1)	96N1
Ile57→Val	6.3	25	−0.8	(1)	96N1

Remark:
(1) for the procedure, see Refs. 94D3 and 95I1

Chymotrypsinogen

Chymotrypsinogen

pH	T	ΔG	Approach	Remark	Ref
3	25	31	heat, optical method		64B2
3	25	31	heat, optical method		64B3
7	25	63±21	heat, optical method		75B
		52	hydrogen exchange		80C
4	25	40.6*	DSC	(1)	84F4

Remark:
(1) Ref. 84F4 contains the dependence of thermodynamic parameters on protein concentration. The given value refers to zero protein concentration

α-Chymotrypsinogen, ΔG determined by various denaturants

Denaturant	T	ΔG	Approach	Ref
urea	25	37.5	linear extrapolation	94P3
methylurea	25	34.9	linear extrapolation	94P3
N,N′-dimethylurea	25	33.1	linear extrapolation	94P3
ethylurea	25	37.8	linear extrapolation	94P3
propylurea	25	27.1	linear extrapolation	94P3

Chymotrypsinogen A, modified by reductive alkylation

Form	pH	T	$\Delta(\Delta G)$	Approach	Ref
native protein	3.0	40	18.2	DSC	91F2
methylated (92%)	3.0	40	20.6	DSC	91F2
ethylated (56%)	3.0	40	20.5	DSC	91F2
n-butylated (49%)	3.0	40	20.6	DSC	91F2

Chymotrpysinogen, guadinated

ΔG	Approach	Ref
59	hydrogen exchange	80C

Coiled-Coils, see α-helix

Cold-Shock Protein

Cold-shock protein, CspB from *Bacillus subtilis*

pH	T	ΔG	Approach	Remarks	Ref
7.0	25	8.9	heat	(1,2)	96S4
7.0	25	8.9	kinetics	(2,3)	96S4

Remarks:
(1) thermal denaturation in the presence of urea, ΔG refers to zero denaturant concentration
(2) buffer: 0.1 M sodium cacodylate/HCl
(3) from the difference of activation parameters for unfolding and refolding

Colicin

Colicin E1 channel peptide, Trp mutants

Mutant	pH	T	$\Delta G_{(N \to I)}$	$\Delta G_{(N \to I)}$	$\Delta G_{(N \to I)}$	Approach	Remarks	Ref
wild-type	6.0	20	35±6	25±2	60±8	GuHCl	(1,2,5)	95S12
Trp355	6.0	20			17±3	GuHCl	(1,4,5)	95S12
Trp367	6.0	20			17±1	GuHCl	(1,4,5)	95S12
Trp404	6.0	20			11±2	GuHCl	(1,4,5)	95S12
Trp413	6.0	20			32±7	GuHCl	(1,4,5)	95S12
Trp424	6.0	20			14±2	GuHCl	(1,3,5)	95S12
Trp431	6.0	20	17±5	10±2	27±7	GuHCl	(1,4,5)	95S12
Trp460	6.0	20	24±4	11±7	35±11	GuHCl	(1,3,5)	95S12
Trp484	6.0	20	18±8	22±7	40±15	GuHCl	(1,4,5)	95S12
Trp495	6.0	20	28±6			GuHCl	(1,3,5)	95S12
Trp507	6.0	20	26±3	43±11	69±14	GuHCl	(1,4,5)	95S12
wild-type, Trp355 and Trp495								
	7.0	20			29±4	GuHCl	(1–4,6)	95S12

Remarks:
(1) linear extrapolation to zero denaturant concentration by a method that includes the pre- and postdenaturational baselines for a non-linear regression of the data; the transitions of biphasic unfolding profiles were fit independently
(2) wild-type colicin channel peptide contains three Trp residues in position 424, 460, and 495
(3) two of the three Trp residues of the wild-type protein were replaced by Phe, indicated is the remaining Trp
(4) all three Trp residues of the wild-type protein were replaced by Phe, indicated is the single Trp mutant replacing either Phe or Tyr
(5) the transition was monitored by fluorescence, buffer: 10 mM dimethylglutaric acid, 100 mM NaCl, pH 6.0
(6) the transition was monitored by CD, non-linear fit through the average of the data, buffer: 10 mM sodium phosphate, 100 mM sodium flouride, pH 7

Collagen

Collagen from various species the following table contains T_{trs} and ΔG_{res} (kJ/mol), see also 82P2

Species	T_{trs}	ΔG_{res}	Approach	Ref
cod skin	15	−0.183	DSC	79P6
halibut	18	−0.12	DSC	79P6
frog skin (Rana temp.)	25	0	DSC	79P6
pike skin	27	0.035	DSC	79P6
carp swim bladder	30	0.090	DSC	79P6
rat skin	37	0.257	DSC	79P6
sheep skin	37	0.230	DSC	79P6

Collagen-like triple helical peptides

Peptide	pH	T	ΔG	Approach	Remarks	Ref
Peptide I	1	25	115	heat	(1,2)	94V1
Peptide I	7	25	110	heat	(1,2)	94V1
Peptide I	13	25	120	heat	(1,2)	94V1
Peptide II	1	25	89	heat	(1,3)	94V1
Peptide II	7	25	92	heat	(1,3)	94V1
Peptide II	13	25	90	heat	(1,3)	94V1

Remarks:
(1) peptide I = (Pro-Hyp-Gly)$_{10}$
(2) ΔG refers to a two-state trimer to monomer transition
(3) peptide II = (Pro-Hyp-Gly)$_4$-(Glu-Lys-Gly)-(Pro-Hyp-Gly)$_5$

Complement Protein

Human complement C1S, fragment (24 kDa)

pH	T	ΔG	Approach	Ref
7.2	25	44*	DSC	89M6

Complement Receptor

Human complement receptor 1, SCR1-3 Domain

Form	pH	T	ΔG	Approach	Remarks	Ref
oxidized	7.4	25	19.5	GuHCl	(1,2)	96C12
oxidized	7.4	25	19.5	GuHCl	(1,3)	96C12
oxidized	7.4	25	19.9	GuHCl	(1,4)	96C12
oxidized	7.4	25	19.6±0.2	GuHCl	(1,5)	96C12
reduced	7.4	25	9.8	GuHCl	(1,2)	96C12
reduced	7.4	25	11.2	GuHCl	(1,3)	96C12

Remarks:
(1) linear extrapolation
(2) transition monitored by fluorescence
(3) transition monitored by near-UV CD
(4) transition monitored by far-UV CD
(5) average value for oxidized (-S-S-) protein

Creatine Kinase

Creatine kinase, mitochondiral, wild-type and mutants

Mutant	pH	T	ΔG	Approach	Remarks	Ref
wild-type	7.4	22	112.2	GuHCl	(1)	94G7
Trp206→Phe	7.4	22	102.9	GuHCl	(1)	94G7
Trp206→Cys	7.4	22	97.4	GuHCl	(1,2)	94G7
Trp213→Cys	7.4	22	102.2	GuHCl	(1)	94G7
Trp223→Phe	7.4	22	108.2	GuHCl	(1)	94G7
Trp223→Cys	7.4	22	108.8	GuHCl	(1)	94G7
Trp264→Cys	7.4	22	85.7	GuHCl	(1)	94G7
Trp268→Cys	7.4	22	107.2	GuHCl	(1)	94G7

Remarks:
(1) ΔG is the octamer stability
(2) partly ($\leq$5%) monomeric

Cro Repressor

λ Cro repressor, covalently linked dimer Val55→Cys

	pH	T	ΔG	Approach	Remarks	Ref
	3.0	25	3.7	DSC	(1)	96F2
	4.0	25	21.1	DSC	(1)	96F2
	4.5	25	27.5	DSC	(1)	96F2
	5.0	25	31.3	DSC	(1)	96F2
	7.0	25	39.8	DSC	(1)	96F2

Remark:
(1) ΔG refers to the $N_2 \rightarrow U_2$ unfolding equilibrium, the unfolding transition, however, is more complex for further details, see Ref. 96F2)

CRP, cAMP Receptor Protein

CRP, cAMP receptor protein

Transition	pH	T	ΔG	Approach	Remarks	Ref
CRP, the intact protein						
trans. (1)	7.9	20	50.2±2.5	GuHCl	(1–5)	93C2
trans. (2)	7.9	20	30.1±0.4	GuHCl	(1–5)	93C2
S-CRP, fragment of cAMP receptor protein, obtained by subtilisin treatment						
trans. (1)	7.9	20	42.3±1.7	GuHCl	(1–5)	93C2
trans. (2)	7.9	20	36.4±1.7	GuHCl	(1–5)	93C2
CH-CRP, fragment of cAMP receptor protein, obtained by chymotrypsin treatment						
trans. (1)	7.9	20	39.7±1.7	GuHCl	(1–5)	93C2
trans. (2)	7.9	20	36.4±1.7	GuHCl	(1–5)	93C2

Remarks:
(1) linear extrapolation
(2) buffer: 50 mM Tris, 0.1 M KCl, 1 mM EDTA, pH 7.9
(3) transition monitored by fluorescence emission at 345 nm, anisotropy, and CD at 222 nm
(4) mechanism: native CRP dimer→2 native CRP monomers,
     2 native CRP monomers→2 denatured monomers
(5) trans. (1): dissociation trans. (2): unfolding

Crystallin

α-Crystallin

	pH	T	ΔG	Approach	Remarks	Ref
	7.4	25	24±5	DSC	(1)	96G4

Remark:
(1) based on thermodynamic quantities that were extrapolated to infinite scan rate

γ-Crystallin, various fractions and glutathione-modified protein

Fraction	pH	T	ΔG	Approach	Remarks	Ref
γ-II	7.0	25	21.8	GuHCl	(1)	90K6
γ-IIIA	7.0	25	17.6	GuHCl	(1)	90K6
γ-IIIB	7.0	25	27.3	GuHCl	(1)	90K6
γ-IVA	7.0	25	34.9	GuHCl	(1)	90K6
Glutathione-modified protein:						
γ-II	7.0	25	16.0	GuHCl	(1)	90K6
γ-IIIA	7.0	25	15.1	GuHCl	(1)	90K6
γ-IIIB	7.0	25	23.1	GuHCl	(1)	90K6
γ-IVA	7.0	25	24.8	GuHCl	(1)	90K6

Remark:
(1) linear extrapolation

Crystallin, modified

Form	pH	T	ΔG	Approach	Remarks	Ref
γ, unmodified	6.8	27	15.9	GuHCl	(1)	93L9
γ, glycated	6.8	27	7.5	GuHCl	(1)	93L9
α, unmodified	6.8	27	4.6	GuHCl	(1)	93L9
α, glycated	6.8	27	5.9	GuHCl	(1)	93L9

Remark:
(1) linear extrapolation

γB-Crystallin, C-terminal domain

	pH	T	ΔG	Approach	Ref
	5	20	35.4	urea, linear extrapol.	94M10

CspB

CspB, cold-shock protein from *Bacillus subtilis*

pH	T	ΔG	Approach	Remarks	Ref
7.0	25	12.4±0.4	urea	(1,2)	95S2
7.0	11	13.0	urea	(1,2)	95S2
7.0	25	11.0±1.4	urea	(2,3)	95S2
7.0	25	11.1±0.6	urea	(2,4)	95S2

Remarks:
(1) equilibrium unfolding, linear extrapolation
(2) buffer: 0.1 M Na cacodylate/HCl
(3) folding kinetics
(4) from final fluorescence values observed in the kinetic experiments

Cytochrome b₅

Cytochrome b₅

pH	T	ΔG	Approach	Remarks	Ref
7.4	25	21	DSC	(1)	83B1
8	25	16	DSC	(1)	83B1
7.4	25	13.6	heat	(2)	82S1
7.4	25	28.4	heat	(3)	82S1

Remarks:
(1) cytochrome b₅ from rabbit, reconstituted with dimyristoyllecithin
(2) optical methods, the data refer to the lower temperature range specified in Ref. 82S1
(3) optical methods, the data refer to the higher temperature range specified in 82S1

Cytochrome b₅, rabbit, fragments

Residues	pH	T	ΔG	Approach	Remarks	Ref
1–90	5.5	25	22.5±2.1	urea	(1)	93P5
1–90	7	25	25.0±2	DSC		80P
12–97	7	25	27.6±2	DSC		80B

Remark:
(1) linear extrapolation to zero denaturant concentration by a method that includes the pre- and postdenaturational baselines for a non-linear regression of the data

Cytochrome b_5, bovine, tryptic fragments

Mutant	pH	T_{trs}	ΔT	$\Delta(\Delta G)$	Approach	Remarks	Ref
Tryptic fragment, variant (Ala7-Lys90):							
oxidized	7.0	67.4±0.7	0.0	0.0	heat		92N1
reduced	7.0	73.2±0.8	5.8	3.79	heat	(1)	92N1
Tryptic fragment, variant (Ala1-Lys90):							
oxidized	7.0	73.0±1.1	5.6	5.01	heat	(1)	92N1
reduced	7.0	79.2±1.5	11.8	10.08	heat	(1)	92N1
Tryptic fragment, variant (Ala1-Ser104):							
oxidized	7.0	73.1	5.7		heat		92N1

Remark:
(1) $\Delta(\Delta G)$ was obtained using $\Delta(\Delta G) = \Delta T \times \Delta S$

Cytochrome b_5, bovine recombinant

Mutant	pH	T	ΔT	$\Delta(\Delta G)$	Approach	Remarks	Ref
Oxidized form:							
Ala7-Lys90	7.0	67.4±0.7	0.0	0.0	heat	(1)	93H6
Ala1-Ser104	7.0	67.4±0.7	5.7	4.35	heat		93H6
Ala1-Lys90	7.0	67.4±0.7	5.8	3.79	heat		93H6
Reduced form:							
Ala7-Lys90	7.0	67.4±0.7	5.6	5.01	heat		93H6
Ala1-Ser104	7.0	67.4±0.7	11.3	12.38	heat		93H6
Ala1-Lys90	7.0	67.4±0.7	11.8	10.08	heat		93H6

Remark:
(1) the mutant Ala7-Lys90, oxidized, serves as the reference protein

Apocytochrome b_5 from rabbit liver

pH	T	ΔG	Approach	Ref
7.4	25	7.1±1.1	DSC	93P3

Cytochrome b_{562}

Cytochrome b_{562}

Form	pH	T	ΔG	Approach	Remarks	Ref
ferri	7.0	20±0.2	27.6±2.1	urea	(1)	91F1
apo	7.0	20±0.2	13.4±2.1	urea	(1)	91F1

Remark:
(1) linear extrapolation

Cytochrome b$_{562}$, wild-type and mutants

Mutant	pH	T	ΔT	Δ(ΔG)	Approach	Remarks	Ref
wild-type	7.4	66	0.0	0.0	heat	(1,2)	93R1
Glu18→Arg	7.4	66	−1.87	−2.09	heat	(3)	93R1
Glu18→Gln	7.4	66	−0.97	−1.09	heat	(3)	93R1
Glu18→Leu	7.4	66	0.26±0.1	0.29±0.08	heat	(3)	93R1
Gln25→Glu	7.4	66	−0.90	−1.00	heat	(3)	93R1
Gln25→Lys	7.4	66	−1.19	−1.34	heat	(3)	93R1
double mutant (Glu18→Arg and Gln25→Lys)							
	7.4	66	−6.53	−7.32	heat	(3)	93R1
double mutant (Glu18→Gln and Gln25→Lys)							
	7.4	66	−5.04	−5.65	heat	(3)	93R1
double mutant (Glu18→Leu and Gln25→Lys)							
	7.4	66	−2.01	−2.26	heat	(3)	93R1
double mutant (Glu18→Leu and Gln25→Glu)							
	7.4	66	1.02	1.13	heat	(3)	93R1

Remarks:
(1) the wild-type has the positions Glu18 and Gln25 buffer: 20 mM potassium phosphate, 1 mM EDTA
(2) ΔG of the wild-type protein amounts to 27.6 kJ/mol at 25°C, pH 7.4
(3) Δ(ΔG) was obtained using Δ(ΔG)= ΔT×ΔS

Cytochrome c

The data entries are arranged as follows:

a) data for various species obtained by different approaches,
b) various forms, mutants and chemically modified protein,
c) various states and transitions.

a) Data for various species obtained by different approaches

Cytochrome c, ferric form, from various species

Species	pH	T	ΔG	Approach	Remarks	Ref
bovine	4.8	25	37.7±2.5	DSC	74P1	
bovine	6.5	25	64.4	GuHCl	(1)	74K
bovine	6.5	25	35.1	GuHCl	(2)	74K
bovine	7	25	32.1	GuHCl	(2)	78M1
candida krusei	7	25	58.6	GuHCl	(1)	74K
candida krusei	7	25	30.3	GuHCl	(2)	74K
chicken	7	25	32.1	GuHCl	(2)	78M1
dog	7	25	30.4	GuHCl	(2)	78M1
donkey	7	25	32.1	GuHCl	(2)	78M1
frog	7	25	47	GuHCl	(2)	82B
horse	1.7–1.9	20	13.8	pH, salt	(4)	90G3
horse	4	16±1	41.9±5.9	urea	(3)	82H
horse	5.6	23	17	urea	(2)	81H1
horse			39	hydrogen exchange		80C
horse	6.5	25	53.1	GuHCl	(1)	74K
horse	6.5	25	30.4	GuHCl	(2)	74K
horse	7	25	24.9	GuHCl and heat		77K
horse	7	25	30.9	GuHCl	(2)	78M1

Cytochrome c, ferric form, from various species (continued)

Species	pH	T	ΔG	Approach	Remarks	Ref
horse	7	25	35.6	GuHCl	(2)	86S1
horse	7.0	25	36.1±1.5	GuHCl	(5)	93P5
horse	7.6	25	43.9	GuHCl	(2)	83S
horse	7.6	25	80.3	GuHCl	(1)	83S
rabbit	7	25	32.1	GuHCl	(2)	78M1
tuna	7	25	32.7	GuHCl	(2)	78M1

Remarks:
(1) denaturant binding model
(2) linear extrapolation
(3) urea gradient electrophoresis, linear extrapolation
(4) acid and salt-induced subtransition of the A state
(5) linear extrapolation to zero denaturant concentration by a method that includes the pre- and postdenaturational baselines for a non-linear regression of the data

Cytochrome c from various species, and recombinant protein

Form	pH	T	ΔG	Approach	Remarks	Ref
bullfrog	7.0	27	30.1	GuHCl	(1,5)	92B4
chicken	7.0		39.8	urea	(1)	92S5
guanaco	7.0		38.2	urea	(1)	92S5
horse	7.0		36.5	urea	(1)	92S5
horse	7.0		33.9	GuHCl	(1)	92S5
horse	7.0	27	31.0	GuHCl	(1,5)	92B4
horse	10.5		27.6	urea	(1)	92S5
horse-CN	7.0		22.0	urea	(1,3)	92S5
horse di-CMc	7.0		18.0	urea	(1,2)	92S5
horse di-CMc	7.0		16.3	GuHCl	(1,2)	92S5
kangaroo	7.0		39.4	urea	(1)	92S5
mink	7.0	37.9		urea	(1)	92S5
penguin	7.0		35.2	urea	(1)	92S5
pigeon	7.0		34.5	urea	(1)	92S5
rat RNc-II	7.0		38.3	urea	(1,4)	92S5
tuna	7.0	27	37.7	GuHCl	(1,5)	92B4

Remarks:
(1) linear extrapolation
(2) di-CM-c = di(met-65-sulfur,met-80-sulfur)- carbamoylcytochrome c
(3) CN complex
(4) RNc-II = recombinant wild-type rat cytochrome c
(5) reconsideration of data from Ref. 82B

Cytochrome c from horse heart

Urea Concentration	pH	T	ΔG	Approach	Remarks	Ref
	6.0	25	31.6±1.3	GuHCl	(1)	94A1
	6.0	25	31.3±0.2	urea	(1)	94A1
GuHCl-induced unfolding in the presence of urea:						
2.0 M urea	6.0	25	22.3±2.5	GuHCl	(1,2)	94A1
4.0 M urea	6.0	25	14.2±0.6	GuHCl	(1,2)	94A1
6.0 M urea	6.0	25	3.2±0.5	GuHCl	(1,2)	94A1

Remarks:
(1) linear extrapolation
(2) the results support the linear extrapolation procedure

Cytochrome c from horse heart

pH	T	ΔG	Approach	Remarks	Ref
4.9	10	34.3±2.5	GuHCl	(1)	96S9
4.9	10	36.8±1.7		(2)	96S9

Remarks:
(1) equilibrium unfolding, linear extrapolation
(2) activation parameters obtained from kinetic experiments under two-state conditions

Cytochrome c from horse heart

pH	T	ΔG	Approach	Remarks	Ref
4.75	23	46*	urea	(1,2)	92J2

Remarks:
(1) linear extrapolation
(2) the data were taken from Fig. 6 in Ref. 92J2

Cytochrome c from horse heart

pH	T	ΔG	Approach	Remarks	Ref
7.0	25	28.7±0.5	GuHCl	(1–3)	96A1

Remarks:
(1) linear extrapolation
(2) measured in 0.03 M cacodylic buffer containing 0.1 M KCl
(3) Ref. 96A1 contains a systematic study of ΔG of cytochrome c in the presence of various salts

b) Various forms, mutants and chemically modified protein

Cytochrome c from horse heart, oxidized form

pH	T	ΔG	Approach	Remarks	Ref
5.0	10	42.3*	GuHCl	(1)	94E3
7.0	10	43.6*	GuHCl	(1)	94E3

Remark:
(1) linear extrapolation

Cytochrome c, ferrous form

Species	pH	T	ΔG	Approach	Remarks	Ref
bovine	7	25	34.6	GuHCl	(1)	78M1
horse	7	25	33.4	GuHCl	(1)	78M1

Remark:
(1) linear extrapolation

Reduced cytochrome c from horse heart, CO complex

	pH	T	ΔG	Approach	Remarks	Ref
	6.5	40	40.3*	GuHCl	(1)	93J4

Remark:
(1) linear extrapolation

Cytochrome c, various forms and species

Form/Species	pH	T	$\Delta(\Delta G)$	Approach	Remarks	Ref
horse	7	25	0.0	GuHCl	(1)	86S1
horse, cyanide complex						
	7	25	0.0	GuHCl	(2)	86S1
candida	7	25	−10.9	GuHCl	(2)	86S1
saccharomyces, monomer						
	7	25	−16.5	GuHCl	(2)	86S1
saccharomyces, dimer						
	7	25	−25	GuHCl	(2)	86S1

Remarks:
(1) reference value is ΔG = 35.6 kJ/mol obtained by GuHCl induced unfolding and linear extrapolation
(2) linear extrapolation

Cytochrome c from rat, recombinant, wild-type and mutants

Mutant	pH	T	ΔG	Approach	Remarks	Ref
wild-type	7.0	25	38.2	urea	(1,2)	94K2
wild-type	7.2	25	39.7	urea	(1,2)	95Q2
His26→Val	7.2	25	23.9	urea	(1,2)	95Q2
Pro30→Ala	7.0	15.7		urea	(1,2)	92S5
Pro30→Ala	7.0	25	15.7	urea	(1,2)	94K2
Pro30→Ala	7.2	25	15.7	urea	(1,2)	95Q2
Pro30→Val	7.2	25	18.8	urea	(1,2)	95Q2
His33→Phe	7.2	25	42.6	urea	(1,2)	95Q2
Asn52→Ile	7.0	25	26.6	urea	(1,2)	94K2
Asn52→Ile	7.2	25	26.6	urea	(1,2)	95Q2
Tyr67→Phe	7.0	25	30.1	urea	(1,2)	94K2
Tyr67→Phe	7.0		30.1	urea	(1,2)	92S5
Tyr67→Phe	7.2	25	30.1	urea	(1,2)	95Q2
double mutant (His26→Val and His33→Phe)						
	7.2		30.8	urea	(1,2)	95P2

Cytochrome c from rat, recombinant, wild-type and mutants (continued)

Mutant	pH	T	ΔG	Approach	Remarks	Ref
double mutant (His26→Val and Asn52→Ile)						
	7.2		32.0	urea	(1,2)	95P2
double mutant (Pro30→Ala and Tyr67→Phe)						
	7.0		28.2	urea	(1,2)	92S5
	7.0	25	28.2	urea	(1,2)	94K2
	7.2	25	28.2	urea	(1,2)	95P2

Remarks:
(1) linear extrapolation
(2) recombinant rat cytochrome c, called RNc, see also Refs. 89L5, 92S5, and 94K9

Cytochrome c from horse, wild-type and mutants

Mutant	pH	T	ΔG	Approach	Remarks	Ref
wild-type	7.0	10	40.6±4.6	GuHCl	(1,2)	96C14
Leu94→Ile	7.0	10	41.4±2.9	GuHCl	(1,2)	96C14
Leu94→Val	7.0	10	37.2±2.9	GuHCl	(1,2)	96C14
Leu94→Ala	7.0	10	25.9±2.1	GuHCl	(1,2)	96C14

Remarks:
(1) linear extrapolation to zero denaturant concentration by a method that includes the pre- and post-denaturational baselines for a non-linear regression of the data
(2) measured in 0.1 M sodium phosphate

Cytochrome c from horse heart, ferric form, chemically modified

Modification	pH	T	ΔG	Approach	Remarks	Ref
succinylated	4	16±1	38.9±8.4	urea	(1)	82H
guanidinated		44		hydrogen exchange		80C

Remark:
(1) urea gradient electrophoresis, linear extrapolation

Cytochrome c from horse heart, native and carboxymethylated (modified) at Met65 and Met80

Form	pH	T	ΔG	Approach	Remarks	Ref
native	7.0	25	63.2	GuHCl	(1)	89S1
native	7.0	25	25.9	GuHCl	(2,4)	89S3
native	7.0	25	24.3	GuHCl	(3,4)	89S3
native	7.0	25	21.3	urea	(3,4)	89S3
native	7.0	25	28.0	urea	(2,4)	89S3
modified	7.0	25	11.5	urea	(3,4)	89S3
modified	7.0	25	17.6	urea	(2,4)	89S3
modified	7.0	25	11.7	GuHCl	(3,4)	89S3
modified	7.0	25	14.6	GuHCl	(2,4)	89S3

Remarks:
(1) denaturant binding model, k = 1.2
(2) monitored by CD at 222 nm
(3) monitored by Soret absorption at 408 nm
(4) linear extrapolation

Table 1. Gibbs Energy Change – Molar Values

Cytochrome c from horse heart, native and modified protein

Form	pH	T	ΔG	Approach	Remarks	Ref
native	7.0		31.0±0.4	GuHCl	(1,2)	95J3
native	7.0		35.6±2.1	GuHCl	(1,3)	95J3
native	7.0		35.1±0.8	GuHCl	(1,4)	95J3
modified	7.0		17.2±0.4	GuHCl	(1,2,5)	95J3
modified	7.0		16.7±2.9	GuHCl	(1,3,5)	95J3
modified	7.0		13.8±2.1	GuHCl	(1,4,5)	95J3

Remarks:
(1) linear extrapolation
(2) transition monitored by fluorescence
(3) transition monitored by absorbance
(4) transition monitored by CD at 222 nm
(5) modified protein: S-(carboxymethyl)methionine cytochrome c

Porphyrin-cytochrome c from horse heart

pH	T	ΔG	Approach	Ref
3.1	20	−3.6	heat, v.H.	96H1
3.4	20	−0.9	heat, v.H.	96H1
3.9	20	2.4	heat, v.H.	96H1
4.8	20	6.4	heat, v.H.	96H1
6.0	20	5.8	heat, v.H.	96H1
7.0	20	7.1	heat, v.H.	96H1

c) Various states and transitions

Cytochrome c from horse heart, various folded states of porphyrin-cytochrome c, apo-cytochrome c, and holo-cytochrome c

Form	pH	T	ΔG	Approach	Remarks	Ref
holo-cytochrome c:						
native	7.0	20	37.5	urea	(1)	96H1
native	7.0	20	34.2	GuHCl	(1)	96H1
m.g. (2)	2.0	20	10.8	GuHCl	(1,3)	96H1
porphyrin-cytochrome c:						
m.g.-like	7.0	20	6.1	urea	(1)	96H1
m.g.-like	7.0	20	5.4	GuHCl	(1)	96H1
m.g.-like	2.0	20	3.4	urea	(1,3)	96H1
apo-cytochrome c:						
m.g.-like	2.0	20	2.3	urea	(1,3)	96H1

Remarks:
(1) linear extrapolation
(2) m.g. = molten globule
(3) measured in the presence of 200 mM $NaClO_4$

Apocytochrome c from chicken, wild-type and mutant

Mutant	pH	T	$\Delta(\Delta G)$	Approach	Remarks	Ref
wild-type	7.0	25	0.0	$NaClO_4$	(1)	96T4
Val92→Ala	7.0	25	1.1	$NaClO_4$	(1)	96T4

Remark:
(1) $NaClO_4$ titration treated by a denaturant binding model

Ferricytochrome c from horse heart, high- and low-temperature transitions

Transition	pH	T	ΔG	Approach	Remarks	Ref
trans. (1)	7.0	20	2.1	heat	(1,2)	93M18
trans. (2)	7.0	20	9.2	heat	(1,2)	93M18
trans. (1)	7.0	20	1.7	heat	(1,3)	93M18
trans. (2)	7.0	20	15.1	heat	(1,3)	93M18
trans. (1)	7.0	20	3.3	heat	(1,4)	93M18
trans. (2)	7.0	20	13.4	heat	(1,4)	93M18
trans. (1)	7.0	20	4.6	heat	(1,5)	93M18
trans. (2)	7.0	20	15.9	heat	(1,5)	93M18
trans. (1)	7.0	20	2.5	heat	(1,6)	93M18
trans. (2)	7.0	20	18.8	heat	(1,6)	93M18
trans. (1)	7.0	20	0.8	heat	(1,7)	93M18
trans. (2)	7.0	20	16.3	heat	(1,7)	93M18

Remarks:
(1) the thermal transition was resolved into a low-temperature component and a high-temperature component, for details of the approach, see Ref. 93M18
(2) protein dissolved in water, without buffer
(3) protein in 0.2 M KCl
(4) protein in 0.1 M $NaClO_4$
(5) protein in 0.1 M cacodylate buffer
(6) protein in 0.1 M sodium phosphate buffer
(7) buffer: mixture of 0.1 M cacodylate + 0.1 M $NaClO_4$

Cytochrome c from horse heart

Salt Concentration	pH	T	ΔG	Approach	Remarks	Ref
0	1.8	20	8.54	GuHCl	(1,2)	93H1
0.4 M NaCl	1.8	20	10.25	GuHCl	(1,2)	93H1

Remarks:
(1) linear extrapolation
(2) the data refer to GuHCl-induced refolding

Cytochrome c, wild-type and mutants, salt-induced transition of the A-state

Mutant	pH	T	ΔG	Approach	Remarks	Ref
wild-type	2.0	10	14.8±3.3	KCl titration	(1)	96C15
wild-type	2.0	10	17.3±5.3	KCl titration	(2)	96C15
Leu94→Ala	2.0	10	3.1±0.8	KCl titration	(1)	96C15
Leu94→Ala	2.0	10	0.4±0.7	KCl titration	(2)	96C15
Leu94→Val	2.0	10	11.3±1.5	KCl titration	(1)	96C15
Leu94→Val	2.0	10	10.8±3.8	KCl titration	(2)	96C15

Remarks:
(1) transition monitored by fluorescence
(2) transition monitored by CD

Cytochrome c, native and A-state

Mutant/Transition	pH	T	$\Delta(\Delta G)$	Approach	Remarks	Ref
wild-type, A→U	2.0	65±0.3	0.0	heat	(1,2)	96C15
wild-type, N→U	2.0	83±0.2	0.0	heat	(1,2)	96C15
Leu94→Val, A→U	2.0	51±0.6	−7.5±0.4	heat	(1,2)	96C15
Leu94→Val, N→U	2.0	79±0.3	−5.0±0.4	heat	(1,2)	96C15

Remarks:
(1) buffer: 20 mM phosphate, 1.0 M KCl
(2) $\Delta(\Delta G)$ was calculated at T_{trs} of the wild-type using $\Delta C_p = 4.6$ kJ//mol/K for the A→U transition
(3) $\Delta(\Delta G)$ was calculated at T_{trs} of the wild-type using $\Delta C_p = 5.86$ kJ//mol/K for the N→U transition

Cytochrome c, calcium chloride-induced unfolding

Transition	pH	T	ΔG	Approach	Remarks	Ref
N→X	7.0	25	34.8±0.8	$CaCl_2$	(1–3)	94A2
X→D	7.0	25	6.1±0.3	$CaCl_2$	(1–3)	94A2
N→X	7.0	25	29.2±0.4	$CaCl_2$	(1,2,4)	94A2
X→D	7.0	25	6.9±0.3	$CaCl_2$	(1,2,4)	94A2

Remarks:
(1) calcium chloride-induced unfolding, resolved into two subtransitions
(2) linear extrapolation
(3) transition monitored by difference absorption at 405 nm
(3) transition monitored by difference absorption at 290 nm

Cytochrome c, from horse heart, unfolding in the presence of calcium chloride

Salt Concentration	pH	T	ΔG	Approach	Remarks	Ref
0.50 M $CaCl_2$	7.0	25	26.6±0.5	GuHCl	(1)	94A2
1.00 M $CaCl_2$	7.0	25	23.3±0.6	GuHCl	(1)	94A2
1.40 M $CaCl_2$	7.0	25	13.5±1.0	GuHCl	(1)	94A2
2.25 M $CaCl_2$	7.0	25	2.8±1.0	GuHCl	(1)	94A2
extrapolated	7.0	25	36.4±0.5	GuHCl	(2,3)	94A2
extrapolated	7.0	25	28.9±0.5	GuHCl	(2,3)	94A2

Remarks:
(1) linear extrapolation, given is the apparent ΔG value
(2) combined approach, for details see Ref. 94A2
(3) extrapolation to zero GuHCl and $CaCl_2$ concentration, for the details, see Ref. 94A2

Cytochrome c from horse heart, in the presence of osmolytes (amino acids)

Concentration	pH	T	ΔG	Approach	Remarks	Ref
control	6.0	25	35.11±0.54	GuHCl	(1)	94T5
0.60 M Ser	6.0	25	34.70±0.54	GuHCl	(1)	94T5
0.45 M Ala	6.0	25	36.30±1.25	GuHCl	(1)	94T5

Remark:
(1) linear extrapolation

Cytochrome c from horse heart, in the presence of osmolytes (amino acids)

Amino Acid	pH	T_{trs}	ΔT	Δ(ΔG)	Remarks	Ref
1 M Ala			9.0	0	(1)	94T5
1 M Arg			−5.2	−0.253	(1)	94T5
1 M Gly			3.3	0	(1)	94T5
1 M His			−2.6	−0.148	(1)	94T5
1 M Ile			0	0	(1)	94T5
1 M Leu			0	0	(1)	94T5
1 M Lys			6.0	0	(1)	94T5
1 M Met			2.6	0	(1)	94T5
1 M Phe			0	0	(1)	94T5
1 M Pro			3.1	0	(1)	94T5
1 M Ser			3.7	0	(1)	94T5
1 M Thr			3.4	0	(1)	94T5
1 M Val			1.8	0	(1)	94T5

Remark:
(1) the results are based on extrapolation of experimental results to 1 M amino acid concentration; Δ(ΔG) values are given for 25°C

Cytochrome c₂

Cytochrome c₂ from *Rhodobacter capsulatus*

Mutant	pH	T	ΔG	Approach	Remarks	Ref
Oxidized form:						
wild-type	7.5	20	15.5±1.3	GuHCl	(1)	91C1
Tyr75→Cys	7.5	20	15.5±1.3	GuHCl	(1)	91C1
Tyr75→Phe	7.5	20	18.5±1.3	GuHCl	(1)	91C1
Reduced form:						
wild-type	7.5	20	31.2±3.6	GuHCl	(1)	91C1
Tyr75→Cys	7.5	20	18.4±3.6	GuHCl	(1)	91C1
Tyr75→Phe	7.5	20	19.9±3.6	GuHCl	(1)	91C1

Remark:
(1) linear extrapolation

Cytochrome c$_2$ from *Rhodobacter capsulatus*

Mutant	pH	T	$\Delta(\Delta G)$	Approach	Remarks	Ref
Oxidized form:						
wild-type	7.5	20	0.0	GuHCl	(1)	91C1
Tyr75→Cys	7.5	20	−1.3±0.1	GuHCl	(1)	91C1
Tyr75→Phe	7.5	20	1.5±0.1	GuHCl	(1)	91C1
Reduced form:						
wild-type	7.5	20	0.0	GuHCl	(1)	91C1
Tyr75→Cys	7.5	20	−17.2±1.8	GuHCl	(1)	91C1
Tyr75→Phe	7.5	20	−7.7±1.8	GuHCl	(1)	91C1

Remark:
(1) $\Delta(\Delta G)$ is the difference in ΔG of wild-type and mutant at GuHCl concentrations of the half transition

Cytochrome c$_2$ from *Rhodobacter capsulatus* wild-type and mutants

Mutant	pH	T	ΔG	Approach	Remarks	Ref
wild-type	7.5	25	18.4	GuHCl	(1)	91C2
Lys12→Asp	7.5	25	13.8	GuHCl	(1)	91C2
Lys14→Glu	7.5	25	14.6	GuHCl	(1)	91C2
Lys32→Glu	7.5	25	20.1	GuHCl	(1)	91C2
double mutant (Lys14→Glu and Lys32→Glu)						
	7.5	25	15.1	GuHCl	(1)	91C2

Remark:
(1) linear extrapolation

Cytochrome c$_2$ from *Rhodobacter capsulatus*

Mutant/Form	pH	T	ΔG	$\Delta(\Delta G)$	Approach	Remarks	Ref
wild-type/ox.	7.5	25	18.4±1.3	0.0	GuHCl	(1)	93C1
Trp67→Tyr, ox.	7.5	25	7.9±1.3	−10.5±1.3	GuHCl	(1)	93C1
wild-type/red.	7.5	25	31.4±2.5	0.0	GuHCl	(1)	93C1
Trp67→Tyr, red.	7.5	25	18.8±2.5	−12.6±2.5	GuHCl	(1)	93C1

Remark:
(1) linear extrapolation

Cytochrome c-552

Cytochrome c-552 from various species

Species	pH	T	ΔG	Approach	Remarks	Ref
Thermus th.	7.5	25	119.2±0.6	GuHCl	(1)	78N1
Thermus th.	7.5	25	121	GuHCl	(2)	79H
Hydro. th.	7.0	25	91.6	GuHCl	(1)	89S1

Abbreviations (species):
Thermus th. = *Thermus thermophilus*
Hydro. th.   = *Hydrogenobacter thermophilus*

Remarks:
(1) ligand binding model, k = 1.2
(2) ligand binding model

Cytochrome P450

Apocytochrome P450 from *Pseudomonas putida*

pH	T	ΔG	Approach	Remarks	Ref
8.0	25	7.5±1.0	DSC	(1)	93P4

Remark:
(1) buffer: 20 mM HEPES

Iso-1-Cytochrome c

Iso-1-cytochrome c mutants with blocked Cys107 (–SCH₃)

Mutant	pH	T	ΔG	Approach	Remarks	Ref
wild-type	7.2	25	12.1±2.1	GuHCl	(1)	88H
wild-type	7.2	25	11.7±2.1	GuHCl	(2)	88H
Lys32→Leu	7.2	25	11.3±2.1	GuHCl	(1)	88H
Lys32→Leu	7.2	25	12.6±2.1	GuHCl	(2)	88H
Lys32→Gln	7.2	25	12.6±2.1	GuHCl	(1)	88H
Lys32→Gln	7.2	25	7.1±2.1	GuHCL	(2)	88H
Lys32→Tyr	7.2	25	5.0±2.1	GuHCl	(1)	88H
Lys32→Tyr	7.2	25	2.9±2.1	GuHCl	(2)	88H
Lys32→Trp	7.2	25	5.0±2.1	GuHCl	(1)	88H
Lys32→Trp	7.2	25	3.3±2.1	GuHCl	(2)	88H

Remarks:
(1) monitored by CD, linear extrapolation
(2) monitored by fluorescence, linear extrapolation

Iso-1-cytochrome c mutants, with blocked Cys107 (–SCH₃)

Mutant	pH	T	ΔG	Approach	Remarks	Ref
w.t.(Asn57)	6.0	25	15.9	heat	(1)	89D
Asn57→Ile	6.0	25	33.5	heat	(1)	89D

Remark:
(1) measured in 100 mM sodium phosphate

Iso-1-cytochrome c mutants, with blocked Cys107 (–SCH₃)

Mutant	pH	T_{trs}	ΔT	$\Delta(\Delta G)$	Approach	Remarks	Ref
wild-type	6	46.5±1	0.0	0.0	heat	(1)	88H
Lys32→Leu	6	44.9±1	−1.6	−1.7	heat	(2)	88H
Lys32→Gln	6	41.5±1	−5	−5	heat	(2)	88H
Lys32→Tyr	6	36.9±1	−9.6	−10	heat	(2)	88H
Lys32→Trp	6	37.4±1	−9.1	−9.2	heat	(2)	88H

Remarks:
(1) reference value, obtained by van't Hoff treatment
(2) van't Hoff treatment

Iso-1-cytochrome c (Cys107→Ala)

Mutant	pH	ΔT	Δ(ΔG)		Remarks	Ref
Cys107→Ala		0.0	0.0		(1–3)	92H4
double mutant (Thr74→Ala and Cys107→Ala)						
		0.5	0.4		(2,4)	92H4
double mutant (Thr74→Glu and Cys107→Ala)						
		3.6	3.3		(2,5)	92H4

Remarks:
(1) reference protein for the following entries
(2) measured at 2 mM ionic strength
(3) $T_{trs} = 51°C$
(4) $T_{trs} = 51.5°C$
(5) $T_{trs} = 54.6°C$

Iso-1-ferricytochrome c from *Saccharomyces cerevisiae*, Cys102→Thr variant

Mutant	pH	T	ΔG	Approach	Remarks	Ref
Cys102→Thr	4.6	26.8	21.3±1.3	heat		94C7
Cys102→Thr	7.2	27	26.4±1.3	GuHCl	(1,2)	93A3
(Phe10→Met and Cys102→Thr)						
	7.2	27	14.2±1.3	GuHCl	(1)	93A3
(Phe10→Cys, S-methyl cysteine derivative, and Cys102→Thr)						
	7.2	27	17.2±1.3	GuHCl	(1)	93A3

Remarks:
(1) linear extrapolation
(2) reference value for the following mutants

Iso-1-cytochrome c, cysteine mutants, oxidized form

Mutant	pH	T	ΔG	Approach	Remarks	Ref
wild-type	7.0		10.0±1.2	urea	(1)	94H10
Cys102→Thr	7.0		27.1±1.2	urea	(1)	94H10
double mutant (Arg13→Cys and Cys102→Thr)						
	7.0		27.6±1.2	urea	(1)	94H10
double mutant (Asp90→Cys and Cys102→Thr)						
	7.0		13.8±1.2	urea	(1)	94H10
multiple mutant (Arg13→Cys, Asp90→Cys and Cys102→Thr)						
	7.0		3.8±1.2	urea	(1)	94H10

Remark:
(1) linear extrapolation

Iso-1-cytochrome c from yeast, mutations in positions 6 and 52

Mutant	pH	T	ΔG	Approach	Remarks	Ref
wild-type	7.0	22	24.5	GuHCl	(1,2)	95L6
Gly6→Ala	7.0	22	9.9	GuHCl	(2)	95L6
Asn52→Ile	7.0	22	41.7	GuHCl	(2)	95L6
double mutant (Gly6→Ala and Asn52→Ile)						
	7.0	22	24.5	GuHCl	(2)	95L6
double mutant (Gly6→Ser and Asn52→Ile)						
	7.0	22	23.2	GuHCl	(2)	95L6
	4.6	27	11.6	DSC		95L6

Remarks:
(1) wild-type = Asn52
(2) linear extrapolation to zero denaturant concentration by a method that includes the pre- and postdenaturational baselines for a non-linear regression of the data

Iso-1-cytochrome c from yeast, mutations in position 52

Mutant	pH	T	ΔG	Approach	Remarks	Ref
wild-type	7.0	22	24.5	GuHCl	(1,2)	95L5
wild-type	5.0	27	23.5	DSC	(1)	95L5
wild-type	4.6	27	16.9	DSC	(1)	95L5
wild-type	4.0	27	10.5	DSC	(1)	95L5
wild-type	3.5	27	6.9	DSC	(1)	95L5
Asn52→Ala	7.0	22	36.9	GuHCl	(2)	95L5
Asn52→Ala	4.6	27	19.5	DSC		95L5
Asn52→Gln	7.0	22	21.0	GuHCl	(2)	95L5
Asn52→Gln	4.6	27	16.5	DSC		95L5
Asn52→His	7.0	22	17.4	GuHCl	(2)	95L5
Asn52→His	4.6	27	11.5	DSC		95L5
Asn52→Ile	7.0	22	41.7	GuHCl	(2)	95L5
Asn52→Ile	5.0	27	32.0	DSC		95L5
Asn52→Ile	4.6	27	30.0	DSC		95L5
Asn52→Ile	4.0	27	23.2	DSC		95L5
Asn52→Ile	3.5	27	17.8	DSC		95L5
Asn52→Leu	7.0	22	44.3	GuHCl	(2)	95L5
Asn52→Leu	4.6	27	27.6	DSC		95L5
Asn52→Met	7.0	22	41.4	GuHCl	(2)	95L5
Asn52→Met	4.6	27	26.3	DSC		95L5
Asn52→Met	4.0	27	22.7	DSC		95L5
Asn52→Phe	7.0	22	24.3	GuHCl	(2)	95L5
Asn52→Ser	7.0	22	23.0	GuHCl	(2)	95L5
Asn52→Ser	5.0	27	18.1	DSC		95L5
Asn52→Ser	4.6	27	16.7	DSC		95L5
Asn52→Ser	4.0	27	13.6	DSC		95L5
Asn52→Thr	7.0	22	32.5	GuHCl	(2)	95L5
Asn52→Thr	5.0	27	21.3	DSC		95L5
Asn52→Thr	4.6	27	19.3	DSC		95L5
Asn52→Thr	4.0	27	14.8	DSC		95L5
Asn52→Thr	3.5	27	11.0	DSC		95L5
Asn52→Val	7.0	22	33.0	GuHCl	(2)	95L5
Asn52→Val	4.6	27	23.8	DSC		95L5
Asn52→Val	4.0	27	20.8	DSC		95L5

Remarks:
(1) wild-type = Asn52
(2) linear extrapolation to zero denaturant concentration by a method that includes the pre- and postdenaturational baselines for a non-linear regression of the data

Iso-1-cytochrome c from baker yeast, recombinant, wild-type and mutants

Mutant	pH	T	ΔG	Approach	Remarks	Ref
wild-type	7.0	25	6.1	urea	(1,2)	94K9
wild-type	7.0	25	6.6	GuHCl	(1,2)	94K9
Asn52→Ile	7.0	25	12.1	urea	(1,3)	94K9
Asn52→Ile	7.0	25	12.6	GuHCl	(1,3)	94K9

Remarks:
(1) linear extrapolation
(2) called SC-iso1c
(3) called SC-iso1c-N52I

Iso-1-cytochrome c, mutants

Mutant	pH	T	$\Delta(\Delta G)$	Approach	Remarks	Ref
wild-type	6	46.5	0.0	heat	(1,2)	91H1
Asn52→Ile	6	46.5	18.0	heat	(2)	91H1
double mutant (Gly29→Ser and Asn52→Ile)						
	6	46.5	−6.3	heat	(2)	91H1
double mutant (His33→Pro and Asn52→Ile)						
	6	46.5	−3.3	heat	(2)	91H1
Cys102→Ala	6	46.5	12.1	heat		91H1
Cys102→Ser	6	46.5	11.7	heat		91H1
double mutant (Asn52→Ala and Cys102→Ala)						
	6	46.5	16.3	heat		91H1
double mutant (Asn52→Gly and Cys102→Ala)						
	6	46.5	3.8	heat		91H1
double mutant (Asn52→Ile and Cys102→Ala)						
	6	46.5	25.9	heat		91H1

Remarks:
(1) recombinat protein containing Gly29, His33, Asn52, Cys102 as the wild-type = reference protein for the following mutants
(2) blocked Cys102-SCH$_3$

Iso-1-cytochrome c, mutants

Mutant	pH	T	ΔG	Approach	Remarks	Ref
wild-type	7		18.0	GuHCl	(1,2)	91H1
wild-type	7		13.4	GuHCl	(1–3)	91H1
double mutant (Gly29→Ser and Asn52→Ile)						
	7		9.2	GuHCl	(2,3)	91H1
Cys102→Ala	7		19.7	GuHCl	(2)	91H1
double mutant (Asn52→Ala and Cys102→Ala)						
	7		23.8	GuHCl	(2)	91H1
double mutant (Asn52→Gly and Cys102→Ala)						
	7		15.9	GuHCl	(2)	91H1
double mutant (Asn52→Ile and Cys102→Ala)						
	7		29.3	GuHCl	(2)	91H1

Remarks:
(1) recombinat protein containing Gly29, His33, Asn52, Cys102 as the wild-type = reference protein for the following mutants
(2) linear extrapolation
(3) blocked Cys102-SCH$_3$

Iso-1-cytochrome c mutants with blocked Cys102 (–SCH₃), wild-type and mutants in position 71

Mutant	pH	T	ΔG	Approach	Ref
wild-type	6	20	15.1	GuHCl	86R
Pro71→Val	6	20	10.9	GuHCl	86R
Pro71→Thr	6	20	7.9	GuHCl	86R
Pro71→Ile	6	20	7.9	GuHCl	86R

Iso-1-cytochrome c (Cys102→Ser) form, oxidized, wild-type and mutants in position 73

Mutant	pH	T	ΔG	Approach	Remarks	Ref
wild-type	7.5	25.0	23.7±1.7	GuHCl	(1,2)	93B6
Lys73→Ala	7.5	25	21.8±1.5	GuHCl	(1,2)	95H6
Lys73→Arg	7.5	25	23.6±0.5	GuHCl	(1,2)	95H6
Lys73→Gly	7.5	25	20.6±1.7	GuHCl	(1,2)	95H6
Lys73→Ile	7.5	25	18.7±0.5	GuHCl	(1,2)	95H6
Lys73→Leu	7.5	25	23.6±0.3	GuHCl	(1,2)	95H6
Lys73→Met	7.5	25.0	19.5±0.7	GuHCl	(1,2)	93B6
Lys73→Phe	7.5	25.0	18.9±1.3	GuHCl	(1,2)	93B6
Lys73→Trp	7.5	25.0	17.7±1.5	GuHCl	(1,2)	93B6
Lys73→Tyr	7.5	25.0	19.3±1.2	GuHCl	(1,2)	93B6
Lys73→Val	7.5	25	17.9±1.3	GuHCl	(1,2)	95H6

Remarks:
(1) linear extrapolation
(2) buffer: 20 mM Tris with 40 mM NaCl, pH 7.5

Iso-1-cytochrome c, Cys102→Thr variant (wild-type) and mutants, N→D and A→D transitions

Mutant	pH	T	Δ(ΔG)	Approach	Remarks	Ref
Thermodynamic parameters of the (N→D) transition:						
wild-type	4.6	52.6	0.0	heat	(1–3)	95M4
Ala7→Leu	4.6	52.6	0.9±1.2	heat	(1–3)	95M4
Ala7→Tyr	4.6	52.6	1.8±1.1	heat	(1–3)	95M4
Phe10→Cys	4.6	52.6	−14.0±1.5	heat	(1–3,5)	95P3
Phe10→Ile	4.6	52.6	−4.8±1.3	heat	(1–3)	95P3
Phe10→Met	4.6	52.6	−14.7±1.6	heat	(1–3)	95P3
Phe10→Trp	4.6	52.6	−5.4±1.3	heat	(1–3)	95P3
Phe10→Tyr	4.6	52.6	−2.0±1.1	heat	(1–3)	95P3
Leu94→Ala	4.6	52.6	−15.0±1.6	heat	(1–3)	95P3
Leu94→Ile	4.6	52.6	1.4±1.2	heat	(1–3)	95P3
Leu94→Thr	4.6	52.6	−12.8±1.4	heat	(1–3)	95P3
Leu94→Val	4.6	52.6	−4.4±1.3	heat	(1–3)	95P3
Tyr97→Ala	4.6	52.6	−20.4±1.9	heat	(1–3)	95P3
Tyr97→Phe	4.6	52.6	1.0±1.2	heat	(1–3)	95P3
double mutant (Leu94→Ile and Tyr97→Phe)						
	4.6	52.6	−3.2±1.2	heat	(1–3)	95P3
double mutant (Leu94→Ala and Tyr97→Phe)						
	4.6	52.6	−18.8±1.8	heat	(1–3)	95P3
Thermodynamic parameters of the (A→D) transition:						
wild type	2.1	35.2	0.0	heat	(1–4)	95M4
Ala7→Leu	2.1	35.2	−0.9±0.6	heat	(1–4)	95M4
Ala7→Tyr	2.1	35.2	0.3±0.6	heat	(1–4)	95M4
Phe10→Trp	2.1	35.2	−6.4±0.6	heat	(1–4)	95M4
Phe10→Tyr	2.1	35.2	−3.6±0.6	heat	(1–4)	95M4
Leu94→Ala	2.1	35.2	<−18	heat	(1–4)	95M4

Iso-1-cytochrome c, Cys102→Thr variant (wild-type) and mutants, N→D and A→D transitions (continued)

Mutant	pH	T	Δ(ΔG)	Approach	Remarks	Ref
Leu94→Ile	2.1	35.2	−0.9±0.6	heat	(1–4)	95M4
Leu94→Thr	2.1	35.2	←−13	heat	(1–4)	95M4
Tyr97→Ala	2.1	35.2	←−18	heat	(1–4)	95M4
Tyr97→Phe	2.1	35.2	−1.1±0.6	heat	(1–4)	95M4

Remarks:
(1) the Cys102→Thr variant is referred to as the wild-type, all variants also contain the Cys102→Thr mutation
(2) the considered states and the corresponding buffer conditions are:
   A state – 0.33 M Na_2SO_4/H_2SO_4, pH 2.1 acid denatured state – 0.01 M Na_2SO_4/H_2SO_4, pH 2.1
   N state – 0.05 M potassium phosphate, pH 7.0 or 0.05 M acetate for the pos. 7 mutants
(3) Δ(ΔG) was determined from $\Delta(\Delta G) = (\Delta H/T_{trs})_{w.t.} \times \Delta T$
(4) for the transition temperature, see Table 2 containing enthalpy and heat capacity changes

Iso-1-cytochrome c, mutants, oxidized [ΔG(III)] and reduced [ΔG(II)] forms

Mutant	pH	T	ΔG(III)	ΔG(II)	Approach	Remarks	Ref
Cys102→Ala	7.0	25	20.9	67.4	GuHCl	(1)	94K8
double mutant (Phe82→Ser and Cys102→Ala)							
	7.0	25	14.6	57.3	GuHCl	(1)	94K8
double mutant (Arg38→Ala and Cys102→Ala)							
	7.0	25	15.9	57.7	GuHCl	(1)	94K8
double mutant (Asn52→Ile and Cys102→Ala)							
	7.0	25	37.7	76.6	GuHCl	(1)	94K8
multiple mutant (Arg38→Ala, Asn52→Ile and Cys102→Ala)							
	7.0	25	34.3	73.6	GuHCl	(1)	94K8
multiple mutant (Arg38→Ala, Phe82→Ser and Cys102→Ala)							
	7.0	25	9.6	48.1	GuHCl	(1)	94K8
multiple mutant (Asn52→Ile, Phe82→Ser and Cys102→Ala)							
	7.0	25	23.0	60.2	GuHCl	(1)	94K8
multiple mutant (Arg38→Ala, Asn52→Ile, Phe82→Ser and Cys102→Ala)							
	7.0	25	25.9	60.2	GuHCl	(1)	94K8

Remark:
(1) ΔG(III) was determined by linear extrapolation to zero denaturant concentration using a method that includes the pre- and postdenaturational baselines for a non-linear regression of the data. ΔG(II) was determined combining ΔG(III) with an electrochemical approach

Iso-1-cytochrome c from *Sacchararomyces cerevisiae*, wild-type and mutant, oxidized and reduced forms

Mutant	pH	T	ΔG	Approach	Remarks	Ref
wild-type, ox.	4.6	27	23.4±3.3	GuHCl	(1,2)	96D2
Asp52→Ile, ox.	4.6	27	28.5±5.9	GuHCl	(1,2)	96D2
wild-type, red.	4.6	27	63±8	GuHCl	(1,2)	96D2
Asp52→Ile, red.	4.6	27	75±17	GuHCl	(1,2)	96D2

Remarks:
(1) linear extrapolation to zero denaturant concentration by a method that includes the pre- and postdenaturational baselines for a non-linear regression of the data
(2) for the electrochemistry, see S. Hilgen-Willis, E.F. Bowden, and G.J. Pielak: J. Inorg. Biochem. 51 (1993) 649

Iso-1 cytochrome c, chemically modified forms

Form	pH	T	ΔG	Approach	Remarks	Ref
monomer (1)	7	25	12.6±1.7	GuHCl	(1)	85B
monomer (2)	7	20	8.8	GuHCl	(2)	83Z
monomer (3)	7	25	7.9±0.8	GuHCl	(1)	85B
dimer	7	25	4.6±1.7	GuHCl	(1)	85B

Explanation of the forms:
monomer (1) – without modification
monomer (2) – iodacetamide blocked protein at Cys102
monomer (3) – alkylated protein
dimer   – linked by interchain disulfide bond

Remarks:
(1) mean value of various approaches
(2) linear extrapolation

Iso-1-cytochrome c, variants

Mutant	pH	T	ΔG	Approach	Ref
(1) w.t.	7.2	27	17.2	GuHCl, linear extrapol.	92B4
(2)	7.2	27	17.2	GuHCl, linear extrapol.	92B4
(4)	4.6	27	26.8	GuHCl, linear extrapol.	92B4
(4)	7.2	27	26.8	GuHCl, linear extrapol.	92B4
(4)	4.6	27	23.8	DSC	92B4
(4)	6.0	27	24.7	DSC	92B4
(5)	4.6	27	12.1	GuHCl, linear extrapol.	92B4
(5)	7.2	27	12.6	GuHCl, linear extrapol.	92B4
(5)	4.6	27	17.2	DSC	92B4
(5)	6.0	27	18.4	DSC	92B4

Mutants:
(1) position 20: Val, position 102: Cys methylated
(2) position 20: Val20→Cys methylated, position 102: Thr
(3) position 20: Val20→Cys methylated, position 102: Cys methylated
(4) position 20: Val, position 102: Cys→Thr
(5) position 20: Val20→Cys, position 102: Cys

Iso-1-cytochrome c with engineered disulfide

Mutant	pH	T	ΔG	Approach	Remarks	Ref
wild-type	4.6	25	22.2±1.7	heat	(1,5)	96B6
disulfide	4.6	25	17.2±1.7	heat	(2–5)	96B6

Remarks:
(1) on transfer from H_2O to D_2O, ΔG increases by 4.6±1.7 kJ/mol
(2) disulfide connects position 20 (usually Val) with position 102 (usually Thr)
(3) on transfer from H_2O to D_2O, ΔG increases by 4.2±1.7 kJ/mol
(4) the stability of the disulfide mutant is decreased by Δ(ΔG) = −5.0±2.5 kJ/mol, see also Ref. 92B4 (Δ(ΔG)
    = −6.7 kJ/mol)
(5) Ref. 96B6 contains additional data on local unfolding

Iso-1-cytochrome c, ISO-2-Cytochrome c and composite forms, oxidized

Mutant	pH	T	ΔG	Approach	Remarks	Ref
iso-1 Cys102-meth.						
	6.0	25	16.0*	heat	(1,2)	94L4
iso-2	6.0	25	18.3*	heat	(1)	94L4
comp1 Cys102-meth.						
	6.0	25	11.1*	heat	(1–3)	94L4
comp2 Cys102-meth.						
	6.0	25	11.5*	heat	(1,2,4)	94L4
comp3 Cys102-meth.						
	6.0	25	11.4*	heat	(1,2,5)	94L4

Remarks:
(1) see also Fig. 8 in Ref. 94L4
(2) S-methylated at Cys102
(3) composite protein from the allele CYC1–136-B
(4) composite protein from the allele CYC1–158-B
(5) composite protein from the allele CYC1–136-C

Iso-1-cytochrome c, ISO-2-Cytochrome c and composite forms, oxidized

Mutant	pH	T	Δ(ΔG)	Approach	Remarks	Ref
iso-1 Cys102→Ala						
	6.0	50.7	4.6±0.3	heat		94L4
iso-1 Cys102-meth.						
	6.0	50.7	0	heat	(1,2)	94L4
iso-2	6.0	50.7	3.1±0.1	heat		94L4
comp1 Cys102-meth.						
	6.0	50.7	−2.2±0.2	heat	(2,3)	94L4
comp2 Cys102-meth.						
	6.0	50.7	−0.6±0.04	heat	(2,4)	94L4
comp3 Cys102-meth.						
	6.0	50.7	−1.6±0.04	heat	(2,5)	94L4

Remarks:
(1) reference protein having at pH 6.0 T_{trs} = 50.7°C
(2) S-methylated at Cys102
(3) composite protein from the allele CYC1–136-B
(4) composite protein from the allele CYC1–158-B
(5) composite protein from the allele CYC1–136-C

Iso-2-Cytochrome c

Iso-2-cytochrome c, wild-type and mutants

Mutant	pH	T	ΔG	Approach	Remarks	Ref
wild-type	7.2	20	13	GuHCl	(1)	81N
wild-type (Pro71)	7.2	20	15.9±2.1	GuHCl	(2)	87W
Pro71→Thr	7.2	20	12.6±2.1	GuHCl	(2)	87W
wild-type (Pro76)	6	20	15.9	GuHCl	(1)	88W4
Pro76→Gly	6	20	10.9	GuHCl	(1)	88W4

Remarks:
(1) linear extrapolation
(2) fluorescence measurement, linear extrapolation

Iso-2-cytochrome c, wild-type and mutants, oxidized and reduced forms

Mutant	pH	T	ΔG	Approach	Ref
wild-type					
oxidized	6.0	25	18.4±1.3	DSC	96M3
reduced	6.0	25	43.9±5.0	DSC	96M3
mutant (Asn52→Ile)					
oxidized	6.0	25	26.8±2.5	DSC	96M3
reduced	6.0	25	55.2±5.9	DSC	96M3

Iso-2-cytochrome c from yeast, wild-type and mutants

Mutant	pH	T	ΔG	Approach	Remarks	Ref
wild-type	6.0	20	20.5±1.3	GuHCl	(1)	95V3
wild-type	6.0	20	17.6±0.8	DSC, T_{trs}=55.4°C	(2)	95V3
Pro25→Gly	6.0	20	17.2±0.8	GuHCl	(1)	95V3
Pro25→Gly	6.0	20	19.7±0.4	DSC, T_{trs}=54.3°C	(2)	95V3
His33→Asn	6.0	20	15.1±0.8	GuHCl,	(1)	95V3
His33→Asn	6.0	20	14.2±0.4	DSC, T_{trs}=50.0°C	(2)	95V3
Asn52→Ile	6.0	20	35.1±2.1	GuHCl	(1)	95V3
Asn52→Ile	6.0	20	27.6±0.8	DSC, T_{trs}=64.6°C	(2)	95V3
Pro76→Gly	6.0	20	15.9±0.8	GuHCl	(1)	95V3
Pro76→Gly	6.0	20	5.4±0.8	DSC, T_{trs}=51.4°C	(2)	95V3
double mutant (Pro30→Ala and Asn52→Ile)						
	6.0	20	19.7±1.3	GuHCl	(1)	95V3
	6.0	20	18.8±0.4	DSC, T_{trs}=48.0°C	(2)	95V3
double mutant (Asn52→Ile and Ile75→Met)						
	6.0	20	25.1±0.8	GuHCl	(1)	95V3
	6.0	20	16.3±0.8	DSC, T_{trs}=60.0°C	(2)	95V3

Remarks:
(1) linear extrapolation to zero denaturant concentration by a method that includes the pre- and postdenaturational baselines for a non-linear regression of the data
(2) ΔG was calculated using ΔC_p = 6.61±0.29 kJ/mol/K from Ref. 95M11

Iso-2-cytochrome c from yeast, stability of slow-folding intermediates detected by absorbance and fluorescence

Mutant	pH	T	ΔG	Approach	Remarks	Ref
wild-type	6.0	20	19.7±0.4	GuHCl, absorbance	(1)	95V3
wild-type	6.0	20	16.7±0.2	GuHCl, fluorescence	(2)	95V3
Pro25→Gly	6.0	20	16.7±1.7	GuHCl, absorbance	(1)	95V3
Pro25→Gly	6.0	20	13.0±0.4	GuHCl, fluorescence	(2)	95V3
His33→Asn	6.0	20	16.7±0.4	GuHCl, absorbance	(1)	95V3
His33→Asn	6.0	20	18.8±1.7	GuHCl, fluorescence	(2)	95V3
Asn52→Ile	6.0	20	17.8±0.2	GuHCl, absorbance	(1)	95V3
Asn52→Ile	6.0	20	19.2±1.7	GuHCl, fluorescence	(2)	95V3
Pro76→Gly	6.0	20	18.8±0.4	GuHCl, fluoescence	(2)	95V3
double mutant (Pro30→Ala and Asn52→Ile)						
	6.0	20	18.8±0.8	GuHCl, absorbance	(1)	95V3
	6.0	20	15.5±0.8	GuHCl, fluorescence	(2)	95V3
Pro76→Gly	6.0	20	18.8±0.4	GuHCl, fluorescence	(2)	95V3
double mutant (Asn52→Ile and Ile75→Met)						
	6.0	20	28.0±2.9	GuHCl, absorbance	(1)	95V3
	6.0	20	15.9±0.4	GuHCl, fluorescence	(2)	95V3

Remarks:
(1) kinetic approach, absorbance-detected slow folding intermediate
(2) kinetic approach, fluorescence-detected slow folding intermediate

Cytotoxin, see Toxins

De Novo Synthesized Proteins/Peptides

70-Residue synthetic two-stranded α-helical coiled-coil

Mutant	pH	T	ΔG	Approach	Remarks	Ref
	7.0	20	40.6	GuHCl	(1,2)	92Z4
Leu5→Ala	7.0	20	35.6	GuHCl	(1,2)	92Z4
Leu9→Ala	7.0	20	28.0	GuHCl	(1,2)	92Z4
Leu12→Ala	7.0	20	28.5	GuHCl	(1,2)	92Z4
Leu16→Ala	7.0	20	28.5	GuHCl	(1,2)	92Z4
Leu19→Ala	7.0	20	28.0	GuHCl	(1,2)	92Z4
Leu23→Ala	7.0	20	28.5	GuHCl	(1,2)	92Z4
Leu26→Ala	7.0	20	30.5	GuHCl	(1,2)	92Z4
Leu30→Ala	7.0	20	32.2	GuHCl	(1,2)	92Z4
Leu33→Ala	7.0	20	36.8	GuHCl	(1,2)	92Z4

Remarks:
(1) linear extrapolation
(2) buffer: 50 mM phosphate, 0.1 M KCl

Designed heterodimeric coiled-coil

Concentration	pH	T	ΔG	Approach	Remarks	Ref
acid-p1/base-p1 heterodimer						
1 M	7.0	20	42.3	urea	(1,2)	93O3
3 mM	7.0	20	28.0	urea	(1,2)	93O3
base-p1 homodimer						
1 M	7.0	20	17	urea	(1,2)	95L12
BASE-pLL homodimer						
1 M	7.0	25	34	urea	(1,2)	95L12
acid-pLL/base-pLL heterotetramer						
1 M	7.0	25	113	urea	(1,2)	95L12

Remarks:
(1) linear extrapolation, the data were calculated for standard concentration of 1 M
(2) for the sequence, see Ref. 93O3
(3) pLL differs from p1 in Asn14→Leu replacement

Two-stranded α-helical coiled-coils

Form	pH	T	ΔG	Approach	Remarks	Ref
parallel oriented α-helical chains with interchain attractions (P/A):						
P/A	7	20	10.5	urea	(1–3)	94M18
parallel oriented α-helical chains with interchain repulsions (P/R):						
P/R	7	20	4.5	urea	(1–3)	94M18
antiparallel oriented α-helical chains with interchain attractions (AP/A):						
AP/A	7	20	14	urea	(1–3)	94M18
antiparallel oriented α-helical chains with interchain repulsions (AP/R):						
AP/R	7	20	7.5	urea	(1–3)	94M18

Remarks:
(1) prepared from two 35-residue peptides with heptad repeat: Leu-Glu-Ala-Leu-Glu-Gly-Lys or Leu-Ala-Glu-Leu-Lys-Gly-Glu
(2) linear extrapolation
(3) ΔG was taken from Fig. 5D in Ref. 94M18

Newly designed coiled-coil analogs

Mutant	pH	T	ΔG	Approach	Remarks	Ref
20A	7.0	20	0	GuHCl	(1–3)	94M19
20A	7.0	20	0	urea	(1–3)	94M19
15A5R	7.0	20	0.8	GuHCl	(1–3)	94M19
15A5R	7.0	20	6.3	urea	(1–3)	94M19
10A10R	7.0	20	0.8	GuHCl	(1–3)	94M19
10A10R	7.0	20	15.5	urea	(1–3)	94M19
20R	7.0	20	0.8	GuHCl	(1–3)	94M19
20R	7.0	20	24.3	urea	(1–3)	94M19

Remarks:
(1) the coiled-coil analogs were designed as to change the number of intrachain and interchain electrostatic attractions (A) systematically to repulsions (R); for the sequences, the reader is referred to Ref. 94M19
(2) linear extrapolation
(3) GuHCl and urea give different data for ΔG since the ionic nature of GuHCl masks electrostatic interactions in contrast to the uncharged urea

De novo designed coiled-coil with mutations of amino acids that are involved in interchain interactions

Mutant	pH	T	Δ(ΔG)	Approach	Remarks	Ref
Glu$^{(-)}$→Gln	7	20	0.9	urea	(1)	94Z6
Lys$^{(+)}$→Gln	7 and 3	20	1.1	urea	(1)	94Z6
Gln→Glu$^{(o)}$	3	20	1.9	urea	(1)	94Z6
Glu$^{(-)}$→Glu$^{(o)}$	7 and 3	20	2.7	urea	(1)	94Z6

Remark:
(1) linear extrapolation, for further details see Ref. 94Z6

De novo synthesized coiled-coils

Mutant	pH	T	Δ(ΔG)	Approach	Remarks	Ref
native(o)	7	20	0.0	GuHCl	(1,3,4)	95K11
E$_1$(20)(o)	7	20	−0.8	GuHCl	(1–4)	95K11
E$_2$(13,22)(o)	7	20	−0.8	GuHCl	(1–4)	95K11
E$_2$(20,22)(o)	7	20	−0.8	GuHCl	(1–4)	95K11
E$_2$(15,20)(o)	7	20	−0.8	GuHCl	(1–4)	95K11
native(o)	3	20	0.0	GuHCl	(1,3,4)	95K11
E$_1$(20)(o)	3	20	4.2	GuHCl	(1–4)	95K11
E$_2$(13,22)(o)	3	20	8.8	GuHCl	(1–4)	95K11
E$_2$(20,22)(o)	3	20	8.4	GuHCl	(1–4)	95K11
E$_2$(15,20)(o)	3	20	9.6	GuHCl	(1–4)	95K11
native(o)	7	20	0.0	urea	(1,3,4)	95K11
E$_1$(20)(o)	7	20	−1.3	urea	(1–4)	95K11
E$_2$(13,22)(o)	7	20	−2.1	urea	(1–4)	95K11
E$_2$(20,22)(o)	7	20	−2.1	urea	(1–4)	95K11
E$_2$(15,20)(o)	7	20	−5.9	urea	(1–4)	95K11
native(o)	3	20	0.0	urea	(1,3,4)	95K11
E$_1$(20)(o)	3	20	3.3	urea	(1–4)	95K11
E$_2$(13,22)(o)	3	20	7.9	urea	(1–4)	95K11
E$_2$(20,22)(o)	3	20	8.4	urea	(1–4)	95K11
E$_2$(15,20)(o)	3	20	7.5	urea	(1–4)	95K11
native(r)	7	20	0.0	urea	(1,3,4)	95K11
E$_1$(20)(r)	7	20	−0.4	urea	(1–4)	95K11
E$_2$(13,22)(r)	7	20	−2.5	urea	(1–4)	95K11
E$_2$(20,22)(r)	7	20	−2.5	urea	(1–4)	95K11

De novo synthesized coiled-coils (continued)

Mutant	pH	T	$\Delta(\Delta G)$	Approach	Remarks	Ref
$E_2(15,20)(r)$	7	20	−6.7	urea	(1–4)	95K11
native(r)	3	20	0.0	urea	(1,3,4)	95K11
$E_1(20)(r)$	3	20	1.7	urea	(1–4)	95K11
$E_2(13,22)(r)$	3	20	8.8	urea	(1–4)	95K11
$E_2(20,22)(r)$	3	20	9.6	urea	(1–4)	95K11
$E_2(15,20)(r)$	3	20	10.0	urea	(1–4)	95K11

Remarks:
(1) the native coiled-coil consists of two identical 35-residue polypeptide chains with a heptad repeat (Gln-Val-Gly-Ala-Leu-Gln-Lys) and a Cys in position 2 to allow formation of an interchain disulfide bridge
(2) one or two Glu substitutions for Gln per chain
(3) index "o" for oxidized, "r" for reduced
(4) $\Delta(\Delta G)$ is based on the difference in denaturant half concentration and average slope

De novo designed antiparallel four-stranded coiled-coils

Form	pH	T	ΔG	Approach	Remarks	Ref
coil-LL	6.9	25	92.0	sedimentation	(1–3)	96B5
coil-LL	6.9	25	86.2	thermal	(1,3)	96B5
coil-LL	6.9	57	84.1	thermal	(1,3)	96B5
coil-VL	6.9	25	75.3	sedimentation	(1–3)	96B5
coil-VL	6.9	25	70.3	thermal	(1,3)	96B5
coil-VL	6.9	57	59.0	thermal	(1,3)	96B5

Remarks:
(1) ΔG refers to the 1 M standard state
(2) from sedimentation equilibrium
(3) peptide design: Gly-Asn-Ala-Asp-Glu-Leu-Tyr-Arg-Met-X-Asp-Ala-Leu-Arg-Glu-His-X-Gln-Ser-Leu-Arg-Arg-Lys-X-Arg-Ser-Gly with X = Val for coil-VL and X = Leu for coil-LL

Trimeric coiled-coil

pH	T	ΔG	Approach	Remarks	Ref
7.5	25	77.0±0.8	GuHCl	(1–3)	96B7

Remarks:
(1) coiled-coil designed by placing Val at each a position and Leu at each d position of the heptad repeating unit
(2) treatment by a monomer/trimer equilibrium
(3) buffer: 10 mM MOPS

Four-chain coiled-coil, synthetic

Peptide	pH	T	ΔG	Approach	Remarks	Ref
Lac-21	7.5	25	66	GuHCl	(1–3)	95F1
Lac-28	7.5	25	126	GuHCl	(1–3)	95F1
Lac-35	7.5	25	217	GuHCl	(1–3)	95F1
Lac21Glu/Lys	7.5	25	93.7±5.4	GuHCl	(1–4)	96F1

Remarks:
(1) synthetic peptides derived from sequences contained in the tetramerization domain of Lac repressor
(2) bundles that contain α-helical chains of 21, 28, and 35 amino acids
(3) ΔG is the tetramer stability obtained by linear extrapolation
(4) Lac-21 based peptide with either Glu or Lys at all b and c heptad positions

Four-helix bundle protein from de novo synthesis

pH	T	ΔG	Approach	Remark	Ref
7	–	94	GuHCl	(1)	88R2

Remark:
(1) linear extrapolation

Four-helix bundle, redesigned Zn-binding synthetic protein

Form	pH	T	ΔG	Approach	Remarks	Ref
Ac-α_4-CONH$_2$	7.0		64.4	GuHCl	(1)	93H2
H3-α_4 in the absence of Zn^{2+}:						
	7.0		26.0	GuHCl	(1)	93H2
H3-α_4 in the presence of Zn^{2+}:						
	7.0		40.8	GuHCl	(1)	93H2
H6-α_4 in the absence of Zn^{2+}:						
	7.0		10.5	GuHCl	(1)	93H2
H6-α_4 in the presence of Zn^{2+}:						
	7.0		43.1	GuHCl	(1)	93H2

Remark:
(1) linear extrapolation

Four-helix bundles consisting of two disulfide-linked helix-loop-helix parts with variable loops

Mutant	pH	T	ΔG	Approach	Remarks	Ref
GCPRG	7		51.9±6.3	GuHCl	(1–3)	94R2
GPCRG	7		64.9±7.5	GuHCl	(1–3)	94R2
GCRRG	7		51.0±6.3	GuHCl	(1–3)	94R2
GPRCG	7		60.7±5.4	GuHCl	(1–3)	94R2
GPRRC	7		34.7±3.3	GuHCl	(1–3)	94R2

Remarks:
(1) linear extrapolation to zero denaturant concentration by a method that includes the pre- and postdenaturational baselines for a non-linear regression of the data
(2) mutant description: indicated is the loop that connects two α-helices, e.g., (α-helix)-(Gly-Cys-Pro-Arg-Gly)-(α-helix); the four-helix bundle is formed by a disulfide bridge between Cys residues that belong to the loop region
(3) buffer: 10 mM MOPS, 100 mM NaCl, pH 7

Helichrome, artificial hemeprotein

pH	T	ΔG	Approach	Remark	Ref
7.5	27	18.4	GuHCl	(1)	90S

Remark:
(1) linear extrapolation

Helix forming model peptide
α-helix propensity derived from folding studies on two model peptides normalized so that the helix propensity of Gly is zero

Amino Acid	$\Delta(\Delta G)(1)$	Ref	$\Delta(\Delta G)(2)$	Ref
Aib	2.89	9O2		
Ala	3.22	9O2	3.31	90L4
Arg	2.85	9O2		
Asn	0.29	9O2	0.75	90L4
Asp	0.63	9O2		
Cys	0.96	9O2		
Gln	1.38	9O2	2.01	90L4
Glu	1.13	9O2		
Gly	0.00	9O2	0.00	90L4
His	0.25	9O2		
Ile	0.96	9O2	1.63	90L4
Leu	2.59	9O2	2.59	90L4
Lys	2.72	9O2		
Met	2.09	9O2	2.38	90L4
Phe	1.72	9O2		
Pro	ca.–12	9O2		
Ser	1.46	9O2	1.17	90L4
Thr	0.46	9O2	0.96	90L4
Trp	1.88	9O2		
Tyr	0.71	9O2		
Val	0.59	9O2	1.42	90L4

Remarks:
(1) model peptide: 4x(Leu-Glu-Ala-Leu-Glu-X-Lys), with guest position X; measured at pH 7.5 and 23±2°C by urea denaturation with correction for the dimer/monomer equilibrium
(2) model peptide: (Tyr-Ser-Glu$_4$-Lys$_4$-X$_3$-Glu$_4$-Lys$_4$) with guest position X; measured at pH 7 at 4°C by CD, the approach is given in Ref. 90L4
(3) for comparison with amino acid replacements in proteins, see: LYSOZYME phage T4, position 44, Ref. 93B5, BARNASE, position 9, Ref. 91H3

De novo synthesized greek key jellyroll protein

pH	T	ΔG	Approach	Remark	Ref
8.5	20	0.4	urea	(1)	95S10

Remark:
(1) linear extrapolation

Leucine zippers, synthetic model peptides

pH	T	ΔG	Approach	Remarks	Ref
7.0	20	31.4	GuHCl	(1–4)	93Z4

Remarks:
(1) linear extrapolation
(2) measured in 50 mM phosphate buffer with 0.1 M KCl
(3) the data were taken from Fig. 8 in Ref. 93Z4
(4) data for further model peptides are contained in Ref. 93Z4

Minibody = designed metal-binding Protein

	pH	T	ΔG	Approach	Remarks	Ref
	5.8	21	10.5	urea	(1,2)	93P2

Remarks:
(1) linear extrapolation
(2) measured in the presence of 1 mM Zn^{2+}

De novo designed Ru_2Cu_2-metalloprotein

Form	pH	T	ΔG	Approach	Remarks	Ref
holo	6.0	25	14.6±0.8	GuHCl	(1)	93G4
apo	6.0	25	8.3	GuHCl	(1)	93G4

Remark:
(1) non-linear fit of the denaturation equilibrium curve

QLRb-4, 80-residues protein recovered from a random sequence library, predominantly composed of Gln, Leu and Arg

Protein Concentration	pH	T	ΔG	Approach	Remarks	Ref
2.5 μM	5.8	25	188±9	GuHCl	(1,2)	95D1
10.0 μM	5.8	25	183±5	GuHCl	(1,2)	95D1
19.0 μM	5.8	25	184±5	GuHCl	(1,2)	95D1

Remarks:
(1) linear extrapolation
(2) ΔG refers to dissociation and unfolding of the native tetramer into unfolded monomers at a standard state of 1 M

Protein B, de novo designed protein

	pH	T	ΔG	Approach	Remarks	Ref
	7.5	20	15.5	urea	(1,2)	93K2

Remarks:
(1) linear extrapolation
(2) buffer: 50 mM sodium phosphate with 100 mM NaCl

Protein F, de novo designed protein

	pH	T	ΔG	Approach	Remarks	Ref
	7.5	20	18.4	urea	(1,2)	93K2

Remarks:
(1) linear extrapolation
(2) buffer: 50 mM sodium phosphate with 100 mM NaCl

Desoxyribonuclease

Desoxyribonuclease I

	pH	T	ΔG	Approach	Remarks	Ref
	7	24	39	GuHCl	(1)	82O
	7	24	37	GuHCl	(2)	82O

Remarks:
(1) in the presence of 1 mM calcium, linear extrapolation
(2) in the absence of calcium, linear extrapolation

Dihydrofolate Reductase

Dihydrofolate reductase, wild-type and mutants (*E. coli*), ΔG values

Mutant	pH	T	ΔG	Approach	Remarks	Ref
wild-type	7.8	15	24.7±1.3	urea	(1)	87P2
wild-type	7.8	15	25.6±1.5	urea	(2)	86T3
wild-type	7.8	15	22.3±2.8	urea	(3)	86T3
wild-type	7.8	15	22.7±4.4	urea	(4)	86T3
wild-type	7.8	15	24.7±1.3	urea	(1)	87P3
wild-type	7.8	15	24.8±1.4	urea	(1)	89P6
wild-type	7.8	15	30.6±3.3	urea	(1)	91K6
Asp27→Asn	7.8	15	30.6±5.4	urea	(1)	87P2
Asp27→Asn	7.8	15	30.5±5.4	urea	(1)	87P3
Leu28→Arg	7.8	15	31.8±2.9	urea	(1)	87P2
Leu28→Arg	7.8	15	31.8±2.9	urea	(1)	87P3
Leu28→Arg	7.8	15	32.0±2.8	urea	(1)	89P6
Phe31→Val	7.8	15	18.4±2.1	urea	(1)	87P2
Phe31→Val	7.8	15	18.4±2.1	urea	(1)	87P3
Arg44→Leu	7.8	15	31.0±7.1	urea	(1)	87P2
Arg44→Leu	7.8	15	31.0±7.1	urea	(1)	87P3
Thr113→Val	7.8	15	19.7±2.9	urea	(1)	87P2
Thr113→Val	7.8	15	19.7±2.9	urea	(1)	87P3
Gln139→Lys	7.8	15	20.5±4.2	urea	(1)	87P2
Glu139→Lys	7.8	15	20.5±4.2	urea	(1)	87P3
Glu139→Lys	7.8	15	20.7±4.3	urea	(1)	89P6
Glu139→Gln	7.8	15	23.0±1.8	urea	(1)	89P6
double mutant (Leu28→Arg and Glu139→Gln)						
	7.8	15	26.7±2.3	urea	(1)	89P6
wild-type (Val75)						
	7.8	15	24.7±1.3	urea	(1)	89G1
Val75→Ala	7.8	15	25.1±2.1	urea	(1)	89G1
Val75→Cys	7.8	15	25.1±3.3	urea	(1)	89G1
Val75→Ser	7.8	15	25.9±2.5	urea	(1)	89G1
Val75→His	6.5	15	19.2±2.5	urea	(1)	89G1
Val75→His	7.8	15	16.3±1.3	urea	(1)	89G1
Val75→Ile	7.8	15	19.2±2.1	urea	(1)	89G1

Dihydrofolate reductase, wild-type and mutants (*E. coli*), ΔG values (continued)

Mutant	pH	T	ΔG	Approach	Remarks	Ref
Val75→Arg	7.8	15	19.7±2.1	urea	(1)	89G1
Val75→Tyr	7.8	15	14.6±1.3	urea	(1)	89G1
Val75→Tyr	7.8	15	19.2±2.9	urea	(3)	89G1
Val75→Tyr	7.8	15	18.0±1.3	urea	(5)	89G1
Val75→Tyr	7.8	15	38.9±20.9	urea	(6)	89G1

Remarks:
(1) linear extrapolation
(2) UV spectroscopy, linear extrapolation
(3) CD, linear extrapolation
(4) fluorescence, linear extrapolation
(5) transition N→I, three-state unfolding model
(6) transition I→U, three-state unfolding model

Dihydrofolate reductase, wild-type and mutants, difference values

Mutant	ΔT	$\Delta(\Delta G)$	Approach	Remarks	Ref
wild-type (Asp27)	0	0	DSC		87S2
Asp27→Asn	3.8	5.4	DSC	(1)	87S2
Asp27→Ser	5.2	4.2	DSC	(2)	87S2
wild-type (Trp74)	0	0	urea	(3)	89G2
Trp74→Phe		−5.0	urea	(4)	89G2

Remarks:
(1) the enthalpy change is increased by 63 kJ/mol
(2) the enthalpy change is decreased by 29 kJ/mol
(3) transition midpoint of the wild-type protein at pH 7.8 and 15°C is 3.1 M urea
(4) the ΔG value refers to the conditions of remark (3)

Dihydrofolate reductase, wild-type and cysteine mutants

Mutant	pH	T	ΔG	Approach	Remarks	Ref
wild-type	7.8	15	25.5±1.3	urea	(1,2)	93I3
Cys85→Ser	7.8	15	24.3±1.7	urea	(1)	93I3
Cys152→Glu	7.8	15	24.3±0.8	urea	(1)	93I3
double mutant (Cys85→Ser and Cys152→Glu)						
	7.8	15	18.8±0.8	urea	(1,3)	93I3
double mutant (Cys85→Ser and Cys152→Glu)						
	7.8	15	18.4±0.8	urea	(1,4)	93I3

Remarks:
(1) linear extrapolation, data treatment by a method that includes the pre- and postdenaturational baselines for a non-linear regression of the data
(2) data from Ref. 86T2
(3) transition monitored by UV spectroscopy
(4) transition monitored by CD

Dihydrofolate reductase from *E. coli*, wild-type and mutants

Mutant	pH	T	ΔG	Approach	Remarks	Ref
wild-type	7.2	20	21.6±0.5	urea	(1)	91A1
Val88→Ala	7.2	20	23.2±1.4	urea	(1)	91A1
Val88→Ile	7.2	20	24.7±2.1	urea	(1)	91A1
Val88 deleted	7.2	20	19.5±3.3	urea	(1)	91A1

Remark:
(1) linear extrapolation

Dihydrofolate reductase, wild-type and mutants

Mutant	pH	T	ΔG	Approach	Remarks	Ref
wild-type	7.8	15	24.7±1.3	urea	(1,2)	92T2
Arg44→Leu	7.8	15	30.5±1.7	urea	(1,3)	92T2
Pro66→Ala	7.8	15	30.1±2.1	urea	(1,4)	92T2
double mutant (Arg44→Ala and Pro66→Ala)						
	7.8	15	28.5±2.8	urea	(1,5)	92T2

Remarks:
(1) linear extrapolation
(2) slope m = (−7.9±0.4) kJ/(mol×M_D)
(3) slope m = (−9.6±0.4) kJ/(mol×M_D)
(4) slope m = (−10.0±0.8) kJ/(mol×M_D)
(5) slope m = (−9.6±0.8) kJ/(mol×M_D)

Dihydrofolate reductase mutants, comparison of reduced and oxidized form

Mutant	pH	T	ΔG	Approach	Remarks	Ref
Pro39→Cys (red.)	9	25	49.4	GuHCl	(1)	87V2
Pro39→Cys (ox.)	9	25	56.9	GuHCl	(1)	87V2

Remark:
(1) UV spectroscopy, linear extrapolation

Dihydrofolate reductase, wild-type and circularly permuted mutant, in the presence and in the absence of ligands

Mutant/Ligand	T	ΔG	Approach	Remarks	Ref
wild-type	23	20.1	urea	(1)	94P5
wild-type + ligand 1	23	57.7	urea	(1,2)	94P5
Pr20	23	10.9	urea	(1,3)	94P5
Pr20 + ligand 1	23	19.7	urea	(1–3)	94P5
Pr20 + ligand 2	23	54.8	urea	(1,3,4)	94P5

Remarks:
(1) linear extrapolation
(2) ligand 1: 250 μM NADPH, 25 μM methotrexate
(3) Pr20 = circularly permuted mutant protein
(4) ligand 2: 250 μM NADPH, 200 μM methotrexate

Dihydrofolate reductase from *E. coli*, wild-type and mutants in the presence of ligands

Mutant/Ligand	pH	T	ΔG	Approach	Remarks	Ref
wild-type, no ligand	7.0	15	26.5	urea	(1,2)	93G3
wild-type, no ligand	7.0	15	37.2	GuHCl	(1,2)	93G3
wild-type, folic acid	7.0	15	30.1	urea	(1–3)	93G3
wild-type, NADPH	7.0	15	37.2	urea	(1–3)	93G3
wild-type, 0.1 M KCl	7.0	15	36.1	urea	(1,2)	93G3
Gly121→Leu, no ligand	7.0	15	24.9	urea	(1,2)	93G3
Gly121→Val, no ligand	7.0	15	21.4	urea	(1,2)	93G3
Gly121→Val, no ligand	7.0	15	25.7	GuHCl	(1,2)	93G3
Gly121→Val, folic ac.	7.0	15	28.0	urea	(1–3)	93G3
Gly121→Val, NADPH	7.0	15	26.7	urea	(1–3)	93G3
Gly121→Val, 0.1 M KCl	7.0	15	30.7	urea	(1,2)	93G3

Remarks:
(1) linear extrapolation
(2) buffer: 10 mM potassium phosphate, 0.1 mM EDTA, 1.4 mM 2-mercaptoethanol
(3) folic acid and NADPH were added in 1:1 molar ratio with respect to enzyme

Dihydrofolate reductase from *E. coli*

Transition	pH	T	ΔG	Approach	Remarks	Ref
N→I	7.0	15	7.5	heat	(1–3)	96O2
I→U	7.0	15	7.9	heat	(1–3)	96O2
N→U	7.0	15	15.5	heat	(1–3)	96O2

Remarks:
(1) for details of the mathematical treatment, see Ref. 96O2
(2) T_{trs} amounts to 45.0, 53.0, and 49.4°C, respectively
(2) measured in 10 mM potassium phosphate, 0.1 mM EDTA, 0.1 mM DTT

R67 Dihydrofolate Reductase

General remark: R67 DHFR, encoded by an R plasmid, does not show any homology with chromosomal DHFR. R67 DHFR is a polypeptide chain that is 78 residues long. R67 DHFR is tetrameric at high pH (8.0) and dimeric at low pH (5.0).

R67 dihydrofolate reductase (DHFR), truncated and dimeric protein

Form	pH	T	ΔG	Approach	Remarks	Ref
Complete and truncated protein:						
complete	5.0	30	56.5±2.2	GuHCl	(1,2)	91R2
complete	5.0	30	64.9±2.4	GuHCl	(1,3)	91R2
complete	5.0	30	55.2±2.1	GuHCl	(1,4)	91R2
complete	5.0	30	55.6±2.7	GuHCl	(1,5)	91R2
truncated	5.0	30	44.4±0.9	GuHCl	(1,2,6,7)	91R2
truncated	5.0	30	50.2±2.4	GuHCl	(1,4,6)	91R2

R67 dihydrofolate reductase(DHFR), truncated and dimeric protein (continued)

Form	pH	T	ΔG	Approach	Remarks	Ref
Native and dimeric form:						
native	5.0	23	55.2±2.1	GuHCl	(1,8)	93Z5
native	5.0	23	55.6±2.7	GuHCl	(1,9)	93Z5
double	5.0	23	30.5±1.3	GuHCl	(1,8,10)	93Z5
double	5.0	23	28.6±1.5	GuHCl	(1,9,10)	93Z5
trans. (1)	8.0	30	40.3	GuHCl	(11,12)	94Z7
trans. (2)	8.0	30	51.7	GuHCl	(11,13)	94Z7

Remarks:
(1) linear extrapolation to zero denaturant concentration by a method that includes the pre- and postdenaturational baselines for a non-linear regression of the data according to Santoro & Bolen, Refs. 88B2, 88S4
(2) measured by difference UV absorption at 290 nm (94 μM DHFR)
(3) measured by CD at 217 nm (12 μM DHFR)
(4) measured by tryptophan fluorescence, excitation 290 nm, emission 340 nm, 12 μM DHFR
(5) see (4), 1.2 μM DHFR
(6) 16 N-terminal residues were cleaved off by chymotrypsin
(7) 140 μM DHFR
(8) 6.0 μM DHFR
(9) 0.6 μM DHFR
(10) dimeric R67 DHFR, constructed by gene duplication
(11) mechanism: tetramer→2 dimer→4 unfolded
(12) ΔG refers to the tetramer→2 dimer transition, the transition was monitored by enzyme acitvity and CD
(13) ΔG refers to the dimer→unfolded transition

DNA-Binding Protein

DNA-Binding Protein HU from *Bacillii*

Form	pH	T	ΔG	Approach	Remarks	Ref
Bac. cald.	6.0	20	23.4	heat	(1)	90W4
Bac. stea.	6.0	20	20.9	heat	(2)	90W4
Bac. subt.	6.0	20	10.5	heat	(3)	90W4
Bac. glob.	6.0	20	9.2	heat	(4)	90W4

Remarks:
(1) Bac. cald. = *Bacillus caldolyticus* having optimum growth temperature above 70°C, T_{trs} of the protein at 68°C
(2) Bac. stea. = *Bacillus stearothermophilus* having optimum growth temperature at 65°C, T_{trs} of the protein at 64°C
(3) Bac. subt. = *Bacillus subtilis* having optimum growth tempeature at 37°C, T_{trs} of the protein at 43°C
(4) Bac. glob. = *Bacillus globigii* having optimum growth temperature at 30°C, T_{trs} of the protein at 41°C

Table 1. Gibbs Energy Change – Molar Values

Papillomavirus strain-16 E2 DNA-binding domain in various buffers

Buffer	pH	T	ΔG	Approach	Remarks	Ref
acetate	5.6	25	46.1±1.0	urea	(1–3)	96M9
acetate	5.6	25	45.4±2.1	urea	(1,2,4)	96M9
acetate	5.6	25	41.0±1.2	urea	(1,3,5)	96M9
MES	6.1	25	41.4±1.4	urea	(1,3,5)	96M9
Bis-Tris	7.0	25	53.3±1.6	urea	(1,3,5)	96M9
citrate-phosphate buffer						
	7.0	25	67.7±2.6	urea	(1,3,5)	96M9
Tris-HCl	8.0	25	53.3±2.5	urea	(1,3,5)	96M9

Remarks:
(1) treatment by model in which the dimer is dissociated and unfolded simultaneously, assuming linear dependence of ΔG on denaturant concentration
(2) transition monitored by fluorescence
(3) protein concentration 10 μM
(4) protein concentration 1 μM
(5) transition monitored by far-UV CD

DNA-binding protein Sso7d from the hyperthermophile *Sulfolobus solfataricus*

pH	T	ΔG	Approach	Remarks	Ref
2.5	25	14.2	DSC	(1,2)	96K5
2.6	25	16.3	DSC	(1,2)	96K5
3.0	25	18.8	DSC	(1,2)	96K5
3.0	25	19.7	DSC	(1,3)	96K5
3.1	25	22.6	DSC	(1,2)	96K5
3.5	25	25.9	DSC	(1,2)	96K5
4.0	25	28.5	DSC	(1,2)	96K5
4.5	25	32.2	DSC	(1,2)	96K5
4.75	25	30.5	DSC	(1,2)	96K5
5.5	25	31.0	DSC	(1,2)	96K5
6.0	25	30.5	DSC	(1,2)	96K5
6.0	25	32.2	DSC	(1,3)	96K5
6.5	25	31.0	DSC	(1,2)	96K5
7.0	25	32.6	heat	(1,2)	96K5
2.5	25	13.0	heat	(1,2)	96K5
2.5	25	13.0	heat	(1,3)	96K5
2.7	25	14.2	heat	(1,2)	96K5
2.8	25	16.3	heat	(1,3)	96K5
2.9	25	16.7	heat	(1,2)	96K5
3.0	25	19.2	heat	(1,3)	96K5
3.1	25	19.7	heat	(1,2)	96K5
3.3	25	22.6	heat	(1,3)	96K5
3.5	25	24.3	heat	(1,2)	96K5
	9	~30	DSC, heat	(4)	96K5
	77	~12	DSC, heat	(5)	96K5
	90	~4	DSC, heat	(6)	96K5

Remarks:
(1) ΔG was calculated using ΔCp = 2.59 kJ/mol/K
(2) cloned non-methylated protein (c-Sso7d)
(3) ε-mono methylated protein purified from *Sulfolobus solfataricus* (m-Sso7d). No significant differences between c-Sso7d and m-Sso7d were reported in Ref. 96K5
(4) 9°C is the temperature of maximum stability
(5) 77°C is the temperature of optimal growth
(6) 90°C is the maximum growth temperature

Sac7d DNA-binding protein from *Sulfolobus acidocaldarius*, recombinant

pH	T	ΔG	Approach	Remarks	Ref
7.0	20	32.2±1.3	DSC	(1)	96M2
7.0	20	22.2±1.3	GuHCl	(2)	96M2
7.0	20	25.5±0.8	GuHCl	(3)	96M2
7.0	28	21.3±0.8	DSC	(4,5)	96M2
4.0	20	26.4±0.8	DSC	(6)	96M2
4.0	20	14.2±1.3	urea	(2)	96M2
4.0	20	20.1	urea	(3)	96M2
4.0	23	18.4±0.8	DSC	(7,8)	96M2

Remarks:
(1) using T_{trs} = 90.7±0.8°C, ΔH = 244±4 kJ/mol and ΔCp = 2.08±0.08 kJ/mol/K
(2) linear extrapolation to zero denaturant concentration by a method that includes the pre- and postdenaturational baselines for a non-linear regression of the data
(3) denaturant binding model
(4) using the data from (1) except the more reliable ΔCp = 3.59±0.08 kJ/mol/K from global non-linear regression of the chemical denaturation data constrained by DSC determined values of ΔH and T_{trs}
(5) the protein achieves maximum stability at 28°C (pH 7)
(6) using T_{trs} = 80.2°C, ΔH = 222 kJ/mol and ΔCp = 2.08±0.08 kJ/mol/K
(7) using the data from (6) except the more reliable ΔCp = 3.59±0.08 kJ/mol/K from global non-linear regression of the chemical denaturation data constrained by DSC determined values of ΔH and T_{trs}
(9) the protein achieves maximum stability at 23°C (pH 4)

DsbA, see also protein disulfide isomerase

DsbA, 21-kDa soluble periplasmic protein which facilitates disulfide bond formation

Form	pH	T	ΔG	Approach	Remarks	Ref
oxidized	7.5	25	36.4±3.3	urea	(1,2)	93Z1
reduced	7.5	25	51.5±2.5	urea	(1,2)	93Z1

Remarks:
(1) linear extrapolation
(2) buffer: 0.1 M Tris, 0.2 M KCl, 1mM EDTA

DsbA from *E. coli*, wild-type and cysteine mutants

Mutant	pH	T	ΔG	Approach	Remarks	Ref
(Cys30SH and Cys33SH)						
	7.5	25	54.0	urea	(1–3)	94Z3
(Cys30SH and Cys33→Ser)						
	7.5	25	49.4	urea	(1–3)	94Z3
(Cys30→Ser and Cys33SH)						
	7.5	25	46.4	urea	(1–3)	94Z3
(Cys30→Ser and Cys33→Ser)						
	7.5	25	45.6	urea	(2,3)	94Z3
(Cys30SSG and Cys33→Ser)						
	7.5	25	43.1	urea	(2–4)	94Z3
(Cys30-Cys33 disulfide bond)						
	7.5	25	40.6	urea	(2,3)	94Z3

Remarks:
(1) Cys30SH, Cys33SH – free thiol groups
(2) linear extrapolation
(3) buffer: 0.1 M Tris-HCl, 0.2 M KCl, 1 mM EDTA
(4) Cys30SSG – mixed disulfide with glutathione

Dystrophin

Dystrophin rod domain, recombinant, fragments

Fragment	pH	T	ΔG	Approach	Remarks	Ref
F 117	7.6	20	6.3	urea	(1,2)	95K1
F 117	7.6	20	6.3	heat, v.H.	(2)	95K1
F 119	7.6	20	14.2	urea	(1,2)	95K1
F 119	7.6	20	14.6	heat, v.H.	(2)	95K1

Remarks:
(1) linear extrapolation
(2) fragments F 117 and F 119 are composed of 117 and 119 residues, respectively

Eglin

Eglin c, proteinase inhibitor

	pH	T	ΔG	Approach	Ref
	7.00	25	33.3*	DSC	95B1

Fatty Acid Binding Protein

Intestinal fatty acid binding protein

Mutant	pH	T	ΔG	Approach	Remarks	Ref
wild-type	7.2	20	21.8±1.4	GuHCl	(1)	93J3
wild-type	9.6	20	23.4±2.5	GuHCl	(1)	93J3
Ile23→Cys	7.2	20	25.6±1.7	GuHCl	(1)	93J3
Ile23→Cys	9.6	20	31.3±1.8	GuHCl	(1)	93J3
Ser53→Cys	7.2	20	24.3±1.1	GuHCl	(1)	93J3
Ser53→Cys	9.6	20	29.7±3.2	GuHCl	(1)	93J3
Val60→Cys	7.2	20	25.2±1.2	GuHCl	(1)	93J3
Val60→Cys	9.6	20	25.9±1.6	GuHCl	(1)	93J3
Leu72→Cys	7.2	20	21.3±1.3	GuHCl	(1)	93J3
Leu72→Cys	9.6	20	22.9±1.5	GuHCl	(1)	93J3
Leu89→Cys	7.2	20	19.3±0.8	GuHCl	(1)	93J3
Leu89→Cys	9.6	20	16.3±2.0	GuHCl	(1)	93J3
Ala104→Cys	7.2	20	32.5±3.8	GuHCl	(1)	93J3
Ala104→Cys	9.6	20	29.7±3.1	GuHCl	(1)	93J3

Remark:
(1) linear extrapolation, data treatment by a method that includes the pre- and postdenaturational baselines for a non-linear regression of the data, see also Ref. 90R4

Human muscle fatty acid binding protein, wild-type and mutants

Mutant	pH	T	ΔG	Approach	Remarks	Ref
apo	8.0	25	45.5±5.6	urea	(1)	96P3
holo	8.0	25	49.1±0.9	urea	(1)	96P3
apo	8.0	25	28.8±0.6	GuHCl	(1)	96P3
holo	8.0	25	37.0±2.7	GuHCl	(1)	96P3
Phe16→Tyr	8.0	25	31.1±1.6	urea	(1)	96P3
Thr40→Glu	8.0	25	31.8±6.0	urea	(1)	96P3
Thr40→Gln	8.0	25	27.2±1.7	urea	(1)	96P3
Thr40→Val	8.0	25	37.8±3.7	urea	(1)	96P3
Phe57→Ser	8.0	25	30.3±2.5	urea	(1)	96P3
Arg106→Thr	8.0	25	30.7±1.7	urea	(1)	96P3
Arg106→Lys	8.0	25	26.0±4.3	urea	(1)	96P3
Arg106→Gln	8.0	25	42.9±6.9	urea	(1)	96P3
Tyr128→Phe	8.0	25	29.5±0.6	urea	(1)	96P3

Remark:
(1) linear extrapolation

Rat intestinal fatty acid binding protein

pH	T	ΔG	Approach	Remarks	Ref
7.2	20	40.6±1.8	dimethylurea	(1,2)	90R4
7.2	20	23.2±2.1	GuHCl	(1,3)	90R4
7.2	20	24.4±1.0	GuHCl	(1,4)	90R4
7.2	20	21.8±1.4	GuHCl	(1,5)	90R4
7.2	20	23.9±1.7	GuHCl	(1,6)	90R4
7.2	20	42.7±2.4	GuHSCN	(1,5)	90R4
7.2	20	19.6±1.0	KSCN	(1,5)	90R4
7.2	20	42.9±2.1	urea	(1,3)	90R4
7.2	20	40.3±3.4	urea	(1,4)	90R4
7.2	20	42.0±1.0	urea	(1,5)	90R4
7.2	20	40.6±2.3	urea	(1,5,6)	90R4

Remarks:
(1) linear extrapolation to zero denaturant concentration by a method that includes the pre- and postdenaturational baselines for a non-linear regression of the data
(2) monitored by fluorescence at 352 nm
(3) monitored by absorbance at 292 nm
(4) monitored by ellipticity at 216 nm
(5) monitored by fluorescence at 338 nm
(6) stopped flow, monitored by fluorescence at 305 nm
(7) in the presence of 1 M NaCl

Rat intestinal fatty acid binding protein, with bound ligand

pH	T	ΔG	Approach	Remarks	Ref
7.2	20	32.9±1.2	GuHCl	(1–3)	90R4
7.2	20	52.5±1.2	urea	(1–3)	90R4

Remarks:
(1) linear extrapolation to zero denaturant concentration by a method that includes the pre- and postdenaturational baselines for a non-linear regression of the data
(2) in the presence of twofold molar excess of oleate
(3) monitored by fluorescence at 338 nm

E. coli-derived rat intestinal fatty acid binding protein

Mutant	pH	T	ΔG	Approach	Remarks	Ref
wild-type	7.4	20	21.8±1.4	GuHCl	(1)	96K4
Δ17-SG	7.4	20	16.9±0.6	GuHCl	(1,2)	96K4

Remarks:
(1) linear extrapolation to zero denaturant concentration by a method that includes the pre- and postdenaturational baselines for a non-linear regression of the data
(2) Δ17-SG is a variant missing a helix, residues 15–31 deleted and a Ser-Gly linker after residue 14 inserted

Ferredoxin

Ferredoxin from *Anabaena*, wild-type and mutants

Mutant	pH	T	ΔG	Approach	Remarks	Ref
wild-type	7.5	25±0.2	26.4±1.7	GuHCl	(1–3)	95H9
Arg42→Ala	7.5	25±0.2	16.3±0.8	GuHCl	(1–3)	95H9
Arg42→His	7.5	25±0.2	20.5±0.8	GuHCl	(1–3)	95H9
Thr48→Ala	7.5	25±0.2	26.8±1.3	GuHCl	(1–3)	95H9
Phe65→Ala	7.5	25±0.2	22.2±1.3	GuHCl	(1–3)	95H9
Phe65→Ile	7.5	25±0.2	22.6±1.3	GuHCl	(1–3)	95H9
Phe65→Trp	7.5	25±0.2	23.8±0.8	GuHCl	(1–3)	95H9
Phe65→Tyr	7.5	25±0.2	24.3±1.3	GuHCl	(1–3)	95H9
Glu94→Asp	7.5	25±0.2	21.8±0.8	GuHCl	(1–3)	95H9
Glu94→Gln	7.5	25±0.2	28.9±2.1	GuHCl	(1–3)	95H9
Glu94→Lys	7.5	25±0.2	24.7±0.8	GuHCl	(1–3)	95H9
Glu95→Lys	7.5	25±0.2	30.5±1.3	GuHCl	(1–3)	95H9
wt*	7.5	25±0.2	18.8±0.8	GuHCl	(1,3,4)	95H9
His42→Arg*	7.5	25±0.2	35.6±0.8	GuHCl	(1,3,4)	95H9

Remarks:
(1) linear extrapolation
(2) *Anabaena* 7120 vegetative ferredoxin
(3) buffer: 20 mM Tris-HCl, 40 mM NaCl, pH 7.5
(4) *Anabaena* 7120 heterocyst ferredoxin

High-potential [4Fe-4S] iron-sulfur protein from *Ectothiorhodospira halophila*

Mutant	pH	T	Δ(ΔG)	Approach	Remarks	Ref
wild-type	7	25	0	DSC	(1)	95I2
Tyr12→His	7	25	−107	DSC		95I2
Tyr12→His	9	25	−115	DSC		95I2
Tyr12→Phe	7	25	−54	DSC		95I2
Tyr12→Phe	9	25	−50	DSC		95I2
Tyr12→Phe	10	25	−51	DSC		95I2
Tyr12→Phe	10.8	25	−82	DSC		95I2
Tyr12→Trp	7	25	−127	DSC		95I2

Remark:
(1) negative ΔH values were observed in unfolding of wild-type and mutant proteins

Ferric Enterobactin Receptor

Ferric enterobactin receptor

pH	T	ΔG	Approach	Remarks	Ref
7.2	22	27.0±4.5	GuHCl	(1–3)	95K9
7.2	22	24.9±2.3	urea	(1–3)	95K9

Remarks:
(1) linear extrapolation
(2) transition was monitored by ESR
(3) measured in 2% Triton X-100, 20 mM MOPS, pH 7.2

Fibrinogen

Fibrinogen, D-fragment

pH	T	ΔG	Approach	Ref
3.5	25	23.6*	DSC	82M1

Fibrinogen, bovine, proteolytic fragments derived from the C-terminal regions

Fragment	pH	T	ΔG	Approach	Remarks	Ref
TSD	3.5	37	42	DSC	(1)	95L7
TSD_1	3.5	37	21	DSC	(1,2)	95L7
TSD_2	3.5	37	17	DSC	(1,2)	95L7

Remarks:
(1) TSD thermostable region of the D fragment
(2) ΔG was taken from Fig. 6 in Ref. 95L7

Fibroblast Growth Factor, acidic, see growth factor

Fibronectin

Fibronectin, gelatin binding fragment

pH	T	ΔG	Approach	Remark	Ref
–	25	9.6	GuHCl	(1)	89I2

Remark:
(1) linear extrapolation

FK Binding Protein

Human FK binding protein, prolyl isomerase

pH	T	ΔG	Approach	Remarks	Ref
7.2	25.0	24.7	urea	(1–3)	93E1
7.2	25.0	27.2	urea	(1,2,4)	93E1
7.2	25.0	24.3	urea	(1,2,5)	93E1
7.2	25.0	25.9	urea	(1,2,6)	93E1

Remarks:
(1) linear extrapolation
(2) buffer: 75 mM sodium phosphate, 25 mM glycine, 1.0 mM DTT, 0.04% sodium azide, pH 7.2
(3) measured by fluorescence
(4) measured by second derivative spectroscopy
(5) measured by CD
(6) measured by NMR

FK 506 binding protein, prolyl isomerase

pH	T	ΔG	Approach	Remarks	Ref
8.0	20	19.2±1.3	GuHCl	(1,2)	96V3

Remarks:
(1) linear extrapolation to zero denaturant concentration by a method that includes the pre- and postdenaturational baselines for a non-linear regression of the data
(2) buffer: 0.02 M Tris, 1 mM EDTA, 1 mM β-mercaptoethanol

Flagellin

Flagellin, monomeric, normal type

pH	T	ΔG	Approach	Ref
7	25	72*	DSC	84F1

Flavodoxin

Apoflavodoxin from *Anabaena* PCC7119, in the presence of KCl

KCl (M)	pH	T	ΔG	Approach	Remarks	Ref
0.0	7.0	25	17.1±0.5	urea	(1,2,6)	96G3
0.0	7.0	25	16.3	DSC	(3,4)	96G3
0.0	7.0	25	14.4	DSC	(3,5)	96G3
0.1	7.0	25	17.7	DSC	(3,4)	96G3
0.1	7.0	25	15.6	DSC	(3,5)	96G3
0.2	7.0	25	18.6	DSC	(3,4)	96G3
0.2	7.0	25	16.7	DSC	(3,5)	96G3
0.3	7.0	25	19.2	DSC	(3,4)	96G3
0.3	7.0	25	17.6	DSC	(3,5)	96G3
0.5	7.0	25	20.1	DSC	(3,4)	96G3
0.5	7.0	25	18.2	DSC	(3,5)	96G3
0.75	7.0	25	21.9	DSC	(3,4)	96G3
0.75	7.0	25	20.0	DSC	(3,5)	96G3
1.0	7.0	25	23.4	DSC	(3,4)	96G3
1.0	7.0	25	21.3	DSC	(3,5)	96G3

Remarks:
(1) linear extrapolation
(2) measured in 50 mM MOPS buffer, transition monitored by fluorescence
(3) buffer: 50 mM sodium phosphate + X M KCl
(4) using temperature invariant ΔCp = 5.8 kJ/mol/K
(5) using temperature dependent ΔCp, $\Delta Cp(T) = -34.33 + 0.3256 \times T - 0.000614 \times T^2$
(6) apoflavodoxin is not a monomer at pH 2.0 and the transition is not two-state

Four Helix Bundle, see de novo synthesized proteins/peptides

Galactoside-Binding Protein

Galactoside-binding protein in the presence of Ca^{2+}

pH	T	ΔG	Approach	Remarks	Ref
7.8	25	25*	DSC	(1)	81S2

Remark:
(1) calculated per cooperative unit AB_2

Gene V Protein

Gene V protein, for engeneering and stability of gene V protein – see also Ref. 95T9

Gene V protein from bacteriophage F1, wild-type and mutants

Mutant	pH	T	ΔG	Approach	Remarks	Ref
wild-type	7	25	68.9±2.5	GuHCl	(1)	89S2
wild-type	7.0	25.0	68.2±2.9	GuHCl	(1)	91L6
Val35→Ile	7	25	67.1±2.1	GuHCl	(1)	89S2
Ile47→Val	7	25	58.7±2.5	GuHCl	(1)	89S2
double mutant (Val35→Ile and Ile47→Val)						
	7	25	57.3±2.1	GuHCl	(1)	89S2

Remark:
(1) linear extrapolation, ΔG refers to the unfolding of the dimer into two monomers, i.e., $N_2 \rightarrow 2U$

Gene V protein, wild-type and mutants

Mutant	pH	T	ΔG	Approach	Remarks	Ref
wild-type			68.2±4.2	GuHCl	(1)	91S2
Val35→Ala			61.1±4.2	GuHCl	(1)	91S2
Val35→Cys			66.1±4.2	GuHCl	(1)	91S2
Val35→Ile			65.7±4.2	GuHCl	(1)	91S2
Val35→Leu			64.0±4.2	GuHCl	(1)	91S2
Val35→Met			63.6±4.2	GuHCl	(1)	91S2
Val35→Phe			56.9±4.2	GuHCl	(1)	91S2
Ile47→Ala			43.9±4.2	GuHCl	(1)	91S2
Ile47→Cys			53.6±4.2	GuHCl	(1)	91S2
Ile47→Leu			66.5±4.2	GuHCl	(1)	91S2
Ile47→Met			61.1±4.2	GuHCl	(1)	91S2
Ile47→Phe			63.2±4.2	GuHCl	(1)	91S2
Ile47→Val			59.8±4.2	GuHCl	(1)	91S2
double mutants:						
(Val35→Ala and Ile47→Leu)			63.6±4.2	GuHCl	(1)	91S2
(Val35→Ala and Ile47→Met)			54.0±4.2	GuHCl	(1)	91S2
(Val35→Ala and Ile47→Phe)			59.4±4.2	GuHCl	(1)	91S2
(Val35→Ala and Ile47→Val)			56.9±4.2	GuHCl	(1)	91S2
(Val35→Cys and Ile47→Cys)			48.5±4.2	GuHCl	(1)	91S2
(Val35→Leu and Ile47→Leu)			59.0±4.2	GuHCl	(1)	91S2
(Val35→Leu and Ile47→Met)			57.3±4.2	GuHCl	(1)	91S2
(Val35→Leu and Ile47→Phe)			57.3±4.2	GuHCl	(1)	91S2
(Val35→Leu and Ile47→Val)			51.5±4.2	GuHCl	(1)	91S2
(Val35→Ile and Ile47→Leu)			67.8±4.2	GuHCl	(1)	91S2

Gene V protein, wild-type and mutants (continued)

Mutant	pH	T	ΔG	Approach	Remarks	Ref
(Val35→Ile and Ile47→Met)			61.5±4.2	GuHCl	(1)	91S2
(Val35→Ile and Ile47→Phe)			66.1±4.2	GuHCl	(1)	91S2
(Val35→Ile and Ile47→Val)			59.4±4.2	GuHCl	(1)	91S2
(Val35→Met and Ile47→Leu)			64.9±4.2	GuHCl	(1)	91S2
(Val35→Met and Ile47→Met)			54.8±4.2	GuHCl	(1)	91S2
(Val35→Met and Ile47→Phe)			67.4±4.2	GuHCl	(1)	91S2
(Val35→Phe and Ile47→Leu)			53.1±4.2	GuHCl	(1)	91S2

Remark:
(1) for the procedure see W.Sandberg and T.C. Terwilliger, Ref. 89S2

Gene V protein from bacteriophage F1, wild-type and mutants

Mutant	pH	T	Δ(ΔG)	Approach	Remarks	Ref
wild-type	7.0	25	0	GuHCl	(1)	91Z1
Glu30→Phe	7.0	25	8.8	GuHCl	(1)	91Z1
Glu30→Met	7.0	25	2.9	GuHCl	(1)	91Z1
Cys33→Ser	7.0	25	−17.6	GuHCl	(1)	91Z1
Asp36→Cys	7.0	25	−8.4	GuHCl	(1)	91Z1
Asp36→Asn	7.0	25	−4.2	GuHCl	(1)	91Z1
Ile47→Thr	7.0	25	−31	GuHCl	(1)	91Z1
Asp50→His	7.0	25	−6.3	GuHCl	(1)	91Z1
Phe68→Leu	7.0	25	−17.6	GuHCl	(1)	91Z1
Lys69→His	7.0	25	−5.0	GuHCl	(1)	91Z1
Lys69→Met	7.0	25	0.8	GuHCl	(1)	91Z1
Val70→Cys	7.0	25	−13.0	GuHCl	(1)	91Z1
Val70→Pro	7.0	25	−21.0	GuHCl	(1)	91Z1

Remark:
(1) the data refer to the difference stability compared with the wild-type protein in 2 M GuHCl

Gene V protein

Mutant	pH	T	ΔG	Δ(ΔG)	Remarks	Ref
wild-type	7	25	37.8±1.3	0	(1–3)	93S1
Ile6→Val	7	25		−2.8	(1,2,4)	93S1
Phe13→Thr	7	25		−2.8	(1,2,4)	93S1
Leu28→Val	7	25		4.6	(1,2,4)	93S1
Glu30→Phe	7	25		8.2	(1,2,4)	93S1
Leu32→Tyr	7	25		4.4	(1,2,4)	93S1
Cys33→Met	7	25		−14.6	(1,2,4)	93S1
Cys33→Val	7	25		−0.8	(1,2,4)	93S1
Val35→Ala	7	25		−9.4	(1,2,4)	93S1
Val35→Cys	7	25		−6.1	(1,2,4)	93S1
Val35→Phe	7	25		−13.4	(1,2,4)	93S1
Val35→Ile	7	25		−2.8	(1,2,4)	93S1
Val35→Leu	7	25		−11.4	(1,2,4)	93S1
Val35→Met	7	25		−4.6	(1,2,4)	93S1
Tyr41→Phe	7	25		−2.6	(1,2,4)	93S1
Val45→Cys	7	25		−0.2	(1,2,4)	93S1
Ile47→Cys	7	25		−22.2	(1,2,4)	93S1
Ile47→Phe	7	25		−8.5	(1,2,4)	93S1
Ile47→Leu	7	25		−2.8	(1,2,4)	93S1
Ile47→Met	7	25		−9.2	(1,2,4)	93S1

120   Table 1. Gibbs Energy Change – Molar Values

Gene V protein (continued)

Mutant	pH	T	ΔG	Δ(ΔG)	Remarks	Ref
Ile47→Val	7	25		−11.0	(1,2,4)	93S1
His64→Cys	7	25		2.1	(1,2,4)	93S1
Leu65→Pro	7	25		−6.2	(1,2,4)	93S1
Phe73→Trp	7	25		3.2	(1,2,4)	93S1
Met77→Ile	7	25		6.8	(1,2,4)	93S1
Met77→Val	7	25		5.1	(1,2,4)	93S1
Arg82→Cys	7	25		−6.3	(1,2,4)	93S1
Ala86→Thr	7	25		−2.8	(1,2,4)	93S1
Ala86→Val	7	25		2.0	(1,2,4)	93S1
Phe68→Leu	7	25		−18.0	(1,2,4)	93S1

Double mutants:

Mutant	pH	T	ΔG	Δ(ΔG)	Remarks	Ref
(Ile6→Val and Glu30→Phe)	7	25		3.0	(1,2,4)	93S1
(Ile6→Val and Met77→Ile)	7	25		1.4	(1,2,4)	93S1
(Ile6→Val and Met77→Val)	7	25		−0.2	(1,2,4)	93S1
(Phe13→Thr and Glu30→Phe)	7	25		6.7	(1,2,4)	93S1
(Leu28→Val and Phe68→Leu)	7	25		−14.5	(1,2,4)	93S1
(Glu30→Phe and Ala86→Thr)	7	25		4.9	(1,2,4)	93S1
(Glu30→Phe and Ala86→Val)	7	25		8.6	(1,2,4)	93S1
(Leu32→Tyr and Arg82→Cys)	7	25		0.6	(1,2,4)	93S1
(Cys33→Val and Val35→Cys)	7	25		−6.6	(1,2,4)	93S1
(Cys33→Met and Ile47→Cys)	7	25		−24.0	(1,2,4)	93S1
(Val35→Ala and Ile47→Phe)	7	25		−15.4	(1,2,4)	93S1
(Val35→Ala and Ile47→Leu)	7	25		−12.4	(1,2,4)	93S1
(Val35→Ala and Ile47→Met)	7	25		−18.9	(1,2,4)	93S1
(Val35→Ala and Ile47→Val)	7	25		−18.9	(1,2,4)	93S1
(Val35→Cys and Ile47→Cys)	7	25		−30.1	(1,2,4)	93S1
(Val35→Phe and Ile47→Leu)	7	25		−17.7	(1,2,4)	93S1
(Val35→Ile and Ile47→Phe)	7	25		−8.7	(1,2,4)	93S1
(Val35→Ile and Ile47→Leu)	7	25		−4.9	(1,2,4)	93S1
(Val35→Ile and Ile47→Met)	7	25		−11.9	(1,2,4)	93S1
(Val35→Ile and Ile47→Val)	7	25		−13.1	(1,2,4)	93S1
(Val35→Leu and Ile47→Phe)	7	25		−16.7	(1,2,4)	93S1

Gene V protein (continued)

Mutant	pH	T	ΔG	Δ(ΔG)	Remarks	Ref
(Val35→Leu and Ile47→Leu)						
	7	25		−15.0	(1,2,4)	93S1
(Val35→Leu and Ile47→Met)						
	7	25		−22.6	(1,2,4)	93S1
(Val35→Leu and Ile47→Val)						
	7	25		−21.3	(1,2,4)	93S1
(Val35→Met and Ile47→Phe)						
	7	25		−9.8	(1,2,4)	93S1
(Val35→Met and Ile47→Leu)						
	7	25		−7.1	(1,2,4)	93S1
(Val35→Met and Ile47→Met)						
	7	25		−15.1	(1,2,4)	93S1
(Tyr41→Phe and Phe73→Trp)						
	7	25		−2.8	(1,2,4)	93S1
(Val35→Cys and Arg82→Cys)						
	7	25		−4.4	(1,2,4)	93S1
(His64→Cys and Phe68→Leu)						
	7	25		−17.0	(1,2,4)	93S1
(Leu65→Pro and Phe68→Leu)						
	7	25		−17.8	(1,2,4)	93S1

Remarks:
(1) GuHCl–induced unfolding, data treatment by linear extrapolation, for the procedure see also Ref. 91L6
(2) the data refer to the dimeric protein
(3) mean of 10 measurements
(4) Δ(ΔG) was determined at 2 M GuHCl, the estimated error of Δ(ΔG) of the mutants amounts to ±1.7 kJ/mol

Gene 32 Protein

Gene 32 protein of phage T4, native Zn^{2+}, and reconstituted forms

Form	pH	T_{trs}	ΔT	Δ(ΔG)	Approach	Remarks	Ref
Zn^{2+}	8	55.4±0.2	0	0	DSC	(1)	88K1
apo	8	49.3±0.4	−6.1	−7.1±0.4	DSC	(2)	88K1
Cd^{2+}	8	53.3±0.2	−2.1	−3.8±0.4	DSC	(2)	88K1
Co^{2+}	8	56.4±0.2	1	1.7±0.4	DSC	(2)	88K1
Zn^{2+}-reco	8	54.9	0	−0.8	DSC	(2–3)	88K1

Remarks:
(1) native Zn^{2+} form
(2) ΔT and Δ(ΔG) refer to the native Zn^{2+} form
(3) reconstituted Zn^{2+} form

Table 1. Gibbs Energy Change – Molar Values

Gene 32 protein of phage T4, complexed with polynucleotide native Zn(II), and reconstituted forms

Form	pH	T_{trs}	ΔT	$\Delta(\Delta G)$	Approach	Remarks	Ref
Zn^{2+}	8	59.9±0.2	4.6	7.9	DSC	(1–2)	88K1
apo	8	49.8+0.2	0.5	0	DSC	(1–2)	88K1
Cd^{2+}	8	57.3+0.1	4	7.1	DSC	(1–2)	88K1
Co^{2+}	8	60.9+0.6	4.5	8.8	DSC	(1–2)	88K1

Remarks:
(1) complexed with stoichiometric amount of poly(dT)
(2) ΔT and $\Delta(\Delta G)$ refer to the indicated protein in the absence of poly(dT)

Glucanase

(1,3–1,4)-β-glucanase, various forms

Form	pH	T	ΔG	Approach	Remarks	Ref
MAC	6.0	25	54.0±4.5	GuHCl	(1,2)	94K2
AMY	6.0	25	20.8±0.4	GuHCl	(1,3)	94K2
H(A16-M)Na	6.0	25	58.1±3.1	GuHCl	(1,4,5)	94K2
H(A16-M)Ca	6.0	25	62.3±3.9	GuHCl	(1,4,6)	94K2

Remarks:
(1) linear extrapolation
(2) glucanase from *Bacillus macerans*
(3) glucanase from *Bacillus amyloliquefaciens*
(4) hybrid from consisting of 16 N-terminal residues of AMY and 198 C-proximal residues of MAC
(5) sodium form, measured in the presence of 22 mM Na^+
(6) calcium form, measured in the presence of 5 mM Ca^{2+}

(1–3,1–4)-β-glucanase, hybrid forms

Mutant	pH	T	ΔG	Approach	Remarks	Ref
measurements in the presence of $CaCl_2$:						
H(A12-M)	6.0	25	10.4	DSC	(1)	94W3
measurements in the presence of EDTA:						
H(A12-M)	6.0	25	9.9	DSC	(2)	94W3
H(A12-M)ΔY13	6.0	25	8.4	DSC	(2)	94W3
H(A16-M)	6.0	25	10.8	DSC	(2)	94W3

Remarks:
(1) buffer: 2 mM cacodylate, 1 mM $CaCl_2$, 1.5 M GuHCl
(2) buffer: 2 mM cacodylate, 1 mM EDTA, 1.5 M GuHCl

Glucanase

(1,3–1,4)-β-glucanase, hybrid forms

Mutant	pH	T	ΔG	Approach	Remarks	Ref
MAC	6.0	25	72.0	GuHCl	(1,2,4)	96W2
MAC	6.0	25	51.9	GuHCl	(1–3)	96W2
MAC	6.0	25	37.7	DSC	(1,5,6)	96W2
MAC	6.0	25	25.5	DSC	(1,6,7)	96W2
hybrid 1	6.0	25	73.3	GuHCl	(1–3)	95W4
hybrid 1	6.0	25	73.2	GuHCl	(1–3)	96W2
hybrid 1	6.0	25	43.1	DSC	(1,6,7)	96W2
hybrid 2	6.0	25	67.4	GuHCl	(1–3)	95W4
hybrid 2	6.0	25	89.6	GuHCl	(1,2,4)	95W4
hybrid 2	6.0	25	89.5	GuHCl	(1,2,4)	96W2
hybrid 2	6.0	25	67.4	GuHCl	(1–3)	96W2
hybrid 2	6.0	25	43.5	DSC	(1,5,6)	96W2
hybrid 2	6.0	25	37.7	DSC	(1,6,7)	96W2
hybrid 3	6.0	25	64.5	GuHCl	(1–3)	95W4
hybrid 3	6.0	25	64.4	GuHCl	(1–3)	96W2
hybrid 3	6.0	25	36.0	DSC	(1,6,7)	96W2
hybrid 4	6.0	25	70.3	GuHCl	(1,2,4)	96W2
hybrid 4	6.0	25	68.6	GuHCl	(1–3)	96W2
hybrid 4	6.0	25	42.7	DSC	(1,5,6)	96W2
hybrid 4	6.0	25	34.3	DSC	(1,6,7)	96W2
hybrid 5	6.0	25	74.5	GuHCl	(1–3)	96W2
hybrid 5	6.0	25	42.7	DSC	(1,6,7)	96W2

Remarks:
(1) the hybrids are composed of N-terminal segments of glucanase from *Bacillus amyloliquefaciens* (AMY) and of C-proximal residues of glucanase from *Bacillus macerans* (MAC)
   hybrid 1: (1–16)AMY.MAC(17–214)
   hybrid 2: (1–12)AMY.MAC(13–214)
   hybrid 3: (1–12)AMY.des-Tyr13MAC(14–214)
   hybrid 4: H(A12-M)ΔY13FMA
   hybrid 5: H(A12-M)ΔF14
(2) linear extrapolation, relative error in ΔG about 10%
(3) buffer: 2 mM sodium cacodylate, 1 mM EDTA, pH 6.0
(4) buffer: 2 mM sodium cacodylate, 1 mM CaCl$_2$, pH 6.0
(5) measured in 2 mM sodium cacodylate, 1.5 M GuHCl, 1 mM CaCl$_2$, pH 6.0
(6) ΔG was calculated using ΔCp = 7.9±1.3 kJ/mol/K
(7) measured in 2 mM sodium cacodylate, 1.5 M GuHCl, 1 mM EDTA, pH 6.0

Glucoamylase

Glucoamylase from *Aspergillus niger*, fragments

Form	pH	T	ΔG	Approach	Remarks	Ref
G2	7.8	20	125±30	DSC	(1–3)	92W3
G1C	7.8	20	16±4	DSC	(3,4)	92W3
G1C	7.8	20	32±2	GuHCl	(4,7,8)	92W3
G1C499	7.8	20	13±10	DSC	(3,5)	92W3
G1C499	7.8	20	35±3	GuHCl	(5,7,8)	92W3
G1C509	7.8	20	19±7	DSC	(3,6)	92W3
G1C509	7.8	20	23±2	GuHCl	(6–8)	92W3

Remarks:
(1) G2 = glucoamylase 2
(2) ΔG results from a formal thermodynamic treatment
(3) protein conc. was 0.122 mM
(4) G1C = glucoamylase 1 fragment (471–616)
(5) G1C499 = glucoamylase 1 fragment (499–616)
(6) G1C509 = glucoamylase 1 fragment (509–616)
(7) linear extrapolation
(8) ΔG values obtained by DSC and GuHCl may be different due to the formal application of thermodynamic equations to a multidomain protein

Glucoamylase from *Aspergillus awamori*, wild-type and alanine mutants

Mutant	pH	T	ΔG	Approach	Remarks	Ref
wild-type	4.5	25	28.1±1.0	GuHCl	(1,2)	96C9
Gly137→Ala	4.5	25	21.5±0.7	GuHCl	(1,2)	96C9
Gly139→Ala	4.5	25	14.5±0.5	GuHCl	(1,2)	96C9
double mutant (Gly137→Ala and Gly139→Ala)						
	4.5	25	21.1±1.2	GuHCl	(1,2)	96C9
Gly251→Ala	4.5	25	28.0±1.4	GuHCl	(1,2)	96C9
Gly383→Ala	4.5	25	26.3±1.0	GuHCl	(1,2)	96C9

Remarks:
(1) linear extrapolation
(2) Ref. 96C9 contains additional data for irreversible thermoinactivation

Glucose Dehydrogenase

Glucose dehydrogenase

pH	T	ΔG	Approach	Remark	Ref
–	–	3.9	enzyme activity		83T
–	–	6.9	GuHCl	(1)	83T

Remark:
(1) linear extrapolation

Glutamine Synthetase

Glutamine synthetase from *Bacillus caldolyticus*, isoforms

Form	pH	T	ΔG	Approach	Remarks	Ref
apo E-I	6.7	25	58.2±14.2	GuHCl	(1,2)	91W2
apo E-I	6.7	25	16.3±4.2	GuHCl	(1,3)	91W2
E-II	6.7	25	8.4±2.5	GuHCl	(1,2)	91W2
E-II	6.7	25	20.5±3.3	GuHCl	(1,3)	91W2

Remarks:
(1) three-state unfolding pathway
(2) ΔG refers to the N→I transition
(3) ΔG refers to the I→D transition

Glutaredoxin

Glutaredoxin from *E. coli*, oxidized and reduced form

Form	pH	T	ΔG	Approach	Remarks	Ref
oxidized	7.0	25	19.7	heat		91S3
oxidized	7.0	30	17.2	heat		91S3
oxidized	8.0	30	17.6	GuHCl	(1)	91S3
reduced	7.0	25	23.4	heat		91S3
reduced	7.0	30	20.9	heat		91S3
reduced	8.0	30	25.5	GuHCl	(1)	91S3

Remark:
(1) linear extrapolation

Glutathione S-Transferase

Porcine glutathione S-transferase, native and modified enzyme

Form	pH	T	ΔG	Approach	Remarks	Ref
unmodified	6.5	20	97.5	GuHCl	(1–3)	91D8
unmodified	6.5	20	113.8	urea	(1–3)	91D8
unmodified	6.5	20	94.1	GuHCl	(1,2,4)	95S9
alkylated	6.5	20	94.1	GuHCl	(1,2,4,5)	95S9
oxidized	6.5	20	48.5	GuHCl	(1,2,4,6)	95S9

Remarks:
(1) linear extrapolation
(2) the data refer to the transition of the folded dimer into unfolded monomers (N$_2$→2U)
(3) buffer: 20 mM MES, 100 mM NaCl, 1 mM EDTA, 5 mM DTT
(4) buffer: 20 mM MES, 100 mM NaCl, 1 mM EDTA, 0.02% NaN$_3$
(5) alkylated by IAEDANS
(6) H$_2$O$_2$ oxidized protein and oxidized protein treated with 5 mM DTT

Glycoprotein

α1-Acid glycoprotein

pH	T	ΔG	Approach	Remarks	Ref
5.0	20	21.3	GuHCl	(1)	90R3
5.0	20	28.9	GuHCl	(2)	90R3

Remarks:
(1) linear extrapolation
(2) transfer model, $\varepsilon = 0.12$

Glycinin, 11 S Globulin from Soybean, see plant seed proteins

Glycosidase

1,3–1,4-β-D-Glucan 4-Glucanohydrolase from *Bacillus licheniformis*, wild-type and mutants

Mutant	pH	T	ΔG	Approach	Remarks	Ref
wild-type	7.3	37	25.5	urea	(1)	95P5
reduced w.t.	7.3	37	26.6	urea	(1)	95P5
Cys61→Ala	7.3	37	9.6	urea	(1)	95P5
Cys90→Ala	7.3	37	15.1	urea	(1)	95P5
double mutant (Cys61→Ala and Cys90→Ala)						
	7.3	37	5.9	urea	(1)	95P5

Remark:
(1) linear extrapolation

Granulocyte Macrophage Stimulating Factor

Granulocyte macrophage stimulating factor, species

Species	pH	T	ΔG	Approach	Remarks	Ref
human	7.7	25	25.14±6.4	GuHCl	(1)	88W3
human	7.7	25	20.35±4.73	GuHCl	(2)	88W3
human	9	4	24.2 ±1.7	urea	(3)	88W3
murine	7.7	25	23.43±2.26	GuHCl	(4)	88W3
murine	7.7	25	26.37±2.35	GuHCl	(2)	88W3
murine	9	4	21.3 ±0.8	urea	(3)	88W3

Remarks:
(1) CD at 294 nm, linear extrapolation
(2) CD at 220 nm, linear extrapolation
(3) urea gradient gel electrophoresis
(4) CD at 278 nm, linear extrapolation

GroEL

Chaperonin GroEL, dissociation and unfolding

Form	pH	T	ΔG	Approach	Remarks	Ref
GroEL	7.8	25	54.6±6.4	urea	(1,2)	95G8
GroEL	7.8		13.3±1.3	urea	(1,3)	95G8
GroEL-Mg complex						
	7.8	25	75.9±13.8	urea	(1,2)	95G8
	7.8		63.4±6.4	urea	(1,3)	95G8
GroEL-MgAMP-PNP complex						
	7.8	25	32.0±4.9	urea	(1,2)	95G8
GroEL-MgADP complex						
	7.8	25	16.9±2.5	urea	(1,2)	95G8
	7.8		10.7±0.9	urea	(1,3)	95G8

Remarks:
(1) linear extrapolation; buffers: 50 mM Tris, pH 7.8 and 50 mM Tris, 10 mM $MgCl_2$, 10 mM mercaptoethanol
(2) transition monitored by tryptophan fluorescence
(3) transition monitored by light scattering

GroES

Chaperonin GroES

pH	T	ΔG	Approach	Remarks	Ref
7.8	23	18.5±1.7	urea	(1–3)	96S5
7.8	23	24.1±5.4	urea	(1,2,4)	96S5
7.8	23	14.8±1.7	urea	(1,2,5)	96S5
7.8	23	20.7±2.5	urea	(1,2,6)	96S5

Remarks:
(1) linear extrapolation
(2) buffer: 10 mM Tris-HCl, 100 mM NaCl
(3) transition monitored by intrinsic fluorescence
(4) transition monitored by fluorescence (reversibility)
(5) transition monitored by bis-ANS binding
(6) transition monitored by dansyl-labeled GroES

Growth Factor

Acidic fibroblast growth factor

pH	T	ΔG	Approach	Remark	Ref
7	20	27.2±1.3	urea	(1)	93M1

Remark:
(1) linear extrapolation

Table 1. Gibbs Energy Change – Molar Values

Acidic fibroblast growth factor in the presence of polyanions

Conditions	pH	T	ΔG	Approach	Remarks	Ref
In the presence of heparin:						
$0.0 \times$ heparin	7.2	4	27.2±0.8	urea	(1)	93B8
$0.1 \times$ heparin	7.2	4	18.8±0.4	urea	(1)	93B8
$0.3 \times$ heparin	7.2	4	29.7±8	urea	(1)	93B8
$1.0 \times$ heparin	7.2	4	36.0±0.8	urea	(1)	93B8
$3.0 \times$ heparin	7.2	4	38.5±2.1	urea	(1)	93B8
$10.0 \times$ heparin	7.2	4	37.2±3.3	urea	(1)	93B8
In the presence of NaCl at 3-fold excess heparin:						
0.0 M NaCl	7.2	4	32.2±1.3	urea	(1)	93B8
0.1 M NaCl	7.2	4	38.5±2.1	urea	(1)	93B8
0.25 M NaCl	7.2	4	41.0±1.3	urea	(1)	93B8
0.5 M NaCl	7.2	4	35.1±0.4	urea	(1)	93B8
1.0 M NaCl	7.2	4	37.2±0.8	urea	(1)	93B8
2.0 M NaCl	7.2	4	46.4±3.8	urea	(1)	93B8
In the presence of sulfated heparin (S heparin):						
1.5% S heparin	7.2	4	30.1±0.8	urea	(1)	93B8
4–7% S heparin	7.2	4	28.5±0.8	urea	(1)	93B8
8–10% S heparin	7.2	4	30.1±0.8	urea	(1)	93B8
13.5% S heparin	7.2	4	38.5±2.1	urea	(1)	93B8
In the presence of low MW heparin:						
	7.2	4	37.2±2.1	urea	(1)	93B8
In the presence of heparin sulfate:						
	7.2	4	40.6±2.5	urea	(1)	93B8
In the presence of β-cyclodextrin (BCD, S – sulfated):						
BCD	7.2	4	29.3±0.8	urea	(1)	93B8
8% S BCD	7.2	4	27.2±0.4	urea	(1)	93B8
14% S BCD	7.2	4	38.1±0.8	urea	(1)	93B8
In the presence of inositol:						
inositol	7.2	4	37.2±0.4	urea	(1)	93B8
inositol triphosphate	7.2	15	31.4±0.8	urea	(1)	93B8

Remark:

(1) linear extrapolation, unless otherwise specified, ligands are present in 3-fold weight excess, buffer: 50 mM sodium phosphate, 0.1 M NaCl

Acidic fibroblast growth factor, temperature dependence of ΔG measured in the presence of a threefold weight excess of heparin

pH	T	ΔG	Approach	Remarks	Ref
7.2	4	38.5±2.1	urea	(1–3)	93B8
7.2	15	43.8±1.3	urea	(1–3)	93B8
7.2	25	38.9±0.8	urea	(1–3)	93B8
7.2	35	37.7±3.8	urea	(1–3)	93B8
7.2	40	31.4±0.8	urea	(1–3)	93B8

Remarks:
(1) linear extrapolation
(2) thermal transition in the absence of heparin at $T_{trs} = 45°C$ (Ref. 93V2)
(3) thermal transition in the presence of a threefold excess of heparin at $T_{trs} = 64°C$ (Ref. 93V2)

Growth factor, epidermal, cyanogen bromide derivative

Modification	pH	T	ΔG	Approach	Remarks	Ref
epidermal	7.5	25	67±29	GuHCl	(1)	76H
epidermal, cyanogen bromide derivative						
	7.5	25	18	GuHCl	(1)	76H

Remark:
(1) ΔG is the mean value obtained by several approaches

Growth factor from mouse β-nerve

pH	T	ΔG	Approach	Remarks	Ref
7	23	80.8±4.2	GuHCl	(1,2)	92T3
4	23	61.5±4.6	urea	(1,2)	92T3

Remarks:
(1) linear extrapolation
(2) ΔG is the mean of several measurements at varying protein concentrations

Growth Hormone

Growth hormone, species

Species	pH	T	ΔG	Approach	Remarks	Ref
bovine	7.9–8.5	25	58.6	GuHCl	(1)	74H
ovine	7.9–8.5	25	40.6	GuHCl	(1)	74H
rat	7.9–8.5	25	33.5	GuHCl	(1)	74H

Remark:
(1) transfer model

Table 1. Gibbs Energy Change – Molar Values

Porcine growth hormone (PGH), recombinant, native and truncated protein

Mutant	pH	T	ΔG	Approach	Remarks	Ref
Met(1–190)	8.0	26	16.0±0.7	urea	(1,2)	95C1
P-band	8.0	26	15±2	urea	(1,3)	95C1

Remarks:
(1) linear extrapolation
(2) Met(1–190) is the full-length PGH with 2 intact disulfides
(3) P-band is the 20 kDa variant lacking the C-terminal disulfide loop

Human growth hormone

Protein Concentration	pH	T	ΔG	Approach	Remarks	Ref
0.1 mg/ml	2.5	23	43.9	GuHCl	(1–3)	95D2
0.1 mg/ml	2.5	23	43.5	GuHCl	(1,2,5)	95D2
1.0 mg/ml	1.5	23	22.6	GuHCl	(1–3)	95D2
1.0 mg/ml	1.5	23	33.5	GuHCl	(1,2,4)	95D2
1.0 mg/ml	2.0	23	20.9	GuHCl	(1–3)	95D2
1.0 mg/ml	2.0	23	35.1	GuHCl	(1,2,4)	95D2
1.0 mg/ml	2.5	23	20.1	GuHCl	(1–3)	95D2
1.0 mg/ml	2.5	23	35.1	GuHCl	(1,2,4)	95D2
1.0 mg/ml	3.0	23	24.3	GuHCl	(1–3)	95D2
1.0 mg/ml	3.0	23	33.5	GuHCl	(1,2,4)	95D2
0.1 mg/ml	7.5	23	60.7	GuHCl	(1–3)	95D2
0.1 mg/ml	8.0	23	65.3	GuHCl	(1,2,5)	95D2
1.0 mg/ml	8.0	23	44.8	GuHCl	(1–3)	95D2
1.0 mg/ml	8.0	23	45.2	GuHCl	(1,2,4)	95D2
1.0 mg/ml	8.0	23	43.9	GuHCl	(1,2,6)	95D2

Remarks:
(1) linear extrapolation
(2) estimated error in ΔG ±4.2 kJ/mol
(3) transition monitored by CD at 222 nm
(4) transition monitored by CD at 295 nm
(5) transition monitored by fluorescence at 350 nm
(6) transition monitored by optical absorption at 295 nm

Human growth hormone, native and reduced forms

Form	pH	T	ΔG	Approach	Remarks	Ref
native	7.5		60.7±4.2	GuHCl	(1)	90B8
tetra-S-modified	7.5		22.2±2.1	GuHCl	(1,2)	90B8
di-S-modified	7.5		46.9±3.3	GuHCl	(1,2)	90B8

Remarks:
(1) linear extrapolation
(2) reduced and alkylated protein

Human growth hormone (hGH)

Form	pH	T	ΔG	Approach	Remarks	Ref
native	7.5	22±1	57.7	GuHCl	(1,2)	95Y6
native	7.5	22±1	51.0	GuHCl	(1,3)	95Y6
2-RCAM	7.5	22±1	18.4	GuHCl	(1,4)	95Y6

Remarks:
(1) linear extrapolation; for the dependence of the apparent ΔG value on the protein concentration, see 93D2
(2) transition monitored by fluorescence intensity
(3) transition monitored by absorbance
(4) 2-RCAM = tetra-S-carbamidomethylated hGH

Guanidino Kinases

Mitochondrial creatine kinase compared with muscle-type creatine kinase and arginine kinase

Protein/ Transition	pH	T	ΔG	Approach	Remarks	Ref
Mi-CK, N→I	7.4	22	22.3±2.6	GuHCl	(1–3)	95G11
Mi-CK, I→U	7.4	22	9.3±2.6	GuHCl	(1–3)	95G11
Mi-CK, N→U	7.4	22	31.6	GuHCl	(1–3)	95G11
Mi-CK, N→I	7.4	22	37.6±2.2	urea	(1–3)	95G11
Mi-CK, I→U	7.4	22	9.2±0.2	urea	(1–3)	95G11
Mi-CK, N→U	7.4	22	46.8	urea	(1–3)	95G11
M-CK, N→I	7.4	22	18.7±0.9	GuHCl	(1,3,4)	95G11
M-CK, I→U	7.4	22	10.3±0.6	GuHCl	(1,3,4)	95G11
M-CK, N→U	7.4	22	29.0	GuHCl	(1,3,4)	95G11
Arg-K, N→I	7.4	22	15.8±8.7	GuHCl	(1,3,5)	95G11
Arg-K, I→U	7.4	22	9.2±2.3	GuHCl	(1,3,5)	95G11
Arg-K, N→U	7.4	22	25.0	GuHCl	(1,3,5)	95G11

Remarks:
(1) data were obtained fitting the biphasic curves to a unimolecular three-state denaturation model based on linear dependence of ΔG on denaturant concentration
(2) Mi-CK: chicken sarcomeric mitochondrial creatine kinase, recombinant
(3) buffer: 100 mM sodium phosphate, 5 mM mercaptoethanol, 0.1 mM EDTA, pH 7.4
(4) M-CK: rabbit muscle creatine kinase
(5) Arg-K: lobster tail arginine kinase

Heat Shock Protein, see also GroEL, GroES, and cold-shock protein

α-Helix Propensity

α-Helix propensity, see also de novo sytheszied proteins, helix forming model peptide
see also lysozyme phage T4, site 44 mutants

α-Helix propensity, based on site 44 mutants of lysozyme phage T4 the data refer to the Ser44→Gly mutant of the pseudo wild-type as the reference protein

Amine Acid	ΔT	$\Delta(\Delta H)$	$\Delta(\Delta G)$	Remarks	Ref
Ala	2.76	30	4.02	(1)	94B6
Arg	2.23	25	3.22	(1)	94B6
Asn	1.15	0	1.63	(1)	94B6
Asp	1.23	0	1.76	(1)	94B6
Cys	1.20	−37	1.76	(1)	94B6
Gln	2.30	25	3.35	(1)	94B6
Glu	1.55	4	2.22	(1)	94B6
Gly	0.0	0	0.00	(1)	94B6
His	1.67	30	2.38	(1)	94B6
Ile	2.46	9	3.51	(1)	94B6
Leu	2.64	30	3.85	(1)	94B6
Lys	2.12	25	3.05	(1)	94B6
Met	2.47	21	3.60	(1)	94B6
Phe	1.73	25	2.47	(1)	94B6
Pro	−8.77	−150	−10.5	(1)	94B6
Ser	1.55	17	2.22	(1)	94B6
Thr	1.58	13	2.26	(1)	94B6
Trp	2.70	9	2.43	(1)	94B6
Tyr	2.09	21	3.01	(1)	94B6
Val	1.84	13	2.64	(1)	94B6

Remark:
(1) measured by heat denaturation, for details, see also lysozyme phage T4, site 44 mutants

Helix propensity of amino acids ($\Delta(\Delta G)_{Helix}$) and helix N-cap propensity ($\Delta(\Delta G)_{N\text{-}Cap}$) measured in alanine-based peptides without helix-stabilizing side-chain interactions (at 273 K)

Residue	$\Delta(\Delta G)_{Helix}$	Remark	$\Delta(\Delta G)_{N\text{-}CAP}$	Ref
Ala	7.87	(1)	0.0	94C4
Arg$^+$	6.99	(1)		94C4
Asn	4.18	(1)	5.90	94C4
Asp^o	4.18	(1)		94C4
Asp$^-$	4.18	(1)		94C4
Cys	4.44	(1)		94C4
Gln	5.48	(1)	−3.89	94C4
Gluø	5.86	(1)		94C4
Glu$^-$	5.02	(1)		94C4
Gly	0.0	(1)	4.52	94C4
His^o	4.60	(1)		94C4
His$^+$	0.42	(1)		94C4
Ile	4.94	(1)	2.43	94C4
Leu	6.69	(1)	2.97	94C4
Lys$^+$	6.36	(1)	−2.09	94C4
Met	5.73	(1)	1.63	94C4
Phe	3.97	(1)		94C4
Pro	<−21	(1)	1.38	94C4
Ser	4.60	(1)	4.69	94C4

Helix propensity of amino acids ($\Delta(\Delta G)_{Helix}$) and helix N-cap propensity ($\Delta(\Delta G)_{N-Cap}$) measured in alanine-based peptides without helix-stabilizing side-chain interactions (at 273 K) (continued)

Residue	$\Delta(\Delta G)_{Helix}$	Remark	$\Delta(\Delta G)_{N-Cap}$	Ref
Thr	2.34	(1)	2.68	94C4
Trp	(4.06)–4.60	(1,2)		94C4
Tyr	4.64–(5.36)	(1,2)		94C4
Val	3.47	(1)	0.42	94C4
acetyl group			5.15	94C4

Remarks:
(1) given is $\Delta(\Delta G)$ relative to Gly at 0°C, 0.1 M NaCl for uncharged residues and Lys, 10 mM NaCl for Arg, Asp, Glu, and His, the data from Ref. 94C4 are figured out as positive values
(2) data in brackets represent values corrected for error in fraction helix measurement caused by aromatic contribution

Helix propagation and N-cap propensities of the amino acids measured in alanine-based peptides

Residue	T	ΔG_{Helix}	$\Delta(\Delta G)_{N-Cap}$	Remarks	Ref
Ala	0	−1.13		(1,2)	96R3
Glu°	0	0.88		(1,2)	96R3
Cys°	0	2.68		(1,2)	96R3
Cys	0	−3.85		(1,2)	96R3
Asp⁻	0	2.26	−4.18	(1,2)	96R3
Glu⁻	0	1.46	−1.63	(1,2)	96R3
Phe	0	3.05	−1.63	(1,2)	96R3
Gly	0	7.1	−3.10	(1,2)	96R3
His⁺	0	3.51		(1,2)	96R3
Ile	0	1.84	−1.05	(1,2)	96R3
His°	0	2.38	−1.72	(1,2)	96R3
Lys⁺	0	0.08	0.75	(1,2)	96R3
Leu	0	0.40	−1.63	(1,2)	96R3
Met	0	1.05	−0.63	(1,2)	96R3
Asn	0	2.89	−4.18	(1,2)	96R3
Asp°	0	2.18		(1,2)	96R3
Pro	0	>15.9	−0.67	(1,2)	96R3
Gln	0	1.17	5.02	(1,2)	96R3
Arg⁺	0	−0.22	0.0	(1,2)	96R3
Ser	0	2.18	−3.10	(1,2)	96R3
Thr	0	3.97	−1.84	(1,2)	96R3
Val	0	3.22	+0.09	(1,2)	96R3
Trp	0	2.89	−2.93	(1,2)	96R3
Tyr	0	1.76	−4.02	(1,2)	96R3
acetyl	0	−3.60		(1,2)	96R3

Remarks:
(1) the data are based on the Lifson-Roig statistical weights, for details of the peptide synthesis and and CD measurements, see Ref. 96R3
(2) the data refer to the amino acid residues in water at 0°C and have the same sign as in Ref. 96R3

HBsu

HBsu, DNA-binding histone-like protein

Mutant	pH	T	ΔG	Approach	Ref
Phe50→Trp	7.5	20	36	urea, linear extrapol.	93W3
Phe79→Trp	7.5	20	40	urea, linear extrapol.	93W3

Helichrome, see de novo synthesized proteins/peptides

Hemolysin

Hemolysin of *Vibrio parahemolyticus*

	pH	T	ΔG	Approach	Ref
	8.15	25	32	DSC	83U

Histone

Histone GH1 and GH5 fragment

Fragment	pH	T	ΔG	Approach	Ref
GH1	7	25	10*	DSC	82T1
GH5	7	25	14*	DSC	82T1

Hirudin

Hirudin

	pH	T	ΔG	Approach	Ref
	7.0	25	21	heat	91O

Histocompatibility Complex

Class I major histocompatibility complex, empty and filled heterodimer (HD)

Form	pH	T	ΔG	Approach	Ref
empty HD	7	37	5.0	heat	92F1
filled HD	7	37	23.4	heat	92F1

HIV-1 Capsid Protein

HIV-1 Capsid protein rp24

pH	T	ΔG	Approach	Remarks	Ref
5.8	25	18.1	GuHCl	(1,2)	95M16
5.8	25	22.4	urea	(1,2)	95M16
5.8	25	19.7	GuHCl	(1,3)	95M16
5.8	25	24.1	urea	(1,3)	95M16
2.0	25	12.4	GuHCl	(1,2)	95M16
2.0	25	12.8	urea	(1,3)	95M16

Remarks:
(1) linear extrapolation
(2) the transition was monitored by CD at 222 nm
(3) the transition was monitored by fluorescence intensity ratio I_{370nm}/I_{370nm}

HIV-1 Capsid protein p24, recombinant, modified by maleylation

Form	pH	T	ΔG	Approach	Remarks	Ref
rp24	5.8	25	18.1	GuHCl	(1,2)	95M16
rp24F	5.8	25	15.4	GuHCl	(1,3)	96E3
rp24I	5.8	25	12.1	GuHCl	(1,4)	96E3

Remarks:
(1) linear extrapolation
(2) unmodified protein
(3) 9 amino groups substituted
(4) 12 amino groups substituted

HIV-1 Protease

Human immunodeficiency virus type 1 (HIV-1) protease

Form	pH	T	ΔG	Approach	Remarks	Ref
free	6.0	25	59.4±5.9	urea	(1–3)	92G3
inhibited	6.0	25	80.8±−2.9	urea	(1,3,4)	92G3

Remarks:
(1) linear extrapolation, ΔG refers to the unfolding of the dimer into two monomers, i.e., $N_2 \rightarrow 2U$
(2) average value from measurements conducted from 0.11 to 1.1 μM protein
(3) buffer: 50 mM MES, 0.2 M NaCl, 1 mM EDTA, 1 mM DTT, and 2% glycerol
(4) 10 μM inhibitor

Human immunodeficiency virus protease (HIV-1), wild-type and mutant, salt dependence

NaCl Concentration	pH	T	ΔG	Approach	Remarks	Ref
HIV-1:						
0.1 M	7.0	25	9.02±0.29	urea	(1,2)	96S13
1.0 M	7.0	25	13.13±0.53	urea	(1,2)	96S13
0.1 M	5.0	25	11.58±0.41	urea	(1,2)	96S13
1.0 M	5.0	25	14.22±0.95	urea	(1,2)	96S13
Mutant (Gln7→Lys, Leu33→Ile and Leu63→Ile):						
0.1 M	7.0	25	10.40±0.41	urea	(1,2)	96S13
1.0 M	7.0	25	13.55±0.52	urea	(1,2)	96S13
0.1 M	5.0	25	12.63±0.64	urea	(1,2)	96S13
1.0 M	5.0	25	14.35±0.84	urea	(1,2)	96S13

Remarks:
(1) linear extrapolation to zero denaturant concentration by a method that includes the pre- and postdenaturational baselines for a non-linear regression of the data
(2) buffer: 25 mM formic acid, 25 mM acetic acid, 25 mM MES, 75 mM Tris, 1 mM EDTA, 5% (v/v) glycerol

SIV Protease simian immunodeficiency virus (SIV)

pH	T	ΔG	Approach	Remarks	Ref
6.0	25	55.6±5.4	urea	(1–3)	92G3

Remarks:
(1) linear extrapolation, ΔG refers to the unfolding of the dimer into two monomers, i.e., $N_2 \rightarrow 2U$
(2) average value from measurements conducted from 0.20 to 0.54 µM protein
(3) buffer: 50 mM MES, 0.2 M NaCl, 1 mM EDTA, 1 mM DTT, and 2% glycerol

HPr

HPr, histidine-containing phosphocarrier protein from *Bacillus subtilis*

pH	T	ΔG	Approach	Remarks	Ref
7.0	0	10.8	urea	(1)	95S3
7.0	5	12.8	urea	(1)	95S3
7.0	10	14.4	urea	(1)	95S3
7.0	15	15.6	urea	(1)	95S3
7.0	20	16.6	urea	(1)	95S3
7.0	25	17.6	urea	(1)	95S3
7.0	30	16.9	urea	(1)	95S3
7.0	35	16.6	urea	(1)	95S3
7.0	40	15.9	urea	(1)	95S3

Remark:
(1) linear extrapolation

Histidine-containing protein (HPr), influence of phosphorylation on protein stability

Mutant	pH	T	$\Delta(\Delta G)$	Approach	Remarks	Ref
HPr	7.0	30	0.0	urea	(1)	95P10
HPr(Ser-P)	7.0	30	3.3±0.6	urea	(1,2)	95P10
HPr(Ser-P)	7.0	73.6	2.9±0.8	heat	(2,3)	95P10
Ser46→Asp	7.0	30	3.1±0.6	urea	(1)	95P10
Ser46→Asp	7.0	73.6	2.9±0.8	heat	(2,3)	95P10

Remarks:
(1) linear extrapolation
(2) Ser46 phosphorylated
(3) $\Delta(\Delta G)$ was determined from $\Delta(\Delta G) = (\Delta H/T_{trs})_{w.t.} \times \Delta T$

Histidine-containing protein, phosphocarrier protein from the bacterial phosphoenolpyruvate: sugar phosphotransferase system

Mutant	pH	T	ΔG	Approach	Remarks	Ref
wild-type	7.0	30	20.7±0.3	urea	(1)	95H2
Ser31→Ala	7.0	30	19.3±0.3	urea	(1)	95H2

Remark:
(1) linear extrapolation

HPr, histidine-containing phosphocarrier protein from *E. coli*

pH	T	ΔG	Approach	Remarks	Ref
7.0	0	19.4±0.4	urea	(1)	96N2
7.0	5	21.3±0.2	urea	(1)	96N2
7.0	10	22.2±0.2	urea	(1)	96N2
7.0	15	22.6±0.6	urea	(1)	96N2
7.0	20	22.8±0.2	urea	(1)	96N2
7.0	25	22.2±0.3	urea	(1)	96N2
7.0	30	20.9±0.2	urea	(1)	96N2
7.0	35	19.6±0.2	urea	(1)	96N2
7.0	40	17.3±0.3	urea	(1)	96N2
7.0	45	14.1±0.2	urea	(1)	96N2
7.0	50	10.1±0.9	urea	(1)	96N2

Remark:
(1) linear extrapolation to zero denaturant concentration by a method that includes the pre- and postdenaturational baselines for a non-linear regression of the data

HPr, histidine-containing phosphocarrier protein from *E. coli*

Urea Concentration	pH	T	ΔG	Approach	Remarks	Ref
0.0 M	7.0	16.9	22.8	heat	(1)	96N2
0.0 M	7.0	16.9	22.6	heat	(2)	96N2
1.0 M	7.0	16.9	17.8	heat	(1)	96N2
2.0 M	7.0	16.9	13.1	heat	(1)	96N2
3.0 M	7.0	16.9	8.2	heat	(1)	96N2
4.0 M	7.0	16.9	4.1	heat	(1)	96N2

Remarks:
(1) thermal unfolding in the presence of urea, transition monitored by CD at 222 nm
(2) from ΔG versus urea concentration, extrapolated to zero denaturant concentration

HPr, histidine-containing phosphocarrier protein from *E. coli*, wild-type and N-Cap mutations

Mutant	pH	T	$\Delta(\Delta G)$	Approach	Remarks	Ref
wild-type	7.0	30	0.0	urea	(1)	96T2
Ser46→Ala	7.0	30	−4.6±0.8	heat		96T2
Ser46→Ala	7.0	30	−3.8±0.8	urea	(1)	96T2
Ser46→Ala	7.0	30	−(0.8–3.8)	NH exchange		96T2
Ser46→Asn	7.0	30	1.3±0.8	heat		96T2
Ser46→Asn	7.0	30	2.1±0.8	urea	(1)	96T2
Ser46→Asn	7.0	30	(3.3–6.3)	NH exchange		96T2
Ser46→Asp	7.0	30	5.9±0.8	heat		96T2
Ser46→Asp	7.0	30	6.3±0.8	urea	(1)	96T2
Ser46→Asp	7.0	30	(7.5–7.9)	NH exchange		96T2

Remark:
(1) linear extrapolation using the m value of the wild-type protein according to $\Delta(\Delta G) = m \times (\Delta C_m)$

Immunoglobulin

Immunoglobulin, Bence-Jones protein

pH	T	ΔG	Approach	Ref
7.4	25	19*	DSC, data per monomer	77Z

Immunoglobulin, C_L fragment, various forms

Form	pH	T	ΔG	Approach	Remarks	Ref
NAG, λ:						
intact	7.5	25	23.8	GuHCl	(1)	86G
intact	7.5	25	29.7	GuHCl	(2)	86K2
reduced	7.5	25	6.7	GuHCl	(1)	86G
Hg form	7.5	25	5.9	GuHCl	(1)	86G
OKU, κ:						
intact	7.5	25	21.4	GuHCl	(1)	87G1
reduced	7.5	25	1.7	GuHCl	(1)	87G1
reduced	7.5	25	12.6	GuHCl	(1,3)	87G1
Light chain (OKU):						
intact	7.5	25	21.3	GuHCl	(1,4)	88G3
intact	7.5	25	19.7	GuHCl	(1,5)	88G3
intact	7.5	25	15.9	GuHCl	(4,6)	88G3
intact	7.5	25	14.6	GuHCl	(5,6)	88G3
intact	7.5	25		GuHCl	(7)	88G3
intact	7.5	25	25.9	GuHCl	(1)	88G3
intact	7.5	25	19.7	GuHCl	(6)	88G3

Remarks:
(1) denaturant binding model, k = 0.6
(2) unfolding kinetics
(3) in the presence of 0.5 M ammonium sulfate
(4) monitored by fluorescence
(5) monitored by CD
(6) linear extrapolation
(7) dependence of ΔG on ammonoium sulfate concentration gives a linear dependence with
   $d(\Delta G)/dc = 15.1$ kJ/mol

Immunoglobulin, C_L fragment

Fragment	pH	T	ΔG	Approach	Remarks	Ref
C_L intact	7.5	25	26	GuHCl	(1)	79G
C_L intact	7.5	25	24	GuHCl	(2)	79G
C_L reduced	7.5	25	7	GuHCl	(2)	79G
C_L reduced	7.5	25	8	GuHCl	(1)	79G
C_L intact	8	25	22	GuHCl	(2)	85A
C_L intact	10.3	25	22	GuHCl	(2)	85A

Remarks:
(1) denaturant binding model, k = 1.2
(2) denaturant binding model, k = 0.6

Immunoglobulin, C_L fragment Type κ reduced and modified with IAEDANS

pH	T	ΔG	Approach	Remark	Ref
7.5	25	12.6±1.3	GuHCl	(1)	91K3

Remark:
(1) denaturant binding model, k = 0.6

Immunoglobulin, C_L fragment, native and modified form

Form	pH	T	ΔG	Approach	Remarks	Ref
native	7.5	25	15.1±7.1	heat		90O1
native	7.5	25	23.8	GuHCl	(1)	90O1
native	7.5	25	17.2±0.4	GuHCl	(2)	90O1
NFK-Trp187	7.5	25	8.8±3.3	heat	(3)	90O1
NFK-Trp187	7.5	25	10.0±0.4	GuHCl	(1,3)	90O1
NFK-Trp187	7.5	25	7.9±0.4	GuHCl	(2,3)	90O1
Kyn-Trp187	7.5	25	9.2±5.0	heat	(4)	90O1
Kyn-Trp187	7.5	25	11.3±0.8	GuHCl	(1,4)	90O1
Kyn-Trp187	7.5	25	8.8±0.8	GuHCl	(2,4)	90O1

Remarks:
(1) binding model, k = 0.6
(2) linear extrapolation
(3) modified by ozone oxidation of Trp187 to the N-formylkynurenine derivative
(4) modified by ozone oxidation of Trp187 to the kynurenine derivative

Immunoglobulin, light chain, C_L fragment (105–214)

Residues	pH	T	ΔG	Approach	Remarks	Ref
105–214	7.5	25	23.8	GuHCl	(1,2)	87G2
105–214	7.5	25	23.8	GuHCl	(1,3)	87G2
109–214	7.5	25	23.8	GuHCl	(1,2)	87G2
109–214	7.5	25	23.4	GuHCl	(1,3)	87G2
113–214	7.5	25	19.2	GuHCl	(1,2)	87G2
113–214	7.5	25	18.8	GuHCl	(1,3)	87G2

Remarks:
(1) denaturant binding model, k = 0.6
(2) transition monitored by CD
(3) transition monitored by fluorescence

C_L fragment of type κ immunoglobulin, fragments

Residues	pH	T	ΔG	Approach	Remarks	Ref
109–214	7.5	25	23.8±0.4	GuHCl	(1)	91I
109–214	7.5	25	17.2±0.4	GuHCl	(2)	91I
109–211	7.5	25	25.1±0.8	GuHCl	(1)	91I
109–211	7.5	25	18.0±0.4	GuHCl	(2)	91I
109–207	7.5	25	2.5±0.4	GuHCl	(1)	91I
109–207	7.5	25	2.5±0.4	GuHCl	(2)	91I

Remarks:
(1) denaturant binding model, k = 0.6
(2) linear extrapolation

Immunoglobulin fragments, light chain (OKU, κ)

Fragment	pH	T	ΔG	Approach	Remarks	Ref
C_L	7.5	25	21.8±0.8	GuHCl	(1)	87T
V_L	7.5	25	26.4±0.8	GuHCl	(1)	87T

Remark:
(1) denaturant binding model, k = 0.6

Immunoglobulin, fab fragment, regions

Region	pH	T	ΔG	Approach	Remarks	Ref
C-region	–	–	62.8	GuHCl	(1,2)	76R1
V-region	–	–	50.2	GuHCl	(1,3)	76R1

Remarks:
(1) transfer model
(2) with 16.7 kJ/Mol for domain-domain interaction
(3) with 5.4 kJ/Mol for domain-domain interaction

Immunoglobulin, single-chain F_v fragment, modified

Form	pH	T	ΔG	Approach	Remarks	Ref
4.4.20/212	8.0	25	18.1±1.3	GuHCl	(1,2)	91P1
4.4.20/212	8.0	25	18.1±1.5	GuHCl	(1)	95G13
4.4.20/205	8.0	25	20.3±1.5	GuHCl	(1,3)	91P1
4.4.20/205	8.0	25	20.3±1.3	GuHCl	(1)	95G13
4.4.20/202	8.0	25	16.2±1.1	urea	(1,4)	91P1
4.4.20/212	8.0	25	21.4±1.3	urea	(1,2)	91P1
4.4.20/205	8.0	25	22.7±1.2	urea	(1,3)	91P1
04-01/212	8.0	25	6.0±0.5	GuHCl	(1)	95G13

Remarks:
(1) linear extrapolation, fit of the entire transition curve
(2) linker peptide, called 4.4.20/212: Gly-Ser-Thr-Ser- Gly-Ser-Gly-Lys-Ser-Ser-Glu-Gly-Lys-Gly
(3) linker peptide, called 4.4.20/205: Ser-Ser-Ala-Asp-Asp-Ala-Lys-Lys-Asp-Ala-Ala-Lys-Lys-Asp-Asp-Ala-
    Lys-Lys-Asp-Asp-Ala-Lys-Lys-Asp-Gly
(4) linker peptide, called 4.4.20/202: Gly-Lys-Ser-Ser-Gly-Ser-Gly-Ser-Glu-Ser-Lys-Ser

HCDR (heavy chain complementary region) transplants

Antibody	pH	T	ΔG	Approach	Remarks	Ref
HCDR1[4-4-20]	8.0		12.05	GuHCl	(1)	95G13
HCDR2[4-4-20]	8.0		12.89	GuHCl	(1)	95G13
HCDR3[4-4-20]	8.0		8.45	GuHCl	(1)	95G13
HCDR1-2[4-4-20]	8.0		12.01	GuHCl	(1)	95G13
HCDR1-3[4-4-20]	8.0		10.17	GuHCl	(1)	95G13
HCDR2-3[4-4-20]	8.0		7.87	GuHCl	(1)	95G13
HCDR1-2-3[4-4-20]	8.0		12.18	GuHCl	(1)	95G13

Remark:
(1) linear extrapolation

Human immunoglobulin G, domains

Domain	pH	T	ΔG	Approach	Remarks	Ref
CH2	7.5	25	24	GuHCl	(1)	82S2
PFC (CH3)	7.5	25	23	GuHCl	(1)	82S2
R.A.FC(T)	7.5	25	24	GuHCl	(1)	82S2

Remark:
(1) denaturant binding model, k = 0.6

Immunoglobulin, Wes L chain

	pH	T	ΔG	Approach	Remark	Ref
	7	25	23	GuHCl	(1)	73R

Remark:
(1) transfer model

Immunoglobulin variable domain in McPC603 V_K

Mutant	pH	T	ΔG	Approach	Remarks	Ref
wild-type	7.4	20	13.5	GuHCl	(1,2)	94S8
Ala15→Leu	7.4	20	19.2	GuHCl	(1,2)	94S8
Met21→Ile	7.4	20	14.5	GuHCl	(1,2)	94S8
Met21→Leu	7.4	20	12.2	GuHCl	(1,2)	94S8
Lys24→Arg	7.4	20	12.8	GuHCl	(1,2)	94S8
Phe32→Tyr	7.4	20	15.1	GuHCl	(1,2)	94S8
Pro43→Ser	7.4	20	12.8	GuHCl	(1–3)	94S8
Thr63→Ser	7.4	20	14.7	GuHCl	(1,2)	94S8
Gln79→Glu	7.4	20	11.8	GuHCl	(1,2,4)	94S8
Asn90→Gln	7.4	20	17.9	GuHCl	(1,2)	94S8
Ala100→Gly	7.4	20	13.6	GuHCl	(1–3)	94S8
Leu106→Ile	7.4	20	15.0	GuHCl	(1,2)	94S8

Remarks:
(1) linear extrapolation
(2) buffer: phosphate buffered saline solution (PBS)
(3) interacts with V_H domain
(4) interacts with C_K domain

Immunoglobulin V_L domain, deletion mutants with various newly synthesized loops inserted into CDRs of immunoglobulin V_L domain

Mutant	pH	T	$\Delta(\Delta G)$	Approach	Remarks	Ref
wild-type REI	7.4	25	0	GuHCl	(1)	95H4
RGD17	7.4	25	>−28	GuHCl	(1)	95H4
RGD4	7.4	25	>−28	GuHCl	(1)	95H4
RGD18	7.4	25	−26	GuHCl	(1)	95H4
RGD12	7.4	25	−21	GuHCl	(1)	95H4
RGD26	7.4	25	>−28	GuHCl	(1)	95H4
RGD23	7.4	25	−23	GuHCl	(1)	95H4
RGD21	7.4	25	>−28	GuHCl	(1)	95H4
RGD22	7.4	25	>−28	GuHCl	(1)	95H4
RGD14	7.4	25	>−28	GuHCl	(1)	95H4
RGD15	7.4	25	−28	GuHCl	(1)	95H4
RGD35	7.4	25	−25	GuHCl	(1)	95H4
RGD32	7.4	25	−23	GuHCl	(1)	95H4
RGD34	7.4	25	−21	GuHCl	(1)	95H4
RGD1	7.4	25	−19	GuHCl	(1)	95H4

Remark:
(1) linear extrapolation to zero denaturant concentration by a method that includes the pre- and postdenaturational baselines for a non-linear regression of the data

REI_V, the variable domain of human immunoglobulin κ light chain

Mutant/Form	pH	T	ΔG	Approach	Remarks	Ref
wild-type[S–S]	7.0	25	24.6±0.4	urea	(1,2)	96F5
wild-type[S–S]	7.0	25	26.5	urea	(1,2,4)	94F3
wild-type[S–S]	7.0	25	26.1±1.9	urea	(1,2)	94K7
Tyr32→His[S–S]	7.0	25	30.2±0.5	urea	(1,2)	96F5
Tyr32→His[S–S]	7.0	25	31	urea	(1,2,4)	94F3
Tyr32→His[(SH)₂]	7.0	25	11.7±1.3	urea	(1,3)	96F5
Tyr32→His[(SH)₂]	7.0	25	14	urea	(1,3,4)	94F3
Thr39→Lys[S–S]	7.0	25	29.9±0.5	urea	(1,2)	96F5
Thr39→Lys[(SH)₂]	7.0	25	12.1±0.9	urea	(1,3)	96F5
Ser67→His[S–S]	7.0	25	22.6±1.3	urea	(1,2)	94K7
Tyr71→Phe[S–S]	7.0	25	30.0±0.5	urea	(1,2)	96F5
Tyr71→Phe[(SH)₂]	7.0	25	11.5±1.8	urea	(1,3)	96F5
Phe73→Leu[S–S]	7.0	25	28.7±0.5	urea	(1,2)	96F5
Leu94→His[S–S]	7.0	25	27.7±1.5	urea	(1,2)	94K7
double mutant (Tyr32→His and Thr39→Lys)[S–S]						
	7.0	25	35.5±0.6	urea	(1,2)	96F5
[(SH)₂]	7.0	25	17.4±0.8	urea	(1,3)	96F5
double mutant (Thr39→Lys and Tyr71→Phe)[S–S]						
	7.0	25	36.0±0.6	urea	(1,2)	96F5
[(SH)₂]	7.0	25	18.3±0.9	urea	(1,3)	96F5

REI$_V$, the variable domain of human immunoglobulin κ light chain (continued)

Mutant/Form	pH	T	ΔG	Approach	Remarks	Ref
double mutant (Thr39→Lys and Phe73→Leu)$^{(S-S)}$						
	7.0	25	34.0±0.8	urea	(1,2)	96F5
$^{(SH)}$$_2$	7.0	25	14.6±0.9	urea	(1,3)	96F5
double mutant (Tyr71→Phe and Phe73→Leu)$^{(S-S)}$						
	7.0	25	34.5±1.0	urea	(1,2)	96F5
$^{(SH)}$$_2$	7.0	25	14.2±0.9	urea	(1,3)	96F5
triple mutant (Tyr39→Lys, Tyr71→Phe and Phe73→Leu)$^{(S-S)}$						
	7.0	25	40.6±0.7	urea	(1,2)	96F5
$^{(SH)}$$_2$	7.0	25	17.9±1.0	urea	(1,3)	96F5
double mutant (Cys23→Val and Tyr32→His)						
	7.0	25	12.6±0.5	urea	(1)	96F5
	7.0	25	14	urea	(1,4)	94F3
triple mutant (Cys23→Val, Tyr32→His and Thr39→Lys)						
	7.0	25	13.5±0.6	urea	(1)	96F5

Remarks:
(1) linear extrapolation to zero denaturant concentration by a method that includes the pre- and postdenaturational baselines for a non-linear regression of the data
(2) oxidized form (S-S)
(3) reduced form (SH)$_2$
(4) the data were taken from Fig. 3 in Ref. 94F3

Immunoglobulin superfamily domain 18' from twitchin

pH	T	ΔG	Approach	Remarks	Ref
4.5	20	12.3±1.0	urea	(1–3)	96F3
5.0	20	13.1±1.0	urea	(1–3)	96F3
6.0	20	14.9±1.1	urea	(1–3)	96F3
7.0	20	16.6±1.3	urea	(1–3)	96F3
7.0	20	15.4±3.6	urea	(1,2,4)	96F3

Remarks:
(1) linear extrapolation
(2) buffer: 450 mM sodium acetate, pH 4.5 or 5.0; 450 mM potassium phosphate, pH 6.0 or 7.0
(3) transition monitored by fluorescence
(4) transition monitored by CD

Immunoglobulin Binding Protein, see Protein G

Insulin

Insulin

pH	T	ΔG	Approach	Remarks	Ref
3	20	27 (>=27)	hydrogen exchange		66H
7.5	23	18.8±2.1	GuHCl	(1,2)	90B7

Remarks:
(1) linear extrapolation
(2) Zn free form in the presence of 20% (v./v.) ethanol for eliminating problems of protein aggregation

Insulin, variant B-chain (Pro28→Trp and Lys29→Pro)

	pH	T	ΔG	Approach	Remarks	Ref
	7.5	25	16.3	GuHCl	(1,2)	92B8

Remark:
(1) linear extrapolation
(2) buffer: 20 mM HEPES, 1 mM EDTA, 20% (v./v.) ethanol

Insulin modified, native and crosslinked at Lys-A1 and Lys-B29

Cross Link	pH	T	ΔG	Approach	Remarks	Ref
A or B chain	7.5	23	18.8	GuHCl	(1–3)	91B4
acetylated	7.5	23	16.3	GuHCl	(1–3)	91B4
oxaloyl	7.5	23	16.7	GuHCl	(1,2)	91B4
succinoyl	7.5	23	28.9	GuHCl	(1,2)	91B4
adipoyl	7.5	23	33.5	GuHCl	(1,2)	91B4
suberoyl	7.5	23	26.4	GuHCl	(1,2)	91B4
ethyleneglycolbissuccinoyl						
	7.5	23	26.8	GuHCl	(1,2)	91B4

Remarks:
(1) linear extrapolation
(2) the measurements were made in the presence of 20% (v./v.) ethanol
(3) reference protein without cross-links

Human insulin, wild-type and mutants

Mutant	pH	T	ΔG	Approach	Remarks	Ref
wild-type	7.5	23	18.4	GuHCl	(1)	92B5
(His10→Asp)/B-chain	7.5	23	21.3	GuHCl	(1)	92B5
(Pro28→Ala)/B-chain	7.5	23	16.7	GuHCl	(1)	92B5
(Lys29→Gly)/B-chain	7.5	23	18.0	GuHCl	(1)	92B5
des-(Thr30)/B-chain	7.5	23	19.2	GuHCl	(1)	92B5
des-(23–30)/B-chain	7.5	23	15.9	GuHCl	(1)	92B5
(Pro28→Ala and Lys29→Pro)/B-chain						
	7.5	23	20.1	GuHCl	(1)	92B5
(Pro28→Gln and Lys29→Pro)/B-chain						
	7.5	23	20.1	GuHCl	(1)	92B5
(Pro28→Glu and Lys29→Pro)/B-chain						
	7.5	23	19.2	GuHCl	(1)	92B5
(Pro28→Gly and Lys29→Pro)/B-chain						
	7.5	23	19.2	GuHCl	(1)	92B5
(Pro28→Leu and Lys29→Pro)/B-chain						
	7.5	23	17.6	GuHCl	(1)	92B5
(Pro28→Lys and Lys29→Pro)/B-chain						
	7.5	23	15.1	GuHCl	(1)	92B5
(Pro28→Phe and Lys29→Pro)/B-chain						
	7.5	23	20.5	GuHCl	(1)	92B5
(Pro28→Ser and Lys29→Pro)/B-chain						
	7.5	23	16.7	GuHCl	(1)	92B5

Human insulin, wild-type and mutants (continued)

Mutant	pH	T	ΔG	Approach	Remarks	Ref
(Pro28→Trp and Lys29→Pro)/B-chain						
	7.5	23	16.3	GuHCl	(1)	92B5
(Pro28→Val and Lys29→Pro)/B-chain						
	7.5	23	19.7	GuHCl	(1)	92B5
(Pro28→Aba and Lys29→Pro)/B-chain						
	7.5	23	22.6	GuHCl	(1)	92B5
(His10→Asp, Pro28→Asp and Lys29→Pro)/B-chain						
	7.5	23	23.8	GuHCl	(1)	92B5
(His10→Asp, Pro28→Gln and Lys29→Pro)/B-chain						
	7.5	23	22.6	GuHCl	(1)	92B5
(His10→Asp, Pro28→Glu and Lys29→Pro)/B-chain						
	7.5	23	21.8	GuHCl	(1)	92B5
(His10→Asp, Pro28→Lys and (Lys29→Pro)/B-chain						
	7.5	23	19.7	GuHCl	(1)	92B5
(His10→Asp, Pro28→Val and Lys29→Pro)/B-chain						
	7.5	23	20.5	GuHCl	(1)	92B5

Remark:
(1) linear extrapolation, the measurements were made in the presence of 20% ethanol to prevent aggregation

Human insulin, biosynthetic, wild-type and mutants

Mutant	pH	T	ΔG	Approach	Remarks	Ref
wild-type	2.0	23	24.3	GuHCl	(1–3)	93B7
wild-type	8.0	23	18.4	GuHCl	(2,3)	93B7
HisB10→Asp	2.0	23	20.7	GuHCl	(1,3,4)	93B7
HisB10→Asp	6.0	23	27.2	GuHCl	(3,4)	93B7
HisB10→Asp	8.0	23	20.5	GuHCl	(3,4)	93B7
OrnB5→Asp	3.5	23	18.4	GuHCl	(3,5,6)	93B7
OrnB5→Asp	6.0	23	18.4	GuHCl	(3,5,6)	93B7
OrnB5→Asp	8.0	23	17.8	GuHCl	(3,5,6)	93B7

Remarks:
(1) acid stabilization of insulin
(2) the data were taken from Fig. 1C in Ref. 93B7, the paper contains the pH dependence of ΔG from pH 2–9
(3) buffer: 50 mM Tris, 25 mM MES, 25 mM acetate, 1 mM EDTA, 20% (v./v.) ethanol, pH 8.0
(4) the data were taken from Fig. 1D in Ref. 93B7, the paper contains the pH dependence of ΔG from pH 2–9
(5) Orn: ornithine derivative
(6) the data were taken from Fig. 1D in Ref. 93B7, the paper contains the pH dependence of ΔG from pH 3.5–8.5

Human insulin

Form/Mutant	pH	T	ΔG	Δ(ΔG)	Approach	Remarks	Ref
monomers (3 μM)	8.0	25	15.7		GuHCl	(1)	93K1
dimers (37 μM)	8.0	25	16.2±0.2		GuHCl	(1)	93K1
oligomers (250 μM)							
	8.0	25	16.2		GuHCl	(1)	93K1
SerB9→Asp	8.0	25	20.8		GuHCl	(1,2)	93K1
SerB9→Asp	8.0	25	16.2	0.4	GuHCl	(1,3)	93K1
HisB10→Asp	8.0	25	21.1	5.4	GuHCl	(1)	93K1
HisB10→Glu	8.0	25	21.8	6.0	GuHCl	(1)	93K1
HisB10→Thr	8.0	25	14.8	−0.9	GuHCl	(1)	93K1
ThrA8→Ala	8.0	25	16.4	0.7	GuHCl	(1)	93K1
ThrA8→His	8.0	25	23.1	7.4	GuHCl	(1)	93K1
ThrA8→Arg	8.0	25	21.0	5.2	GuHCl	(1)	93K1
des-ThrB27	8.0	25	16.6	0.8	GuHCl	(1)	93K1
des-TyrB26	8.0	25	14.3	−1.4	GuHCl	(1)	93K1
TyrB26→Thr	8.0	25	14.5	−1.3	GuHCl	(1)	93K1
des-(B26–B30)	8.0	25	14.1	−1.6	GuHCl	(1)	93K1
PheB25→His	8.0	25	19.9	4.2	GuHCl	(1)	93K1
PheB25→Asp	8.0	25	18.2	2.5	GuHCl	(1)	93K1
triple mutant (ThrA8→His, HisB10→Asp and PheB25→His)							
	8.0	25	28.9	13.1	GuHCl	(1)	93K1

Remarks:
(1) linear extrapolation to zero denaturant concentration by a method that includes the pre- and postdenaturational baselines for a non-linear regression of the data
(2) measured by CD in the near UV spectrum at 270 nm
(3) measured by CD in the far UV spectrum at 224 nm

Interleukin

Interleukin-1β

	pH	T	ΔG	Approach	Remark	Ref
	6.5	25	29.4	GuHCl	(1)	87C

Remark:
(1) linear extrapolation

Interleukin-1 β, wild-type and mutants

Mutant	pH	T	ΔG	Δ(ΔG)	T_A	Remarks	Ref
wild-type	6.5	25	38.1±6.3	0.0	58.6	(1–3)	93C4
Lys97→Arg	6.5	25	25.1±2.1	−2.1±0.4	56.8	(1–3)	93C4
Lys97→Gly	6.5	25	29.3±2.9	−5.0±0.8	54.4	(1–3)	93C4
Lys97→Val	6.5	25	44.4±7.5	3.4±0.8	53.4	(1–4)	93C4
Lys97→Val	6.5	25	38.9±11.3	1.3±0.8	52.4	(1–3,5)	93C4
Thr9→Ala	6.5	25	29.3±2.5	−3.4±0.8	58.6	(1–3)	93C4
Thr9→Leu	6.5	25	25.1±2.9	−2.9±0.4	57.4	(1–3)	93C4
Thr9→Gln	6.5	25	26.4±5.9	−8.0±1.7	53.0	(1–3)	93C4
Thr9→Gly	6.5	25	22.6±3.4	−10.9±1.7	52.6	(1–3)	93C4

Remarks:
(1) ΔG was determined by GuHCl denaturation, linear extrapolation to zero denaturant concentration by a method that includes the pre- and postdenaturational baselines for a non-linear regression of the data
(2) Δ(ΔG) refers to C_m of the wild-type protein at 1.35 M GuHCl
(3) T_A is the temperature of aggregation, see Ref. 93C3
(4) the protein was isolated from the soluble fraction
(5) the protein was isolated from inclusion bodies

Interleukin-4

pH	T	ΔG	Approach	Remark	Ref
7.2	22	24.7	GuHCl	(1)	91W3

Remark:
(1) linear extrapolation

Murine interleukin-6, recombinant

pH	T	ΔG	Approach	Remarks	Ref
4.0	25	37.7±2.9	urea	(1,2,5,6)	95W3
4.0	25	28.8±2.1	urea	(1,2,4,6)	95W3
7.4	25	14.2±2.5	urea	(1,3,4,6)	95W3
4.0	25	7.5±1.7	GuHCl	(1,2,4,7)	95W3
7.4	25	7.5±1.3	GuHCl	(1,3,4,7)	95W3

Remarks:
(1) transition monitored by CD at 222 nm
(2) buffer: 10 mM sodium acetate, pH 4.0
(3) buffer: 10 mM Tris-HCl, pH 7.4
(4) linear extrapolation, no pre- and posttransitional dependence of the signal
(5) linear extrapolation assuming dependence of the pre- and postdenaturational signal
(6) at urea-induced unfolding a two-state transition is observed, CD and fluorescence monitored unfolding transitions are superimposable
(7) in the presence of GuHCl a biphasic transition is observed

Interleukin-6, murine (mIL-6)

Form	pH	T	ΔG	Approach	Remarks	Ref
mIL-6	4.0	25	28.5	urea	(1,2)	93W1
pMC5	4.0	25	25.1	urea	(1,3)	93W1
pMC5H	4.0	25	34.3	urea	(1,4)	93W1

Remarks:
(1) linear extrapolation
(2) mIL-6: murine interleukin-6
(3) pMC5: mIL-6 lacking residues 183–187 (Ser-Thr-Arg-Gln-Thr)
(4) pMC5H: mIL-6 containing residues 183–188 from human interleukin-6 (Ala-Leu-Arg-Gln-Met)

Interleukin-6, recombinant, N-terminally truncated form

Form	pH	T	ΔG	Approach	Remarks	Ref
	7.5	25	20.9±0.4	urea	(1–3)	95B7

Remarks:
(1) 163 residue protein lacking 22 N-terminal residues of the 185-residue chain of human interleukin-6 with Cys45→Ser and Cys51→Ser replacements
(2) linear extrapolation
(3) buffer: 50 mM potassium phosphate, 0.1 M NaCl, pH 7.5

Interleukin-6, murine, Trp36 and Trp160 sulfenylated

Form	pH	T	ΔG	Δ(ΔG)	Remarks	Ref
mIL-6 unmodified	4.0	25	25.1±1.8		(1,2)	93Z2
mIL-6 modified	4.0	25		–3.98	(1,3)	93Z2

Remarks:
(1) urea gradient electrophoresis
(2) see also, Zhang et al., Eur. J. Biochem. 207 (1992) 903
(3) Trp36 and Trp160 sulfenylated

Recombinant human interleukin-6

Mutant	pH	T	ΔG	Approach	Remarks	Ref
wild-type	8.6	25	19.2	GuHCl	(1)	94R4
reduced w.t.	8.6	25	19.2	GuHCl	(1)	94R4
mutant	8.6	25	19.2	GuHCl	(1,2)	94R4

Remarks:
(1) linear extrapolation
(2) mutant protein, Cys(45,51,74,84)→Ala

Kanamycin

Kanamycin nucleotidyltransferase, wild-type and mutants

Mutant	Transition	pH	T	ΔG	Approach	Remarks	Ref
wild-type	N→I	7	25	13	urea	(1)	88M3
wild-type	N→D	7	25	46	urea	(1)	88M3
Asp80→Ala	N→I	7	25	21	urea	(1)	88M3
Asp80→Ala	N→D	7	25	84	urea	(1)	88M3
Asp80→Phe	N→I	7	25	25	urea	(1)	88M3
Asp80→Phe	N→D	7	25	126	urea	(1)	88M3
Asp80→Tyr	N→I	7	25	31	urea	(1)	88M3
Asp80→Tyr	N→D	7	25	130	urea	(1)	88M3

Remark:
(1) linear extrapolation, the data were taken from Fig. 5 in Ref. 88M3

Lac-Repressor

Lac-Repressor, headpiece

pH	T	ΔG	Approach	Remarks	Ref
7.25	20	10	urea	(1)	81S1
	20	8.4±3.3	heat		81S1
7.25	20	12.1	urea	(2)	81S1
8	25	10.8	DSC		81H2

Remarks:
(1) linear extrapolation
(2) transfer model

Lac-Repressor, substitutions that alter the monomer-monomer interface

Mutant	pH	T	ΔG	Approach	Remarks	Ref
−11 deletion	7.5		80.3±3.3	urea	(1–3,4)	94C6
−11 deletion	7.5		80.8±5.9	urea	(1–3,5)	94C6
R3	7.5		100.0±4.2	urea	(2,3,5,6)	94C6
Tyr282→Asp	7.5		20.1±1.3	urea	(3,5,7)	94C6

Remarks:
(1) −11 deletion = 11 C-terminal residues deleted
(2) ΔG refers to transition of a dimer to unfolded monomers, data treatment assuming linear dependence of ΔG on denaturant concentration
(3) buffer: 10 mM Tris-HCl, 0.1 M potassium sulfate, pH 7.5
(4) transition monitored by CD at 222 nm
(5) transition monitored by fluorescence
(6) R3 = mutant with the C-terminal Leu heptad repeats replaced by the GCN4 dimerization sequence
(7) Tyr282→Asp = monomeric mutant, equilibrium treatment by a monomer unfolding model

α-Lactalbumin

The data are arranged as follows:

a) holo α-lactalbumin, various approaches and different species,
b) apo α-lactalbumin,
c) various transitions,
d) modified α-lactalbumin.

a) Holo α-lactalbumin, various approaches and different species

Bovine α-lactalbumin, various approaches

pH	T	ΔG	Approach	Remarks	Ref
5.5	25	15.9	GuHCl	(1a)	76K1
5.5	25	17.6	GuHCl	(2)	81C2
6.3	25	21.9	DSC		81P1
6.65	25	27.2	GuHCl	(3)	76K2
7	25	18.4	GuHCl	(1)	76K1
7	25	28.5	GuHCl	(1b)	73S
7	25	18.4	urea	(2)	82A
7	25	17.6	GuHCl, GuHSCN(2)		82A
7.3	25	26.4	GuHSCN	(1b)	74T1
7.3	25	18	GuHSCN	(4)	74T1
7.3	25	23.8	GuHSCN, anion binding		74T1
7.3	25	17.6	GuHSCN	(5a)	74T1
7.3	25	22.2	GuHSCN	(5b)	74T1
		18	hydrogen exchange		80C

Remarks:
(1a) denaturant binding model, pH function
(1b) denaturant binding model
(2) linear extrapolation
(3) three-state model
(4) eq. (11) Ref. 69A
(5a) transfer model, ε = 0.14
(5b) transfer model, ε = 0.16

α-Lactalbumin from various species

Species	pH	T	ΔG	Approach	Remarks	Ref
bovine	7.0	25	29*	heat	(1)	89H1
goat	7.0	25	35*	heat	(1)	89H1
human	6–7	25	26.4	GuHCl	(2)	78N2
human	7.0	25	29*	heat	(1)	89H1
guinea pig	7.0	25	28*	heat	(1)	89H1

Remarks:
(1) enthalpy change and transition temperature were taken from Fig. 1 of Ref. 89H1
(2) denaturant binding model

Bovine α-lactalbumin in the presence of calcium chloride

CaCl$_2$ Concentration	pH	T	ΔG	Approach	Remarks	Ref
0.0 mM	7.0	25.0	15.7	GuHCl	(1,2)	86I
0.1 mM	7.0	25.0	25.4	GuHCl	(1,3)	86I
1.0 mM	7.0	25.0	30.2	GuHCl	(1,3)	86I
12.0 mM	7.0	25.0	37.1	GuHCl	(1,3)	86I

Remarks:
(1) linear extrapolation
(2) N→D transition in the absence of CaCl$_2$
(3) N→D transition in the presence of CaCl$_2$

b) Apo α-lactalbumin

α-Lactalbumin from various species, calcium free form

Species	pH	T	ΔG	Approach	Remarks	Ref
bovine	8	25	12.5±0.3	DSC, I=0.1		85P3
bovine	7.0	25	9.6*	heat	(1)	89H1
goat	7.0	25	12.3*	heat	(1)	89H1
human	7.0	25	6.5*	heat	(1)	89H1
human	8	25	6.1±0.1	DSC, I=0.1		85P3
guinea pig	7.0	25	5.4*	heat	(1)	89H1

Remark:
(1) enthalpy change and transition temperature were taken from Fig. 1 of Ref. 89H1.

Goat α-lactalbumin, apo form

pH	T	ΔG	Approach	Remarks	Ref
7.5	25	−0.5±0.1	heat	(1)	91D7
7.5	25	−0.5±0.1	calorimetry	(1,2)	91D7

Remarks:
(1) apo form in the absence of metal ions
(2) calorimetry, special approach to resolve conformational and Mn^{2+} binding process

c) Various transitions

α-Lactalbumin, A→D transition

pH	T	ΔG	Approach	Remarks	Ref
7.0	25.0	5.94	GuHCl	(1,2)	86I

Remarks:
(1) linear extrapolation
(2) A→D transition

α-Lactalbumin, bovine, molten globule intermediate

Transition	pH	T	ΔG	Approach	Remarks	Ref
I→U	8.0	25	7.86±0.17	GuHCl	(1,2)	96A2

Remarks:
(1) linear extrapolation
(2) from burst hase kinetics and equilibrium unfolding

α-Lactalbumin, bovine, various transitions in the presence and in the absence of calcium

Concentration (mM)	pH	T	ΔG	Approach	Remarks	Ref
Bovine α-lactalbumin:						
0, N→U	7.0	4.5	15.3±1.0	GuHCl	(1,2)	92I1
1, N→U	7.0	4.5	29.2±1.6	GuHCl	(1,2)	92I1
N→A	7.0	4.5	5.4±0.8	GuHCl	(1,2)	92I1
Form 2CM-3SS-BLA					(3)	
0, N→U	7.0	4.5	5.3±1.2	GuHCl	(1–3)	92I1
1, N→U	7.0	4.5	16.2±5.0	GuHCl	(1–3)	92I1
A→U	7.0	4.5	0.6±0.3	GuHCl	(1–3)	92I1

Remarks:
(1) limear extrapolation
(2) based on data from Ref. 86I
(3) 2CM-3SS-BLA = modified protein with 3 disulfide bonds, i.e., Cys6 and Cys120 selectively reduced and carboxymethylated

α-Lactalbumin, various species and mutants, molten globule unfolding transition

Mutant	pH	T	ΔG	Approach	Remarks	Ref
Authentic wild-type forms:						
bovine	2.0	20	6.07±0.46	urea	(1)	95U
guinea pig	2.0	20	12.97±0.29	urea	(1)	95U
goat	2.0	20	2.76±0.21	urea	(1)	95U
Mutants of recombinant goat α-lactalbumin:						
reco. goat	2.0	20	2.93±0.21	urea	(1)	95U
Thr29→Ile	2.0	20	3.22±0.29	urea	(1)	95U
Ala30→Thr	2.0	20	2.34±0.21	urea	(1)	95U
Ala30→lle	2.0	20	4.18±0.25	urea	(1)	95U
Thr33→Ile	2.0	20	4.27±0.88	urea	(1)	95U
(Ala30→Ile and Thr33→Ile)	2.0	20	5.77±0.29	urea	(1)	95U

Remark:
(1) extrapolation to zero denaturant concentration by a method that includes the pre- and postdenaturational baselines for a non-linear regression of the data

α-Lactalbumin, transition A→U of the molten globule state into the unfolded protein monitoreed by ^{1}H NMR of individual residues

Residue	pH	T	ΔG	Approach	Remarks	Ref
Tyr103 C3,5H	2.0	35	4.85±0.17	urea	(1,2)	93S7
Tyr103 C2,6H	2.0	35	5.52±0.25	urea	(1,2)	93S7
Trp104 C4H	2.0	35	8.24±0.92	urea	(1,2)	93S7
Trp104 C5H	2.0	35	5.86±0.63	urea	(1,2)	93S7
TrpU2 C4H	2.0	35	10.21±0.96	urea	(1–3)	93S7
TrpU2 C5H	2.0	35	5.23±0.33	urea	(1–3)	93S7
TrpU3 C4H	2.0	35	3.35±0.75	urea	(1–3)	93S7
His107 C2H	2.0	35	5.56±0.38	urea	(1–3)	93S7

Remarks:
(1) linear extrapolation
(2) given is the pD value
(3) U = unassigned Trp

d) Modified α-lactalbumin

Bovine α-lactalbumin, modified forms

Form	pH	T	ΔG	Approach	Ref
guanidinated			26	hydrogen exchange	80C

α-Lactalbumin modified, three-disulfide derivative, Cys6-Cys120 disulfide bond reduced and carboxy-methylated

pH	T	Δ(ΔG)	Approach	Remark	Ref
8.5	25.0	−10.0	urea	(1)	90K8

Remark:
(1) Δ(ΔG) refers to the native protein

α-Lactalbumin, native protein and disulfide derivatives

Solvent	pH	T	ΔG	Approach	Remarks	Ref
native α-lactalbumin:						
Tris, Ca^{2+}	8.0	25	28	DSC	(1)	96H2
Tris	8.0	25	21	DSC	(2)	96H2
Tris, EDTA	8.0	25	2	DSC	(3)	96H2
3SS α-lactalbumin					(4)	
Tris, Ca^{2+}	8.0	25	18	DSC	(1)	96H2
Tris	8.0	25	15	DSC	(2)	96H2
2SS α-lactalbumin					(5)	
Tris, Ca^{2+}	8.0	25	2	DSC	(1)	96H2

Remarks:
(1) 10 mM Tris, 2 mM CaCl$_2$, pH 8.0
(2) 10 mM Tris, pH 8.0
(3) 10 mM Tris, 1 mM EDTA, pH 8.0
(4) Cys6 and Cys120 reduced and blocked by carboxymethylation
(5) Cys6, Cys120, Cys28 and Cys111 reduced and blocked by carboxymethylation

β-Lactamase

β-Lactamase

pH	T	ΔG	Approach	Remarks	Ref
6.6	24	9.1±0.8	GuHCl	(1)	83M1
6.6	24	8.8±1.2		(1,3)	83M1
6.6	24	8.8±0.7		(1,4)	83M1
6.6	24	19.2±1.3		(1,5)	83M1
6.6	24	19.2		(1,6)	83M1
7	25	25	GuHCl	(2)	76R2
7	20	8.5±0.9	GuHCl	(1)	76R2
7	20	8.5±0.9	GuHCl	(1)	76R2

Remarks:
(1) linear extrapolation
(2) transfer model
(3) denaturant: biguanide hydrochloride
(4) denaturant: propylbiguanide hydrochloride
(5) denaturant: hexylbiguanide hydrochloride
(6) denaturant: N-decylbiguanide hydrochloride

β-Lactamase, three-state unfolding

Transition	pH	T	ΔG	Approach	Remarks	Ref
N→H	6.5	20.5	10.40±2.3	urea	(1)	85M2
N→U	6.5	20.5	9.99±2.51	urea	(1)	85M2
N→H	7	20	8.52±0.9	GuHCl		80A
H→U	7	20	11.19±0.74	GuHCl		80A
N→H	7	20	10.15±1.5	urea		80A
H→U	7	20	10.09±1.61	urea		80A

Remark:
(1) linear extrapolation

β-Lactamase mutants, transition N→U

Mutant	pH	T	ΔG	Approach	Remarks	Ref
P54	6.5	20	22.13±1.49	GuHCl	(1)	85C2
PC1	6.5	20	23.95±2.0	GuHCl	(1)	85C2

Remark:
(1) linear extrapolation

β-Lactamase, precursor and mature lactamase

Transition	pH	T	ΔG	Approach	Remarks	Ref
Pre-β-lactamase:						
N→H	7.0	25	−2.1±0.4	urea	(1–4)	94Z1
N→H	5.5	25	12.6±1.7	urea	(1–4)	94Z1
H→U	7.0	25	28.9±3.8	urea	(1–4)	94Z1
Mature β-lactamase:						
N→H	7.0	25	36.4±2.5	urea	(1–4)	94Z1
N→H	7.0	25	31.0±2.1	urea	(1–3,5)	94Z1
N→H	7.0	25	26.8±2.1	urea	(1–3,5,6)	94Z1
H→U	7.0	25	17.6±1.7	urea	(1–4)	94Z1
N→H	7.0	25	37.7	heat	(2,3,5)	94Z2
H→U	7.0	25	16.7	heat	(2,3,5)	94Z2

Remarks:
(1) data analysis in terms of a three-state model in which ΔG is a linear function of the denaturant concentration
(2) the model U→H→N involves unfolded (U), native (N) and enzymatically inactive β-lactamase with native-like fluorescence properties (H)
(3) folding buffer containing 100 mM potassium phosphate (pH 7.0), 100 mM urea, 100 mM ammonium sulfate, 0.01% Tween, 10 mM DTT
(4) folding was started from protein denatured in 8 M urea
(5) unfolding was started from the N state
(6) measured in the presence of GroEL

TEM-1 β-lactamase, three-state transition

Transition	pH	T	ΔG	Approach	Remarks	Ref
N→I	7	25	21.8±1.7	GuHCl	(1,2)	95V2
N→I	7	25	18.8±1.7	GuHCl	(1,3)	95V2
I→U	7	25	23.8±0.8	GuHCl	(1,3)	95V2

Remarks:
(1) data analysis by a non-linear fit according to a three-state folding model that is based on linear dependence of ΔG on the denaturant concentration
(2) the transition was monitored by fluorescence
(3) the transition was monitored by CD

TEM β-lactamase, wild-type and mutants

Mutant	pH	T	Δ(ΔG)	Approach	Remarks	Ref
wild-type	7.0	50.6	0.0	heat, v.H.	(1)	95R2
Glu104→Lys	7.0	50.6	+2.7	heat, v.H.	(1,2)	95R2
Gly238→Ser	7.0	50.6	−5.2	heat, v.H.	(1,2)	95R2
double mutant (Glu104→Lys and Gly104→Lys)						
	7.0	50.6	−3.2	heat, v.H.	(1,2)	95R2
Arg164→His	7.0	50.6	−0.7	heat, v.H.	(1,2)	95R2
Arg164→Ser	7.0	50.6	−2.6	heat, v.H.	(1,2)	95R2
double mutant (Arg164→Ser and Glu240→Lys)						
	7.0	50.6	+3.1	heat, v.H.	(1,2)	95R2
Ser235→Ala	7.0	50.6	+2.4	heat, v.H.	(1,2)	95R2

Remarks:
(1) Δ(ΔG) was calculated assuming ΔCp = 8.8±1.7 kJ/mol/K
(2) reference temperature at which Δ(ΔG) = 0 for the wild-type protein

TEM β-lactamase, wild-type and mutants

Transition	pH	T	ΔG	Approach	Remarks	Ref
Wild-type:						
trans. (1)	7.0	25	21.8±1.7	GuHCl	(1)	95R2
trans. (2)	7.0	25	23.8±0.8	GuHCl	(1)	95R2
Mutant (Glu104→Lys):						
trans. (1)	7.0	25	25.5±1.7	GuHCl	(1)	95R2
trans. (2)	7.0	25	22.6±1.3	GuHCl	(1)	95R2
Mutant (Gly238→Ser):						
trans. (1)	7.0	25	8.4±0.8	GuHCl	(1)	95R2
trans. (2)	7.0	25	20.9±2.5	GuHCl	(1)	95R2

Remark:
(1) transition curves monitored by fluorescence were analyzed by a two-state model, transition curves monitored by far UV CD were analyzed by a three-state model (N→I→U)

TEM-1 β-lactamase, wild-type and mutant, three-state transition

Transition	pH	T	ΔG	Approach	Remarks	Ref
Wild-type:						
N→I	7	25	21.8±1.7	GuHCl	(1,2)	95V2
N→I	7	25	18.8±1.7	GuHCl	(1,3)	95V2
I→U	7	25	23.8±0.8	GuHCl	(1,3)	95V2
Mutant (cis-Pro167→Thr):						
N→I	7	25	12.1±1.3	GuHCl	(1,2)	96V2
N→I	7	25	13.8±1.7	GuHCl	(1,3)	96V2
I→U	7	25	22.6±2.9	GuHCl	(1,3)	96V2

Remarks:
(1) data analysis by a non-linear fit according to a three-state folding model that is based on linear dependence of ΔG on the denaturant concentration
(2) the transition was monitored by fluorescence
(3) the transition was monitored by CD

RTEM β-lactamase from *E. coli*, transitions

Transition	pH	T	ΔG	Approach	Remarks	Ref
	7.0	25	8.8	GuHCl	(1)	96S2
trans. (1)	7.0	25	11.0	GuHCl	(2)	96S2
trans. (2)	7.0	25	15.0	GuHCl	(2)	96S2
trans. (1)	7.0	25	10.0	GuHCl	(3)	96S2
trans. (2)	7.0	25	13.0	GuHCl	(3)	96S2
N→U	7.0	25	24.5	GuHCl		96S2
I→U	7.0	25	14.0	GuHCl		96S2

Remarks:
(1) from equilibrium inactivation experiments
(2) from non-linear least square fitting of fluorescence intensity change at 340 nm and from fluorescence intensity maximum shift,
(3) from non-linear least square fitting of CD data at 222 nm

RTEM1 β-lactamase from *E. coli*, transitions, oxidized (ox.) and reduced (red.) protein

Form/ Transition	pH	T	ΔG	Approach	Remarks	Ref
ox., N→I	7.0	25	24.2±3.0	GuHCl	(1)	96F4
ox., N→I	7.0	25	22.8±3.1	GuHCl	(2)	96F4
ox., I→U	7.0	25	12.2±1.9	GuHCl	(1)	96F4
ox., I→U	7.0	25	0.0	GuHCl	(1,3)	96F4
red., N→I	7.0	25	23.1±3.9	GuHCl	(1)	96F4
red., N→I	7.0	25	23.2±1.3	GuHCl	(2)	96F4
red., I→U	7.0	25	12.1±2.3	GuHCl	(1)	96F4
red., I→U	7.0	25	−3.5	GuHCl	(1,3)	96F4

Remarks:
(1) ΔG from equilibrium unfolding monitored by CD, analysis by a three-state model
(2) ΔG from equilibrium unfolding monitored by fluorescence, analysis by a two-state model
(3) in the presence of 2 M GuHCl

Lactate Dehydrogenase

Porcine lactate dehydrogenase

pH	T	ΔG	Approach	Ref
7.6	20	110	pressure denaturation and dissociation	82M2

Lactate dehydrogenase from *Bacillus stearothermophilus*, wild-type and mutant

Mutant	pH	T	ΔG	Approach	Remarks	Ref
Wild-type	7.4	25		GuHCl	(1,2)	94N2
trans. (1)			33±10			
trans. (2)			56±2			
trans. (3)			94±47			
Asp143→Asn	7.4	25		GuHCl	(1,2)	94N2
trans. (1)			16±4			
trans. (2)			34±5			
trans. (3)			67±36			

Remarks:
(1) wild-type (Asp143)
(2) non-linear regression of the transition by a three-state equilibrium, assuming linear dependence of ΔG on the denaturant concentration

β-Lactoglobulin

Bovine and goat β-lactoglobulin

Species	pH	T	ΔG	Approach	Remarks	Ref
bovine	2	25	40	DSC		89L2
bovine A	3.15	25	42.7	urea	(1)	75P
bovine A	3.15	25	55.2	GuHCl	(1)	75P
bovine B	3.15	25	49	urea	(1)	75P
bovine	5.5	0	33 (>33)	hydrogen exchange		66H
bovine	5.5	25	41	DSC		89L2
bovine A,B	7.2	16±1	31.4±3.3	urea, electrophoresis		82H
goat		25	60.7		(2)	64T
goat	3.2	25	109	GuHCl	(3a)	79P1
goat	3.2	25	93	GuHCl	(3b)	79P1
goat	3.2	25	52.3	GuHCl	(1)	74G1
goat	3.2	25	43.9	urea	(1)	74G1
goat	3.2	25	77	GuHCl	(4)	74G1
goat	3.2	25	50.6	urea	(4)	74G1

Remarks:
(1) linear extrapolation
(2) estimation, minimal value
(3a) denaturant binding model, k = 1.2
(3b) denaturant binding model, k = 0.8
(4) transfer model

Bovine β-lactoglobulin in the presence of kosmotropic salts, salt concentration 0.2 M

Salt	pH	T	ΔG	Δ(ΔG)	Approach	Remarks	Ref
	6.85	25	27.8	0.0	urea	(1)	88K2
chloride	6.85	25	29.3	1.5	urea	(1)	88K2
tartrate	6.85	25	31.7	3.9	urea	(1)	88K2
sulfate	6.85	25	36.8	9.0	urea	(1)	88K2
phosphate	6.85	25	38.4	10.6	urea	(1)	88K2
citrate	6.85	25	40.9	13.1	urea	(1)	88K2

Remark:
(1) linear extrapolation

β-Lactoglobulin B mixed disulfide derivates

Derivate	T	ΔG	Approach	Remarks	Ref
β-LG-SH (1)	25	46.4	urea	(2,3)	83C
β-LG-S-S-CH_2CH_2COOH	25	27.4	urea	(2,3)	83C
β-LG-S-S-CH_2CH_2OH	25	22.2	urea	(2,3)	83C
β-LG-S-S-$CH_2CH_2CH_3$	25	16.3	urea	(2,3)	83C
β-LG-S-S-$CH_2CH_2NH_2$	25	3.0	urea	(2,3)	83C

Remarks:
(1) abbreviation: β-LG = β-lactoglobulin, β-LG-SH = the native protein
(2) linear extrapolation
(3) the data refer to pH 2.83

β-Lactoglobulin, wild-type and mutant

Mutant	pH	T	ΔG	Approach	Remarks	Ref
wild-type	7.1		46.9	GuHCl	(1)	94K1
Trp19→Tyr	7.1		18.0	GuHCl	(1)	94K1

Remark:
(1) denaturant binding model, k = 1.0

λ Cro Protein

λ Cro Protein, wild-type and mutants

Mutant	pH	T_{trs}	ΔT	Δ(ΔG)	Approach	Remarks	Ref
wild-type	7	40	0	0	heat, v.H.		89P3
wild-type	7	39.5	0	0	heat, v.H.		90P
Gln16→Leu	7	54	14	11.7±0.8	heat, v.H.		89P3
Tyr26→Asp	7	54.0	14.5	11.3	heat	(1)	90P
Tyr26→Cys	7	51	11	9.2±0.8	heat, v.H.		89P3
Tyr26→Cys	7	51.0	11.5	9.2	heat	(1)	90P
Tyr26→Gln	7	47	7.5	5.9	heat	(1)	90P
Tyr26→His	7	49.5	10	7.9	heat	(1)	90P
Tyr26→Leu	7	46	6.5	4.6	heat	(1)	90P
Tyr26→Phe	7	41.5	2	0.4	heat	(1)	90P
Tyr26→Trp	7	37.5	−2	−0.4	heat	(1)	90P
Tyr26→Val	7	44.5	5	3.8	heat	(1)	90P
Ala36→Ser	7	41	1	1.7±0.8	heat, v.H.		89P3

Remark:
(1) Δ(ΔG) refers to the wild-type, and determination of the equilibrium constant at 45°C

λ Cro Protein, wild-type and double mutants

Mutant	pH	T_{trs}	ΔT	Δ(ΔG)	Remarks	Ref
Ile30→Leu	7	35	0	0.0	(1,2)	89P3
(Ile30→Leu and Tyr26→Cys)						
	7	47	12	7.5±0.8	(2)	89P3
(Ile30→Leu and Gln27→Pro)						
	7	38	3	1.3±0.8	(2)	89P3
(Ile30→Leu and Ala36→Ser)						
	7	36	1	1.3±0.8	(2)	89P3

Remarks:
(1) protein Ile30 is the reference protein for the following double mutants
(2) heat denaturation, van't Hoff treatment

λ-Repressor

λ-Repressor, recombinant

pH	T	ΔG	Approach	Remarks	Ref
5.6	37	13.7±0.3	urea	(1–3)	95H7
5.6	37	11.3±1.7	urea	(1,2,4)	95H7
5.6	37	13.0±1.3	urea	(1,2,5)	95H7

Remarks:
(1) linear extrapolation to zero denaturant concentration by a method that includes the pre- and postdenaturational baselines for a non-linear regression of the data
(2) buffer: ~99% D_2O, 10 mM CD_3COOD, 100 mM NaCl, 17 μg/ml 3-(trimethylsilyl)propionic acid, 1 mM NaN_3, pH 5.6
(3) the transition was monitored by CD
(4) the transition was monitored by NMR, chemical shift of Tyr60 δH
(5) the transition was monitored by NMR, chemical shift of Tyr60 εH

λ-Repressor, N-terminal domain, wild-type and mutants

Mutant	pH	T_{trs}	ΔT	Δ(ΔG)	Approach	Ref
wild-type	8	53.4±0.1	0	0	DSC	86H1
Lys4→Gln	8	53.5	2	1.8	DSC	84H2
Tyr22→His	8	28.8	22.7	−9.4	DSC	84H2
Gln33→Tyr	8	57.4	5.9	5.5	DSC	84H2
Gln44→Tyr	8	51.4	−0.1	−0.1	DSC	84H2
Gly46→Ala	8	56.5±0.1	3.1	2.8±0.5	DSC	86H1
Gly48→Ala	8	58.1±0.1	4.7	3.6±0.7	DSC	86H1
Gly48→Asn	8	57.5±0.1	4.1	3.3±0.6	DSC	86H1
Gly48→Ser	8	57.4±0.1	4	2.8±0.5	DSC	86H1
Ala49→Val	8	38.5	−13	−5.1	DSC	84H2
Ala66→Thr	8	29	−22.5	−12.5	DSC	84H2
Ile84→Ser	8	37.2	−14.3	−9.4	DSC	84H2
Tyr88→Cys	8	62.7±0.1	8.8	10	DSC	88S10
multiple mutants:						
(Gly46→Ala and Gly48→Ala)						
	8	59.6±0.1	6.2	4.6±0.7	DSC	86H1
(Gly46→Ala and Gly48→Ala)						
	8	62±0.1	8.1	7.9	DSC	88S10
(Gly46→Ala and Gly48→Ala and Tyr88→Cys)						
	8	70.3±0.1	16.4	17.6	DSC	88S10

λ-Repressor (Res. 1–102), wild-type

Mutant	pH	T	ΔG	ΔT	Δ(ΔG)	Remarks	Ref
wild-type	7	37	12.2*	0	0.0	(1)	89P5
Leu57→Pro	7	37	0.7*	−15	−11.5*	(1)	89P5
Leu57→Cys	7	37	−0.6*	−19	−12.8*	(1)	89P5
Leu57→Ala	7	37	−8.2*	−34	−20.4*	(1)	89P5

Remark:
(1) heat denaturation, van't Hoff treatment

λ-Repressor, monomeric, residues 6-85 of the N-terminal domain, heat and cold denaturation

	pH	T	ΔG	Approach	Remarks	Ref
	8.0	13	19.2±0.8	heat, urea	(1–3)	96H7
	8.0	25	18.0±0.8	heat, urea	(1,2)	96H7
	8.0	37	13.4±0.8	heat, urea	(1,2)	96H7

Remarks:
(1) from a three-dimensional thermal-urea denaturation profile monitored by CD and NMR
(2) buffer: 20 mM KD_2PO_4, 100 mM NaCl in 99% D_2O
(3) temperature of maximum stability

λ-Repressor, truncated monomeric form of the N-terminal domain

Mutant	pH	T	Δ(ΔG)	Approach	Remarks	Ref
λ (residues 6-85)	8.0	37	0	urea	(1–3)	96B10
double mutant (Gly46→Ala and Gly 48→Ala) of λ (residues 6-85)						
	8.0	37	7.4±0.1	urea	(1–3)	96B10

Remarks:
(1) linear extrapolation
(2) kinetic and equilibrium data
(3) buffer: 20 mM K_2HPO_4, 100 mM NaCl

λ-Repressor, dimer formed by linkage between two cysteine residues

Mutant	pH	T	ΔG	ΔT	Remarks	Ref
wild-type	7.5	22	0	0	(1,2)	86S2
Tyr85→Cys (SH)						
	7.5	22	0	1	(1,2)	86S2
Cys85-Cys85'-dimer						
	7.5	22	–2*	–11	(1,2)	86S2
Cys88-Cys88'-dimer						
	7.5	22	5*	8	(1,2)	86S2

Remarks:
(1) ΔG is based on the urea-induced transition curves in Fig. 4 of Ref. 86S2, roughly determined by linear extrapolation
(2) ΔT was taken from Fig. 7 of Ref. 86S2.

λ-Repressor, wild-type (Pro78) and mutants

Mutant	pH	T	ΔG	Approach	Remarks	Ref
wild-type	6.6–6.8	20	21.8	GuHCl	(1–3)	90R2
Pro78→Ala	6.6–6.8	20	15.1	GuHCl	(1,2,4)	90R2
double mutant (Gly46→Ala and Gly48→Ala)						
	6.6–6.8	20	28.9	GuHCl	(1,2,5)	90R2
triple mutant (Gly46→Ala, Gly48→Ala and Pro78→Ala)						
	6.6–6.8	20	22.6	GuHCl	(1–3)	90R2

Remarks:
(1) linear extrapolation
(2) measurement in the presence of 50 mM sodium phosphate with 100 mM KCl
(3) thermal transition at 55°C
(4) thermal transition at 48°C
(5) thermal transition at 62°C

λ-Repressor, wild-type and mutants in positions 14, 17 and 77

Mutant	pH	T	ΔG	Approach	Remarks	Ref
Low salt concentration:						
wild-type	7	25	20.2	urea	(1–3)	94M6
ARS	7	25	15.3	urea	(1,3,4)	94M6
DAS	7	25	15.4	urea	(1,3,4)	94M6
DRA	7	25	14.5	urea	(1,3,4)	94M6
AAS	7	25	13.9	urea	(1,3,4)	94M6
ARA	7	25	16.0	urea	(1,3,4)	94M6
DAA	7	25	10.7	urea	(1,3,4)	94M6
AAA	7	25	14.4	urea	(1,3,4)	94M6
High salt concentration:						
wild-type	7	25	24.7	urea	(1,2,5)	94M6
ARS	7	25	19.8	urea	(1,2,5)	94M6
DAS	7	25	20.8	urea	(1,2,5)	94M6
AAS	7	25	18.5	urea	(1,2,5)	94M6

Remarks:
(1) linear extrapolation to zero denaturant concentration by a method that includes the pre- and postdenaturational baselines for a non-linear regression of the data
(2) wild-type contains Asp14, Arg17, and Ser77
(3) buffer: 50 mM potassium phosphate, 0.1 M KCl, pH 7
(4) mutant description: indicated are the amino acid residues in one letter code that occupy the positions 14, 17, and 77; wild-type corresponds to DRS
(5) buffer with 1 M KCl

λ-Repressor, N-terminal domain (res. 1–102), wild-type and mutants

Mutant	pH	T	ΔG	Approach	Remarks	Ref
wild-type	7.0	25	20.1±0.4	GuHCl	(1)	92L4
wild-type	7.0	5	25.1±0.4	GuHCl	(1)	92L4
Val36→Ile	7.0	25	24.3±0.4	GuHCl	(1)	92L4
Val36→Ile	7.0	5	27.6±0.4	GuHCl	(1)	92L4
Met40→Ala	7.0	25	11.7±0.4	GuHCl	(1)	92L4
Met40→Ala	7.0	5	16.3±0.4	GuHCl	(1)	92L4
double mutant (Met40→Val and Val47→Leu)						
	7.0	25	15.5±0.4	GuHCl	(1)	92L4
double mutant (Val36→Phe and Met40→Leu)						
	7.0	25	14.2±0.4	GuHCl	(1)	92L4
double mutant (Leu18→Ala, Met40→Ala)						
	7.0	25	–0.8±0.4	GuHCl	(1)	92L4
triple mutant (Val36→Leu, Met40→Val and Val47→Ile)						
	7.0	25	15.5±0.4	GuHCl	(1)	92L4
triple mutant (Val36→Ile, Met40→Val and Val47→Leu)						
	7.0	25	16.3±0.4	GuHCl	(1)	92L4
triple mutant (Val36→Ile, Met40→Val and Val47→Ile)						
	7.0	25	16.7±0.4	GuHCl	(1)	92L4
triple mutant (Val36→Leu, Met40→Leu and Val47→Ile)						
	7.0	25	22.2±0.4	GuHCl	(1)	92L4
triple mutant (Val36→Phe, Met40→Ala and Val47→Ile)						
	7.0	5	24.7±0.4	GuHCl	(1)	92L4
	7.0	25	10.9±0.4	GuHCl	(1)	92L4

λ-Repressor, N-terminal domain (res. 1–102), wild-type and mutants (continued)

Mutant	pH	T	ΔG	Approach	Remarks	Ref
triple mutant (Val36→Phe, Met40→Phe and Val47→Phe)						
	7.0	25	7.5±0.4	GuHCl	(1)	92L4
	7.0	5	12.1±0.4	GuHCl	(1)	92L4
quadruple mutant (Val36→Phe, Met40→Phe, Val47→Ile and Leu65→Phe)						
	7.0	25	12.6±0.4	GuHCl	(1)	92L4
	7.0	5	18.4±0.4	GuHCl	(1)	92L4

Remark:
(1) linear extrapolation to zero denaturant concentration by a method that includes the pre- and postdenaturational baselines for a non-linear regression of the data

λ-Repressor and its C-terminal domain

Form	pH	T	ΔG	Approach	Remarks	Ref
complete	8.0	25	8.74	urea	(1)	92B2
C-term.	8.0	25	4.18	urea	(1,2)	92B2

Remarks:
(1) linear extrapolation, the transition was monitored by tryptophane fluorescence
(2) residues 132–236

Laminin

Laminin, C-terminal α-helical coiled-coil region, recombinant, isolated reduced fragments and disulfide-linked fragments

Fragment, Concentration	pH	T	ΔG	Approach	Remarks	Ref
Reduced and alkylated fragments:						
β fragment, 54 μM	7.4	25	37.8	heat	(1,2)	95K3
γ fragment, 65 μM	7.4	25	17.4	heat	(1,2)	95K3
β + γ fragment, 25 μM	7.4	25	42.8	heat	(1,2)	95K3
β fragment, 12 μM	7.4	25	34.9	heat	(1,3)	95K3
β + γ fragment, 22 μM	7.4	25	39.1	heat	(1,3)	95K3
β + γ fragment, 76 μM	7.4	25	30.5	heat	(1,4)	95K3
Disulfide-linked fragments:						
γ-γ	7.4	25	2.3	heat	(1,2)	95A5
β-γ	7.4	25	31.8	heat	(1,2)	95A5
β-γ	7.4	25	32.5	heat	(1,3)	95A5
β-γ	7.4	25	30	heat	(1,4)	95A5

Remarks:
(1) fragments:
    α – residues 2044 to 2146
    β – residues 1700 to 1786
    γ – residues 1506 to 1607
(2) buffer: 5 mM phosphate, pH 7.4
(3) buffer: 5 mM phosphate, 0.1 M NaCl, pH 7.4
(4) buffer: 5 mM phosphate, 1 M urea, pH 7.4

Leucine Zippers, synthetic model peptides, see also de novo synthesized proteins

Leucine zipper peptide GCN4-p1, synthetic 33-residue peptide from the yeast transcriptional activator GCN4

Concentration	pH	T	ΔG	Approach	Remarks	Ref
6.0 µM	7.0	5	41.3±3.0	GuHCl	(1–3)	95Z2
12.4 µM	7.0	5	43.8±3.0	GuHCl	(1–3)	95Z2
12.6 µM	7.0	5	44.1±2.5	GuHCl	(1–3)	95Z2
17.6 µM	7.0	5	44.7±1.8	GuHCl	(1–3)	95Z2
37.5 µM	7.0	5	43.9±1.5	GuHCl	(1–3)	95Z2
	7.0	5	43.9±1.0	kinetics, local fit		95Z2
	7.0	5	42.8±1.6	kinetics, global fit		95Z2
	7.0	5	43.1±0.8	kinetics	(4)	95Z2

Remarks:
(1) peptide concentrations are expressed in terms of total concentration of monomer
(2) ΔG refers to unfolding of the dimer into two monomers
(3) linear extrapolation
(4) simulation of equilibrium unfolding from the kinetic data

Leucine zipper, 29-residue peptide

Mutant	pH	T	ΔG	Approach	Remarks	Ref
LZ	7.0	20	33.6±0.3	urea	(1,2)	95L3
LZ(Glu16→Ala)	7.0	20	23.6±0.9	urea	(1,2)	95L3
LZ(Ala14→Pro)	7.0	20	11.7±0.2	urea	(1,2)	95L3

Remarks:
(1) LZ = Ac-Glu-Tyr-Glu-Ala-Leu-Glu-Lys-Lys-Leu-Ala-Ala-Leu-Glu-Ala-Lys-Leu-Gln-Ala-Leu-Glu-Lys-Lys-Leu-Glu-Ala-Leu-Glu-His-Gly
(2) linear extrapolation

Leucine zipper peptide GCN4-p1, 33-residue peptide derived from the bZIP transcriptional activator GCN4

Variant	pH	T	ΔG	Approach	Remarks	Ref
GCN4-p1	5.5	10	43.9±0.4	GuHCl	(1,2)	96S8
GCN4-p1	5.5	10	42.7±0.4	GuHCl	(3,4)	96S8
Asp7→Gly	5.5	10	36.4±0.8	GuHCl	(1,2)	96S8
Asp7→Gly	5.5	10	36.4±0.4	GuHCl	(3,4)	96S8
Asp7→Ala	5.5	10	39.3±0.4	GuHCl	(1,2)	96S8
Asp7→Ala	5.5	10	41.4±0.8	GuHCl	(3,4)	96S8
Ser14→Gly	5.5	10	35.1±0.4	GuHCl	(1,2)	96S8
Ser14→Gly	5.5	10	37.7±0.4	GuHCl	(3,4)	96S8
Ser14→Ala	5.5	10	44.4±2.5	GuHCl	(1,2)	96S8
Ser14→Ala	5.5	10	45.6±0.4	GuHCl	(3,4)	96S8
Asn21→Ala	5.5	10	43.9±0.8	GuHCl	(3,4)	96S8
Ala24→Gly	5.5	10	35.6±0.8	GuHCl	(1,2)	96S8
Lys28→Gly	5.5	10	37.3±0.4	GuHCl	(3,4)	96S8

Remarks:
(1) equilibrium unfolding, linear extrapolation, the data refer to the $N_2 \to 2U$ unfolding transition of the homodimer, ΔG is extrapolated to 1 M standard peptide concentration
(2) measured in 100 mM sodium acetate, pH 5.5, at peptide concentration from 20 to 50 µM
(3) kinetics, difference between unfolding and folding activation parameters
(4) measured in 100 mM sodium acetate, pH 5.5, at peptide concentration from 5 to 8 µM

Leucine zipper, heterodimeric coiled coil

Mutant	pH	T	ΔG	Approach	Remarks	Ref
AB	7.2	10	41.0±0.8	ITC	(1,2)	96J1
AB12	7.2	10	42.3±0.8	ITC	(1–3)	96J1
A12B	7.2	10	39.8±0.5	ITC	(1,2,4)	96J1
AB19	7.2	10	38.9±1.2	ITC	(1,2,5)	96J1
A19B	7.2	10	40.1±0.4	ITC	(1,2,6)	96J1
A12B12	7.2	10	36.5±0.4	ITC	(1,2,7)	96J1
AB	7.2	20	44.1±0.5	ITC	(1,2)	96J1
AB12	7.2	20	43.6±0.4	ITC	(1–3)	96J1
A12B	7.2	20	39.7±0.5	ITC	(1,2,4)	96J1
AB19	7.2	20	38.2±0.5	ITC	(1,2,5)	96J1
A19B	7.2	20	40.8±0.4	ITC	(1,2,6)	96J1
A12B12	7.2	20	35.9±0.4	ITC	(1,2,7)	96J1
A19B19	7.2	20	30.3±0.4	ITC	(1,2,8)	96J1
A12B19	7.2	20	31.2±0.4	ITC	(1,2,9)	96J1
A19B12	7.2	20	32.4±0.2	ITC	(1,2,10)	96J1
AB	7.2	30	44.7±0.4	ITC	(1,2)	96J1
AB12	7.2	30	44.0±0.4	ITC	(1–3)	96J1
A12B	7.2	30	40.7±0.4	ITC	(1,2,4)	96J1
AB19	7.2	30	37.9±0.4	ITC	(1,2,5)	96J1
A19B	7.2	30	38.9±0.2	ITC	(1,2,6)	96J1
A12B12	7.2	30	35.3±1.3	ITC	(1,2,7)	96J1
AB	7.2	40	45.3±0.4	ITC	(1,2)	96J1
AB12	7.2	40	42.5±0.4	ITC	(1–3)	96J1
A12B	7.2	40	37.7±0.4	ITC	(1,2,4)	96J1
AB19	7.2	40	35.9±0.4	ITC	(1,2,5)	96J1
A19B	7.2	40	36.8±0.3	ITC	(1,2,6)	96J1
A12B12	7.2	40	33.6±0.4	ITC	(1,2,7)	96J1

Remarks:
(1) heterodimer composed of the acidic peptide (A):
     Ac-Glu-Tyr-Gln-Ala-Leu-Glu-Lys-Glu-Val-Ala-Gln-(Leu/Ala)-Glu-Ala-Glu-Asn-Gln-Ala-(Leu/Ala)-
     Glu-Lys-Glu-Val-Ala-Gln-Leu-Glu-His-Glu-Gly-amide, and the basic peptide (B): Ac-Glu-Tyr-Gln-
     Ala-Leu-Lys-Lys-Lys-Val-Ala-Gln-(Leu/Ala)-Lys-Ala-Lys-Asn-Gln-Ala-(Leu/Ala)-Lys-Lys-Lys-Val-
     Ala-Gln-Leu-Lys-His-Lys-Gly-amide
(2) buffer: 10 mM sodium phosphate
(3) peptide B Leu12→Ala
(4) peptide A Leu12→Ala
(5) peptide B Leu19→Ala
(6) peptide A Leu19→Ala
(7) double mutant, peptides A and B Leu12→Ala
(8) double mutant, peptides A and B Leu19→Ala
(9) double mutant, peptide A Leu12→Ala and B Leu19→Ala
(10) double mutant, peptide A Leu12→Ala and B Leu19→Ala

Lipase

Lipase from *Aspergillus* and *Rhizopus*

Species	pH	T	ΔG	Approach	Remarks	Ref
aspergillus	7		46	heat	(1)	82T3
rhizopus	7		16.8	heat	(1)	82T3

Remark:
(1) transition monitored by CD

Lipoprotein, see apolipoprotein

Luciferase

Luciferase, bacterial, recombinant

Transition	pH	T	ΔG	Approach	Remarks	Ref
trans. (1)	7.0	18	18.9±1.3	urea	(1,2,4)	93C6
trans. (2)	7.0	18	84.2±0.8	urea	(1,3,4)	93C6
trans. (1)	7.0	18	18.4±3.3	urea	(1,2,5)	93C6
effect of Tween 20:						
trans. (1)	7.0	18	18.3±1.9	urea	(1,2,6,7)	93C6
trans. (2)	7.0	18	83.3±0.6	urea	(1,3,6,7)	93C6
trans. (1)	7.0	18	21.0±4.6	urea	(1,2,6,8)	93C6
trans. (2)	7.0	18	80.3±6.7	urea	(1,3,6,8)	93C6

Remarks:
(1) linear extrapolation to zero denaturant concentration by a method that includes the pre- and postdenaturational baselines for a non-linear regression of the data
(2) transition of the active herterodimer into the inactive heterodimer
(3) transition of the inactive heterodimer into unfolded subunits
(4) mean of 11 measurements at protein conc. 5-25 µg/ml monitored by fluorescence at 280 and 295 nm, and CD at 222 nm
(5) monitored by enzymatic activity
(6) mean of 3 measurements monitored by fluorescence at 280 and 295 nm, and CD at 222 nm
(7) measured in the presence of TWEEN 20
(8) measured in the absence of TWEEN 20

Lysozyme HEW = HEN Egg White/chicken lysozyme

The data entries are arranged as follows,
a) data obtained by various approaches,
b) wild-type and mutants,
c) modified lysozyme,
d) data obtained in the presence of solutes

a) Data obtained by various approaches

HEN Egg White lysozyme, ΔG determined by various approaches

pH	T	ΔG	Approach	Remarks	Ref
1 (<1)	25	37.7	GuHCl	(2d)	74P1
2	25	24.3	DSC	(4)	79V
2.9	25	29.7	GuHCl	(2a)	75P
2.9	25	28.9	urea	(2a)	75P
2.9	25	38.9	GuHCl	(2b)	79P1
2.9	25	50.6	urea	(2c)	75P
2.9	25	44.8	GuHCl	(2d)	79P1
2.9	25	24.3	GuHCl	(1)	74G1
2.9	25	24.3	urea	(1)	74G1
2.9	25	27.2	urea	(3)	74G1
2.9	25	38.1	GuHCl	(3)	74G1
4	60	17.6	DSC		76P2
4	60	18	acid denaturation		76P2
5.4	20	31	hydrogen exchange		72N

HEN Egg White lysozyme, ΔG determined by various approaches (continued)

pH	T	ΔG	Approach	Remarks	Ref
5.5	25	37.6	GuHCl	(1)	83A1
5.5	25	37.2	GuHCl	(1)	82A
5.5	25	38.1	GuHSCN	(1)	82A
5.5	25	36.8	urea	(1)	82A
7.0	25.0	43.1	GuHCl	(1)	86I
7	25	44.4	GuHCl	(5a)	69A
7	25	48.1	GuHCl	(5b)	69A
7	25	59.4	GuHCl	(5c)	69A
7	25	73.6	GuHCl	(5d)	69A
7.3	25	52.7	GuHCl	(3)	74T1
7	25	60.7±3.3	DSC		74P1
7	25	60.7±3.3	DSC		76P3
4.5–8.70		42 (≥42)	hydrogen exchange		66H
4.5–8.70		42 (≥42)	hydrogen exchange		64H
9	25	54.4	hydrogen exchange		80K
		45.2	hydrogen exchange		80C
	25	40.6	heat, viscosity		64S

Remarks:
(1) linear extrapolation
(2a) denaturant binding model, k = 0.1
(2b) denaturant binding model, k = 0.6
(2c) denaturant binding model, k = 1.0
(2d) denaturant binding model, k = 1.2
(3) transfer model
(4) measurement in aqueous mixtures of methanol, ethanol, and 1-propanol, results are extrapolated to zero alcohol concentration
(5a) denaturant binding model with GuHCl activity, i.e., eq. (11) in Ref. 69A, k = 3
(5b) denaturant binding model with GuHCl activity and water replacement, i.e.,eq. (13) in Ref. 69A, k = 5.1
(5c) denaturant binding model with independent ion binding, i.e., eq. (15) in Ref. 69A, k = 1.2
(5d) denaturant binding model with independent ion binding and water replacement, i.e., eq. (16) in Ref. 69A, k = 7

Lysozyme, HEW

pH	T	ΔG	Approach	Remarks	Ref
0.34	25.0	10.90±0.37	GuHCl	(1)	92A1
0.64	25.0	11.40±0.50	GuHCl	(1)	92A1
2.03	25.0	17.05±0.79	GuHCl	(1)	92A1
6.00	25.0	35.91±3.11	GuHCl	(1)	92A1
7.00	25.0	37.18	GuHCl	(1)	92A1

Remark:
(1) linear extrapolation

Lysozyme, HEW

	pH	T	ΔG	Approach	Remarks	Ref
	3.0	25	24.8±0.3	GuHCl	(1)	94A1
	3.0	25	24.8±0.4	urea	(1)	94A1
GuHCl-induced unfolding in the presence of urea:						
1.97 M urea	3.0	25	19.2±2.0	GuHCl	(1,2)	94A1
4.36 M urea	3.0	25	3.9±0.2	GuHCl	(1,2)	94A1

Remark:
(1) linear extrapolation
(2) the results support the linear extrapolation procedure

Lysozyme, HEW, intermediate states

Transition	pH	T	ΔG	Approach	Remarks	Ref
N→I	2.9	20	23.8±1.3	urea	(1–3)	96C10
N→U	2.9	20	48.5±2.1	urea	(1–3)	96C10
N→U	2.9	20	33.8±0.8	urea	(1,2,4)	96C10

Remarks:
(1) linear extrapolation to zero denaturant concentration by a method that includes the pre- and postdenaturational baselines for a non-linear regression of the data
(2) transition monitored by CD at 222 nm and X-ray scattering
(3) analyzed by two two-state transitions
(4) analyzed by two- and three-state transitions

b) Wild-type and mutants

Avian lysozymes and mutants in chicken lysozyme that have relevance for the avian lysozymes

Mutant	pH	T_{trs}	Δ(ΔG)	Approach	Remarks	Ref
chicken, wild-type	6.4	74.0	0	heat	(1–4)	95S7
Bob. quail	6.4	73.2		heat		95S7
Cal. quail	6.4	72.1		heat		95S7
Jpn. quail	6.4	71.8		heat		95S7
Turkey	6.4	74.7		heat		95S7
Rn. pheasant	6.4	76.8		heat		95S7
S. pheasant	6.4	77.0		heat		95S7
G. pheasant	6.4	74.2		heat		95S7
Re. pheasant	6.4	76.2		heat		95S7
Phe3→Tyr	6.4	72.8	−1.9	heat	(1–3)	95S7
His15→Leu	6.4	76.0	+3.2	heat	(1–3)	95S7
Asn19→Lys	6.4	71.2	−4.4	heat	(1–3)	95S7
Arg21→Gln	6.4	73.6	−0.6	heat	(1–3)	95S7
Phe34→Tyr	6.4	74.5	+0.8	heat	(1–3)	95S7
Arg68→Lys	6.4	73.9	−0.2	heat	(1–3)	95S7
Arg73→Lys	6.4	74.6	+1.0	heat	(1–3)	95S7
Asn77→His	6.4	73.0	−1.6	heat	(1–3)	95S7

Avian lysozymes and mutants in chicken lysozyme that have relevance for the avian lysozymes (continued)

Mutant	pH	T_{trs}	$\Delta(\Delta G)$	Approach	Remarks	Ref
Asp101→Gly	6.4	75.2	+1.9	heat	(1–3)	95S7
Gly102→Arg	6.4	75.0	+1.6	heat	(1–3)	95S7
Gly102→Val	6.4	73.9	–0.2	heat	(1–3)	95S7
Arg114→His	6.4	75.8	+2.8	heat	(1–3)	95S7
Gln121→His	6.4	72.8	–1.9	heat	(1–3)	95S7

Remarks:
(1) $\Delta(\Delta G)$ was determined from $\Delta(\Delta G) = \Delta T \times \Delta S$, where ΔS of the wild-type is 1.586 kJ/mol/K and T_{trs} = 74°C, the experimental error in T_{trs} was ±0.3°C
(2) measured in 66 mM potassium phosphate at pH 6.4
(3) the transition was monitored by optical absorption at 292 nm
(4) chicken, wild-type is the reference protein

Chicken lysozyme, mutants concerning position 31 (wild-type Ala31)

Mutant	pH	T_{trs}	$\Delta(\Delta G)$	Approach	Remarks	Ref
wild-type	6.4	74.0	0.0	heat	(1–3)	95S7
Ala31→Val	6.4	77.1	5.0	heat	(1–3)	95S7
Ala31→Ile	6.4	77.6	5.9	heat	(1–3)	95S7
Ala31→Leu	6.4	78.7	7.5	heat	(1–3)	95S7

Remarks:
(1) $\Delta(\Delta G)$ was determined from $\Delta(\Delta G) = \Delta T \times \Delta S$, where ΔS of the wild-type is 1.586 kJ/mol/K and T_{trs} = 74°C, the experimental error in T_{trs} was ±0.3°C
(2) measured in 66 mM potassium phosphate at pH 6.4
(3) the transition was monitored by optical absorption at 292 nm

Lysozyme from chicken, wild-type and mutants

Mutant	pH	T	ΔG	Approach	Remarks	Ref
wild-type	3.5	35	29.4	GuHCl	(1,2)	92I2
wild-type	4.0	35	35.3	GuHCl	(1,2)	92I2
wild-type	4.5	35	40.0	GuHCl	(1,2)	92I2
wild-type	5.0	35	39.3	GuHCl	(1,2)	92I2
wild-type	5.5	35	39.1	GuHCl	(1,2)	92I2
Glu35→Gln	3.5	35	27.2	GuHCl	(1,2)	92I2
Glu35→Gln	4.0	35	33.3	GuHCl	(1,2)	92I2
Glu35→Gln	4.5	35	38.2	GuHCl	(1,2)	92I2
Glu35→Gln	5.0	35	40.0	GuHCl	(1,2)	92I2
Glu35→Gln	5.5	35	41.1	GuHCl	(1,2)	92I2
Glu35→Ala	3.5	35	35.5	GuHCl	(1,2)	92I2
Glu35→Ala	4.0	35	40.1	GuHCl	(1,2)	92I2
Glu35→Ala	4.5	35	43.6	GuHCl	(1,2)	92I2
Glu35→Ala	5.0	35	45.8	GuHCl	(1,2)	92I2
Glu35→Ala	5.5	35	46.5	GuHCl	(1,2)	92I2
Glu35→His	3.5	35	15.0	GuHCl	(1,2)	92I2
Glu35→His	4.0	35	22.2	GuHCl	(1,2)	92I2
Glu35→His	4.5	35	28.5	GuHCl	(1,2)	92I2
Glu35→His	5.0	35	32.8	GuHCl	(1,2)	92I2
Glu35→His	5.5	35	36.6	GuHCl	(1,2)	92I2

Remarks:
(1) linear extrapolation
(2) 0.1 M sodium acetate buffer

Table 1. Gibbs Energy Change – Molar Values

Lysozyme from chicken, wild-type and mutant

Mutant	pH	T	Δ(ΔG)	Approach	Remarks	Ref
wild-type	3.5	73.0	0.0	DSC		92I2
Glu35→Ala	3.5	75.6	5.8	DSC	(1)	92I2
wild-type	5.5	76.4	0.0	DSC		92I2
Glu35→Ala	5.5	81.7	10.0	DSC	(2)	92I2

Remarks:
(1) Δ(ΔG) refers to T_{trs} = 73.0°C of wild-type
(2) Δ(ΔG) refers to T_{trs} = 76.4°C of wild-type

HEN Egg White lysozyme, wild-type and mutants

Mutant	pH	T	ΔG	Approach	Remarks	Ref
wild-type	5.5	35	43.5	GuHCl	(1)	95T10
Gly49→Ala	5.5	35	35.5	GuHCl	(1)	95T10
Gly67→Ala	5.5	35	38.2	GuHCl	(1)	95T10
double mutant (Gly102→Ala and Gly104→Ala)						
	5.5	35	35.1	GuHCl	(1)	95T10
45–49hu	5.5	35	37.9	GuHCl	(1,2)	95T10
multiple	5.5	35	24.6	GuHCl	(1,3)	95T10

Remarks:
(1) linear extrapolation
(2) five residues in position 45–49 (Arg-Asn-Thr-Asp-Gly) were replaced by six (Tyr-Asn-Ala-Gly-Asp-Arg) in human lysozyme
(3) multiple mutant 45–49hu, Gly67→Ala and (Gly102→Ala and Gly104→Ala)

Chicken lysozyme, wild-type and mutants

Mutant	pH	T	ΔG	Approach	Remarks	Ref
wild-type	6.4	25	42.7±2.9	GuHCl	(1,2)	95S6
Ile55→Thr	6.4	25	28.9±3.3	GuHCl	(1,2)	95S6
hs mutant	6.4	25	55.6±3.3	GuHCl	(1–3)	95S7

Remarks:
(1) linear extrapolation to zero denaturant concentration by a method that includes the pre- and postdenaturational baselines for a non-linear regression of the data
(2) measured by fluorescence intensity at 360 nm
(3) hs mutant contains the following six mutations: His15→Leu, Ala31→Val, Ile55→Leu, Ser91→Thr, Asp101→Ser and Arg114→His

Lysozyme, HEW, recombinant

Mutant	pH	T	ΔG	Approach	Remarks	Ref
wild-type	5.5	35	42.3	GuHCl	(1,2)	93U
Asp101→Pro	5.5	35	32.2	GuHCl	(1)	93U
Gly102→Pro	5.5	35	41.8	GuHCl	(1)	93U
double mutant (Asp101→Gly and Gly102→Pro)						
	5.5	35	46.0	GuHCl	(1)	93U

Remarks:
(1) linear extrapolation
(2) Pos. Asp101 and Gly102 unaltered

Chicken lysozyme, mutants concerning position 101 (wild-type Asp101)

Mutant	pH	T_{trs}	$\Delta(\Delta G)$	Approach	Remarks	Ref
wild-type	6.4	74.0	0.0	heat	(1–3)	95S7
Asp101→Ala	6.4	76.0	3.2	heat	(1–3)	95S7
Asp101→Arg	6.4	74.7	1.1	heat	(1–3)	95S7
Asp101→Asn	6.4	74.1	0.2	heat	(1–3)	95S7
Asp101→Gln	6.4	73.8	−0.3	heat	(1–3)	95S7
Asp101→Glu	6.4	74.0	0.0	heat	(1–3)	95S7
Asp101→Gly	6.4	75.2	1.9	heat	(1–3)	95S7
Asp101→Lys	6.4	74.5	0.8	heat	(1–3)	95S7
Asp101→Phe	6.4	75.9	3.0	heat	(1–3)	95S7
Asp101→Ser	6.4	76.3	3.6	heat	(1–3)	95S7

Remarks:
(1) $\Delta(\Delta G)$ was determined from $\Delta(\Delta G) = \Delta T \times \Delta S$, where ΔS of the wild-type is 1.586 kJ/mol/K and $T_{trs} = 74°C$, the experimental error in T_{trs} was ±0.3°C
(2) measured in 66 mM potassium phosphate at pH 6.4
(3) the transition was monitored by optical absorption at 292 nm

HEN Egg White lysozyme, wild-type and mutants

Mutant	pH	T	ΔG	Approach	Ref
wild-type	3.0	20	52.7	heat	92K2
Asn103→Asp	3.0	20	53.6	heat	92K2
Asn106→Asp	3.0	20	53.6	heat	92K2

Lysozyme of chicken, wild-type and mutants

Mutant	pH	T	ΔG	Approach	Remarks	Ref
wild-type	3.5	35	29.4	GuHCl	(1,2)	92I3
wild-type	4.0	35	35.3	GuHCl	(1,2)	92I3
wild-type	4.5	35	40.0	GuHCl	(1,2)	92I3
wild-type	5.0	35	39.3	GuHCl	(1,2)	92I3
wild-type	5.5	35	39.1	GuHCl	(1,2)	92I3
Trp108→Tyr	3.5	35	23.5	GuHCl	(1,3)	92I3
Trp108→Tyr	4.0	35	29.4	GuHCl	(1,3)	92I3
Trp108→Tyr	4.5	35	34.3	GuHCl	(1,3)	92I3
Trp108→Tyr	5.0	35	34.0	GuHCl	(1,3)	92I3
Trp108→Tyr	5.5	35	33.6	GuHCl	(1,3)	92I3
Trp108→Gln	3.5	35	14.4	GuHCl	(1,4)	92I3

Lysozyme of chicken, wild-type and mutants (continued)

Mutant	pH	T	ΔG	Approach	Remarks	Ref
Trp108→Gln	4.0	35	20.2	GuHCl	(1,4)	92I3
Trp108→Gln	4.5	35	23.3	GuHCl	(1,4)	92I3
Trp108→Gln	5.0	35	24.5	GuHCl	(1,4)	92I3
Trp108→Gln	5.5	35	24.7	GuHCl	(1,4)	92I3

Remarks:
(1) linear extrapolation
(2) slope m = 10.84±0.75 kJ/(mol×M_D)
(3) slope m = 13.35±0.88 kJ/(mol×M_D)
(4) slope m = 12.18±1.17 kJ/(mol×M_D)

Chicken lysozyme, core triplet mutants that affect Thr40, Ile55 and Ser91

Mutant	pH	T_{trs}	Δ(ΔG)	Approach	Remarks	Ref
wild-type	5.0	76.4±0.3		heat	(5)	95S6
	6.4	74.0±0.3	0.0	heat	(1–3)	95S6
	6.5	73.9±0.1		heat	(3,4)	90M2
Thr40→Ile	6.4	68.2	−9.2	heat	(1–3)	95S6
Thr40→Ser	6.4	73.3	−1.1	heat	(1–3)	95S6
	6.5	73.0±0.1		heat	(3,4)	90M2
Ile55→Ala	6.4	62.4	−18.4	heat	(1–3)	95S6
Ile55→Leu	6.4	72.8	−1.9	heat	(1–3)	95S6
Ile55→Met	6.4	68.0	−9.5	heat	(1–3)	95S6
Ile55→Phe	6.4	67.5	−10.3	heat	(1–3)	95S6
Ile55→Thr	5.0	63.3±0.4		heat	(5)	95S6
	6.4	60.9	−20.8	heat	(1,6)	95S6
Ile55→Val	6.4	71.6	−3.8	heat	(1–3)	95S6
	6.5	71.2±0.1		heat	(3,4)	90M2
Ser91→Ala	6.4	73.6	−0.6	heat	(1–3)	95S6
Ser91→Asp	6.4	67.9	−9.7	heat	(1–3)	95S6
Ser91→Thr	6.4	76.6	+4.1	heat	(1–3)	95S6
	6.4	77.5±0.1		heat	(3,4)	90M2
Ser91→Tyr	6.4	65.9	−12.8	heat	(1–3)	95S6
Ser91→Val	6.4	73.8	−0.3	heat	(1–3)	95S6

Double mutants:
(Thr40→Ser and Ile55→Val)

Mutant	pH	T_{trs}	Δ(ΔG)	Approach	Remarks	Ref
	6.4	70.5	−5.6	heat	(1–3)	95S6
	6.5	70.6±0.6		heat	(3,4)	90M2

(Thr40→Ser and Ser91→Ala)

| | 6.4 | 71.7 | −3.6 | heat | (1–3) | 95S6 |

(Thr40→Ser and Ser91→Thr)

| | 6.4 | 75.6 | +2.6 | heat | (1–3) | 95S6 |
| | 6.5 | 75.5±0.1 | | heat | (3,4) | 90M2 |

(Thr40→Ser and Ser91→Val)

| | 6.4 | 72.6 | −2.2 | heat | (1–3) | 95S6 |

(Ile55→Ala and Ser91→Thr)

| | 6.4 | 66.7 | −11.6 | heat | (1–3) | 95S6 |

(Ile55→Leu and Ser91→Thr)

| | 6.4 | 77.3 | +5.2 | heat | (1–3) | 95S6 |

(Ile55→Val and Ser91→Ala)

| | 6.4 | 70.7 | −5.2 | heat | (1–3) | 95S6 |

(Ile55→Val and Ser91→Thr)

| | 6.4 | 74.0 | 0 | heat | (1–3) | 95S6 |
| | 6.5 | 74.5±0.2 | | heat | (3,4) | 90M2 |

Chicken lysozyme, core triplet mutants that affect Thr40, Ile55 and Ser91 (continued)

Mutant	pH	T_{trs}	$\Delta(\Delta G)$	Approach	Remarks	Ref
(Ile55→Val and Ser91→Val)						
	6.4	70.7	−5.2	heat	(1–3)	95S6
Triple mutants:						
(Thr40→Ser, Ile55→Val and Ser91→Ala)						
	6.4	69.6	−7.0	heat	(1–3)	95S6
(Thr40→Ser, Ile55→Val and Ser91→Thr)						
	6.4	73.2	−1.3	heat	(1–3)	95S6
	6.5	73.4±0.1		heat	(3,4)	90M2
hs mutant	6.4	84.5±0.3	+16.7	heat	(1–3,7)	95S7
	5.0	86.4±0.5	+16.7	DSC	(7)	95S7

Remarks:
(1) $\Delta(\Delta G)$ was determined from $\Delta(\Delta G) = \Delta T \times \Delta S$, where ΔS of the wild-type is 1.586 kJ/mol/K and $T_{trs} = 74°C$, the experimental error in T_{trs} was ±0.3°C
(2) measured in 66 mM potassium phosphate at pH 6.4
(3) the transition was monitored by optical absorption at 292 nm
(4) measured in 100 mM potassium phosphate, pH 6.5
(5) measured by fluorescence intensity in 0.1 M sodium acetate, pH 5.0
(6) measured at pH 5.0, T_{trs} at pH 6.4 is an extrapolated value, see Ref. 95S6
(7) hs mutant contains the following six mutations: His15→Leu, Ala31→Val, Ile55→Leu, Ser91→Thr, Asp101→Ser and Arg114→His

HEN Egg White lysozyme, wild-type and mutants

Mutant	pH	T	ΔG	Approach	Remarks	Ref
wild-type	3.0	20	52.7	heat		92K2
Asn103→Asp	3.0	20	53.6	heat		92K2
Asn106→Asp	3.0	20	53.6	heat		92K2
native	9.5	20	56	heat	(1)	95K5
wild-type	9.5	20	53	heat	(1)	95K5
Lys13→Asp	9.5	20	25	heat	(1)	95K5
Gly49→Asn	9.5	20	49	heat	(1)	95K5
Pro70→Asn	9.5	20	44	heat	(1)	95K5
Cys94→Ala	9.5	20	33	heat	(1)	95K5
Asn103→Asp	9.5	20	54	heat	(1)	95K5
Met105→Thr	9.5	20	52	heat	(1)	95K5

Remark:
(1) ΔG was taken from Fig. 5 in Ref. 95K5

Table 1. Gibbs Energy Change – Molar Values

HEN Lysozyme mutants containing Gly-Pro and Pro-Gly sequences

Mutant	pH	T	ΔG	Approach	Remarks	Ref
wild-type	5.5	35	41.0	GuHCl	(1–3)	95M18
Pro47-Gly47'	5.5	35	35.6	GuHCl	(1,2)	95M18
Gly47-Pro47'	5.5	35	31.0	GuHCl	(1,2)	95M18
Pro70-Gly71	5.5	35	41.0	GuHCl	(1–3)	95M18
Gly70-Pro70'	5.5	35	36.8	GuHCl	(1,2)	95M18
Pro117-Gly118	5.5	35	30.1	GuHCl	(1,2)	95M18
Gly117-Pro118	5.5	35	32.6	GuHCl	(1,2)	95M18
Pro121-Gly122	5.5	35	29.3	GuHCl	(1,2)	95M18
Gly121-Gly122	5.5	35	33.9	GuHCl	(1,2)	95M18

Remarks:
(1) linear extrapolation
(2) measured in 0.1 M acetate buffer, pH 5.5, at 35°C by fluorescence intensity
(3) wild-type = Pro70, Gly71

c) Modified lysozyme

HEN Egg White lysozyme, recombinant wild-type and mutants with N-glycosylation site

Mutant	pH	T	ΔG	Approach	Remarks	Ref
wild-type	5.5	35	42.3±0.4	GuHCl	(1)	96U1
Asp48→Asn	5.5	35	39.3±0.5	GuHCl	(1)	96U1
Asp48→Asn-HM	5.5	35	41.0±0.5	GuHCl	(1,2)	96U1
Asp87→Asn	5.5	35	38.1±0.6	GuHCl	(1)	96U1
Asp87→Asn-HM	5.5	35	44.8±0.5	GuHCl	(1,2)	96U1

Remarks:
(1) linear extrapolation
(2) AsnXX-HM = Asn in position XX is modified with a sugar of high mannose type

HEN Egg White lysozyme, guanidinated

pH	T	ΔG	Approach	Ref
		45.2	hydrogen exchange	80C

Lysozyme, crosslinked between Glu35 and Trp108

pH	T_{trs}	ΔT	Δ(ΔG)	Approach	Remark	Ref
3.7	40–80	18	28	DSC	(1)	89S6

Remark:
(1) Δ(ΔG) is the difference in ΔG between crosslinked and native lysozyme. The data were obtained in the presence of various concentrations of 1-propanol

Lysozyme HEW, modified

Form	pH	T	ΔG	Approach	Remarks	Ref
unmodified	8.0	40	37.7	GuHCl	(1)	93Y1
101-β	8.0	40	33.9	GuHCl	(1,2)	93Y1
(13–129)cl.	8.0	40	36.8	GuHCl	(1,3)	93Y1

Remarks:
(1) linear extrapolation
(2) the α-aspartylglycyl sequence at Asp101-Gly102 is converted to β
(3) the ε-amino group of Lys13 and the α-carboxylic acid group of Leu129 are cross-linked in an amide bond

Lysozyme HEW, native and modified forms

Form	pH	T	ΔG	Approach	Remarks	Ref
native	7.5	25	20.1±0.4	heat		9001
NFK-Trp62	7.5	25	20.5±5.0	heat	(1)	9001
Kyn-Trp62	7.5	25	20.5±7.5	heat	(2)	9001

Remarks:
(1) modified by ozone oxidation of Trp187 to the N-formylkynurenine derivative
(2) modified by ozone oxidation of Trp187 to the kynurenine derivative

Lysozyme HEW, crosslinked

pH	T	ΔG	Δ(ΔG)	Approach	Remarks	Ref
3	25	52.5*	9.2	DSC	(1–3)	91U1
3	78.1	0	17.2	DSC	(1–3)	91U1

Remarks:
(1) Crosslinked between Aps101 and Trp62 by -CH_2-CO-CH_2-S-CH_2-CH_2-NH-
(2) Protein concentration 40 mg/ml
(3)The data refer to lysoyzme HEW with T_{trs} = 71°C , and ΔH = 485 kJ/mol (Y. Fujita & Y. Noda, Bull. Chem. Soc. Jpn., 56 (1983) 233) and ΔCp = 6.7 kJ/K/mol (Ref. 79P4)

HEN Lysozyme, modified, three-disulfide derivative

Form	pH	T	ΔG	Approach	Remarks	Ref
unmodified	3.8	25.0	54	DSC	(1)	92C4
modified	3.8	25.0	23	DSC	(1,2)	92C4

Remarks:
(1) ΔG was taken from Fig. 4 of Ref. 92C4
(2) Cys6 and Cys127 were specifically reduced and carboxymethylated

HEN Egg White lysozyme, modified, alkylated

Form	pH	T	ΔG	Approach	Remark	Ref
native	3.0	25	46	DSC		92F5
ethylated	3.0	25	38	DSC		92F5
butylated	3.0	25	37	DSC		92F5
hexylated	3.0	25	27	DSC		92F5
2,2-DMP	3.0	25	35	DSC	(1)	92F5
benzylated	3.0	25	38	DSC		92F5

Remark:
(1) 2,2-DMP: 2,2-dimethylpropylated lysozyme

d) Data obtained in the presence of solutes

HEN Egg White lysozyme, in the presence of glycerol and sorbitol

	pH	T	ΔG	Approach	Remarks	Ref
	2	25	19.1	GuHCl	(1,2a,2b)	90G1

Remarks:
(1) linear extrapolation
(2a) increase in ΔG of 2.2 kJ per mole glycerol
(2b) increase in ΔG of 6.5 kJ per mole sorbitol

HEN Egg White lysozyme, in the presence of various added solutes

Added Solute	pH	T	ΔG	Approach	Remarks	Ref
none	3.0	25	41.0	DSC		96L3
none	3.0	40	30.4	DSC		96L3
none	3.0	55	14.7	DSC		96L3
none	3.0	70	−5.9	DSC		96L3
none	4.0	25	51.5	DSC		96L3
none	4.0	40	41.4	DSC		96L3
none	4.0	55	26.2	DSC		96L3
none	4.0	70	6.2	DSC		96L3
0.5 M sucrose	3.0	25	27.4	DSC		96L3
0.5 M sucrose	3.0	40	26.4	DSC		96L3
0.5 M sucrose	3.0	55	17.2	DSC		96L3
0.5 M sucrose	3.0	70	0.1	DSC		96L3
0.5 M sucrose	4.0	25	36.6	DSC		96L3
0.5 M sucrose	4.0	40	36.0	DSC		96L3
0.5 M sucrose	4.0	55	27.2	DSC		96L3
0.5 M sucrose	4.0	70	10.6	DSC		96L3
1.0 M sucrose	4.0	25	27.8	DSC		96L3
1.0 M sucrose	4.0	40	30.5	DSC		96L3
1.0 M sucrose	4.0	55	25.4	DSC		96L3
1.0 M sucrose	4.0	70	12.8	DSC		96L3
1.0 M sucrose	5.0	25	29.3	DSC		96L3
1.0 M sucrose	5.0	40	32.1	DSC		96L3
1.0 M sucrose	5.0	55	27.1	DSC		96L3
1.0 M sucrose	5.0	70	14.6	DSC		96L3
10% glycerol	3.0	25	36.5	DSC	(1)	96L3
10% glycerol	3.0	40	30.2	DSC	(1)	96L3
10% glycerol	3.0	55	16.9	DSC	(1)	96L3
10% glycerol	3.0	70	−3.2	DSC	(1)	96L3

HEN Egg White lysozyme, in the presence of various added solutes (continued)

Added Solute	pH	T	ΔG	Approach	Remarks	Ref
10% glycerol	4.0	25	46.8	DSC	(1)	96L3
10% glycerol	4.0	40	41.0	DSC	(1)	96L3
10% glycerol	4.0	55	28.2	DSC	(1)	96L3
10% glycerol	4.0	70	8.7	DSC	(1)	96L3
10% glycerol	5.0	25	48.5	DSC	(1)	96L3
10% glycerol	5.0	40	42.8	DSC	(1)	96L3
10% glycerol	5.0	55	30.0	DSC	(1)	96L3
10% glycerol	5.0	70	10.5	DSC	(1)	96L3
1.0 M GuHCl	3.0	10	20.8	DSC		96L3
1.0 M GuHCl	3.0	25	12.0	DSC		96L3
1.0 M GuHCl	3.0	40	−2.1	DSC		96L3
1.0 M GuHCl	3.0	55	−21.6	DSC		96L3
1.0 M GuHCl	4.0	10	32.6	DSC		96L3
1.0 M GuHCl	4.0	25	24.4	DSC		96L3
1.0 M GuHCl	4.0	40	10.8	DSC		96L3
1.0 M GuHCl	4.0	55	−8.1	DSC		96L3
1.0 M GuHCl	5.0	10	35.0	DSC		96L3
1.0 M GuHCl	5.0	25	27.0	DSC		96L3
1.0 M GuHCl	5.0	40	13.5	DSC		96L3
1.0 M GuHCl	5.0	55	−5.3	DSC		96L3
1.0 M GuHCl	6.0	10	34.9	DSC		96L3
1.0 M GuHCl	6.0	25	26.9	DSC		96L3
1.0 M GuHCl	6.0	40	13.3	DSC		96L3
1.0 M GuHCl	6.0	55	−5.4	DSC		96L3
2.0 M GuHCl	4.0	10	21.9	DSC		96L3
2.0 M GuHCl	4.0	25	13.7	DSC		96L3
2.0 M GuHCl	4.0	40	0.4	DSC		96L3
2.0 M GuHCl	4.0	55	−17.8	DSC		96L3
2.0 M GuHCl	5.0	10	24.3	DSC		96L3
2.0 M GuHCl	5.0	25	16.2	DSC		96L3
2.0 M GuHCl	5.0	40	3.1	DSC		96L3
2.0 M GuHCl	5.0	55	−15.0	DSC		96L3
2.0 M GuHCl	6.0	10	24.2	DSC		96L3
2.0 M GuHCl	6.0	25	16.1	DSC		96L3
2.0 M GuHCl	6.0	40	2.9	DSC		96L3
2.0 M GuHCl	6.0	55	−15.1	DSC		96L3

Remark:
(1) 10% (v/v) glycerol

Lysozyme, Equine

Equine lysozyme, holo and apo protein

Transition	pH	T	ΔG	Approach	Remarks	Ref
Holo equine lysozyme in the presence of 1.5 mM Ca^{2+}:						
trans. (1)	4.5	25	12	DSC	(1)	95G10
trans. (2)	4.5	25	8.5	DSC	(1)	95G10
Apo equine lysozyme:						
trans. (1)	4.5	25	6	DSC	(1)	95G10
trans. (2)	4.5	25	8.5	DSC	(1)	95G10

Remark:
(1) ΔG was taken from Fig. 8 in Ref. 95G10

Lysozyme, Human

Human lysozyme, wild-type and mutant with non-native disulfide bond

Mutant	pH	T	ΔG	Approach	Remarks	Ref
wild-type	3.0	25	n.d.	GuHCl	(1)	92K1
double mutant (Trp64→Cys and Cys65→Ala) with non-native disulfide bond Cys64-Cys81						
	3.0	25	n.d.	GuHCl	(2)	92K1

Remarks:
(1) GuHCl transition midpoint $m_{1/2}$ = 2.9 M
(2) GuHCl transition midpoint $m_{1/2}$ = 2.7 M

Human lysozyme, wild-type and calcium binding mutant in the presence and in the absence of Ca^{2+}

Mutant/Form	T	ΔG	Approach	Remarks	Ref
wild-type +Ca^{2+}	85	−7.9	DSC	(1)	92K12
Double mutant (Gln86→Asp and Ala92→Asp):					
apo form	85	−13.8	DSC	(2)	92K12
holo form	85	0.0	DSC	(3)	92K12

Remarks:
(1) based on ΔH = 602 kJ/mol and ΔCp = 6.44 kJ/mol/K extrapolated to 85°C, T_{trs} was 80.3°C
(2) based on ΔH = 602 kJ/mol and ΔCp = 6.44 kJ/mol/K extrapolated to 85°C, T_{trs} was 76.5°C
(3) based on ΔH = 573 kJ/mol and ΔCp = 4.06 kJ/mol/K extrapolated to 85°C, T_{trs} was 85.2°C

Human lysozyme, wild-type and mutants

Mutant	pH	T	ΔG	ΔT	Approach	Remarks	Ref
wild-type	2.80	68.8	0.0	0.0	DSC	(1)	92H3
Pro71→Gly	2.80	68.8	−6.7	−4.7	DSC	(1)	92H3
Pro103→Gly	2.80	68.8	−0.4	−0.1	DSC	(1)	92H3
double mutant (Pro71→Gly and Pro103-Gly)							
	2.80	68.8	−6.7	−4.5	DSC	(1)	92H3
Asp91→Pro	2.80	68.8	−1.7	−1.1	DSC	(1)	92H3
Ala47→Pro	2.80	68.8	0.4	0.3	DSC	(1)	92H3
Val110→Pro	2.80	68.8	2.1	1.2	DSC	(1)	92H3

Remark:
(1) ΔT is the shift in thermal transition temperature compared with the wild-type protein

Human lysozyme, wild-type and mutants

Mutant	pH	T	ΔG	Approach	Remark	Ref
wild-type	2.1	57	19.2	DSC	(1)	92K11
Cys77→Ala	3.0	57	0	DSC		92K11
double mutant (Cys77→Ala and Cys95→Ala)						
	3.0	57	0	DSC		92K11

Remark:
(1) the pH value was taken from Fig. 2 in Ref. 92K11

Human lysozyme, wild-type and mutants

Mutant	pH	T_{trs}	ΔT	$\Delta(\Delta G)$	Approach	Remarks	Ref
wild-type	3.0	69.9	0.0	0.0	heat	(1)	90F
Val110→Pro	3.0	71.0	1.1	2.3	heat	(1,2)	90F
double mutant (Cys77→Ala and Cys95→Ala)							
	3.0	55.6	−14.3	−21.0	heat	(1,2)	90F

Remarks:
(1) the data were obtained by multidimensional spectroscopy
(2) the data refer to T_{trs} of the wild-type

Human lysozyme, wild-type and Ile→Val mutants

Mutant	pH	T_{trs}	ΔT	$\Delta(\Delta G)$	Approach	Remarks	Ref
wild-type	2.7	64.9±0.5	0.0	0.0	DSC	(1)	95T1
Ile23→Val	2.7	63.8±0.4	−1.1	−1.5±0.4	DSC	(1)	95T1
Ile56→Val	2.7	61.3±0.3	−3.6	−5.0±0.4	DSC	(1)	95T1
Ile59→Val	2.7	61.5±0.4	−3.4	−4.6±0.4	DSC	(1)	95T1
Ile89→Val	2.7	63.5±0.6	−1.4	−2.0±0.8	DSC	(1)	95T1
Ile106→Val	2.7	62.7±0.3	−2.2	−3.0±0.4	DSC	(1)	95T1

Remark:
(1) T_{trs} = 64.9°C of wild-type is the reference temperature for $\Delta(\Delta G)$ of the mutants

Human lysozyme and the amyloidogenic mutant (Ile56→Thr)

Mutant	pH	T	ΔG	Approach	Remarks	Ref
wild-type	3.0	10	49.4	GuHCl	(1,5)	92T1
wild-type	4.0	10	58.5	GuHCl	(1)	96F6
wild-type	4.0	10	64.9	DSC	(2)	96F6
wild-type	4.0	10	58.6	GuHCl	(1,4)	92T1
wild-type	4.0	10	57.3	GuHCl	(1,5)	92T1
Ile56→Thr	4.0	10	30.6	GuHCl	(1)	96F6
Ile56→Thr	4.0	10	46.5	DSC	(3)	96F6
double mutant (Cys77→Ala and Cys95→Ala)						
	3.0	10	25.5	GuHCl	(1,5)	92T1
	4.0	10	34.3	GuHCl	(1,4)	92T1
	4.0	10	33.5	GuHCl	(1,5)	92T1

Remarks:
(1) linear extrapolation
(2) using T_{trs} = 79.5°C, ΔH = 575 kJ/mol and ΔCp = 6.6 kJ/mol/K from Ref. 92K12
(3) using T_{trs} = 67.0°C, ΔH = 435 kJ/mol and ΔCp = 5.2 kJ/mol/K
(4) transition monitored by CD at 222 nm
(5) transition monitored by fluorescence

Lysozyme Phage T4

The data entries are arranged as follows:

a) data obtained by various approaches,
b) data obtained predominantly on single mutations,
c) data on single and multiple mutations,
d) proline mutations,
e) insertion mutants,
f) mutations affecting disulfide bonds

a) Data obtained by various approaches

Lysozyme phage T4, for studies on lysozyme phage T4 stability along with structural data – see also Ref. 95M8

Lysozyme, phage T4, wild-type and mutants, ΔG values

Mutant	pH	T	ΔG	Approach	Remarks	Ref
wild-type	2	0	30	heat	(1)	77E
wild-type	2	25	14.4*	DSC		89K4
Arg96→His	2	25	−1.7*	DSC		89K4
Trp138→Tyr	2	0	18	heat	(3)	77E
multiple mutant (Trp126,138,158→Tyr)						
	2	0	18	heat	(1)	77E
wild-type	3	25	34.2*	DSC		89K4
wild-type	3	25	15*	heat, optical method		84H1
wild-type	3	49	8.8	heat		80S1
wild-type	3	49	6.7	GuHCl	(2)	80S1
Arg96→His	3	25	12.7*	heat, optical method		84H1
Arg96→His	3	25	20.9*	DSC		89K4
Met102→Val	3	25	11.2*	heat, optical method		84H1
Glu128→Lys	3	25	15.2*	heat, optical method		84H1
Ala146→Thr	3	25	8.2*	heat, optical method		84H1
double mutant (Ile3→Cys and Cys54→Thr, with disulfide bond Cys3-Cys97):						
	5	−9.5	41	GuHCl, kinetics		89C
	5	−2	58	GuHCl, kinetics		89C
	5	5	55.4±4	GuHCl	(2)	89C
	5	12	61.4±1.9	GuHCl	(2)	89C
	5	15	61.5	GuHCl	(2)	89C
	5	20	55.2	GuHCl	(2)	89C
	5	25	53.8	GuHCl	(2)	89C
wild-type	–	25	25	heat		80S1
Arg96→His	–	25	8	heat	(3)	80S1
Met102→Val	–	25	7.5	heat	(3)	80S1
(Glu128→Lys or Ala)		25	17	heat	(3)	80S1
Ala146→Thr	–	25	10	heat	(3)	80S1

Remarks:
(1) optical methods, data are taken from ΔG versus T in Fig. 6 of Ref. 77E
(2) linear extrapolation
(3) data are taken from ΔG versus T in Fig. 1 of Ref. 80S1

Lysozyme phage T4, wild-type and mutants

Mutant	pH	T	ΔG	pH	T_{trs}	Approach	Ref
wild-type	5.7	12	76.1	2.0	41.9	GuHCl (1)	92C1
Ile3→Ala	5.7	12	65.7	2.0	38.1	GuHCl (1)	92C1
Ile3→Gly	5.7	12	66.5	2.0	34.7	GuHCl (1)	92C1
Ile3→Phe	5.7	12	70.3	2.0	37.9	GuHCl (1)	92C1
Ile3→Val	5.7	12	75.7	2.0	39.8	GuHCl (1)	92C1
Pro37→Ala	5.7	12	74.9	2.0	41.1	GuHCl (1)	92C1
Leu39→Pro	5.7	12	64.4	2.0	36.4	GuHCl (1)	92C1
Ala82→Pro	5.7	12	77.8	2.0	42.7	GuHCl (1)	92C1
Pro86→Ala	5.7	12	74.5	2.0	40.6	GuHCl (1)	92C1
Pro143→Ala	5.7	12	72.0	2.0	36.9	GuHCl (1)	92C1
double mutant (Ile3→Cys – Cys97 and Cys54→Thr)							
	5.0	12	61.5	6.5	69	GuHCl (1)	92C1

Remark:
(1) ΔG obtained at 12°C and pH 5.7 by linear extrapolation, T_{trs} was determined at pH 2 in the absence of GuHCl

Lysozyme phage T4, wild-type and mutants, data determined by urea gradient electrophoresis ($C_{1/2}$ = urea concentration at the transition midpoint)

Mutant	pH	T	Δ(ΔG)	$C_{1/2}$	Approach	Remarks	Ref
wild-type	7.0	22	0	6.3	urea	(1,2)	91K5
Met6→Ile	7.0	22	−20	3.9	urea	(1,2)	91K5
Tyr25→Gly	7.0	22	−19	4.0	urea	(1,2)	91K5
Pro37→Ala	7.0	22	0	6.3	urea	(1,2)	91K5
Ser38→Asp	7.0	22	−5	5.6	urea	(1,2)	91K5
Gly51→Asp	7.0	22	−11	5.0	urea	(1,2)	91K5
Cys54→Tyr	7.0	22	−21	3.8	urea	(1,2)	91K5
Ile58→Tyr	7.0	22	−13	4.8	urea	(1,2)	91K5
Gly77→Ala	7.0	22	−5	5.65	urea	(1,2)	91K5
Ala82→Pro	7.0	22	−3	5.9	urea	(1,2)	91K5
Pro86→Gly	7.0	22	−8	5.3	urea	(1,2)	91K5
Leu91→Pro	7.0	22	−34	2.2	urea	(1,2)	91K5
Ala98→Thr	7.0	22	−16	4.4	urea	(1,2)	91K5
Ala98→Val	7.0	22	−21	3.8	urea	(1,2)	91K5
Val103→Ala	7.0	22	−8	5.3	urea	(1,2)	91K5
Gln105→Gly	7.0	22	−13	4.7	urea	(1,2)	91K5
Trp126→Arg	7.0	22	−24	3.4	urea	(1,2)	91K5
Trp138→Tyr	7.0	22	−12	4.9	urea	(1,2)	91K5
Ala146→Ile	7.0	22	−18	4.2	urea	(1,2)	91K5
Ala146→Thr	7.0	22	−12	4.9	urea	(1,2)	91K5
Ala146→Val	7.0	22	−18	4.1	urea	(1,2)	91K5
Val149→Ala	7.0	22	−12	4.9	urea	(1,2)	91K5
Phe153→Cys	7.0	22	−13	4.7	urea	(1,2)	91K5
Gly156→Asp	7.0	22	−17	4.3	urea	(1,2)	91K5
Thr157→Asp	7.0	22	−10	5.1	urea	(1,2)	91K5
Thr157→Phe	7.0	22	−17	4.3	urea	(1,2)	91K5
Ala160→Thr	7.0	22	−19	4.0	urea	(1,2)	91K5
double mutant (Ala98→Val and Thr152→Ser)							
	7.0	22	−23	3.6	urea	(1,2)	91K5
double mutant (Thr155→Ala and Thr157→Ile)							
	7.0	22	−18	4.2	urea	(1,2)	91K5

Remarks:
(1) the urea concentration at half conversion was taken from Fig. 4 in Ref. 91K5
(2) Δ(ΔG) was calculated from the urea dependence of of ΔG, m = −8.4 kJ/(mol×M_D) as given in Ref. 91K5

b) Data obtained predominantly on single mutations

Lysozyme phage T4, substitutions in the substrate binding site

Mutant	pH	ΔT	Δ(ΔG)	Approach	Remarks	Ref
wild-type	5.42	0.0	0.0	heat, v.H.	(1)	95S5
pseudo wild-type	5.42	0.0	0.0	heat, v.H.	(2,3)	95S5
Glu11→Phe	5.42	4.3	7.1	heat, v.H.	(4)	95S5
Glu11→Met	5.42	4.1	6.7	heat, v.H.	(4)	95S5
Glu11→Ala	5.42	2.6	4.6	heat, v.H.	(4)	95S5
Glu11→His	5.42	0.1	0.4	heat, v.H.	(5)	95S5
Glu11→Asn	5.42	−0.6	−0.4	heat, v.H.	(5)	95S5
Asp20→Asn	5.42	3.1	5.4	heat, v.H.	(5)	95S5
Asp20→Thr	5.42	2.2	3.8	heat, v.H.	(5)	95S5
Asp20→Ser	5.42	1.6	2.9	heat, v.H.	(5)	95S5
Asp20→Ala	5.42	−0.8	−1.3	heat, v.H.	(5)	95S5
Gly30→Ala	5.42	0.1	0.4	heat, v.H.	(4)	95S5
Gly30→Phe	5.42	−4.9	−6.3	heat, v.H.	(4)	95S5
Ser117→Val	5.42	5.1	8.4	heat, v.H.	(4)	95S5
Ser117→Ile	5.42	4.2	7.1	heat, v.H.	(4)	95S5
Ser117→Phe	5.42	2.8	4.6	heat, v.H.	(4,6)	95S5
Asn132→Met	5.42	3.6	6.3	heat, v.H.	(4)	95S5
Asn132→Phe	5.42	3.3	5.4	heat, v.H.	(4)	95S5
Asn132→Ile	5.42	3.0	5.0	heat, v.H.	(4)	95S5
double mutant (Ser117→Ala and Asn132→Ile)						
	5.42	5.3	8.4	heat, v.H.	(5)	95S5
double mutant (Ser117→Ala and Asn132→Met)						
	5.42	4.7	7.5	heat, v.H.	(5)	95S5
double mutant (Ser117→Ile and Asn132→Met)						
	5.42	5.5	8.4	heat, v.H.	(4)	95S5
double mutant (Ser117→Ile and Asn132→Ile)						
	5.42	3.6	5.9	heat, v.H.	(4)	95S5

Remarks:
(1) wild-type, T_{trs} = 64.48°C
(2) pseudo wild-type, T_{trs} = 65.10°C
(3) $\Delta G_{(w.t.)} - \Delta G_{(wt^*)}$ = −2.2 kJ/mol at 64.48°C
(4) reference protein is the wild-type, see (1)
(5) reference protein is the pseudo wild-type, see (2,3)
(6) see also mutations concerning position 117 of lysozyme phage T4 and Ref. 93A2

Lysozyme phage T4, mutations in position 3, wild-type (Ile3) and mutants, difference values

Mutant	pH	T_{trs}	ΔT	Δ(ΔG)	Approach	Ref
wild-type	2	41.9	0	0	heat, v.H.	88M2
Ile3→Ala	2		−3.8	−4.6	heat, v.H.	88M2
Ile3→Asp	2		−6.5	−7.5	heat, v.H.	88M2
Ile3→Cys (S-H)						
	2		−1.9	−1.7	heat, v.H.	88M2
Ile3→Cys (S-S)						
	2		4.8	4.2	heat, v.H.	88M2
Ile3→Glu	2		−4.1	−4.6	heat, v.H.	88M2
Ile3→Gly	2		−7.2	−7.5	heat, v.H.	88M2
Ile3→Leu	2		3	3.8	heat, v.H.	88M2
Ile3→Met	2		−0.9	−1.3	heat, v.H.	88M2
Ile3→Phe	2		−4	−4.2	heat, v.H.	88M2

Lysozyme phage T4, mutations in position 3, wild-type (Ile3) and mutants, difference values (continued)

Mutant	pH	T_{trs}	ΔT	$\Delta(\Delta G)$	Approach	Ref
Ile3→Ser	2		−7	−7.9	heat, v.H.	88M2
Ile3→Thr	2		−6.1	−7.1	heat, v.H.	88M2
Ile3→Trp	2		−16.4	−15.1	heat, v.H.	88M2
Ile3→Tyr	2		−9.5	−11.3	heat, v.H.	88M2
Ile3→Val	2		−2.1	−2.5	heat, v.H.	88M2
wild-type	6.5	64.7	0	0	heat, v.H.	88M2
Ile3→Ala	6.5		−1.8	−2.9	heat, v.H.	88M2
Ile3→Asp	6.5		−8.5	−13.4	heat, v.H.	88M2
Ile3→Cys (S-H)						
	6.5		−3.7	−5	heat, v.H.	88M2
Ile3→Cys (S-S)						
	6.5		3.3	5	heat, v.H.	88M2
Ile3→Glu	6.5		−5.7	−8.4	heat, v.H.	88M2
Ile3→Gly	6.5		−5.8	−8.8	heat, v.H.	88M2
Ile3→Leu	6.5		0.9	1.7	heat,-v.H.	88M2
Ile3→Met	6.5		−2.3	−3.8	heat, v.H.	88M2
Ile3→Phe	6.5		−3	−4.6	heat, v.H.	88M2
Ile3→Ser	6.5		−4.6	−7.1	heat, v.H.	88M2
Ile3→Thr	6.5		−6	−9.6	heat, v.H.	88M2
Ile3→Trp	6.5		−8	−11.7	heat, v.H.	88M2
Ile3→Tyr	6.5		−5.9	−9.6	heat, v.H.	88M2
Ile3→Val	6.5		−1.2	−1.7	heat, v.H.	88M2

Lysozyme phage T4, wild-type and mutants

Mutant	pH	T_{trs}	ΔT	$\Delta(\Delta G)$	Approach	Remarks	Ref
wild-type	2.00	38.75	0	0	DSC	(1)	92L1
Ile3→Glu	2.00	38.75	−5.69	−7.4	DSC	(2)	92L1
Ile3→Leu	2.00	38.75	2.66	3.6	DSC	(4)	92L1
Ile3→Phe	2.00	38.75	−4.68	−6.2	DSC	(3)	92L1
Ile3→Pro	2.00	38.75	−11.99	−12.4	DSC	(5)	92L1
Ile3→Thr	2.00	38.75	−8.48	−10.3	DSC	(6)	92L1
wild-type	2.50	46.16	0	0	DSC	(1)	92L1
Ile3→Glu	2.50	46.16	−4.68	−7.2	DSC	(2)	92L1
Ile3→Leu	2.50	46.16	2.75	4.2	DSC	(4)	92L1
Ile3→Phe	2.50	46.16	−3.00	−4.6	DSC	(3)	92L1
Ile3→Pro	2.50	46.16	−9.20	−12.4	DSC	(5)	92L1
Ile3→Thr	2.50	46.16	−6.40	−9.5	DSC	(6)	92L1
wild-type	3.00	53.56	0	0	DSC	(1)	92L1
Ile3→Glu	3.00	53.56	−3.66	−6.4	DSC	(2)	92L1
Ile3→Leu	3.00	53.56	2.84	4.9	DSC	(4)	92L1
Ile3→Phe	3.00	53.56	−1.31	−2.2	DSC	(3)	92L1
Ile3→Pro	3.00	53.56	−6.41	−10.6	DSC	(5)	92L1
Ile3→Thr	3.00	53.56	−4.32	−7.6	DSC	(6)	92L1

Remarks:

T is the reference temperature of the wild-type protein. General expressions for the pH dependence of the transition temperature:

(1) $T_{trs} = A + B \times pH = 9.13 + 14.81 \times pH$ (for pH 1.60 to 2.84)
(2) $T_{trs} = A + B \times pH = (-0.62 \pm 0.13) + (-16.84 \pm 0.05) \times pH$ (for pH 1.98 to 3.05)
(3) $T_{trs} = A + B \times pH = (-2.29 \pm 0.23) + (-18.18 \pm 0.10) \times pH$ (for pH 1.96 to 3.01)
(4) $T_{trs} = A + B \times pH = (11.43 \pm 0.28) + (14.99 \pm 0.12) \times pH$ (for pH 2.0 to 3.19)
(5) $T_{trs} = A + B \times pH = (-14.02 \pm 0.24) + (20.39 \pm 0.10) \times pH$ (for pH 2.0 to 3.01)
(6) $T_{trs} = A + B \times pH = (-7.67 \pm 0.16) + (18.97 \pm 0.06) \times pH$ (for pH 2.02 to 2.97)

Table 1. Gibbs Energy Change – Molar Values

Lysozyme phage T4, mutant

Mutant	pH	T_{trs}	ΔT	$\Delta(\Delta G)$	Approach	Remark	Ref
Ile3→Pro	3.01	46.2±0.4	−7.3±0.6	−11.7±1.7	heat	(1)	92D3

Remark:
(1) the paper refers to data from Ericksson et al. (Ref. 92E3)
(2) $\Delta(\Delta G)$ was obtained using $\Delta(\Delta G) = \Delta T \times \Delta S$

Lysozyme phage T4, mutations in position 6, wild-type (Met6) and mutants, difference values

Mutant	pH	T_{trs}	ΔT	$\Delta(\Delta G)$	Approach	Ref
wild-type	2	42	0	0	heat, v.H.	87B1
Met6→Ile	2		−13		heat, v.H.	87B1
wild-type	6.5	65	0	0	heat, v.H.	87B1
Met6→Ile	6.5		−6		heat, v.H.	87B1

Lysozyme Phage T4, mutations in position 31, wild-type (His31) and mutants, difference values

Mutant	pH	T_{trs}	ΔT	$\Delta(\Delta G)$	Approach	Remarks	Ref
wild-type	5.5	66	0	0	heat		90A3
His31→Asn	5.5	55	−11	−17	heat	(1)	90A3
double mutant (His31→Asn and Asp70→Asn)							
	5.5	55	−11	−17	heat	(1,2)	90A3

Remarks:
(1) the data refer to the transition temperature of the wild-type protein at 66°C
(2) estimated value

Lysozyme phage T4, mutations in position 38, wild-type (Ser38) and mutants, difference values

Mutant	pH	T_{trs}	ΔT	$\Delta(\Delta G)$	Approach	Remarks	Ref
wild-type	3–5		0	0			
Ser38→Asp	3–5		2		heat	(1)	88N
double mutant (Ser38→Asp and Asn144→Asp)							
	3–5		4	6.7	heat	(1)	88N

Remark:
(1) heat capacity change $\Delta Cp = 8.4±0.8$ kJ/mol/K

Lysozyme phage T4, mutations in positon 41, wild-type (Ala41) and mutants, difference values

Mutant	pH	T_{trs}	ΔT	$\Delta(\Delta G)$	Approach	Remarks	Ref
wild-type	2.03		0	0	heat	(1)	90D1
wild-type	2.83		0	0	heat	(1)	90D1
Ala41→Val	2.03		1.6±0.1	1.8	heat	(1)	90D1
Ala41→Val	2.83		1.33±0.2	1.9	heat	(1)	90D1
double mutant (Ala41→Val and Val131→Ala)							
	2.03		2.2±0.1	2.3	heat	(1)	90D1
	2.83		1.81±0.2	2.5	heat	(1)	90D1

Remark:
(1) in the presence of 0.2 M KCl, monitored by CD at 223 nm

Lysozyme phage T4, mutants in position 44, the data serve as basis of amino acid α-helix propensity $\Delta(\Delta G)$ is normalized so that the helix propensity of Gly is zero

Mutant	pH	T	$\Delta(\Delta G)$	Approach	Remarks	Ref
Ala	3.01		4.02	heat		93B5
Arg	3.01		3.22	heat		93B5
Asn	3.01		1.63	heat		93B5
Asp	3.01		1.76	heat		93B5
Cys	3.01		1.76	heat		93B5
Gln	3.01		3.35	heat		93B5
Glu	3.01		2.22	heat		93B5
Gly	3.01		0.00	heat	(1)	93B5
His	3.01		2.38	heat		93B5
Ile	3.01		3.51	heat		93B5
Leu	3.01		3.85	heat		93B5
Lys	3.01		3.05	heat		93B5
Met	3.01		3.60	heat		93B5
Phe	3.01		2.47	heat		93B5
Pro	3.01		−10.5	heat		93B5
Ser	3.01		2.22	heat	(2)	93B5
Thr	3.01		2.26	heat		93B5
Trp	3.01		2.43	heat		93B5
Tyr	3.01		3.01	heat		93B5
Val	3.01		2.64	heat		93B5

Remarks:
(1) reference amino acid
(2) Ser44 = wild-type

Lysozyme phage T4, site 44 mutants

Mutant	pH	T	ΔT	$\Delta(\Delta G)$	Approach	Remarks	Ref
wt*	3.01	51.68	0	0	heat	(1–4)	94B6
Ser44→Ala	3.01	51.68	1.21	1.80	heat	(2–5)	94B6
Ser44→Arg	3.01	51.68	0.68	1.00	heat	(2–5)	94B6
Ser44→Asn	3.01	51.68	−0.40	−0.59	heat	(2–5)	94B6
Ser44→Asp	3.01	51.68	−0.32	−0.46	heat	(2–5)	94B6
Ser44→Cys	3.01	51.68	−0.35	−0.46	heat	(2–5)	94B6
Ser44→Gln	3.01	51.68	0.75	1.13	heat	(2–5)	94B6
Ser44→Glu	3.01	51.68	0.00	0.00	heat	(2–5)	94B6

Lysozyme phage T4, site 44 mutants (continued)

Table 1. Gibbs Energy Change – Molar Values

Mutant	pH	T	ΔT	Δ(ΔG)	Approach	Remarks	Ref
Ser44→Gly	3.01	51.68	−1.55	−2.22	heat	(2–5)	94B6
Ser44→His	3.01	51.68	0.12	0.17	heat	(2–5)	94B6
Ser44→Ile	3.01	51.68	0.91	1.30	heat	(2–5)	94B6
Ser44→Leu	3.01	51.68	1.09	1.63	heat	(2–5)	94B6
Ser44→Lys	3.01	51.68	0.57	0.84	heat	(2–5)	94B6
Ser44→Met	3.01	51.68	0.92	1.38	heat	(2–5)	94B6
Ser44→Phe	3.01	51.68	0.18	0.25	heat	(2–5)	94B6
Ser44→Pro	3.01	51.68	−10.32	−12.7	heat	(2–5)	94B6
Ser44→Thr	3.01	51.68	0.03	0.04	heat	(2–5)	94B6
Ser44→Trp	3.01	51.68	0.15	0.21	heat	(2–5)	94B6
Ser44→Tyr	3.01	51.68	0.54	0.79	heat	(2–5)	94B6
Ser44→Val	3.01	51.68	0.29	0.42	heat	(2–5)	94B6

Remarks:
(1) wt* = pseudo wild-type, containing (Cys54→Thr and Cys97→Ala)
(2) wt* is the reference protein, T_{trs} = 51.68°C
(3) estimated error: T_{trs} ±0.1°C, ΔH ±(4–6)%, Δ(ΔG) ±0.21 (except mutant Ser44→Pro: ±1.5)
(4) buffer: 25 mM KCl, 3 mM H_3PO_4, 17 mM KH_2PO_4
(5) Δ(ΔG) was obtained using Δ(ΔG) = ΔT × ΔS ΔCp = 10.04 was taken from Ref. 92H10, and assumed to
   have the same value for all mutants

Lysozyme phage T4, mutations in position 54, wild-type (Cys54) and mutants, difference values

Mutant	pH	T	ΔT	Δ(ΔG)	Approach	Ref
wild-type	6.5	64.7	0	0	heat, v.H.	88M3
Cys54→Val	6.5		−2	−2.9	heat, v.H.	88M3
Cys54→Thr	6.5		1	1.3	heat, v.H.	88M3
double mutant (Cys54→Val and Cys97→Ser)						
	2.0	38.8		−1.42	DSC	92H7
	2.5	46.2		−3.22	DSC	92H7
	3.0	53.6		−5.31	DSC	92H7
	6.5		−6	−8.8	heat, v.H.	88M3

Lysozyme phage T4, mutations in positon 55, wild-type (Asn55) and mutants, difference values

Mutant	pH	T	ΔT	Δ(ΔG)	Approach	Ref
wild type	2	41.9	0	0	heat, v.H.	89N
Asn55→Gly	2	40.2	−1.7	−2.1	heat, v.H.	89N
wild-type	6.5	64.7	0	0	heat, v.H.	89N
Asn55→Gly	6.5	63.1	−1.6	−2.5	heat, v.H.	89N

Lysozyme phage T4, pseudo wild-type (Cys54→Thr and Cys97→Ala) and mutations in position 59

Mutant	pH	T	ΔT	Δ(ΔG)	Approach	Remarks	Ref
wt*	2.0	41.0	0	0	heat		92B3
wt*	6.5	63.0	0	0	heat		92B3
Thr59→Ala	2.0	41.0	−10.1	−11.7	heat	(1)	92B3
Thr59→Ala	6.5	63.0	−4.0	−6.3	heat	(1)	92B3
Thr59→Asn	2.0	41.0	−2.1	−2.5	heat	(1)	92B3
Thr59→Asn	6.5	63.0	−2.8	−4.6	heat	(1)	92B3

Lysozyme phage T4, pseudo wild-type (Cys54→Thr and Cys97→Ala) and mutations in position 59 (continued)

Mutant	pH	T	ΔT	Δ(ΔG)	Approach	Remarks	Ref
Thr59→Asp	2.0	41.0	−3.1	−3.8	heat	(1)	92B3
Thr59→Asp	6.5	63.0	−3.1	−5.0	heat	(1)	92B3
Thr59→Gly	2.0	41.0	−7.7	−9.2	heat	(1)	92B3
Thr59→Gly	6.5	63.0	−4.1	−6.7	heat	(1)	92B3
Thr59→Ser	2.0	41.0	−2.6	−2.9	heat	(1)	92B3
Thr59→Ser	6.5	63.0	−0.4	−0.8	heat	(1)	92B3
Thr59→Val	2.0	41.0	−10.0	−11.7	heat	(1)	92B3
Thr59→Val	6.5	63.0	−4.0	−6.3	heat	(1)	92B3

Remark:
(1) Δ(ΔG) was obtained using $\Delta(\Delta G) = \Delta T \times \Delta S$

Lysozyme phage T4, mutations in positon 66, wild-type (Leu66) and mutants, difference values

Mutant	pH	T	ΔT	Δ(ΔG)	Approach	Ref
wild-type	2	42	0	0	heat, v.H.	87B1
wild-type	6.5	65	0	0	heat, v.H.	87B1
Leu66→Phe	2		−3		heat, v.H.	87B1
Leu66→Phe	6.5		−1		heat, v.H.	87B1

Lysozyme phage T4, mutations in position 70, wild-type (Asp70) and mutants, difference values

Mutant	pH	T	ΔT	Δ(ΔG)	Approach	Remarks	Ref
wild-type	5.5	66	0	0	heat		90A3
Asp70→Asn	5.5	55	−11	−17	heat	(1,2)	90A3
double mutant (His31→Asn and Asp70→Asn)							
	5.5	55	−11	−17	heat	(1,2)	90A3

Remarks:
(1) the data refer to the transition temperature of the wild-type protein at 66°C
(2) estimated value

Lysozyme phage T4, mutations in position 77, wild-type (Gly77) and mutants, difference values

Mutant	pH	T	ΔT	Δ(ΔG)	Approach	Ref
wild-type	2	41.9±0.4	0	0	heat, v.H.	89S11
Gly77→Ala	2	40.5±0.7	−1.4	−1.7	heat, v.H.	89S11
wild-type	6.5	64.7±0.5	0	0	heat, v.H.	89S11
Gly77→Ala	6.5	65.6±0.2	0.9	1.7	heat, v.H.	89S11

Lysozyme phage T4, mutations in position 82, wild-type (Ala82) and mutants, difference values

Mutant	pH	T	ΔT	Δ(ΔG)	Approach	Ref
wild-type	2	41.9±0.4	0	0	heat, v.H.	87M
wild-type	6.5	64.7±0.5	0	0	heat, v.H.	87M
Ala82→Pro	2	42.7±0.1	0.8	1.3	heat, v.H.	87M
Ala82→Pro	6.5	66.8±0.2	2.1	3.3	heat, v.H.	87M
Ala82→Pro	2.0	38.8		−0.29	DSC	92H7
Ala82→Pro	2.5	46.2		2.13	DSC	92H7
Ala82→Pro	3.0	53.6		5.31	DSC	92H7

Lysozyme phage T4, mutations in position 86

Mutant	pH	T	Δ(ΔG)	Approach	Remarks	Ref
wild-type	2	42	0	heat	(1)	88A
Pro86→Ala	2	40	−1 to −2	heat		88A
Pro86→Arg	2	40	−1 to −2	heat		88A
Pro86→Asp	2	42	0	heat		88A
Pro86→Cys	2	41	−1 to −2	heat		88A
Pro86→Gly	2	40	−1 to −2	heat		88A
Pro86→His	2	40	−1 to −2	heat		88A
Pro86→Ile	2	40	−1 to −2	heat		88A
Pro86→Leu	2	40	−1 to −2	heat		88A
Pro86→Ser	2	41	−1 to −2	heat		88A
Pro86→Thr	2	41	−1 to −2	heat		88A
wild-type	4	63	0	heat	(1)	88A
Pro86→Arg	4	62	−1 to −2	heat		88A
Pro86→Asp	4	63	0	heat		88A
Pro86→Gly	4	62	−1 to −2	heat		88A
wild-type	6	66	0	heat	(1)	88A
Pro86→Arg	6	63	−1 to −2	heat		88A
Pro86→Asp	6	65	0	heat		88A
Pro86→His	6	62	−1 to −2	heat		88A

Remark:
(1) the given Δ(ΔG) value serves as the reference value

Lysozyme phage T4, mutations in position 93, wild-Type (Ala93) and mutants, difference values

Mutant	pH	T	ΔT	Δ(ΔG)	Approach	Ref
Ala93→Pro	2.0	38.8		0.13	DSC	92H7
Ala93→Pro	2.5	46.2		2.13	DSC	92H7
Ala93→Pro	3.0	53.6		4.60	DSC	92H7

Lysozyme phage T4, mutations in positon 96, wild-type (Arg96) and mutants, difference values

Mutant	pH	T	ΔT	Δ(ΔG)	Approach	Ref
wild-type	2	42	0	0	heat, v.H.	87B1
Arg96→His	2		−13		heat, v.H.	87B1
wild-type	6.5	65	0	0	heat, v.H.	87B1
wild-type	6.5	65	0	0	heat, v.H.	88W2
Arg96→His	6.5		−6		heat, v.H.	87B1
Arg96→His	6.5		−6	−11.7	heat, v.H.	88W2

Lysozyme phage T4, pseudo wild-type (wt*) and mutants in position 99

Mutant	pH	T	ΔT	Δ(ΔG)	Approach	Remarks	Ref
wt*	3.01	51.76	0.0	0.0	heat	(1)	93E2
Leu99→Ala	3.01	51.76	−15.7	−20.9	heat	(1)	93E2
Leu99→Ile	3.01	51.76	−4.0	−5.9	heat	(1)	93E2
Leu99→Met	3.01	51.76	−2.0	−2.9	heat	(1)	93E2
Leu99→Phe	3.01	51.76	−1.1	−1.7	heat	(1)	93E2
Leu99→Val	3.01	51.76	−6.6	−9.6	heat	(1)	93E2
double mutant (Leu99→Ala and Phe153→Ala)							
	3.01	51.76	−41.8	−34.7	heat	(1)	93E2
wt*	5.71	64.88	0.0	0.0	heat	(1)	93E2
Leu99→Ala	5.7	64.88	−11.4	−18.8	heat	(1)	93E2
Leu99→Ile	5.7	64.88	−3.7	−6.3	heat	(1)	93E2
Leu99→Met	5.7	64.88	−1.5	−2.5	heat	(1)	93E2
Leu99→Phe	5.7	64.88	−0.7	−1.3	heat	(1)	93E2
Leu99→Val	5.7	64.88	−5.2	−8.4	heat	(1)	93E2
double mutant (Leu99→Ala and Phe153→Ala)							
	5.7	64.88	−22.8	−28.9	heat	(1)	93E2

Remark:
(1) reference protein is the cysteine-free pseudo wild-type (Cys54→Thr and Cys97→Ala)

Lysozyme phage T4, pseudo wild-type (Cys54→Thr and Cys97→Ala) and mutants in positions 102 and 133

Mutant	pH	T_{trs}	ΔT	Δ(ΔG)	Approach	Remarks	Ref
wt*	3.0	51.6	0.0	0.0	heat	(1)	91D3
Met102→Lys	3.0	16.6	−35.0	−37.2	heat	(1–3)	91D3
Leu133→Asp	3.0	36.7	−14.9	−20.5	heat	(1–3)	91D3
wild-type*	4.4	64.6	0.0	0.0	heat	(1)	91D3
Leu133→Asp	4.4	49.6	−15.0	−23.0	heat	(1–3)	91D3
wt*	5.3	65.3	0.0	0.0	heat	(1)	91D3
Met102→Lys	5.3	45.0	−20.3	−28.9	heat	(1–3)	91D3
wt*	6.5	61.9	0.0	0.0	heat	(1)	91D3
Leu133→Asp	6.5	44.0	−17.9	−23.8	heat	(1–3)	91D3
wt*	10.4	47.9	0.0	0.0	heat	(1,4)	91D3
Met102→Lys	10.4	37.8	−10.1	−9.2	heat	(1–4)	91D3
Leu133→Asp	10.4	28.9	−19.0	−15.5	heat	(1–4)	91D3

Remarks:
(1) measurements in 20 mM potassium phosphate with 25 mM KCl
(2) the mutant protein contains additionally the replacements (Cys54→Thr and Cys97→Ala) of the template
    (pseudo wild-type)
(3) Δ(ΔG) values refer at each pH to T_{trs} of the wt*
(4) buffer (1) with 0.1 mM EDTA

Table 1. Gibbs Energy Change – Molar Values

Lysozyme phage T4, wild-type and mutants in position 105

Mutant	pH	T	ΔT	Δ(ΔG)	Approach	Remarks	Ref
wild-type	2.1	42.0	0.0	0.0	heat	(1,2)	93P10
Gln105→Ala	2.1	42.0	−3.5	−3.8	heat	(1,2)	93P10
Gln105→Glu	2.1	42.0	−2.1	−2.1	heat	(1,2)	93P10
Gln105→Gly	2.1	42.0	−7.2	−7.1	heat	(1,2)	93P10
wild-type	5.8	66.3	0.0	0.0	heat	(1,2)	93P10
Gln105→Ala	5.8	66.3	−1.6	−2.5	heat	(1,2)	93P10
Gln105→Glu	5.8	66.3	−3.0	−4.6	heat	(1,2)	93P10
Gln105→Gly	5.8	66.3	−3.9	−6.3	heat	(1,2)	93P10

Remark:
(1) Δ(ΔG) was obtained using $\Delta(\Delta G) = \Delta T \times \Delta S$
(2) the estimated error in Δ(ΔG) is ±0.6 kJ/mol

Lysozyme phage T4, wild-type and mutants in position 113

Mutant	pH	T	ΔT	Δ(ΔG)	Approach	Remarks	Ref
Gly113→Ala	2.0	38.8		2.09	DSC	(1)	92H7
Gly113→Ala	2.5	46.2		2.26	DSC	(1)	92H7
Gly113→Ala	3.0	53.6		2.26	DSC	(1)	92H7

Remark:
(1) Δ(ΔG) refers to T_{trs} of the wild-type protein

Lysozyme phage T4, wild-type and mutants in position 117

Mutant	pH	T	ΔT	Δ(ΔG)	Approach	Ref
wild-type	3.0	51.70±0.14	0	0	heat	93A2
Ser117→Phe	3.0	51.70±0.14	4.8±0.2	5.9±0.4	heat	93A2
wild-type	5.4	65.09±0.22	0	0	heat	93A2
Ser117→Phe	5.4	65.09±0.22	2.8±0.3	4.6±0.4	heat	93A2

Lysozyme phage T4, mutations in position 124, wild-type (Lys124) and mutants, difference values

Mutant	pH	T	ΔT	Δ(ΔG)	Approach	Ref
wild-type	2	41.9	0	0	heat, v.H.	89N
Lys124→Gly	2	41.9	0	0	heat, v.H.	89N
wild-type	6.5	64.7	0	0	heat, v.H.	89N
Lys124→Gly	6.5	64.5	−0.2	−0.4	heat, v.H.	89N

Lysozyme phage T4, wild-type and mutants, alanine replacements within α-helix 115–123

Mutant	pH	T_{trs}	ΔT	Δ(ΔG)	Approach	Remarks	Ref
wt*	3.01	51.65	0.0	0.0	heat	(1–3)	95B4
Thr115→Ala	3.01	51.22	−0.43	−0.6±0.3	heat	(1–3)	95B4
Asn116→Ala	3.01	52.14	0.49	0.7±0.3	heat	(1–3)	95B4
Ser117→Ala	3.01	55.29	3.64	5.3±0.3	heat	(1–3)	95B4
Leu118→Ala	3.01	39.6	−12.05	−14.6±0.5	heat	(1–3)	92E3
Arg119→Ala	3.01	51.12	−0.53	−0.8±0.3	heat	(1–3)	95B4
Met120→Ala	3.01	51.07	−0.58	−0.8±0.3	heat	(1–3)	95B4

Lysozyme phage T4, wild-type and mutants, alanine replacements within α-helix 115–123 (continued)

Mutant	pH	T_{trs}	ΔT	$\Delta(\Delta G)$	Approach	Remarks	Ref
Leu121→Ala	3.01	42.5	−9.15	−11.3±0.5	heat	(1–3)	92E3
Gln122→Ala	3.01	50.94	−0.71	−1.0±0.3	heat	(1–3)	95B4
Gln123→Ala	3.01	51.01	−0.64	−0.9±0.3	heat	(1–3)	95B4
double mutant (Thr115→Ala and Ser117→Ala)							
	3.01	54.49	2.84	4.0±0.3	heat	(1–3)	95B4
double mutant (Ser117→Ala and Arg119→Ala)							
	3.01	55.02	3.37	4.9±0.3	heat	(1–3)	95B4
double mutant (Met120→Ala and Gln122→Ala)							
	3.01	50.90	−0.75	−1.1±0.3	heat	(1–3)	95B4
double mutant (Thr115→Ala and Arg119→Ala)							
	3.01	51.13	−0.52	−0.7±0.3	heat	(1–3)	95B4
double mutant (Asn116→Ala and Met120→Ala)							
	3.01	52.23	0.58	0.9±0.3	heat	(1–3)	95B4
double mutant (Arg119→Ala and Gln123→Ala)							
	3.01	51.14	−0.51	−0.7±0.3	heat	(1–3)	95B4
multiple mutant (residues 115–123→Ala except Leu118 and Leu121)							
	3.01	54.84	3.19	4.2±0.3	heat	(1–3)	95B4

Remarks:
(1) $\Delta(\Delta G)$ was determined from $\Delta(\Delta G) = \Delta T \times \Delta S$, where ΔS and ΔT are the values for unfolding of the mutant
(2) wt* is the reference protein
(3) buffer: 0.0029 M H_3PO_4, 0.017 M KH_2PO_4, 0.025 M KCl, pH 3.01

Lysozyme phage T4, wild-type and mutants, alanine replacements within α-helix 126–134

Mutant	pH	T_{ref}	ΔT	$\Delta(\Delta G)$	Approach	Remark	Ref
wild-type	2.0	40.77±0.2	0	0	heat, v.H.		91Z2
Glu128→Ala	2.0	40.8	0.6±0.25	0.7	heat, v.H.		91Z2
Val131→Ala	2.0	40.8	1.0±0.25	1.1	heat, v.H.		91Z2
Leu133→Ala	2.02	40.75	−17.0±2.0	−17.5	heat, v.H.	(1)	92Z2
Double mutants:							
(Asp127→Ala and Glu128→Ala)							
	2.02	40.75	0.9±0.25	1.0	heat, v.H.		92Z2
(Glu128→Ala and Val131→Ala)							
	2.0	40.8	1.5±0.25	1.7	heat, v.H.		91Z2
(Val131→Ala and Asn132→Ala)							
	2.0	40.8	2.3±0.25	2.6	heat, v.H.		91Z2
Triple mutant:							
(Glu128→Ala, Val121→Ala and Asn132→Ala)							
	2.0	40.8	3.4±0.22	3.9	heat, v.H.		91Z2
Multiple mutants:							
(Asp127→Ala, Glu128→Ala, Val131→Ala and Asn132→Ala)							
	2.02	40.75	4.0±0.25	4.2	heat, v.H.		92Z2
(Glu128→Ala, Val131→Ala, Asn132→Ala and Leu133→Ala)							
	2.02	40.75	−10.3±0.5 −	10.8	heat, v.H.		92Z2
(Asp127→Ala, Glu128→Ala, Val131→Ala, Asn132→Ala and Leu133→Ala)							
	2.02	40.75	−9.4±0.5	−9.5	heat, v.H.		92Z2

Remark:
(1) see also Ref. 92E3 for $\Delta(\Delta G)$ obtained under slightly different conditions

Lysozyme phage T4, wild-type and mutants in position 129

Mutant	pH	T	Δ(ΔG)	Approach	Remarks	Ref
wild-type	1.99	41.4±0.3	0.0	heat, v.H.	(1)	89K1
wild-type	2.86	52.1±0.4	0.0	heat, v.H.	(1)	89K1
Ala129→Val	1.99	38.5±0.5	−3.3	heat, v.H.	(1)	89K1
Ala129→Val	2.8	50.2±0.6	−2.9	heat, v.H.	(1)	89K1

Remark:
(1) measured in 0.2 M KCl

Lysozyme phage T4, mutations in position 131, wild-type (Val131) and mutants, difference values

Mutant	pH	T	ΔT	Δ(ΔG)	Approach	Remarks	Ref
wild-type	2.03		0	0	heat	(1)	90D1
wild-type	2.83		0	0	heat	(1)	90D1
Val131→Ala	2.03		0.56±0.2	0.6	heat	(1)	90D1
Val131→Ala	2.83		0.90±0.3	1.3	heat	(1)	90D1
Val131→Thr	2.03		−0.41±0.1	−0.5	heat	(1)	90D1
Val131→Thr	2.83		−0.10±0.2	−0.2	heat	(1)	90D1
double mutant (Ala41→Val and Val131→Ala)							
	2.03		2.2±0.1	2.3	heat	(1)	90D1
	2.83		1.81±0.2	2.5	heat	(1)	90D1

Remark:
(1) in the presence of 0.2 M KCl, monitored by CD at 223 nm

Lysozyme phage T4, site 131 mutants

Mutant	pH	T	ΔT	Δ(ΔG)	Approach	Remarks	Ref
wild-type	3.01	53.08	0.0	0.0	heat	(1–3)	94B6
Val131→Ala	3.01	53.08	0.66	1.09	heat	(1–4)	94B6
Val131→Asp	3.01	53.08	0.22	0.33	heat	(1–4)	94B6
Val131→Glu	3.01	53.08	0.52	0.84	heat	(1–4)	94B6
Val131→Gly	3.01	53.08	−1.80	−0.68	heat	(1–4)	94B6
Val131→Ile	3.01	53.08	0.41	0.67	heat	(1–4)	94B6
Val131→Leu	3.01	53.08	0.23	0.38	heat	(1–4)	94B6
Val131→Met	3.01	53.08	0.32	0.50	heat	(1–4)	94B6
Val131→Ser	3.01	53.08	−0.12	−0.21	heat	(1–4)	94B6
Val131→Thr	3.01	53.08	−0.33	−0.50	heat	(1–4)	94B6

Remarks:
(1) wild-type (Val131) is the reference protein, T_{trs} = 53.08°C
(2) estimated error: T_{trs} ±0.1°C, ΔH ±(4–6)%, Δ(ΔG) ±0.21
(3) buffer: 25 mM KCl, 3 mM H_3PO_4, 17 mM KH_2PO_4
(4) Δ(ΔG) was obtained using Δ(ΔG) = ΔT × ΔS ΔCp = 10.04 kJ/mol/K was taken from Ref. 92H10, and assumed to have the same value for all mutants

Mutations in position 133 – see also position 102,
Lysozyme phage T4, wild-type and mutants, position 133

Mutant	pH	T	$\Delta(\Delta G)$	Approach	Remarks	Ref
wild-type	1.99	41.4±0.3	0.0	heat, v.H.	(1)	89K1
wild-type	2.86	52.1±0.4	0.0	heat, v.H.	(1)	89K1
Leu133→Phe	1.99	40.7±0.5	−0.8	heat, v.H.	(1)	89K1
Leu133→Phe	2.85	51.3±0.6	−1.3	heat, v.H.	(1)	89K1

Remark:
(1) measured in 0.2 M KCl

Lysozyme phage T4, mutants with unnatural amino acids

Mutant	pH	T	ΔT	$\Delta(\Delta G)$	Approach	Remarks	Ref
wild-type		43.5±0.2	0	0.0	heat	(1)	92M3
Leu133→[1]		43.5	4.3	5.2±0.4	heat	(2,3)	92M3
Leu133→[2]		43.5	1.9	2.5±0.5	heat	(2,4)	92M3
Leu133→[3]		43.5	−4.0	−4.5±0.4	heat	(2,5)	92M3
Leu133→[4]		43.5	−10.6	−11.0±2.3	heat	(2,6)	92M3
Leu133→[5]		31	−12.5	−13.8	heat	(2,7,8)	92M3

Remarks:
(1) van't Hoff treatment
(2) the data refer to T_{trs} of the wild-type $\Delta(\Delta G)$ was calculated using an estimated value for ΔCp of 10.5 kJ/K/mol
(3) mutant [1]: Leu133→S-2-amino-3-cyclopentylpropanoic acid
(4) mutant [2]: Leu133→S,S-2-amino-4-methyl hexanoic acid
(5) mutant [3]: Leu133→norvaline
(6) mutant [4]: Leu133→O-methylserine
(7) mutant [5]: Leu133→ethylglycine, $\Delta(\Delta G)$ was calculated at the mutant T_{trs} = 31°C with ΔH = 492 kJ/mol and ΔCp = 7.5 kJ/mol/K of the wild-type (data from Ref. 91D3)
(8) approximate ΔT and $\Delta(\Delta G)$ values

Lysozyme phage T4, mutations in position 144, wild-type (Asn144) and mutants, difference values

Mutant	pH	T	ΔT	$\Delta(\Delta G)$	Approach	Remarks	Ref
wild-type	3–5		0	0			
Asn144→Asp	3–5		1.5		heat	(1)	88N
double mutant (Ser38→Asp and Asn144→Asp)							
	3–5		4	6.7	heat	(1)	88N

Remark:
(1) heat capacity change ΔCp = 8.4±0.8 kJ/K/mol

Lysozyme phage T4, mutations in position 146, wild-type (Ala146) and mutants, difference values

Mutant	pH	T	ΔT	$\Delta(\Delta G)$	Approach	Ref
wild-type	6.5	65	0	0	heat, v.H.	88W2
Ala146→Thr	6.5	−6		−6.3	heat, v.H.	88W2

Table 1. Gibbs Energy Change – Molar Values

Lysozyme phage T4, pseudo wild-type (wt*) and mutants in position 153

Mutant	pH	T	ΔT	Δ(ΔG)	Approach	Remarks	Ref
wt*	3.01	51.76	0.0	0.0	heat	(1)	93E2
Phe153→Ala	3.01	51.76	−12.3	−14.6	heat	(1)	93E2
Phe153→Val	3.01	51.76	−6.1	−7.5	heat	(1)	93E2
Phe153→Ile	3.01	51.76	−1.5	−2.1	heat	(1)	93E2
Phe153→Leu	3.01	51.76	+0.6	+0.8	heat	(1)	93E2
Phe153→Met	3.01	51.76	−2.4	−3.3	heat	(1)	93E2
double mutant (Leu99→Ala and Phe153→Ala)							
	3.01	51.76	−41.8	−34.7	heat	(1)	93E2
wt*	5.71	64.88	0.0	0.0	heat	(1)	93E2
Phe153→Ala	5.7	64.88	−9.3	−15.9	heat	(1)	93E2
Phe153→Ile	5.7	64.88	−0.5	−0.8	heat	(1)	93E2
Phe153→Leu	5.7	64.88	+0.8	+1.3	heat	(1)	93E2
Phe153→Met	5.7	64.88	−1.6	−2.5	heat	(1)	93E2
Phe153→Val	5.7	64.88	−4.5	−7.5	heat	(1)	93E2
double mutant (Leu99→Ala and Phe153→Ala)							
	5.7	64.88	−22.8	−28.9	heat	(1)	93E2

Remark:
(1) reference protein is the cysteine-free pseudo wild-type (Cys54→Thr and Cys97→Ala)

Lysozyme phage T4, mutations in position 156, wild-type (Gly156) and mutants, difference values

Mutant	pH	T	ΔT	Δ(ΔG)	Approach	Ref
wild-type	2	41.9±0.4	0	0	heat, v.H.	87G3
Gly156→Asp	2		−5.1	−5.9	heat, v.H.	87G3
wild-type	6.5	64.7±0.5	0	0	heat, v.H.	87G3
Gly156→Asp	6.5		−6.1	−9.6	heat, v.H.	87G3

Lysozyme phage T4, mutations in position 157, wild-type (Thr157) and mutants, difference values

Mutant	pH	T	ΔT	Δ(ΔG)	Approach	Ref
wild-type	2	42	0	0	heat, v.H.	87A
wild-type	2	42±0.5	0		heat, v.H.	87G4
Thr157→Ala	2		−5.4	−5.9	heat, v.H.	87A
Thr157→Arg	2		−5.1	−5.4	heat, v.H.	87A
Thr157→Asn	2		−1.7	−1.9	heat, v.H.	87A
Thr157→Asp	2		−4.2	−4.6	heat, v.H.	87A
Thr157→Cys	2		−4.9	−5.4	heat, v.H.	87A
Thr157→Glu	2		−5.8	−6.3	heat, v.H.	87A
Thr157→Gly	2		−4.2	−4.6	heat, v.H.	87A
Thr157→His	2		−7.9	−8.8	heat, v.H.	87A
Thr157→Ile	2		−11	−12.1	heat, v.H.	87A
Thr157→Ile	2		−11	−12.9	heat, v.H.	87G4
Thr157→Leu	2		−5	−5.4	heat, v.H.	87A
Thr157→Phe	2		−9.2	−2.4	heat, v.H.	87A
Thr157→Ser	2		−2.5	−2.8	heat, v.H.	87A
Thr157→Val	2		−6	−6.7	heat, v.H.	87A
wild-type	4	62.8±1		0	heat, v.H.	87G4
Thr157→Ile	4		−5.8	−8.8	heat, v.H.	87G4
wild-type	6	64.5±1		0	heat, v.H.	87G4
Thr157→Ile	6		−3	−5	heat, v.H.	87G4

Lysozyme phage T4, wild-type and mutants in position 157

Mutant	pH	T	Δ(ΔG)	Approach	Ref
wild-type	2.0	38.8	0	DSC	91C7
wild-type	2.5	46.2	0	DSC	91C7
wild-type	3.0	53.6	0	DSC	91C7
Thr157→Ala	2.0	38.8	−4.6	DSC	91C7
Thr157→Ala	2.5	46.2	−3.8	DSC	91C7
Thr157→Ala	3.0	53.6	−2.1	DSC	91C7
Thr157→Asn	2.0	38.8	−4.2	DSC	91C7
Thr157→Asn	2.5	46.2	−4.6	DSC	91C7
Thr157→Asn	3.0	53.6	−5.0	DSC	91C7
Thr157→Arg	2.0	38.8	−4.6	DSC	91C7
Thr157→Arg	2.5	46.2	−2.5	DSC	91C7
Thr157→Arg	3.0	53.6	1.3	DSC	91C7
Thr157→Glu	2.0	38.8	−6.3	DSC	91C7
Thr157→Glu	2.5	46.2	−5.4	DSC	91C7
Thr157→Glu	3.0	53.6	−3.3	DSC	91C7
Thr157→Ile	2.0	38.8	−8.8	DSC	91C7
Thr157→Ile	2.5	46.2	−7.9	DSC	91C7
Thr157→Ile	3.0	53.6	−6.3	DSC	91C7
Thr157→Leu	2.0	38.8	−7.1	DSC	91C7
Thr157→Leu	2.5	46.2	−7.1	DSC	91C7
Thr157→Leu	3.0	53.6	−7.1	DSC	91C7
Thr157→Val	2.0	38.8	−7.1	DSC	91C7
Thr157→Val	2.5	46.2	−6.7	DSC	91C7
Thr157→Val	3.0	53.6	−5.0	DSC	91C7

Lysozyme phage T4, mutations in position 160, wild-type (Ala160) and mutants, difference values

Mutant	pH	T	ΔT	Δ(ΔG)	Approach	Ref
wild-type	2	42	0	0	heat, v.H.	87B1
Ala160→Thr	2		−12		heat	87B1
wild-type	6.5	65	0	0	heat, v.H.	87B1
Ala160→Thr	6.5		−7		heat	87B1

Lysozyme phage T4, mutants concerning helix dipole interactions

Mutant	pH	ΔT	Δ(ΔG)	Approach	Remarks	Ref
Ser38→Asp	2.0	−0.3±0.5	−0.4±0.6	heat	(1)	91N1
Ser38→Asp	6.9	1.5±0.5	2.5±0.8	heat	(1)	91N1
Ser38→Asn	2.0	−0.3±0.4	−0.4±0.4	heat	(2)	91N1
Ser38→Asn	6.7	−0.1±0.6	0.0±0.8	heat	(2)	91N1
Asp92→Asn	2.0	−0.3±0.4	−0.4±0.4	heat	(2)	91N1
Asp92→Asn	6.7	−3.7±0.5	−5.9±0.8	heat	(2)	91N1
Thr109→Asp	2.0	−0.8±0.4	−1.3±0.4	heat	(2)	91N1
Thr109→Asp	6.7	1.5±0.6	2.5±0.8	heat	(2)	91N1
Thr109→Asn	2.0	−0.1±0.4	0.0±0.4	heat	(2)	91N1
Thr109→Asn	6.7	0.3±0.5	0.4±0.8	heat	(2)	91N1
Thr115→Glu	2.0	−1.7±0.5	−2.1±0.6	heat	(1)	91D1
Thr115→Glu	6.5	0.7±0.5	1.3±0.8	heat	(1)	91D1

Lysozyme phage T4, mutants concerning helix dipole interactions (continued)

Mutant	pH	ΔT	Δ(ΔG)	Approach	Remarks	Ref
Asn116→Asp	2.0	−0.2±0.3	−0.4±0.4	heat	(1)	91N1
Asn116→Asp	5.7	1.6±0.4	2.5±0.6	heat	(1)	91N1
Arg119→Met	2.0	−0.9±0.3	−1.3±0.4	heat	(1)	91N1
Arg119→Met	5.7	0.3±0.5	0.4±0.8	heat	(1)	91N1
Double mutant (Asn116→Asp and Arg119→Met)						
	2.0	−0.8±0.3	−1.3±0.4	heat	(1)	91N1
	5.7	1.6±0.4	2.5±0.6	heat	(1)	91N1
Asn144→Asp	2.0	−0.2±0.5	−0.4±0.6	heat	(1)	91N1
Asn144→Asp	6.9	1.4±0.5	2.1±0.8	heat	(1)	91N1
Asn144→His	2.0	−2.3±0.4	−2.5±0.4	heat	(2)	91N1
Asn144→His	6.7	0.7±0.6	1.3±0.8	heat	(2)	91N1

Remarks:
(1) ΔT and Δ(ΔG) refer to wild-type having T_{trs} = 40.9±0.1°C at pH 2.0 and 66.1±0.2°C at pH 5.7
(2) ΔT and Δ(ΔG) refer to the pseudo wild-type Cys54→Thr and Cys97→Ala (wt*) having T_{trs} = 39.5±0.2°C
    at pH 2.0 and 62.4±0.2°C at pH 6.7

Lysozyme phage T4, pseudo wild-type (wt*) and mutants

Mutant	pH	T	ΔT	Δ(ΔG)	Approach	Remarks	Ref
wt*	3.0	51.55±0.15	0.0	0.0	heat	(1–3)	93B4
Ala41→Ser	3.0	51.55	−1.77	−2.5	heat	(1–3)	93B4
Ala42→Ser	3.0	51.55	−7.49	−9.6	heat	(1–3)	93B4
Ala49→Ser	3.0	51.55	−1.53	−2.1	heat	(1–3)	93B4
Ala73→Ser	3.0	51.55	−1.27	−1.7	heat	(1–3)	93B4
Val75→Thr	3.0	51.55	−3.70	−5.4	heat	(1–3)	93B4
Ala82→Ser	3.0	51.55	−0.99	−1.3	heat	(1,3)	93B4
Val87→Thr	3.0	51.55	−4.55	−6.7	heat	(1–3)	93B4
Ala93→Ser	3.0	51.55	−0.52	−0.8	heat	(1–3)	93B4
Ala98→Ser	3.0	51.55	−7.47	−10.5	heat	(1–3)	93B4
Ala130→Ser	3.0	51.55	−2.89	−4.2	heat	(1–3)	93B4
Ala134→Ser	3.0	51.55	−0.44	−0.4	heat	(1–3)	93B4
Val149→Thr	3.0	51.55	−10.08	−11.7	heat	(1–3)	93B4

Remarks:
(1) wt* = (Cys54→Thr and Cys97→Ala) is the reference protein according to Matsumura & Matthews:
    Science 243 (1989) 792.
(2) ΔC_p = 10.46 kJ/mol/K was taken from 91Z2, and assumed to be identically for the mutants
(3) buffer: 25 mM KCl, 20 mM KH_2PO_4, pH 3.0

Lysozyme phage T4, mutants with unnatural and natural amino acids

Mutant	pH	T_{trs}	ΔT	$\Delta(\Delta G)$	Approach	Remarks	Ref
Ser44→Ala	2.5	49.4±0.1	0.0	0.0	heat	(1–3)	94C11
wild-type	2.5	47.8±0.1	−1.6±0.1	−1.8±0.1	heat	(1,2)	94C11
Ser44→AHA	2.5	49.2±0.1	−0.2±0.1	0.2±0.1	heat	(1,2,4)	94C11
Ser44→ADBA	2.5	46.9±0.1	−2.5±0.1	−2.9±0.1	heat	(1,2,5)	94C11
Ans68→Ala	2.5	48.0±0.1	0	0	heat	(1–3)	94C11
wild-type	2.5	47.8±0.1	−0.2±0.1	−0.2±0.1	heat	(1,2)	94C11
Asn68→AHA	2.5	48.1±0.1	0.1±0.1	−0.1±0.1	heat	(1,2,4)	94C11
Asn68→ADBA	2.5	49.0±0.1	1.0±0.1	1.1±0.1	heat	(1,2,5)	94C11

Remarks:
(1) wild-type contains Ser44 and Asn68
(2) $\Delta(\Delta G)$ was calculated from $\Delta(\Delta G) = \Delta T \times \Delta S$
(3) the alanine mutants were taken as reference proteins
(4) AHA = L-2-aminohexanoic acid
(5) ADBA = L-2-amino-3,3-dimethylbutanoic acid

c) Data on single and multiple mutations

Lysozyme phage T4, wild-type and mutants

Mutant	pH	ΔT	$\Delta(\Delta G)$	Approach	Remarks	Ref
wild-type	2.0	(40.4)	0.0	heat	(1)	91D1
wild-type	3.0	(52.6)	0.0	heat	(1)	91D1
wild-type·	4.0	(63.6)	0.0	heat	(1)	91D1
wild-type	5.5	(66.7)	0.0	heat	(1)	91D1
wild-type	6.5	(64.6)	0.0	heat	(1)	91D1
Lys60→His	2.0	−3.4	−3.3	heat, v.H.		91D1
Lys60→His	6.5	−0.4	−0.4	heat, v.H.		91D1
Lys83→His	2.0	−2.2	−2.1	heat, v.H.		91D1
Lys83→His	6.5	−1.0	−1.7	heat, v.H.		91D1
Ser90→His	2.0	−4.1	−4.2	heat, v.H.		91D1
Ser90→His	5.5	−2.7	−4.6	heat, v.H.		91D1
Ser90→His	6.5	−2.9	−4.6	heat, v.H.		91D1
Thr115→Glu	2.0	−1.7	−2.1	heat, v.H.		91D1
Thr115→Glu	3.0	−0.6	−0.8	heat, v.H.		91D1
Thr115→Glu	4.0	−0.3	−0.4	heat, v.H.		91D1
Thr115→Glu	5.5	0.1	0.2	heat, v.H.		91D1
Thr115→Glu	6.5	0.7	1.3	heat, v.H.		91D1
Gln123→Glu	2.0	1.0	1.3	heat, v.H.		91D1
Gln123→Glu	5.5	0.2	0.4	heat, v.H.		91D1
Gln123→Glu	6.5	1.2	1.7	heat, v.H.		91D1
Asn144→Glu	2.0	0.9	0.8	heat, v.H.		91D1
Asn144→Glu	5.5	1.4	2.1	heat, v.H.		91D1
Asn144→Glu	6.5	1.5	2.1	heat, v.H.		91D1
double mutant (Lys60→His and Leu13→Asp)						
	2.0	−6.1	−6.3	heat, v.H.		91D1
	5.5	−7.1	−11.7	heat, v.H.		91D1
	6.5	−7.1	−11.7	heat, v.H.		91D1
double mutant (Lys83→His and Ala112→Asp)						
	2.0	−4.8	−5.0	heat, v.H.		91D1
	5.5	−3.4	−5.4	heat, v.H.		91D1
	6.5	−3.9	−6.3	heat, v.H.		91D1

Table 1. Gibbs Energy Change – Molar Values

Lysozyme phage T4, wild-type and mutants (continued)

Mutant	pH	ΔT	Δ(ΔG)	Approach	Remark	Ref
double mutant (Ser90→His and Gln122→Asp)						
	2.0	−5.8	−5.9	heat, v.H.		91D1
	5.5	−4.1	−6.7	heat, v.H.		91D1
	6.5	−5.7	−9.2	heat, v.H.		91D1
double mutant (Thr115→Glu and Lys83→Met)						
	2.0	−1.5	−2.1	heat, v.H.		91D1
	6.5	0.1	0.2	heat, v.H.		91D1
double mutant (Asn144→Glu and Lys147→Met)						
	2.0	1.1	1.3	heat, v.H.		91D1
	5.5	1.2	2.1	heat, v.H.		91D1
	6.5	1.2	1.7	heat, v.H.		91D1

Remark:
(1) for the wild-type protein the transition temperature is given in brackets instead of ΔT, Δ(ΔG) refers to the wild-type protein at identical pH value

Lysozyme phage T4, wild-type and mutants

Mutant	pH	T	ΔT	Δ(ΔG)	Approach	Remarks	Ref
wild-type	3.0	40	0.0	0.0	heat	(1–4)	91D4
Ala98→Val	3.0	40	−14.8	−20.5	heat	(1–4)	91D4
Val149→Cys	3.0	40	−5.1	−9.2	heat	(1–4)	91D4
Thr152→Ser	3.0	40	−6.6	−10.9	heat	(1–4)	91D4
double mutant (Ala98→Val and Thr152→Ser)							
	3.0	40	−13.9	−20.1	heat	(1–4)	91D4
triple mutant (Ala98→Val, Val149→Cys and Thr152→Ser)							
	3.0	40	−12.1	−18.4	heat	(1–4)	91D4
triple mutant (Ala98→Val, Val149→Ile and Thr152→Ser)							
	3.0	40	−12.0	−18.4	heat	(1–4)	91D4

Remarks:
(1) Δ(ΔG) was extrapolated to 40°C
(2) ΔCp was taken from Ref. 89K4
(3) ΔT refers to the transition temperature of the wild-type protein (T_{trs} = 54.1°C)
(4) buffer: 20 mM potassium phosphate and 25 mM KCl

Lysozyme phage T4, wild-type and mutants

Mutant	pH	T	ΔT	Δ(ΔG)	Approach	Remarks	Ref
wild-type	2.83	50.9	0.0	0.0	heat	(1)	91D2
	5.3	66.7	0.0	0.0	heat	(1)	91D2
Lys16→Glu	2.83	49.4	−1.5	−2.1	heat	(1)	91D2
	5.3	67.8	1.1	2.1	heat	(1)	91D2
Arg119→Glu	2.83	50.3	−0.6	−0.8	heat	(1)	91D2
	5.3	66.6	0.1	0.2	heat	(1)	91D2
Lys135→Glu	2.83	50.2	−0.7	−0.8	heat	(1)	91D2
	5.3	64.4	−2.3	−4.2	heat	(1)	91D2
Lys147→Glu	2.83	51.1	0.2	0.4	heat	(1)	91D2
	5.3	65.1	−1.6	−2.9	heat	(1)	91D2
Arg154→Glu	2.83	51.5	0.6	0.8	heat	(1)	91D2
	5.3	64.1	−2.6	−4.6	heat	(1)	91D2

Lysozyme phage T4, wild-type and mutants (continued)

Mutant	pH	T	ΔT	Δ(ΔG)	Approach	Remarks	Ref
double mutant (Lys16→Glu and Arg119→Glu)							
	2.83	49.1	−1.8	−2.5	heat	(1)	91D2
	5.3	67.6	0.9	1.7	heat	(1)	91D2
double mutant (Lys16→Glu and Lys135→Glu)							
	2.83	48.3	−2.6	−3.8	heat	(1)	91D2
	5.3	65.1	−1.6	−2.9	heat	(1)	91D2
double mutant (Lys16→Glu and Arg154→Glu)							
	2.83	51.1	0.2	0.4	heat	(1)	91D2
	5.3	64.8	−1.9	−3.3	heat	(1)	91D2
double mutant (Arg119→Glu and Lys135→Glu)							
	2.83	49.4	−1.5	−2.1	heat	(1)	91D2
	5.3	64.1	−2.6	−4.6	heat	(1)	91D2
double mutant (Lys135→Glu and Lys147→Glu)							
	2.83	50.0	−0.9	−1.3	heat	(1)	91D2
	5.3	63.0	−3.7	−6.3	heat	(1)	91D2
triple mutant (Lys16→Glu, Lys135→Glu and Lys147→Glu)							
	2.83	48.4	−2.5	−3.3	heat	(1)	91D2
	5.3	62.7	−4.0	−7.1	heat	(1)	91D2
triple mutant (Arg119→Glu, Lys135→Glu and Lys147→Glu)							
	2.83	50.0	−0.9	−1.3	heat	(1)	91D2
	5.3	62.9	−3.8	−6.7	heat	(1)	91D2
multiple mutant (Lys16→Glu, Arg119→Glu, Lys135→Glu and Lys147→Glu)							
	2.83	48.3	−2.6	−0.9	heat	(1)	91D2
	5.3	64.2	−2.5	−4.2	heat	(1)	91D2

Remark:
(1) buffer: 20 mM potassium phosphate and 25 mM KCl

Lysozyme phage T4, pseudo wild-type (wt*) and mutants, alanine replacements within α-Helix 39-50

Mutant	pH	T_{ref}	ΔT	Δ(ΔG)	Approach	Remarks	Ref
wt*	3.0	51.65	0.0	0.0	heat	(1–4)	92H2
wt*	6.7	62.2	0.0	0.0	heat	(1–4)	92H2
Asn40→Ala	3.0	51.65	1.2	1.8	heat	(1–4)	92H2
Asn40→Ala	6.7	62.2	0.88	1.3	heat	(1–4)	92H2
Lys43→Ala	3.0	51.65	−2.95	−4.0	heat	(1–4)	92H2
Lys43→Ala	6.7	62.2	−3.08	−4.3	heat	(1–4)	92H2
Ser44→Ala	3.0	51.65	1.2	1.8	heat	(1–4)	92H2
Ser44→Ala	6.7	62.2	0.98	1.4	heat	(1–4)	92H2
Glu45→Ala	3.0	51.65	1.47	2.3	heat	(1–4)	92H2
Glu45→Ala	6.7	62.2	0.04	0.04	heat	(1–4)	92H2
Leu46→Ala	3.0	51.65	−8.39	−11.0	heat	(1–4)	92H2
Leu46→Ala	6.7	62.2	−6.4	−7.8	heat	(1–4)	92H2
Asp47→Ala	3.0	51.65	−0.81	−1.2	heat	(1–4)	92H2
Asp47→Ala	6.7	62.2	−2.75	−4.0	heat	(1–4)	92H2
Lys48→Ala	3.0	51.65	−0.93	−1.3	heat	(1–4)	92H2
Lys48→Ala	6.7	62.2	−1.68	−2.3	heat	(1–4)	92H2
double mutant (Glu45→Ala and Lys48→Ala)							
	3.0	51.65	1.04	1.6	heat	(1–4)	92H2
	6.7	62.2	0.03	0.04	heat	(1–4)	92H2

Lysozyme phage T4, pseudo wild-type (wt*) and mutants, alanine replacements within α-Helix 39-50 (continued)

Mutant	pH	T_{ref}	ΔT	$\Delta(\Delta G)$	Approach	Remarks	Ref
multiple Ala replacements in positions 40–49							
	3.0	51.65	−8.47	−8.8	heat	(1–4)	92H2
	6.7	62.2	−10.7	−10.6	heat	(1–4)	92H2
multiple Ala replacements in positions 40–42, 44–45, 47–49							
	3.0	51.65	3.09	4.4	heat	(1–4)	92H2
	6.7	62.2	−1.56	−2.3	heat	(1–4)	92H2

Remarks:
(1) ΔT and $\Delta(\Delta G)$ refer to the pseudo wild-type Cys54→Thr and Cys97→Ala (wt*) having T_{trs} = 51.65°C at pH 3.0 and 62.2°C at pH 6.7
(2) van't Hoff analysis using ΔCp = 10.5 kJ/mol/K
(3) error in T_{trs} ±0.15°C and in $\Delta(\Delta G)$ ±1.0 kJ/mol
(4) measured in 20 mM phosphate with 25 mM KCl at pH 3 and 10 mM phosphate with 0.15 M KCl at pH 6.7

Lysozyme Phage T4, core repacking variants

Mutant	pH	T_{trs}	ΔT	$\Delta(\Delta G)$	Approach	Remarks	Ref
wt*	5.4	65.3	0.0	0.0	heat	(1–3)	96B1
Leu121→Ala	5.4	59.2	−6.1	−9.6	heat	(1–3)	96B1
Ala129→Leu	5.4	62.1	−3.2	−5.4	heat	(1–3)	96B1
Ala129→Met	5.4	60.1	−5.2	−7.9	heat	(1–3)	96B1
Phe153→Ala	5.4	55.6	−9.3	−14.2	heat	(1,2,4)	93E2
double mutant (Leu121→Ala and Ala129→Leu)							
	5.4	62.5	−2.8	−4.6	heat	(1–3)	96B1
double mutant (Leu121→Ala and Ala129→Met)							
	5.4	62.6	−2.7	−4.2	heat	(1–3)	96B1
double mutant (Leu129→Met and Phe153→Ala)							
	5.4	52.6	−12.7	−18.0	heat	(1–3)	96B1

Remarks:
(1) wt* is the cysteine free pseudo wild-type, reference protein
(2) $\Delta(\Delta G)$ refers to T_{trs} of wt* at 59.0°C, calculated using ΔCp = 10.46 kJ/mol/K
(3) buffer: 0.10 M sodium chloride, 1.4 mM acetic acid, 8.6 mM sodium acetate
(4) data from Ref. 93E2, measured in 0.2 M KCl at pH 5.7, T_{trs} of the wt* amounts to 64.9°C in the buffer

Lysozyme phage T4, pseudo wild-type and mutants

Mutant	pH	T	ΔT	$\Delta(\Delta G)$	Approach	Remarks	Ref
wt*	3.01	51.76	0.0	0.0	heat	(1,2)	92H10
wt*	5.70	64.89	0.0	0.0	heat	(1,3)	92H10
Met102→Leu	3.01	51.76	−2.54	−3.0±0.8	heat	(4)	92H10
Met102→Leu	3.01	51.76	−2.54	−3.1±0.8	heat	(5)	92H10
Met102→Leu	5.70	64.89	−2.31	−3.3±0.8	heat	(4)	92H10
Met102→Leu	5.70	64.89	−2.31	−3.4±0.8	heat	(5)	92H10
Val111→Ile	3.01	51.76	−2.32	−2.8±0.8	heat	(4)	92H10
Val111→Ile	3.01	51.76	−2.32	−2.9±0.8	heat	(5)	92H10
Val111→Ile	5.70	64.89	−2.60	−3.4±0.8	heat	(4)	92H10
Val111→Ile	5.70	64.89	−2.60	−3.6±0.8	heat	(5)	92H10
Val111→Phe	3.01	51.76	−4.77	−5.6±0.8	heat	(4)	92H10
Val111→Phe	3.01	51.76	−4.77	−6.0±0.8	heat	(5)	92H10

Lysozyme phage T4, pseudo wild-type and mutants (continued)

Mutant	pH	T	ΔT	Δ(ΔG)	Approach	Remarks	Ref
Val111→Phe	5.70	64.89	−4.63	−6.6±0.8	heat	(4)	92H10
Val111→Phe	5.70	64.89	−4.63	−7.1±0.8	heat	(5)	92H10
double mutant (Leu99→Phe and Met102→Leu)							
	3.01	51.76	−2.34	−2.9±0.8	heat	(4)	92H10
	3.01	51.76	−2.34	−3.0±0.8	heat	(5)	92H10
	5.70	64.89	−2.21	−3.1±0.8	heat	(4)	92H10
	5.70	64.89	−2.21	−3.1±0.8	heat	(5)	92H10
double mutant (Leu99→Phe and Val111→Ile)							
	3.01	51.76	−2.70	−3.3±0.8	heat	(4)	92H10
	3.01	51.76	−2.70	−3.4±0.8	heat	(5)	92H10
	5.70	64.89	−2.51	−3.7±0.8	heat	(4)	92H10
	5.70	64.89	−2.51	−3.9±0.8	heat	(5)	92H10
double mutant (Leu99→Phe and Phe153→Leu)							
	3.01	51.76	+0.09	+0.1±0.8	heat	(4)	92H10
	3.01	51.76	+0.09	+0.1±0.8	heat	(5)	92H10
	5.70	64.89	+0.03	+0.9±0.8	heat	(4)	92H10
	5.70	64.89	+0.03	+0.9±0.8	heat	(5)	92H10
double mutant (Met102→Leu and Val111→Phe)							
	3.01	51.76	−5.49	−5.8±0.8	heat	(4)	92H10
	3.01	51.76	−5.49	−6.3±0.8	heat	(5)	92H10
	5.70	64.89	−6.19	−8.0±0.8	heat	(4)	92H10
	5.70	64.89	−6.19	−8.8±0.8	heat	(5)	92H10
double mutant (Val111→Ile and Phe153→Leu)							
	3.01	51.76	−3.52	−4.4±0.8	heat	(4)	92H10
	3.01	51.76	−3.52	−4.6±0.8	heat	(5)	92H10
	5.70	64.89	−3.24	−4.3±0.8	heat	(4)	92H10
	5.70	64.89	−3.24	−4.6±0.8	heat	(5)	92H10
triple mutant (Leu99→Phe, Met102→Leu and Val111→Ile)							
	3.01	51.76	−4.02	−4.6±0.8	heat	(4)	92H10
	3.01	51.76	−4.02	−4.8±0.8	heat	(5)	92H10
	5.70	64.89	−3.83	−4.7±0.8	heat	(4)	92H10
	5.70	64.89	−3.83	−5.0±0.8	heat	(5)	92H10
triple mutant (Leu99→Phe, Met102→Leu and Phe153→Leu)							
	3.01	51.76	−0.68	−0.9±0.8	heat	(4)	92H10
	3.01	51.76	−0.68	−0.9±0.8	heat	(5)	92H10
	5.70	64.89	−0.74	−0.9±0.8	heat	(4)	92H10
	5.70	64.89	−0.74	−0.9±0.8	heat	(5)	92H10
triple mutant (Leu99→Phe, Val111→Ile and Phe153→Leu)							
	3.01	51.76	−1.73	−2.3±0.8	heat	(4)	92H10
	3.01	51.76	−1.73	−2.3±0.8	heat	(5)	92H10
	5.70	64.89	−1.25	−1.8±0.8	heat	(4)	92H10
	5.70	64.89	−1.25	−1.8±0.8	heat	(5)	92H10
mutliple mutant (Leu99→Phe, Met102→Leu, Val111→Ile and Phe153→Leu)							
	3.01	51.76	−1.82	−2.2±0.8	heat	(4)	92H10
	3.01	51.76	−1.82	−2.3±0.8	heat	(5)	92H10
	5.70	64.89	−2.07	−2.6±0.8	heat	(4)	92H10
	5.70	64.89	−2.07	−2.7±0.8	heat	(5)	92H10

Remarks:
(1) pseudo wild-type (Cys54→Thr and Cys97→Ala)
(2) reference value for pH 3.01
(3) reference value for pH 5.70
(4) Δ(ΔG) was obtained using Δ(ΔG) = ΔT × ΔS
(5) Δ(ΔG) was calculated by means of the constant ΔCp model using ΔCp = 10.46 kJ/mol/K at pH 3.01, and
ΔCp = 14.64 kJ/mol/K at pH 5.70 from Ref. 89K4

Table 1. Gibbs Energy Change – Molar Values

Lysozyme phage T4, wild-type and core packing variants concerning positions 121, 129, 133, 149, and 153

Mutant	pH	T_{ref}	$\Delta(\Delta G)$	Approach	Remarks	Ref
wild-type	3.0	51.7	0.0	heat	(1,2)	93B1
multiple mutants:						
(Leu121→Ala, Ala129→Met and Phe153→Leu)						
	3.0	51.7	−4.6	heat	(2)	93B1
(Leu121→Met, Ala129→Leu, Leu133→Met, Val149→Ile and Phe153→Trp)						
	3.0	51.7	−5.4	heat	(2)	93B1
(Leu121→Ile, Ala129→Leu, Leu133→Met and Phe153→Trp)						
	3.0	51.7	−5.4	heat	(2)	93B1
(Leu121→Ala, Ala129→Met and Val149→Ile)						
	3.0	51.7	−5.9	heat	(2)	93B1
(Leu121→Ile, Ala129→Trp and Leu133→Met)						
	3.0	51.7	−5.9	heat	(2)	93B1
(Leu121→Ala, Ala129→Val, Leu133→Met and Phe153→Leu)						
	3.0	51.7	−9.6	heat	(2)	93B1
(Leu121→Met, Leu133→Val and Phe153→Leu)						
	3.0	51.7	−10.5	heat	(2)	93B1
(Leu121→Ala, Ala129→Val, Leu133→Ala and Phe153→Leu)						
	3.0	51.7	−14.6	heat	(2)	93B1

Remarks:
(1) wild-type containing Leu121, Ala129, Leu133, Val149, and Phe153
(2) van't Hoff treatment using $\Delta C_p = 10.5$ kJ/mol/K

Lysozyme phage T4, pseudo wild-type (wt*) and mutants

Mutant	pH	T	ΔT	$\Delta(\Delta G)$	Approach	Remarks	Ref
wt*	5.4	65.12±0.2	0.0	0.0	heat	(1–3)	93P9
wt*	6.78	62.19±0.2	0.0	0.0	heat	(1–3)	93P9
Arg14→Lys	5.4	65.12	−0.08	−0.1	heat	(1–3)	93P9
Arg14→Lys	6.78		−0.53		heat	(1–3)	93P9
Glu22→Lys	5.4	65.12	1.37	2.4	heat	(1–3)	93P9
Glu22→Lys	6.78		1.58		heat	(1–3)	93P9
Thr26→Ser	5.4	65.12	1.35	2.4	heat	(1–3)	93P9
Thr26→Ser	6.78		1.35		heat	(1–3)	93P9
Asn40→Asp	5.4	65.12	1.14	1.8	heat	(1–3)	93P9
Asn40→Asp	6.78		1.28		heat	(1–3)	93P9
Ala41→Asp	5.4	65.12	0.71	1.2	heat	(1–3)	93P9
Ala41→Asp	6.78		1.10		heat	(1–3)	93P9
Ala41→Val	5.4		0.58	1.1	heat	(1–4)	93P9
Asp70→Lys	5.4	57.2	1.7	1.5	heat	(1–3,5)	93P9
Arg80→Lys	5.4	65.12	−0.43	−0.7	heat	(1–3)	93P9
Arg80→Lys	6.78		−0.34		heat	(1–3)	93P9
double mutant (Arg80→Lys and Arg119→His):							
	5.4	65.12	−1.20	−2.0	heat	(1–3)	93P9
	6.78		−1.25		heat	(1–3)	93P9
Ala93→Thr	5.4	65.12	0.13	0.3	heat	(1–3)	93P9
Ala93→Thr	6.78		0.16		heat	(1–3)	93P9
Gly113→Glu	5.4	65.12	0.79	1.3	heat	(1–3)	93P9
Gly113→Glu	6.78		1.03		heat	(1–3)	93P9
Arg119→His	5.4	65.12	−0.74	−1.2	heat	(1–3)	93P9
Arg119→His	6.78		−0.83		heat	(1–3)	93P9
Thr151→Ser	5.4	65.12	0.93	1.6	heat	(1–3)	93P9

Lysozyme phage T4, pseudo wild-type (wt*) and mutants (continued)

Mutant	pH	T	ΔT	Δ(ΔG)	Approach	Remarks	Ref
Thr151→Ser	6.78		0.97		heat	(1–3)	93P9
Phe153→Leu	5.4	65.12	0.88	1.5	heat	(1–3)	93P9
Phe153→Leu	6.78		0.51		heat	(1–3)	93P9
Asn163→Asp	5.4	65.12	−0.50	−0.9	heat	(1–3)	93P9
Asn163→Asp	6.78		−0.49		heat	(1–3)	93P9

Remarks:
(1) the data refer to the pseudo wild-type protein (wt*) if not otherwise indicated
(2) buffer pH 5.4: 10 mM sodium acetate, 0.1 M NaCl, buffer pH 6.78: 10 mM potassium phosphate, 0.15 M KCl
(3) Δ(ΔG) was obtained using Δ(ΔG) = ΔT × ΔS
(4) the data refer to the wild-type protein
(5) the data refer to the mutant His31→Asn

Lysozyme phage T4, combination of stability enhancing point mutations

Mutant	pH	ΔT	Δ(ΔG)	Approach	Remarks	Ref
wild-type	3.0		0.0	heat	(1–4)	95Z1
	5.4		0.0	heat	(1–4)	95Z1
Ile3→Leu	3.0	2.11	3.5	heat	(1–3)	95Z1
	5.4	1.69	2.8	heat	(1–3)	95Z1
Ser38→Asp	3.0	0.46	0.9	heat	(1–3)	95Z1
	5.4	1.36	2.5	heat	(1–3)	95Z1
Ala41→Val	3.0	1.10	1.7	heat	(1–3)	95Z1
	5.4	0.70	1.3	heat	(1–3)	95Z1
Ala82→Pro	3.0	1.08	1.9	heat	(1–3)	95Z1
	5.4	1.34	2.4	heat	(1–3)	95Z1
Asn116→Asp	3.0	0.90	1.4	heat	(1–3)	95Z1
	5.4	1.36	2.5	heat	(1–3)	95Z1
Val131→Ala	3.0	0.72	1.4	heat	(1–3)	95Z1
	5.4	0.87	1.6	heat	(1–3)	95Z1
Asn144→Asp	3.0	0.94	1.7	heat	(1–3)	95Z1
	5.4	0.99	1.7	heat	(1–3)	95Z1
Val131→Ala	3.0	0.72	1.4	heat	(1–3)	95Z1
	5.4	0.87	1.6	heat	(1–3)	95Z1
double mutant (Val131→Ala and Ala41→Val)						
	3.0	1.83	3.2	heat	(1–3)	95Z1
	5.4	1.51	2.8	heat	(1–3)	95Z1
double mutant (Ser38→Asp and Asn144→Asp)						
	3.0	1.35	2.5	heat	(1–3)	95Z1
	5.4	2.36	4.3	heat	(1–3)	95Z1
triple mutant (Ser38→Asp, Ala82→Pro and Asn144→Asp)						
	3.0	2.58	4.4	heat	(1–3)	95Z1
	5.4	3.75	6.7	heat	(1–3)	95Z1
multiple mutant (Ser38→Asp, Ala82→Pro, Asn144→Asp and Ile3→Leu)						
	3.0	4.87	8.2	heat	(1–3)	95Z1
	5.4	5.38	9.7	heat	(1–3)	95Z1
multiple mutant (Ser38→Asp, Ala82→Pro, Asn144→Asp, Ile3→Leu and Val131→Ala)						
	3.0	5.56	9.4	heat	(1–3)	95Z1
	5.4	6.06	10.8	heat	(1–3)	95Z1

Lysozyme phage T4, combination of stability enhancing point mutations (continued)

Mutant	pH	ΔT	$\Delta(\Delta G)$	Approach	Remarks	Ref
multiple mutant (Ser38→Asp, Ala82→Pro, Asn144→Asp, Ile3→Leu, Val131→Ala and Ala41→Val)						
	3.0	6.50	11.0	heat	(1–3)	95Z1
	5.4	6.91	12.0	heat	(1–3)	95Z1
multiple mutant (Ser38→Asp, Ala82→Pro, Asn144→Asp, Ile3→Leu, Val131→Ala, Ala41→Val and Asn116→Asp)						
	3.0	7.26	12.2	heat	(1–3)	95Z1
	5.4	8.32	14.9	heat	(1–3)	95Z1

Remarks:
(1) buffer: 25 mM KCl, 3 mM H_3PO_4, 17 mM KH_2PO_4 pH 3.01, and 100 mM NaCl, 1.4 mM acetic acid, 8.6 mM sodium acetate, pH 5.42
(2) $\Delta(\Delta G)$ was calculated at 57°C at pH 3.0 and 70°C at pH 5.4 taking ΔCp to be 7.5 kJ/mol/K at pH 3.0 and 10.5 kJ//mol/K at pH 5.4; ΔH was determined by van't Hoff treatment
(3) estimated errors in ΔT ±25°C and in $\Delta(\Delta G)$ ±0.4 kJ/mol
(4) T_{trs} of wild-type amounts to 53.42°C at pH 3.0 and 66.51°C at pH 5.4

Lysozyme phage T4, spin-labeled mutants

Mutant	pH	ΔT	$\Delta(\Delta G)$	Approach	Remarks	Ref
Asp22→R1	2.95	0.0	0.0	heat	(1–3)	96M4
Asn40→R1	2.95	−1.2	−1.7	heat	(1–3)	96M4
Ser44→R1	2.95	0.7	1.0	heat	(1–3)	96M4
Lys48→R1	2.95	−3.2	−4.6	heat	(1–3)	96M4
Gly51→R1	2.95	−2.5	−3.3	heat	(1–3)	96M4
Asp61→R1	2.95	−1.0	−1.3	heat	(1–3)	96M4
Lys65→R1	2.95	−1.0	−1.3	heat	(1–3)	96M4
Asp72→R1	2.95	0.4	0.6	heat	(1–3)	96M4
Ala74→R1	2.95	0.0	0.0	heat	(1–3)	96M4
Val75→R1	2.95	−3.2	−4.6	heat	(1–3)	96M4
Arg80→R1	2.95	−4.5	−5.0	heat	(1–3)	96M4
Asn81→R1	2.95	−2.9	−4.2	heat	(1–3)	96M4
Val87→R1	2.95	−11.0	−16.7	heat	(1–3)	96M4
Leu99→R1	2.95	−13.3	−17.6	heat	(1–3)	96M4
Ala130→R1	2.95	−3.1	−4.2	heat	(1–3)	96M4
Val131→R1	2.95	0.5	0.7	heat	(1–3)	96M4
Asn132→R1	2.95	1.5	4.2	heat	(1–3)	96M4
Leu133→R1	2.95	−9.5	−13.0	heat	(1–3)	96M4
Lys135→R1	2.95	−1.8	−2.5	heat	(1–3)	96M4
Ile150→R1	2.95	−4.8	−6.3	heat	(1–3)	96M4
Phe153→R1	2.95	−13.0	−17.2	heat	(1–3)	96M4
double mutant (Lys65→R1 and Asn68→Ala)						
	2.95	−2.0	−2.9	heat	(1–3)	96M4
double mutant (Lys65→R1 and Asn68→Gly)						
	2.95	−4.0	−5.4	heat	(1–3)	96M4
multiple mutant (Asp72→R1, Asn68→Ala, Gln69→Ala and Arg76→Ala)						
	2.95	−2.2	−2.6	heat	(1–3)	96M4

Remarks:
(1) reference protein is the pseudo wild-type Cys54→Thr and Cys97→Ala
(2) R1 = the corresponding position of the protein was replaced by cysteine and labeled with a sulfhydryl-specific nitroxide reagent
(3) measured in 20 mM potassium phosphate and 25 mM KCl, the transition was monitored by CD at 223 nm

d) Proline mutations

Lysozyme phage T4, mutants X→Pro

Mutant	pH	T	ΔT	Δ(ΔG)	Approach	Remarks	Ref
Gln69→Pro	2.0	38.5	−12.9	−13.8	heat	(1,2)	92S4
	6.5	63.2	−7.6	−12.1	heat	(1,2)	92S4
Asp72→Pro	2.0	38.5	−10.2	−10.9	heat	(1,3)	92S4
	6.5	63.4	−7.1	−11.3	heat	(1,3)	92S4
Ala74→Pro	3.5	39.3	−18.2	−23.8	heat	(1,4)	92S4
	4.0	62.2	−15.3	−20.9	heat	(1,4)	92S4
	5.5	65.7	−12.4	−18.8	heat	(1,4)	92S4
	6.5	62.9	−12.1	−19.2	heat	(1,4)	92S4

Remarks:
(1) the data refer to the Cys-free pseudo wild-type, T is the thermal transition temperature of the reference
    protein
(2) T_{trs} of the mutant is 25.6°C at pH 2.0 and 55.6°C at pH 6.5
(3) T_{trs} of the mutant is 28.3°C at pH 2.0 and 56.3°C at pH 6.5
(4) at the given pH values is T_{trs} of the mutant 39.3, 46.9, 53.3, and 62.9°C, respectively

Lysozyme phage T4, wild-type and mutants Ala82→Pro and Ala93→Pro

Mutant	pH	T	Δ(ΔG)	Approach	Remarks	Ref
Ala82→Pro	2.0	38.8	−0.29	DSC	(1)	92H7
Ala82→Pro	2.5	46.2	2.13	DSC	(1)	92H7
Ala82→Pro	3.0	53.6	5.31	DSC	(1)	92H7
Ala93→Pro	2.0	38.8	0.13	DSC	(1)	92H7
Ala93→Pro	2.5	46.2	2.13	DSC	(1)	92H7
Ala93→Pro	3.0	53.6	4.60	DSC	(1)	92H7

Remark:
(1) Δ(ΔG) refers to T_{trs} of the wild-type protein

Lysozyme phage T4, mutations Pro86→X

Mutant	pH	T	Δ(ΔG)	Approach	Remarks	Ref
wild-type	2	42	0	heat	(1)	88A
Pro86→Ala	2	40	−1 to −2	heat		88A
Pro86→Arg	2	40	−1 to −2	heat		88A
Pro86→Asp	2	42	0	heat		88A
Pro86→Cys	2	41	−1 to −2	heat		88A
Pro86→Gly	2	40	−1 to −2	heat		88A
Pro86→His	2	40	−1 to −2	heat		88A
Pro86→Ile	2	40	−1 to −2	heat		88A
Pro86→Leu	2	40	−1 to −2	heat		88A
Pro86→Ser	2	41	−1 to −2	heat		88A
Pro86→Thr	2	41	−1 to −2	heat		88A
wild-type	4	63	0	heat	(1)	88A
Pro86→Arg	4	62	−1 to −2	heat		88A
Pro86→Asp	4	63	0	heat		88A
Pro86→Gly	4	62	−1 to −2	heat		88A
wild-type	6	66	0	heat	(1)	88A
Pro86→Arg	6	63	−1 to −2	heat		88A
Pro86→Asp	6	65	0	heat		88A
Pro86→His	6	62	−1 to −2	heat		88A

Remark:
(1) the indicated Δ(ΔG) value serves as the reference value for the mutants at the same pH

Lysozyme phage T4, wild-type and mutations X→Pro

Mutant	pH	T	ΔT	Δ(ΔG)	Approach	Remarks	Ref
wild-type	2.0	41.9	0.0	0.0	heat	(1)	92N2
Lys60→Pro	2.0	41.9	0.3±0.5	0.4±0.6	heat	(1)	92N2
Gly77→Ala	2.0	41.9	−1.4±0.8	−1.7±0.8	heat	(1)	92N2
Ala82→Pro	2.0	41.9	0.8±0.4	1.3±0.6	heat	(1)	92N2
Ala93→Pro	2.0	41.9	0.4±0.4	0.4±0.4	heat	(1)	92N2
Gly113→Ala	2.0	41.9	0.4±0.6	0.4±0.6	heat	(1)	92N2
wild-type	6.5	64.7	0.0	0.0	heat	(2)	92N2
Lys60→Pro	6.5	64.7	−0.1±0.7	0.0±1.0	heat	(2)	92N2
Gly77→Ala	6.5	64.7	0.9±0.5	1.7±0.8	heat	(2)	92N2
Ala82→Pro	6.5	64.7	2.1±0.5	3.3±0.8	heat	(2)	92N2
Ala93→Pro	6.5	64.7	0.2±0.6	0.4±0.8	heat	(2)	92N2
Gly113→Ala	6.5	64.7	0.8±0.6	1.3±0.8	heat	(2)	92N2

Remarks:
(1) buffer: 0.2 M KCl, 10 mM HCl, and 10 mM H_3PO_4
(2) buffer: 0.15 M KCl, 20 mM potassium phosphate

Lysozyme phage T4, wild-type and proline mutations X→Pro

Mutant	pH	T	ΔG	Approach	Remarks	Ref
wild-type	7.0	25	69.2	urea	(1)	96G7
Leu91→Pro	7.0	25	23.3	urea	(1,2)	96G7
Leu99→Pro	7.0	25	18.1	urea	(1,3)	96G7

Remarks:
(1) linear extrapolation
(2) thermal transition temperature that amounts $T_{trs} = 65°C$ for wild-type is reduced by $\Delta T = -34.3°C$
(3) thermal transition temperature that amounts $T_{trs} = 65°C$ for wild-type is reduced by $\Delta T = -38.2°C$

e) Insertion mutants

Lysozyme phage T4, insertion mutants

Mutant	pH	T	ΔT	Δ(ΔG)	Approach	Remarks	Ref
(Asn40-Ala-Ala41)							
	5.4	53	−7.9	−11.7	heat	(1–3)	93H4
(Asn40→Leu and Leu40-Ala-Ala41)							
	5.4	53	−3.9	−5.9	heat	(1–3)	93H4
(Asn40-Ala-Ala-Ala41)							
	5.4	53	−3.0	−5.0	heat	(1–3)	93H4
(Asn40-Ala-Ala-Ala-Ala41)							
	5.4	53	−4.8	−7.1	heat	(1–3)	93H4
(Asn40-Ala-Ala-Ala-Ala-Ala41)							
	5.45	53	−6.4	−9.2	heat	(1–3)	94H7
(Asn40-Ser-Leu-Asp-Ala41)							
	5.4	53	−2.0	−3.3	heat	(1–3)	93H4
(Ser44-Ala-Glu45)							
	5.45	53	−11.7	−15.9	heat	(1–3)	93H4
(Ser44-Ala-Ala-Glu45)							
	5.4	53	−9.9	−13.8	heat	(1–3)	93H4
(Ser44-Ala-Ala-Ala-Glu45)							
	5.45	53	−7.3	−10.5	heat	(1–3)	94H7

Lysozyme phage T4, insertion mutants

Mutant	pH	T	ΔT	Δ(ΔG)	Approach	Remarks	Ref
(Ser44→Ala and Ala44-Ala-Ala-Glu45)							
	5.4	53	−4.4	−7.1	heat	(1–3)	93H4
(Ser44-Ala-Ala-Ala-Glu45)							
	5.4	53	−6.0	−9.2	heat	(1–3)	93H4
(Lys48-Ala-Ala49)							
	5.4	53	−11.4	−14.6	heat	(1–3)	93H4
(Lys48-Ala-Ala-Ala49)							
	5.4	53	−10.5	−14.2	heat	(1–3)	93H4
(Lys48-His-Pro-Ala49)							
	5.4	53	−7.1	−10.5	heat	(1–3)	93H4
(Lys48-Ala-Ala-Ala-Ala49)							
	5.4	53	−14.8	−17.6	heat	(1–3)	93H4
(Lys48-Ala-Ala-Ala-Ala-Ala49)							
	5.45	53	−12.5	−16.7	heat	(1–3)	94H7
(Asn40→Leu, Lys43→Ala and Ser44-Ala-Glu45)							
	5.45	53	−5.4	−7.5	heat	(1–3)	94H7
(Asn40-Glu-Ser-Ala41)							
	5.45	53	−1.6	−2.9	heat	(1–3)	94H7
(Asn40-Ser-Leu-Asp-Ala41 and Leu46→Ala)							
	5.45	53	−8.1	−11.7	heat	(1–3)	94H7
(Ser44-Ala-Ala-Glu45 and Leu46→Ala)							
	5.45	53	−11.2	−15.5	heat	(1–3)	94H7
(Lys48-Leu-Pro-Ala49)							
	5.45	53	−7.3	−10.5	heat	(1–3)	94H7
(Asn40-Ala-Ala-Ala41 and Lys48-Leu-Pro-Ala49)							
	5.45	53	−9.8	−13.8	heat	(1–3)	94H7
(Leu39-Ala-Asn40)							
	5.45	53	−2.5	−3.8	heat	(1–3)	94H7
(Ala42-Lys-Lys43)							
	5.45	53	−11.0	−15.5	heat	(1–3)	94H7

Remarks:
(1) the data refer to the pseudo wild-type having T_{trs} = 65.15 °C and ΔH = 569 kJ/mol
(2) Δ(ΔG) was calculated for 53°C using ΔCp = 18.8 kJ/mol/K
(3) buffer: 10 mM sodium acetate with 0.1 M NaCl

Lysozyme phage T4, pseudo wild-type (wt*), extended, and permuted form

Mutant	pH	T	Δ(ΔG)	Approach	Remarks	Ref
wt*	7.4	22	0.0	GuHCl	(1,2)	93Z3
extended	7.4	22	0	GuHCl	(1–3)	93Z3
permuted	7.4	22	−4.2	GuHCl	(1,2,4)	93Z3
wt*	7.0	22	0	urea	(5)	93Z3
extended	7.0	22	0.8	urea	(3,5)	93Z3
permuted	7.0	22	−4.6	urea	(4,5)	93Z3
wt*	2.5	45	0	DSC	(6,7)	93Z3
extended	2.5	45	−1.3±0.4	DSC	(3,7)	93Z3
permuted	2.5	45	−4.6±0.4	DSC	(4,7)	93Z3

Remarks:
(1) linear extrapolation
(2) buffer: 50 mM Tris-HCl, 1 mM MgCl$_2$, 0.1 mM DTT
(3) the protein contains at the C-terminus additionally Ser-4*(Gly)-Ala
(4) the permuted protein begins at residue 37 of the pseudo wild-type and ends at residue 36; the chain termini are joined by the linker peptide Ser-4*(Gly)-Ala; additional amino acid replacements are: Pro37→Met, Ser38→Asp, and Asn2→Asp
(5) urea gradient electrophoresis, buffer: 50 M MOPS/imidazole
(6) data from Ref. 89K4
(7) buffer: 20 mM potassium phosphate, 25 mM KCl, 0.1 mM DTT

Lysozyme phage T4, insertion, deletion, and extension mutants

Mutant	pH	T	Δ(ΔG)	Approach	Remarks	Ref
Insertions:						
(Phe4-Ala-Glu5)	3.0	42	−15.5	heat	(1–6)	96V4
(Phe4-Ala-Glu5)	5.4	53	−13.0	heat	(1–6)	96V4
(Arg8-Ala-Ile9)	3.0	42	n.d.	heat	(1–6)	96V4
(Arg8-Ala-Ile9)	5.4	53	−23.4	heat	(1–6)	96V4
(Asn40-Ala-Ala41)	3.0	42	−10.9	heat	(1–6)	96V4
(Asn40-Ala-Ala41)	5.4	53	−11.7	heat	(1–7)	96V4
(Ser44-Ala-Glu45)	3.0	42	−13.4	heat	(1–6)	96V4
(Ser44-Ala-Glu45)	5.4	53	−15.9	heat	(1–7)	96V4
(Lys48-Ala-Ala49)	3.0	42	−12.1	heat	(1–6)	96V4
(Lys48-Ala-Ala49)	5.4	53	−15.5	heat	(1–7)	96V4
(Glu64-Ala-Lys65)	3.0	42	−16.7	heat	(1–6)	96V4
(Glu64-Ala-Lys65)	5.4	53	−15.5	heat	(1–6)	96V4
(Asn68-Ala-Gln69)	3.0	42	−19.7	heat	(1–6)	96V4
(Asn68-Ala-Gln69)	5.4	53	−18.4	heat	(1–6)	96V4
(Ala73-Ala-Ala74)	3.0	42	−22.2	heat	(1–6)	96V4
(Ala73-Ala-Ala74)	5.4	53	−17.6	heat	(1–6)	96V4
(Val75-Ala-Arg76)	3.0	42	−14.6	heat	(1–6)	96V4
(Val75-Ala-Arg76)	5.4	53	−11.3	heat	(1–6)	96V4
(Tyr88-Ala-Asp89)	3.0	42	−17.6	heat	(1–6)	96V4
(Tyr88-Ala-Asp89)	5.4	53	−16.3	heat	(1–6)	96V4
(Arg96-Ala-Ala97)	3.0	42	n.d.	heat	(1–6)	96V4
(Arg96-Ala-Ala97)	5.4	53	−27.6	heat	(1–6)	96V4
(Glu108-Ala-Thr109)	3.0	42	−11.7	heat	(1–6)	96V4
(Glu108-Ala-Thr109)	5.4	53	−11.3	heat	(1–6)	96V4
(Thr115-Ala-Asn116)	3.0	42	−8.4	heat	(1–6)	96V4
(Thr115-Ala-Asn116)	5.4	53	−7.5	heat	(1–6)	96V4
(Arg119-Ala-Met120)	3.0	42	−10.9	heat	(1–6)	96V4
(Arg119-Ala-Met120)	5.4	53	−10.9	heat	(1–6)	96V4

Lysozyme phage T4, insertion, deletion, and extension mutants (continued)

Mutant	pH	T	Δ(ΔG)	Approach	Remarks	Ref
(Asp127-Ala-Glu128)	3.0	42	−13.8	heat	(1–6)	96V4
(Asp127-Ala-Glu128)	5.4	53	−14.6	heat	(1–6)	96V4
(Val131-Ala-Asn132)	3.0	42	−12.6	heat	(1–6)	96V4
(Val131-Ala-Asn132)	5.4	53	−12.1	heat	(1–6)	96V4
(Asn140-Ala-Gln141)	3.0	42	−9.2	heat	(1–6)	96V4
(Asn140-Ala-Gln141)	5.4	53	−8.8	heat	(1–6)	96V4
(Asn144-Ala-Arg145)	3.0	42	−3.3	heat	(1–6)	96V4
(Asn144-Ala-Arg145)	5.4	53	−2.9	heat	(1–6)	96V4
(Lys147-Ala-Arg148)	3.0	42	−7.9	heat	(1–6)	96V4
(Lys147-Ala-Arg148)	5.4	53	−8.8	heat	(1–6)	96V4
(Arg148-Ala-Val149)	3.0	42	−16.7	heat	(1–6)	96V4
(Arg148-Ala-Val149)	5.4	53	−19.2	heat	(1–6)	96V4
(Ile150-Ala-Thr151)	3.0	42	−16.7	heat	(1–6)	96V4
(Ile150-Ala-Thr151)	5.4	53	−25.0	heat	(1–6)	96V4
Deletions:						
Arg8 deleted	3.0	42	n.d.	heat	(1–6)	96V4
Arg8 deleted	5.4	53	−22.6	heat	(1–6)	96V4
Ser44 deleted	3.0	42	−12.6	heat	(1–6)	96V4
Ser44 deleted	5.4	53	−11.7	heat	(1–6)	96V4
Ala73 deleted	3.0	42	−19.2	heat	(1–6)	96V4
Ala73 deleted	5.4	53	−16.3	heat	(1–6)	96V4
Arg119 deleted	3.0	42	−15.9	heat	(1–6)	96V4
Arg119 deleted	5.4	53	−15.1	heat	(1–6)	96V4
Asp127 deleted	3.0	42	−11.3	heat	(1–6)	96V4
Asp127 deleted	5.4	53	−13.0	heat	(1–6)	96V4
Extension:						
(Leu164-Ala-Ala-Ala-Ala)						
	3.0	42	+1.3	heat	(1–6)	96V4
	5.4	53	+0.8	heat	(1–6)	96V4

Remarks:
(1) the data refer to the cysteine-free pseudo wild-type (wt*, Cys54→Thr and Cys97→Ala)
(2) the thermal stabilities were determined in 0.10 M NaCl, 1.4 mM acetic acid, 8.6 mM sodium acetate, pH 5.42, and in 0.025 M KCl, 2.95 mM H_3PO_4, 17.0 mM KH_2PO_4, pH 3.01
(3) van't Hoff treatment of transitions monitored by CD at 223 nm
(4) ΔCp was taken as 10.46 kJ/mol/K at 53°C for pH 5.4 buffer, and as 7.53 kJ/mol/K at 42°C for pH 3.0 buffer according to Ref. 95Z1
(5) the error in T_{trs} is ±0.15°C near T_{trs} of wt* and increased to approximately ±0.35°C as the ΔH value decreased
(6) the error in Δ(ΔG) is ±0.6 kJ/mol for the most stable mutants and the value increased to approximately ±4.2 kJ/mol for the least stable mutants
(7) data from Refs. 93H4 and 94H7

Lysozyme phage T4, insertion mutants following Ala73 and Arg148

Mutant	pH	ΔT	Δ(ΔG)	Approach	Remarks	Ref
(Ala73-Ala)	3.0	−19.7	−22.2	heat	(1–7)	96V4
(Ala73-Ala)	5.4	−12.5	−17.6	heat	(1–6)	96V4
(Ala73-Ala-Ala)	3.0	−29.2	−25.9	heat	(1–7)	96V4
(Ala73-Ala-Ala)	5.4	−18.0	−23.8	heat	(1–7)	96V4
(Ala73-Ala-Ala-Ala)	3.0	−21.6	−20.1	heat	(1–7)	96V4
(Ala73-Ala-Ala-Ala)	5.4	−15.9	−20.9	heat	(1–7)	96V4
(Ala73-Leu)	3.0	−14.2	−18.0	heat	(1–6)	96V4
(Ala73-Leu)	5.4	−9.0	−13.0	heat	(1–6)	96V4
(Ala73-Arg)	3.0	−17.6	−21.3	heat	(1–7)	96V4
(Ala73-Arg)	5.4	−8.2	−12.1	heat	(1–6)	96V4
(Ala73-Val-Leu)	3.0	−5.8	−8.4	heat	(1–6)	96V4
(Ala73-Val-Leu)	5.4	−5.2	−7.1	heat	(1–6)	96V4
(Arg148-Ala)	3.0	−14.4	−16.7	heat	(1–6)	96V4
(Arg148-Ala)	5.4	−13.7	−19.2	heat	(1–6)	96V4
(Arg148-Ala-Ala)	3.0	−19.5	−18.8	heat	(1–6)	96V4
(Arg148-Ala-Ala)	5.4	−17.5	−22.6	heat	(1–7)	96V4
(Arg148-Ala-Ala-Ala)	3.0	−20.0	−18.8	heat	(1–6)	96V4
(Arg148-Ala-Ala-Ala)	5.4	−15.9	−20.9	heat	(1–7)	96V4
(Arg148-Ala-Ala-Ala-Ala)						
	3.0			heat	(8)	96V4
	5.4	−23.2	−23.4	heat	(1–7)	96V4
(Arg148-Asp)	3.0	−9.1	−12.6	heat	(1–6)	96V4
(Arg148-Asp)	5.4	−14.7	−20.1	heat	(1–7)	96V4
(Arg148-Ser)	3.0	−17.0	−18.0	heat	(1–6)	96V4
(Arg148-Ser)	5.4	−15.7	−20.9	heat	(1–6)	96V4
(Arg148-Asp-Ser)	3.0	−20.7	−18.0	heat	(1–6)	96V4
(Arg148-Asp-Ser)	5.4	−19.0	−23.4	heat	(1–7)	96V4
(Arg148-Thr-Thr)	3.0	−19.9	−17.2	heat	(1–6)	96V4
(Arg148-Thr-Thr)	5.4	−20.5	−24.7	heat	(1–7)	96V4
(Arg148-Val-Pro)	3.0	−18.0	−15.9	heat	(1–6)	96V4
(Arg148-Val-Pro)	5.4	−22.0	−26.4	heat	(1–7)	96V4

Remarks:
(1) the data refer to the cysteine-free pseudo wild-type (wt*, Cys54→Thr and Cys97→Ala) having $T_{trs} = 65.22°C$ at pH 5.4 and $T_{trs} = 51.66°C$ at pH 3.0
(2) the thermal stabilities were determined in 0.10 M NaCl, 1.4 mM acetic acid, 8.6 mM sodium acetate, pH 5.42, and in 0.025 M KCl, 2.95 mM H_3PO_4, 17.0 mM KH_2PO_4, pH 3.01
(3) van't Hoff treatment of transitions monitored by CD at 223 nm
(4) ΔCp was taken as 10.46 kJ/mol/K at 53°C for pH 5.4 buffer, and as 7.53 kJ/mol/K at 42°C for pH 3.0 buffer according to Ref. 95Z1
(5) the error in T_{trs} is ±0.15°C near T_{trs} of wt* and increased to approximately ±0.35°C as the ΔH value decreased
(6) the error in Δ(ΔG) is ±0.6 kJ/mol for the most stable mutants and the value increased to approximately ±4.2 kJ/mol for the least stable mutants
(7) estimated error in Δ(ΔG) in excess of ±2.1 kJ/mol
(8) not two-state

f) Mutations affecting disulfide bonds

Lysozyme phage T4, wild-type, pseudo wild-type (wt*), and mutants

Mutant	pH	T	Δ(ΔG)	Approach	Remarks	Ref
pseudo wild-type (wt*)	3.01	51.8	0.0	heat	(1)	92E3
wild-type (w.t.)	3.01	51.8	2.9	heat	(1,2)	92E3
Leu46→Ala of wt*	3.01	51.8	−11.3	heat	(1,2)	92E3
Leu99→Ala of wt*	3.01	51.8	−20.9	heat	(1,2)	92E3
Leu118→Ala of wt*	3.01	51.8	−14.6	heat	(1,2)	92E3
Leu121→Ala of wt*	3.01	51.8	−11.3	heat	(1,2)	92E3
Leu133→Ala of w.t.	3.01	51.8	−15.1	heat	(1–3)	92E3
Phe153→Ala of wt*	3.01	51.8	−14.6	heat	(1,2)	92E3
(Leu99→Ala and Phe153→Ala) of wt*	3.01	51.8	−34.7	heat	(1,2)	92E3

Remarks:
(1) wt* = (Cys54→Thr and Cys97→Ala), wt* is the reference protein for the following mutants
(2) Δ(ΔG) was calculated at 51.8°C using a constant value for the heat capacity change of ΔCp = 10.5 kJ/K/mol
(3) the mutant protein was created using the gene for wild-type lysozyme T4; all other mutants were constructed with the gene of pseudo wild-type lysozyme (wt*)

Lysozyme phage T4, mutant (Ala146→Cys), modified with cystamine (Ala146-SSCH₂CH₂NH₃⁺)

pH	T	ΔG	Δ(ΔG)	Approach	Remarks	Ref
2.44		−7.8	−5.0	heat	(1)	92L6
2.85		−7.1	−6.7	heat	(1)	92L6
5.13		−3.3	−4.6	heat	(2)	92L6
5.32		−3.4	−4.6	heat	(1)	92L6

Remarks:
(1) Δ(ΔG) was obtained using Δ(ΔG) = ΔT × ΔS, which refers to the pseudo wild-type Ala146→Cys buffer: 20 mM potassium phosphate, 25 mM KCl
(2) see (1) except buffer: 100 mM potassium acetate

Lysozyme phage T4, mutations which affect disulfide bonds, mutants and reference proteins, difference values

Mutant	pH	T	ΔT	Δ(ΔG)	Approach	Ref
wild-type	6.5	6.5	0	0	heat, v.H.	88W2
Cys54→Thr	6.5		1	1.3	heat, v.H.	88W2
Cys54→Val	6.5		−2	−2.9	heat, v.H.	88W2
Arg96→His	6.5		−7	−11.7	heat, v.H.	88W2
Ala146→Thr	6.5		−6	−6.3	heat, v.H.	88W2
Ile3→Cys, with disulfide bond Cys3-Cys97						
	6.5		3	5	heat, v.H.	88W2
double mutant (Cys54→Val and Cys97→Ser)						
	6.5		−6	−8.8	heat, v.H.	88W2
double mutant (Ile3→Cys-Cys97 and Cys54→Val)						
	6.5		1	1.7	heat, v.H.	88W2
double mutant (Ile3→Cys-Cys97 and Cys54→Thr)						
	6.5		4	6.3	heat, v.H.	88W2
triple mutant (Ile3→Cys-Cys97, Cys54→Thr and Arg96→His)						
	6.5		−5	−10.5	heat, v.H.	88W2
triple mutant (Ile3→Cys-Cys97, Cys54→Thr and Ala146→Thr)						
	6.5		−3	−2.1	heat, v.H.	88W2

Lysozyme phage T4, extended (ext.) and circularly permuted (per.) mutants, oxidized (ox.) and reduced (red.) forms

Disulfides	Template	Form	N	pH	T	$\Delta(\Delta G)_{rel}$	Approach	Remarks	Ref
Cys9-Cys164	ext.	ox.	155	2.5	50	9.2±0.8	heat	(1,2)	94Z5
Cys9-Cys164	ext.	red.		2.5	50	−4.6±0.8	heat	(1,2)	94Z5
Cys9-Cys164	per.	ox.	15	2.5	50	3.8±0.8	heat	(1,2)	94Z5
Cys9-Cys164	per.	red.		2.5	50	−2.9±0.8	heat	(1,2)	94Z5
Cys21-Cys142	ext.	ox.	121	2.5	50	14.6±0.4	heat	(1,2)	94Z5
Cys21-Cys142	ext.	red.		2.5	50	1.3±0.4	heat	(1,2)	94Z5
Cys21-Cys142	per.	ox.	49	2.5	50	5.0±0.4	heat	(1,2)	94Z5
Cys21-Cys142	per.	red.		2.5	50	1.3±0.4	heat	(1,2)	94Z5
Cys3-Cys97	ext.	ox.	94	2.5	50	7.1±0.4	heat	(1,2)	94Z5
Cys3-Cys97	ext.	red.		2.5	50	−1.3±0.4	heat	(1,2)	94Z5
Cys3-Cys97	per.	ox.	76	2.5	50	7.5±0.4	heat	(1,2)	94Z5
Cys3-Cys97	per.	red.		2.5	50	1.3±0.4	heat	(1,2)	94Z5

Remarks:
(1) $\Delta(\Delta G)_{rel}$ is the stabilitiy relative to the cysteine-free templates
(2) N is the length of the loops made by disulfides in the extended and permuted proteins

Lysocyme phage T4, pseudo wild-type (Cys54→Thr and Cys97→Ala) and mutants, effect of benzene on the stability

Benzene Concentration	pH	T	ΔT	$\Delta(\Delta G)$	Approach	Remarks	Ref
wt* (Cys54→Thr and Cys97→Ala)							
0.0	3.01	51.6	0.0	0.0	heat	(1)	92E2
10.4	3.01	51.5	−0.1	0.0	heat	(1)	92E2
Mutant (Leu99→Ala)							
0.0	3.01	36.2	0.0	0.0	heat	(1,2)	92E2
2.6	3.01	40.7	4.5	5.9	heat	(1,2)	92E2
5.6	3.01	41.6	5.4	7.1	heat	(1,2)	92E2
7.5	3.01	42.2	6.0	7.9	heat	(1,2)	92E2
Mutant (Phe153→Ala)							
0.0	3.01	39.5	0.0	0.0	heat	(1,2)	92E2
3.0	3.01	39.5	0.0	0.0	heat	(1,2)	92E2
7.2	3.01	39.4	−0.1	0.0	heat	(1,2)	92E2
Double mutant (Leu99→Ala and Phe153→Ala)							
0.0	4.46	41.7	0.0	0.0	heat	(1,2)	92E2
10.4	4.46	44.7	3.0	2.5	heat	(1.2)	92E2

Remarks:
(1) measurements in 20 mM potassium phosphate and 25 mM KCl in the presence of varying amounts of benzene. Reference state is the mutant in the presence of 0.0 mM benzene
(2) the mutant proteins contain additionally the replacements (Cys54→Thr and Cys97→Ala) of the template (pseudo wild-type)

Malate Dehydrogenase

Malate dehydrogenase, mitochondrial (mMDH) and cytosolic (cMDH)

Form	pH	T	$\Delta G(1)$	$\Delta G(2)$	Approach	Remarks	Ref
mMDH	7.5	25	15.1	19.2	GuHCl	(1,2)	94S7
mMDH + citrate	7.5	25	22.6	15.9	GuHCl	(1,2)	94S7
cMDH	7.5	25	24.7	26.8	GuHCl	(1,2)	94S7

Remarks:
(1) the unfolding profiles contain two unfolding transitions
(2) for details of the data treatment, see 93S9

Malate dehydrogenase, wild-type and mutant

Mutant	pH	T	ΔG	Approach	Remarks	Ref
wild-type	7.2	20	14.3	GuHCl	(1)	94G2
Arg102→Gln	7.2	20	13.1	GuHCl	(1)	94G2

Remark:
(1) linear extrapolation

Maltose-Binding Protein

Maltose-binding protein, wild-type

	pH	T	ΔG	Approach	Remarks	Ref
	7.6–7.8	25	53.6	GuHCl	(1)	95D5
	7.6–7.8	25	46	urea	(1)	95D5

Remark:
(1) linear extrapolation

Maltose binding protein, mutants

Mutant	pH	T	$\Delta(\Delta G)$	Approach	Remarks	Ref
Trp10→Ala	7.6–7.8	25	−18.0	GuHCl	(1)	95D5
Ala276→Gly	7.6–7.8	25	−7.9	GuHCl	(1)	95D5

Remark:
(1) linear extrapolation

Maltose binding protein from *E. coli*, wild-type and mutant

Mutant	pH	T	ΔG	$c_{1/2}$	m	Approach	Remarks	Ref
wild-type	7.5	25	39.7±6.3	0.98±0.02	41.8±8.4	GuHCl	(1,2)	96B4
double mutant (Gly32→Asp and Ile33→Pro)								
	7.5	25	23.0±6.3	0.62±0.02	37.7±8.4	GuHCl	(1,2)	96B4

Remarks:
(1) linear extrapolation
(2) transition monitored by fluorescence

Maltose binding protein from *E. coli*, wild-type and mutants

Mutant	pH	T	$\Delta(\Delta G)$	$c_{1/2}$	m	Approach	Remarks	Ref
wild-type	7.6	25	0.0	3.50	11.3±1.3	urea	(1,2)	93C5
Val8→Gly	7.6	25	−4.6	3.14	15.1±2.5	urea	(1,2)	93C5
Gly19→Cys	7.6	25	−9.6	2.66	11.7±0.8	urea	(1,2)	93C5
Asp55→Asn	7.6	25	0.0	3.49	13.0±1.7	urea	(1,2)	93C5
Ala276→Gly	7.6	25	−6.3	2.98	13.4±1.3	urea	(1,2)	93C5
Tyr283→Asp	7.6	25	−13.4	2.44	13.8±1.7	urea	(1,2)	93C5
Thr345→Ile	7.6	25	+2.9	3.79	10.0±1.7	urea	(1,2)	93C5

Remarks:
(1) linear extrapolation
(2) $\Delta(\Delta G)$ is based on $\Delta(\Delta G) = <m> \times (\Delta c_{1/2})$ where $(\Delta c_{1/2})$ is the difference between $c_{1/2}$ of mutant and wild-type and $<m>$ is the average value of m_i

Mannose transporter

Mannose transporter from *E. coli*, hydrophilic subunit, IIB domain

pH	T	ΔG	Approach	Remarks	Ref
7.4	22	27.2	GuHCl	(1)	94M5

Remark:
(1) linear extrapolation

Melittin

Melittin

pH	T	ΔG	Remarks	Ref
4.7	20	39.9	(1,3)	92G2
5	20	96.2	(1,2)	92H1

Remarks:
(1) concentration dependent unfolding transition with $F_4 \rightarrow 4U$
(2) the data refer to ΔG at zero net charge of melittin (see also Ref. 92G2)
(3) the data refer to ΔG at net charge = 6 of melittin (see also Ref. 92H1)

Metalloproteinases

Metalloproteinases-1 and -2 of tissue inhibitor (TIMP), N-terminal domain

Form	pH	T	ΔG	Approach	Remarks	Ref
ΔTIMP-1	7.5	25	49.4	GuHCl	(1,2)	94W4
ΔTIMP-1 A21	7.5	25	34.3	GuHCl	(1,3)	94W4
ΔTIMP-2 T21	7.5	25	22.2	GuHCl	(1,4)	94W4
ΔCHOΔTiMP-1	7.5	25	43.5	GuHCl	(1,5)	94W4

Remarks:
(1) linear extrapolation
(2) truncated TIMP molecule, TIMP-1 residues 1–126 TIMP-2 residues 1–127
(3) obtained by site directed mutagesis to generate the sequence 18–22 Val-Ile-Arg-Ala-Lys
(4) derived from the full-length cDNA clone
(5) obtained by site directed mutagesis of the ΔTIMP-1 template to produce Asn30→Gln and Asn78→Gln

Methionine Repressor Protein

Methionine repressor protein (MetJ) from *E. coli*

pH	T	ΔG	Approach	Remarks	Ref
6.0	25	42	DSC	(1)	92J1
7.0	25	34	DSC	(1)	92J1

Remark:
(1) buffer pH 7.0: 25 mM potassium phosphate, 100 mM KCl, Ref. 92J1 contains further data for the dependence of ΔG on pH and ionic strength.

Minibody – see de novo synthesized proteins

Myoglobin

The data entries are arranged as follows:

a) data for various species and approaches,
b) data for holo and apomyoglobin,
c) amino acid replacements,
d) complexes, solvents, and chemically modified myoglobin.

a) Data for various species and approaches

Metmyoglobin of various species, ΔG values

Species	pH	T	ΔG	Approach	Remarks	Ref
alligator	8	25	27.6±0.8	GuHCl	(1a)	87K3
alligator	8	25	24.7±0.8	GuHCl	(1e)	87K3
alligator	8	25	27.6±0.8	GuHCl	(2)	87K3
armadillo	8	25	27	GuHCl	(2d)	90K2
armadillo	8	25	41	GuHCl	(2c)	90K2
armadillo	8	25	26	GuHCl	(1a)	90K2
armadillo	8	25	16	GuHCl	(3)	90K2
armadillo	8	25	38±6	DSC		90K2

Metmyoglobin of various species, ΔG values (continued)

Species	pH	T	ΔG	Approach	Remarks	Ref
carp	8	25	25±3	DSC		90K2
chicken	8	25	34.3±1.3	GuHCl	(1a)	85H
chicken	8	25	34.7±1.3	GuHCl	(1b)	85H
cow	7	25	47.3	GuHCl	(2)	73P2
cow	7.1	25	35.1	GuHCl	(1)	77M
dog	7.1	25	26.4	GuHCl	(1)	77M
horse	4.8	25	10.5	heat, optical method		82C
horse	6	25	44.8	GuHCl	(1c)	79P1
horse	6	25	36.8	GuHCl	(1d)	79P1
horse	6	25	39.7	GuHCl	(1a)	79P1
horse	6	25	40.6	GuHCl	(2)	79P1
horse	6.6	25	31.8±0.8	GuHCl	(3)	82A
horse	6.6	25	33.9±0.8	urea	(3)	82A
horse	7	25	49.4	GuHCl	(1)	73P3
horse	7	25	43.5	GuHCl	(2)	73P3
horse	7	25	41.3	acid denaturation		73P3
horse	7	25	49	acid denaturation		73P3
horse	7	25	46	GuHCl	(2)	73P2
horse	7	25	40.6	GuHCl	(1a)	83B2
horse	7	25	38.5	GuHCl	(2a)	83B2
horse	7	25	77.4	GuHCl	(2b)	83B2
horse	7	25	50.2	acid denaturation		83B2
horse	7	25	66.9	GuHCl	(4)	83B2
horse	7.1	25	31.8	GuHCl	(1)	77M
horse	7.6	25	62.8	urea	(2)	68S
horse	8	25	50±3	DSC		90K2
horse	11	25	21	heat, optical method		82C
human	7	25	36.8	GuHCl	(2)	73P2
opossum	8	25	52.3±2.1	GuHCl	(1a)	88K4
opossum	8	25	46.4±2.1	GuHCl	(1e)	88K4
opossum	8	25	53.6±2.5	GuHCl	(2)	88K4
opossum	8	25	43±6	DSC		90K2
porpoise	7	25	34.7	acid denaturation		73P2
rabbit	8	25	69.5±2.1	GuHCl	(1a)	88K4
rabbit	8	25	67.4±1.7	GuHCl	(1e)	88K4
rabbit	8	25	69.5±2.1	GuHCl	(2)	88K4
raccoon	8	25	61.9±1.7	GuHCl	(1a)	88K4
raccoon	8	25	55.6±1.7	GuHCl	(1e)	88K4
raccoon	8	25	63.6±2.1	GuHCl	(2)	88K4
raccoon	8	25	42±3	DSC		90K2
rat	8	25	46±5	DSC		90K2
rat	8	25	65.7±1.3	GuHCl	(1a)	88K4
rat	8	25	60.7±1.3	GuHCl	(1e)	88K4
rat	8	25	67.4±1.3	GuHCl	(2)	88K4
seal	7	25	56.9	acid denaturation		73P2
sperm whale	6.2–9.4	21	50.2	hydrogen exchange		71O
sperm whale	6.2–9.4	21	50.2	hydrogen exchange		70A2
sperm whale	7	25	48.7	acid denaturation		73P3
sperm whale	7	25	58.6	acid denaturation		73P3
sperm whale	7	25	59.8	GuHCl	(1)	73P3
sperm whale	7	25	55.3	GuHCl	(2)	73P3
sperm whale	7	25	56.9	GuHCl	(2)	73P2
sperm whale	7.1	25	44.4	GuHCl	(1)	77M

Metmyoglobin of various species, ΔG values (continued)

Species	pH	T	ΔG	Approach	Remarks	Ref
sperm whale	7.1	25	56.5	GuHCl	(1)	78M2
sperm whale	7.1	25	43.1	GuHCl	(3)	78M2
sperm whale	7.6	25	62.8	urea	(2)	68S
sperm whale	8	25	57.7±2.5	GuHCl	(1a)	85H
sperm whale	8	25	55.6±2.5	GuHCl	(2c)	85H
sperm whale	9	25	58.6	pH denaturation		67H
turtle	7	25	30.1	GuHCl	(2)	73P2

Remarks:
(1)   denaturant binding model
(1a)   denaturant binding model, k = 0.6
(1b)   denaturant binding model, k = 0.2
(1c)   denaturant binding model, k = 1.2
(1d)   denaturant binding model, k = 0.8
(1e)   denaturant binding model, modified
(2)   transfer model
(2a)   transfer model, ε = 0.24
(2b)   transfer model, ε = 0.4
(2c)   transfer model, ε = 0.25
(2d)   transfer model, ε = 0.16
(3)   linear extrapolation
(4)   three-state model

Metmyoglobin of various species, difference ΔG values

Species	pH	T	Δ(ΔG)	Remarks	Ref
sperm whale	4	25	0.0	(1,2)	83F1
bottlenosed whale	4	25	9.8	(2)	83F1
dwarf sperm whale	4	25	1	(2)	83F1
goose-beaked whale	4	25	10.3	(2)	83F1
Hubbs beaked whale	4	25	15	(2)	83F1
humpback whale	4	25	14.1	(2)	83F1
killer whale	4	25	10.8	(2)	83F1
minke whale	4	25	13.3	(2)	83F1
pilot whale	4	25	10.8	(2)	83F1
pygmy sperm whale	4	25	−1.6	(2)	83F1
sei whale	4	25	14.6	(2)	83F1

Remarks:
(1) reference value, I = 0.01
(2) acid denaturation

Table 1. Gibbs Energy Change – Molar Values

Myoglobin from horse heart

pH	T	ΔG	Approach	Remarks	Ref
4.60	25.0	-2.06±0.18	GuHCl	(1)	92A1
4.92	25.0	5.10±0.12	GuHCl	(1)	92A1
5.26	25.0	10.47±0.02	GuHCl	(1)	92A1
5.46	25.0	12.44±0.02	GuHCl	(1)	92A1
5.52	25.0	16.39±1.13	GuHCl	(1)	92A1
5.88	25.0	22.41±0.43	GuHCl	(1)	92A1
7.00	25.0	31.57	GuHCl	(2)	92A1

Remarks:
(1) linear extrapolation
(2) ΔG was adjusted to pH 7 by calculations outlined in Ref. 92A1

Myoglobin from horse heart

pH	T	ΔG	Approach	Remarks	Ref
6.0	25	21.9±0.5	GuHCl	(1–3)	96A1

Remarks:
(1) linear extrapolation
(2) measured in 0.03 M cacodylic buffer containing 0.1 M KCl
(3) Ref. 96A1 contains a systematic study of ΔG of myoglobin in the presence of various salts

Myoglobin from horse heart, GuHCl denaturation measured in the presence of urea

Concentration	pH	T	ΔG	Approach	Remarks	Ref
	6.0	25	22.4±0.4	GuHCl	(1–3)	96G10
0.1–1 M KCl	6.0	25	22.4±0.4	GuHCl	(1–3)	96G10
0.49 M urea	6.0	25	22.5±0.5	GuHCl	(1–3)	96G10
1.00 M urea	6.0	25	21.6±0.3	GuHCl	(1–3)	96G10
2.30 M urea	6.0	25	10.0±0.2	GuHCl	(1–3)	96G10
3.67 M urea	6.0	25	0.7±0.2	GuHCl	(1–3)	96G10

Remarks:
(1) linear extrapolation
(2) measured in 0.03 M cacodylic acid with 0.1 M KCl
(3) from the results it is concluded that the ΔG dependence on urea conc. is linear and the ΔG dependence on GuHCl conc. is curved

Myoglobin from horse heart, urea denaturation measured in the presence of GuHCl

Concentration	pH	T	ΔG	Approach	Remarks	Ref
	6.0	25	31.8±1.4	urea	(1–3)	96G10
0.16 M KCl	6.0	25	31.7±1.5	urea	(1–3)	96G10
0.33–1 M KCl	6.0	25	33.4±0.9	urea	(1–3)	96G10
0.16 M GuHCl	6.0	25	26.7±0.1	urea	(1–3)	96G10
0.33 M GuHCl	6.0	25	17.8±0.2	urea	(1–3)	96G10
0.50 M GuHCl	6.0	25	11.8±0.7	urea	(1–3)	96G10
0.67 M GuHCl	6.0	25	11.1±0.3	urea	(1–3)	96G10
1.00 M GuHCl	6.0	25	2.9±0.1	urea	(1–3)	96G10

Remarks:
(1) linear extrapolation
(2) measured in 0.03 M cacodylic acid with 0.1 M KCl
(3) from the results it is concluded that the ΔG dependence on urea conc. is linear and the ΔG dependence on GuHCl conc. is curved

b) Data for holo- and apomyoglobin

Apomyoglobin of various species, ΔG values

Species	pH	T	ΔG	Approach	Remark	Ref
horse	2–2.2	20	15.1	pH, salt	(1)	90G3
sperm whale	5	31	11	DSC		88G4
sperm whale	8.1–9.8	25	37.7	hydrogen exchange		71O
sperm whale	8.1–9.8	25	37.7	hydrogen exchange		70A1

Remark:
(1) acid- and salt-induced subtransitions of the A state

Apomyoglobin from horse

	pH	T	ΔG	Approach	Remarks	Ref
	2.0	20	11.5	GuHCl	(1,2)	94S5
	8.0	20	13.2	GuHCl	(1,3)	94S5

Remarks:
(1) linear extrapolation
(2) salt refolded apomyoglobin in 10 mM sodium phosphate, 5 mM sodium acetate, 0.15 M NaCl
(3) folded apomyoglobin

Myoglobin from horse, holo and apoprotein, native and molten globule states

Form	pH	T	ΔG	Approach	Remarks	Ref
holo, native	6.0	20	33.3	GuHCl	(1)	95N3
apo, native	6.0	20	7.97	GuHCl	(1)	95N3
apo, m.g.	2.0	20	12.6	GuHCl	(1,2)	95N3
holo, native	6.0	20	53.0	urea	(1)	95N3
apo, native	6.0	20	20.3	urea	(1)	95N3
apo, m.g.	2.0	20	9.73	urea	(1,2)	95N3
apo, m.g.	2.0	20	13.3	urea	(1,3)	95N3

Remarks:
(1) linear extrapolation
(2) chloride-stabilized molten globule (m.g.)
(3) trichloroacetate-stabilized molten globule (m.g.)

Apomyoglobin, effect of mutations on intermediate stability, I→U transition (for N→U, see the following table)

Mutant	pH	T	ΔG	Approach	Remarks	Ref
wild-type	4.2	4	12.1	urea	(1,2)	96K2
Trp7→Phe	4.2	4	7.5	urea	(1,2)	96K2
Trp14→Phe	4.2	4	10.0	urea	(1,2)	96K2
His36→Gln	4.2	4	11.7	urea	(1,2)	96K2
Val68→Thr	4.2	4	11.3	urea	(1,2)	96K2
Phe123→Lys	4.2	4	9.2	urea	(1,2)	96K2
Ala130→Lys	4.2	4	8.4	urea	(1,2)	96K2
Ala130→Leu	4.2	4	12.6	urea	(1,2,3)	96K2
Met131→Ala	4.2	4	8.4	urea	(1,2)	96K2

Remarks:
(1) linear extrapolation to zero denaturant concentration by a method that includes the pre- and postdenaturational baselines for a non-linear regression of the data
(2) transition monitored by CD and fluorescence, buffer: 4 mM citrate buffer pH 4.2
(3) average for ΔG obtained by fluorescence (12.1 kJ/mol) and CD (13.0 kJ/mol)

Apomyoglobin, effect of mutations on the protein stability, N→U transition (for I→U, see the preceeding table)

Mutant	pH	T	ΔG	Approach	Remarks	Ref
wild-type	7.8	4	23.4	urea	(1,2)	96K2
Trp7→Phe	7.8	4	19.7	urea	(1,2)	96K2
Trp14→Phe	7.8	4	18.8	urea	(1,2)	96K2
His36→Gln	7.8	4	20.1	urea	(1,2)	96K2
Val68→Thr	7.8	4	20.9	urea	(1,2)	96K2
Phe123→Lys	7.8	4	14.6	urea	(1,2)	96K2
Ala130→Lys	7.8	4	14.6	urea	(1,2)	96K2
Ala130→Leu	7.8	4	19.7	urea	(1,2,3)	96K2
Met131→Ala	7.8	4	14.2	urea	(1,2)	96K2

Remarks:
(1) linear extrapolation to zero denaturant concentration by a method that includes the pre- and postdenaturational baselines for a non-linear regression of the data
(2) transition monitored by CD and fluorescence, buffer: 10 mM HEPES buffer pH 7.8

Apomyoglobin from horse

Salt Concentration	pH	T	ΔG	Approach	Remarks	Ref
0.0 M NaCl	1.8	20	8.70	GuHCl	(1,2)	93H1
0.4 M NaCl	1.8	20	11.46	GuHCl	(1,2)	93H1

Remarks:
(1) linear extrapolation
(2) the data refer to GuHCl-induced refolding

Apomyoglobin from sperm whale, recombinant, molten globule folding intermediates

Transition	pH	T	ΔG	Approach	Remarks	Ref
N→U	1–9	0	27.6	pH and urea		93B2
N→I$_1$	1–9	0	18.8	pH and urea		93B2
I$_1$→U	1–9	0	8.8	pH and urea		93B2
I$_1$→U	6.1	5	10.5±2.1	urea	(1)	93J2
I$_1$→U	4.2	4	10.5	urea	(2,3)	95L8
I$_2$→U	4.2	4	24.7	urea	(3,4)	95L8

Remarks:
(1) two-state fit for the intermediate form
(2) linear extrapolation to zero denaturant concentration by a method that includes the pre- and postdenaturational baselines for a non-linear regression of the data
(3) intermediate, low salt form, maximally populated at pH 4.2
(4) intermediate, trichloroacetate form (20 mM CCl$_3$COONa present)

Apomyoglobin from sperm whale, recombinant

Transition	pH	T	ΔG	Approach	Ref
N→I	1–9	0	18.8	pH and urea	93B2
I→U	1–9	0	8.8	pH and urea	93B2

Apomyoglobin, sperm whale, recombinant, molten globule form

Mutant	pH	T	ΔG	Approach	Remarks	Ref
wild-type	7.8	0	19.5	urea	(1,2)	95K8
double mutant (Gly 23→Ala and Gly25→Ala)						
	7.8	0	21.6	urea	(1,2)	95K8
wild-type	4.0	0	14.8	urea	(1,3)	95K8
double mutant (Gly 23→Ala and Gly25→Ala)						
	4.0	0	10.3	urea	(1,3)	95K8

Remarks:
(1) linear extrapolation
(2) buffer: 10 mM HEPES
(3) buffer: 10 mM citrate, 0.5 M KCl

Apomyoglobin from sperm whale, mutant (Asp122→Asn), transitions

Transition	pH	T	ΔG	Approach	Remarks	Ref
N→I	7.0	25	15.1±1.7	GuHCl	(1,2)	95R1
N→I	7.0	25	15.5±1.3	GuHCl	(1,3)	95R1
N→I	7.0	25	8.4±2.5	GuHCl	(1,4)	95R1
I→U	7.0	25	13.0±2.5	GuHCl	(1,2)	95R1
I→U	7.0	25	14.6±2.9	GuHCl	(1,3)	95R1
I→U	7.0	25	17.2±10.0	GuHCl	(1,4)	95R1

Remarks:
(1) measured by a multidimensional experimental approach, data treatment by linear extrapolation
(2) transition monitored by CD at 222 nm
(3) transition monitored by CD at 235 nm
(4) transition monitored by fluorescence at 285 nm

Apomyoglobin, recombinant sperm whale, wild-type and mutants

Mutant	pH	T	ΔG	Approach	Remarks	Ref
wild-type	7.5	0	24.3±1.3	urea	(1)	91H4
Phe123→Thr	7.5	0	9.6±0.4	urea	(1)	91H4
Ala130→Leu	7.5	0	14.6±0.8	urea	(1)	91H4
Ala130→Lys	7.5	0	8.8±0.4	urea	(1)	91H4

Remark:
(1) linear extrapolation

Apomyoglobin, histidine mutants

Mutant	Transition	pH	T	ΔG	Remarks	Ref
wild-type	N→I	>7	0	19.0±1.0	(1,2)	94B4
wild-type	I→U	>7	0	8.6±0.7	(1,2)	94B4
His24→Val,	N→I	>7	0	16.8±1.1	(1,2)	94B4
His24→Val,	I→U	>7	0	8.7±0.4	(1,2)	94B4
His36→Gln,	N→I	>7	0	14.1	(1,2)	94B4
His36→Gln,	I→U	>7	0	8.1	(1,2)	94B4
His48→Gln,	N→I	>7	0	16.9±1.0	(1,2)	94B4
His48→Gln,	I→U	>7	0	8.2±0.3	(1,2)	94B4
His64→Gln,	N→I	>7	0	17.6±0.5	(1,2)	94B4
His64→Gln,	I→U	>7	0	8.2±0.3	(1,2)	94B4
His82→Gln,	N→I	>7	0	19.7±0.8	(1,2)	94B4
His82→Gln,	I→U	>7	0	7.8±0.4	(1,2)	94B4
His93→Gly,	N→I	>7	0	19.2±0.5	(1,2)	94B4
His93→Gly,	I→U	>7	0	8.6±0.2	(1,2)	94B4
His97→Gln,	N→I	>7	0	19.1±0.8	(1,2)	94B4
His97→Gln,	I→U	>7	0	8.1±0.4	(1,2)	94B4
His113→Gln,	N→I	>7	0	17.5±0.7	(1,2)	94B4
His113→Gln,	I→U	>7	0	9.0±0.3	(1,2)	94B4
His119→Phe,	N→I	>7	0	16.1±0.8	(1,2)	94B4
His119→Phe,	I→U	>7	0	8.7±0.3	(1,2)	94B4
double mutant (His24→Val and His119→Phe)						
	N→I	>7	0	22.5±0.8	(1,2)	94B4
	I→U	>7	0	10.0±0.5	(1,2)	94B4

Remarks:
(1) acid and urea-induced unfolding
(2) three-state fit of the transitions

Human myoglobin, wild-type and mutants, apo and holo protein, ΔG values

Mutant	pH	T	ΔG	Approach	Remarks	Ref
Apo protein:						
wild-type	7.6	3	21.3	urea	(1)	89H3
Cys110→Ala	7.6	3	25.5	urea	(1)	89H3
Cys110→Asp	7.6	3	6.3	urea	(1)	89H3
Cys110→Ser	7.6	3	20.5	urea	(1)	89H3
Holo protein:						
wild-type	7.6	25	32.6	urea	(1)	89H3
Cys110→Ala	7.6	25	44.4	urea	(1)	89H3
Cys110→Asp	7.6	25	34.3	urea	(1)	89H3
Cys110→Leu	7.6	25	25.1	urea	(1)	89H3
Cys110→Ser	7.6	25	34.7	urea	(1)	89H3

Remark:
(1) linear extrapolation

c) *Amino acid replacements*

Myoglobin sperm whale, wild-type and mutants, cyanide complex

Mutant	pH	T	ΔT	Δ(ΔG)	Approach	Remarks	Ref
wild-type	9.6		0.0	0.0	heat	(1–3)	93L5
Ile28→Leu	9.6		−1.3	−2.6	heat	(1–3)	93L5
Ile28→Met	9.6		−0.9	−1.8	heat	(1–3)	93L5
Leu29→Ile	9.6		−4.5	−8.3	heat	(1–3)	93L5
Leu29→Met	9.6		0.7	1.4	heat	(1–3)	93L5
Leu69→Ile	9.6		−1.1	−2.2	heat	(1–3)	93L5
Leu69→Met	9.6		−0.4	−0.8	heat	(1–3)	93L5
Ile111→Leu	9.6		−1.9	−3.7	heat	(1–3)	93L5
Leu135→Ile	9.6		−4.3	−8.0	heat	(1–3)	93L5
Leu135→Met	9.6		−2.0	−3.8	heat	(1–3)	93L5
Ile142→Leu	9.6		1.7	3.5	heat	(1–3)	93L5
Ile142→Met	9.6		2.5	5.2	heat	(1–3)	93L5

Remarks:
(1) buffer: 20 mM potassium phosphate, 0.1 M KCl
(2) measured in the presence of 0.5 mM KCN
(3) Δ(ΔG) was calculated using the expression $\Delta(\Delta G) = -\Delta T/T_{trs}(\text{w.t.}) \times [\Delta H(\text{w.t.}) + \Delta C p \times \Delta T]$

Table 1. Gibbs Energy Change – Molar Values

Myoglobin sperm whale, wild-type and mutants, cyanide complex

Mutant	pH	T_{trs}	ΔT	$\Delta(\Delta G)$	Approach	Remarks	Ref
wild-type	9.5	82.2	0.0	0.0	heat	(1,2)	93P7
Leu9→Ala	9.5	81.4	−0.8	−1.7	heat	(1,2)	93P7
Leu11→Ala	9.5	81.4	−0.8	−1.8	heat	(1,2)	93P7
Val13→Ala	9.5	80.9	−1.3	−2.8	heat	(1,2)	93P7
Val66→Ala	9.5	83.6	1.4	3.1	heat	(1,2)	93P7
Thr67→Ala	9.5	81.7	−0.5	−1.1	heat	(1,2)	93P7
Pro88→Ala	9.5	83.3	1.1	2.5	heat	(1,2)	93P7
Val114→Ala	9.5	79.3	−2.9	−6.1	heat	(1,2)	93P7
Gly129→Ala	9.5	84.2	2.0	4.4	heat	(1,2)	93P7
Leu137→Ala	9.5	78.6	−3.6	−7.4	heat	(1,2)	93P7
Ile142→Ala	9.5	78.3	−3.9	−8.0	heat	(1,2)	93P7

Remarks:
(1) measured in the presence of KCN
(3) $\Delta(\Delta G)$ was calculated using the expression $\Delta(\Delta G) = -\Delta T/T_{trs}(w.t.)\times[\Delta H(w.t.)+\Delta Cp\times\Delta T]$

Myoglobin, sperm whale (recombinant), wild-type and mutants, cyanide complex

Mutant	pH	T_{trs}	ΔT	$\Delta(\Delta G)$	Approach	Remarks	Ref
wild-type	9.6	82.2	0.0	0.0	heat	(1,2)	93P8
Leu9→Ala	9.6	82.2	−0.8	−1.7	heat	(1,2)	93P8
Leu11→Ala	9.6	82.2	−0.8	−1.8	heat	(1,2)	93P8
Val13→Ala	9.6	82.2	−1.3	−2.8	heat	(1,2)	93P8
Gly23→Ala	9.6	82.2	−2.2	−4.7	heat	(1,2)	93P8
Thr51→Ala	9.6	82.2	−2.8	−5.9	heat	(1,2)	93P8
Val66→Ala	9.6	82.2	1.4	3.1	heat	(1,2)	93P8
Thr67→Ala	9.6	82.2	−0.5	−1.1	heat	(1,2)	93P8
Pro88→Ala	9.6	82.2	1.1	2.5	heat	(1,2)	93P8
Val114→Ala	9.6	82.2	−2.9	−6.1	heat	(1,2)	93P8
His116→Ala	9.6	82.2	0.3	0.7	heat	(1,2)	93P8
Ser117→Ala	9.6	82.2	−0.5	−1.1	heat	(1,2)	93P8
Gly129→Ala	9.6	82.2	2.0	4.4	heat	(1,2)	93P8
Leu137→Ala	9.6	82.2	−3.6	−7.4	heat	(1,2)	93P8
Ile142→Ala	9.6	82.2	−3.9	−8.0	heat	(1,2)	93P8
Leu149→Ala	9.6	82.2	−3.2	−6.7	heat	(1,2)	93P8

Remarks:
(1) thermal transition monitored by heme absorbance at 420 nm and CD at 222 nm
(2) buffer: 20 mM potassium phosphate, 100 mM KCl, and 0.5 mM KCN

Myoglobin from sperm whale, wild-type and mutants

Mutant	pH	T	ΔT	$\Delta(\Delta G)$	Approach	Remarks	Ref
wild-type	11.0	76.5	0.0	0.0	heat	(1)	93L6
Ile28→Ala	11.0	76.5	−4.4	−5.6	heat	(1,2)	93L6
Ile28→Leu	11.0	76.5	−1.7	−2.3	heat	(1,2)	93L6
Ile28→Met	11.0	76.5	−1.7	−2.3	heat	(1,2)	93L6
Ile28→Val	11.0	76.5	+0.1	+0.2	heat	(1,2)	93L6
Leu29→Ala	11.0	76.5	−8.9	−9.9	heat	(1,2)	93L6
Leu29→Ile	11.0	76.5	−3.6	−4.7	heat	(1,2)	93L6
Leu29→Met	11.0	76.5	+0.3	+0.5	heat	(1,2)	93L6
Leu29→Val	11.0	76.5	−6.0	−7.2	heat	(1,2)	93L6
Leu49→Ile	11.0	76.5	−2.5	−3.3	heat	(1,2)	93L6

Myoglobin from sperm whale, wild-type and mutants (continued)

Mutant	pH	T	ΔT	Δ(ΔG)	Approach	Remarks	Ref
Leu69→Ala	11.0	76.5	−3.8	−4.9	heat	(1,2)	93L6
Leu69→Ile	11.0	76.5	−0.1	−0.1	heat	(1,2)	93L6
Leu69→Met	11.0	76.5	0.0	0.0	heat	(1,2)	93L6
Leu69→Val	11.0	76.5	−0.3	−0.4	heat	(1,2)	93L6
Ile111→Ala	11.0	76.5	−6.4	−7.7	heat	(1,2)	93L6
Ile111→Leu	11.0	76.5	−2.0	−2.7	heat	(1,2)	93L6
Ile111→Met	11.0	76.5	−3.7	−4.8	heat	(1,2)	93L6
Leu135→Ile	11.0	76.5	−5.2	−6.4	heat	(1,2)	93L6
Leu135→Met	11.0	76.5	−3.3	−2.5	heat	(1,2)	93L6
Leu135→Val	11.0	76.5	−8.3	−9.4	heat	(1,2)	93L6
Ile142→Ala	11.0	76.5	−3.5	−4.6	heat	(1,2)	93L6
Ile142→Leu	11.0	76.5	+1.8	+2.6	heat	(1,2)	93L6
Ile142→Met	11.0	76.5	+2.6	+3.9	heat	(1,2)	93L6
Ile142→Val	11.0	76.5	−0.3	−0.5	heat	(1,2)	93L6

Remarks:
(1) measured in the presence of cyanide, buffer: 40 mM glycine/KOH, 0.5 mM KCN
(1) Δ(ΔG) was obtained using $\Delta(\Delta G) = \Delta T[\Delta H_{wt} + \Delta C_p \times \Delta T]/T_{trs.wt}$ using $\Delta H_{wt} = 494$ kJ/mol
and $\Delta C_p = 11.7$ kJ/mol/K

Myoglobin, human cyanomet myoglobin, recombinant, wild-type and mutants

Mutant	pH	T	ΔG	Approach	Remarks	Ref
wild-type	5.0	23	19.1±1.16	urea	(1)	93K10
Leu29→Ala	5.0	23	26.0±1.68	urea	(1)	93K10
Leu72→Ala	5.0	23	16.0±1.11	urea	(1)	93K10
Leu104→Ala	5.0	23	16.6±2.41	urea	(1)	93K10

Remark:
(1) linear extrapolation

Myoglobin, sperm whale, recombinant, multiple alanine substitutions

Mutant	pH	T	Δ(ΔG)	Approach	Remarks	Ref
Multiple mutants (N=2):						
(Leu11, Val66)→Ala	9.6	81.1	−2.4	heat	(1,2)	94L6
(Val13, Val66)→Ala	9.6	83.8	3.6	heat	(1,2)	94L6
(Val23, Gly66)→Ala	9.6	83.0	1.9	heat	(1,2)	94L6
(Val66, Val114)→Ala	9.6	81.0	−2.6	heat	(1,2)	94L6
(Gly66, Val129)→Ala	9.6	84.7	5.7	heat	(1,2)	94L6
(Val66, Leu149)→Ala	9.6	80.3	−4.1	heat	(1,2)	94L6
(Val66, Leu137)→Ala	9.6	81.4	−1.8	heat	(1,2)	94L6
(Val66, Leu135)→Ala	9.6	71.0	−20.0	heat	(1,2)	94L6
(Gly129, Val114)→Ala	9.6	83.9	3.8	heat	(1,2)	94L6
Multiple mutants (N=3):						
(Val13, Val66, Val114)→Ala						
	9.6	80.3	−4.1	heat	(1,2)	94L6
(Leu149, Val114, Gly129)→Ala						
	9.6	79.4	−5.9	heat	(1,2)	94L6
(Val66, Gly129, Gly23)→Ala						
	9.6	84.2	4.6	heat	(1,2)	94L6
(Val66, Gly129, Leu149)→Ala						
	9.6	82.5	0.7	heat	(1,2)	94L6

Myoglobin, sperm whale, recombinant, multiple alanine substitutions (continued)

Mutant	pH	T	$\Delta(\Delta G)$	Approach	Remarks	Ref
(Val66, Val114, Gly129)→Ala						
	9.6	83.4	−2.7	heat	(1,2)	94L6
(Val114, His116, Ser117)→Ala						
	9.6	83.5	2.9	heat	(1,2)	94L6
Multiple mutants (N=4):						
(Gly23, Val66, Gly129, Val13)→Ala						
	9.6	83.6	3.1	heat	(1,2)	94L6
(Gly23, Val114, His116, Ser117)→Ala						
	9.6	83.1	2.0	heat	(1,2)	94L6
(Val66, Val114, Val13, Gly23)→Ala						
	9.6	84.9	6.2	heat	(1,2)	94L6
(Val66, Val114, Gly129, Leu149)→Ala						
	9.6	80.6	−3.0	heat	(1,2)	94L6
Mutliple mutants (N=5):						
(Val66, Val114, Gly129, Leu149, Gly23)→Ala						
	9.6	79.6	−1.3	heat	(1,2)	94L6
Mutliple mutants (N=6):						
(Gly23, Val66, Val114, Gly129, Leu149, Val13)→Ala						
	9.6	82.9	1.5	heat	(1,2)	94L6
(Val66, Val114, His116, Ser117, Gly129, Leu149)→Ala						
	9.6	83.9	3.8	heat	(1,2)	94L6

Remarks:
(1) reference protein is the wild-type with T_{trs} = 82.2°C, ΔH = 780 kJ/mol, ΔC_p = 11.7 kJ/mol/K
(2) buffer: 20 mM potassium phosphate, 100 mM KCl, 0.5 mM KCN, pH 9.6

d) Complexes, solvents, and chemically modified myoglobin:

Complexes of horse metmyoglobin

Ligand	pH	T	ΔG	Approach	Ref
azide	4.8	25	13	heat, optical method	82C
azide	11	25	30	heat, optical method	82C
cyanide	4.8	25	16	heat, optical method	82C
cyanide	11	25	33	heat, optical method	82C
fluoride	4.8	25	11	heat, optical method	82C
fluoride	11	25	24	heat, optical method	82C

Complexes of sperm whale metmyoglobin

Ligand	pH	T	ΔG	Approach	Remarks	Ref
azide	7.1	25	59	GuHCl	(1)	78M2
azide	7.1	25	44.8	GuHCl	(2)	78M2
cyanide	7.1	25	61.9	GuHCl	(1)	78M2
cyanide	7.1	25	47.3	GuHCl	(2)	78M2
cyanide	10	25	50.2±3.3	DSC		74P1

Remarks:
(1) denaturant binding model
(2) linear extrapolation

Chemically modified metmyoglobin

Species	pH	T	Δ(ΔG)		Remarks	Ref
sperm whale	–	25	0		(1,2)	87R
sperm whale with oxidized Trp7 and Trp14	–	25	–22		(2)	87R

Remarks:
(1) reference value
(2) acid denaturation

Myoglobin from horse heart in the presence of glycerol

Solvent	pH	T	ΔG	Approach	Remarks	Ref
water	7.4	25.0±0.1	32.0±0.2	GuHCl	(1)	96B3
glycerol	7.4	25.0±0.1	15.0±0.3	GuHCl	(1,2)	96B3

Remarks:
(1) linear extrapolation
(2) measured in 60% (v/v) glycerol-Tris buffer

Myosin

Myosin rod, subtransition 1

pH	T	ΔG	Approach	Ref
7.02	37	14.2	DSC	90B4

Light meromyosin (=LMM), deconvolution

Transition	pH	T	ΔG	Approach	Remarks	Ref
trans. (1)	7.03	37	9.6	DSC	(1,2)	90B4
trans. (2)	7.03	37	8.4	DSC	(1,3)	90B4
trans. (3)	7.03	37	15.9	DSC	(1,4)	90B4
trans. (4)	7.03	37	11.3	DSC	(1,5)	90B4
trans. (5)	7.03	37	34.7	DSC	(1,6)	90B4

Remarks:
(1) further data for other pH values are contained in Tab. 3 of Ref. 90B4
(2) the data refer to the transition at 42.4°C
(3) the data refer to the transition at 48.8°C
(4) the data refer to the transition at 49.0°C
(5) the data refer to the transition at 53.7°C
(6) the data refer to the transition at 54.8°C

Myosin subfragment 2 (=S-2), deconvolution

Transition	pH	T	ΔG	Approach	Remarks	Ref
trans. (1)	7.02	37	7.1	DSC	(1,2)	90B4
trans. (2)	7.02	37	15.1	DSC	(1,3)	90B4
trans. (3)	7.02	37	14.2	DSC	(1,4)	90B4

Remarks:
(1) further data for other pH values are contained in Tab. 4 of Ref. 90B4
(2) the data refer to the transition at 42.1°C
(3) the data refer to the transition at 44.7°C
(4) the data refer to the transition at 49.8°C

N-(5'-Phosphoribosyl) Anthranilate Isomerase, see anthranilate isomerase

Neurotoxin, see Toxins

Neurotrophins

Neurotrophins

Protein	pH	T	ΔG	Approach	Remarks	Ref
BDNF	3.5	23±2	71.1±5.9	GuHCl	(1,2)	94T6
BDNF	7	23±2	110.5±10.0	GuHCl	(1,2)	94T6
hNGF	7	23±2	96.2±12.6	GuHCl	(1,2)	94T6
mNGF	3.5	23±2	53.1±2.9	GuHCl	(1,2)	94T6
mNGF	4	23±2	65.7	GuHCl	(1,2)	94T6
mNGF	4.5	23±2	67.4	GuHCl	(1,2)	94T6
mNGF	5	23±2	72.0	GuHCl	(1,2)	94T6
mNGF	5.5	23±2	65.3	GuHCl	(1,2)	94T6
mNGF	6	23±2	76.1	GuHCl	(1,2)	94T6
mNGF	7	23±2	73.2	GuHCl	(1,2)	94T6
mNGF	7	23±2	80.8±4.6	GuHCl	(1,2)	94T6
NT-3	7	23±2	95.0±5.4	GuHCl	(1,2)	94T6
NT-4/5	3.5	23±2	79.5±4.6	GuHCl	(1,2)	94T6
NT-4/5	7	23±2	87.0±5.0	GuHCl	(1,2)	94T6

Remarks:
(1) abbreviations:
   BDNF – brain-derived neurotrophic factor
   hNGF – human nerve growth factor
   mNGF – mouse nerve growth factor
   NT-3 – neurotrophin 3
   NT-4/5 – neurotrophin 4/5
(2) approach based on linear extrapolation for dimeric proteins (transition $N_2 \rightarrow 2D$) using non-linear regression of the transition curve

Nuclease from *Staphylococcus Aureus*

The data are arranged as follows

a) wild-type and mutants
b) insertions and deletion mutants
c) hybrid forms
d) chemically modified and ligated forms

a) wild-type and mutants

Staphylococcal nuclease, for m-value effects at denaturant-induced unfolding of Staphylococcal nuclease – see also Ref. 95S8

Staphylococcal nuclease, wild-type and mutants

Mutant	pH	T	ΔG	Approach	Remarks	Ref
wild-type	4	25	4.5	DSC	(1)	88G5
wild-type	4.5	25	10	DSC	(1)	88G5
wild-type	7	2	23.4	GuHCl	(3,4)	86S3
wild-type	7	2	25.5	urea	(3,4)	86S3
wild-type	7	25	17.5	DSC	(1)	88G5
wild-type	7	25	15.8*	DSC	(5)	85C1
wild-type	7	25	25.8*	DSC	(2)	85C1
wild-type	7	25	34.8*	DSC	(6)	85C1
Ile18→Met	7	2	20.9	GuHCl	(3)	86S3
Ile18→Met	7	2	22.6	urea	(3)	86S3
Ala69→Thr	7	2	12.1	GuHCl	(3)	86S3
Ala69→Thr	7	2	13.4	urea	(3)	86S3
Gly79→Ser	7	2	12.6	GuHCl	(3)	86S3
Gly79→Ser	7	2	13.4	urea	(3)	86S3
Gly88→Val	7	2	19.3	GuHCl	(3)	86S3
Gly88→Val	7	2	21.8	urea	(3)	86S3
Ala90→Ser	7	2	15.1	GuHCl	(3)	86S3
Ala90→Ser	7	2	16.7	urea	(3)	86S3
Met98→Ile	8.0	20	3.5	GuHCl	(7)	89S9
His121→Pro	8.0	20	1	GuHCl	(7)	89S9
Val66→Leu	7	2	22.6	GuHCl	(3)	86S3
Val66→Leu	7	2	26.4	urea	(3)	86S3
double mutant (Ile18→Met and Ala69→Thr)						
	7	2	9.2	GuHCl	(3)	86S3
	7	2	9.6	urea	(3)	86S3
double mutant (Ile18→Met and Ala90→Ser)						
	7	2	11.7	GuHCl	(3)	86S3
	7	2	13.4	urea	(3)	86S3
double mutant (Ala69→Thr and Ala90→Ser)						
	8.0	20	2.7	GuHCl	(7)	89S9
double mutant (Val66→Leu and Gly79→Ser)						
	7	2	9.6	GuHCl	(3)	86S3
	7	2	10.5	urea	(3)	86S3

Table 1. Gibbs Energy Change – Molar Values

Staphylococcal nuclease, wild-type and mutants (continued)

Mutant	pH	T	ΔG	Approach	Remarks	Ref
double mutant (Val66→Leu and Gly88→Val)						
	7	2	14.7	GuHCl	(3)	86S3
	7	2	19.7	urea	(3)	86S3
triple mutant (Val66→Leu and Gly79→Ser and Gly88→Val)						
	7	2	10.9	GuHCl	(3)	86S3
	7	2	11.3	urea	(3)	86S3

Remarks:
(1) the data were taken from Fig. 6 of Ref. 88G5
(2) in the presence of calcium
(3) linear extrapolation, measurements monitored by intrinsic fluorescence and CD
(4) reference value for the following mutants
(5) without ligands
(6) in the presence of thymidine-3',5'-diphosphate
(7) linear extrapolation

Staphylococcal nuclease

pH	T	ΔG	$c_{1/2}$	m	Approach	Remarks	Ref
8.0	10	26.4±1.7	2.5	10.5	urea	(1)	94C12
7.0	2	25.5	2.56	10.0	urea	(2)	86S3

Remarks:
(1) urea gradient electrophoresis
(2) linear extrapolation, reference value from Ref. 86S3

Staphylococcal nuclease

Mutant	pH	T	ΔG	$c_{1/2}$	m	Approach	Remarks	Ref
wild-type	7.0	20	23.0	0.82	28.0	GuHCl	(1–3)	86S4
	7.0	20	23.4	0.82	28.6	GuHCl	(1–3)	86S4
	6.8	20	22.6	0.82	27.4	GuHCl	(1,2,4)	86S4
Val66→Leu	7.0	20	22.6	0.96	23.5	GuHCl	(1–3)	86S4
Phe76→Val	7.0	20	10.9	0.39	28.0	GuHCl	(1–3)	86S4
Gly79→Ser	7.0	20	13.4	0.53	25.8	GuHCl	(1–3)	86S4
His124→Arg	7.0	20	27.2	1.00	27.2	GuHCl	(1–3)	86S4
His124→Leu	7.0	20	28.0	1.08	26.0	GuHCl	(1–3)	86S4
	7.0	20	28.9	1.08	26.6	GuHCl	(1–3)	86S4
double mutant (Phe76→Val and His124→Leu)								
	7.0	20	17.6	0.60	29.4	GuHCl	(1–3)	86S4
double mutant (Gly79→Ser and Val66→Leu)								
	7.0	20	9.2	0.55	17.1	GuHCl	(1–3)	86S4
	7.0	20	9.6	0.56	17.4	GuHCl	(1–3)	86S4
double mutant (Gly79→Ser and His124→Leu)								
	7.0	20	18.0	0.70	25.5	GuHCl	(1–3)	86S4

Staphylococcal nuclease (continued)

Mutant	pH	T	ΔG	$c_{1/2}$	m	Approach	Remarks	Ref
double mutant (His124→Arg and Val66→Leu)								
	7.0	20	28.5	1.17	24.4	GuHCl	(1–3)	86S4
double mutant (His124→Leu and Val66→Leu)								
	7.0	20	30.1	1.25	24.4	GuHCl	(1–3)	86S4
	7.0	20	30.1	1.25	24.1	GuHCl	(1–3)	86S4

Remarks:
(1) linear extrapolation
(2) transition monitored by fluorescence intensity, excitation at 295 nm, emission at 325 nm
(3) buffer: 25 mM potassium phosphate, 100 mM NaCl, pH 7.0
(4) buffer: 25 mM sodium cacodylate, 100 mM NaCl, pH 6.8

Staphylococcal nuclease, wild-type

pH	T	ΔG	Approach	Remarks	Ref
7.0	20	23.35±0.17	GuHCl	(1–4)	95S14
7.0	20	23.18±0.17	GuHCl	(1–3,5)	95S14

Remarks:
(1) the measurements were made on an automated device
(2) mean of 10 determinations
(3) buffer: 25 mM sodium phosphate, 100 mM NaCl, pH 7.0
(4) linear extrapolation
(5) linear extrapolation to zero denaturant concentration by a method that includes the pre- and postdenaturational baselines for a non-linear regression of the data

Staphylococcal nuclease, recombinant

pH	T	ΔG	Approach	Remarks	Ref
5.5	21	10.9±1.3	pressure jump	(1)	95V4
5.5	21	13.4±1.3	pressure	(2)	95V4

Remarks:
(1) from the ratio of the rate constants, calculated for $p° = 1$ atm
(2) from the pressure dependent equilibrium unfolding profile, monitored by fluorescence intensity

Staphylococcal nuclease, recombinant

pH	T	ΔG	Approach	Remarks	Ref
7.0	5	18.0±0.2	urea	(1–3)	95R1

Remarks:
(1) measured by a multidimensional experimental approach using CD and fluorescence
(2) linear extrapolation, global fit
(3) measured in 0.05 M Tris

Staphylococcal nuclease, wild-type and mutants

Mutant	pH	T	ΔG	Approach	Remarks	Ref
wild-type	7.0	20	22.6	GuHCl	(1)	94S9
Ile15→Ala	7.0	20	11.7	GuHCl	(1)	94S9
Ile15→Gly	7.0	20	8.8	GuHCl	(1)	94S9
Lys16→Ala	7.0	20	21.8	GuHCl	(1)	94S9
Lys16→Gly	7.0	20	19.7	GuHCl	(1)	94S9
Asp19→Ala	7.0	20	22.2	GuHCl	(1)	94S9
Asp19→Gly	7.0	20	21.3	GuHCl	(1)	94S9
Asp21→Ala	7.0	20	25.5	GuHCl	(1)	94S9
Asp21→Gly	7.0	20	23.8	GuHCl	(1)	94S9
Lys28→Ala	7.0	20	19.7	GuHCl	(1)	94S9
Lys28→Gly	7.0	20	19.7	GuHCl	(1)	94S9
Gly29→Ala	7.0	20	17.6	GuHCl	(1)	94S9
Leu38→Ala	7.0	20	15.9	GuHCl	(1)	94S9
Leu38→Gly	7.0	20	20.5	GuHCl	(1)	94S9
Lys45→Ala	7.0	20	23.8	GuHCl	(1)	94S9
Lys45→Gly	7.0	20	23.4	GuHCl	(1)	94S9
Lys49→Ala	7.0	20	21.3	GuHCl	(1)	94S9
Lys49→Gly	7.0	20	21.8	GuHCl	(1)	94S9
Gly50→Ala	7.0	20	23.8	GuHCl	(1)	94S9
Val51→Ala	7.0	20	21.8	GuHCl	(1)	94S9
Val51→Gly	7.0	20	21.3	GuHCl	(1)	94S9
Glu52→Ala	7.0	20	22.2	GuHCl	(1)	94S9
Glu52→Gly	7.0	20	20.9	GuHCl	(1)	94S9
Tyr54→Ala	7.0	20	13.8	GuHCl	(1)	94S9
Tyr54→Gly	7.0	20	15.1	GuHCl	(1)	94S9
Asp77→Ala	7.0	20	9.6	GuHCl	(1)	94S9
Asp77→Gly	7.0	20	13.4	GuHCl	(1)	94S9
Gly79→Ala	7.0	20	13.0	GuHCl	(1)	94S9
Gly86→Ala	7.0	20	20.5	GuHCl	(1)	94S9
Asp95→Ala	7.0	20	8.8	GuHCl	(1)	94S9
Asp95→Gly	7.0	20	11.3	GuHCl	(1)	94S9
Gly96→Ala	7.0	20	15.5	GuHCl	(1)	94S9
Gly107→Ala	7.0	20	4.2	GuHCl	(1)	94S9
Tyr113→Ala	7.0	20	23.0	GuHCl	(1)	94S9
Tyr113→Gly	7.0	20	21.8	GuHCl	(1)	94S9
Asn118→Ala	7.0	20	14.2	GuHCl	(1)	94S9
Asn118→Gly	7.0	20	15.1	GuHCl	(1)	94S9
Asn119→Ala	7.0	20	17.6	GuHCl	(1)	94S9
Asn119→Gly	7.0	20	17.6	GuHCl	(1)	94S9
Lys136→Ala	7.0	20	19.7	GuHCl	(1)	94S9
Lys136→Gly	7.0	20	23.0	GuHCl	(1)	94S9
Asn138→Ala	7.0	20	18.4	GuHCl	(1)	94S9
Asn138→Gly	7.0	20	23.4	GuHCl	(1)	94S9

Remark:
(1) linear extrapolation, estimated error in ΔG ±0.6, for further details see Ref. 90S4

Staphylococcal nuclease, substitution mutants, difference ∆G values

Mutant	pH	T	∆(∆G)	Approach	Remarks	Ref
wild-type	7.0	20	0.0	GuHCl	(1)	90S7
Ile18→Gly	7.0	20	−10.9	GuHCl	(1)	90S7
Thr33→Ser	7.0	20	−5.9	GuHCl	(1)	90S7
Leu36→Ala	7.0	20	−15.1	GuHCl	(1)	90S7
Leu36→Gly	7.0	20	−22.6	GuHCl	(1)	90S7
Leu37→Ala	7.0	20	−7.5	GuHCl	(1)	90S7
Leu37→Gly	7.0	20	−16.3	GuHCl	(1)	90S7
Ala60→Gly	7.0	20	−6.3	GuHCl	(1)	90S7
Ala60→Val	7.0	20	−12.1	GuHCl	(1)	90S7
Phe61→Ala	7.0	20	−10.0	GuHCl	(1)	90S7
Phe61→Gly	7.0	20	−19.7	GuHCl	(1)	90S7
Thr62→Ala	7.0	20	−10.0	GuHCl	(1)	90S7
Thr62→Gly	7.0	20	−14.2	GuHCl	(1)	90S7
Glu73→Ala	7.0	20	−6.7	GuHCl	(1)	90S7
Phe76→Ala	7.0	20	−17.2	GuHCl	(1)	90S7
Phe76→Gly	7.0	20	−19.7	GuHCl	(1)	90S7
Asp77→Ala	7.0	20	−13.4	GuHCl	(1)	90S7
Asp77→Gly	7.0	20	−9.2	GuHCl	(1)	90S7
Gly79→Ser	7.0	20	−10.0	GuHCl	(1)	90S7
Asp83→Ala	7.0	20	−16.3	GuHCl	(1)	90S7
Asp83→Gly	7.0	20	−11.7	GuHCl	(1)	90S7
Ala90→Ser	7.0	20	−8.4	GuHCl	(1)	90S7
Tyr91→Ser	7.0	20	−22.2	GuHCl	(1)	90S7
Asp95→Ala	7.0	20	−15.1	GuHCl	(1)	90S7
Asp95→Gly	7.0	20	−13.0	GuHCl	(1)	90S7
Leu103→Ala	7.0	20	−19.2	GuHCl	(1)	90S7
Asn118→Asp	7.0	20	−10.5	GuHCl	(1)	90S7

Remark:
(1) linear extrapolation, measurement in 25 mM phosphate buffer with 100 mM NaCl

Staphylococcal nuclease, substitution mutants, difference ∆G values

Mutant	pH	T	∆(∆G)	Approach	Remarks	Ref
wild-type	7.0	20	0.0	GuHCl	(1,2)	90S4
Leu7→Ala	7.0	20	−6.7	GuHCl	(2)	90S4
Leu7→Gly	7.0	20	−6.3	GuHCl	(2)	90S4
Leu14→Ala	7.0	20	−9.6	GuHCl	(2)	90S4
Leu14→Gly	7.0	20	−15.5	GuHCl	(2)	90S4
Ile15→Val	7.0	20	−3.3	GuHCl	(2)	90S4
Ile15→Ala	7.0	20	−11.3	GuHCl	(2)	90S4
Ile15→Gly	7.0	20	−13.8	GuHCl	(2)	90S4
Ile18→Val	7.0	20	−4.6	GuHCl	(2)	90S4
Ile18→Ala	7.0	20	−10.5	GuHCl	(2)	90S4
Ile18→Gly	7.0	20	−10.5	GuHCl	(2)	90S4
Val23→Ala	7.0	20	−12.1	GuHCl	(2)	90S4
Val23→Gly	7.0	20	−23.4	GuHCl	(2)	90S4
Leu25→Ala	7.0	20	−11.3	GuHCl	(2)	90S4
Leu25→Gly	7.0	20	−18.8	GuHCl	(2)	90S4
Met26→Ala	7.0	20	−6.3	GuHCl	(2)	90S4
Met26→Gly	7.0	20	−9.2	GuHCl	(2)	90S4

Staphylococcal nuclease, substitution mutants, difference ∆G values (continued)

Mutant	pH	T	∆(∆G)	Approach	Remarks	Ref
Tyr27→Ala	7.0	20	−17.6	GuHCl	(2)	90S4
Tyr27→Gly	7.0	20	−24.3	GuHCl	(2)	90S4
Met32→Ala	7.0	20	−7.1	GuHCl	(2)	90S4
Met32→Gly	7.0	20	−10.0	GuHCl	(2)	90S4
Phe34→Ala	7.0	20	−15.5	GuHCl	(2)	90S4
Phe34→Gly	7.0	20	−25.9	GuHCl	(2)	90S4
Leu36→Ala	7.0	20	−14.6	GuHCl	(2)	90S4
Leu36→Gly	7.0	20	−22.2	GuHCl	(2)	90S4
Leu37→Ala	7.0	20	−7.1	GuHCl	(2)	90S4
Leu37→Gly	7.0	20	−15.9	GuHCl	(2)	90S4
Leu38→Ala	7.0	20	−7.1	GuHCl	(2)	90S4
Leu38→Gly	7.0	20	−2.5	GuHCl	(2)	90S4
Val39→Ala	7.0	20	−9.2	GuHCl	(2)	90S4
Val39→Gly	7.0	20	−19.7	GuHCl	(2)	90S4
Val51→Ala	7.0	20	−1.3	GuHCl	(2)	90S4
Val51→Gly	7.0	20	−1.7	GuHCl	(2)	90S4
Tyr54→Ala	7.0	20	−9.2	GuHCl	(2)	90S4
Tyr54→Gly	7.0	20	−7.9	GuHCl	(2)	90S4
Phe61→Ala	7.0	20	−9.6	GuHCl	(2)	90S4
Phe61→Gly	7.0	20	−20.1	GuHCl	(2)	90S4
Met65→Ala	7.0	20	−8.4	GuHCl	(2)	90S4
Met65→Gly	7.0	20	−19.2	GuHCl	(2)	90S4
Val66→Ala	7.0	20	−9.2	GuHCl	(2)	90S4
Val66→Gly	7.0	20	−18.4	GuHCl	(2)	90S4
Ile72→Val	7.0	20	−7.5	GuHCl	(2)	90S4
Ile72→Ala	7.0	20	−21.3	GuHCl	(2)	90S4
Ile72→Gly	7.0	20	−27.2	GuHCl	(2)	90S4
Val74→Ala	7.0	20	−13.0	GuHCl	(2)	90S4
Val74→Gly	7.0	20	−27.6	GuHCl	(2)	90S4
Phe76→Ala	7.0	20	−16.7	GuHCl	(2)	90S4
Phe76→Gly	7.0	20	−19.7	GuHCl	(2)	90S4
Tyr85→Ala	7.0	20	−1.7	GuHCl	(2)	90S4
Tyr85→Gly	7.0	20	−4.2	GuHCl	(2)	90S4
Leu89→Ala	7.0	20	−10.9	GuHCl	(2)	90S4
Leu89→Gly	7.0	20	−13.4	GuHCl	(2)	90S4
Tyr91→Ala	7.0	20	−22.2	GuHCl	(2)	90S4
Tyr91→Gly	7.0	20	−28.0	GuHCl	(2)	90S4
Ile92→Val	7.0	20	−2.1	GuHCl	(2)	90S4
Ile92→Ala	7.0	20	−16.7	GuHCl	(2)	90S4
Ile92→Gly	7.0	20	−27.6	GuHCl	(2)	90S4
Tyr93→Ala	7.0	20	−27.2	GuHCl	(2)	90S4
Tyr93→Gly	7.0	20	−31.4	GuHCl	(2)	90S4
Met98→Ala	7.0	20	−19.2	GuHCl	(2)	90S4
Met98→Gly	7.0	20	−18.8	GuHCl	(2)	90S4
Val99→Ala	7.0	20	−13.4	GuHCl	(2)	90S4
Val99→Gly	7.0	20	−20.9	GuHCl	(2)	90S4
Leu103→Ala	7.0	20	−19.2	GuHCl	(2)	90S4
Leu103→Gly	7.0	20	−27.6	GuHCl	(2)	90S4
Val104→Ala	7.0	20	−12.1	GuHCl	(2)	90S4
Val104→Gly	7.0	20	−27.2	GuHCl	(2)	90S4
Leu108→Ala	7.0	20	−24.3	GuHCl	(2)	90S4
Leu108→Gly	7.0	20	−30.1	GuHCl	(2)	90S4
Val111→Ala	7.0	20	−19.7	GuHCl	(2)	90S4

Staphylococcal nuclease, substitution mutants, difference ΔG values (continued)

Mutant	pH	T	$\Delta(\Delta G)$	Approach	Remarks	Ref
Val111→Gly	7.0	20	−20.5	GuHCl	(2)	90S4
Tyr113→Ala	7.0	20	0	GuHCl	(2)	90S4
Tyr113→Gly	7.0	20	−1.3	GuHCl	(2)	90S4
Val114→Ala	7.0	20	0	GuHCl	(2)	90S4
Val114→Gly	7.0	20	−0.8	GuHCl	(2)	90S4
Tyr115→Ala	7.0	20	−1.3	GuHCl	(2)	90S4
Tyr115→Gly	7.0	20	−2.9	GuHCl	(2)	90S4
Leu125→Ala	7.0	20	−20.5	GuHCl	(2)	90S4
Leu125→Gly	7.0	20	−29.3	GuHCl	(2)	90S4
Leu137→Ala	7.0	20	−9.6	GuHCl	(2)	90S4
Leu137→Gly	7.0	20	−19.2	GuHCl	(2)	90S4
Ile139→Val	7.0	20	−6.3	GuHCl	(2)	90S4
Ile139→Ala	7.0	20	−14.6	GuHCl	(2)	90S4
Ile139→Gly	7.0	20	−18.4	GuHCl	(2)	90S4

Remarks:
(1) ΔG of the wild-type = 22.9 kJ/mol
(2) linear extrapolation, buffer used: 25 mM sodium phosphate with 100 mM NaCl (see Ref. 91S6)

Staphylococcal nuclease, wild-type and mutants

Mutant	pH	ΔT	$\Delta(\Delta G)$	Approach	Ref
wild-type	7.0	0.0	0.0	DSC	95C2
Leu7→Ala	7.0	−3.2	−2.0	DSC	95C2
Val23→Ala	7.0	−12.5	−7.2	DSC	95C2
Lys24→Gly	7.0	−6.3	−1.7	DSC	95C2
Ala69→Thr	7.0	−13.7	−8.7	DSC	95C2
Glu75→Ala	7.0	−5.6	0.3	DSC	95C2
Glu75→Gly	7.0	−15.2	−5.7	DSC	95C2
Gly79→Ser	7.0	−4.4	−2.7	DSC	95C2
Gly88→Trp	7.0	4.6	8.8	DSC	95C2
Leu137→Ala	7.0	−8.7	−8.0	DSC	95C2
triple mutant (Val66→Leu, Gly88→Val and Gly79→Ser)					
	7.0	3.5	3.2	DSC	95C2

Staphylococcal nuclease, mutants and cross-linked dimers

Mutant	pH	T	ΔG	Approach	Remarks	Ref
A. untreated cysteine standard						
wild-type	7.0	20	23.0	GuHCl	(1–3)	95B11
Gly29→Cys	7.0	20	18.4	GuHCl	(1–3)	95B11
Gly50→Cys	7.0	20	19.7	GuHCl	(1–3)	95B11
Glu57→Cys	7.0	20	20.1	GuHCl	(1–3)	95B11
Ala60→Cys	7.0	20	18.0	GuHCl	(1–3)	95B11
Lys70→Cys	7.0	20	20.9	GuHCl	(1–3)	95B11
Lys78→Cys	7.0	20	21.3	GuHCl	(1–3)	95B11
Arg105→Cys	7.0	20	11.7	GuHCl	(1–3)	95B11
Ala112→Cys	7.0	20	19.7	GuHCl	(1–3)	95B11
Lys134→Cys	7.0	20	20.1	GuHCl	(1–3)	95B11

Staphylococcal nuclease, mutants and cross-linked dimers (continued)

Mutant	pH	T	ΔG	Approach	Remarks	Ref
B. phenylalanine substitution mutants						
Gly29→Phe	7.0	20	18.4	GuHCl	(1–3)	95B11
Gly50→Phe	7.0	20	21.3	GuHCl	(1–3)	95B11
Glu57→Phe	7.0	20	19.2	GuHCl	(1–3)	95B11
Ala60→Phe	7.0	20	n.d.	GuHCl	(1–3)	95B11
Lys70→Phe	7.0	20	23.0	GuHCl	(1–3)	95B11
Lys78→Phe	7.0	20	23.4	GuHCl	(1–3)	95B11
Arg105→Phe	7.0	20	10.0	GuHCl	(1–3)	95B11
Ala112→Phe	7.0	20	17.6	GuHCl	(1–3)	95B11
Lys134→Phe	7.0	20	21.8	GuHCl	(1–3)	95B11
C. MCA-treated protein					(4)	
wild-type	7.0	20	22.6	GuHCl	(1–3)	95B11
Gly29→Cys	7.0	20	18.0	GuHCl	(1–3)	95B11
Gly50→Cys	7.0	20	19.2	GuHCl	(1–3)	95B11
Glu57→Cys	7.0	20	20.5	GuHCl	(1–3)	95B11
Ala60→Cys	7.0	20	17.6	GuHCl	(1–3)	95B11
Lys70→Cys	7.0	20	21.3	GuHCl	(1–3)	95B11
Lys78→Cys	7.0	20	20.1	GuHCl	(1–3)	95B11
Arg105→Cys	7.0	20	10.0	GuHCl	(1–3)	95B11
Ala112→Cys	7.0	20	20.5	GuHCl	(1–3)	95B11
Lys134→Cys	7.0	20	20.9	GuHCl	(1–3)	95B11
D. mustard-treated monomer					(5)	
wild-type	7.0	20	22.6	GuHCl	(1–3)	95B11
Gly29→Cys	7.0	20	18.0	GuHCl	(1–3)	95B11
Gly50→Cys	7.0	20	17.6	GuHCl	(1–3)	95B11
Glu57→Cys	7.0	20	19.2	GuHCl	(1–3)	95B11
Lys70→Cys	7.0	20	20.1	GuHCl	(1–3)	95B11
Lys78→Cys	7.0	20	20.5	GuHCl	(1–3)	95B11
Arg105→Cys	7.0	20	10.5	GuHCl	(1–3)	95B11
Ala112→Cys	7.0	20	18.8	GuHCl	(1–3)	95B11
Lys134→Cys	7.0	20	20.1	GuHCl	(1–3)	95B11
E. DBP-treated monomer					(6)	
wild-type	7.0	20	3.8	GuHCl	(1–3)	95B11
Ala60→Cys	7.0	20	7.5	GuHCl	(1–3)	95B11
F. BMH-treated monomer					(7)	
wild-type	7.0	20	22.6	GuHCl	(1–3)	95B11
Gly29→Cys	7.0	20	15.5	GuHCl	(1–3)	95B11
Gly50→Cys	7.0	20	17.6	GuHCl	(1–3)	95B11
Glu57→Cys	7.0	20	18.0	GuHCl	(1–3)	95B11
Ala60→Cys	7.0	20	15.9	GuHCl	(1–3)	95B11
Lys70→Cys	7.0	20	18.4	GuHCl	(1–3)	95B11
Lys78→Cys	7.0	20	18.0	GuHCl	(1–3)	95B11
Arg105→Cys	7.0	20	11.3	GuHCl	(1–3)	95B11
Ala112→Cys	7.0	20	16.7	GuHCl	(1–3)	95B11
Lys134→Cys	7.0	20	17.6	GuHCl	(1–3)	95B11
G. BMH dimer					(7)	
Gly29→Cys						
trans. (1)	7.0	20	15.1	GuHCl	(2,3,8)	95B11
trans. (2)	7.0	20	14.6	GuHCl	(2,3,8)	95B11
Gly50→Cys						
trans. (1)	7.0	20	20.5±0.8	GuHCl	(2,3,8)	95B11
trans. (2)	7.0	20	18.8±1.3	GuHCl	(2,3,8)	95B11

Staphylococcal nuclease, mutants and cross-linked dimers (continued)

Mutant	pH	T	ΔG	Approach	Remarks	Ref
Glu57→Cys						
trans. (1)	7.0	20	17.2	GuHCl	(2,3,8)	95B11
trans. (2)	7.0	20	11.3±1.3	GuHCl	(2,3,8)	95B11
Ala60→Cys						
trans. (1)	7.0	20	16.3	GuHCl	(2,3,8)	95B11
trans. (2)	7.0	20	12.1	GuHCl	(2,3,8)	95B11
Lys70→Cys						
trans. (1)	7.0	20	17.6±0.8	GuHCl	(2,3,8)	95B11
trans. (2)	7.0	20	11.3±2.5	GuHCl	(2,3,8)	95B11
Lys78→Cys						
trans. (1)	7.0	20	13.8	GuHCl	(2,3,8)	95B11
trans. (2)	7.0	20	19.7	GuHCl	(2,3,8)	95B11
Arg105→Cys						
trans. (1)	7.0	20	4.2	GuHCl	(2,3,8)	95B11
trans. (2)	7.0	20	13.8	GuHCl	(2,3,8)	95B11
Ala112→Cys						
trans. (1)	7.0	20	17.6	GuHCl	(2,3,8)	95B11
trans. (2)	7.0	20	5.4	GuHCl	(2,3,8)	95B11
Lys134→Cys						
trans. (1)	7.0	20	18.4±0.8	GuHCl	(2,3,8)	95B11
trans. (2)	7.0	20	15.5±1.7	GuHCl	(2,3,8)	95B11
H. mustard dimer					(5)	
Lys70→Cys						
trans. (1)	7.0	20	18.8±0.8	GuHCl	(2,3,8)	95B11
trans. (2)	7.0	20	9.2±2.9	GuHCl	(2,3,8)	95B11

Remarks:
(1) linear extrapolation, transition was monitored by fluorescence
(2) buffer: 25 mM sodium phosphate, 100 mM NaCl, 0.5 mM tricarboxyethyl phosphine (TCEP), pH 7.0
(3) error in ΔG is estimated to be ±0.4 kJ/mol if not otherwise indicated
(4) MCA = ε-maleimidocapronic acid
(5) mustard = bis(2-chloroethyl)sulfide
(6) DBP = 1,3-dibromo-2-propanol
(7) BMH = 1,6-bismaleimidohexane
(8) treatment by a three-state thermodynamic unfolding model

Staphylococcal nuclease wild-type and mutant, position 28

Mutant	pH	T	ΔG	Approach	Remarks	Ref
wild-type	7.0	20	19.2±2.1	GuHCl	(1,2)	94E1
wild-type	7.0	20	20.9±0.3	GuHCl	(1,2)	94E1
wild-type	7.0	20	19.2±0.2	GuHCl	(1,2)	94E1
wild-type	7.0	20	23.7±0.1	GuHCl	(1,3)	94E1
wild-type	7.0	20	25.0±0.1	GuHCl	(1,4)	94E1
Ser28→Gly	7.0	20	4.1±0.6	GuHCl	(1,2)	94E1
Ser28→Gly	7.0	20	3.9±0.1	GuHCl	(1,2)	94E1

Remarks:
(1) for the approach see Ref. 94E1
(2) from lifetime of fluorescence decay
(3) from fluorescence intensity
(4) from fluorescence anisotropy

Staphylococcal nuclease, wild-type and mutants position 32

Mutant	pH	T	ΔG	Approach	Remarks	Ref
wild-type	7.0	20	23.35±0.17	GuHCl	(1–4)	95S14
wild-type	7.0	20	23.18±0.17	GuHCl	(1–3,5)	95S14
Met32→Leu	7.0	20	19.7±0.4	GuHCl	(1–4)	96S10
Met32→Ile	7.0	20	20.5±0.4	GuHCl	(1–4)	96S10

Remarks:
(1) the measurements were made on an automated device
(2) mean of 10 determinations
(3) buffer: 25 mM sodium phosphate, 100 mM NaCl, pH 7.0
(4) linear extrapolation
(5) linear extrapolation to zero denaturant concentration by a method that includes the pre- and postdenaturational baselines for a non-linear regression of the data

Staphylococcal nuclease, wild-type and mutants position 66

Mutant	pH	T	ΔG	Approach	Remarks	Ref
wild-type	7.0	20	23.4	GuHCl	(1,2)	91S6
wild-type	7.0	20	27.7	GuHCl	(1,5)	91S6
wild-type	7.5	20	22.2	GuHCl	(1,2)	91S6
wild-type	8.0	20	22.2	GuHCl	(1,2)	91S6
wild-type	8.4	20	22.2	GuHCl	(1,2)	91S6
wild-type	8.8	20	21.8	GuHCl	(1,2)	91S6
wild-type	9.2	20	20.5	GuHCl	(1,3)	91S6
wild-type	9.6	20	18.8	GuHCl	(1,3)	91S6
wild-type	10.0	20	13.4	GuHCl	(1,4)	91S6
wild-type	10.4	20	8.8	GuHCl	(1,4)	91S6
Val66→Lys	7.0	20	−8.0	GuHCl	(1,2)	91S6
Val66→Lys	7.5	20	−6.7	GuHCl	(1,2)	91S6
Val66→Lys	8.0	20	−3.8	GuHCl	(1,2)	91S6
Val66→Lys	8.4	20	−2.1	GuHCl	(1,2)	91S6
Val66→Lys	8.8	20	0.0	GuHCl	(1,2)	91S6
Val66→lYS	9.2	20	0.0	GuHCl	(1,3)	91S6
Val66→Lys	9.6	20	0.4	GuHCl	(1,3)	91S6
Val66→Lys	10.0	20	−1.3	GuHCl	(1,3)	91S6
Val66→Lys	10.4	20	−5.9	GuHCl	(1,3)	91S6
Val66→Met	7.0	20	22.6	GuHCl	(1,2)	91S6
Val66→Met	7.0	20	22.7	GuHCl	(1,6)	91S6
Val66→Met	8.8	20	21.3	GuHCl	(1,2)	91S6
Val66→Met	9.6	20	18.4	GuHCl	(1,3)	91S6

Remarks:
(1) linear extrapolation, unfolding in the absence of calcium and inhibitor
(2) buffer: 25 mM Tris-HCl with 100 mM NaCl
(3) buffer: 25 mM ethanolamine with 100 mM NaCl
(4) buffer: 50 mM ethanolamine with 100 mM NaCl
(5) reference value, buffer: 25 mM Tris-HCl
(6) reference value, buffer: 25 mM sodium phosphate with 100 mM NaCl

Staphylococcal nuclease, mutant Asp77→Ala, in the presence of 100 mM NaCl

Transition	pH	T	ΔG	Approach	Ref
trans. (1)	7.0	20	10±2	DSC, deconvolution	94C3
trans. (2)	7.0	20	9.2±2	DSC, deconvolution	94C3

Nuclease from *Staphylococcus aureus*, mutant His124→Leu (H124L), oxidized (ox.) and reduced (red.) forms of its engineered disulfide mutants

Mutant/Cys Form	pH	T	ΔG	Approach	Remarks	Ref
H124L	5.5	21.5	25.0±0.9	GuHCl	(1)	96H4
H124L, Cys80-Cys116 ox.						
	5.5	21.5	20.8±1.1	GuHCl	(1)	96H4
H124L, Cys80-Cys116 red.						
	5.5	21.5	22.6±1.2	GuHCl	(1)	96H4
Double mutant (Gly79→Ser and His124→Leu):						
Cys80-Cys116 ox.						
	5.5	21.5	23.0±0.6	GuHCl	(1)	96H4
Cys80-Cys116 red.						
	5.5	21.5	18.8±1.1	GuHCl	(1)	96H4
H124L, Cys79-Cys118 ox.						
	5.5	21.5	23.2±0.1	GuHCl	(1)	96H4
H124L, Cys79-Cys118 red.	5.5					
		21.5	13.7±1.2	GuHCl	(1)	96H4
H124L, Cys77-Cys118 ox.						
	5.5	21.5	24.9±0.6	GuHCl	(1)	96H4
H124L, Cys77-Cys118 red.						
	5.5	21.5	11.5±1.0	GuHCl	(1)	96H4
H124L	5.5	40	18.4±1.8	heat	(2)	96H4
H124L, Cys80-Cys116 ox.						
	5.5	40	12.6±1.3	heat	(2)	96H4
H124L, Cys80-Cys116 red.						
	5.5	40	7.7±1.7	heat	(2)	96H4
Double mutant (Gly79→Ser and His124→Leu):						
Cys80-Cys116 ox.						
		40	11.9±1.5	heat	(2)	96H4
Cys80-Cys116 red.	5.5					
	5.5	40	1.3±0.3	heat	(2)	96H4
H124L, Cys79-Cys118 ox.						
	5.5	40	24.4±1.8	heat	(2)	96H4
H124L, Cys79-Cys118 red.						
	5.5	40	8.0±1.4	heat	(2)	96H4
H124L, Cys77-Cys118 ox.						
	5.5	40	8.3±1.0	heat	(2)	96H4
H124L, Cys77-Cys118 red.						
	5.5	40	−5.2±0.9	heat	(2)	96H4

Remarks:
(1) linear extrapolation
(2) van't Hoff treatment, from $[\Delta G = \Delta H - T\Delta S]_{thermal}$ at 313 K

Staphylococcal nuclease, wild-type and mutants

Mutant	pH	T	$\Delta(\Delta G)_{total}$	$\Delta(\Delta G)_{trs.1}$	Approach	Remarks	Ref
wild-type	7.0	20	0	0	DSC	(1)	94C3
Val66→Leu	7.0	20	3.7	0.1	DSC	(1)	94C3
Val66→Trp	7.0	20	−4.9	−10	DSC	(1)	94C3
Val66→Ala	7.0	20	−9.0	−9.0	DSC	(1)	94C3
Gly88→Val	7.0	20	3.6	−1.0	DSC	(1)	94C3
Glu75→Val	7.0	20	−2.3	−7.7	DSC	(1)	94C3
Asp77→Ala	7.0	20	−1.5	−11	DSC	(1)	94C3

Remark:
(1) see also the thermodynamic quantities in Table 2

Staphylococcal nuclease, wild-type and mutants

Mutant	pH	T	ΔG	Approach	Remarks	Ref
wild-type	5.31	40	6.0±0.6	heat	(1)	90A1
Ile18→Met	5.42	40	5.1±0.7	heat	(1)	90A1
His46→Tyr	5.38	40	6.0±0.6	heat	(1)	90A1
Gly79→Ser	5.28	40	−1.5±0.2	heat	(1)	90A1
Leu89→Phe	5.38	40	0.9±0.2	heat	(1)	90A1
Val23→Phe	5.35	40	−0.47±0.04	heat	(1)	90A1
His124→Leu	5.20	40	8.6±1.6	heat	(1)	90A1
double mutant (Phe76→Val and His124→Leu)						
	5.21	40	6±1	heat	(1)	90A1

Remark:
(1) transition monitored by NMR

Staphylococcal nuclease, wild-type and mutants

Mutant	pH	T	ΔG	Approach	Remarks	Ref
wild-type	7.0	20	24.1	heat, v.H.		91E1
wild-type	7.0	20	25.6±2.1	urea	(1)	91E1
wild-type	7.0	20	23.0±1.3	GuHCl	(1)	91E1
wild-type	7.0	20	25.5	heat, v.H.		88S8
wild-type	7.0	20	24.7*	DSC	(3)	88S8
Val66→Leu	7.0	20	23.4	heat, v.H.		88S8
Gly88→Val	7.0	20	21.8	heat, v.H.		88S8
Ala69→Thr	7.0	20	13.8	heat, v.H.		88S8
double mutant (Ile18→Met and Ala90→Ser)						
	7.0	20	11.3	heat, v.H.		88S8
double mutant (Val66→Leu and Gly88→Val)						
	7.0	20	16.3	heat, v.H.		88S8
triple mutant (Val66→Leu, Gly88→Val and Gly79→Ser)						
	7.0	20	13.4	heat, v.H.		88S8
NC	7.0	20	7.4	heat	(2)	91E1
NC	7.0	20	5.3±0.7	pressure	(2)	91E1

Staphylococcal nuclease, wild-type and mutants (continued)

Mutant	pH	T	ΔG	Approach	Remarks	Ref
NC-Ser28→Gly	7.0	20	5.7	heat	(2)	91E1
NC-Ser28→Gly	7.0	20	3.6±0.8	urea	(1,2)	91E1
NC-Ser28→Gly	7.0	20	5.3±1.3	GuHCl	(1,2)	91E1
NC-Ser28→Gly	7.0	20	5.2±0.7	pressure	(2)	91E1

Remarks:
(1) linear extrapolation
(2) NC = nuclease-conA is a mutant protein containing a six amino acid residue β-turn substitute from concanavalin A, i.e., Ser-Ser-Asn-Gly-Ser-Pro instead of the residues 27-31 Tyr-Lys-Gly-Gln-Pro in the wild-type Staphylococcal nuclease
(3) ΔG was calculated from other thermodynamic quantities given in Ref. 88S8

Staphylococcal nuclease, wild-type and mutants

Mutant	pH	T	ΔG	Approach	Remarks	Ref
wild-type	7.0	0	19.2±1.7	GuHSCN	(1,2)	91A2
Ser28→Gly	7.0	0	4.0±0.6	GuHSCN	(1,2)	91A2
Ser28→Gly	7.0	0	14.6	GuHSCN	(1,3)	91A2

Remarks:
(1) linear extrapolation
(2) 10 mM cacodylate buffer
(3) 10 mM cacodylate buffer with 10 mM $CaCl_2$ and 0.1 mM pdTp (3′,5′-deoxythymidine biphosphate)

Staphylococcal nuclease, double substitution mutants

Mutant	pH	T	Δ(ΔG)	Approach	Remarks	Ref
(Leu-25-Ala and Met-26-Ala)						
	7.0	20	−18.4	GuHCl	(1,2)	92S13
(Met-32-Ala and Thr-33-Ala)						
	7.0	20	−13.8	GuHCl	(1,2)	92S13
(Leu-37-Ala and Leu-38-Ala)						
	7.0	20	−7.5	GuHCl	(1,2)	92S13
(Leu-38-Ala and Val-39-Ala)						
	7.0	20	−16.7	GuHCl	(1,2)	92S13
(Phe-61-Ala and Thr-62-Ala)						
	7.0	20	−19.2	GuHCl	(1,2)	92S13
(Met-65-Ala and Val-66-Ala)						
	7.0	20	−18.8	GuHCl	(1,2)	92S13
(Glu-73-Ala and Val-74-Ala)						
	7.0	20	−21.8	GuHCl	(1,2)	92S13

Remarks:
(1) linear extrapolation
(2) buffer: 25 mM sodium phosphate with 100 mM NaCl

Staphylococcal nuclease, wild-type and mutants

Mutant	pH	T	ΔG	Approach	Remarks	Ref
wild-type	7.0	20.0	23.0	GuHCl	(1)	92G4
Thr4→Ala	7.0	20.0	22.2	GuHCl	(1)	92G4
Thr4→Gly	7.0	20.0	22.2	GuHCl	(1)	92G4
Pro11→Ala	7.0	20.0	21.3	GuHCl	(1)	92G4
Pro11→Gly	7.0	20.0	18.8	GuHCl	(1)	92G4
Ala12→Gly	7.0	20.0	13.0	GuHCl	(1)	92G4
Ala12→Val	7.0	20.0	18.4	GuHCl	(1)	92G4
Thr13→Ala	7.0	20.0	20.1	GuHCl	(1)	92G4
Thr13→Gly	7.0	20.0	18.4	GuHCl	(1)	92G4
Ala17→Gly	7.0	20.0	21.8	GuHCl	(1)	92G4
Ala17→Val	7.0	20.0	15.1	GuHCl	(1)	92G4
Gly20→Ala	7.0	20.0	21.8	GuHCl	(1)	92G4
Gly20→Val	7.0	20.0	13.4	GuHCl	(1)	92G4
Thr22→Ala	7.0	20.0	16.3	GuHCl	(1)	92G4
Thr22→Gly	7.0	20.0	13.0	GuHCl	(1)	92G4
Gly29→Ala	7.0	20.0	17.6	GuHCl	(1)	92G4
Gly29→Val	7.0	20.0	10.5	GuHCl	(1)	92G4
Gln30→Ala	7.0	20.0	21.8	GuHCl	(1)	92G4
Gln30→Gly	7.0	20.0	19.2	GuHCl	(1)	92G4
Pro31→Ala	7.0	20.0	20.9	GuHCl	(1)	92G4
Pro31→Gly	7.0	20.0	16.3	GuHCl	(1)	92G4
Thr33→Ala	7.0	20.0	17.2	GuHCl	(1)	92G4
Thr33→Gly	7.0	20.0	12.6	GuHCl	(1)	92G4
Thr41→Ala	7.0	20.0	23.0	GuHCl	(1)	92G4
Thr41→Gly	7.0	20.0	14.6	GuHCl	(1)	92G4
Pro42→Ala	7.0	20.0	23.4	GuHCl	(1)	92G4
Pro42→Gly	7.0	20.0	21.3	GuHCl	(1)	92G4
Thr44→Ala	7.0	20.0	21.3	GuHCl	(1)	92G4
Thr44→Gly	7.0	20.0	20.5	GuHCl	(1)	92G4
Pro47→Ala	7.0	20.0	20.5	GuHCl	(1)	92G4
Pro47→Gly	7.0	20.0	22.6	GuHCl	(1)	92G4
Gly50→Ala	7.0	20.0	23.8	GuHCl	(1)	92G4
Gly50→Val	7.0	20.0	18.4	GuHCl	(1)	92G4
Gly55→Ala	7.0	20.0	20.5	GuHCl	(1)	92G4
Gly55→Val	7.0	20.0	15.5	GuHCl	(1)	92G4
Pro56→Ala	7.0	20.0	23.0	GuHCl	(1)	92G4
Pro56→Gly	7.0	20.0	18.8	GuHCl	(1)	92G4
Ala58→Gly	7.0	20.0	12.1	GuHCl	(1)	92G4
Ala58→Val	7.0	20.0	11.3	GuHCl	(1)	92G4
Ser59→Ala	7.0	20.0	24.7	GuHCl	(1)	92G4
Ser59→Gly	7.0	20.0	18.4	GuHCl	(1)	92G4
Ala60→Gly	7.0	20.0	17.2	GuHCl	(1)	92G4
Ala60→Val	7.0	20.0	11.3	GuHCl	(1)	92G4
Thr62→Ala	7.0	20.0	13.0	GuHCl	(1)	92G4
Thr62→Gly	7.0	20.0	8.4	GuHCl	(1)	92G4
Asn68→Ala	7.0	20.0	20.9	GuHCl	(1)	92G4
Asn68→Gly	7.0	20.0	20.9	GuHCl	(1)	92G4
Ala69→Gly	7.0	20.0	14.6	GuHCl	(1)	92G4
Ala69→Val	7.0	20.0	13.4	GuHCl	(1)	92G4
Gly79→Ala	7.0	20.0	13.0	GuHCl	(1)	92G4
Gly79→Val	7.0	20.0	13.0	GuHCl	(1)	92G4
Gln80→Ala	7.0	20.0	22.6	GuHCl	(1)	92G4
Gln80→Gly	7.0	20.0	17.2	GuHCl	(1)	92G4
Thr82→Ala	7.0	20.0	19.2	GuHCl	(1)	92G4

Staphylococcal nuclease, wild-type and mutants (continued)

Mutant	pH	T	ΔG	Approach	Remarks	Ref
Thr82→Gly	7.0	20.0	14.6	GuHCl	(1)	92G4
Gly86→Ala	7.0	20.0	20.5	GuHCl	(1)	92G4
Gly86→Val	7.0	20.0	7.5	GuHCl	(1)	92G4
Gly88→Ala	7.0	20.0	22.6	GuHCl	(1)	92G4
Gly88→Val	7.0	20.0	19.2	GuHCl	(1)	92G4
Ala90→Gly	7.0	20.0	14.6	GuHCl	(1)	92G4
Ala90→Val	7.0	20.0	22.2	GuHCl	(1)	92G4
Ala94→Gly	7.0	20.0	13.0	GuHCl	(1)	92G4
Ala94→Val	7.0	20.0	17.6	GuHCl	(1)	92G4
Gly96→Ala	7.0	20.0	15.5	GuHCl	(1)	92G4
Gly96→Val	7.0	20.0	7.5	GuHCl	(1)	92G4
Asn100→Ala	7.0	20.0	1.3	GuHCl	(1)	92G4
Asn100→Gly	7.0	20.0	1.7	GuHCl	(1)	92G4
Ala102→Gly	7.0	20.0	17.6	GuHCl	(1)	92G4
Ala102→Val	7.0	20.0	15.1	GuHCl	(1)	92G4
Gln106→Ala	7.0	20.0	23.4	GuHCl	(1)	92G4
Gln106→Gly	7.0	20.0	16.7	GuHCl	(1)	92G4
Gly107→Ala	7.0	20.0	4.2	GuHCl	(1)	92G4
Gly107→Val	7.0	20.0	<–8.4	GuHCl	(1)	92G4
Ala109→Gly	7.0	20.0	18.8	GuHCl	(1)	92G4
Ala109→Val	7.0	20.0	10.0	GuHCl	(1)	92G4
Ala112→Gly	7.0	20.0	23.0	GuHCl	(1)	92G4
Ala112→Val	7.0	20.0	16.7	GuHCl	(1)	92G4
Pro117→Ala	7.0	20.0	26.4	GuHCl	(1)	92G4
Pro117→Gly	7.0	20.0	26.7	GuHCl	(1)	92G4
Asn118→Ala	7.0	20.0	14.2	GuHCl	(1)	92G4
Asn118→Gly	7.0	20.0	15.1	GuHCl	(1)	92G4
Asn119→Ala	7.0	20.0	17.6	GuHCl	(1)	92G4
Asn119→Gly	7.0	20.0	17.6	GuHCl	(1)	92G4
Thr120→Ala	7.0	20.0	18.0	GuHCl	(1)	92G4
Thr120→Gly	7.0	20.0	14.2	GuHCl	(1)	92G4
Gln123→Ala	7.0	20.0	21.3	GuHCl	(1)	92G4
Gln123→Gly	7.0	20.0	20.5	GuHCl	(1)	92G4
Ser128→Ala	7.0	20.0	25.9	GuHCl	(1)	92G4
Ser128→Gly	7.0	20.0	16.3	GuHCl	(1)	92G4
Ala130→Gly	7.0	20.0	18.4	GuHCl	(1)	92G4
Ala130→Val	7.0	20.0	18.0	GuHCl	(1)	92G4
Gln131→Ala	7.0	20.0	22.2	GuHCl	(1)	92G4
Gln131→Gly	7.0	20.0	13.0	GuHCl	(1)	92G4
Ala132→Gly	7.0	20.0	7.5	GuHCl	(1)	92G4
Ala132→Val	7.0	20.0	2.9	GuHCl	(1)	92G4
Asn138→Ala	7.0	20.0	18.4	GuHCl	(1)	92G4
Asn138→Gly	7.0	20.0	23.4	GuHCl	(1)	92G4
Ser141→Ala	7.0	20.0	21.3	GuHCl	(1)	92G4
Ser141→Gly	7.0	20.0	19.2	GuHCl	(1)	92G4
Asn144→Ala	7.0	20.0	22.6	GuHCl	(1)	92G4
Asn144→Gly	7.0	20.0	20.9	GuHCl	(1)	92G4
Ala145→Gly	7.0	20.0	21.3	GuHCl	(1)	92G4
Ala145→Val	7.0	20.0	21.8	GuHCl	(1)	92G4
Gln149→Gly	7.0	20.0	21.3	GuHCl	(1)	92G4

Remark:
(1) linear extrapolation

Staphylococcal nuclease, proline 117 mutants

Mutant	pH	T	ΔG	Approach	Remarks	Ref
wild-type	5.5		17.7	GuHCl	(1)	94H11
Pro117→Ala	5.5		20.1	GuHCl	(1)	94H11
Pro117→Gly	5.5		24.7	GuHCl	(1)	94H11
Pro117→Thr	5.5		17.6	GuHCl	(1)	94H11

Remark:
(1) linear extrapolation

Staphylococcal nuclease from *Staphylococcus aureus*, mutant (His124→Leu) and proline-to-glycine substitutions, measured in the presence of GuHCl

Mutant	pH	T	ΔG	Approach	Remarks	Ref
His124→Leu	5.5	21	14.2±0.4	GuHCl	(1,4)	96V5
His124→Leu	5.5	21	10.0±0.8	pressure	(2,4)	96V5
His124→Leu	5.5	21	14.6±5.9	pressure	(3,4)	96V5
His124→Leu	5.5	21	7.9±0.4	GuHCl	(1,5)	96V5
His124→Leu	5.5	21	3.3±0.4	pressure	(2,5)	96V5
His124→Leu	5.5	21	4.6±3.8	pressure	(3,5)	96V5
(His124→Leu and Pro42→Gly)						
	5.5	21	10.0±1.7	pressure	(2,5)	96V5
	5.5	21	8.8±2.1	pressure	(3,5)	96V5
(His124→Leu and Pro47→Gly)						
	5.5	21	11.7±1.7	GuHCl	(1,5)	96V5
	5.5	21	11.7±2.1	pressure	(2,5)	96V5
	5.5	21	12.6±8.4	pressure	(3,5)	96V5
(His124→Leu and Pro117→Gly)						
	5.5	21	13.4±0.4	GuHCl	(1,5)	96V5
	5.5	21	13.0±1.7	pressure	(2,5)	96V5
	5.5	21	10.5±6.3	pressure	(3,5)	96V5
(His124→Leu, Pro47→Gly and Pro117→Gly)						
	5.5	21	14.5±0.2	GuHCl	(1,5)	96V5
	5.5	21	15.5±2.9	pressure	(2,5)	96V5
	5.5	21	21.8±8.4	pressure	(3,5)	96V5

Remarks:
(1) from equilibrium unfolding
(2) from equilibrium pressure unfolding
(3) from kinetics of pressure unfolding
(4) measured in the presence of 0.5 M GuHCl
(5) measured in the presence of 0.75 M GuHCl

Staphylococcal nuclease, mutants

Mutant	pH	T	ΔT	Δ(ΔG)	Approach	Remarks	Ref
wild-type	5.0	47.0	0.0	0.0	DSC	(1)	93T1
Leu25→Ala	5.0	47.0	−15.8	−11.7	DSC	(1)	93T1
Val66→Leu	5.0	47.0	3.8	2.5	DSC	(1)	93T1
Gly79→Ser	5.0	47.0	−8.5	−5.0	DSC	(1)	93T1
Gly88→Val	5.0	47.0	3.6	2.5	DSC	(1)	93T1
Ala90→Ser	5.0	47.0	−13.5	−10.9	DSC	(1)	93T1
His124→Leu	5.0	47.0	8.3	7.9	DSC	(1)	93T1
double mutant (Val66→Leu and Gly88→Val)							
	5.0	47.0	10.9	5.0	DSC	(1)	93T1
triple mutant (Val66→Leu, Gly79→Ser and Gly88→Val)							
	5.0	47.0	7.3	2.1	DSC	(1)	93T1
wild-type	7.0	51.4	0.0	0.0	DSC	(1)	93T1
Leu25→Ala	7.0	51.4	−11.5	−10.9	DSC	(1)	93T1
Val66→Leu	7.0	51.4	3.9	3.3	DSC	(1)	93T1
Gly79→Ser	7.0	51.4	−7.6	−5.4	DSC	(1)	93T1
Gly88→Val	7.0	51.4	2.9	2.1	DSC	(1)	93T1
Ala90→Ser	7.0	51.4	−10.7	−10.5	DSC	(1)	93T1
His124→Leu	7.0	51.4	5.1	5.4	DSC	(1)	93T1
double mutant (Val66→Leu and Gly88→Val)							
	7.0	51.4	6.4	4.2	DSC	(1)	93T1
triple mutant (Val66→Leu, Gly79→Ser and Gly88→Val)							
	7.0	51.4	4.7	1.3	DSC	(1)	93T1

Remark:
(1) estimated uncertainties: Δ(ΔG) = ±1.7 kJ/mol

Staphylococcal nuclease, mutants, difference ΔG values

Mutant	pH	T	Δ(ΔG)	Approach	Remarks	Ref
Mutants having m_{GuHCl} similar as AS the wild-type:						
Ile15→Val	7.0	20	−3.39	GuHCl	(1,2)	93G7
Ile72→Val	7.0	20	−7.40	GuHCl	(1,2)	93G7
Tyr85→Ala	7.0	20	−1.72	GuHCl	(1,2)	93G7
Ile92→Val	7.0	20	−2.09	GuHCl	(1,2)	93G7
Tyr113→Ala	7.0	20	0.13	GuHCl	(1,2)	93G7
Ala130→Gly	7.0	20	−4.64	GuHCl	(1,3)	93G7
Mutants having m_{GuHCl} less than the wild-type:						
Leu7→Ala	7.0	20	−6.61	GuHCl	(1,2)	93G7
Val23→Phe	7.0	20	−9.62	GuHCl	(1)	93G7
Leu37→Ala	7.0	20	−7.03	GuHCl	(1,2)	93G7
Thr62→Ala	7.0	20	−10.04	GuHCl	(1,3)	93G7
Val66→Leu	7.0	20	−0.42	GuHCl	(1,4)	93G7
Glu75→Val	7.0	20	−9.66	GuHCl	(1)	93G7
Gly79→Asp	7.0	20	−9.54	GuHCl	(1)	93G7
Gly79→Ser	7.0	20	−11.13	GuHCl	(1,4)	93G7
Gly88→Val	7.0	20	−3.77	GuHCl	(1,4)	93G7
Pro117→Leu	7.0	20	0.84	GuHCl	(1)	93G7
Asn118→Asp	7.0	20	−10.04	GuHCl	(1)	93G7

Staphylococcal nuclease, mutants, difference ΔG values (continued)

Mutant	pH	T	Δ(ΔG)	Approach	Remarks	Ref
Mutants having m_{GuHCl} higher than the wild–type:						
Ile18→Met	7.0	20	−2.09	GuHCl	(1,4)	93G7
Thr33→Ser	7.0	20	−4.18	GuHCl	(1)	93G7
Thr62→Gly	7.0	20	−14.56	GuHCl	(1,2)	93G7
Ala69→Thr	7.0	20	−11.0	GuHCl	(1,4)	93G7
Ala90→Ser	7.0	20	−7.95	GuHCl	(1,4)	93G7

Remarks:
(1) for the procedure and for the groups, see Ref. 92G4; the data refer to Δ(ΔG) = 28.7 kJ/mol of the wild-type protein
(2) see also Ref. 90S4
(3) see also Ref. 92G4
(4) see also Ref. 86S3

Staphylococcal nuclease, wild-type and mutants

Mutant	pH	T	ΔG	Approach	Remarks	Ref
wild-type	7.0	20	23.4	GuHCl	(1,2)	95B10
Thr13→Cys	7.0	20	17.6	GuHCl	(1,2)	95B10
Thr13→Ser	7.0	20	22.2	GuHCl	(1,2)	95B10
Thr13→Val	7.0	20	21.8	GuHCl	(1,2)	95B10
Thr22→Cys	7.0	20	19.7	GuHCl	(1,2)	95B10
Thr22→Ser	7.0	20	20.9	GuHCl	(1,2)	95B10
Thr22→Val	7.0	20	19.7	GuHCl	(1,2)	95B10
Val23→Ser	7.0	20	3.8	GuHCl	(1,2)	95B10
Val23→Thr	7.0	20	10.0	GuHCl	(1,2)	95B10
Tyr27→Leu	7.0	20	<−12	GuHCl	(1,2)	95B10
Tyr27→Phe	7.0	20	20.9	GuHCl	(1,2)	95B10
Thr33→Cys	7.0	20	18.4	GuHCl	(1,2)	95B10
Thr33→Ser	7.0	20	18.0	GuHCl	(1,2)	95B10
Thr33→Val	7.0	20	25.1	GuHCl	(1,2)	95B10
Val39→Ser	7.0	20	13.4	GuHCl	(1,2)	95B10
Val39→Thr	7.0	20	18.0	GuHCl	(1,2)	95B10
Thr41→Cys	7.0	20	25.9	GuHCl	(1,2)	95B10
Thr41→Ser	7.0	20	18.8	GuHCl	(1,2)	95B10
Thr41→Val	7.0	20	26.8	GuHCl	(1,2)	95B10
Thr44→Cys	7.0	20	22.6	GuHCl	(1,2)	95B10
Thr44→Ser	7.0	20	23.4	GuHCl	(1,2)	95B10
Thr44→Val	7.0	20	23.8	GuHCl	(1,2)	95B10
Val51→Ser	7.0	20	23.0	GuHCl	(1,2)	95B10
Val51→Thr	7.0	20	24.3	GuHCl	(1,2)	95B10
Tyr54→Leu	7.0	20	9.2	GuHCl	(1,2)	95B10
Tyr54→Phe	7.0	20	21.3	GuHCl	(1,2)	95B10
Thr62→Cys	7.0	20	18.8	GuHCl	(1,2)	95B10
Thr62→Ser	7.0	20	14.6	GuHCl	(1,2)	95B10
Thr62→Val	7.0	20	22.9	GuHCl	(1,2)	95B10
Val66→Ser	7.0	20	10.5	GuHCl	(1,2)	95B10
Val66→Thr	7.0	20	17.6	GuHCl	(1,2)	95B10
Val74→Ser	7.0	20	0.8	GuHCl	(1,2)	95B10
Val74→Thr	7.0	20	7.5	GuHCl	(1,2)	95B10
Thr82→Cys	7.0	20	23.0	GuHCl	(1,2)	95B10

Staphylococcal nuclease, wild-type and mutants (continued)

Mutant	pH	T	ΔG	Approach	Remarks	Ref
Thr82→Ser	7.0	20	20.5	GuHCl	(1,2)	95B10
Thr82→Val	7.0	20	24.3	GuHCl	(1,2)	95B10
Tyr85→Leu	7.0	20	23.0	GuHCl	(1,2)	95B10
Tyr85→Phe	7.0	20	23.4	GuHCl	(1,2)	95B10
Tyr91→Leu	7.0	20	7.1	GuHCl	(1,2)	95B10
Tyr91→Phe	7.0	20	13.4	GuHCl	(1,2)	95B10
Tyr93→Leu	7.0	20	4.6	GuHCl	(1,2)	95B10
Tyr93→Phe	7.0	20	15.1	GuHCl	(1,2)	95B10
Val99→Ser	7.0	20	2.1	GuHCl	(1,2)	95B10
Val99→Thr	7.0	20	9.6	GuHCl	(1,2)	95B10
Val104→Ser	7.0	20	2.9	GuHCl	(1,2)	95B10
Val104→Thr	7.0	20	13.0	GuHCl	(1,2)	95B10
Val111>Ser	7.0	20	3.3	GuHCl	(1,2)	95B10
Val111→Thr	7.0	20	13.8	GuHCl	(1,2)	95B10
Tyr113→Leu	7.0	20	24.3	GuHCl	(1,2)	95B10
Tyr113→Phe	7.0	20	23.4	GuHCl	(1,2)	95B10
Val114→Ser	7.0	20	23.4	GuHCl	(1,2)	95B10
Val114→Thr	7.0	20	22.2	GuHCl	(1,2)	95B10
Tyr115→Leu	7.0	20	22.2	GuHCl	(1,2)	95B10
Tyr115→Phe	7.0	20	23.0	GuHCl	(1,2)	95B10
Thr120→Cys	7.0	20	16.3	GuHCl	(1,2)	95B10
Thr120→Ser	7.0	20	20.9	GuHCl	(1,2)	95B10
Thr120→Val	7.0	20	15.9	GuHCl	(1,2)	95B10

Remarks:
(1) linear extrapolation
(2) measured in 25 mM sodium phosphate, 100 mM NaCl, pH 7.0

Staphylococcal nuclease, wild-type and mutants

Mutant	pH	T	ΔG	Approach	Remarks	Ref
wild-type	7.0	20	15.9	urea	(1)	95K2
Val66→Leu	7.0	20	26.4	urea	(1)	95K2
Ala69→Thr	7.0	20	7.9	urea	(1)	95K2
Gly88→Val	7.0	20	20.9	urea	(1)	95K2
Ala90→Ser	7.0	20	9.6	urea	(1)	95K2
(Val66→Leu and Gly88→Val)	7.0	20	22.2	urea	(1)	95K2
(Ala69→Thr and Ala90→Ser)	7.0	20	1.3	urea	(1)	95K2

Remark:
(1) linear extrapolation to zero denaturant concentration by a method that includes the pre- and post-denaturational baselines for a non-linear regression of the data

Staphylococcal nuclease, double mutants and triple mutants, difference ΔG values

Mutant	pH	T	$\Delta(\Delta G)$	Approach	Remarks	Ref
Group 1 mutants:						
(Ile15→Val and Ile72→Val)	7.0	20	−8.41	GuHCl	(1)	93G7
(Ile15→Val and Tyr85→Val)	7.0	20	−4.44	GuHCl	(1)	93G7
(Ile15→Val and Tyr113→Ala)	7.0	20	−2.59	GuHCl	(1)	93G7
(Ile72→Val and Tyr85→Ala)	7.0	20	−6.69	GuHCl	(1)	93G7
(Ile72→Val and Tyr113→Ala)	7.0	20	−3.93	GuHCl	(1)	93G7
(Tyr85→Ala and Tyr113→Ala)	7.0	20	−0.96	GuHCl	(1)	93G7
Group 2 mutants:						
(Ile18→Met and Thr33→Ser)	7.0	20	−8.4	GuHCl	(1)	93G7
(Ile18→Met and Ala69→Thr)	7.0	20	−13.8	GuHCl	(1)	93G7
(Ile18→Met and Ala90→Ser)	7.0	20	−13.22	GuHCl	(1)	93G7
(Ala69→Thr and Ala90→Ser)	7.0	20	−19.7	GuHCl	(1)	93G7
Group 3 mutants:						
(Leu7→Ala and Val23→Phe)	7.0	20	−15.27	GuHCl	(1)	93G7
(Leu7→Ala and Leu37→Ala)	7.0	20	−6.90	GuHCl	(1)	93G7
(Leu7→Ala and Glu75→Val)	7.0	20	−15.23	GuHCl	(1)	93G7
(Leu7→Ala and Gly79→Ser)	7.0	20	−11.59	GuHCl	(1)	93G7
(Val23→Phe and Leu37→Ala)	7.0	20	−13.26	GuHCl	(1)	93G7
(Val23→Phe and Glu75→Val)	7.0	20	−20.04	GuHCl	(1)	93G7
(Val23→Phe and Gly79→Ser)	7.0	20	−19.87	GuHCl	(1)	93G7
(Leu37→Ala and Glu75→Val)	7.0	20	−10.33	GuHCl	(1)	93G7
(Leu37→Ala and Gly79→Ser)	7.0	20	−7.66	GuHCl	(1)	93G7
(Leu37→Ala and Pro117→Leu)	7.0	20	−7.07	GuHCl	(1)	93G7
(Leu37→Ala and Ile118→Asp)	7.0	20	−7.20	GuHCl	(1)	93G7
(Thr62→Ala and Val66→Leu)	7.0	20	−10.04	GuHCl	(1)	93G7
(Thr62→Ala and Gly88→Val)	7.0	20	−13.39	GuHCl	(1)	93G7
(Glu75→Val and Gly79→Ser)	7.0	20	−19.54	GuHCl	(1)	93G7
(Glu75→Val and Pro117→Leu)	7.0	20	−6.74	GuHCl	(1)	93G7
(Gly79→Ser and Pro117→Leu)	7.0	20	−8.70	GuHCl	(1)	93G7
(Gly79→Asp and Ile118→Asp)	7.0	20	−10.79	GuHCl	(1)	93G7
(Gly79→Ser and Asn118→Asp)	7.0	20	−11.67	GuHCl	(1)	93G7
(Pro117→Leu and Asn118→Asp)	7.0	20	−4.18	GuHCl	(1)	93G7
(Val66→Leu and Gly88→Val)	7.0	20	−8.37	GuHCl	(1)	93G7
(Leu37→Ala, Gly79→Ser and Asn118→Asp)	7.0	20	−6.86	GuHCl	(1)	93G7
(Leu37→Ala, Gly79→Ser and Pro117→Leu)	7.0	20	−7.66	GuHCl	(1)	93G7
(Gly79→Ser, Pro117→Leu and Asn118→Asp)	7.0	20	−12.3	GuHCl	(1)	93G7
Group 4 mutants:						
(Val23→Phe and Thr33→Ser)	7.0	20	−11.38	GuHCl	(1)	93G7
(Val23→Phe and Ala69→Thr)	7.0	20	−16.7	GuHCl	(1)	93G7
(Val23→Phe and Ala90→Ser)	7.0	20	−13.85	GuHCl	(1)	93G7
(Leu37→Ala and Thr33→Ser)	7.0	20	−9.25	GuHCl	(1)	93G7
(Leu37→Ala and Ala69→Thr)	7.0	20	−17.28	GuHCl	(1)	93G7
(Leu37→Ala and Ala90→Ser)	7.0	20	−13.89	GuHCl	(1)	93G7
(Thr62→Gly and Gly88→Val)	7.0	20	−15.31	GuHCl	(1)	93G7
(Glu75→Val and Thr33→Ser)	7.0	20	−13.43	GuHCl	(1)	93G7
(Glu75→Val and Ala69→Thr)	7.0	20	−18.2	GuHCl	(1)	93G7
(Glu75→Val and Ala90→Ser)	7.0	20	−15.06	GuHCl	(1)	93G7

Staphylococcal nuclease, double mutants and triple mutants, difference ΔG values (continued)

Mutant	pH	T	$\Delta(\Delta G)$	Approach	Remarks	Ref
Group 5 mutants:						
(Leu7→Ala and Ile15→Val)	7.0	20	−8.70	GuHCl	(1)	93G7
(Leu7→Ala and Ile72→Val)	7.0	20	−10.08	GuHCl	(1)	93G7
(Leu7→Ala and Tyr85→Ala)	7.0	20	−6.69	GuHCl	(1)	93G7
(Leu7→Ala and Ile92→Val)	7.0	20	−6.44	GuHCl	(1)	93G7
(Leu7→Ala and Tyr113→Ala)	7.0	20	−4.60	GuHCl	(1)	93G7
(Leu7→Ala and Ala130→Gly)	7.0	20	−10.29	GuHCl	(1)	93G7
(Val23→Phe and Ile15→Val)	7.0	20	−10.84	GuHCl	(1)	93G7
(Val23→Phe and Ile72→Val)	7.0	20	−12.25	GuHCl	(1)	93G7
(Val23→Phe and Tyr85→Ala)	7.0	20	−10.50	GuHCl	(1)	93G7
(Val23→Phe and Ile92→Val)	7.0	20	−8.41	GuHCl	(1)	93G7
(Val23→Phe and Tyr113→Ala)	7.0	20	−7.91	GuHCl	(1)	93G7
(Val23→Phe and Ala130→Gly)	7.0	20	−14.6	GuHCl	(1)	93G7
(Leu37→Ala and Ile15→Val)	7.0	20	−9.08	GuHCl	(1)	93G7
(Leu37→Ala and Ile72→Val)	7.0	20	−11.63	GuHCl	(1)	93G7
(Leu37→Ala and Tyr85→Ala)	7.0	20	−8.28	GuHCl	(1)	93G7
(Leu37→Ala and Ile92→Val)	7.0	20	−7.41	GuHCl	(1)	93G7
(Leu37→Ala and Tyr113→Ala)	7.0	20	−6.11	GuHCl	(1)	93G7
(Leu37→Ala and Ala130→Gly)	7.0	20	−10.59	GuHCl	(1)	93G7
(Glu75→Val and Ile15→Val)	7.0	20	−11.84	GuHCl	(1)	93G7
(Glu75→Val and Ile72→Val)	7.0	20	−13.05	GuHCl	(1)	93G7
(Glu75→Val and Tyr85→Ala)	7.0	20	−10.50	GuHCl	(1)	93G7
(Glu75→Val and Ile92→Val)	7.0	20	−8.58	GuHCl	(1)	93G7
(Glu75→Val and Tyr113→Ala)	7.0	20	−6.49	GuHCl	(1)	93G7
(Glu75→Val and Ala130→Gly)	7.0	20	−14.2	GuHCl	(1)	93G7
(Gly79→Ser and Ile15→Val)	7.0	20	−13.0	GuHCl	(1)	93G7
(Gly79→Ser and Ile72→Val)	7.0	20	−15.10	GuHCl	(1)	93G7
(Gly79→Ser and Tyr85→Ala)	7.0	20	−11.25	GuHCl	(1)	93G7
(Gly79→Ser and Ile92→→Val)	7.0	20	−10.67	GuHCl	(1)	93G7
(Gly79→Ser and Tyr113→Ala)	7.0	20	−9.29	GuHCl	(1)	93G7
(Gly79→Ser and Ala130→Gly)	7.0	20	−15.40	GuHCl	(1)	93G7

Remark:
(1) for the procedure and for the groups, see Ref. 92G4; the data refer to $\Delta(\Delta G) = 28.7$ kJ/mol of the wild-type protein

Staphylococcal nuclease, full length wild-type (res. 1–149), and fragments (res. 1–136) forming compact denatured states

Mutant	pH	T	ΔG	Approach	Remarks	Ref
Full length protein (res. 1–149):						
wild-type	7.0	20	22.3±0.4	GuHCl	(1–3)	93G5
wild-type	7.0	20	21.0±0.4	GuHCl	(1,2,4)	93G5
Gly88→Trp	7.0	20	17.2±0.4	GuHCl	(1–3)	93G5
Gly88→Trp	7.0	20	18.1±0.4	GuHCl	(1,2,4)	93G5
Lys70→Trp	7.0	20	18.8±0.4	GuHCl	(1–3)	93G5
Lys70→Trp	7.0	20	20.4±0.4	GuHCl	(1,2,4)	93G5
Val66→Trp	7.0	20	9.7±0.4	GuHCl	(1–3)	93G5
Val66→Trp	7.0	20	6.9±0.4	GuHCl	(1,2,4,6)	93G5

Staphylococcal nuclease, full length wild-type (res. 1–149), and fragments (res. 1–136) forming compact denatured states (continued)

Mutant	pH	T	ΔG	Approach	Remarks	Ref
Large fragment (res. 1–136):						
wild-type	7.0	20	2.9±0.4	GuHCl	(1,2,4)	93G5
Gly88→Trp	7.0	20	11.7±0.8	GuHCl	(1–3,5)	93G5
Gly88→Trp	7.0	20	10.0±1.3	GuHCl	(1,2,4,5)	93G5
Lys70→Trp	7.0	20	5.4±0.4	GuHCl	(1–3,5)	93G5
Lys70→Trp	7.0	20	1.3±1.3	GuHCl	(1,2,4,5)	93G5
Val66→Trp	7.0	20	10.0±0.4	GuHCl	(1–3,5)	93G5
Val66→Trp	7.0	20	4.2±1.3	GuHCl	(1,2,4–6)	93G5

Remarks:
(1) linear extrapolation
(2) for the procedure, see Ref. 90S4
(3) transition monitored by fluorescence
(4) transition monitored by CD at 222 nm
(5) the protein shares similarities with a molten globule
(6) biphasic transition

Staphylococcal nuclease

pH	T	ΔG	Approach	Remarks	Ref
7.5		18.2	GuHCl	(1)	93S9
7.5		17.6	urea	(1)	93S9

Remark:
(1) for details of the data treatment it is referred to Ref. 93S9

Staphylococcal nuclease

pH	T	ΔG	Approach	Remarks	Ref
7.0	25	18.8	DSC	(1,2)	96S11

Remarks:
(1) measured in the presence of 2.0 M GuHCl
(2) the paper contains an analysis of folding kinetics in terms of a triphasic process

Staphylococcal nuclease, wild-type and mutants

Mutant	pH	T	ΔG	Δ(ΔG)	Approach	Remarks	Ref
wild-type	7.0	0	19.2	0	GuHSCN	(1)	93N1
Pro31→Ala	7.0	0	16.7	−2.5	GuHSCN	(1)	93N1
Pro117→Gly	7.0	0	25.9	6.7	GuHSCN	(1)	93N1
Pro117→Thr	7.0	0	23.8	4.6	GuHSCN	(1)	93N1

Remark:
(1) linear extrapolation

Staphylococcal nuclease, wild-type and tryptophan mutants

Mutant	pH	T	ΔG	Remarks	Ref
wild-type	7.20	20	22.3±0.6	(1,2)	96E2
Val66→Trp*	7.20	20	7.8±0.7	(1–3)	96E2
Val66→Trp	7.20	20	5.4±1.0	(1,2)	96E2
Val66→Trp	7.20	20	8.9±0.5	(1,2,4)	96E2
Val66→Trp	7.20	20	7.9±0.8	(1,2,5)	96E2

Remarks:
(1) linear extrapolation to zero denaturant concentration by a method that includes the pre- and postdenaturational baselines for a non-linear regression of the data
(2) parameters from a global fit of simultaneously acquired data from fluorescence and CD at 222 and 235 nm
(3) mutant Val66→Trp* does not contain Trp140
(4) a three-state model including $\Delta G_2 = 2.1 \pm 0.9$ kJ/mol provides a better fit
(5) results of a three-state fit including a fixed $\Delta G_2 = 7.8$ kJ/mol

b) Insertion and deletion mutants

Staphylococcal nuclease, insertion mutants, difference ΔG values

Mutant	pH	T	Δ(ΔG)	Approach	Remarks	Ref
wild-type	7.0	20	0.0	GuHCl	(1)	90S7
18-Gly-19	7.0	20	−11.7	GuHCl	(1)	90S7
31-Ala-32	7.0	20	−18.8	GuHCl	(1)	90S7
31-Gly-32	7.0	20	−20.1	GuHCl	(1)	90S7
32-Ala-33	7.0	20	−25.9	GuHCl	(1)	90S7
32-Gly-33	7.0	20	−28.5	GuHCl	(1)	90S7
36-Ala-37	7.0	20	−10.9	GuHCl	(1)	90S7
36-Gly-37	7.0	20	−18.0	GuHCl	(1)	90S7
60-Ala-61	7.0	20	−8.4	GuHCl	(1)	90S7
60-Gly-61	7.0	20	−10.9	GuHCl	(1)	90S7
61-Ala-62	7.0	20	−17.2	GuHCl	(1)	90S7
61-Gly-62	7.0	20	−24.7	GuHCl	(1)	90S7
63-Ala-64	7.0	20	−29.3	GuHCl	(1)	90S7
63-Gly-63	7.0	20	−31.0	GuHCl	(1)	90S7
73-Ala-74	7.0	20	−31.8	GuHCl	(1)	90S7
73-Gly-74	7.0	20	−33.1	GuHCl	(1)	90S7
76-Ala-77	7.0	20	−13.0	GuHCl	(1)	90S7
76-Gly-77	7.0	20	−11.3	GuHCl	(1)	90S7
79-Gly-80	7.0	20	−7.9	GuHCl	(1)	90S7
80-Ala-81	7.0	20	−9.2	GuHCl	(1)	90S7
80-Gly-81	7.0	20	−10.0	GuHCl	(1)	90S7
82-Gly-83	7.0	20	−16.7	GuHCl	(1)	90S7
90-Ala-91	7.0	20	−11.3	GuHCl	(1)	90S7
94-Ala-95	7.0	20	−25.1	GuHCl	(1)	90S7
95-Ala-96	7.0	20	−25.1	GuHCl	(1)	90S7
95-Gly-96	7.0	20	−22.2	GuHCl	(1)	90S7
103-Ala-104	7.0	20	<−33	GuHCl	(1)	90S7
118-Ala-119	7.0	20	−7.9	GuHCl	(1)	90S7
118-Gly-119	7.0	20	−6.3	GuHCl	(1)	90S7

Staphylococcal nuclease, insertion mutants, difference ΔG values (continued)

Mutant	pH	T	Δ(ΔG)	Approach	Remarks	Ref
126-Ala-127	7.0	20	−25.5	GuHCl	(1)	90S7
126-Gly-127	7.0	20	−25.5	GuHCl	(1)	90S7
131-Ala-132	7.0	20	−30.5	GuHCl	(1)	90S7
138-Ala-139	7.0	20	<−33	GuHCl	(1)	90S7
138-Gly-139	7.0	20	−25.1	GuHCl	(1)	90S7

Remark:
(1) linear extrapolation, measurement in 25 mM phosphate buffer with 100 mM NaCl

Staphylococcal nuclease, insertion mutants

Mutant	pH	T	Δ(ΔG)	Approach	Remarks	Ref
wild-type	7.0	20	0.0	GuHCl	(1–3)	92S13
11-Ala-12	7.0	20	−16.7	GuHCl	(1,2)	92S13
11-Gln-12	7.0	20	−17.2	GuHCl	(1,2)	92S13
11-Gly-12	7.0	20	−15.9	GuHCl	(1,2)	92S13
11-Leu-12	7.0	20	−16.7	GuHCl	(1,2)	92S13
11-Pro-12	7.0	20	−17.6	GuHCl	(1,2)	92S13
11-Ala-Gly-12	7.0	20	−16.3	GuHCl	(1,2)	92S13
11-Gly-Gly-12	7.0	20	−15.5	GuHCl	(1,2)	92S13
18-Ala-19	7.0	20	−13.8	GuHCl	(1,2)	92S13
18-Gln-19	7.0	20	−13.8	GuHCl	(1,2)	92S13
18-Gly-19	7.0	20	−11.7	GuHCl	(1,2)	92S13
18-Leu-19	7.0	20	−13.4	GuHCl	(1,2)	92S13
18-Pro-19	7.0	20	−16.7	GuHCl	(1,2)	92S13
18-Ala-Gly-19	7.0	20	−11.7	GuHCl	(1,2)	92S13
18-Gly-Gly-19	7.0	20	−10.9	GuHCl	(1,2)	92S13
31-Ala-32	7.0	20	−18.8	GuHCl	(1,2)	92S13
31-Gln-32	7.0	20	−21.3	GuHCl	(1,2)	92S13
31-Gly-32	7.0	20	−20.1	GuHCl	(1,2)	92S13
31-Leu-32	7.0	20	−23.8	GuHCl	(1,2)	92S13
31-Pro-32	7.0	20	−18.0	GuHCl	(1,2)	92S13
31-Ala-Gly-32	7.0	20	−17.6	GuHCl	(1,2)	92S13
31-Gly-Gly-32	7.0	20	−12.6	GuHCl	(1,2)	92S13
36-Ala-37	7.0	20	−10.9	GuHCl	(1,2)	92S13
36-Gln-37	7.0	20	−20.1	GuHCl	(1,2)	92S13
36-Gly-37	7.0	20	−18.0	GuHCl	(1,2)	92S13
36-Leu-37	7.0	20	−8.4	GuHCl	(1,2)	92S13
36-Pro-37	7.0	20	−30.1	GuHCl	(1,2)	92S13
36-Ala-Gly-37	7.0	20	−5.0	GuHCl	(1,2)	92S13
36-Gly-Gly-37	7.0	20	−11.3	GuHCl	(1,2)	92S13
60-Ala-61	7.0	20	−8.4	GuHCl	(1,2)	92S13
60-Gln-61	7.0	20	−9.2	GuHCl	(1,2)	92S13
60-Gly-61	7.0	20	−10.9	GuHCl	(1,2)	92S13
60-Leu-61	7.0	20	−7.1	GuHCl	(1,2)	92S13
60-Pro-61	7.0	20	−16.3	GuHCl	(1,2)	92S13
60-Ala-Gly-61	7.0	20	−6.7	GuHCl	(1,2)	92S13
60-Gly-Gly-61	7.0	20	−11.3	GuHCl	(1,2)	92S13
61-Ala-62	7.0	20	−17.2	GuHCl	(1,2)	92S13
61-Gln-62	7.0	20	−20.1	GuHCl	(1,2)	92S13
61-Gly-62	7.0	20	−24.7	GuHCl	(1,2)	92S13
61-Leu-62	7.0	20	−14.6	GuHCl	(1,2)	92S13

Staphylococcal nuclease, insertion mutants (continued)

Mutant	pH	T	$\Delta(\Delta G)$	Approach	Remarks	Ref
61-Pro-62	7.0	20	−25.9	GuHCl	(1,2)	92S13
61-Ala-Gly-62	7.0	20	−23.0	GuHCl	(1,2)	92S13
61-Gly-Gly-62	7.0	20	−23.0	GuHCl	(1,2)	92S13
76-Ala-77	7.0	20	−13.0	GuHCl	(1,2)	92S13
76-Gln-77	7.0	20	−13.8	GuHCl	(1,2)	92S13
76-Gly-77	7.0	20	−11.3	GuHCl	(1,2)	92S13
76-Leu-77	7.0	20	−19.2	GuHCl	(1,2)	92S13
76-Pro-77	7.0	20	−13.8	GuHCl	(1,2)	92S13
76-Ala-Gly-77	7.0	20	−13.4	GuHCl	(1,2)	92S13
76-Gly-Gly-77	7.0	20	−13.0	GuHCl	(1,2)	92S13
79-Ala-80	7.0	20	−9.6	GuHCl	(1,2)	92S13
79-Gln-80	7.0	20	−10.9	GuHCl	(1,2)	92S13
79-Gly-80	7.0	20	−7.9	GuHCl	(1,2)	92S13
79-Leu-80	7.0	20	−10.0	GuHCl	(1,2)	92S13
79-Pro-80	7.0	20	−10.5	GuHCl	(1,2)	92S13
79-Ala-Gly-80	7.0	20	−11.7	GuHCl	(1,2)	92S13
79-Gly-Gly-80	7.0	20	−3.3	GuHCl	(1,2)	92S13
90-Ala-91	7.0	20	−11.3	GuHCl	(1,2)	92S13
90-Gln-91	7.0	20	−31.4	GuHCl	(1,2)	92S13
90-Gly-91	7.0	20	−15.9	GuHCl	(1,2)	92S13
90-Leu-91	7.0	20	−20.9	GuHCl	(1,2)	92S13
90-Pro-91	7.0	20	−31.4	GuHCl	(1,2)	92S13
90-Ala-Gly-91	7.0	20	−15.5	GuHCl	(1,2)	92S13
90-Gly-Gly-91	7.0	20	−15.1	GuHCl	(1,2)	92S13
126-Ala-127	7.0	20	−26.4	GuHCl	(1,2)	92S13
126-Gln-127	7.0	20	−23.4	GuHCl	(1,2)	92S13
126-Gly-127	7.0	20	−25.5	GuHCl	(1,2)	92S13
126-Leu-127	7.0	20	−22.6	GuHCl	(1,2)	92S13
126-Pro-127	7.0	20	<−33.5	GuHCl	(1,2)	92S13
126-Ala-Gly-127	7.0	20	<−33.5	GuHCl	(1,2)	92S13
126-Gly-Gly-127	7.0	20	<−33.5	GuHCl	(1,2)	92S13

Remarks:
(1) linear extrapolation
(2) buffer: 25 mM sodium phosphate with 100 mM NaCl
(3) the reference value of ΔG of the wild-type protein is ΔG = 23.4 kJ/mol from Ref. 90S7

Staphylococcal nuclease, wild-type, single and deletion mutants

Mutant	pH	T	ΔG	Approach	Remarks	Ref
wild-type	7.0	20	23.0	GuHCl	(1,2)	91P4
Glu43 →Asp	7.0	20	23.8	GuHCl	(1)	91P4
44–49 del.	7.0	20	33.5	GuHCl	(1,3)	91P4
Glu43-Asp and residues (44–49) deleted						
	7.0	20	33.5	GuHCl	(1,3)	91P4

Remarks:
(1) linear extrapolation
(2) the wild-type contains Glu43
(3) deletion mutant without the residues 44–49 (Thr-Lys-His-Pro-Lys-Lys)

c) Hybrid forms

Staphylococcal nuclease, hybrid protein, and various liganded forms

Form	Ligand	pH	T	ΔG	Approach	Remarks	Ref
native	–	7.4	25	19.2		(1)	89H4
native	$CaCl_2$	7.4	25	21.8		(1,2)	89H4
native	$CaCl_2$ + pdTp	7.4	25	26.4		(1,3)	89H4
hybrid	–	7.4	25	5		(1)	89H4
hybrid	$CaCl_2$	7.4	25	7.9		(1,2)	89H4
hybrid	$CaCl_2$ + pdTp	7.4	25	16.3		(1,3)	89H4

Explanation:
hybrid = nuclease hybrid with (Ser-Ser-Asn-Gly-Ser-Pro) from concanavalin instead of residues 27–31 = (Tyr-Lys-Gly-Gln-Pro)

Remarks:
(1) GuHCl, linear extrapolation
(2) in the presence of 10 mM calcium chloride
(3) in the presence of 10 mM calcium chloride and 3′,5′deoxythymidine biphosphate (pdTp)

Staphylococcal nuclease, wild-type and mutants

Mutant	pH	T	ΔG	Approach	Remarks	Ref
wild-type	7.0	0	19.2±1.7	GuHSCN	(1,2)	91A2
hybrid and mutant:						
Ser28→Gly	7.0	0	4.0±0.6	GuHSCN	(1–3)	91A2
Ser28→Gly	7.0	0	14.6	GuHSCN	(1,3,4)	91A2

Remarks:
(1) linear extrapolation
(2) in the absence of calcium and pdTp (3′,5′deoxy-thymidine biphosphate)
(3) hybrid protein in which the res. 27–30 of Staphylococcal nuclease (Tyr-Lys-Gly-Gln) are replaced by the residues 160–164 of concanavalin (Ser-Ser-Asn-Gly-Ser)
(4) in the presence of 10 mM calcium and 0.1 mM pdTp

Staphylococcal nuclease, wild-type and mutants

Mutant	pH	T	ΔG	Approach	Remarks	Ref
wild-type	7.0	20	24.1	heat, v.H.		91E1
wild-type	7.0	20	25.6±2.1	urea	(1)	91E1
wild-type	7.0	20	23.0±1.3	GuHCl	(1)	91E1
wild-type	7.0	20	25.5	heat, v.H.		88S8
wild-type	7.0	20	24.7*	DSC	(3)	88S8
NC	7.0	20	7.4	heat	(2)	91E1
NC	7.0	20	5.3±0.7	pressure	(2)	91E1
NC-Ser28→Gly	7.0	20	5.7	heat	(2)	91E1
NC-Ser28→Gly	7.0	20	3.6±0.8	urea	(1,2)	91E1
NC-Ser28→Gly	7.0	20	5.3±1.3	GuHCl	(1,2)	91E1
NC-Ser28→Gly	7.0	20	5.2±0.7	pressure	(2)	91E1

Remarks:
(1) linear extrapolation
(2) NC = nuclease-conA is a mutant protein containing a six amino acid residue β-turn substitute from concanavalin A, i.e., Ser-Ser-Asn-Gly-Ser-Pro instead of the residues 27–31 Tyr-Lys-Gly-Gln-Pro in the wild-type Staphylococcal nuclease
(3) ΔG was calculated from other thermodynamic quantities given in Ref. 88S8

d) Chemically modified and ligated forms

Staphylococcal nuclease A in the presence of calcium and thymidine 3',5'-diphosphate (pdTp)

Compound	pH	T	ΔG	m	Approach	Remarks	Ref
1 mM EGTA	7.0	4.5	19.4±0.8	9.4±0.4	urea	(1,2)	91S8
10 mM Ca^{2+}	7.0	4.5	17.4±0.8	7.3±0.3	urea	(1,2)	91S8
1 mM pdTp + 1 mM EGTA							
	7.0	4.5	25.4±2.0	9.7±0.7	urea	(1,2)	91S8
1 mM pdTp + 10 mM Ca^{2+}							
	7.0	4.5	25.1±1.0	7.0±0.3	urea	(1,2)	91S8

Remarks:
(1) urea-induced equilibrium unfolding monitored by CD at 225 nm, pre- and postdenaturational slope from kinetic experiments, nonlinear two-state fit with linear dependence of ΔG on denaturant concentration
(2) buffer: 50 mM cacodylate, 50 mM NaCl and the indicated compounds

Staphylococcal nuclease

Form	pH	T	ΔG	Approach	Remarks	Ref
unligated	5.5	37	26.9±1.3	NH exchange	(1)	93L8
unligated	5.5	37	25.6±1.7	urea	(1,2)	93L8
ligated	5.5	37	34.5±1.3	NH exchange	(1,3)	93L8
ligated	5.5	37	30.8±0.8	urea	(1,2)	93L8

Remarks:
(1) monitored by ^{15}N NMR, pH glass electrode reading
(2) linear extrapolation
(3) measured in the presence of 1 mM pdTp and 10 mM CaCl$_2$

Staphylococcal nuclease, mixed disulfide variants produced by disulfide bond formation between Val23→Cys and alkane thiols

Form	pH	T	ΔG	Approach	Remarks	Ref
unmodified protein:						
	8.0	25	17.1	urea	(1–3)	95W7
	8.0	25	17.3	urea	(1,2,4)	95W7
methyl	8.0	25	24.6	urea	(1–3)	95W7
methyl	8.0	25	23.8	urea	(1,2,4)	95W7
ethyl	8.0	25	22.9	urea	(1–3)	95W7
ethyl	8.0	25	22.9	urea	(1,2,4)	95W7
propyl	8.0	25	17.1	urea	(1–3)	95W7
propyl	8.0	25	17.3	urea	(1,2,4)	95W7
butyl	8.0	25	14.1	urea	(1–3)	95W7
butyl	8.0	25	15.3	urea	(1,2,4)	95W7
pentyl	8.0	25	11.6	urea	(1–3)	95W7
pentyl	8.0	25	11.3	urea	(1,2,4)	95W7
ME	8.0	25	13.4	urea	(1–3,5)	95W7
ME	8.0	25	14.7	urea	(1,2,4,5)	95W7

Remarks:
(1) linear extrapolation to zero denaturant concentration by a method that includes the pre- and postdenaturational baselines for a non-linear regression of the data
(2) buffer: 0.05 M Tris, 1 mM EDTA, pH 8.0
(3) transition monitored by fluorescence, the estimated error amounts to ±0.6 kJ/mol
(4) transition monitored by CD, the estimated error amounts to ±1.3 kJ/mol
(5) ME = β-mercaptoethanol

Nucleoside Diphosphate Kinase

Nucleoside diphosphate kinase from *Dictyostelium discoideum*

pH	T	ΔG	Approach	Remark	Ref
8.0	25	22.7	urea	(1)	93L3

Remark:
(1) linear extrapolation

Opsin

Opsin, from various species

Species	pH	T	ΔG	Approach	Ref
bovine	6.8	25	42.4*	DSC	88S7
frog	6.8	25	7.9*	DSC	88S7
rat	6.8	25	14*	DSC	88S7

Ovalbumin

Ovalbumin

pH	T	ΔG	Approach	Remark	Ref
7	25	24.8	GuHCl	(1)	76A
		56.5	hydrogen exchange		71O

Remark:
(1) denaturant binding model, k = 1.2

Ovomucoid

Ovomucoid from chicken egg white

Transition	pH	T	ΔG	Approach	Remarks	Ref
Intact protein:						
trans. (1)	7.0	25	23.9	urea	(1)	91D5
trans. (2)	7.0	25	32.0	urea	(1)	91D5
Domains I and II:						
	7.0	25	19.5	urea	(1,2)	91D5
Domain I:						
	7.0	25	14.2	urea	(1,2)	91D5
Domains II and III:						
trans. (1)	7.0	25	14.2	urea	(1)	91D5
trans. (2)	7.0	25	31.0	urea	(1)	91D5
Domain III:						
	7.0	25	31.3	urea	(1)	91D5

Remarks:
(1) linear extrapolation
(2) single transition

Turkey ovomucoid third domain

Mutant	pH	T	$\Delta(\Delta G)$	Approach	Ref
Gly32→Ala			3.3		90O3
Asn28→Ser			−2.1		90O3
double mutant (Gly32→Ala and Asn28→Ser)			0.8		90O3
Tyr20→His			−3.3		90O3
Asn45→Glu			1.3		90O3
double mutant (Tyr20→His and Asn45→Glu)			−2.5		90O3

Ovomucoid third domain, effect of pH and solvent on ΔG

Solvent	pH	T	ΔG	Approach	Ref
D_2O	1.5	30	12.6±0.4	DSC/heat, v.H.	95S18
H_2O	1.5	30	10.5±0.4	DSC/heat, v.H.	95S18
200 mM KCl	1.5	30	13.4±0.4	DSC/heat, v.H.	95S18
D_2O	4.5	30	26.4±0.8	DSC/heat, v.H.	95S18
H_2O	4.5	30	25.1±0.8	DSC/heat, v.H.	95S18
200 mM KCl	4.5	30	26.4±0.8	DSC/heat, v.H.	95S18

Ovotransferrin

Ovotransferrin, N-terminal half-molecule, oxidized (S-S) and reduced (-SH) form

Form	pH	T	ΔG	Approach	Remarks	Ref
oxidized	8.5	0	20.9	GuHCl	(1)	91H2
oxidized	8.5	6	22.2	GuHCl	(1)	91H2
oxidized	8.5	0	21.3	GuHCl	(2)	91H2
oxidized	8.5	6	23.0	GuHCl	(2)	91H2
reduced	8.5	0	4.2	GuHCl	(1)	91H2
reduced	8.5	6	9.2	GuHCl	(1)	91H2
reduced	8.5	0	5.0	GuHCl	(2)	91H2
reduced	8.5	6	11.7	GuHCl	(2)	91H2

Remarks:
(1) linear extrapolation
(2) denaturant binding model, k = 1.0

Pancreatic Polypeptide

Pancreatic polypeptide from chicken, monomeric state

Form	pH	T	ΔG	Approach	Ref
monomeric	6	25	5.9	heat, optical methods	86K1
dimeric	6	25	15.1	heat, optical methods	86K1

Table 1. Gibbs Energy Change – Molar Values

Papain

Papain

pH	T	ΔG	Approach	Ref
3.8	25	76*	DSC	78T

Parvalbumin

Parvalbumin, various species

Species	pH	T	ΔG	Approach	Remarks	Ref
carp	7	25	46	pH	(1,2)	78F
carp	7	25	83.7	pH	(1,3)	78F
carp	7	4.5	20	GuHCl	(4,5)	88K5
carp	7	4.5	44	GuHCl	(4,6)	88K5
rat	7.74	25	4.3	heat	(7,8)	86W

Remarks:
(1) parvalbumin carp, component III
(2) apparent Gibbs energy change, in the presence of 0.1 mM calcium chloride
(3) apparent Gibbs energy change, in the presence of 33 mM calcium chloride
(4) linear extrapolation
(5) in the presence of 10 mM EGTA and 1 mM calcium chloride
(6) in the presence of 1 mM calcium chloride
(7) parvalbumin rat, apo-form
(8) monitored by NMR and CD

Penicillin G Acylase

Penicillin G acylase, α peptide

pH	T	ΔG	Approach	Remark	Ref
7.5	22	11.8	urea	(1)	90L2

Remark:
(1) linear extrapolation

Pepsin

Pepsin

pH	T	ΔG	Approach	Remarks	Ref
6.5	25	45.2	DSC	(1)	82P3
6.5	25	45.2	DSC	(1)	81P2
7.5	25	16±1.5	urea	(2)	89M5
7.5	25	17.4±1.8	urea	(2,3)	89M5

Remarks:
(1) multidomain protein with the following contributions, 7.9 and 12.5 kJ/mol for the N-terminal lobe, 16.3
    and 8.4 kJ/mol for the C-terminal lobe, 26.8 kJ/mol for the N-terminal lobe in the presence of pepstatin
(2) linear extrapolation
(3) pepsin inactivated with diazoacetylglycine ethyl ester

Pepsinogen

Pepsinogen

pH	T	ΔG	Approach	Remarks	Ref
6–8	25	31.8	urea	(1)	78A
6–8	25	49	urea	(2)	78A
6–8	25	27.2	urea	(3)	78A
6.5	25	65.7	DSC	(4)	82P3
6.5	25	65.7	DSC	(4)	81P2

Remarks:
(1) denaturant binding model, k = 0.1
(2) denaturant binding model, k = 1
(3) linear extrapolation
(4) multidomain protein with 25.9 kJ/mol and 39.7 kJ/mol, respectively, for the two cooperative blocks of pepsinogen

Pepsinogen

pH	T	ΔG	Approach	Remarks	Ref
7.5	25	26.6±1.0	urea	(1)	90S9
9.0	25	11.4±0.7	urea	(1)	90S9

Remark:
(1) linear extrapolation

Peroxidase

Peroxidase, horseradish, apoprotein

Conditions	pH	T	ΔG	Approach	Remarks	Ref
	7.4	21	9.2	GuHCl	(1,2)	93P1
	7.4	21	7.6	GuHCl	(1,3)	93P1
with protoporphyrin IX, in the presence of 1 mM calcium:						
	7.4	21	16.7	GuHCl	(1,2)	93P1
with protoporphyrin IX, in the absence of calcium:						
	7.4	21	9.2	GuHCl	(1,2)	93P1

Remarks:
(1) linear extrapolation
(2) monitored by tryptophan fluorescence

Horseradish peroxidase type II

pH	T	ΔG	Approach	Remarks	Ref
3.2	22	26.857	urea	(1)	95M17
3.2	22	28.227	GuHCl	(1)	95M17
3.2	25	24.842	urea	(1)	95M17
3.2	25	26.167	GuHCl	(1)	95M17
3.2	27	23.557	urea	(1)	95M17
3.2	27	23.760	GuHCl	(1)	95M17
3.2	30	21.683	urea	(1)	95M17
3.2	30	20.855	GuHCl	(1)	95M17
6.4	22	26.996	urea	(1)	95M17
6.4	22	27.939	GuHCl	(1)	95M17
6.4	25	24.842	urea	(1)	95M17
6.4	25	26.681	GuHCl	(1)	95M17
6.4	27	24.571	urea	(1)	95M17
6.4	27	26.024	GuHCl	(1)	95M17
6.4	30	23.826	urea	(1)	95M17
6.4	30	23.510	GuHCl	(1)	95M17
10.0	22	28.058	urea	(1)	95M17
10.0	22	28.015	GuHCl	(1)	95M17
10.0	25	26.605	urea	(1)	95M17
10.0	25	26.779	GuHCl	(1)	95M17
10.0	27	25.272	urea	(1)	95M17
10.0	27	24.556	GuHCl	(1)	95M17
10.0	30	23.199	urea	(1)	95M17
10.0	30	21.893	GuHCl	(1)	95M17

Remark:
(1) ΔG results of a treatment by linear extrapolation, denaturant binding model and Tanford's transfer model. Additional data are given in Ref. 95M17

Phosphatase

Red kidney bean purple acid phosphatase, in the presence and in the absence of phosphate

Form	pH	T	ΔG	Approach	Remarks	Ref
free	7.4	25	10.5	GuHCl	(1,2)	96C4
liganded	7.4	25	25.1	GuHCl	(1,3)	96C4

Remarks:
(1) linear extrapolation
(2) buffer: 10 mM Tris, 500 mM KCl, pH 7.4
(3) buffer (2) with 10 mM NaH_2PO_4

Phosphoglycerate Kinase

Phosphoglycerate kinase from various species

Species	pH	T	ΔG	Approach	Remarks	Ref
horse	7.4	23	12	GuHCl	(1,2)	84D1
horse	7.4	23	26	GuHCl	(1,3)	84D1
horse	7.5	23	25±2	GuHCl	(3,5,6)	88C
horse	7.5	23	13±2	GuHCl	(5–7)	88C
horse	7.5	23	13±2	GuHCl	(5,6,8)	88C

Phosphoglycerate kinase from various species (continued)

Species	pH	T	ΔG	Approach	Remarks	Ref
horse	7.5	23	21±2	GuHCl	(3,5,9)	88C
horse	7.5	23	14±2	GuHCl	(5,7,9)	88C
horse	7.5	23	10±2	GuHCl	(5,8,9)	88C
thermus th.	7.5	25	49.7±0.9	GuHCl	(4)	77N
yeast	7	25	17.9	GuHCl	(1)	77B
yeast	7	25	20.6	GuHCl	(5)	77B
yeast	7.5	25	22.3±0.5	GuHCl	(4)	77N

Explanation (species):
horse = horse muscle
thermus th. = *thermus thermophilus*

Remarks:
(1) linear extrapolation
(2) monitored by optical methods
(3) monitored by enzyme activity
(4) denaturant binding model
(5) transfer model
(6) in the presence of 100 mM phosphate,
(7) monitored by ellipticity
(8) monitored by fluorescence
(9) in the absence of phosphate

Phosphoglycerate kinase from yeast, wild-type and mutants

Mutant	pH	T	ΔG	Approach	Remarks	Ref
wild-type	7	20	33.5±2.1	GuHCl	(1)	90Y2,90B1
Cys97→Ala	7	20	23.0±2.1	GuHCl	(1)	90Y2,90B1
double mutants:						
(Cys97→Ala and Ala183→Cys)						
	7	20	33.5±2.1	GuHCl	(1)	90Y2,90B1
(Cys97→Ala and Ile285→Cys)						
	7	20	25.1±2.1	GuHCl	(1)	90Y2,90B1
(Cys97→Ala and Thr324→Cys)						
	7	20	20.9±2.1	GuHCl	(1)	90Y2,90B1
(Cys97→Ala and Val376→Cys)						
	7	20	20.9±2.1	GuHCl	(1)	90Y2,90B1

Remark:
(1) linear extrapolation, transition monitored by CD

Phosphoglycerate kinase from yeast, recombinant, tryptophan mutants, the naturally occuring Trp residues Trp308 and Trp333 are replaced by Phe if not otherwise indicated

Transition	pH	T	ΔG	Approach	Remarks	Ref
mutant Tyr48→Trp						
N→I	7.5	25	22.2±1.5	GuHCl	(1,2,5)	95S4
I→U	7.5	25	15.9±1.5	GuHCl	(1,2,5)	95S4
mutant Tyr48→Trp						
N→I	7.5	25	18.0±2.8	GuHCl	(1,3,5)	95S4
I→U	7.5	25	14.8±6.6	GuHCl	(1,3,5)	95S4

Phosphoglycerate kinase from yest, recombinant, tryptophan mutants, the naturally occuring Trp residues Trp308 and Trp333 are replaced by Phe if not otherwise indicated (continued)

Transition	pH	T	ΔG	Approach	Remarks	Ref
mutant Tyr122→Trp						
N→I	7.5	25	20.1±1.0	GuHCl	(1,2,5)	95S4
I→U	7.5	25	14.4±2.4	GuHCl	(1,2,5)	95S4
mutant Tyr122→Trp						
N→I	7.5	25	24.2±3.2	GuHCl	(1,3,5)	95S4
I→U	7.5	25	11.4±5.9	GuHCl	(1,3,5)	95S4
mutant Phe194→Trp						
N→I	7.5	25	26.4±1.2	GuHCl	(1,2,4)	95S4
I→U	7.5	25	11.2±9.0	GuHCl	(1,2,4)	95S4
mutant Phe194→Trp						
N→I	7.5	25	22.8±2.0	GuHCl	(1,3,4)	95S4
I→U	7.5	25	9.7±6.4	GuHCl	(1,3,4)	95S4
mutant Leu399→Trp						
N→U	7.5	25	17.4±0.8	GuHCl	(1,2,5)	95S4
mutant Leu399→Trp						
N→I	7.5	25	14.2±2.3	GuHCl	(1,3,5)	95S4
I→U	7.5	25	21.2±10.1	GuHCl	(1,3,5)	95S4
mutant Tyr48→Trp and Trp308 remaining						
N→I	7.5	25	28.6±1.5	GuHCl	(1,2,6)	95S4
I→U	7.5	25	20.7±2.4	GuHCl	(1,2,6)	95S4
mutant Tyr48→Trp and Trp308 remaining						
N→I	7.5	25	24.8±3.0	GuHCl	(1,3,6)	95S4
I→U	7.5	25	9.8±8.5	GuHCl	(1,3,6)	95S4
mutant Tyr48→Trp and Trp333 remaining						
N→I	7.5	25	28.7±1.3	GuHCl	(1,2,6)	95S4
I→U	7.5	25	18.6±2.4	GuHCl	(1,2,6)	95S4
mutant Tyr48→Trp and Trp333 remaining						
N→I	7.5	25	33.0±4.4	GuHCl	(1,3,6)	95S4
I→U	7.5	25	15.2±6.5	GuHCl	(1,3,6)	95S4
Trp308 remaining						
N→U	7.5	25	39.4±5.8	GuHCl	(1,2,6)	95S4
Trp308 remaining						
N→I	7.5	25	28.7±0.3	GuHCl	(1,3,6)	95S4
I→U	7.5	25	20.4±0.5	GuHCl	(1,3,6)	95S4
Trp333 remaining						
N→U	7.5	25	32.0±1.5	GuHCl	(1,2,6)	95S4
Trp333 remaining						
N→I	7.5	25	31.3±4.3	GuHCl	(1,3,6)	95S4
I→U	7.5	25	8.2±17.1	GuHCl	(1,3,6)	95S4
no Trp remaining						
N→I	7.5	25	22.5±3.1	GuHCl	(1,2,6)	95S4
I→U	7.5	25	18.2±4.9	GuHCl	(1,2,6)	95S4
no Trp remaining						
N→I	7.5	25	25.4±5.0	GuHCl	(1,3,6)	95S4
I→U	7.5	25	35.0±16.3	GuHCl	(1,3,6)	95S4

Remarks:
(1) for the procedure, see Ref. 95M7
(2) the transition was monitored by fluorescence
(3) the transition was monitored by CD
(4) according to the headline, the mutant corresponds to the following triple mutant Phe194→Trp, Trp308→Phe and Trp333→Phe
(5) see also remark (4)
(6) Trp308 and/or Trp333 replaced by Phe

Phosphoglycerate kinase from yeast, single tryptophan mutants

Mutant	pH	T	ΔG	Approach	Remarks	Ref
wild-type	7.5	25	59.8±9.5	urea	(1–3)	96S14
wild-type	7.5	25	68.7±13.9	urea	(1,2,4)	96S14
Trp308	7.5	25	55.4±15.3	urea	(1–3)	96S14
Trp308	7.5	25	60.2±8.8	urea	(1,2,4)	96S14
Trp333	7.5	25	51.6±6.3	urea	(1–3)	96S14
Trp333	7.5	25	49.2±6.3	urea	(1,2,4)	96S14
Trp194	7.5	25	31.1±1.7	urea	(1–3)	96S14
Trp194	7.5	25	20.9±2.6	urea	(1,2,4)	96S14
Trp194						
N→I	7.5	25	29.3	urea	(1,3,5)	96S14
I→U	7.5	25	20.9	urea	(1,3,5)	96S14
Trp194						
N→I	7.5	25	29.3	urea	(1,4,5)	96S14
I→U	7.5	25	20.9	urea	(1,4,5)	96S14
Trp48						
N→I	7.5	25	25.9±2.7	urea	(1,3,5)	96S14
I→U	7.5	25	20.2±3.7	urea	(1,3,5)	96S14
Trp48						
N→I	7.5	25	42.6±5.8	urea	(1,4,5)	96S14
I→U	7.5	25	29.1±6.4	urea	(1,4,5)	96S14
Trp48						
N→I	7.5	25	25.9	urea	(1,4–6)	96S14
I→U	7.5	25	20.1	urea	(1,4–6)	96S14
Trp122						
N→I	7.5	25	32.9±5.2	urea	(1,3,5)	96S14
I→U	7.5	25	21.2±10.5	urea	(1,3,5)	96S14
Trp122						
N→I	7.5	25	21.3±2.4	urea	(1,4,5)	96S14
I→U	7.5	25	63.3±44.6	urea	(1,4,5)	96S14
Trp122						
N→I	7.5	25	33.1	urea	(1,4–6)	96S14
I→U	7.5	25	21.3	urea	(1,4–6)	96S14

Remarks:
(1) single Trp mutants, Trp48, Trp122, Trp194, Trp308, and Trp333 designate single Trp's in (Tyr48→Trp, Trp308→Phe, and Trp333→Phe), (Tyr122→Trp, Trp308→Phe, and Trp333→Phe), (Phe194→Trp, Trp308→Phe, and Trp333→Phe), Trp333→Phe, and Trp308→Phe
(2) two-state fit
(3) transition monitored by fluorescence
(4) transition monitored by CD
(5) three-state fit
(6) three-state fit using fixed ΔG(N→I) and ΔG(I→U) from fluorescence-monitored transitions

Phosphoglycerate kinase from yeast and its domains

Domain	pH	T	ΔG	Approach	Remarks	Ref
intact PGK	7.5	21	32.6±2.1	GuHCl	(1)	90Y2,90M5
C-domain	7.5	21	18.8±2.1	GuHCl	(1)	90Y2,90M5
N-domain	7.5	21	16.7±2.1	GuHCl	(1)	90Y2,90M5

Remark:
(1) linear extrapolation

Phosphoglycerate kinase from yeast, wild-type, mutants, and isolated domains

Mutant	pH	T	ΔG	Approach	Remarks	Ref
wild-type	7.5	22	32.6±2.1	GuHCl	(1–3)	95G2
wild-type	7.5	22	31.1±1.1	urea	(1–3)	95G2
wild-type	7.5	22	32.8±0.5	urea	(1,2,5)	95G2
wild-type	7.5	22	34.5±0.1	urea	(1,2,6)	95G2
Trp308→Tyr	7.5	22	24.3±2.1	GuHCl	(1,2,4)	95G2
Trp308→Tyr	7.5	22	17.6±10.5	GuHCl	(1–3)	95G2
Trp308→Tyr	7.5	22	30.5±2.5	urea	(1–3)	95G2
Trp308→Tyr	7.5	22	22.2±0.4	urea	(1,2,5)	95G2
Trp308→Tyr	7.5	22	37.7±0.8	urea	(1,2,6)	95G2
Trp333→Phe	7.5	22	37.7±4.6	GuHCl	(1,2,4)	95G2
Trp333→Phe	7.5	22	41.8±11.7	GuHCl	(1–3)	95G2
Trp333→Phe	7.5	22	34.7±2.1	urea	(1–3)	95G2
Trp333→Phe	7.5	22	33.9±0.8	urea	(1,2,5)	95G2
Trp333→Phe	7.5	22	31.8±1.1	urea	(1,2,6)	95G2
wild-type	7.5	22	32.6±2.1	GuHCl	(1–3)	95R6
Δ(404–415)	7.5	22	11.3±3.3	GuHCl	(1–3,7)	95R6
Δ(404–415)	7.5	22	11.7±1.3	GuHCl	(1,2,5,7)	95R6
C-domain	7.5	22	18.8±2.1	GuHCl	(1–3,8)	95R6
N-domain	7.5	22	16.7±2.1	GuHCl	(1–3,8)	95R6

Remarks:
(1) linear extrapolation
(2) buffer: 20 mM Tris, 0.5 mM EDTA, 1 mM mercapto-ethanol, pH 7.5
(3) transition was monitored by CD
(4) transition was monitored by enzyme activity
(5) transition monitored by variation in the fluorescence maximum wavelength
(6) transition monitored by fluorescence intensity
(7) deletion mutant des-(404–415)
(8) isolated C-terminal domain and N-terminal domain, respectively

Phosphoglycerate kinase from yeast, wild-type, and deletion mutants

Mutant	pH	T	ΔG	Approach	Remarks	Ref
wild-type	7.5	25	45.6±6.3	GuHCl	(1,3)	95M7
wild-type	7.5	25	39.1±3.6	GuHCl	(1,5)	95M7
Δ(413–415)	7.5	25	31.9±9.5	GuHCl	(1,2,4)	95M7
Δ(413–415)	7.5	25	28.2±3.9	GuHCl	(1–3)	95M7
Δ(413–415)	7.5	25	22.3±1.8	GuHCl	(1,2,5)	95M7
Δ(401–415)	7.5	25	18.7±3.2	GuHCl	(1,2,4)	95M7
Δ(401–415), transition (N→I)						
	7.5	25	14.6±0.9	GuHCl	(1,2,5,6)	95M7
Δ(401–415), transition (I→U)						
	7.5	25	7.5±10.7	GuHCl	(1,2,5,6)	95M7

Remarks:
(1) linear extrapolation to zero denaturant concentration to a two-state or three-state unfolding model by a method that includes the pre- and postdenaturational baselines for a non-linear regression of the data
(2) deletion mutants Δ(413–415) and Δ(401–415) correspond to des-(413–415) and des-(401–415)
(3) transition monitored by total fluorescence intensity
(4) transition monitored by fluorescence λ_{max}
(5) transition monitored by CD
(6) resolved into the transitions (N→I) and (I→U)

Phosphoglycerate kinase from yeast, wild-type and Trp mutants

Mutant	pH	T	ΔG	Approach	Remarks	Ref
wild-type	7.5	25	49.2±1.5	GuHCl	(1,2)	94S10
Trp308→Phe	7.5	25	32.3±1.3	GuHCl	(1–3)	94S10
Trp333→Phe	7.5	25	33.6±2.8	GuHCl	(1,2,4)	94S10
double mutant (Trp333→Phe and Trp308→Phe)						
	7.5	25	12.7±1.2	GuHCl	(1,2)	94S10
wild-type	7.5	25	39.0±3.0	GuHCl	(1,5)	94S10
Trp308→Phe	7.5	25	22.5±2.1	GuHCl	(1,3,5)	94S10
Trp333→Phe	7.5	25	23.8±1.9	GuHCl	(1,4,5)	94S10
double mutant (Trp333→Phe and Trp308→Phe)						
	7.5	25	12.5±1.3	GuHCl	(1,5)	94S10

Remarks:
(1) analysis in terms of denaturant binding model and linear extrapolation
(2) transition monitored by CD
(3) Trp333 remains unaffected
(4) Trp308 remains unaffected
(5) transition monitored by fluorescence

Phosphoglycerate kinase from yeast, recombinant and circularly permuted variants

Mutant	pH	T	ΔG	Approach	Remarks	Ref
wild-type	7.5	22	33.1±2.1	activity	(1,2)	95R5
PGK292p	7.5	22	17.6±0.8	activity	(1–3)	95R5
PGK72p	7.5	22	13.8±0.8	activity	(1,2,4)	95R5

Remarks:
(1) linear extrapolation to zero denaturant concentration by a method that includes the pre- and post-denaturational baselines for a non-linear regression of the data
(2) measured in 20 mM Tris-HCl, 0.5 mM EDTA, 1 mM mercaptoethanol, pH 7.5
(3) the natural chain termini were directly connected, new termini were created between Ala291 and Asp292 that belong to the C-domain of PGK
(4) the natural chain termini were directly connected, new termini were created between Asn71 and Glu72 that belong to the N-domain of PGK

Phosphoglycerate kinase from *Bacillus stearothermophilus*

Mutant/Trans.	pH	T	ΔG	Approach	Remarks	Ref
Mutant "W315" (see remark 1):						
trans. (1)	7.5		–	GuHCl	(3,4)	93S9
trans. (2)	7.5		47.7	GuHCl	(3,5)	93S9
trans. (3)	7.5		34.3	GuHCl	(3,6)	93S9
Mutant "W315" (see remark 1):						
trans. (1)	7.5		–	urea	(3,4)	93S9
trans. (2)	7.5		47.1	urea	(3,5)	93S9
trans. (3)	7.5		34.1	urea	(3,6)	93S9
Mutant "W379" (see remark 2):						
trans. (1)	7.5		20.7	GuHCl	(3,4)	93S9
trans. (2)	7.5		45.4	GuHCl	(3,5)	93S9
trans. (3)	7.5		37.4	GuHCl	(3,6)	93S9

Phosphoglycerate kinase from *Bacillus stearothermophilus* (continued)

Mutant/Trans.	pH	T	ΔG	Approach	Remarks	Ref
Mutant "W379" (see remark 2):						
trans. (1)	7.5		15.8	urea	(3,4)	93S9
trans. (2)	7.5		45.1	urea	(3,5)	93S9
trans. (3)	7.5		34.1	urea	(3,6)	93S9

Remarks:
(1) mutant Trp290→Tyr, Trp315 remains as the only Trp
(2) triple mutant Trp290→Tyr, Trp315→Tyr, and Phe379→Trp, to introduce Trp in a new position
(3) the protein shows three transitions: F→I_1→I_2→U; for details of the data treatment it is referred to Ref. 93S9
(4) F→I_1 is unfolding of the N-terminus (not detected on mutant "W315")
(5) I_1→I_2 is unfolding of the C-terminus to partially ordered structure
(6) I_2→U is the complete loss of structure in the C-terminus

Phosphoglycerate kinase from *Bacillus stearothermophilus*

Transition	pH	T	ΔG	Approach	Remarks	Ref
Wild-type, intact protein, transition of the C-domain:						
N→U	7.2	25	55.2	GuHCl	(1,5)	96P1
N→U	7.2	25	57.3	GuHCl	(1,6)	96P1
I→U	7.2	25	14.2	GuHCl	(1)	96P1
Wild-type, intact protein, transition of the N-domain:						
N→U	7.2	25	38.9	GuHCl	(1,5)	96P1
N→U	7.2	25	38.1	GuHCl	(1,6)	96P1
I→U	7.2	25	21.3	GuHCl	(1)	96P1
Mutant, intact protein, transition of the C-domain:						
N→U	7.2	25	23.8	GuHCl	(2,5)	96P1
N→U	7.2	25	23.8	GuHCl	(2,6)	96P1
I→U	7.2	25	2.9	GuHCl	(2)	96P1
Mutant, intact protein, transition of the N-domain:						
N→U	7.2	25	40.2	GuHCl	(2,5)	96P1
N→U	7.2	25	38.1	GuHCl	(2,6)	96P1
I→U	7.2	25	37.2	GuHCl	(2)	96P1
Transition of the isolated C-domain:						
N→U	7.2	25	31.0	GuHCl	(4,5)	96P1
I→U	7.2	25	13.4	GuHCl	(4)	96P1
Transition of the isolated N-domain:						
N→U	7.2	25	39.7	GuHCl	(3,5)	96P1
N→U	7.2	25	37.2	GuHCl	(3,6)	96P1
I→U	7.2	25	21.3	GuHCl	(3)	96P1

Remarks:
(1) PGK′: single Trp template Trp290→Tyr
(2) mutant Ile217→Ala and Leu221→Ala
(3) isolated N-domain, residues 1–175
(4) isolated C-domain, residues 185–394
(5) from equilibrium data
(6) from kinetics

Phosphoglycerate kinase from *Bacillus stearothermophilus* in the presence of sucrose

pH	T	ΔG	Approach	Remarks	Ref
7.2	25	36.6±1.1	GuHCl	(1–3)	95T7

Remarks:
(1) linear extrapolation
(2) measured in 50 mM triethanolamine buffer, 2 mM DTT in the presence of different sucrose concentrations (0, 10, 20 and 30%)
(3) influence of sucrose on transition midpoint and slope, however, no influence of sucrose on ΔG could be found

Phospholipase A2

Phospholipase A2, porcine, wild-type and mutants

Mutant	pH	T	Δ(ΔG)	Approach	Remarks	Ref
wild-type	7.0	20	0.0	GuHCl	(1)	91G4
Asp89→Asn	7.0	20	−0.8	GuHCl	(1)	91G4
double mutant (Asp89→Asn and Glu92→Lys)						
	7.0	20	−3.2	GuHCl	(1)	91G4

Remark:
(1) linear extrapolation

Phycocyanin

Phycocyanin

pH	T	ΔG	Approach	Remark	Ref
6	25	18.4	urea	(1)	77C1

Remark:
(1) linear extrapolation

Phycocyanins from mesophile and thermophile strains

Species	pH	T	ΔG	Approach	Remarks	Ref
thermophilic	6	25	18.0	urea	(1–3)	94C5
thermophilic	6	25	36.4	urea	(1,4)	94C5
mesophilic	6	25	16.7	urea	(1,3,5)	94C5
mesophilic	6	25	22.2	urea	(1,4,5)	94C5

Remarks:
(1) linear extrapolation
(2) thermophilic – from *Synechococcus lividus*
(3) monitored by CD
(4) monitored by absorption at 620 nm
(5) mesophilic – from *Phormidium luridum*

Plant Seed Proteins

11S Protein from *Vicia faba* and *soybean*

Species	pH	T	ΔG	Approach	Remarks	Ref
vicia faba	7.6	25	36*	DSC	(1)	85D2
soybean	7.6	25	64*	DSC	(1,2)	85D1

Remarks:
(1) data per protomer, measurement in the absence of sodium chloride
(2) 11S protein from soybean = glycinin

Plasminogen Activator

Tissue plasminogen activator kringle-2 domain, mutants

Mutant	pH	T	ΔG	Approach	Ref
wild-type	3	25	13.1*	DSC	89K2
wild-type	4.5	25	29.5*	DSC	89K2
wild-type	6–7.5	25	38.3*	DSC	89K2
Val65→Met	3.5	25	8.3*	DSC	89K2
Val65→Met	4.5	25	19.3*	DSC	89K2
Val65→Met	5.9	25	28.4*	DSC	89K2

Tissue plasminogen activator kringle-2 domain, mutants difference Δ(ΔG) values

Mutant	pH	T_{trs}	ΔT	Δ(ΔG)	Approach	Ref
wild-type	4.5	64.3±0.9	0.0	0.0	DSC	89K2
Val65→Ile	4.5	64.8	0.5	1.3	DSC	89K2
Val65→Thr	4.5	60	−4.3	−4.2	DSC	89K2
Val65→Leu	4.5	55.8	−8.5	−7.9	DSC	89K2
Val65→Ala	4.5	55.5	−8.8	−8.4	DSC	89K2
Val65→Met	4.5	54.4	−9.9	−9.2	DSC	89K2
Val65→Ser	4.5	52.9	−11.4	−10	DSC	89K2

Tissue plasminogen activator, kringle-2, wild-type and mutants

Mutant	pH	T	Δ(ΔG)	Ref
wild-type			0.0	90K1
His64→Tyr			12.1	90K1
(Arg68→Gly)			2.9	90K1
double mutant (His64→Tyr and Arg68→Gly)			14.2	90K1

Plasminogen Activator Inhibitor 1

Plasminogen activator inhibitor 1, wild-type and mutants

Mutant	pH	T	ΔT	Δ(ΔG)	Approach	Remarks	Ref
Active form:							
wild-type	6.6	50.2±0.3	0.0	0.0	heat		94L2
Arg30→Glu	6.6	50.2±0.3	−2.6	−5.0±1.7	heat	(1)	94L2
Glu350→Arg	6.6	50.2±0.3	−0.5	−1.3±2.1	heat	(1)	94L2
Glu350→Pro	6.6	50.2±0.3	+0.5	+1.3±2.1	heat	(1)	94L2
double mutant (Arg30→Glu and Glu350→Arg)							
	6.6	50.2±0.3	−2.2	−3.8±1.3	heat	(1)	94L2
Latent form:							
wild-type	6.6	67.5±1	0.0	0.0	heat	(1)	94L2
Arg30→Glu	6.6	67.5±1	−7.5	−7.1±2.5	heat	(1)	94L2
Glu350→Arg	6.6	67.5±1	−6.5	−5.4±3.8	heat	(1)	94L2
Glu350→Pro	6.6	67.5±1	−3.5	−3.8±4.2	heat	(1)	94L2
double mutant (Arg30→Glu and Glu350→Arg)							
	6.6	67.5±1	−6.5	−6.3±2.9	heat	(1)	94L2

Remark:
(1) Δ(ΔG) is based on ΔH of the mutant protein and ΔT

Serpin plasminogen activator inhibitor, active and latent conformation

Form	pH	T	ΔG	Approach	Remarks	Ref
active	6.5	8	~50	urea	(1–3)	96W1
latent	6.5	8	~88	urea	(1–3)	96W1

Remarks:
(1) the serpin plasminogen acitivator inhibitor folds into an active structure and then converts slowly to a more stable, but low-activity 'latent' conformation
(2) transitions followed by fluorescence emission maximum
(3) buffer: 50 mM sodium phosphate, 0.1 M NaCl, 1% glycerol, pH 6.5

Plastocyanin

Apoplastocyanin

pH	T	ΔG	Approach	Remarks	Ref
7.0	25	24.3±2.5	GuHCl	(1–3)	93K9
7.0	25	22.2±3.3	GuHCl	(1,2,4)	93K9

Remarks:
(1) linear extrapolation, data treatment by a method that includes the pre- and postdenaturational baselines for a non-linear regression of the data
(2) buffer: 50 mM potassium phosphate, 0.5 M Na_2SO_4, 2 mM mercaptoethanol, 0.1 mM EDTA
(3) monitored by CD at 226 nm
(4) monitored by fluorescence at 303 nm

Procarboxypeptidase, see carboxypeptidase

Prochymosin

Prochymosin, natural and recombinant protein

Form	pH	T	ΔG	Approach	Remarks	Ref
natural	7.5	25	27.0±1.0	urea	(1)	90S9
natural	9.0	25	14.9±0.9	urea	(1)	90S9
recombinant	7.5	25	26.9±1.2	urea	(1)	90S9
recombinant	9.0	25	14.5±0.6	urea	(1)	90S9

Remark:
(1) linear extrapolation

Prolyl Isomerase, see also FK binding protein

Protein Disulfide Isomerase, see also DsbA

Protein disulfide isomerase

Transition	pH	T	ΔG	Approach	Remarks	Ref
overall	7.5	23	24.3±1.3	urea	(1–4)	93M15
trans. (1)	7.5	23	22.6±1.3	GuHCl	(1,2,5)	93M15
trans. (2)	7.5	23	31.8±3.8	GuHCl	(1,2,5)	93M15
trans. (1)	7.5	23	21.8±2.1	GuHCl	(1,3,5)	93M15
trans. (2)	7.5	23	16.7±2.9	GuHCl	(1,3,5)	93M15

Remarks:
(1) multiparameter fit of the entire transition curve, based on linear dependence of lnK on the denaturant concentration
(2) monitored by fluorescence
(3) monitored by CD
(4) the urea-induced transition can be described by a two-state model
(5) the GuHCl-induced transition occurs in two distinct steps, trans. (1) = N→I, trans. (2) = I→U

Protein disulfide isomerase (DsbA)

Form	pH	T	ΔG	Approach	Remarks	Ref
reduced	7.0	20	65.0±4.0	GuHCl	(1,2)	93W6
reduced	7.0	30	48.3±2.8	GuHCl	(1,2)	93W6
oxidized	7.0	20	50.3±4.2	GuHCl	(1,3)	93W6
oxidized	7.0	30	33.5±1.2	GuHCl	(1,3)	93W6

Remarks:
(1) linear extrapolation to zero denaturant concentration by a method that includes the pre- and post-denaturational baselines for a non-linear regression of the data
(2) buffer: 100 mM sodium phosphate, 1 mM EDTA, 50 mM DTT, pH 7.0
(3) buffer: 100 mM sodium phosphate, 1 mM EDTA, pH 7.0

Protein disulfide isomerase, urea denaturation in the presence of GuHCl

GuHCl (M)	Transition	pH	T	ΔG	Approach	Remarks	Ref
0.0		7.5	23	24.3±1.3	urea	(1–3)	93M15
0.5	trans. (1)	7.5	23	23.8±8.4	urea	(1–3)	93M15
0.5	trans. (2)	7.5	23	16.7±4.2	urea	(1–3)	93M15
0.9	trans. (1)	7.5	23	14.6±2.1	urea	(1–3)	93M15
0.9	trans. (2)	7.5	23	26.8±3.8	urea	(1–3)	93M15
1.35	trans. (1)	7.5	23	8.8±2.1	urea	(1–3)	93M15
1.35	trans. (2)	7.5	23	21.8±3.3	urea	(1–3)	93M15

Remarks:
(1) multiparameter fit of the entire transition curve, based on linear dependence of lnK on the denaturant concentration
(2) transition monitored by fluorescence
(3) the transition in the absence of GuHCl can be described by a two-state model, whereas in the presence of GuHCl two distinct steps occur (trans. (1) = N→I, trans. (2) = I→U)

Protein G, Immunoglobulin Binding Protein

Streptococcal protein G, IgG-binding domains B1 and B2

	pH	T	ΔG	Approach	Remarks	Ref
B1	5.40	5	30	DSC	(1)	92A2
B2	5.40	5	25	DSC	(1,2)	92A2

Remarks:
(1) at conditions of maximal protein stability
(2) $\Delta(\Delta G) = -5.5$ kJ/mol relativ to the domain B1

Streptococcal protein G, IgG-binding domains

Mutant	pH	T	ΔG	Approach	Remarks	Ref
GB1	11.2	25	6.1	heat	(1–3)	92A3
GB2	11.2	25	3.9	heat	(1–3)	92A3

Remarks:
(1) the transition monitored by CD
(2) ΔC_p was taken from Ref. 92A2
(3) the mutant proteins differ in six positions: Ile6→Val, Leu7→Ile, Glu19→Lys, Ala24→Glu, Val29→Ala, and Glu42→Val in GB1

Immunoglobulin binding domain GB1 of streptococcal protein G

	pH	T	ΔG	Approach	Remarks	Ref
	2	25	17.2	urea	(1)	95F3
	4	5	20.1	GuHCl	(2)	94K11

Remarks:
(1) data analysis by a non-linear fit according to a three-state unfolding model that is based on linear dependence of ΔG on denaturant concentration
(2) linear extrapolation

Streptococcal protein G, B1 domain, mutants in position 53

Mutant	pH	T	ΔG	Approach	Remarks	Ref
Thr53	5.2	20	9.2±0.25	GuHCl	(1)	96S7
Thr53	5.2	20	14.6±5.0	GuHCl	(2)	96S7
Thr53→Ala	5.2	20	8.8±0.4	GuHCl	(1)	96S7
Thr53→Ala	5.2	20	7.9±0.8	GuHCl	(2)	96S7
Thr53→Arg	5.2	20	8.8±0.3	GuHCl	(1)	96S7
Thr53→Asp	5.2	20	5.4±0.4	GuHCl	(1)	96S7
Thr53→Glu	5.2	20	8.8±0.4	GuHCl	(1)	96S7
Thr53→Ile	5.2	20	9.6±0.4	GuHCl	(1)	96S7
Thr53→Phe	5.2	20	10.2±0.4	GuHCl	(1)	96S7

Remarks:
(1) linear extrapolation to zero denaturant concentration by a method that includes the pre- and post-denaturational baselines for a non-linear regression of the data
(2) from kinetics determined by stopped flow fluorescence, ΔG refers to zero denaturant concentration

Staphylococcal IgG binding protein G, B1 domain, mutants

Mutant	pH	T_{trs}	ΔG	Approach	Remark	Ref
Thr2→Gln	5.2	81.89	41.7	heat	(1)	94S6
double mutant (Thr2→Gln and Ile6→Ala)						
	5.2	71.12	29.5	heat		94S6
multiple mutant (Thr2→Gln, Ile6→Ala, Thr44→Ala and Thr53 unchanged)						
	5.2	68.73	27.4	heat		94S6
multiple mutant (Thr2→Gln, Ile6→Ala, Thr44→Ala and Thr53→Ala)						
	5.2	57.05	19.3	heat		94S6

Remark:
(1) the calorimetric enthalpy amounts to $\Delta H^{cal} = 280$ kJ/mol, and the van't Hoff enthalpy to $\Delta H^{v.H.} = 271$ kJ/mol

Mutants of an immunoglobulin (Ig) G binding protein based upon the B domain of protein A from *Staphylococcus aureus*, Trp substitution

Mutant	pH	T	ΔG	Approach	Remarks	Ref
SpA$_B^*$-2, double mutant (Tyr18→Trp and Tyr76→Trp)						
	8.5	25	6.3	GuHCl	(1,2)	94B7
C-SpA$_B^*$-2, double mutant (Tyr18→Trp and Tyr76→Trp)						
	8.5	25	6.7	GuHCl	(1,3)	94B7
SpA$_B^*$-2, double mutant (Ile20→Trp and Ile78→Trp)						
	8.5	25	5.4	GuHCl	(1,2)	94B7
C-SpA$_B^*$-2, double mutant (Ile20→Trp and Ile78→Trp)						
	8.5	25	7.5	GuHCl	(1,3)	94B7
SpA$_B^*$-2, double mutant (Phe34→Trp and Phe92→Trp)						
	8.5	25	10.0	GuHCl	(1,2)	94B7
C-SpA$_B^*$-2, double mutant (Phe34→Trp and Phe92→Trp)						
	8.5	25	12.6	GuHCl	(1,3)	94B7
SpA$_B^*$-2, double mutant (Ile35→Trp and Ile93→Trp)						
	8.5	25	11.3	GuHCl	(1,2)	94B7
SpA$_B^*$-2, double mutant (Leu48→Trp and Leu106→Trp)						
	8.5	25	10.9	GuHCl	(1,2)	94B7
C-SpA$_B^*$-2, double mutant (Leu48→Trp and Leu106→Trp)						
	8.5	25	15.1	GuHCl	(1,3)	94B7
C-SpA$_B^*$-2, mutant Leu48→Trp						
	8.5	25	13.4	GuHCl	(1,3)	94B7

Mutants of an immunoglobulin (Ig) G binding protein based upon the B domain of protein A from *Staphylococcus aureus*, Trp substitution (continued)

Mutant	pH	T	ΔG	Approach	Remarks	Ref
C-SpA$_B^*$-2, mutant Leu106→Trp						
	8.5	25	14.2	GuHCl	(1,3)	94B7
C-SpA$_B^*$-2, mutant Leu113→Trp						
	8.5	25	12.6	GuHCl	(1,3)	94B7
SpA$_B^*$-1, mutant Tyr18→Trp						
	8.5	25	4.6	GuHCl	(1,2)	94B7
C-SpA$_B^*$-1, mutant Tyr18→Trp						
	8.5	25	5.0	GuHCl	(1,3)	94B7
C-SpA$_B^*$-1, mutant Ile20→Trp						
	8.5	25	6.7	GuHCl	(1,3)	94B7
SpA$_B^*$-1, mutant Phe34→Trp						
	8.5	25	9.6	GuHCl	(1,2)	94B7
C-SpA$_B^*$-1, mutant Phe34→Trp						
	8.5	25	11.7	GuHCl	(1,3)	94B7
C-SpA$_B^*$-1, mutant Leu48→Trp						
	8.5	25	11.3	GuHCl	(1,3)	94B7
C-SpA$_B^*$-1, mutant Leu55→Trp						
	8.5	25	14.2	GuHCl	(1,3)	94B7

Remarks:
(1) linear extrapolation
(2) SpA$_B^*$ corresponds to the truncated (53 residues) protein
(3) C-SpA$_B^*$ corresponds to the truncated (61 residues) protein

B1 Immunoglobulin-binding domain from streptococcal protein G, wild-type and mutants

Mutant	pH	T	ΔG	Approach	Remarks	Ref
wild-type	5.5	25	19.2	GuHCl	(1,2)	95O2
Asp31→Gly	5.5	25	14.2	GuHCl	(1,2)	95O2
Ala35→Pro	5.5	25	5.0	GuHCl	(1,2)	95O2
Thr62 deletd	5.5	25	2.9	GuHCl	(1,2)	95O2
double mutant (Thr60→Ser and Asp55→Gly)						
	5.5	25	5.0	GuHCl	(1,2)	95O2
wild-type	5.5	25	24.7	heat	(2)	95O2
Asp31→Gly	5.5	25	18.8	heat	(2)	95O2
Ala35→Pro	5.5	25	7.1	heat	(2)	95O2
Thr62 deletd	5.5	25	5.4	heat	(2)	95O2
double mutant (Thr60→Ser and Asp55→Gly)						
	5.5	25	2.9	heat	(2)	95O2

Remarks:
(1) linear extrapolation to zero denaturant concentration by a method that includes the pre- and post-denaturational baselines for a non-linear regression of the data
(2) buffer: 10 mM MES, 150 mM NaCl, pH 5.5

Table 1. Gibbs Energy Change – Molar Values

Streptococcal protein G, core mutants of the immunoglobulin binding domain

Mutant	pH	T	ΔG	Approach	Remarks	Ref
wild-type	5.4	25	23.5±1.0	GuHCl	(1–3)	96G9
Ile6→Leu	5.4	25	18.6±0.5	GuHCl	(1–4)	96G9
double mutant (Ile6→Val and Phe52→Val)						
	5.4	25	8.5±0.3	GuHCl	(1–4)	96G9
multiple mutant (Ile6→Val, Leu7→Val and Phe52→Leu)						
	5.4	25	7.5±0.3	GuHCl	(1–4)	96G9
multiple mutant (Leu5→Val, Ile6→Val, Leu7→Phe and Val54→Leu)						
	5.4	25	11.7±0.2	GuHCl	(1–4)	96G9
multiple mutant (Leu5→Phe, Ile6→Leu, Leu7→Ile and Phe52→Val)						
	5.4	25	4.9±0.4	GuHCl	(1–4)	96G9
multiple mutant (Leu5→Met, Leu7→Ile Phe52→Leu and Val54→Ile)						
	5.4	25	4.0±0.8	GuHCl	(1–4)	96G9
multiple mutant (Leu5→Met, Ile6→Val, Leu7→Val, Phe52→Val and Val54→Leu)						
	5.4	25	–3.3	GuHCl	(1–5)	96G9

Remarks:
(1) linear extrapolation
(2) buffer: 50 mM sodium phosphate
(3) transition monitored by fluorescence
(4) the series of mutants was generated with Thr2→Gln
(5) approximate ΔG value

Immunoglobulin-binding domain B1 from protein G, β-sheet propensity, see β-sheet

Site 53 mutants (guest site) of the B1 domain of staphylococcal IgG binding protein G, β-sheet propensity, see β-sheet

Protein Kinase

cAMP-Dependent protein kinase I, regulatory subunit in the presence and in the absence of cAMP

pH	T	ΔG	Approach	Remarks	Ref
7.0	23	29.3	urea	(1)	91L5
7.0	23	30.1	GuHCl	(1)	91L5
7.0	23	41.8±4.2	urea	(1,2)	91L5

Remarks:
(1) linear extrapolation
(2) in the presence of 158 μmol cAMP

Proteinase, Aspartic

Aspartic proteinase from *Fungi*

Fungus	pH	T	ΔG	Approach	Remarks	Ref
M. miehei	5.4		55.3±0.8	GuHCl	(1,2)	91B5
	5.4		41.6±0.6	GuHCl	(1,3)	91B5
M. miehei, deglycosylated form						
E. par.	5.4		28.6±1.0	GuHCl	(1,2)	91B5
	5.4		22.0±0.5	GuHCl	(1,2)	91B5
	5.4		29.8±0.7	GuHCl	(1,3)	91B5

Explanation:
(1) M. miehei = *Mucor miehei*
(2) E. par. = *Endothia parasitica*

Remarks:
(1) linear extrapolation
(2) monitored by activity
(3) monitored by UV spectroscopy

Protease, Neutral

Neutral protease from *Bacillus stearothermophilus*, mutant (Ala166→Ser)

Mutant	pH	T	ΔT	$\Delta(\Delta G)$	Approach	Remarks	Ref
wild-type	5.0	68	0.0	0.0	heat	(1)	91V3
Ala166→Ser	5.0	68	1.2	1.8	heat	(1)	91V3

Remark:
(1) monitored by heat inactivation, in the presence of 5 mM CaCl$_2$, 0.5% (v./v.) isopropanol, and 62.5 mM NaCl

Proteinase Inhibitor

Proteinase Inhibitor PMP-C, effect of glycosylation

Mutant	pH	T	$\Delta(\Delta G)$	Approach	Remarks	Ref
nonfucosylated	3	20	0.0	H/D	(1,2)	96M7
fucosylated	3	20	4.6±0.4	H/D	(1–3)	96M7
fucosylated	3	20	4.6	heat	(3,4)	96M7

Remarks:
(1) measured by H/D exchange rate, monitored by NMR
(2) measured in D$_2$O
(3) fucosylated at Thr9
(4) monitored by NMR, T$_{trs}$ increases by about 25°C on fucosylation, ΔH amounts to 75.7 and 80.8 kJ/mol for the unmodified and modified protein, respectively

Pyridoxal Kinase

Pyridoxal Kinase

pH	T	ΔG	Approach	Remarks	Ref
7.0	25	6.5	GuHCl	(1,2)	93P6
7.0	25	6.3	GuHCl	(1,3)	93P6

Remarks:
(1) linear extrapolation
(2) monitored by CD
(2) monitored by fluorescence

Pyruvate Kinase

Pyruvate kinase from rabbit muscle

pH	T	ΔG	Approach	Remarks	Ref
6.5	5	6.9	GuHCl	(1,2)	81D
6.5	5	66.5	GuHCl	(1,3)	81D
6.5	5	65.3	GuHCl	(1,4)	81D

Remarks:
(1) linear extrapolation
(2) transition $A_4 \rightarrow A_4^*$
(3) transition $A_4 \rightarrow 2A_2$
(4) transition $A_2 \rightarrow 2A_1$

Retinoic Acid-Binding Protein

Cellular retinoic acid-binding protein, wild-type and mutants

Mutant	pH	T	ΔT	$\Delta(\Delta G)$	Approach	Remarks	Ref
wild-type	7.0	54.0	0.0	0.0	heat		92Z3
w.t. + RA	7.0	54.0	17.3	23.4	heat	(1,2)	92Z3
Arg111→Gln	7.0	54.0	9.4	12.6	heat	(2)	92Z3
Arg131→Gln	7.0	54.0	11.6	15.5	heat	(2)	92Z3
double mutant (Arg111→Gln and Arg131→Gln)							
	7.0	54.0	14.5	19.7	heat	(2)	92Z3

Remarks:
(1) RA = retinoic acid
(2) $\Delta(\Delta G)$ was obtained using $\Delta(\Delta G) = \Delta T \times \Delta S$

Retinol-Binding Protein

Cellular retinol-binding protein II, wild-type

Mutant	pH	T	ΔT	Δ(ΔG)	Approach	Remark	Ref
without	7.0	61.0	0.0	0.0	heat		92Z3
retinol	7.0	61.0	6.4	8.8	heat	(1)	92Z3

Remark:
(1) Δ(ΔG) was obtained using $\Delta(\Delta G) = \Delta T \times \Delta S$

Rhizopuspepsin

Rhizopuspepsin from *Rhizopus chinensis*, wild-type and mutants

Mutant	pH	T	ΔG	Approach	Remarks	Ref
pI6	3.0	25	71.1±7.5	GuHCl	(1,2)	95L11
wild-type	3.0	25	66.9±7.1	GuHCl	(1,3)	95L11
Asp30→Ile	3.0	25	45.6±3.3	GuHCl	(1)	95L11
Asp77→Thr	3.0	25	63.2±5.9	GuHCl	(1)	95L11
double mutant (Asp30→Ile and Asp77→Thr)						
	3.0	25	46.0±2.9	GuHCl	(1)	95L11

Remarks:
(1) linear extrapolation to zero denaturant concentration by a method that includes the pre- and post-denaturational baselines for a non-linear regression of the data
(2) naturally occurring isoenzyme pI6
(3) wild-type recombinant

Rhodanese

Rhodanese

	pH	T	ΔG	Approach	Remarks	Ref
	7.4	23	14.2	GuHCl	(1,2)	89T1
	7.4	23	11	GuHCl	(1,3)	89T1

Remarks:
(1) linear extrapolation
(2) first of two subsequent transitions at 1.5 M GuHCl with slope of −9.8
(3) second of two subsequent transitions at 3.3 M GuHCl with slope of −3.4

Rhodopsin

Rhodopsin, various species

Species	pH	T	ΔG	Approach	Ref
bovine	6.8	25	37.7*	DSC	88S7
frog	6.8	25	33.8*	DSC	88S7
rat	6.8	25	39.7*	DSC	88S7

Ribonuclease from *Bacillus amyloliquefaciens*, see Barnase

Ribonuclease from *Bacillus intermedius 7P*, see Binase

Ribonuclease A

The data are arranged as follows:

a) various pH values and approaches,
b) wild-type and mutants,
c) ribonuclease A under various solvent conditions,
d) fragments and modified forms of ribonuclease A.

a) Various pH values and approaches

Ribonuclease, bovine pancreatic

pH	T	ΔG	Approach	Remarks	Ref
1.30	25.0	7.96±0.26	GuHCl	(1)	92A1
2.10	25.0	11.01±0.50	GuHCl	(1)	92A1
2.5	30	3.8	heat		69B2
2.5	30	3.8	heat		67B
2.77	30	8.6	heat		69B2
2.77	30	8.6	heat		67B
2.8	17.1	22.6	urea	(1)	89P1
2.8	21.1	20.3	urea	(1)	89P1
2.8	24.9	18.2	urea	(1)	89P1
2.8	27.75	14.6	urea	(1)	89P1
3	25	19.7±0.4	GuHCl	(1)	83A2
3.00	25.0	23.70±0.41	GuHCl	(1)	92A1
3	25	15.9±0.8	LiClO₄	(1)	83A2
3	25	53.1±0.8	LiCl, LiBr		83A2
3.02	25	22.1±0.2	GuHCl	(1)	90P1
3.02	25	21.8±1.8	GuHCl	(1a)	90P1
3.02	25	21.8	urea	(8)	90P1
3.15	30	12.9	heat		69B2
3.15	30	12.9	heat		67B
3.28	25	21.9*	DSC		70T2
3.6	37	29	DSC		76T
4	16±1	67	urea	(2)	82H
4.04	25	31*	DSC		70T2
4.7	38	27.2	hydrogen exchange		66H
4.7	38	27.6	DSC		74P1
4.8	25	31.4±0.4		(1,3)	84A
5	25	44.4*	DSC		70T2
5	25	44	DSC	(4)	80J
5.5	25	45.2±2.5	DSC		74P1
6	25	54.4	GuHCl	(5)	70S
6	25	37.2	GuHCl	(6)	70S
6.5	25	59.4	GuHCl	(5a)	72P1
6.5	25	61.5	GuHCl	(5b)	72P1
6.5	25	59.8	GuHCl	(7)	72P1
6.6	25	38.9	GuHCl	(1)	74G1
6.6	25	32.2	urea	(1)	74G1
6.6	25	61.9	GuHCl	(7)	74G1
6.6	25	50.6	urea	(7)	74G1
6.6	25	79.1	GuHCl	(5c)	79P1

Ribonuclease, bovine pancreatic (continued)

pH	T	ΔG	Approach	Remarks	Ref
6.6	25	67.4	GuHCl	(5d)	79P1
6.6	25	68.2	GuHCl	(5e)	79P1
6.98	25	36.5±0.7	GuHCl	(1)	90P1
6.98	25	38.7±2.7	GuHCl	(1a)	90P1
6.98	25	38.5	urea	(8)	90P1
7	25	58.6*	DSC		70T2
7	25	31.4	GuHCl	(1)	82A
7	25	31	urea	(1)	82A
7.0	25	31.4	GuHCl	(1,9)	93A1
7.0	25	31.4	heat	(9)	93A1
7.00	25.0	36.55±0.67	GuHCl	(1)	92A1
7	25	29.7	GuHSCN	(1)	82A
7.35	25	41.8±2.1	urea	(1)	64B1
7.6	25	43.1	GuHCl	(1)	83S
7.6	25	78.2	GuHCl	(5)	83S
9.89	25	35.2±0.8	GuHCl	(1)	90P1
9.89	25	35.1±3.3	GuHCl	(1a)	90P1
9.89	25	36.0	urea	(8)	90P1
	21	75.7	heat		80S1
	43	37.7	heat		80S1
	21	78.7	GuHCl	(1)	80S1
	43	38.5	GuHCl	(1)	80S1
	–	48	hydrogen exchange		80C
neutral	25	47–102	prediction		64T

Remarks:
(1)   linear extrapolation
(1a) linear extrapolation to zero denaturant concentration by a method that includes the pre- and post-denaturational baselines for a non-linear regression of the data
(2)   electrophoresis
(3)   GuHCl + LiCl, LiBr, NaBr
(4)   extrapolated to zero dimethylsulfoxide concentration
(5)   denaturant binding model
(5a) denaturant binding model, k = 1.0
(5b) denaturant binding model, k = 1.3
(5c) denaturant binding model, k = 1.2
(5d) denaturant binding model, k = 0.6
(5e) denaturant binding model, k = 0.8
(6)   equation (11) in Ref. 69A
(7)   transfer model
(8)   from the pH profile in Fig. 6 of Ref. 90P1.
(9)   ribonuclease A, phosphate-free

Ribonuclease A, bovine pancreatic

pH	T	ΔG	Approach	Remarks	Ref
1.60	15	7.2±0.4	heat, v.H.	(1,2)	95T5
1.90	15	10.4±0.4	heat, v.H.	(1,2)	95T5
2.08	15	13.2±0.3	heat, v.H.	(1,2)	95T5
2.0	15	15.4±0.8	GuHCl	(3)	95T5

Remarks:
(1) the transition temperatures amount to 23.0, 25.8 and 28.7°C, respectively
(2) Ref. 95T5 contains additional data on specific volume and compressibility of ribonuclease A
(3) linear extrapolation

280   Table 1. Gibbs Energy Change – Molar Values

Ribonuclease A, dependence of ΔG on pH

pH	T	ΔG	Approach	Remarks	Ref
3.0	25.0	18.74±0.88	urea	(1,2)	95Y5
3.5	25.0	24.18±1.55	urea	(1,2)	95Y5
6.0	25.0	36.90±2.18	urea	(1,2)	95Y5
7.0	25.0	38.74±1.30	urea	(1,2)	95Y5
8.5	25.0	38.91±2.76	urea	(1,2)	95Y5

Remarks:
(1) linear extrapolation to zero denaturant concentration by a method that includes the pre- and post-denaturational baselines for a non-linear regression of the data
(2) Ref. 95Y5 contains additional data concerning the validity of the linear extrapolation method

Ribonuclease A, dependence of ΔG on ionic strength

NaCl Concentration	pH	T	ΔG	Approach	Remarks	Ref
0.0	3.0	25.0	21.21±0.42	GuHCl	(1,2)	95Y5
0.0	3.0	25.0	16.74±0.79	urea	(1,2)	95Y5
0.071	3.0	25.0	18.16±0.96	urea	(1,2)	95Y5
0.121	3.0	25.0	18.74±0.88	urea	(1,2)	95Y5
0.421	3.0	25.0	21.09±0.63	urea	(1,2)	95Y5
0.671	3.0	25.0	22.97±0.50	urea	(1,2)	95Y5
0.921	3.0	25.0	22.97±0.79	urea	(1,2)	95Y5

Remarks:
(1) linear extrapolation to zero denaturant concentration by a method that includes the pre- and post-denaturational baselines for a non-linear regression of the data
(2) Ref. 95Y5 contains additional data concerning the validity of the linear extrapolation method

Ribonuclease A, GuHCl-induced unfolding in the presence of urea

Urea Concentration	pH	T	ΔG	Approach	Remarks	Ref
0.0 M	3.0	25	22.1±0.5	GuHCl	(1)	94A1
0.0 M	3.0	25	22.2±0.5	urea	(1)	94A1
1.0 M	3.0	25	14.7±0.8	urea	(1,2)	94A1
1.8 M	3.0	25	8.1±0.3	urea	(1,2)	94A1
2.25 M	3.0	25	3.2±0.5	urea	(1,2)	94A1

Remarks:
(1) linear extrapolation
(2) the results support the linear extrapolation procedure

Ribonuclease A, apparent Gibbs energy change monitored by amide proton exchange for the following positions

Residue	pH	T	ΔG		Remarks	Ref
Gln11	5.5	34	29.8		(1)	93M9
His12	5.5	34	33.9		(1)	93M9
Met13	5.5	34	33.5		(1)	93M9
Asp14	5.5	34	24.9		(1)	93M9
Met29	5.5	34	26.7		(1)	93M9
Met30	5.5	34	32.2		(1)	93M9
Asn44	5.5	34	27.6		(1)	93M9
Glu49	5.5	34	31.2		(1)	93M9
Ala56	5.5	34	28.0		(1)	93M9
Cys58	5.5	34	40.0		(1)	93M9
Ser59	5.5	34	28.3		(1)	93M9
Gln60	5.5	34	27.9		(1)	93M9
Lys61	5.5	34	24.4		(1)	93M9
Val63	5.5	34	38.1		(1)	93M9
Cys72	5.5	34	35.8		(1)	93M9
Tyr73	5.5	34	36.7		(1)	93M9
Gln74	5.5	34	36.9		(1)	93M9
Cys84	5.5	34	37.3		(1)	93M9
Lys98	5.5	34	33.3		(1)	93M9
Thr100	5.5	34	34.2		(1)	93M9
Ala102	5.5	34	32.8		(1)	93M9
Glu111	5.5	34	31.0		(1)	93M9
His119	5.5	34	37.1		(1)	93M9

Remark:
(1) ΔG is the apparent Gibbs energy change, pH* is the uncorrected value

Ribonuclease A bovine pancreatic, heat and pressure denaturation

pH	T	ΔG	Approach	Remarks	Ref
2.0	25	10.3±0.5	pressure	(1,2)	96L1
2.0	25	10	heat	(3)	96L1

Remarks:
(1) pressure denaturation monitored by fourth derivative spectroscopy
(2) ΔV amounts to –52±2 ml/mol, transition midpoint at 196 MPa
(3) monitored by fourth derivative spectroscopy, T_{trs} amounts to 36.7°C

b) Wild-type and mutants

Ribonuclease A, bovine, wild-type and mutants (cis-Pro)

Mutant	pH	T_{trs}	ΔT	$\Delta(\Delta G)$	Approach	Remarks	Ref
wild-type	4.2	54.5±0.5			heat	(1)	92S6
pseudo w.t.	4.2	56.0±0.5	0.0	0.0	heat	(1,2)	92S6
Pro93→Ala	4.2	46.5±0.5	−9.5	−11.3	heat	(1,3)	92S6
Pro93→Ser	4.2	47.5±0.5	−8.5	−8.8	heat	(1,3)	92S6
Pro114→Ala	4.2	45.5±0.5	−10.5	−13.4	heat	(1,3)	92S6
Pro114→Gly	4.2	45.5±0.5	−10.5	−11.7	heat	(1,3)	92S6
double mutant (Pro93→Ala and Pro114→Gly)							
	4.2	36.0±0.5	−20	−17.6	heat	(1,3)	92S6

Remarks:
(1) buffer: 0.01 M sodium acetate, 0.1 M NaCl
(2) pseudo wild-type, containing an additional N-terminal Met residue. The protein is the reference protein for the following mutants
(3) $\Delta(\Delta G)$ at reference T_{trs} of the pseudo wild-type = 56.0°C

Ribonuclease A, wild-type and mutations in position 97

Mutant	pH	T_{trs}	ΔT	$\Delta(\Delta G)$	Approach	Remarks	Ref
wild-type	5.0	63.3	0.0	0.0	heat	(1)	96E1
Tyr97→Ala	5.0	29.0	−34.3	−50.2	heat	(1,2)	96E1
Tyr97→Gly	5.0	30.0	−33.3	−49.0	heat	(1,2)	96E1
Tyr97→Phe	5.0	53.2	−10.1	−14.8	heat	(1,2)	96E1

Remarks:
(1) measured in 0.10 M sodium acetate containing 0.10 M NaCl
(2) $\Delta(\Delta G)$ was determined by $\Delta(\Delta G) = \Delta T_{trs} \times \Delta S$ at T_{trs} of the unperturbed protein

Ribonuclease A, conformational free energy of unfolding for breakage of the first disulfide bond

Transition	pH	T	ΔG	Approach	Remarks	Ref
N→I	8.0	25	27.5	kinetics	(1,2)	95L4

Remarks:
(1) kinetics of reductive unfolding
(2) double mutants corresponding to the intermediates (Cys65→Ser and Cys72→Ser) and (Cys40→Ser and Cys95→Ser) show a thermal transition at 44 and 42°C, respectively (pH 6.4) compared with 62°C for native ribonuclease A

c) Ribonuclease A under various solvent conditions

Ribonuclease, bovine pancreatic, in the presence of salts

Salt	pH	T	ΔG	Approach	Remarks	Ref
0.1 M NaCl	7	25	36±2.5	GuHCl	(1,2)	86T2
0.1 M NaCl	7	25	38.5±2.9	GuHCl	(1,3)	86T2
0.5 M NaCl	7	25	36±2.9	GuHCl	(1,3)	86T2
1.0 M NaCl	7	25	36±2.9	GuHCl	(1,3)	86T2
0.1 M KCl	7	25	36.8±0.4	GuHCl	(1,3)	86T2
0.49M CsCl	7	25	37.3±2.5	GuHCl	(1,3)	86T2
0.5 M RbCl	7	25	35.2±2.5	GuHCl	(1,3)	86T2
0.45M NH_4Cl	7	25	35.6±1.7	GuHCl	(1–3)	86T2
0.5 M LiCl	7	25	35±0.17	GuHCl	(1–3)	86T2
0.99M LiCl	7	25	32.66±0.04	GuHCl	(1–3)	86T2
0.15M Na_2SO_4	7	25	39.73±0.13	GuHCl	(1–3)	86T2
0.3 M Na_2SO_4	7	25	42.5±0.21	GuHCl	(1–3)	86T2
0.15M $(NH_4)_2SO_4$	7	25	40.19±0.08	GuHCl	(1–3)	86T2
0.3 M $(NH_4)_2SO_4$	7	25	43.04±0.13	GuHCl	(1–3)	86T2
0.2 M Li_2SO_4	7	25	40.19±0.04	GuHCl	(1–3)	86T2
0.4 M Li_2SO_4	7	25	42.62±0.04	GuHCl	(1–3)	86T2
0.5 M NaBr	7	25	32.24±0.17	GuHCl	(1–3)	86T2
0.99M NaBr	7	25	28.93±0.08	GuHCl	(1–3)	86T2
0.76M LiBr	7	25	27.42±0.13	GuHCl	(1–3)	86T2
1.52M LiBr	7	25	18.88±0.13	GuHCl	(1–3)	86T2
0.25M NaSCN	7	25	30.44±0.04	GuHCl	(1–3)	86T2
0.49M NaSCN	7	25	24.28±0.08	GuHCl	(1–3)	86T2
0.09M LiSCN	7	25	34.62±0.13	GuHCl	(1–3)	86T2
0.34M LiSCN	7	25	26.17±0.17	GuHCl	(1–3)	86T2
0.69M LiSCN	7	25	17.08±0.17	GuHCl	(1–3)	86T2

Remarks:
(1) linear extrapolation
(2) monitored by CD
(3) monitored by difference spectroscopy

Ribonuclease, bovine pancreatic, in the presence of glycerol

	pH	T	ΔG	Approach	Remarks	Ref
	2.8	25	18.8	GuHCl	(1,2)	90G1

Remarks:
(1) linear extrapolation
(2) increase in ΔG of 2.1 kJ per mole glycerol

Ribonuclease A, bovine pancreatic, effect of osmolytes (sarcosine)

Molal Concentration	pH	T	ΔG	Approach	Remarks	Ref
0	4.5	25	35±3	DSC	(1,2)	95P4
1.44	4.5	25	42±4	DSC	(1,2)	95P4
2.79	4.5	25	45±4	DSC	(1,2)	95P4
4.06	4.5	25	49±5	DSC	(1,2)	95P4
5.27	4.5	25	45±6	DSC	(1,2)	95P4
6.61	4.5	25	34±6	DSC	(1,2)	95P4

Remarks:
(1) molal concentration (mol/kg)
(2) ΔG from Fig. 6B in Ref. 95P4

Ribonuclease A in the presence of sarcosine

Concentration	pH	T	ΔT	$\Delta(\Delta G)$	Approach	Remark	Ref
0.0 M	6.0	65	0.0	0.0	DSC		92S3
8.2 M	6.0	65	22	30.1	DSC	(1)	92S3

Remark:
(1) $\Delta H = 481\pm25$ kJ/mol independent of the sarcosine conc. from 0 to 8.2 M

Ribonuclease A, in the presence of various added solutes

Added Solute	pH	T	ΔG	Approach	Remarks	Ref
none	4.0	10	32.6	DSC		96L3
none	4.0	25	27.9	DSC		96L3
none	4.0	40	17.6	DSC		96L3
none	4.0	55	2.1	DSC		96L3
none	5.0	10	36.3	DSC		96L3
none	5.0	25	31.7	DSC		96L3
none	5.0	40	21.7	DSC		96L3
none	5.0	55	6.4	DSC		96L3
none	6.0	10	41.4	DSC		96L3
none	6.0	25	37.1	DSC		96L3
none	6.0	40	27.4	DSC		96L3
none	6.0	55	12.3	DSC		96L3
0.5 M sucrose	4.0	10	14.8	DSC		96L3
0.5 M sucrose	4.0	25	20.0	DSC		96L3
0.5 M sucrose	4.0	40	16.6	DSC		96L3
0.5 M sucrose	4.0	55	5.1	DSC		96L3
0.5 M sucrose	5.0	10	17.4	DSC		96L3
0.5 M sucrose	5.0	25	22.7	DSC		96L3
0.5 M sucrose	5.0	40	19.5	DSC		96L3
0.5 M sucrose	5.0	55	8.1	DSC		96L3
0.5 M sucrose	6.0	10	22.5	DSC		96L3
0.5 M sucrose	6.0	25	28.0	DSC		96L3
0.5 M sucrose	6.0	40	25.1	DSC		96L3
0.5 M sucrose	6.0	55	14.0	DSC		96L3
1.0 M sucrose	4.0	10	–0.3	DSC		96L3
1.0 M sucrose	4.0	25	13.8	DSC		96L3
1.0 M sucrose	4.0	40	17.6	DSC		96L3
1.0 M sucrose	4.0	55	11.6	DSC		96L3
1.0 M sucrose	5.0	10	3.0	DSC		96L3
1.0 M sucrose	5.0	25	17.2	DSC		96L3
1.0 M sucrose	5.0	40	21.2	DSC		96L3
1.0 M sucrose	5.0	55	15.4	DSC		96L3
1.0 M sucrose	6.0	10	2.4	DSC		96L3
1.0 M sucrose	6.0	25	16.6	DSC		96L3
1.0 M sucrose	6.0	40	20.5	DSC		96L3
1.0 M sucrose	6.0	55	14.7	DSC		96L3
1.0 M glycine	4.0	10	26.6	DSC		96L3
1.0 M glycine	4.0	25	27.2	DSC		96L3
1.0 M glycine	4.0	40	21.3	DSC		96L3
1.0 M glycine	4.0	55	9.2	DSC		96L3
1.0 M glycine	5.0	10	31.3	DSC		96L3
1.0 M glycine	5.0	25	32.0	DSC		96L3
1.0 M glycine	5.0	40	26.4	DSC		96L3

Ribonuclease A, in the presence of various added solutes (continued)

Added Solute	pH	T	ΔG	Approach	Remarks	Ref
1.0 M glycine	5.0	55	14.6	DSC		96L3
1.0 M glycine	6.0	10	31.8	DSC		96L3
1.0 M glycine	6.0	25	32.6	DSC		96L3
1.0 M glycine	6.0	40	27.0	DSC		96L3
1.0 M glycine	6.0	55	15.2	DSC		96L3
0.5 M GuHCl	4.0	10	42.1	DSC		96L3
0.5 M GuHCl	4.0	25	33.0	DSC		96L3
0.5 M GuHCl	4.0	40	17.7	DSC		96L3
0.5 M GuHCl	4.0	55	−3.3	DSC		96L3
0.5 M GuHCl	5.0	10	47.9	DSC		96L3
0.5 M GuHCl	5.0	25	39.0	DSC		96L3
0.5 M GuHCl	5.0	40	24.1	DSC		96L3
0.5 M GuHCl	5.0	55	3.3	DSC		96L3
0.5 M GuHCl	6.0	10	49.3	DSC		96L3
0.5 M GuHCl	6.0	25	40.5	DSC		96L3
0.5 M GuHCl	6.0	40	25.6	DSC		96L3
0.5 M GuHCl	6.0	55	5.0	DSC		96L3
1.0 M GuHCl	4.0	10	7.4	DSC		96L3
1.0 M GuHCl	4.0	25	9.3	DSC		96L3
1.0 M GuHCl	4.0	40	4.6	DSC		96L3
1.0 M GuHCl	4.0	55	−6.4	DSC		96L3
1.0 M GuHCl	5.0	10	10.5	DSC		96L3
1.0 M GuHCl	5.0	25	12.6	DSC		96L3
1.0 M GuHCl	5.0	40	8.1	DSC		96L3
1.0 M GuHCl	5.0	55	−2.8	DSC		96L3
1.0 M GuHCl	6.0	10	11.5	DSC		96L3
1.0 M GuHCl	6.0	25	13.7	DSC		96L3
1.0 M GuHCl	6.0	40	9.2	DSC		96L3
1.0 M GuHCl	6.0	55	−1.7	DSC		96L3
10% glycerol	4.0	10	22.2	DSC	(1)	96L3
10% glycerol	4.0	25	23.1	DSC	(1)	96L3
10% glycerol	4.0	40	17.9	DSC	(1)	96L3
10% glycerol	4.0	55	7.1	DSC	(1)	96L3
10% glycerol	5.0	10	25.3	DSC	(1)	96L3
10% glycerol	5.0	25	26.4	DSC	(1)	96L3
10% glycerol	5.0	40	21.3	DSC	(1)	96L3
10% glycerol	5.0	55	10.7	DSC	(1)	96L3
10% glycerol	6.0	10	25.7	DSC	(1)	96L3
10% glycerol	6.0	25	26.7	DSC	(1)	96L3
10% glycerol	6.0	40	21.8	DSC	(1)	96L3
10% glycerol	6.0	55	11.1	DSC	(1)	96L3
0.5 M NaCl	4.0	10	23.1	DSC		96L3
0.5 M NaCl	4.0	25	22.4	DSC		96L3
0.5 M NaCl	4.0	40	15.9	DSC		96L3
0.5 M NaCl	4.0	55	3.8	DSC		96L3
0.5 M NaCl	5.0	10	27.2	DSC		96L3
0.5 M NaCl	5.0	25	26.7	DSC		96L3
0.5 M NaCl	5.0	40	20.4	DSC		96L3
0.5 M NaCl	5.0	55	8.5	DSC		96L3
0.5 M NaCl	6.0	10	27.8	DSC		96L3
0.5 M NaCl	6.0	25	27.3	DSC		96L3
0.5 M NaCl	6.0	40	21.0	DSC		96L3
0.5 M NaCl	6.0	55	9.2	DSC		96L3

Remark:
(1) 10% (v/v) glycerol

d) Fragments, modified forms

Ribonuclease, bovine pancreatic, fragments

Fragment	pH	T	ΔG	Approach	Remarks	Ref
des-(1–20)	7	25	7.8*	DSC	70T2	
des-(1–20) + peptide (1–20)						
	7	25	24.5*	DSC	70T2	
des-(119–124)	7	26	3.3	GuHCl	(1)	72P2
des-(120–124)	7	26	3.6	GuHCl	(1)	72P2
des-(121–124)	7	26	42.3±2.9	GuHCl	(2)	72P1

Remarks:
(1) transfer model
(2) mean value of various approaches

Ribonuclease, bovine pancreatic, modified

Modification	pH	T	ΔG	Approach	Remarks	Ref
glycosylation	7	26	63.2±0.8	GuHCl	(1)	73P4
guadination		54		hydrogen exchange		80C
succinylation	4	16±1	49.6	urea	(2)	82H

Remarks:
(1) mean value of various approaches
(2) electrophoresis

Ribonuclease A, modified by reductive alkylation

Mutant	pH	T	Δ(ΔG)	Approach	Ref
native protein	3.0	40	12.3	DSC	91F2
methylated (94%)	3.0	40	4.6	DSC	91F2
ethylated (54%)	3.0	40	7.8	DSC	91F2
n-butylated (40%)	3.0	40	4.9	DSC	91F2
n-hexylated (37%)	3.0	40	3.0	DSC	91F2

Ribonuclease A, C-peptide analogue, in the presence of sodium dodecylsulfate

Salt Concentration	pH	T	ΔG	Approach	Ref
1 mM		25	4.2	heat	90W5
5 mM		25	5.4	heat	90W5
10 mM		25	5.0	heat	90W5

Ribonuclease F1, see Ribonuclease T1, wild-type and cysteine mutant

Ribonuclease H

Ribonuclease H, from *E. coli*, wild-type and mutants

Mutant	pH	T	ΔG	Approach	Remarks	Ref
wild-type	5.5	25	38.7±0.2	GuHCl	(1)	91K1
His62→Ala	5.5	25	40.3±0.2	GuHCl	(1)	91K1
His83→Ala	5.5	25	38.7±0.2	GuHCl	(1)	91K1
His114→Ala	5.5	25	27.7±0.2	GuHCl	(1)	91K1
His124→Ala	5.5	25	40.7±0.2	GuHCl	(1)	91K1
His127→Ala	5.5	25	36.3±0.3	GuHCl	(1)	91K1

Remark:
(1) linear extrapolation

Ribonuclease H from *E. coli*

pH	T	ΔG	Approach	Remarks	Ref
8.0	25	30.1±2.9	GuHCl	(1–3)	94D1
8.0	25	31.0±4.2	urea	(1–3)	94D1
8.0	25	31.8±5.4	urea	(1,3,4)	94D1

Remarks:
(1) linear extrapolation
(2) transition monitored by CD
(3) the thermal transition of ribonuclease H in the presence of 0.5 M GuHCl takes place at $T_{trs} = 55\pm0.2°C$
(4) transition monitored by Trp fluorescence

Ribonuclease H* (RNaseH*), partly folded species.
Free energy of unfolding $(\Delta G)_{unf.}$ in the absence of denaturant and the free energy of fluctuation $(\Delta G)_{fluct.}$ allowing amide hydrogen exchange

Amino Acid	pH	T	$(\Delta G)_{unf.}$	$(\Delta G)_{fluct.}$	Remarks	Ref
Ile7	5.1	25	30.5		(1,2)	96C7
Phe8	5.1	25	31.8		(1,2)	96C7
Thr9	5.1	25	29.3		(1,2)	96C7
Ser12	5.1	25	35.1	25.9	(1,2)	96C7
Gly18	5.1	25	31.4	23.0	(1,2)	96C7
Gly20	5.1	25	31.8		(1,2)	96C7
Tyr22	5.1	25	31.4		(1,2)	96C7
Ala24	5.1	25	33.5		(1,2)	96C7
Ile25	5.1	25	30.5		(1,2)	96C7
Leu26	5.1	25	30.1		(1,2)	96C7
Arg27	5.1	25	30.1		(1,2)	96C7
Lys33	5.1	25	28.0	25.9	(1,2)	96C7
Phe35	5.1	25	31.0		(1,2)	96C7
Ala37	5.1	25	29.3		(1,2)	96C7
Gly38	5.1	25	n.d.	18.4	(1,2)	96C7
Tyr39	5.1	25	31.4		(1,2)	96C7
Thr40	5.1	25	n.d.	16.3	(1,2)	96C7
Thr42	5.1	25	31.8		(1,2)	96C7
Arg46	5.1	25	38.5	31.0	(1,2)	96C7
Met47	5.1	25	41.0		(1,2)	96C7
Glu48	5.1	25	38.9		(1,2)	96C7

Ribonuclease H* (RNaseH*), partly folded species.
Free energy of unfolding $(\Delta G)_{unf.}$ in the absence of denaturant and the free energy of fluctuation $(\Delta G)_{fluct.}$ allowing amide hydrogen exchange (continued)

Amino Acid	pH	T	$(\Delta G)_{unf.}$	$(\Delta G)_{fluct.}$	Remarks	Ref
Ala51	5.1	25	46.4		(1,2)	96C7
Ala52	5.1	25	45.6		(1,2)	96C7
Val54	5.1	25	43.9		(1,2)	96C7
Ala55	5.1	25	44.8		(1,2)	96C7
Leu56	5.1	25	44.8		(1,2)	96C7
Ala58	5.1	25	37.7	18.8	(1,2)	96C7
Leu59	5.1	25	41.8	18.4	(1,2)	96C7
Val65	5.1	25	n.d.	18.8	(1,2)	96C7
Ile66	5.1	25	38.5		(1,2)	96C7
Leu67	5.1	25	35.1		(1,2)	96C7
Ser68	5.1	25	35.1		(1,2)	96C7
Arg75	5.1	25	38.1	26.4	(1,2)	96C7
Ile78	5.1	25	34.7	24.7	(1,2)	96C7
Trp81	5.1	25	n.d.	15.1	(1,2)	96C7
Ile82	5.1	25	n.d.	21.3	(1,2)	96C7
Lys86	5.1	25	28.9	23.8	(1,2)	96C7
Leu103	5.1	25	36.0	20.1	(1,2)	96C7
Trp104	5.1	25	41.4		(1,2)	96C7
Gln105	5.1	25	42.3	33.9	(1,2)	96C7
Arg106	5.1	25	37.2	23.8	(1,2)	96C7
Leu107	5.1	25	42.3		(1,2)	96C7
Ala109	5.1	25	38.9	26.8	(1,2)	96C7
Ala110	5.1	25	42.7	31.0	(1,2)	96C7
Lys117	5.1	25	32.6	28.0	(1,2)	96C7
Trp118	5.1	25	n.d.	16.7	(1,2)	96C7
Glu119	5.1	25	29.3		(1,2)	96C7
Asp134	5.1	25	n.d.	16.7	(1,2)	96C7
Leu136	5.1	25	27.2		(1,2)	96C7
Ala139	5.1	25	32.2		(1,2)	96C7
Ala140	5.1	25	31.8		(1,2)	96C7
Ala141	5.1	25	31.4		(1,2)	96C7
Met142	5.1	25	28.9		(1,2)	96C7

Explanations:
RNaseH* – *E. coli* ribonuclease H with three cysteines replaced by alanines
$(\Delta G)_{unf.}$ – unfolding events that are linearly dependent on guanidine conc.
$(\Delta G)_{fluct.}$ – local fluctuations that are guanidine independent
n.d.   – not determined

Remarks:
(1) measured by amide hydrogen exchange
(2) ^{15}N-labeled RNaseH* in deuterated solution (100 mM d_3-NaAcetate pD_{read} 5.1) with varying amounts of deuterated guanidine salt

Ribonuclease HI

Ribonuclease HI

pH	T	ΔG	Approach	Remarks	Ref
2.0	10	8.4	heat	(1,2)	95Y3
2.0	25	−4.2	heat	(1,2)	95Y3
3.0	10	34.1	heat	(1,2)	95Y3
3.0	25	25.4	heat	(1,2)	95Y3
4.0	10	50.2	heat	(1,2)	95Y3
4.0	25	41.8	heat	(1,2)	95Y3

Remarks:
(1) based on an extended data set and $\Delta Cp = 5.87\pm0.42$ kJ/mol/K
(2) ref. 95Y3 contains general expressions for the temperature dependence of ΔG

Ribonuclease HI from *E. coli*, wild-type and mutants

Mutant	pH	T	ΔG	Δ(ΔG)	Approach	Remarks	Ref
wild-type	5.5	25	41.0±0.8	0.0	urea	(1)	92K6
wild-type	3.0	49.8		0.0	heat	(2)	92K6
Ser68→Ala	5.5	25	41.8±0.8	0.8	urea	(1)	92K6
Ser68→Ala	3.0	49.8		−1.9	heat	(2)	92K6
Ser68→Gly	5.5	25	31.8±0.8	−9.2	urea	(1)	92K6
Ser68→Gly	3.0	49.8		−10	heat	(2)	92K6
Ser68→Leu	5.5	25	36.8±0.8	−4.2	urea	(1)	92K6
Ser68→Leu	3.0	49.8		−2.0	heat	(2)	92K6
Ser68→Thr	5.5	25	41.4±0.8	0.4	urea	(1)	92K6
Ser68→Thr	3.0	49.8		−0.8	heat	(2)	92K6
Ser68→Val	5.5	25	46.4±0.8	5.4	urea	(1)	92K6
Ser68→Val	3.0	49.8		2.4	heat	(2)	92K6

Remarks:
(1) linear extrapolation
(2) $\Delta(\Delta G)$ was obtained using $\Delta(\Delta G) = \Delta T \times \Delta S$

Ribonuclease HI from *E. coli*, wild-type and mutants

Mutant	pH	T	ΔT	Δ(ΔG)	Approach	Remarks	Ref
wild-type	3.0	49.8	0.0	0.0	heat	(1)	92K7
Lys95→Ala	3.0	49.8	0.3	0.4	heat	(1,2)	92K7
Lys91→Arg	3.0	50.3	0.5	0.8	heat	(1,2)	92K7
Lys95→Asn	3.0	49.8	2.9	3.8	heat	(1,2)	92K7
Asp94→Glu	3.0	49.8	−1.2	−1.7	heat	(1,2)	92K7
Lys95→Gly	3.0	49.8	5.7	7.1	heat	(1,2)	92K7
double mutant (Lys95→Ala and Asp94→Glu)							
	3.0	49.8	−2.0	−2.5	heat	(1,2)	92K7
double mutant (Lys91→Arg and Lys95→Gly)							
	3.0	49.8	5.2	6.7	heat	(1,2)	92K7
double mutant (Asp94→Glu and Lys95→Gly)							
	3.0	49.8	3.6	4.6	heat	(1,2)	92K7
mutant R6	3.0	49.8	4.0	5.0	heat	(1,4)	92K7
wild-type	5.5	52.0	0.0	0.0	heat	(3)	92K7
Lys95→Ala	5.5	52.0	0.4	0.4	heat	(2,3)	92K7
Lys91→Arg	5.5	52.0	0.1	0.0	heat	(2,3)	92K7
Lys95→Asn	5.5	52.0	3.2	3.8	heat	(2,3)	92K7

Ribonuclease HI from *E. coli*, wild-type and mutants (continued)

Mutant	pH	T	ΔT	Δ(ΔG)	Approach	Remarks	Ref
Asp94→Glu	5.5	52.0	−1.6	−1.7	heat	(2,3)	92K7
Lys95→Gly	5.5	52.0	6.8	7.9	heat	(2,3)	92K7
double mutant (Lys95→Ala and Asp94→Glu)							
	5.5	52.0	−1.8	−2.1	heat	(2,3)	92K7
double mutant (Lys91→Arg and Lys95→Gly)							
	5.5	52.0	5.8	6.7	heat	(2,3)	92K7
double mutant (Asp94→Glu and Lys95→Gly)							
	5.5	52.0	5.5	6.3	heat	(2,3)	92K7
mutant R6	5.5	52.0	5.6	6.3	heat	(2–4)	92K7

Remarks:
(1) buffer: 10 mM glycine-HCl, 1 mM DTE
(2) Δ(ΔG) was obtained using Δ(ΔG)= ΔT × ΔS
(3) buffer: 20 mM sodium acetate, 1 M GuHCl, 1 mM DTE
(4) residues 91–95 of ribonuclease HI from *E. coli* (Lys-Thr-Ala-Asp-Lys) were replaced by (Arg-Thr-Ala-Glu-Gly) from *Thermus thermophilus*

Ribonuclease HI from *E. coli*, wild-type and Asp134-mutants

Mutant	pH	T	ΔG	Approach	Remarks	Ref
wild-type	3.0	50.0	23.0	heat	(1,2)	94H6
Asp134→Ala	3.0	50.0	28.1	heat	(1,2)	94H6
Asp134→Asn	3.0	50.0	27.6	heat	(1,2)	94H6
Asp134→Gln	3.0	50.0	26.6	heat	(1,2)	94H6
Asp134→Glu	3.0	50.0	24.2	heat	(1,2)	94H6
Asp134→His	3.0	50.0	30.2	heat	(1,2)	94H6
Asp134→Ile	3.0	50.0	26.2	heat	(1,2)	94H6
Asp134→Leu	3.0	50.0	24.0	heat	(1,2)	94H6
Asp134→Ser	3.0	50.0	27.5	heat	(1,2)	94H6
Asp134→Thr	3.0	50.0	26.9	heat	(1,2)	94H6
Asp134→Val	3.0	50.0	27.0	heat	(1,2)	94H6
wild-type	5.5	50.0	23.0	heat	(2–4)	94H6
Asp134→Ala	5.5	50.0	31.9	heat	(2–4)	94H6
Asp134→Asn	5.5	50.0	27.2	heat	(2–4)	94H6
Asp134→Gln	5.5	50.0	28.9	heat	(2–4)	94H6
Asp134→Glu	5.5	50.0	27.5	heat	(2–4)	94H6
Asp134→His	5.5	50.0	32.5	heat	(2–4)	94H6
Asp134→Ile	5.5	50.0	29.2	heat	(2–4)	94H6
Asp134→Leu	5.5	50.0	29.6	heat	(2–4)	94H6
Asp134→Ser	5.5	50.0	28.7	heat	(2–4)	94H6
Asp134→Thr	5.5	50.0	27.4	heat	(2–4)	94H6
Asp134→Val	5.5	50.0	28.4	heat	(2–4)	94H6

Remarks:
(1) buffer 10 mM glycine-HCl pH 3.0
(2) ΔG was calculated using ΔCp = 5.18 kJ/mol/K (M. Oobatake, unpubl. results, cited in Ref. 94H6)
(3) buffer 20 mM sodium acetate, 1 M GuHCl, 0.1 M NaCl, pH 5.5
(4) the correction for GuHCl is 20.9 kJ/mol at pH 5.5, the increase in ΔG on NaCl is 11.3 kJ/mol at pH 5.5 (M. Oobatake, unpubl. results, cited in Ref. 94H6)

Ribonuclease HI from *E. coli*, wild-type and Asp134-mutants

Mutant	pH	T	ΔT	Δ(ΔG)	Approach	Remarks	Ref
wild-type	3.0	50.0	0.0	0.0	heat	(1,2)	94H6
Asp134→Ala	3.0	50.0	2.9	3.7	heat	(1,2)	94H6
Asp134→Asn	3.0	50.0	−0.3	−0.4	heat	(1,2)	94H6
Asp134→Gln	3.0	50.0	1.6	2.0	heat	(1,2)	94H6
Asp134→Glu	3.0	50.0	2.4	3.0	heat	(1,2)	94H6
Asp134→His	3.0	50.0	1.7	2.2	heat	(1,2)	94H6
Asp134→Ile	3.0	50.0	2.3	2.9	heat	(1,2)	94H6
Asp134→Leu	3.0	50.0	3.8	4.8	heat	(1,2)	94H6
Asp134→Ser	3.0	50.0	0.9	1.1	heat	(1,2)	94H6
Asp134→Thr	3.0	50.0	0.4	0.5	heat	(1,2)	94H6
Asp134→Val	3.0	50.0	1.1	1.3	heat	(1,2)	94H6
wild-type	5.5	53.1	0.0	0.0	heat	(2–4)	94H6
Asp134→Ala	5.5	53.1	5.5	6.3	heat	(2–4)	94H6
Asp134→Asn	5.5	53.1	3.2	3.7	heat	(2–4)	94H6
Asp134→Gln	5.5	53.1	4.8	5.5	heat	(2–4)	94H6
Asp134→Glu	5.5	53.1	3.1	3.6	heat	(2–4)	94H6
Asp134→His	5.5	53.1	7.0	8.1	heat	(2–4)	94H6
Asp134→Ile	5.5	53.1	4.6	5.3	heat	(2–4)	94H6
Asp134→Leu	5.5	53.1	5.5	6.3	heat	(2–4)	94H6
Asp134→Ser	5.5	53.1	3.9	4.5	heat	(2–4)	94H6
Asp134→Thr	5.5	53.1	3.9	4.5	heat	(2–4)	94H6
Asp134→Val	5.5	53.1	4.1	4.7	heat	(2–4)	94H6

Remarks:
(1) buffer 10 mM glycine-HCl, pH 3.0
(2) Δ(ΔG) was calculated using $\Delta(\Delta G) = \Delta T \times \Delta S_{(w.t.)}$
(3) buffer 20 mM sodium acetate, 1 M GuHCl, 0.1 M NaCl, pH 5.5
(4) the correction for GuHCl is 20.9 kJ/mol at pH 5.5, the increase in ΔG on NaCl is 11.3 kJ/mol at pH 5.5
    (M. Oobatake, unpubl. results, cited in Ref. 94H6)

Ribonuclease HI, wild-type and mutants

Mutant	pH	ΔT	Δ(ΔG)	Approach	Remarks	Ref
wild-type	3.0	0.0	0.0	heat	(1,2)	95A3
Gly23→Ala	3.0	2.3	2.9	heat	(1,2)	95A3
His62→Pro	3.0	3.4	4.2	heat	(1,2)	95A3
Val74→Leu	3.0	3.7	4.6	heat	(1,2)	95A3
Lys95→Gly	3.0	5.7	7.1	heat	(1,2)	95A3
Asp134→Asn	3.0	−0.3	−0.4	heat	(1,2)	95A3
Asp134→His	3.0	1.7	2.1	heat	(1,2)	95A3
5N mutant	3.0	12.5	15.9	heat	(1–3)	95A3
5H mutant	3.0	14.2	18.0	heat	(1,2,4)	95A3
wild-type	5.5	0.0	0.0	heat	(1,2)	95A3
Gly23→Ala	5.5	1.8	2.1	heat	(1,2)	95A3
His62→Pro	5.5	4.1	4.6	heat	(1,2)	95A3
Val74→Leu	5.5	3.3	3.8	heat	(1,2)	95A3
Lys95→Gly	5.5	6.8	7.9	heat	(1,2)	95A3

Ribonuclease HI, wild-type and mutants (continued)

Mutant	pH	ΔT	$\Delta(\Delta G)$	Approach	Remarks	Ref
Asp134→Asn	5.5	3.2	3.8	heat	(1,2)	95A3
Asp134→His	5.5	7.0	7.9	heat	(1,2)	95A3
5N mutant	5.5	17.6	20.1	heat	(1–3)	95A3
5H mutant	5.5	20.2	23.4	heat	(1,2,4)	95A3

Remarks:
(1) buffers: 10 mM glycine-HCl (pH 3.0) or sodium acetate (pH 5.5) containing 0.1 M NaCl and 1 M GuHCl
(2) $\Delta(\Delta G)$ was determined from $\Delta(\Delta G) = \Delta T \times \Delta S$, the estimated error in $\Delta(\Delta G)$ amounts to ±0.4 kJ/mol
(3) the 5N mutant is a quintuple mutant with Gly23→Ala, His62→Pro, Val74→Leu, Lys95→Gly, and Asp134→Asn
(4) the 5H mutant is a multiple mutant with Gly23→Ala, His62→Pro, Val74→Leu, Lys95→Gly, and Asp134→His

Ribonuclease HI, wild-type and mutants

Mutant	pH	T	ΔG	$\Delta(\Delta G)$	Approach	Remarks	Ref
wild-type	5.5	25.0	38.3	0	GuHCl	(1)	92K8
R4	5.5	25.0	41.4	3.1	GuHCl	(1)	92K8
R5	5.5	25.0	43.8	5.5	GuHCl	(1)	92K8
R6	5.5	25.0	38.2	−0.1	GuHCl	(1)	92K8
R7	5.5	25.0	36.4	−1.9	GuHCl	(1)	92K8
R4/R6	5.5	25.0	46.6	8.3	GuHCl	(1)	92K8
R4/R7	5.5	25.0	42.9	4.6	GuHCl	(1)	92K8
R5/R6	5.5	25.0	44.7	6.4	GuHCl	(1)	92K8
R5/R7	5.5	25.0	44.6	6.4	GuHCl	(1)	92K8
R4/R5/R7	5.5	25.0	48.9	10.6	GuHCl	(1)	92K8
R5/R6/R7	5.5	25.0	53.1	14.8	GuHCl	(1)	92K8
R4/R5/R6/R7	5.5	25.0	56.6	15.3	GuHCl	(1)	92K8

Mutants – Replacements of certain parts of the sequence of *E. coli* by corresponding parts of *Thermus thermophilus*
R4: His(62) of *E. coli* replaced by Pro(62)
R5: Val(74)-Arg-Gln-Gly-Ile-Thr-Gln(80) of *E. coli* replaced by Leu(74)-Lys-Lys-Ala-Phe-DEL.-Glu-(80)-Gly(80b)
R6: Lys(91)-Thr-Ala-Asp-Lys(95) of *E. coli* replaced by Arg(91)-DEL.-DEL.-Glu-Gly(95)
R7: Leu(111)-Gly-Gln-His-Gln-Ile-Lys-Trp-Glu-Trp(120) of *E. coli* replaced by Met(111)-Ala-Pro-DEL.-Arg-Val-Arg-Phe-His-Phe(120)

Remark:
(1) linear extrapolation

Ribonuclease HI, wild-type and mutants

Mutant	pH	T	ΔT	$\Delta(\Delta G)$	Approach	Remarks	Ref
wild-type	3.0	49.8	0.0	0.0	heat	(1,2)	92K8
R1	3.0	49.8	−2.0	−10.9	heat	(1,2)	92K8
R2	3.0	49.8	−2.3	−2.9	heat	(1,2)	92K8
R4	3.0	49.8	3.4	4.2	heat	(1,2)	92K8
R5	3.0	49.8	2.7	3.3	heat	(1,2)	92K8
Gly(80b)	3.0	49.8	1.2	1.7	heat	(1,2)	92K8
R6	3.0	49.8	4.0	5.0	heat	(1,2)	92K8
R7	3.0	49.8	−5.7	−7.1	heat	(1,2)	92K8
Gln113→Pro	3.0	49.8	−0.6	−0.8	heat	(1,2)	92K8

Ribonuclease HI, wild-type and mutants (continued)

Mutant	pH	T	ΔT	Δ(ΔG)	Approach	Remarks	Ref
R7-E119	3.0	49.8	−1.1	−1.3	heat	(1,2)	92K8
R8	3.0	49.8	0.4	0.4	heat	(1,2)	92K8
R9	3.0	49.8	−18.9	−23.8	heat	(1,2)	92K8
R1/R2	3.0	49.8	−4.5	−5.9	heat	(1,2)	92K8
R1/R4	3.0	49.8	0.4	0.4	heat	(1,2)	92K8
R1/R2/R4	3.0	49.8	−2.1	−2.5	heat	(1,2)	92K8
R4/R6	3.0	49.8	6.7	8.4	heat	(1,2)	92K8
R4/R7	3.0	49.8	−2.0	−2.5	heat	(1,3)	92K8
R5/R6	3.0	49.8	6.0	7.5	heat	(1,2)	92K8
R5/R7	3.0	49.8	−1.9	−2.5	heat	(1,2)	92K8
R4/R5/R7	3.0	49.8	2.0	2.5	heat	(1,2)	92K8
R5/R5/R7	3.0	49.8	1.9	2.5	heat	(1,2)	92K8
R4/R5/R6/R7	3.0	49.8	6.2	7.9	heat	(1,3)	92K8
wild-type	5.5	52.0	0.0	0.0	heat	(1,3)	92K8
R1	5.5	52.0	−1.7	−2.1	heat	(1,3)	92K8
R2	5.5	52.0	−3.2	−3.8	heat	(1,3)	92K8
R4	5.5	52.0	4.1	4.6	heat	(1,3)	92K8
R5	5.5	52.0	4.5	5.0	heat	(1,3)	92K8
Gly(80b)	5.5	52.0	0.8	0.8	heat	(1,3)	92K8
R6	5.5	52.0	5.6	6.3	heat	(1,3)	92K8
R7	5.5	52.0	2.4	2.5	heat	(1,3)	92K8
Gln113→Pro	5.5	52.0	−2.1	−2.5	heat	(1,3)	92K8
R7-E119	5.5	52.0	1.9	2.1	heat	(1,3)	92K8
R8	5.5	52.0	0.0	0.0	heat	(1,3)	92K8
R9	5.5	52.0	−13.8	−15.5	heat	(1,3)	92K8
R1/R2	5.5	52.0	−4.9	−5.4	heat	(1,3)	92K8
R1/R4	5.5	52.0	1.9	2.1	heat	(1,3)	92K8
R1/R2/R4	5.5	52.0	−0.1	0.0	heat	(1,3)	92K8
R4/R6	5.5	52.0	8.8	10.0	heat	(1,3)	92K8
R4/R7	5.5	52.0	6.9	7.9	heat	(1,3)	92K8
R5/R6	5.5	52.0	9.2	10.5	heat	(1,3)	92K8
R5/R7	5.5	52.0	8.2	9.2	heat	(1,3)	92K8
R4/R5/R7	5.5	52.0	12.5	14.2	heat	(1,3)	92K8
R5/R6/R7	5.5	52.0	12.4	13.8	heat	(1,3)	92K8
R4/R5/R6/R7	5.5	52.0	16.7	18.8	heat	(1,3)	92K8

Mutants – Replacements of certain parts of the sequence of E. coli by corresponding parts of Thermus thermophilus

R1:   Met(1)-Leu(2) of E. coli replaced by Met(-4)-Asn-Pro-Ser-Pro-Arg(2)

R2:   Tyr(28)-Arg-Gly-Arg(31) of E. coli replaced by Phe(28)-His-Ala-His(31)

R4:   His(62) of E. coli replaced by Pro(62)

R5:   Val(74)-Arg-Gln-Gly-Ile-Thr-Gln(80) of E. coli replaced by Leu(74)-Lys-Lys-Ala-Phe-DEL.-Glu-(80)-Gly(80b)Gly(80b): insertion 80-Gly-81

R6:   Lys(91)-Thr-Ala-Asp-Lys(95) of E. coli replaced by Arg(91)-DEL.-DEL.-Glu-Gly(95)

R7:   Leu(111)-Gly-Gln-His-Gln-Ile-Lys-Trp-Glu-Trp(120) of E. coli replaced by Met(111)-Ala-Pro-DEL.-Arg-Val-Arg-Phe-His-Phe(120)

R7-E119: Leu(111)-Gly-Gln-His-Gln-Ile-Lys-Trp-Glu-Trp(120) of E. coli replaced by Met(111)-Ala-Pro-DEL.-Arg-Val-Arg-Phe-His-Phe(120) with Glu(119) instead of His(119)

R8:   Ala125→Thr

R9:   Thr(149)-Gly-Tyr-Gln-Val-Glu-Val(155) of E. coli replaced by Cys(149)-Pro-Pro-Arg-Ala-Pro-Thr-Leu-Phe-His-Glu-Glu-Ala(161)

Remarks:
(1) Δ(ΔG) was obtained using Δ(ΔG) = ΔT × ΔS. Reference T_{trs} are 49.8°C and 52.0°C, respectively.
(2) buffer: 10 mM glycine, 1 mM DTT
(3) buffer: 20 mM acetate, 1 M GuHCl

Table 1. Gibbs Energy Change – Molar Values

Ribonuclease HI, wild-type and mutants

Mutant	pH	T	ΔT	$\Delta(\Delta G)$	Approach	Remarks	Ref
wild-type	3.0	49.8	0.0	0.0	heat	(1,2)	93I1
Gly77→Ala	3.0	49.8	−2.9	−3.8	heat	(1)	93I1
Ala80b	3.0	49.8	−1.5	−2.1	heat	(1,3)	93I1
Gly80b	3.0	49.8	1.2	1.7	heat	(1–3)	93I1
double mutant (Gly77→Ala) and insertion (Gly80b)							
	3.0	49.8	2.7	3.3	heat	(1)	93I1

Remarks:
(1) $\Delta(\Delta G)$ was obtained using $\Delta(\Delta G) = \Delta T \times \Delta S$
(2) see also Ref. 92K8
(3) insertion mutant

Ribonuclease HI, wild-type and mutants in position 74

Mutant	pH	T	ΔT	$\Delta(\Delta G)$	Approach	Remarks	Ref
wild-type	3.0	49.8	0.0	0.0	heat		92K8
Val74→Ala	3.0	49.8	−7.6	−9.2	heat	(1)	93I2
Val74→Ile	3.0	49.8	2.4	2.9	heat	(1)	93I2
Val74→Leu	3.0	49.8	3.7	4.6	heat	(1)	93I2
wild-type	5.5	52.0	0.0	0.0	heat		92K8
Val74→Ala	5.5	52.0	−12.7	−14.2	heat	(1)	93I2
Val74→Ile	5.5	52.0	2.1	2.5	heat	(1)	93I2
Val74→Leu	5.5	52.0	3.3	3.8	heat	(1)	93I2

Remark:
(1) $\Delta(\Delta G)$ was obtained using $\Delta(\Delta G) = \Delta T \times \Delta S$

Ribonuclease HI from *E. coli*

Transition	pH	T	ΔG	Approach	Remarks	Ref
I→U	5.5	25	20.0±2.3	GuHCl	(1,2)	95Y4
N→U	5.5	25	39.6±1.5	GuHCl	(1,3)	95Y4

Remarks:
(1) linear extrapolation
(2) from CD values after the initial burst phase in refolding kinetics
(3) based in a three-state mechanism

Ribonuclease HI, wild-type and mutants in the absence of Mg^{2+}

Mutant	pH	T$_{trs}$	ΔT	Δ(ΔG)	Approach	Remarks	Ref
wild-type	9.0	47.0	0.0	0.0	heat	(1,2)	96K1
Asp10→Ala	9.0	60.7	13.7	17.0	heat	(1–3)	96K1
Asp10→Asn	9.0	53.8	6.8	8.2	heat	(1–3)	96K1
Asp10→Glu	9.0	50.4	3.4	4.0	heat	(1–3)	96K1
Asp10→His	9.0	55.2	8.2	9.9	heat	(1–3)	96K1
Asp10→Ser	9.0	56.2	9.2	11.2	heat	(1–3)	96K1
Glu48→Ala	9.0	46.0	−1.0	−1.2	heat	(1–3)	96K1
Glu48→Asp	9.0	46.2	−0.8	−0.9	heat	(1–3)	96K1
Glu48→Gln	9.0	48.0	1.0	1.2	heat	(1–3)	96K1
Asp70→Ala	9.0	50.8	3.0	4.5	heat	(1–3)	96K1
Asp70→Asn	9.0	52.5	5.5	6.6	heat	(1–3)	96K1
Asp70→Glu	9.0	47.4	0.4	0.5	heat	(1–3)	96K1
Asp134→Ala	9.0	53.8	6.8	8.2	heat	(1–3)	96K1
Asp134→Asn	9.0	53.4	6.4	7.7	heat	(1–3)	96K1

Remarks:
(1) van't Hoff treatment
(2) buffer: 50 mM glycine/NaOH containing 2.8 M urea, 20% glycerol
(3) T$_{trs}$ = 47°C of the wild-type protein is the reference value for the determination of Δ(ΔG)

Ribonuclease HI, wild-type and mutants in the presence of Mg^{2+}

Mutant	pH	T$_{trs}$	ΔT	Δ(ΔG)	Approach	Remarks	Ref
wild-type	9.0	57.3	0.0	0.0	heat	(1,2)	96K1
Asp10→Ala	9.0	64.8	7.5	9.8	heat	(1–3)	96K1
Asp10→Asn	9.0	58.2	0.9	1.2	heat	(1–3)	96K1
Asp10→Glu	9.0	57.3	0.0	0.0	heat	(1–3)	96K1
Asp10→His	9.0	56.5	−0.8	−1.0	heat	(1–3)	96K1
Asp10→Ser	9.0	57.6	0.3	0.4	heat	(1–3)	96K1
Glu48→Ala	9.0	56.1	−1.2	−1.5	heat	(1–3)	96K1
Glu48→Asp	9.0	56.6	−0.7	−0.9	heat	(1–3)	96K1
Glu48→Gln	9.0	56.6	−0.7	−0.9	heat	(1–3)	96K1
Asp70→Ala	9.0	56.4	−0.9	−1.1	heat	(1–3)	96K1
Asp70→Asn	9.0	56.1	−1.2	1.5	heat	(1–3)	96K1
Asp70→Glu	9.0	56.5	−0.8	−1.0	heat	(1–3)	96K1
Asp134→Ala	9.0	61.0	3.7	4.8	heat	(1–3)	96K1
Asp134→Asn	9.0	60.8	3.5	4.5	heat	(1–3)	96K1

Remarks:
(1) van't Hoff treatment
(2) buffer: 50 mM glycine/NaOH containing 2.8 M urea, 20% glycerol and 0.2 M MgCl$_2$
(3) T$_{trs}$ = 57.3°C of the wild-type protein is the reference value for the determination of Δ(ΔG)

Table 1. Gibbs Energy Change – Molar Values

Ribonuclease HI, wild-type and mutants in the absence of Mg^{2+}

Mutant	pH	T_{trs}	ΔT	$\Delta(\Delta G)$	Approach	Remarks	Ref
wild-type	3.0	50.0	0.0	0.0	heat	(1,2)	96K1
Asp10→Ala	3.0	58.1	8.1	10.1	heat	(1–3)	96K1
Asp10→Asn	3.0	47.4	−2.6	−3.1	heat	(1–3)	96K1
Asp10→Glu	3.0	53.8	3.8	4.7	heat	(1–3)	96K1
Asp10→His	3.0	51.5	1.5	1.8	heat	(1–3)	96K1
Asp10→Ser	3.0	52.4	2.4	2.9	heat	(1–3)	96K1
Glu48→Ala	3.0	49.8	−0.2	−0.25	heat	(1–3)	96K1
Glu48→Asp	3.0	50.8	0.8	1.0	heat	(1–3)	96K1
Glu48→Gln	3.0	49.8	−0.2	−0.25	heat	(1–3)	96K1
Asp70→Ala	3.0	49.7	−0.3	−0.4	heat	(1–3)	96K1
Asp70→Asn	3.0	48.8	−1.2	−1.5	heat	(1–3)	96K1
Asp70→Glu	3.0	51.8	1.8	2.2	heat	(1–3)	96K1
Asp134→Ala	3.0	52.9	2.9	3.6	heat	(1–3)	96K1
Asp134→Asn	3.0	49.7	−0.3	−0.4	heat	(1–3)	96K1

Remarks:
(1) van't Hoff treatment
(2) buffer: 10 mM glycine/HCl
(3) T_{trs} = 50.0°C of the wild-type protein is the reference value for the determination of $\Delta(\Delta G)$

Ribonuclease T1

Ribonuclease T1 from Aspergillus oryzae and recombinant protein Gln25 form

Ribonuclease T1 stability at various pH values, using different approaches

pH	T	ΔG	Approach	Remarks	Ref
2.92	25	33.6±0.7	GuHCl	(1)	90P1
2.92	25	33.6±1.3	GuHCl	(1a)	90P1
2.92	25	25.9	urea	(5)	90P1
5.95	25.05	29.6	urea	(1)	89P2
6.98	25	31.9±0.7	GuHCl	(1)	90P1
6.98	25	32.2±1.3	GuHCl	(1a)	90P1
6.98	25	23.8	urea	(5)	90P1
7	19.2	26.3	urea	(1)	89P2
7	21.1	24.9	urea	(1)	89P2
7	23.05	24.3	urea	(1)	89P2
7	25	22.9	urea	(1)	89P2
7	25	23.8	urea	(1,6)	89P1
7	25	24.7	heat	(2)	89P1
7	27	20.7	urea	(1)	89P2
7	29	19.7	urea	(1)	89P2
7.9	25	17.2	urea	(1)	89P2
8.05	30	15.1	urea	(1)	86P1
8.05	30	17.2	urea	(3)	86P1

Ribonuclease T1 stability at various pH values, using different approaches (continued)

	pH	T	ΔG	Approach	Remarks	Ref
	8.05	30	19.7	urea	(4)	86P1
	9.91	25	14.2±0.2	GuHCl	(1)	90P1
	9.91	25	14.7±0.2	GuHCl	(1a)	90P1
	9.91	25	12.1	urea	(5)	90P1

Remarks:
(1)   linear extrapolation
(1a) linear extrapolation to zero denaturant concentration by a method that includes the pre- and post-denaturational baselines for a non-linear regression of the data
(2)   assumed heat capacity change of ΔCp = 5.2 kJ/K/mol
(3)   denaturant binding model, k = 0.1
(4)   transfer model
(5)   from the pH profile in Fig. 6 of Ref. 90P1.
(6)   from Tab. 6 in Ref. 89P1

Ribonuclease T1 (Gln25 form), wild-type and mutants, pH profile and various approaches

Mutant	pH	T	ΔG	Approach	Remarks	Ref
Various approaches:						
wild-type	7	25	23.4	urea	(1,4)	89P1
wild-type	7	25	23.4±0.6	urea	(1,2)	89T2
wild-type	7	25	23.4±0.6	heat, v.H.	(2)	89T2
w.t. A.o.	7	25	23.5	urea	(1)	89S8
w.t. A.o.	7	25	28.2	urea	(1,3)	89S8
w.t. A.o.	7	25	23	heat, v.H.		89S8
w.t. T1c	7	25	23.5	urea	(1)	89S8
w.t. T1c	7	25	28	urea	(1,3)	89S8
w.t. T1c	7	25	23	heat, v.H.		89S8
pH profile:						
wild-type	4.97	25	39.3	urea	(1)	90M4
wild-type	5.51	25	38.2	urea	(1)	90M4
wild-type	6.02	25	34.5	urea	(1)	90M4
wild-type	6.51	25	33.2	urea	(1)	90M4
wild-type	7.13	25	31.0	urea	(1)	90M4
wild-type	7.47	25	28.0	urea	(1)	90M4
wild-type	8.02	25	24.8	urea	(1)	90M4
wild-type	8.47	25	23.5	urea	(1)	90M4
wild-type	8.90	25	23.7	urea	(1)	90M4
Various approaches:						
Gln25→Lys	7	25	26.7	urea	(1)	89P1
Gln25→Lys	7	25	26.7	urea	(1)	89S8
Gln25→Lys	7	25	30.6	urea	(1,3)	89S8
Gln25→Lys	7	25	28.9	heat, v.H.		89S8
Glu58→Ala	7	25	19.7	urea	(1)	89P1
Gly58→Ala	7	25	19.5	urea	(1)	89S8
Gly58→Ala	7	25	26	urea	(1,3)	89S8
Glu58→Ala	7	25	18.8	heat, v.H.		89S8
pH profile:						
Glu58→Ala	5.03	25	36.6	urea	(1)	90M4
Glu58→Ala	5.42	25	35.1	urea	(1)	90M4
Glu58→Ala	6.01	25	31.8	urea	(1)	90M4
Glu58→Ala	6.48	25	30.7	urea	(1)	90M4

Ribonuclease T1 (Gln25 form), wild-type and mutants, pH profile and various approaches (continued)

Mutant	pH	T	ΔG	Approach	Remarks	Ref
Glu58→Ala	7.05	25	30.5	urea	(1)	90M4
Glu58→Ala	7.54	25	29.3	urea	(1)	90M4
Glu58→Ala	8.01	25	27.7	urea	(1)	90M4
Glu58→Ala	8.49	25	26.0	urea	(1)	90M4
Glu58→Ala	9.04	25	28.9	urea	(1)	90M4
Glu58→Ala	9.57	25	26.7	urea	(1)	90M4
double mutant(Gln25→Lys and Gly58→Ala)						
	7	25	26.3	urea	(1)	89S8
	7	25	31.4	urea	(1,3)	89S8
	7	25	22.6	heat, v.H.		89S8

Explanations:
w.t. A.o. = ribonuclease T1 from *Aspergillus oryzae*
w.t. T1c = ribonuclease T1 clone

Remarks:
(1) linear extrapolation
(2) mean value obtained using different optical methods
(3) in the presence of 0.15 M NaCl
(4) from Tab. 8 in Ref. 89P1

Ribonuclease T1, isoforms

Form	pH	T	ΔG	Approach	Ref
Gln25	5.0	25	38	DSC	94Y2
Lys25	5.0	25	40	DSC	94Y2

Ribonuclease T1 (Gln25 form), wild-type and mutants stability difference ($\Delta(\Delta G)$ values)

Mutant	pH	T_{trs}	ΔT	$\Delta(\Delta G)$	Approach	Ref
wild-type	7	49.2	0.0	0.0	heat	89P1
Gln25→Lys	7	51.7	2.5	3.1	heat	89P1
Glu58→Ala	7	47.9	−1.3	−1.6	heat	89P1

Ribonuclease T1, in the presence of salts

Salt	pH	T	ΔG	Approach	Remarks	Ref
0.25M NaCl	5	12.5	42.3±1.3	urea	(1,2)	88P2
30 mM MOPS	7	25	23	urea	(1)	88P1
0.1 M NaCl	7	25	26.4	urea	(1)	88P1
0.1 M MgCl$_2$	7	25	30.5	urea	(1)	88P1
0.1 M Na$_2$HPO$_4$	7	25	36.8	urea	(1)	88P1
0.1 M MOPS	7	25	21	urea	(1)	88P2

Remarks:
(1) linear extrapolation
(2) thermal transition at 59.3°C

Ribonuclease T1, Gln25-RNase T1-form, in the presence of NaCl

NaCl Concentration	pH	T	$\Delta(\Delta G)$	Approach	Remarks	Ref
0.00 M	7.0	48.91	0.0	DSC	(1,2)	92H8
0.05 M	7.0	48.91	2.7	DSC	(1,2)	92H8
0.10 M	7.0	48.91	4.0	DSC	(1,2)	92H8
0.20 M	7.0	48.91	6.4	DSC	(1,2)	92H8
0.40 M	7.0	48.91	9.7	DSC	(1,2)	92H8
0.60 M	7.0	48.91	12.4	DSC	(1,2)	92H8
0.80 M	7.0	48.91	14.4	DSC	(1,2)	92H8
1.20 M	7.0	48.91	17.9	DSC	(1,2)	92H8
1.60 M	7.0	48.91	20.8	DSC	(1,2)	92H8

Remarks:
(1) $\Delta(\Delta G)$ was calculated using $\Delta Cp = 6.65$ kJ/K/mol
(2) buffer: 30 mM PIPES and NaCl

Ribonuclease T1, Gln25-RNase T1-form, in the presence of $MgCl_2$

$MgCl_2$ Concentration	pH	T	$\Delta(\Delta G)$	Approach	Remarks	Ref
0.00 M	7.0	48.91	0	DSC	(1,2)	92H8
0.01 M	7.0	48.91	4.3	DSC	(1,2)	92H8
0.02 M	7.0	48.91	5.6	DSC	(1,2)	92H8
0.03 M	7.0	48.91	6.2	DSC	(1,2)	92H8
0.06 M	7.0	48.91	8.1	DSC	(1,2)	92H8
0.10 M	7.0	48.91	9.5	DSC	(1,2)	92H8
0.20 M	7.0	48.91	11.2	DSC	(1,2)	92H8
0.40 M	7.0	48.91	13.9	DSC	(1,2)	92H8
0.60 M	7.0	48.91	16.0	DSC	(1,2)	92H8
0.80 M	7.0	48.91	17.4	DSC	(1,2)	92H8

Remarks:
(1) $\Delta(\Delta G)$ was calculated using $\Delta Cp = 6.65$ kJ/K/mol
(2) buffer: 30 mM PIPES and $MgCl_2$

Ribonuclease T1, native and modified forms

Form	pH	T	ΔG	Approach	Remarks	Ref
native	7.5	25	20.9±9.2	heat		90O1
NFK-Trp59	7.5	25	0.4±1.3	heat	(1)	90O1
Kyn-Trp59	7.5	25	9.2±7.5	heat	(2)	90O1

Remarks:
(1) modified by ozone oxidation of Trp59 to the N-formylkynurenine derivative
(2) modified by ozone oxidation of Trp59 to the kynurenine derivative

Table 1. Gibbs Energy Change – Molar Values

Ribonuclease T1, modified, disulfide bond 2-10 reduced, ΔG values

Form	pH	ΔT	ΔG	Approach	Remarks	Ref
wild-type = reference protein						
	5	12.5	42.3±1.3	urea	(1,2b)	88P2
	7	25	21	urea	(1,2a)	88P2
[(2-10) RCM-T1]						
	5	12.5	28.5±0.8	urea	(1,2b)	88P2
	7	25	7.9	urea	(1,2a)	88P2
[(2-10),(6-103) R-T1]						
	5	12.5	12.6±0.4	urea	(1,2b)	88P2
	5	12.5	23±0.6	urea	(1,2c)	88P2
[(2-10),(6-103) RCAM-T1]						
	5	12.5	6.7±0.2	urea	(1,2b)	88P2
	5	12.5	17.2±0.6	urea	(1,2c)	88P2
[(2-10),(6-103) RCM-T1]						
	5	12.5	4.2±0.2	urea	(1,2b)	88P2
	5	12.5	13.8±0.6	urea	(1,2c)	88P2

Explanation of the forms:
wild-type = reference protein, thermal transition at 59.3°C, remark (2b)
[(2-10) RCM-T1] = disulfide bond 2-10 reduced and carboxymethylated, thermal transition at 53.3°C, remark (2b)
[(2-10), (6-103) R-T1] = disulfide bonds 2-10 and 6-103 reduced, thermal transition at 27.2°C, remark (2b)
[(2-10), (6-103) RCAM-T1] = disulfide bonds 2-10 and 6-103 reduced and carboxamidomethylated, thermal transition at 21.2°C, remark (2b)
[(2-10), (6-103) RCM-T1] = disulfide bonds 2-10 and 6-103 reduced and carboxymethylated, thermal transition at 16.6°C, remark (2b)

Remarks:
(1)   linear extrapolation
(2a) in 0.1 M MOPS
(2b) in the presence of 0.25 M NaCl
(2c) in the presence of 1.5 M NaCl

Ribonuclease T1 from Aspergillus oryzae and recombinant protein Lys25 form

Lys25-Ribonuclease T1, ΔG values

	pH	T	ΔG	Approach	Remarks	Ref
	5.0	25	46	DSC		90K4
	5.0	25	44	GuHCl	(2)	90K4
	5	25	46.6±2.1	DSC		90K5
	5	25	48.7±5	heat	(1)	90K5
	5	25	40.8±1.5	GuHCl	(2)	90K5

Remarks:
(1) van't Hoff treatment of UV melting
(2) linear extrapolation

Ribonuclease T1, isoforms

Form	pH	T	ΔG	Approach	Ref
Gln25	5.0	25	38	DSC	94Y2
Lys25	5.0	25	40	DSC	94Y2

Ribonuclease T1, wild-type and mutant, Lys25 form

Mutant	pH	T	ΔG	Approach	Remarks	Ref
wild-type	5.0	25	45±5	GuHCl	(1,2)	92K5
wild-type	5.0	25	43.0*	GuHCl	(1)	92K3
Trp59→Tyr	5.0	25	50±5	GuHCl	(1,3)	92K5
Trp59→Tyr	5.0	25	50.1*	GuHCl	(1)	92K3
double mutant (Ser54→Gly and Pro55→Asp)						
	5.0	25	38.8*	GuHCl	(1)	92K3
double mutant (Ser54→Gly and Pro55→Asn)						
	5.0	10	46.9±2.5	GuHCl	(1)	92K4

Remarks:
(1) linear extrapolation
(2) T_{trs} in the absence of GuHCl = 61.8±0.5°C
(3) T_{trs} in the absence of GuHCl = 60.6±0.5°C

Ribonuclease T1 (Lys25 form), wild-type and mutants, stability difference (Δ(ΔG) values)

Mutant	pH	T	ΔT	Δ(ΔG)	Approach	Remarks	Ref
wild-type	7	25±0.1		0.0	urea	(1,2)	92S11
wild-type	7	50.9	0.0	0.0	heat	(1)	92S11
Asn9→Ala	7	25±0.1		−3.77	urea	(1,2)	92S11
Asn9→Ala	7	50.9	−2.1	−2.97	heat	(1,3)	92S11
Tyr11→Phe	7	25±0.1		−8.83	urea	(1,2)	92S11
Tyr11→Phe	7	50.9	−6.0	−8.49	heat	(1,3)	92S11
Ser12→Ala	7	25±0.1		−5.15	urea	(1,2)	92S11
Ser12→Ala	7	50.9	−3.2	−4.52	heat	(1,3)	92S11
Ser17→Ala	7	25±0.1		2.80	urea	(1,2)	92S11
Ser17→Ala	7	50.9	1.7	2.38	heat	(1,3)	92S11
Asn36→Ala	7	25±0.1		−0.13	urea	(1,2)	92S11
Asn36→Ala	7	50.9	0.0	0.0	heat	(1,3)	92S11
Tyr42→Phe	7	25±0.1		4.77	urea	(1,2)	92S11
Tyr42→Phe	7	50.9	3.4	4.81	heat	(1,3)	92S11
Asn44→Ala	7	25±0.1		−8.70	urea	(1,2)	92S11
Asn44→Ala	7	50.9	−5.5	−7.78	heat	(1,3)	92S11
Tyr56→Phe	7	25±0.1		−3.26	urea	(1,2)	92S11
Tyr56→Phe	7	50.9	−2.1	−2.97	heat	(1,3)	92S11
Tyr57→Phe	7	25±0.1		−2.09	urea	(1,2)	92S11
Tyr57→Phe	7	50.9	−1.3	−1.84	heat	(1,3)	92S11
Ser64→Ala	7	25±0.1		−6.02	urea	(1,2)	92S11
Ser64→Ala	7	50.9	−4.6	−6.53	heat	(1,3)	92S11
Tyr68→Phe	7	25±0.1		−5.69	urea	(1,2)	92S11
Tyr68→Phe	7	50.9	−4.0	−5.65	heat	(1,3)	92S11
Asn81→Ala	7	25±0.1		−12.01	urea	(1,2)	92S11
Asn81→Ala	7	50.9	−8.6	−12.17	heat	(1,3)	92S11

Remarks:
(1) measurements in 30 mM MOPS buffer
(2) linear extrapolation
(3) Δ(ΔG) was obtained using $\Delta(\Delta G) = \Delta T \times \Delta S_{(w.t.)}$

Table 1. Gibbs Energy Change – Molar Values

Ribonuclease T1, wild-type and mutants (Lys25 form)

Mutant	pH	T	ΔG	Approach	Remarks	Ref
wild-type	5.0	10	41.1	GuHCl	(1)	93M10
wild-type	5.0	25	41.8	GuHCl	(1)	93M10
Pro39→Ala	5.0	10	20.4	GuHCl	(1)	93M10
Pro39→Ala	5.0	25	20.4	GuHCl	(1)	93M10
wild-type	8.0	10	35.3	GuHCl	(1)	93M10
wild-type	8.0	25	24.7	GuHCl	(1)	93M10
Pro39→Ala	8.0	10	17.2	GuHCl	(1)	93M10

Remark:
(1) linear extrapolation, fit of the entire transition curve

Ribonuclease T1, wild-type and Pro73 mutant

Mutant	pH	T	Δ(ΔG)	Approach	Remarks	Ref
Pro73→Val	5.0	10	−6.4	GuHCl	(1–3)	96S3
Pro73→Val	5.0	25	−8.9	GuHCl	(1–3)	96S3
Pro73→Val	8.0	10	−10.2	GuHCl	(1–3)	96S3
Pro73→Val	8.0	25	−8.4	GuHCl	(1–3)	96S3

Remarks:
(1) linear extrapolation to zero denaturant concentration by a method that includes the pre- and post-denaturational baselines for a non-linear regression of the data
(2) reference protein is the wild-type
(3) Δ(ΔG) was determined at [GuHCl]$_{1/2}$ of the wild-type protein

Ribonuclease T1, wild-type (Lys25 form)

Mutant	pH	T	ΔG	ΔT	Δ(ΔG)	Approach	Ref
wild-type	6.0	25	32.7	0.0	0.0	DSC	94S3
Trp59→Tyr	6.0	25	28.8	−2.9	−3.9	DSC	94S3
Tyr24→Trp	6.0	25	37.8	1.6	5.2	DSC	94S3
double mutant (Trp59→Tyr and Tyr24→Trp)							
	6.0	25	33.9	−0.6	1.2	DSC	94S3
Tyr42→Trp	6.0	25	32.1	−0.6	−0.6	DSC	94S3
double mutant (Tyr42→Trp and Trp59→Tyr)							
	6.0	25	28.3	−3.5	−4.3	DSC	94S3
Tyr45→Trp	6.0	25	35.8	−1.3	3.1	DSC	94S3
double mutant (Tyr45→Trp and Trp59→Tyr)							
	6.0	25	28.9	−4.4	−3.8	DSC	94S3
His40→Thr	6.0	25	33.8	−0.5	−1.1	DSC	94S3
double mutant (His40→Thr and Trp59→Tyr)							
	6.0	25	28.8	−3.3	−3.9	DSC	94S3
His92→Ala	6.0	25	30.1	−1.3	−2.6	DSC	94S3
double mutant (His92→Ala and Trp59→Tyr)							
	6.0	25	26.2	−4.0	−6.4	DSC	94S3

Ribonuclease T1, Lys25 form

pH	T	ΔG	Approach	Remarks	Ref
5.0	25	34.0±0.4	urea	(1,2)	93M11
5.0	25	38.0±1.2	urea	(1,3)	93M11
5.0	25	41.1±0.3	GuHCl	(1,4)	93M11

Remarks:
(1) linear extrapolation to zero denaturant concentration by a method that includes the pre- and post-denaturational baselines for a non-linear regression of the data
(2) buffer: 0.01 M sodium acetate, thermal transition at 58.8°C
(3) in the presence of 0.1 M GuHCl, thermal transition at 59.2°C
(4) GuHCl interferes with the determination of ΔG

Ribonuclease T1, wild-type and mutants

Mutant	pH	T	Δ(ΔG)	Approach	Remarks	Ref
wild-type	3.5	55	0.0	heat, pH	(1)	95W2
Glu28→Gln	3.5	55	−0.7	heat, pH	(1)	95W2
Asp29→Asn	3.0	55	−0.8	heat, pH	(1)	95W2
double mutant (Glu28→Gln and Asp29→Asn)						
	3.5	55	−1.0	heat, pH	(1,2)	95W2
wild-type	7.0	55	0.0	heat, pH	(1)	95W2
Glu28→Gln	7.0	55	3.2	heat, pH	(1)	95W2
Asp29→Asn	7.0	55	3.3	heat, pH	(1)	95W2
double mutant (Glu28→Gln and Asp29→Asn)						
	7.0	55	7.0	heat, pH	(1,2)	95W2

Remarks:
(1) combined approach
(2) data were taken from Fig. 6 in Ref. 95W2

Ribonuclease T1, reduced and carboxymethylated

pH	T	ΔG	Approach	Remark	Ref
8.0	15	10.5	NaCl	(1)	92M7

Remark:
(1) linear extrapolation to 2 M NaCl. At pH 8 the protein is largely unfolded. In the presence of NaCl a folding transition is observed with midpoint at about 1 M

Ribonuclease T1, wild-type and mutant, in the absence of disulfide bonds

Mutant	pH	T	ΔG	Approach	Remarks	Ref
RCM-wt	8.0	15	10.9	urea	(1–3,5)	94M21
RCM-(Pro55)	8.0	15	10.2	urea	(1,2,4,5)	94M21

Remarks:
(1) linear extrapolation to zero denaturant concentration by a method that includes the pre- and post-denaturational baselines for a non-linear regression of the data
(2) measured in the presence of 2.5 M NaCl
(3) RCM-wt: both disulfide bonds of wild-type RNase T1 are reduced and the cysteines carboxymethylated
(4) RCM-(Pro55): RNase T1 double mutant (Ser54→Gly and Pro55→Asn), reduced and carboxymethylated, see (3)
(5) RCM-wt and RCM-(Pro55) differ in ΔG by 3.6 kJ/mol at 60°C when the disulfide bonds were left intact

Table 1. Gibbs Energy Change – Molar Values

Ribonuclease T1, wild-type and disulfide-bond cleaved ribonuclease T1

Form	pH	T	ΔG	Approach	Remarks	Ref
wild-type	4	25	31.1±2.4	DSC	(1)	95H3
wild-type	5	25	30.9±2.6	DSC	(1)	95H3
wild-type	6	25	28.9±2.3	DSC	(1)	95H3
wild-type	7	25	19.0±1.9	DSC	(1)	95H3
wild-type	8	25	15.2±1.8	DSC	(1)	95H3
red-cam	4	25	8.0±0.4	DSC	(1,2)	95H3
red-cam	5	25	9.8±0.5	DSC	(1,2)	95H3
red-cam	6	25	12.5±0.5	DSC	(1,2)	95H3
red-cam	7	25	8.6±0.2	DSC	(1,2)	95H3
red-cam	8	25	7.2±0.2	DSC	(1,2)	95H3
complexes with 2'GMP:						
wild-type	5	25	40.0±2.6	DSC	(1)	95H3
red-cam	5	25	16.2±0.5	DSC	(1,2)	95H3

Remarks:
(1) measured in the presence of 2 M NaCl, buffer: 100 mM 3,3'-dimethylglutaric acid, 2 M NaCl, pH 4–8
(2) red-cam, reduced carboxyamidomethyl ribonuclease T1

Ribonuclease mutant (Pro55)-T1 [see remark (1)], disulfide and reduced form

Form	pH	T	ΔG	Approach	Remarks	Ref
2SS	8.0	15	25	urea	(2,3)	94M20
RCM/NaCl	8.0	15	10	urea	(2,4)	94M20

Remarks:
(1) (Pro55)-T1 is the Ser54→Gly and Pro55→Asn double mutant of ribonuclease T1
(2) linear extrapolation
(3) 2SS – both disulfides intact
(4) RCM/NaCl – both disulfide reduced and the Cys residues carboxymethylated; measured in the presence of 2.5 M NaCl since RCM-(Pro55)-T1 is unfolded at pH 8.0 and 15°C

Ribonuclease T1, wild-type and cysteine mutant that corresponds to ribonuclease F1

Mutant	pH	T	ΔG	Approach	Remarks	Ref
wild-type	5.0	10	41.1	GuHCl	(1–3)	94M11
wild-type	5.0	25	41.8	GuHCl	(1–3)	94M11
wild-type	8.0	10	35.3	GuHCl	(1,2)	94M11
wild-type	8.0	25	24.7	GuHCl	(1,2)	94M11
Double mutant (Cys2→Ser and Cys10→Asn) lacking the Cys2-Cys10 disulfide bond:						
mutant	5.0	10	33.8	GuHCl	(1,2,4)	94M11
mutant	5.0	25	30.8	GuHCl	(1,2,4)	94M11
mutant	8.0	10	27.8	GuHCl	(1,2,4)	94M11
mutant	8.0	25	15.6	GuHCl	(1,2,4)	94M11

Remarks:
(1) linear extrapolation to zero denaturant concentration by a method that includes the pre- and post-denaturational baselines for a non-linear regression of the data
(2) buffer pH 5.0: 0.1 M acetate, buffer pH 8.0: 0.1 M Tris
(3) at pH 5.0 is T_{trs} = 60.0 °C (Fig. 6 in Ref. 94M11)
(4) T_{trs} of the double mutant is 7.1°C lower at pH 5.0 than that of the wild-type

Ribonuclease T1, circularly permuted protein

Mutant	pH	T	ΔG	Approach	Remarks	Ref
wild-type	5.0	25	42.3	urea	(1)	96G1
double mutant (Cys2→Ala and Cys10→Ala)						
	5.0	25	26.8	urea	(1,2)	96G1
cp35-Ser1	5.0	25	19.2	urea	(1,3)	96G1
cp35-Ser1	5.0	25	23.8	urea	(1,4)	96G1
cp49-Asp1	5.0	25	12.1	urea	(1,5)	96G1
cp70-Gly1	5.0	25	10.9	urea	(1,6)	96G1
cp96-Ser1	5.0	25	19.2	urea	(1,7)	96G1

Remarks:
(1) linear extrapolation to zero denaturant concentration by a method that includes the pre- and post-denaturational baselines for a non-linear regression of the data
(2) N-terminal sequence: Ala-Ala-Asp-Tyr-Thr-Cys-Gly-Ser-Asn-Ala
(3) N-terminal sequence: Ala-Ser-Asn-Ser-Tyr-Pro-His-Lys-Tyr-Asn
(4) N-terminal sequence: Ala-Ser-Asn-Ser-Tyr-Pro-His-Lys-Tyr-Asn-Asn
(5) N-terminal sequence: Ala-Asp-Phe-Ser-Val-Ser-Ser-Pro-Tyr-Tyr
(6) N-terminal sequence: Ala-Gly-Gly-Ser-Pro-Gly-Ala-Asp-Arg-Val
(7) N-terminal sequence: Ala-Ser-Gly-Asn-Asn-Phe-Val-Glu-Cys-Thr

Ribonuclease T1, wild-type and circularly permuted variants

Mutant	pH	T	ΔG	Approach	Remarks	Ref
wild-type	5.0	10	39.9±3.4	GuHCl	(1)	96J2
double mutant (Cys2-Ala and Cys10→Ala)						
	5.0	10	28.2±1.9	GuHCl	(1)	96J2
cp35	5.0	10	24.6±2.1	GuHCl	(1,2)	96J2
cp49	5.0	10	16.6±1.3	GuHCl	(1,2)	96J2
cp70	5.0	10	17.6±0.8	GuHCl	(1,2)	96J2
cp96	5.0	10	21.8±2.5	GuHCl	(1,2)	96J2

Remarks:
(1) linear extrapolation
(2) cp35, cp49, cp70, and cp96 are circularly permuted mutants of the double mutant (Cys2-Ala and Cys10→Ala) with a new amino-terminus positioned at Ser35, Asp49, Gly70, and Ser96, respectively, and the original termini covalently attached by a Gly-Gly-Gly linker

Ribonuclease T1, Lys-25 form, in the presence of osmolytes

Solute	pH	T_{trs}	ΔH	ΔG	Approach	Remarks	Ref
0.03 M phosphate	7	51.9	464±13	38.1±1.3	heat, v.H.	(1)	94L8
0.50 M urea	7	49.9	481±8	36.8±0.8	heat, v.H.	(1)	94L8
1.0 M urea	7	48.3	469±8	33.5±0.8	heat, v.H.	(1)	94L8
2.0 M urea	7	44.3	435±4	26.4±0.4	heat, v.H.	(1)	94L8
0.25 M TMAO	7	53.0	506±4	43.1±1.3	heat, v.H.	(1)	94L8
0.5 M TMAO	7	54.0	569±8	50.2±0.8	heat, v.H.	(1)	94L8
1.0 M TMAO	7	56.0	523±8	49.4±0.8	heat, v.H.	(1)	94L8
0.5 M urea + 0.25 M TMAO							
	7	51.1	481±4	38.9±0.4	heat, v.H.	(1)	94L8

Ribonuclease T1, Lys-25 form, in the presence of osmolytes (continued)

Solute	pH	T_{trs}	ΔH	ΔG	Approach	Remarks	Ref
1.0 M urea + 0.5 M TMAO							
	7	50.4	485±4	38.5±0.4	heat, v.H.	(1)	94L8
2.0 M urea + 1.0 M TMAO							
	7	49.3	452±4	34.3±0.4	heat, v.H.	(1)	94L8

Remarks:
(1) reference temperature: ΔH is given at T_{trs}, ΔG at 25°C
(2) TMAO – trimethylamine N-oxide

Ribonuclease T1 in buffer and in reverse micelles [AOT/IO, remark (1)]

W_0	pH	T	ΔG	Approach	Remarks	Ref
buffer only	7	25	28.9±0.4	heat	(2)	96S6
4.94	7	25	29.3±0.4	heat	(3)	96S6
6.17	7	25	25.9±0.4	heat	(3)	96S6
7.40	7	25	25.9±0.4	heat	(3)	96S6
12.0	7	25	21.8±0.4	heat	(3)	96S6

Remarks:
(1) AOT = bis(2-ethylhexyl)sodium sulfosuccinate
    IO = isooctane
    W_0 = moles of entrapped water per mole surfactant
(2) 50 mM cacodylate buffer, pH 7
(3) 150 mM AOT/IO at different W_0 values

Ribosomal Protein

Ribosomal protein E-L30

pH	T	ΔG	Approach	Ref
4.5–7	25	22 (<=22)	pH, NMR measurement	87V1

Ribose-Binding Protein

Ribose-binding protein, various forms and mutants

Form	pH	T	ΔG	Approach	Remarks	Ref
mature protein:						
wild-type	7.7	25	21.3	GuHCl	(1)	91T5
Val50→Glu	7.7	25	6.7	GuHCl	(1)	91T5
Ala27→Thr	7.7	25	7.5	GuHCl	(1)	91T5
precursor:						
wild-type	7.7	25	20.1	GuHCl	(1)	91T5
precursor with replacement Leu(-17)→Pro in the leader sequence:						
wild-type	7.7	25	20.5	GuHCl	(1)	91T5
Val50→Glu	7.7	25	4.6	GuHCl	(1)	91T5
Ala27→Thr	7.7	25	10.0	GuHCl	(1)	91T5
precursor with replacements Leu(-17)→Pro and Ser(-18)→Phe in the leader sequence:						
wild-type	7.7	25	17.2	GuHCl	(1)	91T5

Remark:
(1) linear extrapolation

RNA Binding Domain of U1A, see U1A

ROP

ROP, dimeric four-α-helix-bundle protein

pH	T	ΔG	Approach	Remarks	Ref
6.0	25	71.7	DSC	(1,2)	93S10

Remarks:
(1) specific value: $\Delta G = 5.0$ J/g
(2) buffer: 10 M sodium phosphate, 10 mM Na_2SO_4, 1 mM EDTA

ROP, four-helix protein, wild-type and mutants

Mutant	pH	T	ΔG	Approach	Remarks	Ref
wild-type	6.0	25	71.7	DSC	(1)	93H7
wild-type	6.0	25	71.7	DSC	(1)	95S13
Leu41→Ala	6.0	25	46.1	DSC	(1)	95S13
Leu41→Val	6.0	25	61.1	DSC	(1)	95S13
Leu48→Ala	6.0	25	43.4	DSC	(1)	93H7

Remark:
(1) ΔG refers to mol of dimer

ROP, 4-helix bundle, RNA-binding protein, wild-type and mutants

Mutant	pH	T_{trs}	ΔT	$\Delta(\Delta G)$	Approach	Remarks	Ref
wild-type	7.0	67.0	0.0	0.0	heat	(1–3)	95P7
Lys3→Ala	7.0	68.8	1.8	0.8	heat	(1–3)	95P7
Asn10→Ala	7.0	70.2	3.2	2.1	heat	(1–3)	95P7
Phe14→Ala	7.0	68.9	1.9	0.8	heat	(1–3)	95P7
Gln18→Ala	7.0	70.0	3.0	1.7	heat	(1–3)	95P7
Thr21→Ala	7.0	72.6	5.6	3.3	heat	(1–3)	95P7
Lys25→Ala	7.0	68.7	1.7	0.8	heat	(1–3)	95P7

Remarks:
(1) the data treatment accounts for dissociation of a homodimer protein
(2) $\Delta(\Delta G)$ was calculated for 68.7°C using an experimental ΔCp value of 4.2 kJ/mol/K
(3) buffer: 10 mM sodium phosphate, 350 mM NaCl

ROP, 4-helix bundle, wild-type (Asp30) and pos. 30 mutants

Mutant	pH	T_{trs}	$\Delta(\Delta G)$	Approach	Remarks	Ref
wild-type	7.0	68.7	0.0	heat	(1–3)	96P2
Asp30→Ala	7.0	71.2	1.3	heat	(1–3)	96P2
Asp30→Arg	7.0	74.0	3.3	heat	(1–3)	96P2
Asp30→Asn	7.0	73.8	3.3	heat	(1–3)	96P2
Asp30→Cys	7.0	74.0	3.3	heat	(1–3)	96P2
Asp30→Gln	7.0	79.1	7.5	heat	(1–3)	96P2
Asp30→Glu	7.0	74.9	4.2	heat	(1–3)	96P2
Asp30→Gly	7.0	80.3	8.4	heat	(1–3)	96P2
Asp30→His	7.0	74.5	3.8	heat	(1–3)	96P2
Asp30→Ile	7.0	63.4	−3.3	heat	(1–3)	96P2
Asp30→Leu	7.0	68.4	−0.4	heat	(1–3)	96P2
Asp30→Lys	7.0	74.6	3.8	heat	(1–3)	96P2
Asp30→Met	7.0	72.7	2.5	heat	(1–3)	96P2
Asp30→Phe	7.0	68.3	−0.4	heat	(1–3)	96P2
Asp30→Pro	7.0	58.9	−6.7	heat	(1–3)	96P2
Asp30→Ser	7.0	74.7	4.2	heat	(1–3)	96P2
Asp30→Thr	7.0	66.6	−1.7	heat	(1–3)	96P2
Asp30→Trp	7.0	65.9	−1.7	heat	(1–3)	96P2
Asp30→Tyr	7.0	70.4	0.8	heat	(1–3)	96P2
Asp30→Val	7.0	66.6	−1.7	heat	(1–3)	96P2

Remarks:
(1) the data treatment accounts for dissociation of a homodimer protein, for details, see Ref. 95P7
(2) $\Delta(\Delta G)$ was calculated for 68.7°C using an experimental ΔCp value of 4.2 kJ/mol/K
(3) buffer: 10 mM sodium phosphate, 350 mM NaCl

ROP, wild-type and redesigned four-helix bundle

Mutant	pH	T	ΔG	Approach	Remarks	Ref
Rop11	7	25	33.9	GuHCl	(1,2)	94M23
Rop13	7	25	30.5	GuHCl	(1,3)	94M23
Rop21	7	25	31.4	GuHCl	(1,3)	94M23

Remarks:
(1) linear extrapolation
(2) Rop11 corresponds to the wild-type
(3) Rop13 and Rop21 are redesigned four-helix bundles in which idealized layers are created that contain Ala and Leu

ROP variants, hydrophobic core design

Mutant	pH	T_{trs}	ΔG	Approach	Remarks	Ref
wild-type	7	64	32.2	GuHCl	(1–3)	96M11
Ala$_2$Leu$_2$-2	7	72	32.2	GuHCl	(1–4)	96M11
Ala$_2$Leu$_2$-4	7	68	24.3	GuHCl	(1–4)	96M11
Ala$_2$Leu$_2$-6	7	82	33.9	GuHCl	(1–4)	96M11
Ala$_2$Leu$_2$-8	7	91	31.4	GuHCl	(1–4)	96M11
Ala$_2$Leu$_2$-8–rev	7	91	41.4	GuHCl	(1–4)	96M11
Leu$_2$Ala$_2$–8	7	91	53.6	GuHCl	(1–4)	96M11
Ala$_2$Met$_2$-8	7		13.0	GuHCl	(1–5)	96M11

Remarks:
(1) buffer: 100 mM sodium phosphate, 200 mM NaCl, 1 mM DDT for the wild type
(2) T_{trs} from thermal denaturation monitored by CD
(3) ΔG from GuHCl denaturation, linear extrapolation
(4) the mutant description refers to position and number of replacements, see also Ref. 96M11
(5) no transition observable within the temperature range accessible for CD

Serine Hydroxymethyltransferase

Serine hydroxymethyltransferase, wild-type and tryptophan mutants

Transition	pH	T	ΔG	Approach	Remarks	Ref
wild-type						
trans. (1)	7.5		17.2±2.5	urea	(1–3)	96C1
trans. (2)	7.5		18.4±2.9	urea	(1–3)	96C1
mutant Trp16:						
trans. (1)	7.5		17.6±0.8	urea	(1,2,4)	96C1
trans. (2)	7.5		23.0±6.3	urea	(1,2,4)	96C1
mutant Trp385:						
trans. (2)	7.5		24.7±1.7	urea	(1,2,5,6)	96C1

Remarks:
(1) the transitions were monitored by urea gradient electrophoresis and equilibrium unfolding detected by CD
   and fluorescence; the protein exhibits 3 Trp residues in positions 16, 183 and 385; three-state fit except for
   mutant Trp385
(2) buffer: 20 mM Tris, 5 mM mercaptoethanol, 1 mM EDTA
(3) T_{trs} = 67.6°C determined by DSC
(4) Trp16 remaining, the others replaced by Phe, T_{trs} = 63.8°C determined by DSC
(5) Trp385 remaining, the others replaced by Phe, T_{trs} = 60.9°C determined by DSC
(6) two-state fit

Serine hydroxymethyltransferase from *E. coli*, single tryptophan mutants

Mutant	pH	T	ΔG	Approach	Remarks	Ref
wild-type						
N→I	7.5	30	28.9±2.9	urea	(1–3,5)	96C2
I→U	7.5	30	58.6±9.2	urea	(1–3,5)	96C2
wild–type						
N→I	7.5	30	29.3±5.0	urea	(1–3,6)	96C2
I→U	7.5	30	55.6±4.2	urea	(1–3,6)	96C2
Trp183						
N→I	7.5	30	15.1±1.7	urea	(1–3,6)	96C2
I→U	7.5	30	59.8±5.4	urea	(1–3,6)	96C2
Trp16	7.5	30	13.8±0.8	urea	(1,2,4,5)	96C2
Trp16	7.5	30	15.5±0.8	urea	(1,2,4,6)	96C2
Trp385	7.5	30	17.6±0.8	urea	(1,2,4,5)	96C2
Trp385	7.5	30	16.7±0.4	urea	(1,2,4,6)	96C2

Remarks:
(1) serine hydroxymethyltransferase with pyridoxyl 5'-phosphate attached to Lys229
(2) buffer: 20 mM Tris/HCl, 5 mM mercaptoethanol, 1 mM EDTA, pH 7.5
(3) three-state fit
(4) two-state fit
(5) fluorescence signal at 335 nm
(6) fluorescence signal at 380 nm

Serum Albumin

Serum albumin, bovine

Form	pH	T	ΔG	Approach	Ref
native	7.15	25	30 (>30)	DSC	75L
native			32.6	hydrogen exchange	71O
native			39	hydrogen exchange	80C
guanidinated			47	hydrogen exchange	80C

Bovine serum albumin, chemically modified

Modification	pH	T	ΔG	$\Delta(\Delta G)$	Approach	Remarks	Ref
native	7	25	10.6	0.0	urea	(1)	93F2
64% acetylated	7	25	8.1	–0.8	urea	(1)	93F2
96% acetylated	7	25		–5.2	urea	(1)	93F2
66% carbamylated	7	25	8.9	–0.4	urea	(1)	93F2
91% carbamylated	7	25		–4.1	urea	(1)	93F2
62% guanidinated	7	25	10.8	0.5	urea	(1)	93F2
93% guanidinated	7	25	11.0	1.5	urea	(1)	93F2

Remark:
(1) ΔG is based on linear extrapolation, $\Delta(\Delta G)$ is based on ΔG at the half transition

β-**Sheet** (β-sheet propensity, see also chymotrypsin inhibitor)

β-Sheet forming propensity of amino acids based on immunoglobulin-binding domain B1 from protein G Mutant AASS (Ile6→Ala, Thr44→Ala, Thr51→Ser and Thr55→Ser) with a guest site in position 53

Amino acid	T_{trs}	$\Delta(\Delta G)$	Remarks	Ref
Ala	43.8	0.00	(1–3)	94M15
Arg	47.9	1.88	(1–3)	94M15
Asn	42.9	−0.33	(1–3)	94M15
Asp	35.2	−3.93	(1–3)	94M15
Cys	48.5	2.18	(1–3)	94M15
Gln	45.8	0.96	(1–3)	94M15
Glu	44.0	0.04	(1–3)	94M15
Gly	30.2	−5.02	(1–3)	94M15
His	43.3	−0.08	(1–3)	94M15
Ile	53.0	4.18	(1–3)	94M15
Leu	48.4	2.13	(1–3)	94M15
Lys	46.3	1.13	(1–3)	94M15
Met	50.2	3.01	(1–3)	94M15
Pro	<0	<−12.5	(1–3)	94M15
Ser	50.1	2.93	(1–3)	94M15
Thr	53.7	4.60	(1–3)	94M15
Trp	48.7	2.26	(1–3)	94M15
Tyr	52.5	4.02	(1–3)	94M15
Val	51.2	3.43	(1–3)	94M15

Remarks:
(1) mutant AASS-Ala53 is the reference protein
(2) measured by heat denaturation buffer: 50 mM sodium acetate, 150 mM NaCl, pH 5.4
(3) estimated error T_{trs} ±0.5°C, ΔG ±0.25

β-Sheet-forming propensity of amino acids based on immunoglobulin-binding domain of protein G(GB1) with a guest site in position 44 – context specific β-sheet propensity

Amino acid	T_{trs}	$\Delta(\Delta G)$	Remarks	Ref
Ala	54.7	0.00	(1,2)	94M16
Arg	51.2	−1.80	(1,2)	94M16
Asn	52.6	−1.00	(1,2)	94M16
Asp	53.7	−0.42	(1,2)	94M16
Cys	55.1	0.33	(1,2)	94M16
Gln	54.7	0.17	(1,2)	94M16
Glu	57.0	1.30	(1,2)	94M16
Gly	47.6	−3.56	(1,2)	94M16
His	54.6	−0.04	(1,2)	94M16
Ile	54.8	0.08	(1,2)	94M16
Leu	52.8	−1.00	(1,2)	94M16
Lys	51.4	−1.67	(1,2)	94M16
Met	54.2	−0.08	(1,2)	94M16
Phe	56.1	0.67	(1,2)	94M16
Pro	<0	<−16	(1,2)	94M16
Ser	59.4	2.64	(1,2)	94M16
Thr	60.2	3.47	(1,2)	94M16
Trp	53.3	−0.71	(1,2)	94M16
Tyr	55.9	0.46	(1,2)	94M16
Val	56.1	0.71	(1,2)	94M16

Remarks:
(1) Ala44 is the reference protein
(2) measured by heat denaturation, monitored by CD buffer: 10 mM phosphate, 150 mM NaCl, pH 7.3

β-Sheet forming propensity of amino acids based on a zinc-finger host peptide

Amino acid	Δ(ΔG)	Remarks	Ref
Ala	1.46	(1–3)	93K4
Arg	1.84	(1–3)	93K4
Asn	1.59	(1–3)	93K4
Asp	1.72	(1–3)	93K4
Cys	1.97	(1–3)	93K4
Gln	1.67	(1–3)	93K4
Glu	1.72	(1–3)	93K4
Gly	0.0	(1–3)	93K4
His	1.92	(1–3)	93K4
Ile	2.34	(1–3)	93K4
Leu	2.01	(1–3)	93K4
Lys	1.72	(1–3)	93K4
Met	1.92	(1–3)	93K4
Phe	2.30	(1–3)	93K4
Pro	0.96	(1–3)	93K4
Ser	1.63	(1–3)	93K4
Thr	2.01	(1–3)	93K4
Trp	2.01	(1–3)	93K4
Tyr	2.09	(1–3)	93K4
Val	2.22	(1–3)	93K4

Remarks:
(1) peptide: Pro-Tyr-Xaa-Cys-Pro-Glu-Cys-Gly-Lys-Ser-Phe-Ser-Gln-Lys-Ser-Asp-Leu-Val-Lys-His-Gln-Arg-Thr-His-Thr-Gly
(2) measured by Co^{2+} binding; the peptide is unfolded in the absence of Co^{2+} or Zn^{2+}
(3) buffer: 100 mM HEPES, 50 mM NaCl, pH 7.0

β-Sheet propensity, based on site 53 mutants (guest site) of the B1 domain of staphylococcal IgG binding protein G

Guest residue	pH	T_{trs}	Δ(ΔG)	Approach	Remarks	Ref
Ala	5.2	57.05	0.0	heat	(1,2)	94S6
Asn	5.2	61.88	2.18	heat	(1,2)	94S6
Asp	5.2	50.91	−3.56	heat	(1,2)	94S6
Arg	5.2	62.41	1.67	heat	(1,2)	94S6
Cys	5.2	63.99	3.26	heat	(1,2)	94S6
Gln	5.2	60.90	1.59	heat	(1,2)	94S6
Glu	5.2	58.81	0.96	heat	(1,2)	94S6
Gly	5.2	45.95	−5.06	heat	(1,2)	94S6
His	5.2	60.96	1.55	heat	(1,2)	94S6
Ile	5.2	67.78	5.23	heat	(1,2)	94S6
Leu	5.2	62.47	1.88	heat	(1,2)	94S6
Lys	5.2	60.65	1.46	heat	(1,2)	94S6
Met	5.2	64.26	3.77	heat	(1,2)	94S6
Phe	5.2	67.68	4.52	heat	(1,2)	94S6
Pro	5.2	<10	n.d.	heat	(1,2)	94S6
Ser	5.2	64.80	3.64	heat	(1,2)	94S6
Thr	5.2	68.67	5.69	heat	(1,2)	94S6
Trp	5.2	65.73	4.35	heat	(1,2)	94S6
Tyr	5.2	69.22	6.82	heat	(1,2)	94S6
Val	5.2	65.47	3.93	heat	(1,2)	94S6

Remarks:
(1) Δ(ΔG) of the mutant Thr53→Ala is taken as zero
(2) Δ(ΔG) is reported at a temperature of 60°C that is within the transition region of all mutants

Signal Transduction Protein

SEM-5, signal transduction protein from *Caenorhabditis elegans*, SH3 domain

pH	T	ΔG	Approach	Remark	Ref
7.3	25	17.2	GuHCl	(1)	94L5

Remark:
(1) linear extrapolation

Spectrin

Spectrin, SH3 domain

pH	T	ΔG	Approach	Remarks	Ref
2.0	25	2.3	DSC	(1)	94V2
2.25	25	4.2	DSC	(1)	94V2
2.5	25	6.9	DSC	(1)	94V2
2.75	25	9.8	DSC	(1)	94V2
3.0	25	11.6	DSC	(1)	94V2
3.5	25	13.9	DSC	(1)	94V2
3.5	25	14±3.0	DSC		94V2
3.5	25	13±3.0	heat	(2)	94V2
3.5	25	12±0.2	urea	(2)	94V2
3.5	25	12±0.1	urea	(3)	94V2
3.5	25	13±0.7	kinetics	(4)	94V2
4.0	25	15.6	DSC	(1)	94V2
5–7	25	15.5±1.3	GuHCl	(5)	94V2

Remarks:
(1) low ionic strength buffer, 10 mM glycine or 10 mM acetate
(2) transition monitored by fluorescence
(3) transition monitored by CD
(4) kinetics of urea induced unfolding and refolding
(5) ΔG might be lower than the real value by 10–15%

α-Spectrin, SH3 domain, circularly permuted forms

Mutant	pH	T	ΔG	Approach	Remarks	Ref
wt*	7.0	25	17.1	kinetics	(1,2)	95V5
wt*	7.0	25	16.0±0.3	urea	(1,3)	95V5
Ser19-Pro20	7.0	25	9.2	kinetics	(2)	95V5
Ser19-Pro20	7.0	25	8.6±0.3	urea	(3)	95V5
Ser19-Pro20s	7.0	25	10.6	kinetics	(2,4)	95V5
Ser19-Pro20s	7.0	25	10.4±0.2	urea	(3,4)	95V5
Asn47-Asp48	7.0	25	9.5	kinetics	(2)	95V5
Asn47-Asp48	7.0	25	8.7±0.3	urea	(3)	95V5
Asn47-Asp48s	7.0	25	12.3	kinetics	(2,4)	95V5
Asn47-Asp48s	7.0	25	10.5±0.2	urea	(3,4)	95V5
Asn38-Lys39	7.0	25	11.1	kinetics	(2)	95V5
Asn38-Lys39	7.0	25	10.8±0.2	urea	(3)	95V5
wild-type	7.0	25	14.9	kinetics	(2)	95V5
wild-type	7.0	25	14.2±0.1	urea	(3)	95V5
Leu8→Ser	7.0	25	11.1	kinetics	(2,5)	95V5
Leu8→Ser	7.0	25	10.8±0.2	urea	(3,5)	95V5

Remarks:
(1) pseudo wild-type, the first five residues of the protein are substituted by Met-Gly-Thr-Gly
(2) data from kinetics of unfolding and refolding in the presence of urea
(3) linear extrapolation to zero denaturant concentration by a method that includes the pre- and post-denaturational baselines for a non-linear regression of the data
(4) mutant index s: the mutant contains an extra Ser in the connecting loop
(5) mutant of the wild-type protein

SSB

Single stranded DNA binding protein (SSB) from *E. coli*

pH	T	ΔG	Approach	Remarks	Ref
8.1	37	179±1	GuHCl	(1,2)	94F2

Remarks:
(1) linear extrapolation
(2) ΔG refers to a two-state unfolding of tetramer into unfolded monomers

Staphylococcal Protein G, see Protein G, see also β-sheet propensity

Stefin B

Stefin B, low molecular weight cysteine protese inhibitors

pH	T	ΔG	Approach	Ref
8.1	25	24.7	DSC	92Z1

Streptococcal Protein G, see Protein G, see also β-sheet propensity

Subtilisin BPN'

Subtilisin and related proteases

Protease	pH	T	ΔG	Approach	Remarks	Ref
subtilisin DY	7.0	25	(10.0–12.1)±0.8	GuHCl	(1–3)	95G5
subtilisin Carlsberg	7.0	25	(10.0–12.1)±0.8	GuHCl	(1–3)	95G5
subtilisin BPN'	7.0	25	(10.0–12.1)±0.8	GuHCl	(1–3)	95G5
mesentericopeptidase	7.0	25	(10.0–12.1)±0.8	GuHCl	(1–3)	95G5
proteinase K	7.0	25	20.9±1.3	GuHCl	(1–3)	95G5
thermitase	7.0	25	13.4±0.8	GuHCl	(1–3)	95G5

Remarks:
(1) linear extrapolation
(2) phenylmethanesulfonyl (PMS) derivatives were used in these studies
(3) the ΔG values for the four subtilases fall in the range from 10.0 to 12.1 kJ/mol
(4) in the presence of 2×10^{-2} M CaCl$_2$ two transitions are observed; T_{trs} increases by 11–21°C and the stability increases by (28.0–30.1)±1.7 kJ/mol

Subtilisin BPN', wild-type and mutants

Mutant	pH	T_{trs}	ΔT	Δ(ΔG)	Approach	Remark	Ref
wild-type	8	58.9±0.2	0.0	–	DSC		87P1
Thr22→Cys	8	56.4±0.1	–2.5	–	DSC		87P1
Ser87→Cys	8	57.2±0.2	–1.7	–	DSC		87P1
double mutant (Thr22→Cys and Ser87→Cys)							
	8	56.2±0.1	–2.7	0.0	DSC		87P1
double mutant (Thr22→Cys and Ser87→Cys) with disulfide bond Cys22-Cys87							
	8	62±0.1	3.1	5.4	DSC	(1)	87P1

Remark:
(1) Δ(ΔG) is an estimated value which refers to the reduced form

Subtilisin BPN', wild-type and mutants

Mutant	pH	T_{trs}	ΔT	$\Delta(\Delta G)$	Approach	Remarks	Ref
wild-type	8.0	58.5±0.1	0.0	0.0	DSC	(1)	89P4
Met50→Phe	8.0	60.3±0.2	1.8	2.0±0.3	DSC	(1)	89P4
Asn76→Asp	8.0	60.2±0.2	1.7	1.9±0.3	DSC	(1)	89P4
Gly169→Ala	8.0	59.6±0.2	1.1	1.3±0.3	DSC	(1)	89P4
Gln206→Cys	8.0	63.2±0.2	4.7	5.2±0.5	DSC	(1,2)	89P4
Tyr217→Lys	8.0	61.2±0.2	2.7	3.0±0.4	DSC	(1)	89P4
Asn218→Ser	8.0	62.5±0.2	4.0	4.5±0.5	DSC	(1)	89P4

Remarks:
(1) $\Delta(\Delta G)$ was calculated by $\Delta(\Delta G) = \Delta T \times \Delta S_{(w.t.)}$
(2) oxidized form

Subtilisin Inhibitor

Subtilisin Inhibitor

pH	T	ΔG	Approach	Ref
7	25	59*	DSC	81T

Subtilisin inhibitor from *Streptomyces albogriseolus*

pH	T	ΔG	Approach	Remark	Ref
3.21	20	17	heat	(1)	91T2

Remark:
(1) ΔG was taken from Fig. 5 in Ref. 91T2

Streptomyces subtilisin inhibitor, mutants concerning Met73

Mutant	pH	ΔT	$\Delta(\Delta G)$	Approach	Remarks	Ref
wild-type	3.20	0.0	0.0	DSC	(1,2)	95T4
Met73→Ala	3.20	1.66	2.26±0.04	DSC	(1,2)	95T4
Met73→Asp	3.20	3.46	4.52±0.08	DSC	(1,2)	95T4
Met73→Glu	3.20	0.98	1.46±0.04	DSC	(1,2)	95T4
Met73→Gly	3.20	−0.70	−0.88±0.04	DSC	(1,2)	95T4
Met73→Ile	3.20	−2.31	−3.05±0.04	DSC	(1,2)	95T4
Met73→Leu	3.20	−0.43	−0.54±0.04	DSC	(1,2)	95T4
Met73→Lys	3.20	3.71	5.31±0.08	DSC	(1,2)	95T4
Met73→Val	3.20	−1.40	−1.92±0.04	DSC	(1,2)	95T4
wild-type	7.00	0.0	0.0	DSC	(2,3)	95T4
Met73→Ala	7.00	1.16	2.30±0.04	DSC	(2,3)	95T4
Met73→Asp	7.00	3.08	6.69±0.08	DSC	(2,3)	95T4
Met73→Glu	7.00	1.99	4.39±0.04	DSC	(2,3)	95T4
Met73→Gly	7.00	−0.19	−0.25±0.04	DSC	(2,3)	95T4
Met73→Ile	7.00	−1.78	−3.93±0.04	DSC	(2,3)	95T4

Streptomyces subtilisin inhibitor, mutants concerning Met73 (continued)

Mutant	pH	ΔT	Δ(ΔG)	Approach	Remarks	Ref
Met73→Leu	7.00	−0.46	−1.05±0.04	DSC	(2,3)	95T4
Met73→Lys	7.00	0.65	1.46±0.04	DSC	(2,3)	95T4
Met73→Val	7.00	−1.17	−2.76±0.04	DSC	(2,3)	95T4
wild-type	9.50	0.0	0.0	DSC	(2,4)	95T4
Met73→Ala	9.50	0.18	0.54±0.04	DSC	(2,4)	95T4
Met73→Asp	9.50	2.56	5.23±0.08	DSC	(2,4)	95T4
Met73→Glu	9.50	1.06	2.34±0.04	DSC	(2,4)	95T4
Met73→Gly	9.50	−0.87	−2.05±0.04	DSC	(2,4)	95T4
Met73→Ile	9.50	−2.42	−5.36±0.08	DSC	(2,4)	95T4
Met73→Leu	9.50	−1.05	−2.30±0.04	DSC	(2,4)	95T4
Met73→Lys	9.50	0.17	0.21±0.04	DSC	(2,4)	95T4
Met73→Val	9.50	−1.75	−3.97±0.04	DSC	(2,4)	95T4

Remarks:
(1) reference temperature $T_{trs,w.t.}$ = 53.36°C at pH 3.20
(2) for further thermodynamic quantities, see specific values in Table 3
(3) reference temperature $T_{trs,w.t.}$ = 82.21°C at pH 7.00
(4) reference temperature $T_{trs,w.t.}$ = 80.70°C at pH 9.50

Subtilisin inhibitor from *Streptomyces*, effect of intersubunit disulfide bond

Mutant	pH	T	ΔG	Approach	Ref
wild-type	3.0	25	22.73	DSC	94T3
wild-type	3.2	25	23.22	DSC	94T3
wild-type	7.0	25	25.25	DSC	94T3
wild-type	9.5	25	25.29	DSC	94T3
Asp83→Cys	3.0	25	18.68	DSC	94T3
Asp83→Cys	3.2	25	17.89	DSC	94T3
Asp83→Cys	7.0	25	25.36	DSC	94T3
Asp83→Cys	9.5	25	27.29	DSC	94T3
Asp83→Asn	3.0	25	21.62	DSC	94T3
Asp83→Asn	3.2	25	20.36	DSC	94T3
Asp83→Asn	7.0	25	28.92	DSC	94T3
Asp83→Asn	9.5	25	29.39	DSC	94T3

Streptomyces subtilisin inhibitor, mutants concerning Met103

Mutant	pH	ΔT	Δ(ΔG)	Approach	Remarks	Ref
wild-type	3.20	0.0	0.0	DSC	(1,2)	95T3
Met103→Ala	3.20	−4.06	−5.65±0.08	DSC	(1,2)	95T3
Met103→Gly	3.20					
Met103→Ile	3.20	−7.29	−8.42±0.17	DSC	(1,2)	95T3
Met103→Leu	3.20	2.35	3.22±0.13	DSC	(1,2)	95T3
Met103→Val	3.20	−4.34	−5.65±0.08	DSC	(1,2)	95T3
wild-type	7.00	0.0	0.0	DSC	(2,3)	95T3
Met103→Ala	7.00	−4.23	−9.71±0.13	DSC	(2,3)	95T3
Met103→Gly	7.00	−13.57	−28.53±0.33	DSC	(2,3)	95T3
Met103→Ile	7.00	−3.89	−7.90±0.17	DSC	(2,3)	95T3
Met103→Leu	7.00	0.30	0.63±0.04	DSC	(2,3)	95T3
Met103→Val	7.00	−3.56	−7.70±0.08	DSC	(2,3)	95T3
wild-type	9.50	0.0	0.0	DSC	(2,4)	95T3

Streptomyces subtilisin inhibitor, mutants concerning Met103 (continued)

Mutant	pH	ΔT	Δ(ΔG)	Approach	Remarks	Ref
Met103→Ala	9.50	−5.18	−11.72±0.13	DSC	(2,4)	95T3
Met103→Gly	9.50	−14.40	−28.70±0.33	DSC	(2,4)	95T3
Met103→Val	9.50	−4.33	−9.29±0.13	DSC	(2,4)	95T3
Met103→Ile	8.50	−4.77	−9.58±0.13	DSC	(2,4)	95T3
Met103→Leu	9.50	−0.48	−1.21±0.08	DSC	(2,4)	95T3
Met103→Val	9.50	−4.33	−9.29±0.13	DSC	(2,4)	95T3

Remarks:
(1) reference temperature $T_{trs,w.t.}$ = 53.36°C at pH 3.20
(2) for further thermodynamic quantities, see specific values in Table 3
(3) reference temperature $T_{trs,w.t.}$ = 82.21°C at pH 7.00
(4) reference temperature $T_{trs,w.t.}$ = 80.70°C at pH 9.50

Streptomyces subtilisin inhibitor, mutants concerning Val13

Mutant	pH	ΔT	Δ(ΔG)	Approach	Remarks	Ref
wild-type	7.00	0.0	0.0	DSC	(1,2)	95T2
Val13→Ala	7.00	−13.20	−28.36±0.54	DSC	(1,2)	95T2
Val13→Gly	7.00	−27.70	−43.10±0.59	DSC	(1,2)	95T2
Val13→Ile	7.00	−2.39	−3.51±0.17	DSC	(1,2)	95T2
Val13→Leu	7.00	−4.51	−9.25±0.13	DSC	(1,2)	95T2
Val13→Met	7.00	−12.58	−24.77±0.54	DSC	(1,2)	95T2
Val13→Phe	7.00	−13.05	−23.81±0.54	DSC	(1,2)	95T2
wild-type	9.50	0.0	0.0	DSC	(2,3)	95T2
Val13→Ala	9.50	−13.75	−29.00±0.59	DSC	(2,3)	95T2
Val13→Gly	9.50	−25.35	−38.95±0.63	DSC	(2,3)	95T2
Val13→Ile	9.50	−1.83	−3.93±0.13	DSC	(2,3)	95T2
Val13→Leu	9.50	−5.28	−10.54±0.21	DSC	(2,3)	95T2
Val13→Met	9.50	−11.68	−27.72±0.54	DSC	(2,3)	95T2
Val13→Phe	9.50	−12.39	−22.30±0.59	DSC	(2,3)	95T2

Remarks:
(1) reference temperature $T_{trs,w.t.}$ = 82.21°C at pH 7.00
(2) for further thermodynamic quantities, see specific values in Table 3
(3) reference temperature $T_{trs,w.t.}$ = 80.70°C at pH 9.50

Superoxide Dismutase

Superoxide dismutase, wild-type and mutants

Mutant	T	ΔT	Δ(ΔG)	Approach	Remarks	Ref
wild-type, trans. (1)	7.8	74.9	0.0	DSC	(1)	90L1
wild-type, trans. (2)	7.8	83.6	0.0	DSC	(1)	90L1
Cys111→Ser, trans. (1)	7.8	77.7	3.3	DSC	(1,2)	90L1
Cys111→Ser, trans. (2)	7.8	84.5	1.7	DSC	(1,2)	90L1
Cys6→Ala, trans. (1)	7.8	75.7	0.4	DSC	(1,3)	90L1
Cys6→Ala, trans. (2)	7.8	84.1	1.3	DSC	(1,3)	90L1
double mutant (Cys6→Ala and Cys111→Ser)						
trans. (1)	7.8	77.3	2.1	DSC	(1)	90L1
trans. (2)	7.8	82.8	−0.8	DSC	(1)	90L1

Remarks:
(1) two subsequent transitions resolved by deconvolution
(2) Cys6 unchanged
(3) Cys111 unchanged

Fe- superoxide dismutase from *Mycobacterium tuberculosis*, wild-type and mutant

Mutant	pH	T	ΔG	Approach	Remarks	Ref
wild-type	7.9	23	25.1±1.7	GuHCl	(1,2)	96C18
Gly152→Ala	7.9	23	7.1±1.3	GuHCl	(1,2)	96C18

Remarks:
(1) linear extrapolation
(2) buffer: 200 mM Tris-HCl

Tailspike Protein

Tailspike protein of P22 bacteriophage, wild-type and mutants difference values referring to the trimer of the protein

Mutant	pH	T	ΔT	Δ(ΔG)	Approach	Ref
wild-type	7.4	88.4±0.3	0.0	0.0	DSC	89S11
Gly177→Arg	7.4	87.9±0.3	−0.5	−8.8	DSC	89S11
Thr235→Ile	7.4	88.0±0.3	−0.4	−6.7	DSC	89S11
Gly244→Arg	7.4	87.5±0.3	−0.9	−16.7	DSC	89S11
Arg382→Ser	7.4	83.2±0.3	−5.2	−72.8	DSC	89S11
Arg285→Lys	7.4	85.9±0.3	−2.5	−43.1	DSC	89S11
Glu309→Val	7.4	87.8±0.3	−0.6	−9.6	DSC	89S11
Gly323-Asp	7.4	88.5±0.3	0.1	−1.7	DSC	89S11

Tendamistat

Tendamistat, α-amylase inhibitor, wild-type and disulfide mutants

Mutant	pH	T	ΔG	Approach	Remarks	Ref
wild-type	5	4.9	28.5	DSC	(1,2)	92R2
wild-type	7.0	25	29.9	DSC		93H7
wild-type	7.0	55	19.2	DSC		95V8
Double mutants:						
(Cys11→Ala and Cys27→Ala)						
	7.0	55	−5.8	DSC		95V8
(Cys11→Ala and Cys27→Leu)						
	7.0	55	−3.3	DSC		95V8
(Cys11→Ala and Cys27→Ser)						
	7.0	55	−2.7	DSC		95V8
(Cys11→Ala and Cys27→Thr)						
	7.0	55	−1.7	DSC		95V8
(Cys45→Ala and Cys73→Ala)						
	7.0	25	16.8	DSC		93H7
	7.0	55	2.3	DSC		95V8

Remarks:
(1) ΔG was determined using temperature dependent ΔCp
(2) the protein achieves maximum stability at T = 4.9°C

Thermolysin

Thermolysin, apo form

pH	T	ΔG	Approach	Remarks	Ref
7	25	14.2	GuHCl	(1,2)	86C2

Remarks:
(1) linear extrapolation
(2) ΔG was determined from the second of two subsequent transitions at 3 M GuHCl, measurement in the presence of 1 mM EDTA

Thermolysin, fragments

Fragment	pH	T	ΔG	Approach	Remarks	Ref
121-316	7.8	25	28	GuHCl	(1)	85V
121-316	7.8	25	46.8	GuHCl	(2a)	87B1
206-316	7.5	37	13	heat		82V
206-316	7.5	25	23	GuHCl	(1)	82V
206-316	7.5	25	31	GuHCl	(2)	82V
206-316	7.5	25	23	urea	(1)	82V
206-316	7.5	25	27	urea	(2)	82V
206-316	7.8	25	23.4	GuHCl	(1)	85V
206-316	7.8	25	31.3	GuHCl	(2b)	85V
225-316	7.8	25	17.6	GuHCl	(1)	85V
225-316	7.8	25	25.9	GuHCl	(2c)	85V

Remarks:
(1)   linear extrapolation
(2)   transfer model
(2a) transfer model, $\varepsilon = 0.36$
(2b) transfer model, $\varepsilon = 0.28$
(2c) transfer model, $\varepsilon = 0.22$

Thermolysin, fragment 255-316 (dimer)

pH	T	ΔG	Approach	Remarks	Ref
7.5	20	60±6	DSC	(1–3)	94C9

Remarks:
(1) derived from concentration dependent DSC measurements
(2) unfolding equilibrium according to: folded dimer→2 unfolded monomers
(3) ΔG is calculated per dimer with a ΔG of dimerization of –24±2 kJ/mol

Thioredoxin

Thioredoxin from *E. coli*, different forms and mutants

Form/Mutant	pH	T	ΔG	Approach	Remarks	Ref
	6	25	21	GuHCl	(1a)	90B5
	6	25	13	GuHCl	(1b)	90B5
	6	25	27	GuHCl	(1c)	90B5
	7	25	30.5	GuHCl	(1)	84K1
oxidized	7	25	36	GuHCl	(1,2)	87K1
oxidized	7	25	36.4	urea	(1)	87K1
oxidized	7.0	25	39.3	DSC		91B3
reduced	7.0	25	25.9	DSC		91B3
reduced	7	25	25.9	GuHCl	(1)	87K1
red., alkyl.	7	25	19.2	GuHCl	(1,3)	87K1
red., alkyl.	7	25	18	urea	(1)	87K1
reduced	8.7	23	20.9	GuHCl	(1)	89L3
oxidized	8.7	23	33.9	GuHCl	(1)	89L3
oxidized, with cis-Pro76						
	8.7	2	39.7	urea	(1,5a)	89L1
	8.7	25	39.7	urea	(1,5a)	89L1
	8.7	25	41.8	urea	(1,6)	89L1
oxidized, with trans-Pro76						
	8.7	2	27.2	urea	(1,5b)	89L1
reduced, with cis-Pro76						
	8.7	2	25.1	urea	(1,5a)	89L1
	8.7	25	27.2	urea	(1,5a)	89L1
	8.7	25	23	urea	(1,5b)	89L1
	8.7	25	27.2	urea	(1,6)	89L1
reduced, with trans-Pro76						
	8.7	2	18.8	urea	(1,5a)	89L1
	8.7	25	18.8	urea	(1,5a)	89L1
	8.7	2	16.7	urea	(1,5b)	89L1
	8.7	25	16.7	urea	(1,5b)	89L1
mutant Cys32→Ser, with cis-Pro76						
	8.7	2	18.8	urea	(1,5a)	89L1
	8.7	25	25.1	urea	(1,5a)	89L1
	8.7	25	27.2	urea	(1,5b)	89L1
mutant Cys32→Ser, with trans-Pro76						
	8.7	2	10.5	urea	(1,5a)	89L1
	8.7	25	16.7	urea	(1,5a)	89L1
	8.7	2	16.7	urea	(1,5b)	89L1
	8.7	25	16.7	urea	(1,5b)	89L1
double mutant (Cys32→Ser and Cys35→Ser) with cis-Pro76						
	8.7	2	33.5	urea	(1,5a)	89L1
	8.7	25	29.3	urea	(1,5a)	89L1
	8.7	25	29.3	urea	(1,5b)	89L1
double mutant (Cys32→Ser and Cys35→Ser) with trans-Pro76						
	8.7	2	20.9	urea	(1,5a)	89L1
	8.7	2	20.9	urea	(1,5b)	89L1
mutant with an additional Arg between residues 33 and 34, with cis-Pro76, oxidized						
	8.7	25	29.3	urea	(1,5a)	89L1
	8.7	25	27.2	urea	(1,5b)	89L1

Thioredoxin from *E. coli*, different forms and mutants (continued)

Form/Mutant	pH	T	ΔG	Approach	Remarks	Ref
mutant with an additional Arg between residues 33 and 34, with trans-Pro76, oxidized						
	8.7	2	12.6	urea	(1,5b)	89L1
mutant Pro34→Ser						
	7	25	36.8	GuHCl	(1)	87K2
mutant Pro76→Ala						
	7	25	20.9	GuHCl	(1,4)	87K2
mutant Pro76→Ala, oxidized						
	8.7	2	16.7	urea	(1,5a)	89L1
	8.7	25	33.5	urea	(1,5a)	89L1
	8.7	2	18.8	urea	(1,5b)	89L1
	8.7	25	25.1	urea	(1,5b)	89L1

Remarks:
(1)   linear extrapolation
(1a) monitored by fluorescence
(1b) monitored by CD in the far UV
(1b) monitored by CD in the near UV
(2)   thermal transition at 82°C (DSC)
(3)   thermal transition at 64°C (DSC)
(4)   thermal transition at 65°C, after reduction 50°C (DSC)
(5a) urea gradient electrophoresis of the N state
(5b) urea gradient electrophoresis of the D state
(6)   monitored by CD

Thioredoxin from *E. coli*, various forms

Mutant	pH	T	ΔT	Δ(ΔG)	Approach	Remarks	Ref
oxidized	7.0	85.32	0.0	0.0	DSC	(1)	94L1
reduced	7.0	85.32	−12.1±0.1	−12.6±1.7	DSC	(1)	94L1
double mutant (Cys32→Ser and Cys35→Ser)							
	7.0	85.32	−10.5±0.2	−13.0±1.3	DSC	(1)	94L1

Remark:
(1) the data refer to oxidized thioredoxin at pH 7.0, protein conc. 300 µM

Thioredoxin from *Rhodobacter sphaeroides*, wild-type and Phe25 mutants

Mutant	pH	T	ΔG	Approach	Remarks	Ref
wild-type	7.4	23	34	urea	(1,2)	95A8
Phe25→Ile	7.4	23	18	urea	(1,2)	95A8
Phe25→Leu	7.4	23	20	urea	(1,2)	95A8
Phe25→Val	7.4	23	14	urea	(1,2)	95A8
Phe25→Tyr	7.4	23	33	urea	(1,2)	95A8

Remarks:
(1) linear extrapolation
(2) ΔG was taken from Fig. 2B in Ref. 95A8

Thioredoxin from *E. coli*, wild-type and Asp26 mutants

Mutant	pH	T	ΔG	Approach	Remarks	Ref
wild-type	6.0	25	44.4	GuHCl	(1)	91L1
wild-type	6.5	25	41.8	GuHCl	(1)	91L1
wild-type	7.0	25	39.7	GuHCl	(1)	91L1
wild-type	7.5	25	37.7	GuHCl	(1)	91L1
wild-type	8.0	25	36.8	GuHCl	(1)	91L1
wild-type	8.5	25	36	GuHCl	(1)	91L1
wild-type	9.0	25	36	GuHCl	(1)	91L1
Asp26→Ala	7.0	25	55.5	GuHCl	(1)	91L1
Asp26→Ala	7.5	25	55.5	GuHCl	(1)	91L1
Asp26→Ala	8.0	25	55.5	GuHCl	(1)	91L1
Asp26→Ala	8.5	25	55.5	GuHCl	(1)	91L1

Remark:
(1) linear extrapolation

Thioredoxin, wild-type and mutants, oxidized and reduced form

Mutant/Form	pH	T	ΔG	Approach	Remarks	Ref
Wild-type:						
reduced	8.7	23	25.5±1.3	urea	(1)	91L7
oxidized	8.7	23	40.6±1.3	urea	(1)	91L7
Mutant Pro34→Ser:						
reduced	8.7	23	35.6±2.5	urea	(1)	91L7
oxidized	8.7	23	39.3±2.5	urea	(1)	91L7

Remark:
(1) linear extrapolation

Thioredoxin from *E. coli*, wild-type and mutants

Mutant	pH	T_{trs}	ΔT	Δ(ΔG)	Approach	Ref
wild-type	7.0	85.32	0.0	0.0	DSC	95L2
Leu78→Lys	7.0		−13.7	−16.3	DSC	95L2
Leu78→Arg	7.0		−16.1	−16.7	DSC	95L2

Thioredoxin from *E. coli*, oxidized form, in the presence of NaCl

NaCl Concentration	pH	T	ΔG	Approach	Remarks	Ref
0.1 M	7.0	25	37.2±1.3	DSC		92S2
0.5 M	7.0	25	35.1±1.3	DSC		92S2
1.0 M	7.0	25	31.4±1.3	DSC		92S2
1.5 M	7.0	25	32.2±1.3	DSC		92S2
	7.0	25	32.6±0.8	GuHCl	(1,2)	92S2
0.5 M	7.0	25	36.0±3.8	urea	(1,3)	92S2
	7.0	25	33.9±0.4	heat/GuHCl	(4)	92S2

Remarks:
(1) linear extrapolation to zero denaturant concentration by a method that includes the pre- and post-denaturational baselines for a non-linear regression of the data fit of the entire data set
(2) measurement in the presence of 50 mM MOPS buffer
(3) in the presence of 50 mM MOPS buffer with 0.5 M NaCl
(4) thermal/GuHCl combined approach, see Ref. 92S2

Thioredoxin from *E. coli*, mutants, modified triple mutant (Cys32→Ser, Cys35→Ser and Leu78→Cys) modified at position Cys78

Form	pH	T	ΔG	Approach	Remarks	Ref
Cys78 SH	7.0	25	24.4	GuHCl	(1)	93W7
methylated	7.0	25	17.9	GuHCl	(1)	93W7
ethylated	7.0	25	19.4	GuHCl	(1)	93W7
n-propylated	7.0	25	22.2	GuHCl	(1)	93W7
n-butylated	7.0	25	22.3	GuHCl	(1)	93W7
n-pentylated	7.0	25	22.3	GuHCl	(1)	93W7

Remark:
(1) linear extrapolation

Thioredoxin mutant (Cys32→Ser, Cys35→Ser and Leu78→Cys)

Form	pH	T	ΔG	Approach	Remarks	Ref
reduced	8.0	25	21.5	urea	(1)	93W8
modified	8.0	25	24.9	urea	(1,2)	93W8

Remarks:
(1) linear extrapolation
(2) mixed disulfide obtained by reaction with aliphatic thiol

Thymidilate Synthase

Thymidilate synthase

pH	T	ΔG	Approach	Remarks	Ref
7.0	4	19.1	urea	(1,2)	92P2

Remarks:
(1) linear extrapolation
(2) the equilibrium was treated as a two state transition between native dimer and unfolded monomer ($N_2 \rightarrow 2U$)

Thyroid Transcription Factor

Thyroid transcriptor factor 1 (TTF-1) homeodomain

pH	T	ΔG	Approach	Remarks	Ref
7.5	25	6.0	heat, v.H.	(1)	94D2
7.5	25	5.3	urea	(2)	94D2

Remarks:
(1) buffer: 10 mM phosphate, 100 mM NaF, pH 7.5
(2) denaturant binding model

Tissue Factor

Human tissue factor (tTF), recombinant, secreted from yeast, and refolded from *E. coli*

Mutant	pH	T	ΔG	Approach	Remarks	Ref
tTF *E. coli*	7.5	19	21.4±4.6	GuHCl	(1–3)	95S16
tTF double mutant (Thr13→Ala and Asn137→Asp) from yeast						
	7.5	19	21.4±5.0	GuHCl	(1–3)	95S16

Remarks:
(1) linear extrapolation to zero denaturant concentration by a method that includes the pre- and post-denaturational baselines for a non-linear regression of the data
(2) the transition was monitored by fluorescence intensity
(3) buffer: 20 mM phosphate, 0.5 M NaCl, pH 7.5

Titin

Titin from chicken breast muscle

pH	T	ΔG	Approach	Remarks	Ref
7.2	25	14.5	GuHCl	(1–3)	94K3
7.2	25	13.6	GuHCl	(1,2,4)	94K3
7.2	25	12.6	GuHCl	(1,2,5)	94K3

Remarks:
(1) linear extrapolation
(2) buffer: 50 mM sodium phosphate, 0.6 M NaCl, 1 mM DTT, 1 mM EDTA, pH 7.2
(3) transition monitored by fluorescence yield
(4) transition monitored by fluorescence spectral position
(5) transition monitored by fluorescence anisotropy

Toxins

Toxin from different origin

Species	pH	T	ΔG	Approach	Ref
cytotoxin I	4.5	25	24*	DSC	84K2
neurotoxin I	5.7	25	24*	DSC	84K2
neurotoxin II	5	25	36*	DSC	84K2

Transcription Factor

Transcription factor LFB1, dimerization domain (B1-DIM)

pH	T	ΔG	Approach	Remarks	Ref
7.5	0	50.2	GuHCl	(1–4)	91D6
7.5	25	48	heat	(1,2,4)	91D6

Remarks:
(1) the data refer to the the unfolding of the dimer, i.e., $N_2 \rightarrow 2U$
(2) the measurements were made at different protein concentrations
(3) linear extrapolation
(4) measurement in the presence of fluoride

Transferrin

Transferrin N-terminal half-molecule, apo-form

Mutant	pH	T	ΔT	Δ(ΔG)	Approach	Remarks	Ref
wild-type	7.5	25	0.0	0.0	DSC	(1)	93L7
Asp63→Cys	7.5	25	0.4	0.8	DSC	(1)	93L7
Asp63→Ser	7.5	25	−0.8	−1.6	DSC	(1)	93L7
Gly65→Arg	7.5	25	−1.1	−2.4	DSC	(1)	93L7
Lys206→Gln	7.5	25	0.1	0.2	DSC	(1)	93L7
His207→Glu	7.5	25	−0.8	−1.8	DSC	(1)	93L7
N-lobe	7.5	25	1.95	5.4	DSC	(1)	93L7

Remark:
(1) estimated uncertainties: Δ(ΔG) ±0.1 kJ/mol

Triosephosphate Isomerase

Human triosephosphate isomerase (hTIM)

Mutant	pH	T	ΔG	Approach	Remarks	Ref
hTIM	8	25	80.8±1.7	urea	(1,2)	96M1
Gln179→Asp	8	25	87.0±1.7	urea	(1,2)	96M1
Ser105→Asp	8	25	75.3±1.3	urea	(1,2)	96M1
Gln179→Ala	8	25	81.2±0.8	urea	(1,2)	96M1
Lys193→Ala	8	25	84.9±1.7	urea	(1,2)	96M1
Ala215→Pro	8	25	83.3±0.8	urea	(1,2)	96M1
triple mutant (Gln179→Asp, Lys193→Ala and Ala215→Pro)						
	8	25	93.3±1.3	urea	(1,2)	96M1
monomeric double mutant (Met14→Gln and Arg98→Gln)						
	8	25	10.5±0.4	urea	(1,3)	96M1
monomeric multiple mutant(Met14→Gln, Arg98→Gln, Gln179→Asp, Lys193→Ala and Ala215→Pro)						
	8	25	16.3±0.4	urea	(1,3)	96M1

Remarks:
(1) linear extrapolation
(2) ΔG refers to a two-state model based on the transition of the folded dimer to two unfolded monomers, i.e.,
    concerted dissociation/unfolding
(3) the replacements of two interface residues (Met14→Gln and Arg98→Gln) prevents the association

Tropomyosin

Tropomyosin, homodimers and heterodimer

Dimer	pH	T	ΔG	Approach	Remarks	Ref
αα	7.0	39	37.7	heat	(1)	91L4
αβ	7.0	39	51.9	heat	(1)	91L4
ββ	7.0	39	40.2	heat	(1)	91L4

Remark:
(1) in the presence of 0.5 M NaCl, 10 mM sodium phosphate, 1 mM EDTA, and 0.5 mM DTT

α-Tropomyosin mutants

Mutant	pH	T	ΔG	Approach	Remarks	Ref
2a6b9d	7.5	20	17.2±0.4	heat	(1–3)	95G9
2b6b9a	7.5	20	26.4±0.8	heat	(1–3)	95G9
2b6b9d	7.5	20	27.6	heat	(1–3)	95G9
2a6a9d	7.5	20	15.1±5.0	heat	(1–3)	95G9
2a6a9a	7.5	20	20.1	heat	(1–3)	95G9
2a6zip9d	7.5	20	16.3±3.3	heat	(1–3)	95G9
2a6zip9a	7.5	20	21.3±0.4	heat	(1–3)	95G9
2b6b9da	7.5	20	23.5±2.1	heat	(1–3)	95G9
2b6b9ad	7.5	20	31.8	heat	(1–3)	95G9
C2b6b9a	7.5	20	33.5±2.1	heat	(1–3)	95G9
C2zip6b9a	7.5	20	31.0±0.8	heat	(1–3)	95G9
C2rc6b9a	7.5	20	24.3±0.4	heat	(1–3)	95G9

Remarks:
(1) for the detailed mutant description, see Ref. 95G9
(2) buffer: 10 mM sodium phosphate, 500 mM NaCl 1 mM EDTA, 1mM DTT, pH 7.5
(3) ΔG was calculated using a model that accounts for up to three independent helix-coil transitions with dissociation accompanying the helix-coil transition at the highest temperature

Troponin C

Troponin C domain synthetic peptides corresponding to calcium-binding sites III (residues 93-126; SCIII) and IV (residues 129-162; SCIV) of troponin C

Dimer	pH	T	ΔG	Approach	Remarks	Ref
SCIV	7.2	25	51.8	GuHCl	(1,2)	94S4
SCIII	7.2	25	61.5	GuHCl	(1,2)	94S4
SCIII/SCIV	7.2	25	64.8	GuHCl	(1,2)	94S4

Remarks:
(1) linear extrapolation
(2) measured in the presence of 20 mM Ca^{2+}

Trypsin

Trypsin

	pH	T	ΔG	Approach	Remark	Ref
	–	25	67	DSC	(1)	79P4
	4.2	25	54.4±2.5	heat, v.H.		95B8

Remark:
(1) ΔG was taken from Fig. 28 of Ref. 79P4.

Table 1. Gibbs Energy Change – Molar Values

Trypsinogen

Trypsinogen, free and calcium bound trypsinogen

Form	pH	T	ΔG	Approach	Remarks	Ref
free	5.8	25	44.8	GuHCl	(1)	94B9
free	5.8	25	45.8±0.8	GuHCl	(3,4)	96O4
Ca^{2+}	5.8	25	52.3	GuHCl	(1,2)	94B9
Ca^{2+}	5.8	25	51.7±0.9	GuHCl	(3,5)	96O4

Remarks:
(1) linear extrapolation
(2) in the presence of 20 mM Ca^{2+}
(3) linear extrapolation to zero denaturant concentration by a method that includes the pre- and post-denaturational baselines for a non-linear regression of the data
(4) measured in 0.1 M MES in the presence of 2 mM EDTA
(5) measured in 0.1 M MES in the presence of 20 mM $CaCl_2$

Bovine trypsinogen

pH	T	ΔG	Approach	Remarks	Ref
4.2	25	43.5±2.1	heat, v.H.		95B8
7.5	25	44.4±0.8	GuHCl	(1)	95B8
8.3	25	38.9±0.6	GuHCl	(1)	95B8
8.3	25	53.1±0.6	GuHCl	(1,2)	95B8

Remarks:
(1) linear extrapolation
(2) tertiary complex trypsinogen-Ca-Ile-Val

Trypsin Inhibitor

Trypsin inhibitor, bovine pancreatic (BPTI)

pH	T	ΔG	Approach	Ref
2	25	49.3	DSC	87S1
2.0	25	35	DSC	93M4
3.0	25	40	DSC	93M4
3	25	51.7	DSC	83M2
4	25	46*	DSC	79P4
4.9	25	45	DSC	93M4
5	25	59.7	DSC	87S1
7	45	48	hydrogen exchange	79R

Trypsin inhibitor, bovine pancreatic (BPTI), modified

Modification	pH	T	ΔG	Approach	Remarks	Ref
carboxamidomethylation at Cys14 and Cys38						
	2	25	12.6	DSC		87S1
	5	25	23.5	DSC		87S1
	5.3	77	−0.5	heat, NMR	(1,2)	79W
carboxymethylation at Cys14 and Cys38						
	2	25	16.6	DSC		87S1
	5	25	26.2	DSC		87S1
	5.3	77	1.2	heat, NMR	(1,3)	79W
aminoethylation at Cys14 and Cys38						
	5.3	77	−3.0	heat, NMR	(1,4)	79W

Remark:
(1) measurement in heavy water (pD instead of pH)
(2) transition temperature T_{trs} = 76°C
(3) transition temperature T_{trs} = 79°C
(4) transition temperature T_{trs} = 70°C

Trypsin inhibitor, bovine pancreatic (BPTI), wild-type and mutants

Mutant	pH	T	ΔG	Approach	Remarks	Ref
wild-type			29.3		(1,2)	89G3
Tyr23→Leu			0.8		(1,2)	89G3
Tyr35→Gly			7.5		(1,2)	89G3
Asn43→Ala			4.2		(1,2)	89G3
Asn43→Gly			14.6		(1,2)	89G3
double mutants:						
(Cys30→Val and Cys51→Ala)						
	7.8	15	32.6±2.9	GuHCl	(3)	90H2
(Cys30→Ala and Cys51→Ala)						
	7.8	15	25.1±2.9	GuHCl	(3)	90H2
(Cys30→Thr and Cys51→Ala)						
	7.8	15	19.2±0.8	GuHCl	(3)	90H2
(Cys14→Ala and Cys38→Ala)						
	7.8	15	28.9±1.3	GuHCl	(3)	90H2

Remarks:
(1) measured by disulfide reduction equilibrium
(2) the data were taken from Fig. 5a of Ref. 89G3.
(3) linear extrapolation

Bovine pancreatic trypsin inhibitor (BPTI), wild-type and mutant

Mutant	pH	T	ΔG	Approach	Ref
wild-type	1.0	25	42.2	DSC	93H7
wild-type	3.0	25	50.1	DSC	93H7
double mutant (Cys14→Ala and Cys38→Ala)					
	3.0	25	19.5	DSC	93H7

Trypsin inhibitor, pancreatic, wild-type and mutants, $\Delta(\Delta G)$ determined by disulfide reduction

Mutant	pH	T	$\Delta(\Delta G)$	Remarks	Ref
wild-type	8.7	25	0.0	(1)	93G6
Gly12→Asp	8.7	25	−18.0	(1)	93G6
Gly12→Val	8.7	25	−17.2	(1)	93G6
Pro13→Ala	8.7	25	−6.7	(1)	93G6
Pro13→Ser	8.7	25	−5.9	(1)	93G6
Ala16→Thr	8.7	25	−7.1	(1)	93G6
Ala16→Val	8.7	25	−5.4	(1)	93G6
Phe33→Ile	8.7	25	−10.0	(1)	93G6
Phe33→Leu	8.7	25	−5.4	(1)	93G6
Tyr35→Ala	8.7	25	−19.7	(1)	93G6
Tyr35→Asn	8.7	25	−15.5	(1)	93G6
Tyr35→Asp	8.7	25	−15.9	(1)	93G6
Tyr35→Gly	8.7	25	−20.9	(1)	93G6
Tyr35→Leu	8.7	25	−15.5	(1)	93G6
Tyr35→Phe	8.7	25	−2.5	(1)	93G6
Gly36→Asp	8.7	25	−11.7	(1)	93G6
Gly37→Ala	8.7	25	−9.2	(1)	93G6
Gly37→Asp	8.7	25	−7.1	(1)	93G6

Remark:
(1) $\Delta(\Delta G)$ was determined by $\Delta(\Delta G) = - RT \ln(C_{w.t.}/C_{mut.})_{eff.}$ from equilibration with glutathione and DTE

Bovine pancreatic trypsin inhibitor, wild-type and mutants

Mutant	pH	T	ΔG	$\Delta(\Delta G)$	Approach	Remarks	Ref
wild-type	2.0	25	37.7	0.0	DSC	(1)	93K6
Phe22→Ala	2.0	25	32.6	−5.1	DSC	(1)	93K6
Tyr23→Ala	2.0	25	13.0	−24.7	DSC	(1)	93K6
Tyr35→Gly	2.0	25	16.7	−20.9	DSC	(1)	93K6
Asn43→Gly	2.0	25	19.2	−23.8	DSC	(1)	93K6
Phe45→Ala	2.0	25	8.8	−28.9	DSC	(1)	93K6
Phe22→Ala	3.5	30		−8.8	H/D exchange	(1)	93T2
Tyr23→Ala	3.5	30		−29.3	H/D exchange	(1)	93T2
Tyr35→Gly	3.5	30		−23.8	H/D exchange	(1)	93T2
Asn43→Gly	3.5	30		−25.1	H/D exchange	(1)	93T2
Phe45→Ala	3.5	30		−30.1	H/D exchange	(1)	93T2

Remark:
(1) the uncertainties in ΔG derived from DSC = $<\pm3$ kJ/mol

Trypsin inhibitor, bovine pancreatic, wild-type and mutants, comparison of data obained dy DSC and hydrogen exchange (H-Δ(ΔG))

Mutant	pH	T	ΔG	Δ(ΔG)	H-Δ(ΔG)	Remarks	Ref
wild-type	2.0	25	37.7	0.0		(1)	93K5
Asn43→Gly	2.0	25	13.8	−23.8	−25.1	(1)	93K5
Asn44→Gly	2.0	25	18.0	−19.7	−19.7	(1)	93K5
Gly37→Ala	2.0	25	17.2	−20.5	−19.7	(1)	93K5
Phe22→Ala	2.0	25		−5.0	−7.1±1.7	(1)	93T3
Phe45→Ala	2.0	25		−28.9	−28.5±1.3	(1)	93T3
Tyr23→Ala	2.0	25		−24.7	−27.6±1.3	(1)	93T3
Tyr35→Gly	2.0	25	16.7	−20.9	−23.8	(1)	93K5
wild-type	3.5	30	27.6			(2)	93K7

Explanation: H-Δ(ΔG) = Δ(ΔG) determined by hydrogen exchange

Remarks:
(1) approaches: ΔG and Δ(ΔG) were determined by DSC at pH 2.0 H-Δ(ΔG) was determined by hydrogen exchange at pH 3.5 and 30 C
(2) ΔG corresponds to a Δ(ΔG) value for transfer of the protein from 0→8 M urea

Basic pancreatic trypsin inhibitor, intact and cleaved forms

Form	pH	T	ΔG	Approach	Remarks	Ref
intact		25	51	DSC	(1)	96K9
cleaved Lys15,Ala	5.0	25	43.0±2.1	heat	(2,3)	96K9
cleaved Met52,Arg	5.0	25	9.3±2.1	heat	(2,4)	96K9

Remarks:
(1) reference value from 93H7
(2) measured in 50 mM acetate buffer pH 5.0
(3) cleaved between Lys15-Ala16 by hydrolysis
(4) cleaved between Met52-Arg53 by CNBr

Soybean trypsin inhibitor, intact and cleaved forms

Form	pH	T	ΔG	Remarks	Ref
intact	6.4	25	39.1±2.1	(1,2)	96K9
cleaved Arg63, Ile	6.4	25	27.1±2.1	(1-3)	96K9
cleaved Met84, Leu	6.4	25	45.2±2.1	(1,2,4)	96K9

Remarks:
(1) data from thermal denaturation and GuHCl-induced unfolding
(2) measured in 50 mM MES, 0.3 M KCl, pH 6.4
(3) cleaved between Arg63-Ile64 by hydrolysis
(4) cleaved between Met84-Leu85 by CNBr

Tryptophan Repressor

Tryptophan repressor from *E. coli*

pH	T	ΔG	Approach	Remark	Ref
7.5	25	31*	DSC	(1)	88B1

Remark:
(1) the data refer to the transition $A_2 \rightarrow 2B$ of the dimer using $M = 24,700$ Da

Tryptophan aporepressor from *E. coli*

pH	T	ΔG	Approach	Remarks	Ref
7.5	25	97.5±3.8	urea	(1–3)	90G2

Remarks:
(1) linear extrapolation
(2) ΔG is the mean value of independent measurements monitored by difference absorbance, fluorescence, and CD
(3) ΔG is based on a two-state model $D \rightarrow 2U$ (dimer→2 unfolded monomers)

Tryptophan aporepressor from *E. coli*

Form	pH	T	ΔG	Approach	Remarks	Ref
native	7.6	25	93.7±2.9	urea	(1,2)	93M6
intermediate	7.6	25	15.1±1.3	urea	(1.3)	93M6
intermediate	7.6	25	12.6±0.4	urea	(1,4)	93M6

Remarks:
(1) linear extrapolation
(2) based on equilibrium unfolding of the native dimeric protein monitored by CD at 222 nm
(3) based on burst phase kinetics, monitored by CD at 222 nm
(4) based on ANS binding, monitored by fluorescence

Tryptophan aporepressor

Transition	pH	T	ΔG	Approach	Remarks	Ref
	2.5–3	25	8	urea	(1,2)	94E2
trans. (1)	6	25	40	urea	(1,3,4)	94E2
trans. (2)	6	25	13	urea	(1,3,5)	94E2
	7	25	88±13	urea	(1,6)	94E2

Remarks:
(1) linear extrapolation
(2) between pH 2.5 and pH 3 a two-state transition $N \rightarrow U$ (nativ→unfolded) is observed
(3) between pH 3.5 and pH 6 the unfolding follows a $D \rightarrow 2N \rightarrow 2U$ equilibrium (dimer→2 folded monomers→2 unfolded monomers)
(4) ΔG refers to the $D \rightarrow 2N$ transition
(5) ΔG refers to the $N \rightarrow U$ transition
(6) above pH 6 the unfolding can be described by a $D \rightarrow 2U$ (dimer→2 monomers) equilibrium

Tryptophan repressor from *E. coli*, modified

Form	pH	T	ΔG	Approach	Remarks	Ref
native	7.6	80.3±4.6	urea	(1,2)	92F2	
DNS-labeled	7.6	76.6±6.3	urea	(1–3)	92F2	
DNS + Trp	7.6	100.8±10.9	urea	(1–4)	92F2	

Remarks:
(1) linear extrapolation, analysis in terms of a dimer to unfolded monomer model
(2) based on fluorescence measurements
(3) DNS = 5-(dimethylamino)naphthalene-1-sulfonyl chloride
(4) in the presence of 0.4 mM L-tryptophan

Tryptophan repressor, wild-type and superrepressor mutant

Mutant	pH	T	ΔG	Approach	Remarks	Ref
wild-type	7.6		78.7(+4.2,–1.7)	urea	(1,2)	95R3
Ala77→Val	7.6		87.4(+5.0,–3.8)	urea	(1,2)	95R3

Remarks:
(1) linear extrapolation
(2) ΔG refers to a folded dimer to unfolded monomer transition

Tryptophan Synthase

The data are arranged as follows:
a) wild-type and mutants, previous data,
b) various transitions,
c) comparison of tryptophan synthase α subunits from various species and hybrid proteins.

a) Wild-type and mutants, previous data

Tryptophan synthase from *E. coli*, α-subunit, wild-type and mutants, ΔG values

Mutant	pH	T	ΔG	Approach	Remarks	Ref
wild-type	5.5	25	21.8	GuHCl		82Y1
wild-type	6.5	25	52.7	urea	(2c,4)	81M2
wild-type	6.5	25	47.7	urea	(2a,4)	81M2
wild-type	7	25	27.6	heat		84Y2
wild-type	7	25.8	36.8	GuHCl	(2a,3)	79Y
wild-type	7	25	42.7	GuHCl	(2b,3)	79Y
wild-type	7	25	36.8	GuHCl		82Y1
wild-type	7	25	35.1	DSC		82Y2
wild-type	7	25	28	heat		84O
wild-type	7	25	36.8±0.4	GuHCl	(4)	87Y
wild-type	7.2	25	56.9	urea	(2c,4)	81M2
wild-type	7.2	25	50.2	urea	(2a,4)	81M2
wild-type	7.8	25	49	urea	(2c,4)	81M2
wild-type	7.8	25	43.1	urea	(2a,4)	81M2
wild-type	7.8	25	24.1*	DSC		80M2
wild-type	9	25	20.5	GuHCl		82Y1
wild-type	9	25	20.5±1.3	GuHCl	(4)	87Y

Tryptophan synthase from *E. coli*, α-subunit, wild-type and mutants, ΔG values (continued)

Mutant	pH	T	ΔG	Approach	Remarks	Ref
wild-type	9.3	25	24.3	DSC		82Y2
wild-type	9.3	25	16.3	heat		84O
Glu49→Ala	7	25	35.6±0.8	GuHCl	(4)	87Y
Glu49→Ala	9	25	28.5±0.8	GuHCl	(4)	87Y
Glu49→Asn	7	25	34.3±0.8	GuHCl	(4)	87Y
Glu49→Asn	9	25	25.9±0.4	GuHCl	(4)	87Y
Glu49→Asp	7	25	35.6±1.7	GuHCl	(4)	87Y
Glu49→Asp	9	25	29.3±0.4	GuHCl	(4)	87Y
Glu49→Cys	7	25	46±0.4	GuHCl	(4)	87Y
Glu49→Cys	9	25	34.7±1.3	GuHCl	(4)	87Y
Glu49→Gln	5.5	25	11.7	GuHCl		82Y1
Glu49→Gln	7	25.8	26.4	GuHCl	(2a,3)	79Y
Glu49→Gln	7	25	26.4	GuHCl		82Y1
Glu49→Gln	7	25	31	DSC		82Y2
Glu49→Gln	7	25	26.4±1.3	GuHCl	(4)	87Y
Glu49→Gln	7.2	25	25.5	heat		84O
Glu49→Gln	9	25	35.6	GuHCl		82Y1
Glu49→Gln	9	25	35.6±0.8	GuHCl	(4)	87Y
Glu49→Gln	9.3	25	35.1	DSC		82Y2
Glu49→Gln	9.3	25	25.5	heat		84O
Glu49→Gly	7	25	29.7±0.4	GuHCl	(4)	87Y
Glu49→Gly	9	25	26.8±0.4	GuHCl	(4)	87Y
Glu49→His	7	25	42.3±3.3	GuHCl	(4)	87Y
Glu49→His	9	25	38.5±1.7	GuHCl	(4)	87Y
Glu49→Ile	7	25	70.3±2.1	GuHCl	(4)	87Y
Glu49→Ile	9	25	41.8±1.3	GuHCl	(4)	87Y
Glu49→Leu	5.5	25	36	GuHCl	(1,2a)	84Y1
Glu49→Leu	7	25	63	GuHCl	(1,2a)	84Y1
Glu49→Leu	7	25	62.8±2.9	GuHCl	(4)	87Y
Glu49→Leu	9	25	51±1.7	GuHCl	(4)	87Y
Glu49→Leu	9	25	51	GuHCl	(1,2a)	84Y1
Glu49→Lys	7	25	33.1±2.1	GuHCl	(4)	87Y
Glu49→Lys	9	25	31.3±3.8	GuHCl	(4)	87Y
Glu49→Met	5.5	25	24.7	GuHCl		82Y1
Glu49→Met	7	25.8	56.1	GuHCl	(2a,3)	79Y
Glu49→Met	7	25	55.6	GuHCl		82Y1
Glu49→Met	7	25	55.6±0.8	GuHCl	(4)	87Y
Glu49→Met	9	25	35.1±1.3	GuHCl	(4)	87Y
Glu49→Met	9	25	35.1	GuHCl		82Y1
Glu49→Phe	7	25	46.9±0.8	GuHCl	(4)	87Y
Glu49→Phe	9	25	34.7±0.8	GuHCl	(4)	87Y
Glu49→Pro	7	25	34.3±1.3	GuHCl	(4)	87Y
Glu49→Pro	9	25	28.9±3.3	GuHCl	(4)	87Y
Glu49→Ser	5.5	25	16.3	GuHCl		82Y1
Glu49→Ser	7	25	31	GuHCl		82Y1
Glu49→Ser	7	25	32.2	DSC		82Y2
Glu49→Ser	7	25	31±0.8	GuHCl	(4)	87Y
Glu49→Ser	9	25	33.5±3.3	GuHCl	(4)	87Y
Glu49→Ser	9	25	33.5	GuHCl		82Y1
Glu49→Ser	9.3	25	29.7	DSC		82Y2
Glu49→Thr	7	25	36.8±0.8	GuHCl	(4)	87Y
Glu49→Thr	9	25	29.3±3.3	GuHCl	(4)	87Y
Glu49→Trp	7	25	41.4±2.1	GuHCl	(4)	87Y

Tryptophan synthase from *E. coli*, α-subunit, wild-type and mutants, ΔG values (continued)

Mutant	pH	T	ΔG	Approach	Remarks	Ref
Glu49→Trp	9	25	23.8±2.1	GuHCl	(4)	87Y
Glu49→Tyr	5.5	25	20.1	GuHCl		82Y1
Glu49→Tyr	7	25	36.8	GuHCl		82Y1
Glu49→Tyr	7	25	36.8±2.1	GuHCl	(4)	87Y
Glu49→Tyr	9	25	28.5±2.1	GuHCl	(4)	87Y
Glu49→Tyr	9	25	28.5	GuHCl		82Y1
Glu49→Val	5.5	25	25.5	GuHCl		82Y1
Glu49→Val	7	25	50.2	GuHCl		82Y1
Glu49→Val	7	25	50.2±1.3	GuHCl	(4)	87Y
Glu49→Val	9	25	39.3±0.8	GuHCl	(4)	87Y
Glu49→Val	9	25	39.3	GuHCl		82Y1
Gly211→Arg	7.8	25	22.2*	DSC		80M2
Gly211→Glu	7.8	25	26.4*	DSC		80M2

Remarks:
(1) linear extrapolation
(2a) monitored by CD
(2b) monitored by fluorescence
(2c) monitored by UV spectroscopy
(3) three-state fit by means of a modified denaturant binding model
(4) denaturant binding model, three-state fit

b) Various transitions
For a more recent consideration of the transitions of tryptophan synthase α subunit, see Ogasahara, K., Yutani, K.: Biochemistry 36 (1997) 932.

Tryptophan synthase from *E. coli*, α-subunit, wild-type and mutants ΔG values

Mutant	pH	T	ΔG	Transition	Approach	Remarks	Ref
wild-type	7.8	25	23.8±1.7	N→I	urea	(1)	86B
wild-type	7.8	25	20.9±2.5	I→U	urea	(1)	86B
wild-type	7.8	25	23.8±4.2	N→I	urea	(1,2)	86B
wild-type	7.8	25	13.8±4.2	I→U	urea	(1,2)	86B
wild-type	7.8	25	23.8±1.7	N→I	urea	(1)	90T2
wild-type	7.8	25	20.9±2.5	I→U	urea	(1)	90T2
Phe22→Leu	7.8	25	33.1±5.9	N→I	urea	(1)	86B
Phe22→Leu	7.8	25	17.6±5	I→U	urea	(1)	86B
Glu49→Met	7.8	25	39.3±4.6	N→I	urea	(1)	86B
Glu49→Met	7.8	25	29.3±3.8	I→U	urea	(1)	86B
Gly211→Asp	7.8	25	22.6±3.8	N→I	urea	(1)	90T2
Gly211→Asp	7.8	25	17.6±2.1	I→U	urea	(1)	90T2
Gly211→Glu	7.8	25	33.9±2.5	N→I	urea	(1)	86B
Gly211→Glu	7.8	25	22.6±2.1	I→U	urea	(1)	86B
Gly211→Glu	7.8	25	34.3±5	N→I	urea	(1,2)	86B
Gly211→Glu	7.8	25	17.6±3.3	I→U	urea	(1,2)	86B
Gly211→Glu	7.8	25	33.9±2.5	N→I	urea	(1)	90T2
Gly211→Glu	7.8	25	23.0±2.1	I→U	urea	(1)	90T2
Gly211→Ser	7.8	25	24.7±3.3	N→I	urea	(1)	90T2
Gly211→Ser	7.8	25	21.8±3.3	I→U	urea	(1)	90T2
Gly211→Val	7.8	25	22.6±3.8	N→I	urea	(1)	90T2
Gly211→Val	7.8	25	28.9±7.1	I→U	urea	(1)	90T2
Gly211→Trp	7.8	25	17.2±7.5	N→I	urea	(1)	90T2
Gly211→Trp	7.8	25	16.3±8.4	I→U	urea	(1)	90T2

Tryptophan synthase from *E. coli*, α-subunit, wild-type and mutants, ΔG values (continued)

Mutant	pH	T	ΔG	Transition	Approach	Remarks	Ref
Gly234→Asp	7.8	25	29.7±10	N→I	urea	(1)	86B
Gly234→Asp	7.8	25	29.3±12.6	I→U	urea	(1)	86B
Gly234→Lys	7.8	25	25.9±5.9	N→I	urea	(1)	86B
Gly234→Lys	7.8	25	19.7±7.1	I→U	urea	(1)	86B
wild-type	7.8	4	20.9±7.1	N→I	urea	(1)	90T2
wild-type	7.8	4	25.1±3.8	I→U	urea	(1)	90T2
Gly211→Arg	7.8	4	9.2±2.1	N→I	urea	(1)	90T2
Gly211→Arg	7.8	4	30.5±13.8	I→U	urea	(1)	90T2

Remarks:
(1) linear extrapolation
(2) in the presence of 0.5 M NaCl

Tryptophan synthase from *E. coli*, α-subunit, wild-type and mutants, difference Δ(ΔG) values

Mutant	pH	T	Δ(ΔG)	Transition	Approach	Remarks	Ref
wild-type	7.8	25	0.0	N→I	GuHCl	(1)	86H2
wild-type	7.8	25	0.0	I→U	GuHCl	(1)	86H2
Tyr175→Cys	7.8	25	−0.4±0.4	N→I	GuHCl	(1)	86H2
Tyr175→Cys	7.8	25	−2.1±0.4	I→U	GuHCl	(1)	86H2
Gly211→Glu	7.8	25	1.3±0.4	N→I	GuHCl	(1)	86H2
Gly211→Glu	7.8	25	−2.9±0.4	I→U	GuHCl	(1)	86H2
double mutant (Tyr175→Cys and Gly211→Glu)							
	7.8	25	−5.4±0.8	N→I	GuHCl	(1)	86H2
	7.8	25	−4.2±0.4	I→U	GuHCl	(1)	86H2

Remark:
(1) linear extrapolation

Tryptophan synthase from *E. coli*, α-subunit, wild-type and mutants

Transition	pH	T	ΔG	Approach	Remarks	Ref
Wild-type:						
N→I	7.8	23–24	26.8±2.9	urea	(1,2)	95C3
I→U	7.8	23–24	21.8±0.4	urea	(1,2)	95C3
Phe139→Trp						
N→I	7.8	23–24	23.8±5.0	urea	(1,2)	95C3
I→U	7.8	23–24	19.2±3.3	urea	(1,2)	95C3
Phe258→Trp						
N→I	7.8	23–24	20.9±1.7	urea	(1,2)	95C3
I→U	7.8	23–24	17.6±1.7	urea	(1,2)	95C3
double mutant (Phe139→Trp and Phe258→Trp)						
N→I	7.8	23–24	23.4±3.8	urea	(1,2)	95C3
I→U	7.8	23–24	18.4±1.7	urea	(1,2)	95C3

Remarks:
(1) data treatment was made using a three-state model that involves a linear dependence of ΔG on the denaturant concentration
(2) transition monitored by UV absorbance differences at 286 nm

Tryptophan synthase, from *E. coli*, α subunit, wild-type and mutants

Mutant	pH	T	ΔG	Approach	Remarks	Ref
Phe139→Trp						
N→I	7.8	25	23.4±3.3	urea	(1,2)	95C4
I→U	7.8	25	20.1±3.8	urea	(1,2)	95C4
Phe258→Trp	7.8	25	29.3±2.9	urea	(1,3)	95C4

Remarks:
(1) transition monitored by fluorescence
(2) three-state fit assuming linear dependence of ΔG on denaturant concentration
(3) the mutant unfolds according to a two-state model

Tryptophan synthase, from *E. coli*, α subunit

Transition	pH	T	ΔG	Approach	Remarks	Ref
Transitions monitored by optical absorption:						
N→I	7.8	25	28.5±1.7	urea	(1,2)	93S0
I→U	7.8	25	19.2±1.3	urea	(1,2)	93S0
Transitions monitored by CD:						
N→I	7.8	25	25.9±2.1	urea	(1,3)	93S0
I→U	7.8	25	19.2±2.1	urea	(1,3)	93S0
Transitions monitored by NMR:						
N→I	7.8	25	30.1±3.3	urea	(1,4)	93S0
N→I	7.8	25	37.7±3.8	urea	(1,4)	93S0
I→U	7.8	25	34.3±3.8	urea	(1,4)	93S0
I→U	7.8	25	36.4±9.2	urea	(1,4)	93S0
I→U	7.8	25	31.8±5.9	urea	(1,4)	93S0

Remarks:
(1) three-state fit assuming linear dependence of ΔG on denaturant concentration
(2) transition monitored by optical absorbance at 287 nm
(3) transition monitored by CD at 222 nm
(4) transition monitored by NMR using chemical shifts of His92, His146, and His244

Tryptophan synthase from *E. coli*, wild-type and mutants, analysis of the I2 intermediate

Mutant	pH	T	ΔG	Approach	Remarks	Ref
wild-type						
	7.8	25	34.3±3.8	urea, NMR	(1,2)	96S1
	7.8	20	31.4±4.6	urea, NMR	(1,2)	96S1
	7.8	15	25.5±4.2	urea, NMR	(1,2)	96S1
	7.8	10	23.0±3.8	urea, NMR	(1,2)	96S1
	7.8	5	20.1±1.3	urea, NMR	(1,2)	96S1
Pro93→Ser	7.8	25	17.2±3.3	urea, NMR	(1,2)	96S1
Ile95→Ala	7.8	25	33.1±6.3	urea, NMR	(1,2)	96S1
Pro96→Ala	7.8	25	20.5±2.1	urea, NMR	(1,2)	96S1
Leu100→Ala	7.8	25	19.7±2.5	urea, NMR	(1,2)	96S1
Tyr102→Ala	7.8	25	22.6±2.5	urea, NMR	(1,2)	96S1
His244→Ala	7.8	5	16.7±7.1	urea, NMR	(1,2)	96S1
double mutant (Pro93→Ser and His244→Ala)						
	7.8	25	14.2±3.8	urea, NMR	(1,2)	96S1

Remarks:
(1) analysis of the I2 ↔ U transition according to for the general scheme N ↔ I1 ↔ I2 ↔ U
(2) measured in D_2O, the pD is here the uncorrected pH meter reading

Table 1. Gibbs Energy Change – Molar Values

Tryptophan synthase from *E. coli*, α-subunit, wild-type and proline mutants

Mutant	pH	T	ΔG	Approach	Remarks	Ref
wild-type	7	25	19.2	GuHCl	(1,2)	91Y2
wild-type	7	25	17.6	GuHCl	(1,3)	91Y2
wild-type	7	25	36.8	GuHCl	(1,4)	91Y2
Pro28→Ala	7	25	9.2	GuHCl	(1,2)	91Y2
Pro28→Ala	7	25	17.2	GuHCl	(1,3)	91Y2
Pro28→Ala	7	25	26.4	GuHCl	(1,4)	91Y2
Pro28→Gly	7	25	9.6	GuHCl	(1,2)	91Y2
Pro28→Gly	7	25	20.5	GuHCl	(1,3)	91Y2
Pro28→Gly	7	25	30.1	GuHCl	(1,4)	91Y2
Pro132→Ala	7	25	10.0	GuHCl	(1,2)	91Y2
Pro132→Ala	7	25	19.2	GuHCl	(1,3)	91Y2
Pro132→Ala	7	25	29.2	GuHCl	(1,4)	91Y2
Pro132→Gly	7	25	15.1	GuHCl	(1,2)	91Y2
Pro132→Gly	7	25	19.2	GuHCl	(1,3)	91Y2
Pro132→Gly	7	25	34.3	GuHCl	(1,4)	91Y2

Remarks:
(1) the approach is the same as in Ref. 79Y
(2) ΔG refers to the N→I transition
(3) ΔG refers to the I→D transition
(4) ΔG refers to the N→D transition

Tryptophan synthase from *E. coli*, α-subunit, wild-type and proline mutants

Mutant	pH	T_{trs}	ΔT	Δ(ΔG)	Approach	Ref
wild-type	9	54.1	0.0	0.0	DSC	91Y2
Pro57→Ala	9	54.0	−0.1	−0.2	DSC	91Y2
Pro62→Ala	9	52.7	−1.4	−2.2	DSC	91Y2
Pro96→Ala	9	47.8	−6.3	−8.4	DSC	91Y2
Pro132→Ala	9	51.5	−2.6	−3.9	DSC	91Y2
Pro132→Gly	9	51.8	−2.3	−3.3	DSC	91Y2
Pro207→Ala	9	47.4	−6.7	−6.9	DSC	91Y2

Remark:
Δ(ΔG) is given for T_{trs} = 54.1°C of the wild-type protein

Tryptophan synthase, from *E. coli*, α subunit, wild-type and cysteine mutants

Mutant	pH	T	ΔG	Approach	Remarks	Ref
wild-type	9.0	50	4.9±1.1	DSC	(1)	96H6
Cys81→Ala	9.0	50	2.0±0.2	DSC	(1)	96H6
Cys81→Gly	9.0	50	−1.7±0.5	DSC	(1)	96H6
Cys81→Ser	9.0	50	−1.0±0.5	DSC	(1)	96H6
Cys81→Val	9.0	50	−0.6±0.3	DSC	(1)	96H6
Cys118→Ala	9.0	50	−0.8±0.7	DSC	(1)	96H6
Cys118→Ser	9.0	50	−4.6±0.2	DSC	(1)	96H6
Cys118→Val	9.0	50	−0.7±0.3	DSC	(1)	96H6
Cys154→Ala	9.0	50	0.6±0.04	DSC	(1)	96H6
Cys154→Ser	9.0	50	−2.3±0.5	DSC	(1)	96H6
Cys154→Val	9.0	50	0.2±0.2	DSC	(1)	96H6
in the presence of IPP:						
wild-type	9.0	50	6.7±1.0	DSC	(1,2)	96H6
Cys154→Ser	9.0	50	−3.0±0.8	DSC	(1,2)	96H6

Remarks:
(1) for the calorimetric reference values, see Table 2
(2) protein measured in the presence of indole 3-propanol phosphate (IPP) at a 20-fold molar excess

c) Comparison of tryptophan synthase α subunits from various species and hybrid proteins

Tryptophan synthase, α-subunit, species and hybrids

Species	pH	T	ΔG	Transition	Approach	Remarks	Ref
E.coli	7	25	27.6		heat		84Y2
E.coli	7.8	25	23.8±1.7	N→I	urea	(1)	88S9
E.coli	7.8	25	20.9±2.5	I→U	urea	(1)	88S9
S.typh.	7	25	29.3		heat		84Y2
S.typh.	7.8	25	25.5±2.1	N→I	urea	(1)	88S9
S.typh.	7.8	25	25.9±2.9	I→U	urea	(1)	88S9
hybrid (1)	7	25	27.2		heat		84Y2
hybrid (2)	7.8	25	22.6±1.7	N→I	urea	(1)	88S9
hybrid (2)	7.8	25	20.1±2.1	I→U	urea	(1)	88S9
hybrid (3)	7.8	25	22.6±3.3	N→I	urea	(1)	88S9
hybrid (3)	7.8	25	25.9±6.3	I→U	urea	(1)	88S9
hybrid (4)	7.8	25	26.8±6.7	N→I	urea	(1)	88S9
hybrid (4)	7.8	25	16.7±5	I→U	urea	(1)	88S9
hybrid (5)	7.8	25	20.9±4.6	N→I	urea	(1)	88S9
hybrid (5)	7.8	25	26.4±6.7	I→U	urea	(1)	88S9
hybrid (6)	7.8	25	16.3±8.4	N→I	urea	(1)	88S9
hybrid (6)	7.8	25	15.5±10.5	I→U	urea	(1)	88S9

Abbreviations (species/hybrids):
E.coli   – *Escherichia coli*
S.typh.   – *Salmonella typhimurium*
hybrid   – interspecies hybrid from *E.coli* and *S.typh.*
hybrid (1) – hybrid, consisting of residues 1-173 from *S.typh.* and 174-268 from *E.coli*
hybrid (2) – hybrid 6-34
hybrid (3) – hybrid 8-32
hybrid (4) – hybrid 12-28
hybrid (5) – hybrid 14-26
hybrid (6) – hybrid 15-25

Remark:
(1) linear extrapolation

Tryptophan synthase, α subunit, from *E. coli* and *Salmonella typhimurium*

Species/Mutant	pH	T	Δ(ΔG)	Approach	Remarks	Ref
E. coli, wild-type	9	54	0.0	DSC	(1)	96H5
Lys109→Asn, *E. coli*	9	54	−1.80±0.21	DSC		96H5
S. typhimurium	9	54	−8.67	DSC	(2)	96H5

Remarks:
(1) see also Ref. 90S8
(2) see also Ref. 91Y2

Tryptophan synthase, α-subunit, from *Salmonella typhimurium* wild-type and mutants

Transition	pH	T	ΔG	Approach	Remarks	Ref
Wild-type:						
N→I	7.8	25	30.8±2.8	urea	(1)	92C2
I→U	7.8	25	24.3±3.1	urea	(1)	92C2
N→U	7.8	25	55.1±4.2	urea	(1)	92C2
Mutant Phe22→Leu:						
N→I	7.8	25	26.7±3.8	urea	(1)	92C2
I→U	7.8	25	24.0±2.1	urea	(1)	92C2
N→U	7.8	25	50.7±4.3	urea	(1,2)	92C2
Mutant Phe22→Ile:						
N→I	7.8	25	13.2±1.6	urea	(1)	92C2
I→U	7.8	25	22.7±0.9	urea	(1)	92C2
N→U	7.8	25	35.9±1.8	urea	(1,3)	92C2
Mutant Phe22→Val:						
N→I	7.8	25	16.1±1.2	urea	(1)	92C2
I→U	7.8	25	24.6±1.4	urea	(1)	92C2
N→U	7.8	25	40.7±1.8	urea	(1,4)	92C2

Remarks:
(1) three-state model
(2) $\Delta(\Delta G) = -(4.4\pm6.0)$ kJ/mol
(3) $\Delta(\Delta G) = -(19.2\pm4.6)$ kJ/mol
(4) $\Delta(\Delta G) = -(14.4\pm4.6)$ kJ/mol

Tryptophan synthase from *Salmonella typhimurium*, α-subunit, wild-type and mutants

Mutant/Transition	pH	T	ΔG	Δ(ΔG)	Approach	Remarks	Ref4
Wild-type:							
N→I	7.8	25	31.0±5.9	0.0	urea	(1)	93T6
I→U	7.8	25	24.3±6.3	0.0	urea	(1)	93T6
Mutant Ala18→Gly:							
N→I	7.8	25	13.8±2.5	6.3±1.3	urea	(1)	93T6
I→U	7.8	25	24.3±6.3	−0.8±1.7	urea	(1)	93T6
Mutant Ala18→Val:							
N→I	7.8	25	13.8±2.1	3.3±1.3	urea	(1)	93T6
I→U	7.8	25	24.7±6.3	−0.8±1.7	urea	(2)	93T6
Mutant Tyr175→Gln:							
N→I	7.8	25	25.1±7.5	8.8±2.9	urea	(1)	93T6
I→U	7.8	25	33.0±11.3	7.9±0.8	urea	(1)	93T6

Tryptophan synthase from *Salmonella typhimurium*, α-subunit, wild-type and mutants (continued)

Mutant/Transition	pH	T	ΔG	$\Delta(\Delta G)$	Approach	Remarks	Ref
Mutant Leu209→Val:							
N→I	7.8	25	16.7±2.5	3.3±0.4	urea	(1)	93T6
I→U	7.8	25	15.9±3.8	0.8±0.8	urea	(1)	93T6
Mutant Ile232→Val:							
N→I	7.8	25	14.6±3.3	2.9±2.1	urea	(1)	93T6
I→U	7.8	25	18.0±7.9	2.1±1.7	urea	(1)	93T6
Double mutant (Ala18→Gly and Ile232→Val):							
N→I	7.8	25	7.9±2.5	2.5±2.1	urea	(1)	93T6
I→U	7.8	25	23.4±15.9	1.3±3.3	urea	(1)	93T6
Double mutant (Ala18→Val and Ile232→Val):							
N→I	7.8	25	18.4±6.7	5.4±2.5	urea	(1)	93T6
I→U	7.8	25	28.5±10.9	−0.8±2.5	urea	(1)	93T6
Double mutant (Tyr175→Gln and Leu209→Val):							
N→I	7.8	25	15.1±1.7	3.8±0.4	urea	(1)	93T6
I→U	7.8	25	40.6±14.6	−6.7±4.6	urea	(1)	93T6

Remark:
(1) three-state model

Tumor Suppressor

Tumor suppressor p16, recombinant

pH	T	ΔG	Approach	Remarks	Ref
7.5	25	9.6±0.9	urea	(1,2)	96B8

Remarks:
(1) linear extrapolation
(2) measured by CD in 10 mM MOPS, 50 mM NaF, pH 7.5

Twitchin, see Immunoglobulin

Two-Stranded α-Helical Coiled-Coils, see α-helix propensity

Tyrosine Kinase

SH3 Domain of Bruton's Tyrosine kinase

Salt Concen-tration	pH	T	ΔG	Approach	Remarks	Ref
buffer	7.0	25	10.9±1.7	GuHCl	(1,2)	96C11
500 mM	7.0	25	16.7±1.7	GuHCl	(1–3)	96C11

Remarks:
(1) linear extrapolation
(2) buffer: 20 mM sodium phosphate, 50 mM NaCl
(3) buffer (2) with 500 mM sodium sulfate

U1A

RNA binding domain of U1A, N-terminal, wild-type and Tyr mutants

Mutant	pH	T	ΔG	Approach	Remarks	Ref
wild-type	7.0	22	33.9±2.9	GuHCl	(1,2)	96K8
wt*	7.0	22	33.9±2.9	GuHCl	(1–3)	96K8
Tyr13→Phe	7.0	22	31.0±2.5	GuHCl	(1,2)	96K8
Tyr13→Thr	7.0	22	28.5±2.5	GuHCl	(1,2)	96K8
Tyr31→Phe	7.0	22	33.9±2.1	GuHCl	(1,2)	96K8
Tyr31→Ser	7.0	22	24.3±1.7	GuHCl	(1,2)	96K8
Tyr78→Asp	7.0	22	12.6±4.2	GuHCl	(1,2)	96K8
Tyr78→Phe	7.0	22	34.3±2.1	GuHCl	(1,2)	96K8
Tyr86→Phe	7.0	22	23.0±3.3	GuHCl	(1,2)	96K8
Tyr86→Thr	7.0	22	21.8±2.5	GuHCl	(1,2)	96K8

Remarks:
(1) linear extrapolation
(2) measured in 10 mM sodium cacodylate with 50 mM NaCl, pH 7.0
(3) ^{19}F-Tyr labeled protein

Ubiquitin

Ubiquitin

pH	T	ΔG	Approach	Ref
2.0	25	11*	DSC	94W5
3.0	25	24*	DSC	94W5
4.0	25	34*	DSC	94W5

Ubiquitin

pD	T	ΔG	Approach	Remarks	Ref
3.5	25	21	GuHCl	(1,2)	92B6

Remarks:
(1) linear extrapolation
(2) monitored by NMR, pD value without correction.

Ubiquitin, wild-type and mutant (Phe45→Trp)

Mutant	pH	T	ΔG	Δ(ΔG)	Approach	Remarks	Ref
wild-type	5.0	25	31.4	0.0	GuHCl	(1,2)	93K3
wild-type	5.0	25	28.0	0.0	GuHCl	(1,3)	93K3
Phe45→Trp	5.0	25	29.7	1.5	GuHCl	(1,2,4)	93K3
Phe45→Trp	5.0	25	30.5	1.3	GuHCl	(1,3,4)	93K3
Phe45→Trp	2.4	25	26.4		GuHCl	(1,5)	93K3
Phe45→Trp	5.0	25	29.3		GuHCl	(1,5)	93K3

Remarks:
(1) data treatment by linear extrapolation
(2) monitored by NMR, given is the pD value instead of the pH
(3) monitored by CD in the far-UV region
(4) Δ(ΔG) is given for 3.8 M GuHCl
(5) monitored by fluorescence

Ubiquitin, mutants derived from pseudo wild-type (wt* = Phe45→Trp)

Mutant	pH	T	ΔG	Approach	Remarks	Ref
wt*	5.0	25	30.1±0.4	GuHCl	(1–3)	96K3
Val26→Leu	5.0	25	31.4±1.3	GuHCl	(1–3)	96K3
Val26→Ile	5.0	25	34.7±0.8	GuHCl	(1–3)	96K3
Val26→Ala	5.0	25	16.3±0.8	GuHCl	(1–3)	96K3
Val26→Gly	5.0	25	3.8±1.3	GuHCl	(1–3)	96K3
wt*	5.0	25	42.7±1.7	GuHCl	(1–4)	96K3
Val26→Gly	5.0	25	15.5±2.5	GuHCl	(1–4)	96K3

Remarks:
(1) two-state treatment, linear extrapolation
(2) measured by fluorescence
(3) buffer: 25 mM acetate
(4) measured in the presence of 0.4 M Na_2SO_4

Ubiquitin, mutants derived from pseudo wild-type (wt* = Phe45→Trp), apparent Gibbs energies derived from three-state kinetic analysis

Mutant/Trs.	pH	T	ΔG	Approach	Remarks	Ref
wt* (Val26):						
N→I	5.0	25	31.0	GuHCl, kinetics		96K3
I→U	5.0	25	6.4	GuHCl, kinetics		96K3
N→U	5.0	25	37.4	GuHCl, kinetics		96K3
Val26→Leu:						
N→I	5.0	25	27.9	GuHCl, kinetics		96K3
I→U	5.0	25	6.6	GuHCl, kinetics		96K3
N→U	5.0	25	34.5	GuHCl, kinetics		96K3
Val26→Ile:						
N→I	5.0	25	28.7	GuHCl, kinetics		96K3
I→U	5.0	25	7.8	GuHCl, kinetics		96K3
N→U	5.0	25	36.4	GuHCl, kinetics		96K3
Val26→Ala:						
N→I	5.0	25	20.7	GuHCl, kinetics	(1)	96K3
I→U	5.0	25	−2.9	GuHCl, kinetics	(1)	96K3
N→U	5.0	25	34.5	GuHCl, kinetics		96K3

Remark:
(1) this ΔG value is an estimated value

Villin

Villin, F-actin bundling protein, 35 residue subdomain

pH	T	ΔG	Approach	Remarks	Ref
5.0	4.0	13.8±1.7	GuHCl	(1,2)	96M5
5.0	4.0	17.2±1.3	GuHCl	(1,3)	96M5

Remarks:
(1) linear extrapolation
(2) measured in H_2O, the slope of ΔG versus GuHCl conc. amounts to m = 3.35 kJ/mol/M
(3) measured in D_2O, the slope of ΔG versus GuHCl conc. amounts to m = 4.35 kJ/mol/M

Zinc-Finger Peptide, β-sheet propensity, see β-sheet

Table 2.
Enthalpy and Heat Capacity Changes – Molar Values

Acetylcholinesterase

Acetylcholinesterase from *Torpedo californica*

pH	T_{trs}	ΔCp	ΔH	Approach	Remarks	Ref
7.3	42–46		1600±100	DSC	(1,2)	95K13

Remarks:
(1) from scan rate dependent measurements, therefore variation in T_{trs}
(2) measured in 10 mM sodium phosphate, 0.1 M NaCl, pH 7.3

Actin

F-Actin and G-actin, overall heat effect and resolved transitions

Transition	pH	T_{trs}	ΔCp	ΔH	Approach	Ref
F-actin						
overall	8.0	67±0.5		678±42	DSC	90B2
overall			9.6±2.9		DSC	90B3
G-actin:						
overall	7.5	57.1	5.4	197	heat, v.H.	77C3
overall	8.0	57.2±0.5	9.6±2.5	594±21	DSC	90B2
overall			9.6±2.9			90B3
trans. (1)	8.0	52.3		184±21	DSC	90B2
trans. (2)	8.0	56.7		448±42	DSC	90B2

F-Actin vom skeletal muscle

Transition	pH	T_{trs}	ΔCp	ΔH	Approach	Ref
trans. (1)	8.3	59.7±1.4		60.3±1.7	DSC	95L9
trans. (2)	8.3	60.6±1.7		348.2±9.6	DSC	95L9
trans. (3)	8.3	61.4±1.4		104.3±4.6	DSC	95L9

G-Actin from skeletal muscle

Transition	pH	T_{trs}	ΔCp	ΔH	Approach	Ref
trans. (1)	8.3	52.5±1.4		184.4±10.9	DSC	95L9
trans. (2)	8.3	56±1.8		110.2±7.5	DSC	95L9

F-Actin

pH	T_{trs}	ΔC_p	ΔH	Approach	Remark	Ref
		9.6±2.9		DSC		90B3
5.9	74.3±0.2		820±50	DSC	(1)	90B3
6.4	74.0±0.3		791±42	DSC	(1)	90B3
6.9	71.8±0.3		795±54	DSC	(1)	90B3
7.3	70.4±0.2		753±38	DSC	(1)	90B3
7.6	69.1±0.3		749±50	DSC	(1)	90B3
7.9	68.2±0.3		703±50	DSC	(1)	90B3

Remark:
(1) measurement in the presence of 0.2 mM $CaCl_2$, 1 mM Na-ATP, 2 mM $MgCl_2$, 50 mM KCl, and 0.5 mM mercaptoethanol

F-Actin fom chicken breast muscle

pH	T_{trs}	ΔC_p	ΔH	Approach	Remark	Ref
6.5	75.5±0.4		600±40	DSC	(1)	94W1

Remark:
(1) buffer: 50 mM sodium phosphate, 0.6 M NaCl, pH 6.5

F-Actin, in the absence and presence of phalloidin

Molar ratio	pH	T_{trs}	ΔC_p	ΔH	Approach	Remarks	Ref
0.0	8.0	69.5		820	DSC	(1,2)	91L3
1.5	8.0	83.5		940	DSC	(1,2)	91L3

Remarks:
(1) molar ratio actin/phalloidin
(2) ΔH was taken from Fig. 2 of Ref. 91L3

G-Actin, in the presence of varying calcium concentrations and 0.2 mM Na-ATP

Ca^{2+} concentration	pH	T_{trs}	ΔC_p	ΔH	Approach	Ref
0.0 mM	7.0		9.6±2.9			
0.2 mM	7.0	62.7±0.3		611±25	DSC	90B3
0.4 mM	7.0	63.9±0.2		644±29	DSC	90B3
1.0 mM	7.0	65.9±0.3		686±25	DSC	90B3
2.0 mM	7.0	66.9±0.3		682±42	DSC	90B3
4.0 mM	7.0	67.4±0.4		686±25	DSC	90B3
8.0 mM	7.0	67.0±0.4		665±33	DSC	90B3

G-Actin, in the absence and presence of phalloidin

Molar Ratio	pH	T_{trs}	ΔC_p	ΔH	Approach	Remarks	Ref
0.0	8.0	59.5		530	DSC	(1,2)	91L3
1.26	8.0	75		765	DSC	(1,2)	91L3

Remarks:
(1) molar ratio actin/phalloidin
(2) ΔH was taken from Fig. 2 of Ref. 91L3

G-Actin

	pH	T_{trs}	ΔCp	ΔH	Approach	Remark	Ref
	8.0	60.8		664	DSC	(1)	93L4

Remark:
(1) scan rate 37.5 K/h, comparison with kinetics of denaturation, activation energy = 285±2 kJ/mol

Acyl Carrier Protein

Acyl carrier protein

Salt	pH	T_{trs}	ΔCp	ΔH	Approach	Remarks	Ref
low salt	6.1	52.7	3.3	160	DSC	(1,2)	94M12
low (CaCl$_2$)	6.1	64.3	6.4	266	DSC	(1,3)	94M12
high salt	6.1	63.2	2.8	156	DSC	(1,4)	94M12

Remarks:
(1) estimated errors: T_{trs} ±0.2°C, ΔH ±10%, ΔCp ±20%
(2) buffer: 50 mM sodium acetate, 0.5 mM DTT
(3) buffer: 50 mM sodium acetate, 0.5 mM DTT, 8.4 mM CaCl$_2$
(2) buffer: 240 mM sodium acetate, 0.5 mM DTT

Acylphosphatase

Acylphosphatase from horse muscle

	pH	T_{trs}	ΔCp	ΔH	Approach	Remarks	Ref
	3.8	56.8		328	heat, v.H.	(1,2)	94T1
	3.8	54.8		377	heat, v.H.	(1,3)	94T1

Remarks:
(1) measured in D$_2$O
(2) transition monitored by CD
(3) transition monitored by ^{1}H-NMR

Adenylate Kinase

Adenylate kinase from baker's yeast, wild-type and mutants

Mutant	pH	T_{trs}	ΔCp	ΔH	Approach	Remarks	Ref
wild-type	7.5	47.5±1.0		340±40	heat, v.H.		95S11
wild-type	7.5	5–15	8.5±2.3		heat/GuHCl	(1)	95S11
wild-type	4.6	48.4±0.6		240±20	heat, v.H.		95S11
extended	7.5	48.3±0.3		360±10	heat, v.H.	(2)	95S11
Val8→Ile	7.5	46.0±0.6		270±15	heat, v.H.		95S11
Gln48→Glu	7.7	46.2±0.3		280±50	heat, v.H.		95S11
Thr77→His	7.5	46.4±0.7		210±25	heat, v.H.		95S11
Thr77→His	5.0	46.6±2.0		285±10	heat, v.H.		95S11
Thr110→His	7.6	42.7±2.7		175±15	heat, v.H.		95S11
Thr110→His	4.5	40.0±2.1		130±20	heat, v.H.		95S11
Asn169→Asp	7.5	46.9±1.3		200±20	heat, v.H.		95S11
Ile213→Phe	7.5	39.8±0.6		250±20	heat, v.H.		95S11

Remarks:
(1) ΔCp was obtained by a combined approach
(2) extended = adenylate kinase with C-terminal extension Pro-His-His

Adrenodoxin

Adrenodoxin, recombinant, wild-type, in the presence of Na_2S

Na_2S Concentration	pH	T_{trs}	ΔCp	ΔH	Approach	Remarks	Ref
2.0 mM	8.5	46.37±0.37	5.72	324±20	DSC	(1)	95B9
2.5 mM	8.5	46.83±0.11		296±29	DSC	(1)	95B9
5.0 mM	8.5	50.51±0.16	6.48	352±21	DSC	(1)	95B9
10.0 mM	8.5	52.64±0.06	6.01	354±16	DSC	(1)	95B9
10.0 mM	8.5	53.08±0.07	3.41	352±15	DSC	(1)	95B9
10.0 mM	8.5	53.6 ±0.9		351±36	heat, v.H.	(1,2)	95B9
10.0 mM	8.5	51.31±0.09	5.63±2.85	370±6	DSC	(1,2)	95B9
10.0 mM	8.5	51.32±0.10	6.20±1.77	348±6	DSC	(1,2)	95B9
10.0 mM	8.5	51.55±0.17	8.71±1.05	348±12	DSC	(1,2)	95B9
15.0 mM	8.5	54.11±0.13	9.01	380±29	DSC	(1)	95B9
20.0 mM	8.5	55.60±0.07	6.30	379±11	DSC	(1)	95B9
20.0 mM	8.5	55.50±0.07	10.28	366±28	DSC	(1)	95B9
30.0 mM	8.5	57.50±0.06	9.45	376±20	DSC	(1)	95B9
30.0 mM	8.5	57.46±0.05	7.72	376±10	DSC	(1)	95B9
30.0 mM	8.5	55.68±0.07	6.20±1.32	404±19	DSC	(1)	95B9
30.0 mM	8.5	55.57±0.17	4.19±3.07	409±12	DSC	(1,3)	95B9
30.0 mM	8.5	55.27±0.05	4.18±0.81	383±14	DSC	(1,2)	95B9
40.0 mM	8.5	56.56±0.15	2.97±2.14	413±21	DSC	(1)	95B9
50.0 mM	8.5	55.35±0.10		275±10	DSC	(1)	95B9
2–40 mM	8.5	46–56	6.4±2.2		DSC	(4)	95B9
2–40 mM	8.5	46–56	7.5±1.2		DSC	(5)	95B9

Remarks:
(1) buffer: 40 mM glycine, Na_2S as indicated, 1 mM ascorbic acid, pH 8.5
(2) the buffer contains 10 mM mercaptoethanol
(3) the buffer contains 10 mM rather than 1 mM ascorbic acid
(4) mean value from the individual scans
(5) ΔCp from ΔH versus T_{trs}, this value is regarded as the more reliable value

Apoadrenodoxin

pH	T_{trs}	ΔCp	ΔH	Approach	Remarks	Ref
7.4	37.4±3.3		93±14	DSC	(1,2)	95B9

Remarks:
(1) mean of six measurements
(2) buffer: 25 mM potassium phosphate, 1 mM DTT, pH 7.4

Adrenodoxin, recombinant, wild-type, in the presence of Na_2S

pH	T_{trs}	ΔCp	ΔH	Approach	Remarks	Ref
6.5	47.5		376	heat, v.H.	(1,2)	95B9
7.4	53.7		383	heat, v.H.	(1,2)	95B9
8.5	53.6±0.9		351±36	heat, v.H.	(1,2)	95B9
8.5	53.5		316	heat, v.H.	(1,2)	95B9
8.5	53.4		361	heat, v.H.	(1,2)	95B9
9.5	54.0		331	heat, v.H.	(1,2)	95B9
10.5	53.2		270	heat, v.H.	(1,2)	95B9

Remarks:
(1) buffer: 40 mM glycine, 10 mM Na_2S, 1 mM ascorbic acid, 10 mM mercaptoethanol, pH 8.5
(2) the transition was monitored by optical absorption at 415 nm

Adrenodoxin, recombinant, wild-type in the presence of Na_2S

Mutant	pH	T_{trs}	ΔCp	ΔH	Approach	Remarks	Ref
wild-type	8.5	51.7	6.08±0.15	351.3±0.6	DSC	(1)	95B9
(4–108)	8.5	55.3	8.10±0.18	376.0±0.4	DSC	(1,2)	96B11
	8.5	55.1	5.74±0.34	372.2±1.2	DSC	(1,2)	96B11
(4–114)	8.5	50.9	7.49±0.30	313.7±1.1	DSC	(1,3)	96B11
	8.5	51.3	4.27±0.34	325.4±1.1	DSC	(1,3)	96B11
Thr54→Ala	8.5	46.8	6.12±0.28	275.0±0.7	DSC	(1)	96B11
	8.5	47.4	7.47±0.15	295.4±0.5	DSC	(1)	96B11
Thr54→Ser	8.5	51.3	7.04±0.28	350.9±0.9	DSC	(1)	96B11
	8.5	51.9	8.92±0.18	357.4±0.6	DSC	(1)	96B11
His56→Arg	8.5	49.2	4.13±0.27	246.0±0.8	DSC	(1)	96B11
	8.5	48.8	4.21±0.21	244.8±0.6	DSC	(1)	96B11
	8.5	49.5	4.84±0.20	260.9±0.6	DSC	(1)	96B11
	8.5	49.6	4.00±0.24	242.8±0.7	DSC	(1)	96B11
His56→Gln	8.5	46.6	6.99±0.1	249.1±0.8	DSC	(1)	96B11
	8.5	46.3	7.20±0.1	249.7±0.6	DSC	(1)	96B11
His56→Thr	8.5	48.5	6.07±0.18	265.6±0.8	DSC	(1)	96B11
	8.5	49.0	5.78±0.23	276.6±0.7	DSC	(1)	96B11
Asp76→Glu	8.5	53.2	9.17±0.16	366.0±0.7	DSC	(1)	96B11
	8.5	53.6	11.3±0.24	360.8±0.8	DSC	(1)	96B11
Tyr82→Leu	8.5	50.7	8.04±0.1	330.0±0.5	DSC	(1)	96B11

Adrenodoxin, recombinant, wild-type in the presence of Na_2S (continued)

Mutant	pH	T_{trs}	ΔCp	ΔH	Approach	Remarks	Ref
Tyr82→Phe	8.5	51.5	7.54±0.23	350.0±0.7	DSC	(1)	96B11
Tyr82→Ser	8.5	50.6	8.19±0.13	334.3±0.5	DSC	(1)	96B11
Tyr82→Trp	8.5	51.7	7.83±0.30	328.8±0.9	DSC	(1)	96B11
Cys95→Ser	8.5	56.0	7.23±0.11	376.9±0.4	DSC	(1)	96B11
	8.5	55.5	6.34±0.13	371.0±0.4	DSC	(1)	96B11

Remarks:
(1a) see also Ref. 95P1
(1b) buffer: 40 mM glycine, 10 mM Na_2S, 1 mM ascorbic acid, 10 mM mercaptoethanol, pH 8.5
(2)   recombinant short form consisting of residues 4–108
(3)   recombinant short form consisting of residues 4–114

Alanine Peptide

Alanine peptide

pH	T_{trs}	ΔCp	ΔH	Approach	Remarks	Ref
7.0	40	0	168	DSC	(1–3)	91S4

Remarks:
(1) helix forming alanine peptide consisting of 50 residues, i.e.,
    Ac-Tyr-(Ala-Glu-Ala-Ala-Lys-Ala)×8-Phe-NH_2.
(2) the van't Hoff heat amounts to 56.9 kJ/mol
(3) ΔCp was assumed to be zero

Albumin, see also ovalbumin and serum albumin

Aldolase

Aldolase from rabbit muscle, wild-type and Asp128-mutants

Transition	pH	T_{trs}	ΔCp	ΔH	Approach	Remarks	Ref
Wild-type							
trans. (1)	7.4	55.7±1.0		255±113	DSC	(1,2)	94B5
trans. (2)	7.4	58.7±0.08		498±138	DSC	(1,2)	94B5
trans. (3)	7.4	61.2±0.1		142±33	DSC	(1,2)	94B5
Mutant (Asp128→Ala)							
trans. (1)	7.4	45.0±0.3		314±46	DSC	(1,3)	94B5
trans. (2)	7.4	46.8±0.03		297±42	DSC	(1,3)	94B5
Mutant (Asp128→Asn)							
trans. (1)	7.4	45.6±0.3		335±63	DSC	(1,3)	94B5
trans. (2)	7.4	47.0±0.04		347±63	DSC	(1,3)	94B5
Mutant (Asp128→Gln)							
trans. (1)	7.4	45.9±0.3		280±71	DSC	(1,3)	94B5
trans. (2)	7.4	47.1±0.05		322±67	DSC	(1,3)	94B5
Mutant (Asp128→Gly)							
trans. (1)	7.4	44.0±0.3		280±50	DSC	(1,3)	94B5
trans. (2)	7.4	46.0±0.04		305±42	DSC	(1,3)	94B5

Aldolase from rabbit muscle, wild-type and Asp128-mutants (continued)

Transition	pH	T_{trs}	ΔCp	ΔH	Approach	Remarks	Ref
Mutant (Asp128→Val)							
trans. (1)	7.4	43.9±0.2		381±38	DSC	(1,3)	94B5
trans. (2)	7.4	46.5±0.03		326±29	DSC	(1,3)	94B5

Remarks:
(1) results of deconvolution
(2) wild-type unfolds in three transitions
(3) mutants unfold in two transitions

Amylase

Taka-Amylase A

	pH	T_{trs}	ΔCp	ΔH	Approach	Remark	Ref
	7	62	36.4±4.1	2251±40	DSC	(1)	87F

Remark:
(1) measurement in the absence of calcium

Anticoagulant Protein, see Protein C

Apoflavodoxin

Apoflavodoxin from *Anabaena* PCC7119, in the presence of KCl

KCl (M)	pH	T_{trs}	ΔCp	ΔH	Approach	Remarks	Ref
0.0	6.0–9.0	57.3±0.1	5.6±0.1	264±3	DSC	(1)	96G3
0.0	7.0	57.4	5.69	265	DSC	(2)	96G3
0.1	7.0	59.4	5.50	276	DSC	(2)	96G3
0.2	7.0	60.7	5.63	283	DSC	(2)	96G3
0.3	7.0	61.6	5.64	291	DSC	(2)	96G3
0.5	7.0	62.8	5.77	295	DSC	(2)	96G3
0.75	7.0	64.9	5.74	308	DSC	(2)	96G3
1.0	7.0	66.3	5.79	317	DSC	(2)	96G3
2.0	7.0	66.8		160	DSC	(3)	96G3
0.0–1.0	7.0	57–66	5.6±0.1		DSC	(4)	96G3
0.0–1.0	7.0	57–66	5.8		DSC	(5)	96G3
0.0–1.0	7.0	57–66			DSC	(6)	96G3

Remarks:
(1) unfolding of apoflavodoxin is nearly pH independent from pH 6.0 to 9.0
(2) buffer: 50 mM sodium phosphate and KCl as indicated
(3) irreversible transition
(4) average ΔCp from the individual calorimetric curves
(5) ΔCp from ΔH versus T_{trs}
(5) temperature dependent ΔCp from Cp of the folded and unfolded protein, $\Delta Cp(T) = -34.33 + 0.3256 \times T -0.000614 \times T^2$

Apolipoprotein

Apolipoprotein A1

pH	T_{trs}	ΔCp	ΔH	Approach	Ref
9.2	54	10±2	410	DSC	76T

Arabinose-Binding Protein

L-Arabinose-binding protein (*E. coli*)

pH	T_{trs}	ΔCp	ΔH	Approach	Ref
7.4	53.57±0.07	13.2±0.3	635.1±4.6	DSC	83F2

L-Arabinose-binding protein, correlation of thermodynamic quantities with protein concentration (mg/ml) at pH 7.4

Protein Concentration	T_{trs}	ΔCp	ΔH	Approach	Ref
0.49	53.30	12.5	621	DSC	83F2
0.83	53.75	13.2	628	DSC	83F2
0.87	53.38	13.3	643	DSC	83F2
0.94	53.73	11.6	616	DSC	83F2
0.99	53.28	12.5	625	DSC	83F2
1.67	53.60	14.5	624	DSC	83F2
1.98	53.20	12.3	653	DSC	83F2
2.50	53.43	12.6	628	DSC	83F2
2.97	53.19	14.5	666	DSC	83F2
3.34	53.41	14.1	638	DSC	83F2
3.96	52.90	14.3	645	DSC	83F2
4.17	53.33	13.4	636	DSC	83F2

L-Arabinose-binding protein, correlation of thermodynamic quantities with ligand concentration (mM) at pH 7.4

Arabinose Concentration	T_{trs}	ΔCp	ΔH	Approach	Ref
1.0	58.24	10.2	809	DSC	83F2
1.0	58.15	12.3	871	DSC	83F2
5.0	60.60	11.9	833	DSC	83F2
5.0	60.33	11.3	842	DSC	83F2
10.0	61.75	10.2	821	DSC	83F2
10.0	61.18	10.5	864	DSC	83F2
25.0	62.97	13.9	836	DSC	83F2
25.0	62.59	12.3	881	DSC	83F2
50.0	64.02	12.7	813	DSC	83F2
50.0	63.75	10.4	881	DSC	83F2
75.0	64.65	11.4	803	DSC	83F2
75.0	63.65	11.3	834	DSC	83F
100.0	65.10	11.8	815	DSC	83F2
100.0	64.70	10.5	854	DSC	83F2

Arc Repressor

Arc repressor dimer

pH	T_{trs}	ΔCp	ΔH	Approach	Ref
7.3	54	6.7	297	heat, v.H.	89B2

P22 Arc repressor, wild-type and mutants

Mutant	pH	T_{trs}	ΔH	Approach	Remarks	Ref
wild-type	7.5	55.6±0.3	230±8	heat	(1)	93M13
Arc-st5	7.5	54.7±0.3	225±9	heat	(1,2)	93M13
Arc-st6	7.5	57.1±0.3	238±9	heat	(1,3)	93M13
Arc-st11	7.5	56.9±0.3	232±10	heat	(1,4)	93M13
Glu28→Ala-Arc-st11	7.5	53.2±0.5	225±8	heat	(1,4)	93M13
Arg31→Leu-Arc-st11	7.5	35.0±0.5	125±10	heat	(1,4)	93M13
Ser32→Ala-Arc-st11	7.5	33.8±0.2	109±10	heat	(1,4)	93M13

Remarks:
(1) buffer: 50 mM Tris-HCl, 250 mM KCl, 0.2 mM EDTA
(2) Arc-st5: Arc with the C-terminal extension -Lys-Asn-Gln-His-Glu-COOH
(3) Arc-st6: Arc with the C-terminal extension -His-His-His-His-His-His-COOH
(4) Arc-st11: Arc with the C-terminal extension -His-His-His-His-His-His-Lys-Asn-Gln-His-Glu-COOH

Arc repressor, alanine mutants

Mutant	pH	T_{trs}	ΔCp	ΔH	Approach	Remarks	Ref
Arc-st6	7.5	59.0	5.48±0.21	261	heat	(1)	94M14
Arc-st11	7.5	57.9	5.48±0.21	244	heat	(2)	94M14
Met1→Ala-st6	7.5	58.0	5.48±0.21	250	heat	(1)	94M14
Lys2→Ala-st6	7.5	58.7	5.48±0.21	239	heat	(1)	94M14
Gly3→Ala-st6	7.5	58.1	5.48±0.21	251	heat	(1)	94M14
Met4→Ala-st6	7.5	59.2	5.48±0.21	249	heat	(1)	94M14
Ser5→Ala-st6	7.5	57.5	5.48±0.21	233	heat	(1)	94M14
Lys6→Ala-st6	7.5	59.6	5.48±0.21	247	heat	(1)	94M14
Met7→Ala-st6	7.5	55.5	5.48±0.21	220	heat	(1)	94M14
Pro8→Ala-st6	7.5	74.1	5.48±0.21	336	heat	(1)	94M14
Gln9→Ala-st6	7.5	58.4	5.48±0.21	249	heat	(1)	94M14
Phe10→Ala-st6	7.5	40.6	5.48±0.21	146	heat	(1)	94M14
Asn11→Ala-st6	7.5	62.1	5.48±0.21	239	heat	(1)	94M14
Leu12→Ala-st11	7.5	42.3	5.48±0.21	169	heat	(2)	94M14
Arg13→Ala-st6	7.5	57.3	5.48±0.21	215	heat	(1)	94M14
Trp14→Ala-st11	7.5	31.5	5.48±0.21	106	heat	(2)	94M14
Pro15→Ala-st11	7.5	46.6	5.48±0.21	159	heat	(2)	94M14
Arg16→Ala-st6	7.5	59.5	5.48±0.21	251	heat	(1)	94M14
Glu17→Ala-st6	7.5	57.0	5.48±0.21	260	heat	(1)	94M14
Val18→Ala-st6	7.5	56.9	5.48±0.21	264	heat	(1)	94M14
Leu19→Ala-st6	7.5	48.3	5.48±0.21	204	heat	(1)	94M14
Asp20→Ala-st6	7.5	55.3	5.48±0.21	238	heat	(1)	94M14
Leu21→Ala-st11	7.5	39.6	5.48±0.21	154	heat	(2)	94M14
Val22→Ala-st11	7.5	<20			heat	(2,3)	94M14
Arg23→Ala-st11	7.5	56.7	5.48±0.21	236	heat	(2)	94M14
Lys24→Ala-st11	7.5	56.3	5.48±0.21	251	heat	(2)	94M14
Val25→Ala-st6	7.5	59.3	5.48±0.21	244	heat	(1)	94M14
Glu27→Ala-st6	7.5	58.8	5.48±0.21	258	heat	(1)	94M14

Arc repressor, alanine mutants (continued)

Mutant	pH	T_{trs}	ΔC_p	ΔH	Approach	Remarks	Ref
Glu28→Ala-st11	7.5	55.7	5.48±0.21	218	heat	(2)	94M14
Asn29→Ala-st11	7.5	45.3	5.48±0.21	185	heat	(2)	94M14
Gly30→Ala-st11	7.5	47.9	5.48±0.21	204	heat	(2)	94M14
Arg31→Ala-st11	7.5	37.1	5.48±0.21	143	heat	(2)	94M14
Ser32→Ala-st11	7.5	33.5	5.48±0.21	102	heat	(2)	94M14
Val33→Ala-st11	7.5	44.1	5.48±0.21	166	heat	(2)	94M14
Asn34→Ala-st11	7.5	63.0	5.48±0.21	239	heat	(2)	94M14
Ser35→Ala-st6	7.5	63.4	5.48±0.21	239	heat	(1)	94M14
Glu36→Ala-st11	7.5	<20			heat	(2,3)	94M14
Ile37→Ala-st11	7.5	<20			heat	(2,3)	94M14
Tyr38→Ala-st11	7.5	33.0	5.48±0.21	101	heat	(2)	94M14
Gln39→Ala-st11	7.5	61.4	5.48±0.21	258	heat	(2)	94M14
Arg40→Ala-st11	7.5	31.2	5.48±0.21	57	heat	(2)	94M14
Val41→Ala-st11	7.5	<20			heat	(2,3)	94M14
Met42→Ala-st11	7.5	35.6	5.48±0.21	109	heat	(2)	94M14
Glu43→Ala-st6	7.5	56.1	5.48±0.21	249	heat	(1)	94M14
Ser44→Ala-st11	7.5	46.3	5.48±0.21	172	heat	(2)	94M14
Phe45→Ala-st11	7.5	<20			heat	(2,3)	94M14
Lys46→Ala-st11	7.5	57.1	5.48±0.21	221	heat	(2)	94M14
Lys47→Ala-st11	7.5	47.2	5.48±0.21	172	heat	(2)	94M14
Glu48→Ala-st11	7.5	43.2	5.48±0.21	123	heat	(2)	94M14
Gly49→Ala-st11	7.5	48.7	5.48±0.21	192	heat	(2)	94M14
Arg50→Ala-st11	7.5	47.9	5.48±0.21	175	heat	(2)	94M14
Ile51→Ala-st11	7.5	50.9	5.48±0.21	198	heat	(2)	94M14
Gly52→Ala-st11	7.5	60.9	5.48±0.21	241	heat	(2)	94M14

Remarks:
General remark – ΔC_p was determined from ΔH versus T_{trs} of the mutant proteins.
Buffer used: 50 mM Tris, 250 mM KCl, 0.2 mM EDTA, pH 7.5
(1) st6 – carboxy-terminal extension $(His)_6$
(2) st11 – carboxy-terminal extension $(His)_6$-Lys-Asn-Gln-His-Glu
(3) ΔH was not determined

Arc repressor, MYL variant

Mutant	pH	T_{trs}	ΔC_p	ΔH	Approach	Remarks	Ref
MYL	7.5	56.8	–4.5±0.3		heat	(1,2)	96H3

Remarks:
(1) in all mutants, called MYL, hydrophobic interactions between Met31, Tyr36, and Leu40 replace the wild-type salt-bridge interactions between Arg31, Glu36, and Arg40
(2) buffer: 50 mM Tris-HCl and 50 mM KCl

Ascorbate Oxidase

Ascorbate oxidase, various forms

Form	pH	T_{trs}	ΔC_p	ΔH	Approach	Remarks	Ref
holo form	6.0	61.3		1760	DSC	(1)	90S2
	7.0	63.2		2180	DSC	(1)	90S2
type-2 Cu^{2+} depleted form							
	6.0	61.7		1900	DSC	(1)	90S2
	7.0	62.7		2250	DSC	(1)	90S2
apo form	6.0		43.3	990	DSC	(2)	90S2
	7.0	44.9		1410	DSC	(2)	90S2

Remarks:
(1) the transition can be resolved into three subtransitions
(2) the transition can be resolved into two subtransitions

Aspartate Receptor

Aspartate receptor from *E. coli*, cytoplasmic fragment, wild-type and mutants

Mutant	pH	T_{trs}	ΔC_p	ΔH	Approach	Remarks	Ref
wild-type	7	50.6±0.5	4.2	226±29	DSC		95W6
Glu301→Lys	7	51.5		192	DSC		95W6
Thr311→Ile	7	50.8		226	DSC	(1,2)	95W6
Ser325→Leu	7	50.0		160	DSC	(1,2)	95W6
Val346→Met	7	50.3		177	DSC		95W6
Val433→Ile	7	51.2		210	DSC		95W6
Ala436→Val	7	53.7±0.7		159±38	DSC	(2)	95W6
Ser461→Leu	7	59.9±0.5		255±46	DSC	(1,2)	95W6

Remarks:
(1) in size exclusion chromatography cluster formation was observed
(2) the thermogram was resolved into dissociation and unfolding

Aspartate Transcarbamoylase

Aspartate transcarbamoylase

Mutant	pH	T_{trs}	ΔC_p	ΔH	Approach	Remarks	Ref
wild-type	9.0	60.9		2600	DSC	(1)	91P3
Gln288→Ala	9.0	48.2		1215	DSC	(1)	91P3
Gln288→Glu	9.0	44.2		880	DSC	(1)	91P3
Asn291→Asp	9.0	46.0		1130	DSC	(1)	91P3
Arg296→Ala	9.0	64.0		2260	DSC	(1)	91P3
Ala298→Gly	9.0	66.2		2845	DSC	(1)	91P3
Ala298→Val	9.0	49.5		1340	DSC	(1)	91P3

Remark:
(1) ΔH refers to the catalytic trimer of the enzyme

Aspartate transcarbamoylase from *E. coli*, the main heat absorption peaks observable at pH 7

Mutant	T_{trs1}	ΔH_{trs1}	T_{trs2}	ΔH_{trs2}	Approach	Remarks	Ref
wild-type	65.78	2042	72.67	3611	DSC	(1)	90B10
Glu86→Gln	64.27	2021	70.11	2925	DSC	(1,2)	90B10
Asp100→Gly	65.59	2477	–	–	DSC	(3)	90B10
Lys164→Ile	65.18	1879	70.61	3690	DSC	(1)	90B10
Tyr165→Phe	66.77	1950	72.75	4117	DSC	(1)	90B10
Glu239→Ala	65.42	1933	72.58	3736	DSC	(1)	90B10
Glu239→Gln	64.96	1766	72.69	3749	DSC	(1)	90B10
Tyr240→Phe	65.97	2067	72.96	4192	DSC	(1)	90B10
Arg269→Gly	60.57	2615	–	–	DSC	(3)	90B10
Asp271→Ser	65.57	1807	70.52	2607	DSC	(1)	90B10
isolated subunits	50.60	3724	71.40	6025	DSC		90B10

Remarks:
(1) two heat absorption peaks each consisting of at least two subtransitions, deconvolution see Ref. 90B10
(2) ΔCp for the transition is 38±20 kJ/mol/K
(3) single heat absorption peak consisting of at least two subtransitions, deconvolution, see Ref. 90B10

ATPase

Na^+-, K^+-ATPase, free and ligand bound form

Mutant	pH	T_{trs}	ΔCp	ΔH	Approach	Remark	Ref
free enzyme	7.4	57.3		780	DSC		86C1
strophanth.	7.4	63.5		1850	DSC	(1)	86C1

Remark:
(1) in the presence of 10 mM $MgCl_2$ and 1 mM strophanthidin, the data were taken from Fig. 3 in Ref. 86C1

Ca(II), Mg(II)-ATPase from sarcoplasmatic reticulum (SR) of rabbit skeletal muscle

Form	pH	T_{trs}	ΔCp	ΔH	Approach	Remarks	Ref
monomeric	7.0	42.3±0.6		300±60	DSC	(1–3)	94M12
monomeric	7.0	47.4±0.7		386±75	DSC	(1,2,4)	94M12
SR membranes	7.0	52.6±0.1		279±31	DSC	(2,5)	94M12
SR membranes	7.0	54.5±0.4		280±37	DSC	(2,3)	94M12
SR membranes	7.0	53.5±0.5		210±33	DSC	(2,4)	94M12

Remarks:
(1) solubilized in octaethylene glycol dodecyl ether (1 mg/ml)
(2) buffer: 20% glycerol, 100 mM KCl, 20 mM TES/KOH (pH 7.0)
(3) in the presence of 2 mM EGTA/3 mM Mg^{2+}–ADP
(4) in the presence of 0.5 mM Ca^{2+}
(5) in the presence of 2 mM EGTA

ATP Synthase

F1 portion of chloroplast ATP Synthase

Ligand	pH	T_{trs}	ΔCp	ΔH	Approach	Remarks	Ref
2 M ADP	7.5	57.5		3983	DSC	(1–3)	93W2
2 M ADP and 2 M ATP:							
	7.5	65.0		4017	DSC	(1–3)	93W2
4 M ADP	7.5	65.0		5017	DSC	(1–3)	93W2
6 M ADP	7.5	67.7		7786	DSC	(1–3)	93W2

Remarks:
(1) moles ligand per mole protein
(2) buffer: 10 mM HEPES, 200 mM sucrose
(3) the paper contains data for further ligands

Autolysin

LytA amidase, major autolysin from *Streptococcus pneumoniae* and its C-terminal module C-LytA

Transition	pH	T_{trs}	ΔCp	ΔH	Approach	Remarks	Ref
LytA							
trans. (1)	6.9	39.2		247	DSC	(1)	96U2
trans. (2)	6.9	51.5		314	DSC	(1)	96U2
trans. (3)	6.9	43.9		238	DSC	(1)	96U2
trans. (4)	6.9	58.0		845	DSC	(1,2)	96U2
C-LytA							
trans. (1)	6.9	40.4		79	DSC	(1)	96U2
trans. (2)	6.9	54.8		117	DSC	(1)	96U2
trans. (3)	6.9	61.1		481	DSC	(1)	96U2

Remarks:
(1) Ref. 96U2 contains further data for LytA and C-LytA in the presence of choline
(2) $2 \times \Delta H_4$

Azurin

Azurin from *Pseudomonas aeruginosa*

pH	T_{trs}	ΔCp	ΔH	Approach	Remarks	Ref
7.03	85.8	8.5	648	DSC	(1,2)	95L1

Remarks:
(1) measured at a scan rate of 0.5°/min
(2) the data refer to a fit that accounts for the kinetics of irreversible denaturation

Bacteriorhodopsin

Bacteriorhodopsin, white and purple membranes of *Halobacterium halobium*

Form	pH	T_{trs}	ΔC_p	ΔH	Approach	Remarks	Ref
Bacterioopsin in white membrane							
	7.0	79±0.5		209±12	DSC		90M1
Bacteriorhodopsin in purple membrane							
	6.1±0.7	98.4±0.7		385±8	DSC		89B3
	6.5	80			DSC	(1)	92C3
	6.5	96		350	DSC	(2)	92C3
	6.6	101.1±0.2		433±21	DSC	(3)	78J
		79.4±0.8		27.2±2.6	DSC	(1,4)	78J
		95.0±0.7		331±8	DSC	(2,4)	78J
	7.0	74±0.5		71±4	DSC	(1)	90M1
	7.0	97±0.5		360±25	DSC	(2)	90M1
	7.2	96		356	DSC	(5)	87B2
	7.5	97		390	DSC	(3)	92C3
Bleached membrane							
	7.5	82		110	DSC	(3)	92C3
Bacterioopsin reconstituted with all-trans-retinal							
	7.0	101±0.5		435±29	DSC		90M1

Remarks:
(1) pretransition
(2) main transition
(3) measured in 0.1 potassium phosphate
(4) measured in water
(5) from the pH dependence, Figs. 2 and 5 in Ref. 87B2

Bacteriorhodopsin (BR) in the presence and in the absence of retinal (R)

Form	pH	T_{trs}	ΔH	Approach	Remarks	Ref
R(ABCDEFG)$_s$	7	101	418	DSC	(1,2)	92K0
R(ABCDEFG)$_s$	9	83	619	DSC	(1,2)	92K0
R(ABCDEFG)$_s$	7.5	95	346±22	DSC	(1,2,6,7)	93G2
R(ABCDEFG)$_s$	9.5	81	422±10	DSC	(1,2,6,7)	93G2
R(ABCDEFG)$_s$	7.5	95.1	354	DSC	(1,2,6,7)	96A4
R(ABCDEFG)$_s$	9.5	80.8	417	DSC	(1,2,6,7)	96A4
R(ABCDEFG)$_v$	7	95	506	DSC	(1,2)	92K0
R(ABCDEFG)$_v$	7	90	749	heat	(1,5)	92K0
R(ABCDEFG)$_v$	9	87	481	heat	(1,2,8)	92K0
R(AB.CDEFG)$_s$	7	95	234	DSC	(1)	92K0
R(AB.CDEFG)$_s$	9	78	259	DSC	(1)	92K0
R(AB.CDEFG)$_s$	7.5	88.5	290	DSC	(1,3,6)	96A4
R(AB.CDEFG)$_s$	9.5	74.0	259	DSC	(1,3,6,7)	96A4
R(AB.CDEFG)$_v$	7	83	368	DSC	(1)	92K0
R(AB.CDEFG)$_v$	7	80	377	heat	(1,5)	92K0
R(AB.CDEFG)$_v$	9	80	293	heat	(1,8)	92K0
R(ABCDE.FG)$_s$	9.5	75.8	275	DSC	(1,4,6)	96A4
R(A.B.CDEFG)$_v$	7	71	163	DSC	(1)	92K0
R(A.B.CDEFG)$_v$	7	60	142	heat	(1,5)	92K0
R(A.B.CDEFG)$_v$	9	63	142	heat	(1,8)	92K0
(ABCDEFG)$_s$	7	85	96	DSC	(1,2)	92K0
(ABCDEFG)$_s$	9	65	234	DSC	(1,2)	92K0

Bacteriorhodopsin (BR) in the presence and in the absence of retinal (R) (continued)

Form	pH	T_{trs}	ΔH	Approach	Remarks	Ref
(ABCDEFG)$_v$	7	79	84	DSC	(1,2)	92K0
(ABCDEFG)$_v$	9	81	238	heat	(1,2,8)	92K0
(AB.CDEFG)$_s$	7	80	67	DSC	(1)	92K0
(AB.CDEFG)$_s$	9	57	92	DSC	(1)	92K0
(AB.CDEFG)$_s$	7.5		87	DSC	(1,3,6)	96A4
(AB.CDEFG)$_s$	9.5		78	DSC	(1,3,6)	96A4
(AB.CDEFG)$_v$	7	71	84	DSC	(1)	92K0
(AB.CDEFG)$_v$	9	59	109	heat	(1,8)	92K0
(A.B.CDEFG)$_v$	9	53	126	heat	(1,8)	92K0

Remarks:
(1) nomenclature (according to Ref. 92K0):
 – the helices are assigned letters,
 – a dot between letters indicates missing covalent connection between these helices,
 – R indicates the presence of retinal
 – subscript s: sample sheets from membranes
 – subscript v: vesicles
(2) (ABCDEFG) corresponds to uncleaved BR
(3) (AB.CDEFG) chymotrypsin cleaved BR in pos. 71–72
(4) (ABCDE.FG) NaBH$_4$ treated BR, cleaved in pos. 155–156
(5) thermal transition monitored by absorption spectroscopy, van't Hoff heat
(6) measured at different scan rates, T_{trs} for 1 K/min
(7) the scan rate dependent data in Refs. 93G2 and 96A4 follow a two-state kinetic model with an activation energy of 361±15 kJ/mol for intact and 374 kJ/mol for cleaved BR (AB.CDEFG)
(8) thermal transition monitored by UV-CD, van't Hoff heat

Bacteriorhodopsin

pH	T_{trs}	ΔCp	ΔH	Approach	Ref
4.7±0.22	79.9±1.0		393±63	DSC	89B3
5.7±0.06	81.6±1.6		376±33	DSC	89B3
6.3±0.16	74.9±1.4		435±42	DSC	89B3
7.1±0.15	72.4±1.8		397±58	DSC	89B3
7.4±0	67.5±1.8		318±8	DSC	89B3

Table 2. Enthalpy and Heat Capacity Changes – Molar Values

Barnase (ribonuclease from *Bacillus amyloliquefaciens*)

Barnase

pH	T_{trs}	ΔCp	ΔH	Approach	Remarks	Ref
2.2	25.2		272	DSC	(1)	93M2
2.4	27.0		276	DSC	(1)	93M2
2.8	33.5		297	DSC	(1)	93M2
3.2	40.5		372	DSC	(1)	93M2
3.8	47.9	3.6±0.5	448	DSC	(1–3)	93M2
4.5	52.4	3.6±0.5	490	DSC	(1–3)	93M2
4.9	54.6	3.6±0.5	490	DSC	(1–3)	93M2
5.5	55.3	3.6±0.5	485	DSC	(1–3)	93M2
5.9	53.8		448	DSC	(1)	93M2
6.2	53.8		452	DSC	(1)	93M2
8.0	52.2		585	heat	(4)	69H

Remarks:
(1) error less than ±0.1 in T_{trs}, and less than ±30 in ΔH
(2) ΔCp was calculated from the specific value given in Ref. 93M2 (see TABLE 3) using the molar mass of 12,382 Da
(3) ΔCp was found to be proportional to the protein mass, and independent of pH in the range 3.8–5.5; in this range ΔCp from the single calorimetric recordings coincides with $d(\Delta H)/dt$
(4) T_{trs} was found to be unaffected by pH in the range from 5–9

Barnase

pH	T_{trs}	ΔCp	ΔH	Approach	Remark	Ref
	12–56	7.87±2.93	230–630	heat	(1)	94O1
	55	7.87	502	heat		96O3

Remark:
(1) from van't Hoff enthalpies, obtained at different pH values, plotted versus T_{trs}

Barnase

pH	T_{trs}	ΔCp	ΔH	Approach	Remarks	Ref
2.0	23.8		345	DSC		94M7
2.5	31.4		394	DSC		94M7
3.0	39.4		449	DSC		94M7
3.5	46.5		497	DSC		94M7
4.0	50.3		523	DSC		94M7
4.5	52.8		540	DSC		94M7
5.0	53.7		546	DSC		94M7
2–5	24–54	6.2±0.8		DSC	(1)	94M7
2–5	24–54	6.8		DSC	(2)	94M7
2–5	5–125			DSC	(3)	94M7

Remarks:
(1) ΔCp is the average of the calorimetric scans
(2) ΔCp was determined from ΔH versus T_{trs}
(3) ΔCp is temperature dependent according to $\Delta Cp = 0.6 + 0.108 \times T - 0.00028 \times T^2$ (T in K), based on analysis of heat capacity of native and unfolded protein

Barnase

pH	T_{trs}	ΔCp	ΔH	Approach	Remarks	Ref
1.80	22.0		284	DSC		94G6
2.15	22.0		307	DSC		94G6
2.40	25.1		342	DSC		94G6
2.50	20.0		304	DSC		94G6
2.55	30.6		354	DSC		94G6
2.90	36.4		396	DSC		94G6
2.90	36.5		430	DSC		94G6
3.12	39.6		401	DSC		94G6
3.40	42.5		418	DSC		94G6
3.40	41.0		422	DSC		94G6
3.80	47.2		442	DSC		94G6
3.95	48.5		445	DSC		94G6
5.50	55.1		486	DSC		94G6
5.80	54.3		525	DSC		94G6
1.8–5.8	22–55	5.9		DSC	(1)	94G6
	50	5.7		DSC	(2)	94G6
	5	7.1		DSC	(3)	94G6
	25	6.9		DSC	(3)	94G6
	50	5.9		DSC	(3)	94G6
	75	3.9		DSC	(3)	94G6
	100	2.0		DSC	(3)	94G6
	125	0.1		DSC	(3)	94G6

Remarks:
(1) ΔCp was determined from ΔH versus T_{trs}
(2) ΔCp was determined from individual scanning calorimetric recordings
(3) the temperature function of ΔCp was determined from the partial specific heat capacity of folded and unfolded barnase

Barnase, temperature dependence of the heat capacity change at protein unfolding in aqueous solution, and related thermodynamic quantities

T	$\Delta H(exp)$	$\Delta S(exp)$	$\Delta Cp(exp)$	Approach	Remarks	Ref
5	167	379	7.1	DSC	(1–3)	95M3
25	307	866	6.9	DSC	(1–3)	95M3
50	467	1384	5.9	DSC	(1–3)	95M3
75	590	1752	3.9	DSC	(1–3)	95M3
100	664	1959	2.0	DSC	(1–3)	95M3
125	690	2029	0.1	DSC	(1–3)	95M3

Remarks:
(1) the data were taken from Ref. 95M3 with reference to Ref. 94G6
(2) ΔCp was obtained as described in Ref. 90P3
(3) ΔCp in kJ/mol/K, ΔH in kJ/mol and ΔS in J/mol/K

Barnase

pH	T_{trs}	ΔCp	ΔH	Approach	Ref
7.0	54.0±0.5		550±20	DSC	95M5
10.0	48.6±0.5		510±20	DSC	95M5
10.3	48.4±0.5		507±20	DSC	95M5
10.5	47.4±0.5		500±20	DSC	95M5

Barnase, thermodynamic quantities obtained in the presence and in the absence of urea

GuHCl Concentration	pH	T_{trs}	ΔCp	ΔH	Approach	Remarks	Ref
3.0 M	6.3	37.7±0.01		412±1	DSC	(1)	95J1
3.0 M	6.3	37.3±0.1		391±11	heat	(1,2)	95J1
0.0 M	2.85–4.75	36–54	7.15±0.54		DSC	(3)	95J1
1.0 M	2.85–4.75	30–49	7.99±0.50		DSC	(3,4)	95J1
2.0 M	2.85–4.75	24–42	8.45±0.50		DSC	(3,4)	95J1
3.0 M	2.85–4.75	22–36	9.08±0.59		DSC	(3,4)	95J1

Remarks:
(1) measured in 50 mM MES and 3.0 M urea
(2) transition monitored by CD at 230 nm
(3) ΔCp was obtained from ΔH versus T_{trs} varying pH at the given GuHCl conc.
(4) the dependence of ΔCp on urea concentration gives $\Delta Cp = (7.11\pm0.21) + [(0.67\pm0.17)\times c_{urea}]$

Barnase in the presence and absence of GroEL and SecB

Mutant	pH	T_{trs}	ΔH	Approach	Remarks	Ref
wild-type	6.3	53.9	590	heat	(1)	96Z
wild-type + GroEL	6.3	44.7		heat	(1,2)	96Z
wild-type + SecB	6.3	46.8		heat	(1,2)	96Z
wild-type in ^2H_2O	6.9	56.8	577	heat	(1,3)	96Z
wild-type + GroEL in ^2H_2O	6.9	44.4		heat	(1–3)	96Z
Ser91→Ala	6.3	49.6	556	heat	(1)	96Z
Ser91→Ala + GroEL	6.3	40.4		heat	(1,2)	96Z
Ser91→Ala + SecB	6.3	42.3		heat	(1,2)	96Z
Ser91→Ala	6.3	50.3	544	DSC	(4)	96Z
Ser91→Ala + GroEL	6.3	42.5		DSC	(4)	96Z
Ser91→Ala + SecB	6.3	42.6		DSC	(5)	96Z

Remarks:
(1) van't Hoff treatment, transition monitored by CD at 280 nm
(2) measured in 50 mM MES (pH 6.3) at 10 μM barnase in the presence and absence of GroEL (6 μM) or SecB (10 μM)
(3) measured in 20 mM imidazole buffer, ^2H_2O, p^2H 6.9
(4) measured at 14 μM barnase in the absence of GroEL and in the presence of 7 μM GroEL
(5) measured at 18 μM barnase in the presence of 20 μM SecB

Barnase, wild-type and mutants

Mutant	pH	T_{trs}	ΔH	Approach	Ref
wild-type	6.3	53.9	523	heat, v.H.	89K3
Leu14→Ala	6.3	42	452	heat, v.H.	89K3
Ile88→Ala	6.3	42.7	444	heat, v.H.	89K3
Ile88→Val	6.3	51	490	heat, v.H.	89K3
Ile96→Ala	6.3	44.9	469	heat, v.H.	89K3
Ile96→Val	6.3	51.5	506	heat, v.H.	89K3

Barnase, wild-type and mutants

Mutant	pH	T_{trs}	ΔCp	ΔH	Approach	Ref
wild-type	4.40	51.32±0.03	6.7±0.7	538±6	DSC	94M9
Asp8→Ala	4.40	49.56		520	DSC	94M9
Asp12→Gly	4.40	49.52		516	DSC	94M9
Thr16→Arg	4.40	52.8		528	DSC	94M9
Asp22→Met	4.40	49.9		539	DSC	94M9
Thr26→Gly	4.40	47.41		468	DSC	94M9
Ile51→Val	4.40	46.79		492	DSC	94M9
Asp54→Asn	4.40	44.41		488	DSC	94M9
Asp54→Asn	4.40	43.98		481	DSC	94M9
Arg69→Met	4.40	45.7		442	DSC	94M9
Arg83→Gln	4.40	48.18		488	DSC	94M9
Ile88→Val	4.40	47.95		499	DSC	94M9
Ser91→Ala	4.40	46.51		526	DSC	94M9
Ser91→Ala	4.40	46.52		538	DSC	94M9
Ser92→Ala	4.40	44.34		464	DSC	94M9
Ile96→Val	4.40	49.03		502	DSC	94M9
Ile96→Val	4.40	49.02		469	DSC	94M9
double mutant (Thr16→Ser and Glu61→Gly)						
	4.40	47.45		504	DSC	94M9

Barnase, pseudo wild-type and mutants [see remark (3)]

	pH	T_{trs}	ΔCp	ΔH	Approach	Remarks	Ref
wt*	6.3	49.6	5.9	605	heat	(1–4)	96K10
Gln15→Ile	6.3	52.0		530	heat	(1,2)	96K10
His18→Gln	6.3	47.5		614	heat	(1,2)	96K10
Asn58→Ala	6.3	43.8		480	heat	(1,2)	96K10
double mutant (Gln15→Ile and Lys108→Arg)							
	6.3	53.9		740	heat	(1,2)	96K10

Remarks:
(1) vant't Hoff treatment
(2) buffer: 50 mM MES pH 6.3
(3) pseudo wild-type (wt*) containing Val36→Met and His102→Ala
(4) ΔCp = 5.9 kJ/mol/K was taken from Ref. 94G6

Table 2. Enthalpy and Heat Capacity Changes – Molar Values

Barnase, insertion mutants

Mutant	pH	T_{trs}	ΔCp	ΔH	Approach	Remarks	Ref
wild-type	4.4	51.5	6.688	530.0	DSC	(2)	94V3
endo-[RNAseT1-(93–99)] [102a]-barnase							
	4.4	41.8		460.2	DSC	(1,2)	94V3
endo-[RNAseT1-(95–98)] [104a]-barnase							
	4.4	39.1		421.8	DSC	(1,2)	94V3

Remarks:
(1) endo-[RNAseT1-(93–99)][102a]-barnase: the mutant protein contains RNAse T1 residues at positions 93–99 inserted between residues 102 and 103 of barnase endo-[RNAseT1-(95–98)][104a]-barnase: the mutant protein contains RNAse T1 residues at positions 95–98 inserted between residues 104 and 105 of barnase
(2) buffer: 20 mM sodium acetate

Barnase, acid state at pH 2.7

Mutant	T_{trs1}	ΔH_{trs1}	T_{trs2}	ΔH_{trs2}	Approach	Remarks	Ref
wild-type	35.0	340	39.0	120	DSC	(1)	94S1
Ile76→Thr	26.7	337	32.2	69	DSC	(1)	94S1
double mutant (Ile4→Ala and Ile76→Val)							
	28.2	280	33.3	88	DSC	(1)	94S1
double mutant (Ile4→Ala and Tyr78→Phe)							
	25.4	156	29.8	183	DSC	(1)	94S1
double mutant (Ile4→Ala and Ile51→Val)							
	23.9	195	30.0	173	DSC	(1)	94S1
double mutant (Ser8→Cys and His102→Cys)							
	47.4	523			DSC	(2)	94S1
multiple mutant (Ile4→Ala, Ile25→Val, Ile51→Val and Tyr78→Phe)							
	14.5	68	19.9	62	DSC	(1)	94S1

Remarks:
(1) barnase deviates from two-state behaviour below pH 4.0; results of deconvolution
(2) the mutant having an engineered disulfide bond shows a two-state transition

Barnase, various transitions

Transition	pH	T_{trs}	ΔCp	ΔH	Approach	Remarks	Ref
N→D	6.3	25	7.87	353.5	heat	(1–3)	96O3
‡→D	6.3	25	6.28	113		(1,2,4)	96O3
I→D	6.3	25	5.02	130		(1,2,5)	96O3

Remarks:
(1) the values are relative to the thermally denatured state extrapolated to 25°C
(2) at pH 6.3, $\mu = 50$ mM
(3) data from Ref. 94O1
(4) from equilibrium unfolding and unfolding kinetics
(5) from ‡ and refolding kinetics

Barnase and barnase-barstar complex

Transition	pH	T_{trs}	ΔCp	ΔH	Approach	Remarks	Ref
barnase	6.2	53.6	5.7±0.6	565	DSC	(1)	94M2
barnase/NaCl	6.2	54.8		535	DSC	(2)	94M2
Barnase-barstar complex:							
trans. (1)	6.2	72.1		184	DSC	(1,3)	94M2
trans. (2)	6.2	77.9		615	DSC	(1,3)	94M2
Barnase-barstar complex/NaCl:							
trans. (1)	6.2	71.2		427	DSC	(2,3)	94M2
trans. (2)	6.2	74.7		632	DSC	(2,3)	94M2
Barnase-barstar complex/NaCl:							
trans. (1)	8.0	67.5		368	DSC	(2,3)	94M2
trans. (2)	8.0	72.8		485	DSC	(2,3)	94M2

Remarks:
(1) buffer: 10 mM PIPES
(2) buffer: 10 mM PIPES, 50 mM NaCl
(3) subtransitions resolved by deconvolution

Barstar, see also Barnase

Barstar, recombinant, thermal unfolding in the presence of GuHCl

GuHCl Concentration	pH	T_{trs}	ΔCp	ΔH	Approach	Remarks	Ref
0.0 M	8.0	69.8	6.23	292	heat, v.H.		95A1
0.0 M	8.0	69.0	5.90	281	heat, v.H.		95A1
0.2 M	8.0	69.0	6.11	272	heat, v.H.		95A1
0.4 M	8.0	67.1	5.73	244	heat, v.H.		95A1
0.6 M	8.0	64.9	5.65	226	heat, v.H.		95A1
0.8 M	8.0	62.3	5.65	207	heat, v.H.		95A1
1.0 M	8.0	56.8	6.28	196	heat, v.H.		95A1
1.2 M	8.0	52.9	6.36	172	heat, v.H.		95A1
1.4 M	8.0	48.6	6.28	141	heat, v.H.		95A1
1.6 M	8.0	43.3	6.28	106	heat, v.H.		95A1
1.8 M	8.0	37.1	6.28	62	heat, v.H.		95A1
	8.0	0–50	6.11±0.29			(1)	95A1
	8.0	0–75	6.23±0.21			(2)	95A1
	8.0	37–72	6.32±0.36			(3)	95A1
0–1.8	8.0	37–72	5.90±0.17			(4)	95A1

Remarks:
(1) ΔCp was obtained from ΔH versus T_{trs}
(2) ΔCp was obtained from combined heat and GuHCl denaturation
(3) ΔCp was obtained from GuHCl denaturation plotting ΔH versus T_{trs}
(4) from the GuHCl concentration dependence of ΔCp, extrapolated to zero denaturant conc., $\delta(\Delta Cp)/dc$ = 0.222±0.151 (kJ/mol/K)/mol$_{GuHCl}$

Barstar, dependence of thermodynamic quantities on GuHCl concentration

	0°C	24°C	44°C	Ref
ΔG at zero GuHCl concentration	11.72±0.08	18.83±0.08	15.90±0.13	95A1
$\delta(\Delta G)/dc$	−10.46±0.08	−10.04±0.04	−10.04±0.13	95A1
ΔH at zero GuHCl concentration	−139±3		120±4	95A1
$\delta(\Delta H)/dc$	−17.6±3.3		−7.95±3.76	95A1

Barstar, wild-type and double mutant (Cys40→Ala and Cys82→Ala)

Mutant	pH	T_{trs}	ΔCp	ΔH	Approach	Remarks	Ref
wild-type	7	71.5±0.5	5.0±1	251	heat	(1,2)	95K7
Double mutant (Cys40→Ala and Cys82→Ala):							
	7.0	74.0±0.5		265±15	DSC		95M5
	8		5.4±1.3		heat	(1,2)	95K7
	10.0	68.6±0.5		245±15	DSC		95M5
	10.3	67.0±0.5		240±15	DSC		95M5
	10.5	65.7±0.5		235±15	DSC		95M5
	6–10.8	61–76	4.35		DSC	(3)	95M5
		61–76	4.1±2.5		DSC	(4)	95M5
	2.5	25		−95±5	ITC	(5)	95M5

Remarks:
(1) ΔCp was determined from ΔH versus T_{trs} varying GuHCl concentration
(2) the thermal transition was monitored by CD at 220 nm and 275 nm, and absorbance at 287 nm
(3) ΔCp ws obtained from ΔH versus T_{trs} from DSC data of the following mutants: Cys40→Ala, Cys82→Ala, double mutant (Cys40→Ala and Cys82→Ala), triple mutants (Cys40→Ala, Cys82→Ala, and Pro48→Ala) and (Cys40→Ala, Cys82→Ala, and Pro48→Leu) giving $\Delta H(T_{trs})$ = −56 + 4.35×(T_{trs}−273.2) (kJ/mol)
(4) average from individual calorimetric recordings
(5) ΔH was determined by isothermal calorimetric titration changing pH from 7.0 to 2.5

Barstar, double mutant (Cys40→Ala and Cys82→Ala)

NaCl Concentration	pH	T_{trs}	ΔCp	ΔH	Approach	Remarks	Ref
0 mM	6.3	73.6		196.5±10	DSC	(1)	95W5
0 mM	8.0	69.5	5.9±0.3	178.2±9	DSC	(1)	95W5
0 mM	9.5	68.5	5.7±0.3	172.4±9	DSC	(1)	95W5
200 mM	8.0	72.7	6.1±0.3	190.6±9	DSC	(1)	95W5
400 mM	8.0	74.3	5.7±0.3	211.3±9	DSC	(1)	95W5
	6.3–9.5		5.6		DSC	(2)	95W5

Remarks:
(1) ΔCp from the individual scans
(2) ΔCp from ΔH versus T_{trs}

Binase

Ribonuclease from *Bacillus intermedius 7P* (binase)

pH	T_{trs}	ΔCp	ΔH	Approach	Ref
3.5	50.2		475	DSC	87P4
3.6	52		479	DSC	87P4
4	54.2		488	DSC	87P4
5.4	56.5		494	DSC	87P4
6.5	57.5		501	DSC	87P4
7	57.6	3.57±0.98	496	DSC	87P4

Binase, ribonuclease from *Bacillus intermedius*, and its complex with barstar

	pH	T_{trs}	ΔCp	ΔH	Approach	Remarks	Ref
binase	6.2	55.3		397	DSC	(1)	95Y1
binase	8.0	54.6		247	DSC	(1)	95Y1
complex	6.2	81.2		649	DSC	(1,2)	95Y1
complex	8.0	83.4		653	DSC	(1,2)	95Y1

Remarks:
(1) estimated errors are ±6% for ΔH and ±0.3°C for T_{trs}
(2) deconvolution gives:

pH	T_{trs1}	ΔH_{trs1}	T_{trs2}	ΔH_{trs2}
6.2	73.3	293	80.3	414
8.0	75.7	259	82.8	464

estimated errors are ±10% for ΔH and ±1.2°C for T_{trs}

Bromelain

Bromelain

pH	T_{trs}	ΔCp	ΔH	Approach	Remarks	Ref
3.4	59.3		334±17	DSC	(1,2)	95A7

Remarks:
(1) the transition is an irreversible two-state transition
(2) the calorimetric activation enthalpy amounts to 226±11 kJ/mol

Table 2. Enthalpy and Heat Capacity Changes – Molar Values

Calmodulin

Calmodulin from calf brain at varying solvent conditions, transitions resolved by deconvolution

Compounds added	pH	T_{trs}	ΔH	Approach	Remarks	Ref
2.0 mM EDTA						
total	7.28		330±20	DSC	(1)	85T2
	7.0	55		heat	(5,6)	83B3
trans. (1)	7.28	54	200±10	DSC	(1)	85T2
trans. (2)	7.28	36	115±15	DSC	(1)	85T2
2.0 mM EGTA, 2.0 mM MgCl$_2$						
total	7.28		345±20	DSC	(1)	85T2
trans. (1)	7.28	78	200±10	DSC	(1)	85T2
trans. (2)	7.28	61	125±15	DSC	(1)	85T2
2.0 mM EDTA, 150 mM NaCl						
total	7.28		355±20	DSC	(1)	85T2
trans. (1)	7.28	62	210±10	DSC	(1)	85T2
trans. (2)	7.28	53	125±15	DSC	(1)	85T2
10.0 mM EDTA, 9.0 mM CaCl$_2$						
total	7.28		330±20	DSC	(1,2)	85T2
trans. (1)	7.28	63	205±10	DSC	(1,2)	85T2
trans. (2)	7.28	47	120±15	DSC	(1,2)	85T2
10.0 mM EDTA, 9.3 mM CaCl$_2$						
total	7.28		430±20	DSC	(1,3)	85T2
trans. (1)	7.28	101	215±10	DSC	(1,3)	85T2
trans. (2)	7.28	85	125±15	DSC	(1,3)	85T2
trans. (3)	7.28	49	115±15	DSC	(1,3)	85T2
10.0 mM EDTA, 10.0 mM CaCl$_2$						
trans. (1)	7.28	>115	n.d.	DSC	(1,4)	85T2
trans. (2)	7.28	>115	n.d.	DSC	(1,4)	85T2
trans. (3)	7.28	81	130±15	DSC	(1,4)	85T2

Remarks:
(1) buffer: 10 mM cacodylate, pH 7.28, with the indicated substances
(2) free calcium conc. ~$1.8 \cdot 10^{-7}$ M
(3) free calcium conc. ~$2.7 \cdot 10^{-7}$ M
(4) free calcium conc. >10^{-5} M
(5) buffer: 20 mM HEPES, pH 7.0, with the indicated substances
(6) transition monitored by CD at 222 nm

Calmodulin from calf brain, fragments at varying solvent conditions

Components Added	pH	T_{trs}	$\Delta H^{cal.}$	$\Delta H^{v.H.}$	Approach	Remarks	Ref
Fragment TR-1 (residues 1 to 77):							
2.0 mM EDTA	7.25	49	209	201	DSC	(1)	85T2
2.0 mM EDTA	7.0	49			heat	(3,4)	83B3
2.0 mM EGTA	7.25	72	200	198	DSC	(1)	85T2
Fragment TM-1 (residues 1 to 106):							
2.0 mM EDTA	7.25	54	205	201	DSC	(1)	85T2
2.0 mM EDTA	7.0	58			heat	(3,4)	83B3
2.0 mM EGTA	7.25	78	221	212	DSC	(1)	85T2

Calmodulin from calf brain, fragments at varying solvent conditions (continued)

Components Added	pH	T_{trs}	$\Delta H^{cal.}$	$\Delta H^{v.H.}$	Approach	Remarks	Ref
10.0 mM EDTA, 9.3 mM CaCl$_2$							
	7.25	62	213	202	DSC	(1,2)	85T2
0.1 mM CaCl$_2$	7.25	>110	n.d.	n.d.	DSC	(1)	85T2
Fragment TR-2 (residues 78 to 148):							
2.0 mM EDTA	7.25	48	150	146	DSC	(1)	85T2
2.0 mM EDTA	7.0	46			heat	(3,4)	83B3
2.0 mM EGTA	7.25	69	163	158	DSC	(1)	85T2
0.1 mM CaCl$_2$	7.25	97	256	196	DSC	(1)	85T2

Remarks:
(1) buffer: 10 mM cacodylate, pH 7.25, with the indicated substances
(2) free calcium conc. ~$2.7 \cdot 10^{-7}$ M
(3) buffer: 20 mM HEPES, pH 7.0, with the indicated substances
(4) transition monitored by CD at 222 nm

Carbonic Anhydrase

Carbonic anhydrase B

	T_{trs}	ΔH	Approach	Ref
	67	950	DSC	79P4

Carbonic anhydrase from erythrocyte

pH	T_{trs}	ΔCp	ΔH	Approach	Remark	Ref
7.55	62.5	3.9	1075	heat	(1)	91L2

Remark:
(1) ΔCp was estimated from the amino acid content

Carboxypeptidase

Carboxypeptidase B (from porcine pancreas)

pH	T_{trs}	ΔCp	ΔH	Approach	Remarks	Ref
9.0		8–12	414±15	DSC	(1)	91C5
9.0			445±38	DSC	(1,2)	91C5

Remarks:
(1) T_{trs} is scan rate dependent, for the data treatment, see Ref. 91G1
(2) ΔH is the van't Hoff heat from T_{trs} versus logarithm of Zn^{2+} concentration according to Ref. 83F2, for the data treatment, see Ref. 91C5

Carboxypeptidase B from porcine pancreas

pH	T_{trs}	ΔCp	ΔH	Approach	Remarks	Ref
7.5	68.1		716	DSC	(1,2)	91C6
9.0	65.6		710	DSC	(1,3)	91C6

Remarks:
(1) scan rate 1K/min. Ref. 91C6 contains further data for different scan rate
(2) buffer: 1 mM phosphate
(3) buffer: 1mM pyrophosphate

Procarboxypeptidase B from porcine pancreas

pH	T_{trs}	ΔCp	ΔH	Approach	Remarks	Ref
7,5	68.9		840	DSC	(1,2)	91C6
9.0	69.0		858	DSC	(1,3)	91C6

Remarks:
(1) scan rate 1K/min Ref. 91C6 contains further data for different scan rate
(2) buffer: 1 mM phosphate
(3) buffer: 1 mM pyrophosphate

Procarboxypeptidase B from porcine pancreas, fragment, globular activation domain

pH	T_{trs}	ΔCp	ΔH	Approach	Remarks	Ref
3.0	83.8	0.39±0.16	313	DSC	(1)	91C6
7.5	73.6	0.39±0.16	313	DSC	(2)	91C6
7.5	74.2	0.39±0.16	299	DSC	(3)	91C6
9.0	67.8	0.39±0.16	297	DSC	(4)	91C6

Remarks:
(1) buffer: 20 mM glycine
(2) buffer: 1 mM phosphate
(3) buffer: 20 mM phosphate
(4) buffer: 1 mM pyrophosphate

Activation domain of human procarboxypeptidase A2 (ADA2h)

pH	T_{trs}	ΔCp	ΔH	Approach	Ref
7.0	77	3.6±1.4	199	DSC	95V6

Catabolic Activator Protein

Catabolic activator protein from *E. coli*

Transition	pH	T_{trs}	ΔCp	ΔH	Approach	Remarks	Ref
overall	7	66.4±0.1	1.8±1.8	547±25	DSC		88G1
In the presence of cAMP:							
overall	7	67–89	6.7±2.3	603±29	DSC	(1)	88G1
trans. (1)	7	67.1–70.8		153±7	DSC	(1)	88G1
trans. (2)	7	71.5–82.6		264±9	DSC	(1)	88G1
trans. (3)	7	75.6–89.4		142±10	DSC	(1)	88G1

Catabolic activator protein from *E. coli* (continued)

Transition	pH	T_{trs}	ΔCp	ΔH	Approach	Remarks	Ref
In the presence of cGMP:							
overall	7	64–76	3.1±2.8	595±20	DSC	(2)	88G1
trans. (1)	7	64.7–71.8		212±8	DSC	(2)	88G1
trans. (2)	7	68.6–76.8		380±16	DSC	(2)	88G1

Remarks:
(1) cAMP concentration from 0.17 to 12.47 mM
(2) cGMP concentration from 0.17 to 15.0 mM

Cellulase

Cellulase from *Trichoderma reesei*

Enzyme	pH	T_{trs}	ΔCp	ΔH	Approach	Remarks	Ref
CBH I	4.8	64.35		1529	DSC	(1)	92B1
CBH II	4.8	64.05		1430	DSC	(1)	92B1
EG I	4.8	64.55		1212	DSC	(1)	92B1
EG II	4.8	75.05		1966	DSC	(1)	92B1

Remark:
(1) the transition was resolved into two subtransitions by deconvolution

Endoglucanase III, single-domain cellulase from *Trichoderma reesei*

	pH	T_{trs}	ΔCp	ΔH	Approach	Ref
	5.5	59.0	13.4	707	DSC	96A3

Ceruloplasmin

Ceruloplasmin from various species

Form/Species	pH	T_{trs}	ΔCp	ΔH	Approach	Remarks	Ref
sheep	6.8	71.2		2300	DSC	(1)	90D2
sheep, apoprotein							
	6.8	57.2		1565	DSC	(2)	90D2
chicken	7.4	82.1		1435	DSC	(3)	90D2
turtle	7.4	57.8		1505	DSC	(3)	90D2

Remarks:
(1) sheep, native protein: the transition was resolved into four subtransitions by deconvolution
(2) sheep, apo protein: resolved into two subtransitions
(3) chicken and turtle, native proteins: resolved into three subtransitions by deconvolution

Chameleon Sequence, see protein G

Chaperone, see also barnase in the presence of chaperone

Chaperone, see also DnaK

CheY

CheY, globular protein involved in chemotaxis

pH	T_{trs}	ΔCp	ΔH	Approach	Remark	Ref
7	54	7.1	247	heat	(1)	93D3

Remark:
(1) ΔCp was estimated using additionally GuHCl and urea unfolding experiments, see table 1

CheY, chemotactic protein from *E. coli*

pH	T_{trs}	ΔCp	ΔH	Approach	Remarks	Ref
2.5	45±1	3.7±1	245±20	DSC	(1)	93F3
10.3	55±1	3.5±1	280±20	DSC	(1)	93F3
7.0	58±3	2.9±1.5	350±30	DSC	(1)	93F3
7.0	43±3	2.5±1.5	280±20	DSC	(1,2)	93F3

Remarks:
(1) treatment of the calorimetric recording by a model that includes an association-dissociation equilibrium between unfolded monomer and intermediate dimer with the following data

pH	T(ass.)	ΔH(ass.)	T of Cp(max.)
2.5	62	155±20	39
10.3	87	135±20	47
7.0	116	165±30	45
7.0 (2M urea)	49	155±30	40

CheY, bacterial chemotaxis protein from *Salmonella typhimurium*

pH	T_{trs}	ΔCp	ΔH	Remark	Ref
7.0		7.1–9.6		(1)	95D3

Remark:
(1) ΔCp was determined by cold denaturation and thermal denaturation in the absence and in the presence of GuHCl and urea at temperatures ranging from –10 to 80°C

Chromatin

Chromatin, deconvolution

Transition	pH	T_{trs}	ΔCp	ΔH	Approach	Remarks	Ref
trans. (1)	7.8	66.2±1.6		8.4±2.1	DSC	(1,2)	91C3
trans. (2)	7.8	73.9±0.2		7.9±1.3	DSC	(1)	91C3
trans. (3)	7.8	90.6±0.4		15.0±4.2	DSC	(1)	91C3
trans. (4)	7.8	100.8±1.5		18.0±2.9	DSC	(1)	91C3
trans. (5)	7.8	106.6±1.0		27.6±3.8	DSC	(1)	91C3

Remarks:
(1) ΔH in kJ per moles of base pairs
(2) trans. (1) was identified as the histone conformational transition

Chymopapain, see papain

Chymotrypsin

α-Chymotrypsin (mean values)

pH	T_{trs}	ΔCp	ΔH	Approach	Ref
3.8	57	12.6	685	DSC	74P1
3.8	57	12.6	685	DSC	79P4
3.8	57	12.6	685	DSC	74T3
7	49	17.3±2.5	814±18	DSC	85F2

α-Chymotrypsin, pH dependence (data from ref. 74P1)

pH	T_{trs}	ΔCp	ΔH	Approach	Ref
2.21	38	11.5	443	DSC	74P1
2.44	43	12.1	506	DSC	74P1
2.60	45	12.1	548	DSC	74P1
2.82	49	11.6	590	DSC	74P1
3.15	52.5	11.9	622	DSC	74P1
3.30	55	13.7	643	DSC	74P1
3.80	57	13.2	675	DSC	74P1

α-Chymotrypsin, temperature dependence of the heat capacity change at protein unfolding an aqueous solution, and related thermodynamic quantities

T	$\Delta H(exp)$	$\Delta S(exp)$	$\Delta Cp(exp)$	Approach	Remarks	Ref
5	−21	−260	14.9	DSC	(1–3)	95M3
25	268	746	14.1	DSC	(1–3)	95M3
50	598	1813	12.3	DSC	(1–3)	95M3
75	862	2602	8.8	DSC	(1–3)	95M3
100	1039	3099	5.4	DSC	(1–3)	95M3
125	1119	3312	1.0	DSC	(1–3)	95M3

Remarks:
(1) the data were taken from Ref. 95M3 with reference to Ref. 88P3
(2) ΔCp was obtained as described in Ref. 90P3
(3) ΔCp in kJ/mol/K, ΔH in kJ/mol and ΔS in J/mol/K

α-Chymotrypsin, native and immobilized enzyme

Transition	pH	T_{trs}	ΔCp	ΔH	Approach	Remarks	Ref
Free enzyme							
trans. (1)	6.5	51.7		305	DSC		95R4
trans. (2)	6.5	54.7		338	DSC		95R4
Immobilized enzyme							
trans. (1)	5.5	56.4		198	DSC	(1)	95R4
trans. (2)	5.5	64.5		213	DSC	(1)	95R4

Remark:
(1) immobilized α-chymotrypsin on CPC-Silica

Chymotrypsin Inhibitor

Chymotrypsin inhibitor

pH	T_{trs}	ΔCp	ΔH	Approach	Remarks	Ref
2.2	41.4	3.3	182	DSC	(1)	91J1
2.5	47	3.3	206	DSC	(1)	91J1
2.8	55	3.3	221	DSC	(1)	91J1
3.2	63.8	3.3	255	DSC	(1)	91J1
3.5	70.8	3.3	282	DSC	(1)	91J1

Remark:
(1) ΔCp was determined from ΔH versus T_{trs}

Chymotrypsin inhibitor 2, temperature dependence of the heat capacity change at protein unfolding in aqueous solution, and related thermodynamic quantities

T	$\Delta H(exp)$	$\Delta S(exp)$	$\Delta Cp(exp)$	Approach	Remarks	Ref
5	66	119	3.4	DSC	(1–3)	95M3
25	135	360	3.6	DSC	(1–3)	95M3
50	221	636	3.3	DSC	(1–3)	95M3
75	292	850	2.4	DSC	(1–3)	95M3
100	341	986	1.5	DSC	(1–3)	95M3
125	362	1041	0.2	DSC	(1–3)	95M3

Remarks:
(1) the data were taken from Ref. 95M3 with reference to Refs. 91J1 and 91J2
(2) ΔCp was obtained as described in Ref. 90P3
(3) ΔCp in kJ/mol/K, ΔH in kJ/mol and ΔS in J/mol/K

Chymotrypsin inhibitor CI2

pH	T_{trs}	ΔCp	ΔH	Approach	Remarks	Ref
6.3	25	4.06±17	113±13	heat	(1)	96T1
6.3	25	4.06±13	100±3	kinetics	(2)	96T1

Remarks:
(1) from ΔH versus T_{trs}, extrapolated to 25°C
(2) from kinetics of folding and unfolding

Chymotrypsin inhibitor 2, truncated form, wild-type and mutants

Mutant	pH	T_{trs}	ΔCp	ΔH	Approach	Remarks	Ref
wild-type	3.0	64.0±0.2	3.0	255±10	DSC	(1)	93J1
Leu27→Ala	3.0	48.5	3.0	188	DSC	(1)	93J1
Val38→Ala	3.0	61.1	3.0	245	DSC	(1)	93J1
Ile39→Val	3.0	57.7	3.0	225	DSC	(1)	93J1
Val70→Ala	3.0	51.3	3.0	203	DSC	(1)	93J1
Ile76→Val	3.0	65.4	3.0	253	DSC	(1)	93J1
Ile76→Ala	3.0	38.3	3.0	126	DSC	(1)	93J1
wild-type	3.5	73.8	3.0	280	DSC	(1)	93J1
Ile39→Val	3.5	67.3	3.0	257	DSC	(1)	93J1
Ile48→Val	3.5	69.1	3.0	261	DSC	(1)	93J1

Chymotrypsin inhibitor 2, truncated form, wild-type and mutants (continued)

Mutant	pH	T_{trs}	ΔCp	ΔH	Approach	Remarks	Ref
Ile48→Ala	3.5	52.5	3.0	188	DSC	(1)	93J1
Val66→Ala	3.5	45.3	3.0	156	DSC	(1)	93J1
Leu68→Ala	3.5	52.3	3.0	195	DSC	(1)	93J1
Ile76→Ala	3.5	50.3	3.0	172	DSC	(1)	93J1
double mutant (Ile48→Ala and Ile76→Val)							
	3.5	52.8	3.0	147	DSC	(1)	93J1

Remark:
(1) truncated protein: first 19 residues were deleted and Leu20 replaced by Met

Chymotrypsin inhibitor, wild-type and mutant

Mutant	pH	T_{trs}	ΔCp	ΔH	Approach	Ref
wild-type	3.75	72.2	3.3	254	DSC	94G1
Glu45→Ala	3.75	69.9		243	DSC	94G1

Chymotrypsin inhibitor 2 from barley (CI2)

Mutant	pH	T_{trs}	ΔH	Approach	Ref
wild-type	3.0	63.7	251	DSC	94M8
wild-type	3.5	73.6	280	DSC	94M8
Ser31→Ala	3.0	57.9	222	DSC	94M8
Ser31→Gly	3.5	69.0	268	DSC	94M8
double mutant (Glu33→Ala and Glu34→Ala)					
	3.0	60.1	243	DSC	94M8
triple mutant (Ser31→Gly, Glu33→Ala and Glu34→Ala)					
	3.0	56.5	230	DSC	94M8
triple mutant (Ser31→Ala, Glu33→Ala and Glu34→Ala)					
	3.0	56.2	226	DSC	94M8

Barley chymotrypsin inhibitor 2 (CI-2), complexes of the fragments CI-2(20–59) and CI-2(60–83)

Mutant	pH	T_{trs}	ΔH	Approach	Remarks	Ref
wild-type	6.3	46.4	143±4	heat	(1–3)	95R7
Thr22→Ala	6.3	43.5		heat	(1,2)	95R7
Leu27→Ala	6.3	23.9	145±34	heat	(1,2)	95R7
Ser31→Ala	6.3	38.0	148±11	heat	(1,2)	95R7
Glu33→Asn	6.3	42.3	151±20	heat	(1,2)	95R7
Lys36→Ala	6.3	33.8	114±19	heat	(1,2)	95R7
Lys37→Ala	6.3	41.2	146±17	heat	(1,2)	95R7
Lys37→Gly	6.3	35.6	157±17	heat	(1,2)	95R7
Val38→Ala	6.3	41.7	126±6	heat	(1,2)	95R7
Ile39→Val	6.3	31.6	110±2	heat	(1,2)	95R7
Lys43→Ala	6.3	40.6	103±13	heat	(1,2)	95R7
Ile48→Val	6.3	39.2	116±15	heat	(1,2)	95R7
Ile49→Val	6.3	47.8	213±26	heat	(1,2)	95R7
Val53→Ala	6.3	43.6	169±21	heat	(1,2)	95R7
Val53→Thr	6.3	38.1	142±20	heat	(1,2)	95R7
Leu51→Ala	6.3	37.3	178±26	heat	(1,2)	95R7
Leu51→Val	6.3	45.9	156±26	heat	(1,2)	95R7

Barley chymotrypsin inhibitor 2 (CI-2), complexes of the fragments CI-2(20–59) and CI-2(60–83) (continued)

Mutant	pH	T_{trs}	ΔH	Approach	Remarks	Ref
Phe69→Leu	6.3	32.3	115±36	heat	(1,2)	95R7
Val70→Ala	6.3	29.6	140±21	heat	(1,2)	95R7
Ile76→Val	6.3	47.5	145±12	heat	(1,2)	95R7
double mutant (Ser31→Ala and Glu34→Ala)						
	6.3	38.0	163±11	heat	(1,2)	95R7
multiple mutant (Ser31→Ala, Glu33→Ala, and Glu34→Ala)						
	6.3	32.3	222±20	heat	(1,2)	95R7

Remarks:
(1) the fragments associate to give native-like structure
(2) thermal denaturation at 5 µM complex in 10 mM phosphate buffer pH 6.3; the data were fitted to theoretical curves for simultaneous dissociation and unfolding
(3) see also Refs. 93J1 and 94O3

Chymotrypsinogen

Chymotrypsinogen

	pH	T_{trs}	ΔCp	ΔH	Approach	Remarks	Ref
	2	42	15.9	418	DSC		70J
	2	50	8.4	515	DSC		70B
	2.07		15.9±1.1		pressure		71H1
	3	54	11.7	586	DSC		70J
	4	61.9	14	637	DSC	(1)	84F4
	5	62	15.2	620	DSC		71P
cold denaturation:							
	0.8	−11.5	≫0	−284		(2)	85F1
	3	−33		−284			88F

Remarks:
(1) from dependence on protein concentration
(2) undercooled solution without cryoprotectants

Chymotrypsinogen A, modified by reductive alkylation

Form	pH	T_{trs}	ΔCp	ΔH	Approach	Remarks	Ref
native protein	3.0	54.2	10.8	498	DSC	(1)	91F2
methylated (92%)	3.0	56.0	12.0	521	DSC	(1)	91F2
ethylated (56%)	3.0	56.0	16.6	557	DSC	(1)	91F2
n-butylated (49%)	3.0	55.7	16.5	564	DSC	(1)	91F2

Remark:
(1) Ref. 91F2 contains futher data for varying extent of modification and other pH values

Cold-Shock Protein

CspB, cold-shock protein from *Bacillus subtilis*

Buffer	pH	T_{trs}	ΔCp	ΔH	Approach	Remarks	Ref
50 mM HEPES	7.5	52.8	3.7±0.7	110	DSC	(1)	94M4
50 mM ImH	7.5	51.0	3.7±0.7	105	DSC	(2)	94M4
50 mM NaP	7.5	53.4	3.7±0.7	154	DSC	(3)	94M4

Remarks:
(1) unfolding in HEPES buffer follows a dimer→monomer transition (N_2→2U)
(2) unfolding in imidazole (ImH) buffer follows a dimer→monomer transition (N_2→2U)
(3) unfolding in Na phosphate (NaP) buffer follows a two-state transition (N→U)

CspB, cold-shock protein from *Bacillus subtilis*

pH	T	ΔCp	ΔH	Approach	Remarks	Ref
7.0	25	3.8±0.3	58.4±4.3	heat	(1,2)	96S4
7.0	25	3.0±0.5	64.4±4.0	kinetics	(2,3)	96S4

Remarks:
(1) thermal denaturation in the presence of urea, ΔH and ΔCp refer to zero denaturant concentration
(2) buffer: 0.1 M sodium cacodylate/HCl
(3) from the difference of activation parameters for unfolding and refolding

Colicin

Colicin A, pore-forming domain

pH	T_{trs}	ΔCp	ΔH	Approach	Remarks	Ref
3.0	39		85	DSC	(1)	93M17
3.5	49		180	DSC	(1)	93M17
4.0	56		240	DSC	(1)	93M17
5.0	65		385	DSC	(1)	93M17
7.0	69		380	DSC	(1)	93M17
8.0	67		370	DSC	(1)	93M17

Remark:
(1) the data were taken from Fig. 2 in Ref. 93M17 (rounded values)

Collagen

Collagen-like triple helical peptides

Peptide	pH	T_{trs}	$\Delta H°$	Approach	Remarks	Ref
Peptide I	1	61	661	heat	(1,2)	94V1
Peptide I	7	58	652	heat	(1,2)	94V1
Peptide I	13	61	707	heat	(1,2)	94V1
Peptide II	1	44	766	heat	(1,3)	94V1
Peptide II	7	46	736	heat	(1,3)	94V1
Peptide II	13	49	632	heat	(1,3)	94V1

Remarks:
(1) peptide I = (Pro-Hyp-Gly)$_{10}$
(2) $\Delta H°$ refers to 25°C and a two-state trimer to monomer transition
(3) peptide II = (Pro-Hyp-Gly)$_4$-(Glu-Lys-Gly)-(Pro-Hyp-Gly)$_5$

Complement Protein

Human complement C1S, subtransitions and fragments

Transition	pH	T_{trs}	ΔCp	ΔH	Approach	Remarks	Ref
Human complement protein:							
trans. (1)	7.2	37.3		330	DSC	(1)	89M6
trans. (2)	7.2	49.2		582	DSC	(1)	89M6
trans. (3)	7.2	59.5		448	DSC	(1)	89M6
56 kD fragment:							
F56-T1	7.2	36.7		251	DSC	(2)	89M6
F56-T2	7.2	63		435	DSC	(2)	89M6
24 kD fragment:							
F24-T	7.2	36.7	8.4	251	DSC	(3)	89M6

Remarks:
(1) three subtransitions resolved by deconvolution
(2) two subtransitions resolved by deconvolution
(3) two-state transition, ΔCp was estimated from Fig. 6 of Ref. 89M6.

Complement protein C9, human

Transition	pH	T_{trs}	ΔCp	ΔH	Approach	Remarks	Ref
overall	7.2	32–52	29	1104	DSC	(1)	91L9
trans. (1)	7.2	32.3		189	DSC	(2)	91L9
trans. (2)	7.2	48.1		251	DSC	(2)	91L9
trans. (3)	7.2	52.5		675	DSC	(2)	91L9

Remarks:
(1) overall heat except the observed exothermic peak, mean of four independent measurements
(2) subtransitions resolved by deconvolution, mean value

Cro Protein

Phage 434 Cro protein, mutants having changes in the hydrophobic core

Mutant	pH	T_{trs}	ΔCp	ΔH	Approach	Remarks	Ref
C-1	7.0	56	2.9	131	heat, v.H.	(1)	95D4
D-5	7.0	60	2.5	122	heat, v.H.	(2)	95D4
D-8	7.0	50	2.8	100	heat, v.H.	(2)	95D4
D-7	7.0	17	3.1	22	heat, v.H.	(2)	95D4
M-5	7.0	33	2.9	58	heat, v.H.	(3)	95D4

Remarks:
(1) C = control, C-1 variant can be taken as representative of the native sequence
(2) D = designed protein with the indicated (e.g. 5) differences from the native protein
(3) M = minimalist variant

Cro Repressor

Cro Repressor of bacteriophage λ, wild-type and mutant

Mutant	pH	T_{trs}	ΔCp	ΔH	Approach	Remarks	Ref
wild-type	7.0	46.4	7.5*	225	DSC	(1,2)	91G3
Val55→Cys – crosslinked dimer:							
trans. (1)	7.0	54.5		220	DSC	(1,3)	91G3
trans. (2)	7.0	102		250	DSC	(1,3)	91G3

Remarks:
(1) thermodynamic quantities per mole of dimer
(2) ΔCp was taken from Fig. 1 in Ref. 91G3
(3) two separate thermal transitions of the crosslinked dimer

Crystallin

α-Crystallin

Transition	pH	T_{trs}	ΔCp	ΔH	Approach	Remarks	Ref
trans. (1)	7.2	45.6±1.0		410±5	DSC	(1)	91W1
trans. (2)	7.2	60.6±1.3		2510±67	DSC	(2)	91W1

Remarks:
(1) van't Hoff heat $\Delta H^{v.H.} = 381\pm6$ kJ/mol
(2) van't Hoff heat $\Delta H^{v.H.} = 272\pm4$ kJ/mol

α-Crystallin

	pH	T_{trs}	ΔCp	ΔH	Approach	Remark	Ref
	7.4	59.8±1.6	5.7±3.0	336±9	DSC	(1)	96G4

Remark:
(1) ΔH was found to be scan rate dependent; the given value was extrapolated to infinite scan rate

γ-Crystallin, various forms

Form	pH	T_{trs}	ΔCp	ΔH	Approach	Remarks	Ref
γ-II:							
	7.2	71.5		586	heat	(1,2)	92S7
trans. (1)	7.2	50.6±2.8		619±2	DSC	(2,3)	92S7
trans. (2)	7.2	70.4±1.8		527±4	DSC	(2,3)	92S7
	7.2	58.0±1.4		791±3	DSC	(5)	92S7
γ-IIIA:							
	7.2	70.0		336	heat	(1,2)	92S7
	7.2	70.0±1.4		344±15	DSC	(1,2)	92S7
	7.2	52.3±2.9		389±17	DSC	(5)	92S7
γ-IIIB:							
	7.2	73.0		397	heat	(1,2)	92S7
	7.2	68.0±1.2		621±3	DSC	(1,2,4)	92S7
	7.2	52.0±0.7		519±1	DSC	(5)	92S7

γ-Crystallin, various forms (continued)

Form	pH	T_{trs}	ΔCp	ΔH	Approach	Remarks	Ref
γ-IVA:							
	7.2	74.0		451	heat	(1,2)	92S7
	7.2	78.0±1.4		377±4	DSC	(1,2)	92S7
	7.2	55.0±3.0		349±3	DSC	(5)	92S7

Remarks:
(1) transition monitored by spectroscopic methods (CD/fluorescence)
(2) measured in phosphate buffer
(3) from four samples, prepared separately
(4) data from the peak ranging from 37 to 74°C
(5) phosphate buffer with 0.46% tetradecyl-trimethylammonium bromide

α-Crystallin

	pH	T_{trs}	ΔCp	ΔH	Approach	Remarks	Ref
	7.2	61.8		138	DSC	(1,2)	95S17

Remarks:
(1) ΔH refers to the 20 kDa subunit of the 800,000 Da oligomer
(2) measured in 100 mM phosphate pH 7.2

γ-Crystallin, modified

Form	pH	T_{trs}	ΔCp	ΔH	Approach	Ref
unmodified	7.0	72		220	heat, v.H.	93L9
glycated	7.0	62		186	heat, v.H.	93L9

Cytochrome b₅

Cytochrome b₅ from rabbit and fragment 1–90

Form	pH	T_{trs}	ΔCp	ΔH	Approach	Ref
in aqueous solution						
	7.4	59.1±0.1	3.6±0.7	271±36	DSC	83B1
reconstituted with dimyristoyllecithin						
	7.4	63.1±0.5	6.2±1	267±25	DSC	83B1
fragment 1–90						
	7.4	70.7±0.3	6±0.9	336±12	DSC	83B1

Cytochrome b₅, bovine, tryptic fragments

Form	pH	T_{trs}	ΔCp	ΔH	Approach	Ref
Tryptic fragment, variant (Ala7-Lys90):						
oxidized	7.0	67.4±.7		172±52	heat	92N1
reduced	7.0	73.2±.8		227±16	heat	92N1
Tryptic fragment, variant (Ala1-Lys90):						
oxidized	7.0	73.0±1.1		311±12	heat	92N1
reduced	7.0	79.2±1.5		301±59	heat	92N1
Tryptic fragment, variant (Ala1-Ser104):						
oxidized	7.0	73.1				92N1

Cytochrome b$_5$, bovine, recombinant

Mutant	pH	T$_{trs}$	ΔCp	ΔH	Approach	Ref
oxidized form:						
Ala7-Lys90	7.0	67.4±0.7		172±52	heat	93H6
Ala1-Ser104	7.0	73.1±0.4		264±44	heat	93H6
Ala1-Lys90	7.0	73.2±0.8		277±16	heat	93H6
reduced form:						
Ala7-Lys90	7.0	73.0±1.1		311±12	heat	93H6
Ala1-Ser104	7.0	78.7±1.4		420±42	heat	93H6
Ala7-Lys90	7.0	79.2±1.5		301±59	heat	93H6

Apocytochrome b$_5$ from rabbit liver

pH	T$_{trs}$	ΔCp	ΔH	Approach	Remarks	Ref
7.4	48.5±0.5	4.2±.5	149.3±6.7	DSC	(1)	93P3
6–9	44–52	4.2±.6	133–168	DSC	(2)	93P3

Remarks:
(1) ΔH and ΔCp are mean values from seven calorimetric recordings
(2) ΔCp is based on ΔH versus T$_{trs}$

Cytochrome b$_{562}$

Cytochrome b$_{562}$

Form	pH	T$_{trs}$	ΔCp	ΔH	Approach	Ref
ferri	7.0	67.2±0.5	10.0±1.7	393±21	heat	91F1
apo	7.0	52.3±0.9	4.6±1.3	222±17	heat	91F1

Cytochrome c

The data are arranged as folllows:

a) temperature and pH dependence,
b) various transitions,
c) modified forms,
d) in the presemce of ligands, osmolytes etc.

a) Temperature and pH dependence

Bovine cytochrome c, ferric form

pH	T$_{trs}$	ΔCp	ΔH	Approach	Remarks	Ref
4.8	78±0.3	7.5±0.2	444±12	DSC		86P2
	6.61			DSC		74P2
	7.3			DSC		79P4
	7.3			DSC		74P1
specific transitions:						
3.2	72	5	452	DSC	(1)	89P7
2.1	56	0.56	169	DSC	(2)	89P7

Remarks:
(1) the data refer to the N→D transition
(2) the data refer to the (N*)→D transition of the native like protein with bound chloride in the presence of 0.5 M NaCl

Bovine cytochrome c, ferric form, pH dependence (data from ref. 74P1)

pH	T_{trs}	ΔCp	ΔH	Approach	Ref
2.81	52.5		251	DSC	74P1
2.92	54	6.0	272	DSC	74P1
3.05	59	6.2	294	DSC	74P1
3.12	61	6.4	311	DSC	74P1
3.18	62	5.7	322	DSC	74P1
3.40	66	6.2	354	DSC	74P1
3.75	70	7.1	391	DSC	74P1
3.94	72	6.9	403	DSC	74P1
4.52	77	6.2	431	DSC	74P1
4.83	78	6.4	447	DSC	74P1

Cytochrome c

pH	T_{trs}	ΔCp	ΔH	Approach	Remarks	Ref
3.2	60	1.36		DSC	(1–3)	94H8

Remarks:
(1) ΔCp is the experimental value
(2) Ref. 94H8 contains corrections of ΔCp for volume effects
(3) T_{trs} was taken from the figures in Ref. 94H8

Cytochrome c, temperature dependence of the heat capacity change at protein unfolding

T	ΔCp(calc)	ΔCp(exp)	Approach	Remarks	Ref
5	7.5	6.2	DSC	(1–3)	90P3
25	7.4	6.9	DSC	(1–3)	90P3
50	6.7	6.1	DSC	(1–3)	90P3
75	5.0	4.7	DSC	(1–3)	90P3
100	3.3	2.6	DSC	(1–3)	90P3
125	1.1	0.6	DSC	(1–3)	90P3

Remarks:
(1) the calculated value is based on heat capacity of the constituent amino acids minus experimental value for the native protein
(2) the experimental value is based on heat capacity of the heat denatured protein minus experimental value for the native protein
(3) the partial molar heat capacity of hemine is included into the heat capacity change

Cytochrome c, temperature dependence of the heat capacity change at protein unfolding in aqueous solution, and related thermodynamic quantities

T	ΔH(exp)	ΔS(exp)	ΔCp(exp)	Approach	Remarks	Ref
5	−53	−319	6.9	DSC	(1–3)	95M3
25	89	174	6.8	DSC	(1–3)	95M3
50	268	752	6.1	DSC	(1–3)	95M3
75	421	1210	4.4	DSC	(1–3)	95M3
100	532	1520	2.8	DSC	(1–3)	95M3
125	593	1681	0.6	DSC	(1–3)	95M3

Remarks:
(1) the data were taken from Ref. 95M3 with reference to Refs. 93M5 and 93P12
(2) ΔCp was obtained as described in Ref. 90P3
(3) ΔCp in kJ/mol/K, ΔH in kJ/mol and ΔS in J/mol/K

Cytochrome c, horse, deuterated (D) protein in D_2O and undeuterated (H) protein in H_2O

	pH	T_{trs}	ΔCp	ΔH	Approach	Remarks	Ref
D-cyt. c	2.5–5	35–81	4.7±0.2	140–350	DSC in D_2O	(1,2)	95M2
H-cyt. c	2.7–5.5	41–80	5.5±0.3	240–425	DSC in H_2O	(1)	95M2

Remarks:
(1) T_{trs} and ΔH were taken from Fig. 1 in Ref. 95M2
(2) $\Delta(\Delta H)$ between D-cyt. c and H-cyt. c denaturation amounts to $-(43\pm11)$ kJ/mol

Cytochrome c from horse heart

Form	pH	T_{trs}	ΔCp	ΔH	Approach	Remarks	Ref
ferric	4.0	77		250	DSC	(1)	93Y2
ferric	4.0	67		146	heat	(4)	73T
ferric	4.0	67.7	3.8	341	DSC		91V1
ferric	7.0	82			DSC		88L
ferric	7.0	84.4			DSC	(2)	93Y2
ferric	7.0	84.1		410	DSC	(5)	94B1
ferrous	7.0	103			DSC	(3)	93Y2

Remarks:
(1) buffer: 0.125 M sodium acetate
(2) buffer: 25 mM phosphate
(3) buffer: 25 mM phosphate, 1 mM ascorbate
(4) buffer: 15 mM $NaClO_4$, 10 mM sodium acetate
(5) the paper contains further data on cytochrome c denaturation in the presence of heparin

Cytochrome c from horse heart

	pH	T_{trs}	ΔCp	ΔH	Approach	Remarks	Ref
	3.0–5.0	38–80	6.3		DSC	(1)	92F4
	3.0–5.0	101.9±2	6.14±0.67	565±21	DSC	(2,3)	92F4

Remarks:
(1) in the absence of methanol
(2) from measurements in the presence of methanol (0–5% vol/vol) at pH 3.0–5.0
(3) T_{trs} is the convergence temperature for ΔH and ΔS. T_{trs} for ΔS amounts to $T_{trs} = 119\pm2$°C,
    $\Delta S = 1627\pm8$ J/K/mol

b) Various transitions

Ferricytochrome c from horse heart, high- and low-temperature transitions

Transition	pH	T_{trs}	ΔCp	ΔH	Approach	Remarks	Ref
trans. (1)	7.0	25	1.7	134	heat	(1,2)	93M18
trans. (2)	7.0	79	3.3	163	heat	(1,2)	93M18
trans. (1)	7.0	25	0	92	heat	(1,3)	93M18
trans. (2)	7.0	78	7.9	339	heat	(1,3)	93M18
trans. (1)	7.0	29	3.3	134	heat	(1,4)	93M18
trans. (2)	7.0	75	9.2	356	heat	(1,4)	93M18
trans. (1)	7.0	43	0.4	63	heat	(1,5)	93M18
trans. (2)	7.0	77	9.6	393	heat	(1,5)	93M18

Ferricytochrome c from horse heart, high- and low-temperature transitions (continued)

Transition	pH	T_{trs}	ΔC_p	ΔH	Approach	Remarks	Ref
trans. (1)	7.0	34	1.7	67	heat	(1,6)	93M18
trans. (2)	7.0	78	10.0	385	heat	(1,6)	93M18
trans. (1)	7.0	23	0	105	heat	(1,7)	93M18
trans. (2)	7.0	77	8.8	360	heat	(1,7)	93M18

Remarks:
(1) the thermal transition was resolved into a low-temperature component and a high-temperature component, for details of the approach, see Ref. 93M18
(2) protein dissolved in water, without buffer
(3) protein in 0.2 M KCl
(4) protein in 0.1 M $NaClO_4$
(5) protein in 0.1 M cacodylate buffer
(6) protein in 0.1 M sodium phosphate buffer
(7) buffer: mixture of 0.1 M cacodylate and 0.1 M $NaClO_4$

Cytochrome c from horse, various transitions

Transition	pH	T	ΔC_p	ΔH	Approach	Remarks	Ref
trans. (N→D)	2.0–3.2	0	5.76	−1560	heat	(1,2)	92K10
trans. (IIb→D)	2.0–3.2	0	4.69±0.57	−1257±108	heat	(1,3)	92K10
trans. (IIc→D)	2.0–3.2	0	1.68±0.06	−399±37	heat	(1,3)	92K10

Remarks:
(1) the transitions were resolved by multidimensional spectroscopy, T is the reference temperature, T_{trs} was in the range of 20 to 60°C, and cold denaturation of N→D below 0°C
(2) measured in low-salt buffer
(3) measured in high-salt buffer (0.1 to 0.5 M KCl)

Cytochrome c from horse, native→unfolded (N→U) and molten globule→unfolded (I→U) transitions [remarks (1,2)]

Transition	pH	T_{trs}	ΔC_p	ΔH	Approach	Ref
N→U	3–4	20	5.27±0.91	45.5±49.2	heat, CD	94H2
N→U	3–4	20	4.74±0.11	112.3±3.3	DSC, $\Delta H^{v.H.}$	94H2
N→U	3–4	20	5.34±0.10	93.4±3.5	DSC, ΔH^{cal}	94H2
I→U	2	20	1.57±0.14	86.4±4.3	heat, CD	94H2
I→U	2	20	2.67±0.07	69.1±1.8	DSC, $\Delta H^{v.H.}$	94H2
I→U	2	20	1.79±0.54	131.4±13.5	DSC, ΔH^{cal}	94H2

Remarks:
(1) reference temperature 20°C
(2) the molten globule was obtained from intact cytochrome c and variously acetylated forms

Cytochrome c, native and A-state

Mutant/Transition	pH	T_{trs}	ΔH	Approach	Remarks	Ref
wild-type, A→U	2.0	65±0.3	172±4	heat	(1,2)	96C15
wild-type, N→U	2.0	83±0.2	448±25	heat	(1,2)	96C15
Leu94→Val, A→U	2.0	51±0.6	142±8	heat	(1,2)	96C15
Leu94→Val, N→U	2.0	79±0.3	435±29	heat	(1,2)	96C15

Remarks:
(1) buffer: 20 mM phosphate, 1.0 M KCl
(2) transition monitored by CD, two-state treatment

Cytochrome c from horse, salt-induced refolding, determined by isothermal titration calorimetry (ITC), formation of the molten globule

Salt	pH	T	ΔC_p	ΔH	Approach	Remarks	Ref
NaClO$_4$	1.8	20	-2.4 ± 1.0	-122.5	ITC	(1)	94H5
NaClO$_4$	1.8	25		-134.7	ITC	(1)	94H5
NaClO$_4$	1.8	30		-177.7	ITC	(1)	94H5
NaClO$_4$	1.8	35		-160.6	ITC	(1)	94H5
NaClO$_4$	1.8	40		-169.0	ITC	(1)	94H5
Na$_2$SO$_4$	1.8	20	-1.9 ± 0.9	-119.6	ITC	(1)	94H5
Na$_2$SO$_4$	1.8	25		-140.9	ITC	(1)	94H5
Na$_2$SO$_4$	1.8	30		-172.4	ITC	(1)	94H5
Na$_2$SO$_4$	1.8	35		-168.1	ITC	(1)	94H5
Na$_2$SO$_4$	1.8	40		-153.9	ITC	(1)	94H5

Remark:
(1) ΔH is the heat extrapolated to zero salt concentration, ΔC_p is from ΔH versus temperature

Cytochrome c from horse, thermal unfolding of the molten globule in the presence of NaClO$_4$

Salt Concentration	pH	T_{trs}	ΔC_p	ΔH	Approach	Remarks	Ref
20 mM	1.8	34.4	1.1 ± 0.5	135.3	DSC	(1–3)	94H5
30 mM	1.8	40.0		131.5	DSC	(1)	94H5
40 mM	1.8	42.6		145.8	DSC	(1)	94H5
50 mM	1.8	44.1		141.1	DSC	(1)	94H5
60 mM	1.8	46.9		148.1	DSC	(1)	94H5
10 mM	1.8	25.8	1.8 ± 0.1	117.9	heat, v.H.	(4)	94H5
20 mM	1.8	37.0		135.9	heat, v.H.		94H5
50 mM	1.8	47.0		159.1	heat, v.H.		94H5
100 mM	1.8	52.7		164.8	heat, v.H.		94H5

Remark:
(1) ΔH is the heat determined by calorimetry (ΔH^{cal})
(2) ΔC_p is from ΔH^{cal} versus T_{trs}
(3) ΔC_p from $\Delta H^{v.H.}$ versus T_{trs} amounts to $\Delta C_p = 1.3\pm0.6$ kJ/mol/K
(4) ΔC_p is from $\Delta H^{v.H.}$ versus T_{trs}

c) Modified forms

Ferricytochrome c, cysteine mutants

Mutant	pH	T_{trs}	ΔC_p	ΔH	Approach	Ref
Cys102→Thr	4.6	54.7 ± 0.5		374	heat	93A3
Cys102→Thr	6.0	59.6 ± 0.5		455	heat	93A3
(Phe10→Met and Cys102→Thr)						
	4.6	37.6 ± 0.5		264	heat	93A3
	6.0	47.5 ± 0.5		317	heat	93A3
(Phe10→Cys, S-methyl cysteine derivative, and Cys102→Thr)						
	4.6	38.7 ± 0.5		266	heat	93A3
	6.0	48.3 ± 0.5		341	heat	93A3

Ferricytochrome c from horse heart, in the presence and in the absence of lipid

Transition	pH	T_{trs}	ΔCp	ΔH	Approach	Remarks	Ref
trans. (1)	7.4	63.8		159	DSC	(1)	91M2
trans. (2)	7.4	80.7		393	DSC	(1)	91M2
overall	7.4	68.2		293	DSC	(2)	91M2

Remarks:
(1) two transitions detected by DSC and IR on melting of ferricytochrome c in 50 mM HEPES with 100 mM NaCl
(2) in the presence of dimyristoylphosphatidylglycerol and dimyristoylphosphatidylcholine (1:1). Ref. 91M2 contains further data for ferricytochrome c melting in various lipid mixtures.

Porphyrin-cytochrome c from horse heart

pH	T_{trs}	ΔCp	ΔH	Approach	Remarks	Ref
4.8		2.9±0.3	12.1	heat	(1,2)	96H1

Remarks:
(1) reference temperature 20°C
(2) measured by CD

d) In the presemce of ligands, osmolytes etc.

Cytochrome c from horse heart, in the presence of osmolytes (amino acids)

Concentration	pH	T_{trs}	ΔCp	ΔH	Approach	Remark	Ref
none	3.0	45.0	2.81±0.12	67±1.0	heat	(1)	94T5
0.25 M Ala	3.0	45.4		67±2.6	heat, v.H.		94T5
0.50 M Ala	3.0	45.8		65±0.4	heat, v.H.		94T5
0.70 M Ala	3.0	48.0		60±2.3	heat, v.H.		94T5
0.25 M Arg	3.0	41.9		59±1.0	heat, v.H.		94T5
0.50 M Arg	3.0	40.0		51±1.4	heat, v.H.		94T5
0.75 M Arg	3.0	39.8		53±2.0	heat, v.H.		94T5
0.25 M Gly	3.0	45.3		55±1.9	heat, v.H.		94T5
0.50 M Gly	3.0	46.0		50±1.2	heat, v.H.		94T5
0.75 M Gly	3.0	46.3		47±1.3	heat, v.H.		94T5
1.00 M Gly	3.0	48.3		44±2.1	heat, v.H.		94T5
0.09 M His	3.0	43.5		62±2.1	heat, v.H.		94T5
0.17 M His	3.0	43.0		47±1.2	heat, v.H.		94T5
0.08 M Ile	3.0	45.0		67±3.0	heat, v.H.		94T5
0.12 M Ile	3.0	45.2		66±1.3	heat, v.H.		94T5
0.16 M Ile	3.0	45.2		65±2.7	heat, v.H.		94T5
0.05 M Leu	3.0	45.0		70±4.5	heat, v.H.		94T5
0.10 M Leu	3.0	45.0		69±1.7	heat, v.H.		94T5
0.25 M Lys	3.0	49.4		60±1.7	heat, v.H.		94T5
0.50 M Lys	3.0	50.0		54±1.6	heat, v.H.		94T5
0.70 M Lys	3.0	51.5		57±1.0	heat, v.H.		94T5
0.10 M Met	3.0	45.5		63±2.8	heat, v.H.		94T5
0.21 M Met	3.0	45.8		57±3.6	heat, v.H.		94T5
0.05 M Phe	3.0	45.2		67±1.6	heat, v.H.		94T5
0.10 M Phe	3.0	45.2		65±2.4	heat, v.H.		94T5
0.25 M Pro	3.0	45.8		66±3.4	heat, v.H.		94T5
0.50 M Pro	3.0	46.5		65±4.2	heat, v.H.		94T5
0.75 M Pro	3.0	47.5		61±2.7	heat, v.H.		94T5
1.00 M Pro	3.0	49.3		56±2.9	heat, v.H.		94T5

Cytochrome c from horse heart, in the presence of osmolytes (amino acids) (continued)

Concentration	pH	T_{trs}	ΔCp	ΔH	Approach	Remark	Ref
0.25 M Ser	3.0	46.1		54±1.4	heat, v.H.		94T5
0.50 M Ser	3.0	46.9		51±1.0	heat, v.H.		94T5
0.75 M Ser	3.0	48.0		50±1.0	heat, v.H.		94T5
1.00 M Ser	3.0	48.5		49±0.8	heat, v.H.		94T5
0.25 M Thr	3.0	46.0		63±3.4	heat, v.H.		94T5
0.50 M Thr	3.0	46.7		57±2.0	heat, v.H.		94T5
0.10 M Val	3.0	46.3		59±2.2	heat, v.H.		94T5
0.20 M Val	3.0	46.8		54±1.9	heat, v.H.		94T5
0.30 M Val	3.0	46.8		53±1.8	heat, v.H.		94T5

Remark:
(1) ΔCp was determined combining results of thermal and GuHCl denaturation

Ferricytochrome c in the presence of phosphate (P_i) and nucleotides

Ligand	pH	T_{trs}	ΔCp	ΔH	Approach	Remarks	Ref
P_i	7.0	85.6		241	DSC	(1)	95A4
P_i, AMP, ADP	2.5–4.5	35–75	5.8		DSC	(2,3)	95A4
ATP	7.0	75.0		280	DSC	(1,4)	95A4
GTP	7.0	74.6		260	DSC	(1)	95A4
CTP	7.0	75.9		300	DSC	(1)	95A4
UTP	7.0	76.7		275	DSC	(1)	95A4
ADP	7.0	78.5		280	DSC	(1)	95A4
AMP	7.0	84.3		290	DSC	(1)	95A4

Remarks:
(1) conditions: 2 mM/l phosphate buffer, 5 mM/l nucleotides, 110 μM/l cytochrome c
(2) in the presence of P_i, AMP and ADP
(3) general expression: $\Delta H^{cal} = 5.8 \times T_{trs,K} - 1645.5$, negative ΔCp at pH 4.5–7.0
(4) general expression: $\Delta H^{cal} = 3.8 \times T_{trs,K} - 967.9$, negative ΔCp at pH 4.5–7.0

Cytochrome c₁

Cytochrome c₁

Form	pH	T_{trs}	ΔCp	ΔH	Approach	Remarks	Ref
ferrous	7.06	65.3		230	DSC	(1)	93Y2
ferrous	7.0	72		502	DSC	(2)	93Y2
ferrous	7.06	65.3		406	DSC	(3)	93Y2
ferri	7.06	62.3		280	DSC		93Y2
ferri	7.06	65.3		623	DSC	(3)	93Y2

Remarks:
(1) buffer: 25 mM phosphate
(2) buffer: 20 mM phosphate, 1.5 M KCl
(3) in the presence of phospholipid

Cytochrome c$_3$

Cytochrome c$_3$, various proteins

Mutant/Remarks	pH	T$_{trs}$	ΔH	Approach	Remarks	Ref
D.d.N. (1)	7.6	73	268	DSC		95F2
D.v.H. (2)	7.6	121	~710	DSC	(3)	95F2
D.v.H. Hmc (4)	7.6	81	1675	DSC		95F2

Remarks:
(1) from *Desulfovibrio desulfuricans Norway*, tetraheme cytochrome c$_3$, M = 13,000 Da
(2) from *Desulfovibrio desulfuricans Norway*, octaheme cytochrome c$_3$, M = 26,000 Da
(3) extrapolated value
(4) from *Desulfovibrio vulgaris Hildenborough*, high molecular (M = 65,000 Da) cytochrome c$_3$ with 16 hemes

Cytochrome c-552

Cytochrome c-552

pH	T$_{trs}$	ΔCp	ΔH	Approach	Ref
		5.2–7		GuHCl, v.H.	78N1

Iso-1-Cytochrome c

Iso-1-cytochrome c with blocked Cys107 (-SCH$_3$), wild-type and mutants

Mutant	pH	T$_{trs}$	ΔCp	ΔH	Approach	Ref
wild-type	6	46.5±1	7.1	330±21	heat, v.H.	88H
Lys32→Gln	6	41.5±1	7.1	385±92	heat, v.H.	88H
Lys32→Leu	6	44.9±1	7.1	372±25	heat, v.H.	88H
Lys32→Trp	6	37.4±1	7.1	268±25	heat, v.H.	88H
Lys32→Tyr	6	36.9±1	7.1	272±21	heat, v.H.	88H

Iso-1-cytochrome c, mutants

Mutant	pH	T_{trs}	ΔCp	ΔH	Approach	Remarks	Ref
wild-type	6	46.5	8.3	331±13	heat	(1–3)	91H1
Asn52→Ile	6	63.7	8.3	456±25	heat	(2,3)	91H1
double mutant (Gly29→Ser and Asn52→Ile)							
	6	40.3	8.3	243±4	heat	(2,3)	91H1
double mutant (His33→Pro and Asn52→Ile)							
	6	43.1	8.3	255±4	heat	(2,3)	91H1
Cys102→Ala	6	58.1	8.3	372±4	heat	(3)	91H1
Cys102→Ser	6	57.8	8.3	364±13	heat	(3)	91H1
double mutant (Asn52→Ala and Cys102→Ala)							
	6	62.1	8.3	469±21	heat	(3)	91H1
double mutant (Asn52→Gly and Cys102→Ala)							
	6	50.1	8.3	209±13	heat	(3)	91H1
double mutant (Asn52→Ile and Cys102→Ala)							
	6	71.5	8.3	481±4	heat	(3)	91H1

Remarks:
(1) recombinat protein containing Gly29, His33, Asn52, and Cys102 as the wild-type = reference protein for the following mutants
(2) blocked Cys-102-SCH$_3$
(3) ΔCp was assumed to be identical with the value determined for the mutant Cys102→Ala

Iso-1-cytochrome c, variants

Mutant	pH	T_{trs}	ΔCp	ΔH	Approach	Ref
(1) w.-t.	6.0	51.7		354	heat	92B4
(2)	6.0	55.9		386	heat	92B4
(3)	6.0	49.7			not two-state	92B4
(4)	4.6	54.7		374	heat	92B4
(4)	6.0	59.6		455	heat	92B4
(4)	4.6	53.5	5.78	372	DSC	92B4
(4)	6.0	58.5	9.33	416	DSC	92B4
(5)	4.6	53.4		270	heat	92B4
(5)	6.0	58.5		267	heat	92B4
(5)	4.6	52.8	3.87	266	DSC	92B4
(5)	6.0	56.7	3.68	262	DSC	92B4

Mutants:
(1) position 20: Val, position 102: Cys methylated
(2) position 20: Val20→Cys methylated, position 102: Thr
(3) position 20: Val20→Cys methylated, position 102: Cys methylated
(4) position 20: Val, position 102: Cys→Thr
(5) position 20: Val20→Cys, position 102: Cys

Iso-1-cytochrome c with engineered disulfide

Mutant	pH	T_{trs}	ΔCp	ΔH	Approach	Remark	Ref
wild-type	4.6	54.0±1.1	5.9±0.8	333±16	heat		96B6
disulfide	4.6	59.0±1.1	4.2±0.8	258±16	heat	(1)	96B6

Remark:
(1) disulfide connects position 20 (usually Val) with position 102 (usually Thr)

Iso-1-cytochrome c, Cys107-SCH$_3$ blocked

Mutant	pH	T$_{trs}$	ΔCp	ΔH	Approach	Remarks	Ref
w.-t.(Asn57)	6.0	46.5±1.0	8.3	330±21	heat	(1)	89D
Asn57→Ile	6.0	63.7±1.0	8.3	456±38	heat	(1)	89D

Remark:
(1) measured in 100 mM sodium phosphate

Iso-1-ferricytochrome c from *Saccaromyces cerevisiae*, Cys102→Thr variant

	pH	T$_{trs}$	ΔCp	ΔH	Approach	Remark	Ref
	3.00	28.9	5.73±0.25	205	heat, v.H.	(1)	94C7
	3.25	32.2		231	heat, v.H.		94C7
	3.50	38.2		262	heat, v.H.		94C7
	3.75	42.8		285	heat, v.H.		94C7
	4.00	44.0		310	heat, v.H.		94C7
	4.30	49.7		324	heat, v.H.		94C7
	4.60	52.6±1.1		345±16	heat, v.H.		94C7
	5.00	55.3		360	heat, v.H.		94C7

Remark:
(1) ΔCp was determined from ΔH versus T$_{trs}$

Iso-1-cytochrome c and Iso-2-cytochrome c
Iso-1-cytochrome c, Iso-2-cytochrome c and composite forms, oxidized

Mutant	pH	T$_{trs}$	ΔCp	ΔH	Approach	Remarks	Ref
iso-1 Cys102→Ala	6.0	56.2	5.2	292±16	DSC	(1)	94L4
iso-1 Cys102-meth.	6.0	50.7	5.2	270±15	DSC	(1,2)	94L4
iso-1 Cys102-meth.	6.0	50.7	3.6	257	DSC	(1–3)	94L4
iso-2	6.0	54.5	5.2	282±9	DSC	(1)	94L4
comp1 Cys102-meth.	6.0	47.6		218±8	DSC	(4)	94L4
comp2 Cys102-meth.	6.0	49.8		216±11	DSC	(5)	94L4
comp3 Cys102-meth.	6.0	48.4		219±7	DSC	(6)	94L4

Remarks:
(1) ΔCp is from ΔH versus T$_{trs}$ for iso-2-cytochrome c
(2) S-methylated at Cys102
(3) single experimental value
(4) composite protein from the allele CYC1-136-B
(5) composite protein from the allele CYC1-158-B
(6) composite protein from the allele CYC1-136-C

Iso-1-cytochrome c from yeast, mutations in position 52

Mutant	pH	T$_{trs}$	ΔCp	ΔH	Approach	Remarks	Ref
wild-type	5.0	63.3	8.4	448	DSC	(1,2)	95L5
wild-type	4.6	49.2	8.4	330	DSC	(1,2)	95L5
wild-type	4.0	45.4	8.4	269	DSC	(1,2)	95L5
wild-type	3.5	40.1	8.4	231	DSC	(1,2)	95L5
Asn52→Ala	4.6	53.7	8.4	351	DSC	(1,2)	95L5
Asn52→Gln	4.6	51.2	8.4	329	DSC	(1,2)	95L5
Asn52→His	4.6	47.5	8.4	268	DSC	(1,2)	95L5
Asn52→Ile	5.0	64.2	8.4	453	DSC	(1,2)	95L5

Iso-1-cytochrome c from yeast, mutations in position 52 (continued)

Mutant	pH	T_{trs}	ΔCp	ΔH	Approach	Remarks	Ref
Asn52→Ile	4.6	62.2	8.4	424	DSC	(1,2)	95L5
Asn52→Ile	4.0	58.3	8.4	375	DSC	(1,2)	95L5
Asn52→Ile	3.5	52.4	8.4	338	DSC	(1,2)	95L5
Asn52→Leu	4.6	62.1	8.4	408	DSC	(1,2)	95L5
Asn52→Met	4.6	62.6	8.4	402	DSC	(1,2)	95L5
Asn52→Met	4.0	58.9	8.4	372	DSC	(1,2)	95L5
Asn52→Ser	5.0	52.6	8.4	348	DSC	(1,2)	95L5
Asn52→Ser	4.6	51.0	8.4	329	DSC	(1,2)	95L5
Asn52→Ser	4.0	46.7	8.4	296	DSC	(1,2)	95L5
Asn52→Thr	5.0	55.3	8.4	371	DSC	(1,2)	95L5
Asn52→Thr	4.6	54.5	8.4	347	DSC	(1,2)	95L5
Asn52→Thr	4.0	49.2	8.4	315	DSC	(1,2)	95L5
Asn52→Thr	3.5	45.4	8.4	265	DSC	(1,2)	95L5
Asn52→Val	4.6	58.0	8.4	387	DSC	(1,2)	95L5
Asn52→Val	4.0	54.3	8.4	362	DSC	(1,2)	95L5
double mutant (Gly6→Ala and Asn52→Ile)							
	4.6	47.4		271	DSC	(1)	95L6

Remarks:
(1) wild-type = Asn52
(2) ΔCp was determined from ΔH versus T_{trs} for all position-52 replacements at different pH values

Iso-1-cytochrome c, Cys102→Thr variant (wild-type) and mutants, N→D and A→D transitions

Mutant	pH	T_{trs}	ΔCp	ΔH	Approach	Remarks	Ref
Thermodynamic parameters of the (N→D) transition:							
wild-type	4.6	52.6	5.73±0.25	345	heat	(1–3)	95P3
Ala7→Leu	4.6	53.4	5.94±1.13	361	heat	(1–3)	95M4
Ala7→Tyr	4.6	54.4	5.15±0.71	340	heat	(1–3)	95M4
Phe10→Cys	4.6	38.7		266	heat	(1–3,5)	95P3
Phe10→Ile	4.6	48.3	6.57±0.42	334	heat	(1–3)	95P3
Phe10→Met	4.6	37.7	6.28±1.17	251	heat	(1–3)	95P3
Phe10→Trp	4.6	47.4	6.69±0.54	316	heat	(1–3)	95P3
Phe10→Tyr	4.6	49.8	5.27±0.21	316	heat	(1–3)	95P3
Leu94→Ala	4.6	36.4	8.12±0.54	220	heat	(1–3)	95P3
Leu94→Ile	4.6	53.9	7.57±0.46	365	heat	(1–3)	95P3
Leu94→Thr	4.6	38.7	6.23±1.00	242	heat	(1–3)	95P3
Leu94→Val	4.6	48.6	6.44±0.83	341	heat	(1–3)	95P3
Tyr97→Ala	4.6	30.9	7.41±0.63	206	heat	(1–3)	95P3
Tyr97→Phe	4.6	53.6	6.40±0.21	353	heat	(1–3)	95P3
double mutant (Leu94→Ile and Tyr97→Phe)							
	4.6	49.7	7.36±0.42	334	heat	(1–3)	95P3
double mutant (Leu94→Ala and Tyr97→Phe)							
	4.6	33.2	7.45±0.46	225	heat	(1–3)	95P3
	19–61		6.69±0.13		heat	(1,2,6)	95M4/95P3
Thermodynamic parameters of the (A→D) transition:							
wild type	2.1	35.2		159	heat	(1,2,4)	95M4
Ala7→Leu	2.1	33.5		159	heat	(1,2,4)	95M4
Ala7→Tyr	2.1	35.8		174	heat	(1,2,4)	95M4
Phe10→Trp	2.1	22.9		121	heat	(1,2,4)	95M4
Phe10→Tyr	2.1	28.2		170	heat	(1,2,4)	95M4

Iso-1-cytochrome c, Cys102→Thr variant (wild-type) and mutants, N→D and A→D transitions (continued)

Mutant	pH	T_{trs}	ΔCp	ΔH	Approach	Remarks	Ref
Leu94→Ala	2.1	≤ 0			heat	(1,2,4)	95M4
Leu94→Ile	2.1	33.5		150	heat	(1,2,4)	95M4
Leu94→Thr	2.1	≤ 10			heat	(1,2,4)	95M4
Tyr97→Ala	2.1	≤ 0			heat	(1,2,4)	95M4
Tyr97→Phe	2.1	33.1		138	heat	(1,2,4)	95M4

Remarks:
(1) the Cys102→Thr variant is referred to as the wild-type, all variants also contain the Cys102→Thr mutation
(2) the considered states and the corresponding buffer conditions are: A state – 0.33 M Na_2SO_4/H_2SO_4, pH 2.1
acid denatured state – 0.01 M Na_2SO_4/H_2SO_4, pH 2.1 N state – 0.05 M potassium phosphate, pH 7.0 or
0.05 M acetate for the pos. 7 mutants
(3) the uncertainties in T_{trs} are $\pm 1.1°C$ and in ΔH ± 16 kJ/mol
(4) the uncertainties in T_{trs} are $\pm 0.8°C$ and in ΔH ± 5 kJ/mol
(5) Cys10 was modified to form S-methylcysteine
(6) ΔCp was obtained from ΔH versus T_{trs} including data obtained on all the considered mutant proteins

Iso-2-Cytochrome c, see also Iso-1-cytochrome c

Iso-2-cytochrome c

pH	T_{trs}	ΔCp	ΔH	Approach	Remarks	Ref
6.0	6.61 ± 0.29			DSC		95M11
6.0	54.5	5.2	282	DSC	(1)	94L4
5.0	54.9		307	DSC	(2)	94L4
5.0	51.2		288	DSC	(3)	94L4
4.5	48.6		272	DSC	(3)	94L4
4.0	43.7		262	DSC	(3)	94L4
3.5	40.7		236	DSC	(4)	94L4
3.5	37.2		215	DSC	(3)	94L4

Remarks:
(1) ΔCp from ΔH versus T_{trs}
(2) buffer: 0.1 M sodium phosphate
(3) buffer: 0.1 M sodium acetate
(4) buffer: 0.05 M glycine/HCl

Iso-2-cytochrome c, in the presence of guanidine hydrochloride

GuHCl Con-centration (M)	T_{trs}	ΔCp	ΔH	Approach	Ref
0	56.2		292	DSC	94L4
0.25	51.6		279	DSC	94L4
0.5	46.3		240	DSC	94L4
0.75	40.4		185	DSC	94L4
1.0	31.4		145	DSC	94L4

Iso-2-cytochrome c, wild-type and mutants, oxidized and reduced forms

Mutant	pH	T_{trs}	ΔCp	ΔH	Approach	Ref
wild-type						
oxidized	6.0	52.4	6.61±0.29	310±3	DSC	96M3
reduced	6.0	79.4	6.61±0.29	473±12	DSC	96M3
mutant (Asn52→Ile)						
oxidized	6.0	62.3	6.61±0.29	367±6	DSC	96M3
reduced	6.0	85.3	6.61±0.29	540±5	DSC	96M3

Cytochrome c oxidase

Cytochrome c oxidase from *Paracoccus denitrificans*

Transition	pH	T_{trs}	ΔCp	ΔH	Approach	Ref
trans. (1)	8.2	46.7±0.2	22±2	423±42	DSC	94H4
trans. (2)	8.2	67.0±0.2	29±2	1138±84	DSC	94H4

Cytochrome P450

Cytochrome P450 from *Pseudomonas putida*

Transition	pH	T_{trs}	ΔCp	ΔH	Approach	Remarks	Ref
overall	8.0	30–65		980	DSC	(1,2)	93P4
trans. (1)	8.0	41.9±1.0		296±15	DSC	(1)	93P4
trans. (2)	8.0	47.8±0.7		405±38	DSC	(1)	93P4
trans. (3)	8.0	54.3±0.6		279±34	DSC	(1)	93P4

Remarks:
(1) buffer: 20 mM HEPES, 1 mM KCN
(2) the transition could be resolved in three subtransitions

Apocytochrome P450 from *Pseudomonas putida*

	pH	T_{trs}	ΔCp	ΔH	Approach	Remarks	Ref
	8.0	50±2	3.2±0.5	135±10	DSC	(1,2)	93P4

Remarks:
(1) buffer: 20 mM HEPES
(2) mean of five calorimetric scans at pH 8.0

Cytochrome P450 from *Pseudomonas putida*, in the absence and in the presence of the substrate camphor

Transition	pH	T_{trs}	ΔCp	ΔH	Approach	Remarks	Ref
In the absence of the substrate:							
overall	8.0	30–65		980	DSC	(1,2)	93P4
trans. (1)	8.0	41.9±1.0		296±15	DSC	(1)	93P4
trans. (2)	8.0	47.8±0.7		405±38	DSC	(1)	93P4
trans. (3)	8.0	54.3±0.6		279±34	DSC	(1)	93P4
In the presence of camphor:							
overall	8.0	35–70		1310	DSC	(2,3)	94J4
trans. (1)	8.0	48.5±1.6		243±19	DSC	(2,3)	94J4
trans. (2)	8.0	57.8±0.3		480±12	DSC	(2,3)	94J4
trans. (3)	8.0	61.7±0.2		586±24	DSC	(2,3)	94J4

Remarks:
(1) buffer: 20 mM HEPES, 1 mM KCN
(2) the transition could be resolved in three subtransitions
(3) buffer: 20 mM HEPES, 100 mM KCN, 4% camphor, pH 8.0

De Novo Designed Protein

Betabellin 14D

pH	T_{trs}	ΔCp	ΔH	Approach	Ref
5.5	58.4±2.4		106±5	heat, v.H.	94Y1

De novo designed four-helix bundle $\alpha_2 D$

pH	T_{trs}	ΔCp	ΔH	Approach	Ref
6.9	50	0.04±0.004	109±4	heat, v.H.	95R0

Four-helix bundle, redesigned Zn-binding synthetic protein type: α_4 in 6.0 M GuHCl

Transition	T_{trs}	ΔCp	ΔH	Approach	Remarks	Ref
low-temperature	31	0.022	−0.803	heat	(1)	93H2
high-temperature	98	0.022	0.933	heat	(1)	93H2

Remark:
(1) ΔCp is given in kJ/K/(mol residue), ΔG is maximal at 62.5°C, $\Delta G = 2.8$ kJ/mol

De novo designed β/α-barrell protein

pH	T_{trs}	ΔCp	ΔH	Approach	Ref
3.5	52.7±1.3	4.6±1.0	146.2±8.8	DSC	94T4

Peri coil-1, artificially designed peptide

pH	T_{trs}	ΔCp	ΔH	Approach	Remarks	Ref
7.5	−26.9	8.66	−719.9	heat	(1–3)	94K4

Remarks:
(1) NH$_2$-Glu-Glu-Leu-Leu-Pro-Leu-Ala-Glu-Ala-Leu-Ala-Pro-Leu-Leu-Glu-Ala-Leu-Leu-Pro-Leu-Ala-Glu-Ala-Leu-Ala-Pro-Leu-Leu-Lys-Lys-COOH, with prolines at i+7 position
(2) cold denaturation, no thermal transition observable under 90°C
(3) fit of results obtained with peptide concentrations from 23.7 μM to 286.3 μM using a two-state model between pentameric helical state and monomeric denatured state (H$_5$→5D)

De novo designed antiparallel four-stranded coiled-coils

Form	pH	T_{trs}	ΔCp	ΔH	Approach	Remarks	Ref
coil-LL	6.9	57	2.4	139	heat	(1,3)	96B5
coil-VL	6.9	57	4.0	254	heat	(2,3)	96B5

Remarks:
(1) evaluated at 700 μM
(2) evaluated at 40 μM
(3) peptide design: Gly-Asn-Ala-Asp-Glu-Leu-Tyr-Arg-Met-X-Asp-Ala-Leu-Arg-Glu-His-X-Gln-Ser-Leu-Arg-Arg-Lys-X-Arg-Ser-Gly with X = Val for coil-VL and X = Leu for coil-LL

Two-stranded coiled-coil designed for studying electrostatic interactions

Mutant	pH	T_{trs}	ΔH	Approach	Remarks	Ref
$(2A)^{er}(2A)^{ra}$	2.0	55.5	186	DSC	(1,2)	96Y
	3.5	67.0	173	DSC	(1,2)	96Y
	4.5	72.2	165	DSC	(1,2)	96Y
	6.5	86.3	166	DSC	(1,2)	96Y
$(2A)^{er} (1A1R)^{ra}$	6.5	78.2	156	DSC	(1,2)	96Y
$(2A)^{er} (2R)^{ra}$	6.5	70.5	145	DSC	(1,2)	96Y

Remarks:
(1) coiled-coil derived from the sequence Tyr-Lys-Cys-Lys-Ser-Leu-Glu-Ser-Lys-Val-Lys-Ser-Leu-Glu-Ser-Lys-Ala-Lys-Ser-Leu-Glu-Ser-Lys-Val-Lys-Ser-Leu-Glu-Ser-Lys-Val-Lys-Ser-Leu-Glu-Ser
(2) superscripts "er" and "ra" denote inter- and intrachain interactions, respectively. A and R denote attractions and repulsions, respectively

Leucine zipper, de novo synthesized 33 membered polypeptide corresponding to the leucine zipper region of the yeast transcriptional activator GCN4

pH	T_{trs}	ΔCp	ΔH	Approach	Remark	Ref
3.0	70.5		314	DSC	(1)	94P4

Remark:
(1) the data refer to a transition of the type native trimer→3 unfolded monomers

Heterodimeric coiled-coil

Mutant	pH	T	ΔCp	ΔH	Approach	Remarks	Ref
AB	7.2	10	3.0	73.6±1.5	ITC	(1,2)	96J1
AB12	7.2	10	2.6	70.3±0.9	ITC	(1–3)	96J1
A12B	7.2	10	2.7	60.2±1.5	ITC	(1,2,4)	96J1
AB19	7.2	10	2.6	59.4±2.1	ITC	(1,2,5)	96J1
A19B	7.2	10	2.6	66.1±2.1	ITC	(1,2,6)	96J1
A12B12	7.2	10	2.9	58.6±1.3	ITC	(1,2,7)	96J1
AB	7.2	20		103.3±1.8	ITC	(1,2)	96J1
AB12	7.2	20		98.3±1.6	ITC	(1–3)	96J1
A12B	7.2	20		90.0±1.8	ITC	(1,2,4)	96J1
AB19	7.2	20		86.6±2.1	ITC	(1,2,5)	96J1
A19B	7.2	20		90.4±2.9	ITC	(1,2,6)	96J1
A12B12	7.2	20		89.5±1.3	ITC	(1,2,7)	96J1
A19B19	7.2	20		64.0±4.2	ITC	(1,2,8)	96J1
A12B19	7.2	20		64.9±4.2	ITC	(1,2,9)	96J1
A19B12	7.2	20		84.5±1.3	ITC	(1,2,10)	96J1
AB	7.2	30		133.5±1.6	ITC	(1,2)	96J1
AB12	7.2	30		122.6±1.1	ITC	(1–3)	96J1
A12B	7.2	30		113.4±0.9	ITC	(1,2,4)	96J1
AB19	7.2	30		112.1±0.6	ITC	(1,2,5)	96J1
A19B	7.2	30		118.8±0.9	ITC	(1,2,6)	96J1
A12B12	7.2	30		117.2±1.3	ITC	(1,2,7)	96J1

Remarks:
(1a) heterodimer composed of the acidic peptide (A): Ac-Glu-Tyr-Gln-Ala-Leu-Glu-Lys-Glu-Val-Ala-Gln-(Leu/Ala)-Glu-Ala-Glu-Asn-Gln-Ala-(Leu/Ala)-Glu-Lys-Glu-Val-Ala-Gln-Leu-Glu-His-Glu-Gly-amide, and the basic peptide (B): Ac-Glu-Tyr-Gln-Ala-Leu-Lys-Lys-Lys-Val-Ala-Gln- (Leu/Ala)-Lys-Ala-Lys-Asn-Gln-Ala-(Leu/Ala)-Lys-Lys-Lys-Val-Ala-Gln-Leu-Lys-His-Lys-Gly-amide,
(1b) ΔH of protein folding, ΔCp from ΔH versus T_{trs}
(2) buffer: 10 mM sodium phosphate
(3) peptide B Leu12→Ala
(4) peptide A Leu12→Ala
(5) peptide B Leu19→Ala
(6) peptide A Leu19→Ala
(7) double mutant, peptides A and B Leu12→Ala
(8) double mutant, peptides A and B Leu19→Ala
(9) double mutant, peptide A Leu12→Ala and B Leu19→Ala
(10) double mutant, peptide A Leu12→Ala and B Leu19→Ala

Heterodimeric coiled-coil

Mutant	pH	T_{trs}	ΔCp	ΔH	Approach	Remarks	Ref
AB	7.2	77.9	5.9±0.4	224±4	heat	(1–3,5)	96J1
AB	7.2	10–78	4.3±0.4		heat	(1,4,5)	96J1
AB12	7.2	70.2	4.2±1.7	210±4	heat	(1–3,5,6)	96J1
AB12	7.2	10–70	3.6±0.3		heat	(1,2,4–6)	96J1
A12B	7.2	61.5	3.8±2.1	176±5	heat	(1–3,5,7)	96J1
A12B	7.2	10–61	3.6±0.3		heat	(1,2,4,5,7)	96J1
AB19	7.2	52.0	8.8±1.7	185±5	heat	(1–3,5,8)	96J1
AB19	7.2	10–52	4.2±0.4		heat	(1,2,4,5,8)	96J1
A19B	7.2	55.9	5.0±1.3	179±3	heat	(1–3,5,9)	96J1
A19B	7.2	10–56	3.5±0.3		heat	(1,2,4,5,9)	96J1
A12B12	7.2	50.6	9.2±1.3	151±7	heat	(1–3,5,10)	96J1
A12B12	7.2	10–50	4.6±0.4		heat	(1,2,4,5,10)	96J1

Remarks:
(1) heterodimer composed of the acidic peptide (A): Ac-Glu-Tyr-Gln-Ala-Leu-Glu-Lys-Glu-Val-Ala-Gln-(Leu/Ala)-Glu-Ala-Glu-Asn-Gln-Ala-(Leu/Ala)-Glu-Lys-Glu-Val-Ala-Gln-Leu-Glu-His-Glu-Gly-amide, and the basic peptide (B): Ac-Glu-Tyr-Gln-Ala-Leu-Lys-Lys-Lys-Val-Ala-Gln-(Leu/Ala)-Lys-Ala-Lys-Asn-Gln-Ala-(Leu/Ala)-Lys-Lys-Lys-Val-Ala-Gln-Leu-Lys-His-Lys-Gly-amide,
(2) van't Hoff treatment, transition monitored by CD
(3) ΔCp from plotting ΔH versus T_{trs} using the data for different protein concentrations
(4) ΔCp from global fit including data from heat denaturation and ITC
(5) buffer: 10 mM sodium phosphate
(6) peptide B Leu12→Ala
(7) peptide A Leu12→Ala
(8) peptide B Leu19→Ala
(9) peptide A Leu19→Ala
(10) double mutant, peptides A and B Leu12→Ala

Trimeric coiled-coil

pH	T	ΔCp	ΔH	Approach	Remarks	Ref
7.5	25	3.15±0.04	164.2±0.9	heat	(1–3)	96B7

Remarks:
(1) coiled-coil designed by placing Val at each a position and Leu at each d position of the heptad repeating unit
(2) heat denaturation at varying GuHCl conc., treatment by a monomer/trimer equilibrium, data at reference temperature T°
(3) buffer: 10 mM MOPS

Table 2. Enthalpy and Heat Capacity Changes – Molar Values

QLRb-4, 80-residues protein recovered from a random sequence library, predominantly composed of Gln, Leu and Arg

Protein Concentration	pH	T_{trs}	ΔCp	ΔH	Approach	Remarks	Ref
2.3 µM	5.8	59.4±1.1		243±18	heat, GuHCl	(1,2)	95D1
4.6 µM	5.8	66.5±0.9		301±21	heat, GuHCl	(1,2)	95D1
9.2 µM	5.8	73.8±1.3		366±23	heat, GuHCl	(1,2)	95D1
	5.8	59–74	8.4		heat, GuHCl	(1–3)	95D1
	5.8	59–74	5.0–6.7		heat, GuHCl	(1,4)	95D1

Remarks:
(1) van't Hoff treatment, the melting curves were measured in the presence of 3 M GuHCl
(2) the data refer to dissociation and unfolding of the native tetramer into unfolded monomers at a standard state of 1 M
(3) ΔCp from ΔH versus T_{trs}
(4) ΔCp from individual fits of the melting curves

De novo designed protein, see also troponin C

Dihydrofolate Reductase

Dihydrofolate reductase (DHFR) from various species and their complexes

Mutant	pH	T_{trs}	ΔH	Approach	Remarks	Ref
E. coli	6.8	56.3	290	DSC	(1)	95S1
bovine	6.8	49.1	182	DSC	(1)	95S1
E. coli + NADPH	6.8	59.4	313	DSC		95S1
bovine + NADPH	6.8	57.3	318	DSC		95S1
E. coli + TMP	6.8	64.8	379	DSC	(2)	95S1
bovine + TMP	6.8	52.3	394	DSC	(2)	95S1
E. coli + MTX	6.8	70.5	458	DSC	(3)	95S1
bovine + MTX	6.8	62.4	478	DSC	(3)	95S1
E. coli + NADPH + TMP	6.8	73.1	492	DSC	(2)	95S1
bovine + NADPH + TMP	6.8	65.5	490	DSC	(2)	95S1
E. coli + NADPH + MTX	6.8	78.2	542	DSC	(3)	95S1
bovine + NADPH + MTX	6.8	78.1	675	DSC	(3)	95S1

Remarks:
(1) the specific heat capacity change Δcp amounts to 0.66 J/g/K determined from Δh of the two DHFRs and their complexes versus T_{trs}
(2) TMP = trimethoprim
(3) MTX = methotrexate

Dihydrofolate reductase, cysteine free mutant (Cys85→Ser and Cys152→Glu)

Transition	pH	T_{trs}	ΔCp	ΔH	Approach	Remarks	Ref
trans. (1)	7.8	33.1±0.4	5.5±0.6	226±13	heat	(1,2)	95L13
trans. (1)	7.8	33.4±0.7	4.5±1.3	213±25	heat	(1,3)	95L13
trans. (1)	7.8	32.9±0.9	3.6±13.2	255±38	heat	(1,4)	95L13
trans. (1)	7.8	33.1±0.6	5.4±1.0	226±13	heat	(1,5)	95L13
trans. (2)	7.8	45.8±1.9	1.5±1.0	117±8	heat	(1,2)	95L13
trans. (2)	7.8	47.1±3.7	0.0±1.4	146±42	heat	(1,3)	95L13
trans. (2)	7.8	46.2±2.5	1.0±1.1	126±13	heat	(1,5)	95L13

Remarks:
(1) three-state fit assuming ΔCp temperature independent
(2) transition monitored by absorbance at 292 nm, mean of 6 replicate experiments
(3) transition monitored by CD at 222 nm, mean of 5 replicate experiments
(4) transition monitored by CD at 292 nm, single data set, the I→U transition is invisible in CD in the near UV region
(5) global fit of all the data sets indicated above

Dihydrofolate reductase (DHFR) from *E. coli*, wild-type and mutants fused with oligopeptides

Mutant/Transition		pH	T_{trs}	ΔH	Approach	Remarks	Ref
wild-type		7.0	48	152	heat	(1a)	90U2
wild-type		7.0	51	158	heat	(1b)	90U2
wild-type		7.0	49	359	heat	(1c)	90U2
wild-type		7.0	50	329	heat	(1d)	90U2
wild-type		7.0	45	206	heat	(1e)	90U2
G-DHFR,	N→I	7.0	45.59±0.06	195.0±0.9	DSC	(2,3,5)	90U1
G-DHFR	N→D	7.0	51.76±0.07	365.9±1.6	DSC	(2,3,5)	90U1
G-DHFR		7.0	47	273	heat	(1e)	90U2
G-DHFR	N→I	7.7	41.7±0.1	194.4±1.8	DSC	(2,3)	90U1
G-DHFR	N→D	7.7	51.76±0.07	346.0±3.4	DSC	(2,3)	90U1
DHFR-lek,	N→D	7.0	41	142	DSC	(4,5)	90U2
DHFR-lek		7.0	42	315	heat	(1e,4)	90U2

Remarks:
(1) van't Hoff heat, transition monitored by CD (1a) at 203 nm, (1b) at 220 nm, (1c) at 255 nm, (1d) at 260 nm, and (1e) at 290 nm
(2) Cys152 of wild-type replaced by Glu, and (Ile-Gln-Ile) added to the C-terminus
(3) calorimetric enthalpy, three-state model
(4) Cys152 of wild-type replaced by Glu, and (Ile-Arg-Met-Tyr-Gly-Gly-Phe-Leu) added to the C-terminus
(5) see also mutant DHFR-IQI in Ref. 90U2

Dihydrofolate reductase from *E. coli*, wild-type and mutants

Mutant	pH	T_{trs}	ΔCp	ΔH	Approach	Remarks	Ref
wild-type	7.0	49.3		396	DSC	(1)	93G3
Gly121→Leu	7.0	44.1		325	DSC	(1)	93G3
Gly121→Val	7.0	46.9		265	DSC	(1)	93G3
wild-type	7.0	25	7.1–7.5	119	GuHCl	(2)	93G3
Gly121→Val	7.0	25	7.1–7.5	103	GuHCl	(2)	93G3

Remarks:
(1) measured in 10 mM potassium phosphate, 0.1 mM EDTA, 0.1 mM DTT
(2) from the temperature dependence of ΔG measured by GuHCl denaturation

Dihydrofolate reductase, wild-type and circularized mutants

Form	pH	T_{trs}	ΔH	Approach	Remarks	Ref
wild-type	6.8	54.3	183	heat, v.H.		96I
AS protein	6.8	55.2	226	heat, v.H.	(1)	96I
MICGG-AS-C protein					(2)	
oxidized	6.8	58.8	250	heat, v.H.		96I
reduced	6.8	51.9	213	heat, v.H.		96I
MICGG-AS-G2C protein					(3)	
oxidized	6.8	58.6	301	heat, v.H.		96I
reduced	6.8	53.5	221	heat, v.H.		96I
MICGG-AS-G4C protein					(4)	
oxidized	6.8	60.4	316	heat, v.H.		96I
reduced	6.8	53.9	236	heat, v.H.		96I

Remarks:
(1) AS protein: cysteine-free mutant with Cys85→Ala and Cys152→Ser, C-terminal sequence: -Leu-Glu-Arg-Arg-(COOH)
(2) MICGG: N-terminal extension Met-Ile-Cys-Gly-Gly-, Cys residues of C- and N-termini connected by a disulfide bond AS-C: C-terminal sequence extended by Cys, i.e., -Leu-Glu-Arg-Arg-Cys-(COOH), see remark (1)
(3) MICGG: see remark (2) AS-G2C: C-terminal sequence extended by -Gly-Gly-Cys-(COOH), see remark (1)
(4) MICGG: see remark (2) AS-G4C: C-terminal sequence extended by -Gly-Gly-Gly-Gly-Cys-(COOH), see remark (1)

R67 Dihydrofolate Reductase

General remark: R67 DHFR, encoded by an R plasmid, does not show any homology with chromosomal DHFR. R67 DHFR is a polypeptide chain that is 78 residues long. R67 DHFR is tetrameric at high pH (8.0) and dimeric at low pH (5.0).

R67 dihydrofolate reductase (DHFR) and modified forms

Form	pH	T_{trs}	ΔCp	ΔH	Approach	Remarks	Ref
native	5.0	61.25		401	DSC	(1)	93Z5
native	5.0	60.6		397	DSC	(2)	93Z5
double	5.0	55.0		395	DSC	(3,4)	93Z5
double	5.0	56.25		565	DSC	(1,3)	93Z5
double	5.0	55.95		525	DSC	(2,3)	93Z5
native	8.0	70.95		1284	DSC	(5)	94Z7

Remarks:
(1) 296 µM DHFR, buffer without DTT
(2) 296 µM DHFR
(3) dimeric R67 DHFR, constructed by gene duplication
(4) 593 µM DHFR
(5) mechanism: tetramer→2 dimer→4 unfolded, the vant Hoff heat amounts to 613 kJ/mol

DNA-Binding Protein

DNA-binding protein Sso7d from the hyperthermophile *Sulfolobus solfataricus*

pH	T_{trs}	ΔCp	ΔH	Approach	Remarks	Ref
2.5	65.1	1.91	177	DSC	(1,2)	96K5
2.6	68.0	1.82	192	DSC	(1,2)	96K5
3.0	78.7	1.74	202	DSC	(1,2)	96K5
3.0	78.4	4.98	207	DSC	(1,3)	96K5
3.1	83.0	3.50	221	DSC	(1,2)	96K5
3.5	88.9	1.92	240	DSC	(1,2)	96K5
4.0	95.1	3.62	252	DSC	(1,2)	96K5
4.5	97.8	2.31	270	DSC	(1,2)	96K5
4.75	98.4	1.66	262	DSC	(1,2)	96K5
5.5	98.7	1.75	264	DSC	(1,2)	96K5
6.0	97.9	2.21	262	DSC	(1,2)	96K5
6.0	98.4	1.69	270	DSC	(1,3)	96K5
6.5	97.6	4.41	265	DSC	(1,2)	96K5
7.0	98.0	3.59	274	heat	(2,4)	96K5
2.5	64.3		167	heat	(2,4)	96K5
2.5	66.2		167	heat	(3,4)	96K5
2.5	69.3±0.3		166±2	heat	(2,5)	96K5
2.5	69.1		166	heat	(6a)	96K5
2.5	68.7		170	heat	(6b)	96K5
2.7	69.7		174	heat	(2,4)	96K5
2.8	70.1		187	heat	(3,4)	96K5
2.9	76.3		187	heat	(2,4)	96K5
3.0	79.8		202	heat	(3,4)	96K5
3.1	80.7		206	heat	(2,4)	96K5
3.3	86.1		221	heat	(3,4)	96K5
3.5	92.3		230	heat	(2,4)	96K5
2.5	64.9		178		(7a)	96K5
2.5	65.5		176		(7b)	96K5
6.5	97.7		256		(7c)	96K5
6.5	98.3		261		(7d)	96K5
6.5	97.6		252		(7e)	96K5
3.5–7.0		2.43		heat	(8)	96K5
2.5–6.5		2.65		DSC	(9)	96K5
2.5–6.5		2.59		DSC	(10)	96K5

Remarks:
(1) error in T_{trs} ±0.2°C, additional data are reported in Ref. 96K5 on the scan rate dependence of reversibility
(2) cloned unmethylated protein (c-Sso7d)
(3) ε-mono methylated protein purified from *Sulfolobus solfataricus* (m-Sso7d). No significant differences between c-Sso7d and m-Sso7d were reported in Ref. 96K5
(4) van't Hoff heat, transition monitored by CD
(5) van't Hoff heat, joint fit of transitions monitored by CD at 200 and 275 nm
(6) van't Hoff heat, transitions monitored by CD at 200 nm
   (a) upscan
   (b) downscan
(7) measured in various buffers
   (a) acetate adjusted with HCl
   (b) sodium phosphate
   (c) MES/K$^+$
   (d) Tris-HCl
   (e) sodium phosphate
(8) from CD data of c-Sso7d
(9) from DSC data, average of individual scans
(10) DSC, from ΔH versus T_{trs}

Table 2. Enthalpy and Heat Capacity Changes – Molar Values

Sac7d DNA-binding protein from *Solfolobus acidocaldarius*, recombinant

pH	T_{trs}	ΔCp	ΔH	Approach	Remarks	Ref
0.0	63.3		186	DSC	(1)	96M2
0.5	61.3		190	DSC	(1)	96M2
1.0	58.1		195	DSC	(1)	96M2
1.5	58.1		189	DSC	(1)	96M2
2.0	59.1		190	DSC	(1)	96M2
2.5	60.1		188	DSC	(1)	96M2
2.75	62.0		164	DSC	(1)	96M2
3.0	63.2		210	DSC	(1)	96M2
3.2	66.5		188	DSC	(1)	96M2
3.3	68.2		211	DSC	(1)	96M2
3.65	75.2		240	DSC	(1)	96M2
4.0	79.3		235	DSC	(1)	96M2
4.5	86.3		218	DSC	(1)	96M2
4.7	89.0		246	DSC	(1)	96M2
5.0	89.4		268	DSC	(1)	96M2
6.0	90.9		231	DSC	(1)	96M2
7.0	90.7		275	DSC	(1)	96M2
8.0	90.3		254	DSC	(1)	96M2
9.0	90.0		194	DSC	(1)	96M2
10.0	85.2		223	DSC	(1)	96M2
7		3.69±0.25		GuHCl	(2,3)	96M2
4		3.51±0.62		urea	(2,4)	96M2
7		3.61±0.11		GuHCl	(3,5)	96M2
4		3.46±0.18		urea	(4,5)	96M2
		3.59±0.09		GuHCl/urea	(6)	96M2
7		2.97±0.49		GuHCl	(3,7)	96M2
4		3.23±0.79		urea	(4,7)	96M2
7		3.23±0.12		GuHCl	(2,3)	96M2
4		3.11±0.19		urea	(2,4)	96M2
		3.20±0.10		GuHCl/urea	(9)	96M2
1–5	58–89	2.08±0.08		DSC	(10)	96M2
		3.77±0.38		predicted	(11)	96M2
		3.59±0.08			(12)	96M2

Remarks:
(1) buffers: pH 0 to 0.5, HCl; pH 1 to 3.3, glycine; pH 3.65 to 5, potassium acetate; pH 6, PIPES; pH 7, KH_2PO_4; pH 8, BICINE; pH 9 to 10, glycine. All buffers were 0.05 M (except HCl) and contained 0.3 M KCl
(2) multiple independent fit based on the linear extrapolation method
(3) buffer: 0.01 M MOPS pH 7
(4) buffer: 0.05 M potassium acetate, 0.3 M KCl, pH 4
(5) global fit based on the linear extrapolation method
(6) combined fit of GuHCl and urea denaturation data based on the linear extrapolation method
(7) multiple independent fit based on the denaturant binding model
(8) global fit based on the denaturant binding model
(9) combined fit of GuHCl and urea denaturation data based on the denaturant binding model
(10) from ΔH versus T_{trs}
(11) predicted from the change in total accessible surface area
(12) from global non-linear regression of the chemical denaturation data constrained by DSC determined values of ΔH and T_{trs}. The ΔCp value is assumed to be the the most reliable one.

DnaK

DnaK, *E. coli* molecular chaperone, transitions resolved by deconvolution

Transition	pH	T_{trs}	ΔCp	ΔH	Approach	Ref
trans. (1)	6.5	46.4±0.1	15.1±1.3	331±13	DSC	93M14
trans. (2)	6.5	57.0±0.1	9.6±1.3	368±13	DSC	93M14
trans. (3)	6.5	73.4±0.2	7.5±0.4	255±8	DSC	93M14
trans. (1)	7.6	45.2±0.1	10.0±1.3	406±17	DSC	93M14
trans. (2)	7.6	58.0±0.2	16.3±2.1	393±13	DSC	93M14
trans. (3)	7.6	73.3±0.6	7.5±1.7	272±4	DSC	93M14
trans. (1)	7.6	44.8±0.1	13.0±1.7	435±13	DSC	93M14
trans. (2)	7.6	58.0±0.1	20.5±1.3	414±13	DSC	93M14
trans. (3)	7.6	75.0±0.5	3.3±0.4	318±4	DSC	93M14
trans. (1)	9.0	42.6±0.1	14.2±1.3	318±13	DSC	93M14
trans. (2)	9.0	55.8±0.1	15.5±0.8	381±13	DSC	93M14
trans. (3)	9.0	72.2±0.1	3.3±0.4	272±4	DSC	93M14
trans. (1)	10.0	39.9±0.1	7.9±0.8	351±13	DSC	93M14
trans. (2)	10.0	56.2±0.2	18.8±1.7	427±17	DSC	93M14
trans. (3)	10.0	71.3±0.2	7.1±0.8	280±8	DSC	93M14
trans. (1)	10.0	39.9±0.6	10.5±1.7	285±21	DSC	93M14
trans. (2)	10.0	57.2±0.1	8.4±1.7	423±13	DSC	93M14
trans. (3)	10.0	70.4±0.1	14.2±0.8	272±013	DSC	93M14
trans. (1)	6.5–10		11.7±2.5		mean value	93M14
trans. (2)	6.5–10		15.1±4.6		mean value	93M14
trans. (3)	6.5–10		7.1±3.3		mean value	93M14

DnaK, *E. coli* molecular chaperone, fragments

Transition	pH	T_{trs}	ΔCp	ΔH	Approach	Remarks	Ref
N-terminal proteolytic fragment:							
trans. (1)	7.6	47.5±0.1	15.1±1.3	531±17	DSC	(1)	93M14
trans. (2)	7.6	79.4±0.1	0.4±0.4	205±8	DSC	(1)	93M14
N-terminal fragment, recombinant:							
trans. (1)	7.6	48.1±0.1	21.8±1.7	565±63	DSC	(1)	93M14
trans. (2)	7.6	70.4±0.1	0.4±0.4	151±17	DSC	(1)	93M14
C-terminal proteolytic fragment:							
trans. (1)	7.6	50.4±0.6	5.9±1.7	180±17	DSC	(2)	93M14
trans. (2)	7.6	58.2±1.0	6.7±2.5	259±17	DSC	(2)	93M14
trans. (3)	7.6	70.6±0.4	5.0±0.8	251±8	DSC	(2)	93M14

Remarks:
(1) the fragment displays two transitions
(2) the fragment displays three transitions

DnaK, *E. coli* molecular chaperone in the presence of ligands

Transition	pH	T_{trs}	ΔCp	ΔH	Approach	Remarks	Ref
In the presence of ADP:							
trans. (1)	9.0	51.8±.4	10.9±2.5	318±21	DSC	(1,2)	93M14
trans. (2)	9.0	59.0±.1	17.2±1.3	452±13	DSC	(1,2)	93M14
trans. (3)	9.0	72.9±.5	1.7±0.8	259±4	DSC	(1,2)	93M14
In the presence of peptide C:							
trans. (1)	7.6	47.0±.1	4.6±0.8	397±8	DSC	(1,3,4)	93M14
trans. (2)	7.6	60.8±.1	1.7±1.7	385±13	DSC	(1,3,4)	93M14
trans. (3)	7.6	75.7±.2	14.2±0.4	305±8	DSC	(1,3,4)	93M14

Remarks:
(1) the protein displays three transitions resolved by deconvolution
(2) buffer: 25 mM glycine, 2 mM MgCl$_2$, 1 mM ADP
(3) buffer: 25 mM sodium phosphate
(4) peptide C: Lys-Leu-Ile-Gly-Val-Leu-Ser-Ser-Leu-Phe-Arg-Pro-Lys

Eglin

Eglin c, proteinase inhibitor, varying protein concentrations (mg/ml) and pH

Prot. Concentration	pH	T_{trs}	ΔCp	ΔH	Approach	Remarks	Ref
3.16	1.10	41.12	3.6	163	DSC	(1)	95B1
1.43	1.30	41.27	4.1	154	DSC	(1)	95B1
2.14	1.30	42.08	2.8	150	DSC	(1)	95B1
2.58	1.65	46.09	0.9	164	DSC	(1)	95B1
2.47	1.80	47.28	0.5	175	DSC	(1)	95B1
2.18	2.00	46.90	2.4	173	DSC	(1)	95B1
2.70	2.15	51.13	2.1	196	DSC	(1)	95B1
1.56	2.25	51.14	2.1	195	DSC	(1)	95B1
1.80	2.50	51.46	1.7	185	DSC	(1)	95B1
2.16	2.50	54.52	1.0	179	DSC	(1)	95B1
2.56	2.50	55.40	1.3	202	DSC	(1)	95B1
3.37	2.50	54.91	1.2	196	DSC	(1)	95B1
3.78	2.50	51.76	1.6	206	DSC	(1)	95B1
5.51	2.50	55.00	2.1	224	DSC	(1)	95B1
7.08	2.50	51.59	2.3	203	DSC	(1)	95B1
2.65	2.75	57.56	1.5	213	DSC	(1)	95B1
2.78	2.85	59.46	4.3	219	DSC	(1)	95B1
2.91	2.85	60.44	2.7	217	DSC	(1)	95B1
2.30	2.93	63.93	1.8	251	DSC	(1)	95B1
2.32	3.00	63.30	3.5	240	DSC	(1)	95B1
2.37	3.25	68.98	0.9	235	DSC	(1)	95B1
2.21	7.00	85.83	0.8	313	DSC	(1)	95B1
2.45	7.00	85.99	0.8	282	DSC	(1)	95B1
2.91	7.00	85.16	1.8	271	DSC	(1)	95B1
4.42	7.00	85.43	0.9	310	DSC	(1)	95B1
5.01	7.01	85.29	0.7	297	DSC	(1)	95B1
2.54	7.05	85.49	1.5	274	DSC	(1)	95B1
2.20	8.00	86.52	2.1	303	DSC	(1)	95B1
3.00	9.20	86.30	2.9	302	DSC	(1)	95B1

Eglin c, proteinase inhibitor, varying protein concentrations (mg/ml) and pH (continued)

Prot. Concentration	pH	T_{trs}	ΔCp	ΔH	Approach	Remarks	Ref
2.61	10.05	86.55	2.2	303	DSC	(1)	95B1
1.48	10.50	87.05	0.2	308	DSC	(1)	95B1
2.84	10.50	85.45	4.5	282	DSC	(1)	95B1
3.23	10.50	86.58	2.3	303	DSC	(1)	95B1
5.77	10.50	86.37	2.3	300	DSC	(1)	95B1
2.81	10.55	86.02	3.3	287	DSC	(1)	95B1
6.56	10.55	85.68	3.2	304	DSC	(1)	95B1
	1.10–10.55		2.06±0.19		DSC	(2)	95B1
	1.10–10.55		3.05		DSC	(3,4)	95B1

Remarks:
(1) ΔCp from the individual calorimetric scans
(2) ΔCp is the mean value obtained from the individual calorimetric scans
(3) ΔCp from ΔH versus T_{trs}, the value is regarded as the more signifacant one in Ref. 95B1
(4) general expression: $\Delta H = 34.16 + 3.05 \times T_{trs}$

Eglin c, proteinase inhibitor from *Hirudo medicinalis*
temperature dependence of the heat cpacity change at protein unfolding in aqueous solution, and related thermodynamic quantities

T	$\Delta H(exp)$	$\Delta S(exp)$	$\Delta Cp(exp)$	Approach	Remarks	Ref
5	33	−22	4.2	DSC	(1–3)	95M3
25	115	262	3.9	DSC	(1–3)	95M3
50	208	561	3.5	DSC	(1–3)	95M3
75	283	786	2.5	DSC	(1–3)	95M3
100	335	929	1.6	DSC	(1–3)	95M3
125	358	989	0.2	DSC	(1–3)	95M3

Remarks:
(1) the data were taken from Ref. 95M3 with reference to Ref. 95B1
(2) ΔCp was obtained as described in Ref. 90P3
(3) ΔCp in kJ/mol/K, ΔH in kJ/mol and ΔS in J/mol/K

Eglin c, proteinase inhibitor, in the presence of varying GuHCl concentration

GuHCl Concentration	pH	T_{trs}	ΔCp	ΔH	Approach	Remarks	Ref
0.503 M	7.00	77.87	4.1	274	DSC	(1)	95B1
0.945 M	7.00	73.40	1.4	275	DSC	(1)	95B1
1.316 M	7.00	70.14	2.2	262	DSC	(1)	95B1
1.451 M	7.00	67.01	3.2	228	DSC	(1)	95B1
1.493 M	7.00	67.65	1.8	235	DSC	(1)	95B1
1.737 M	7.00	63.69	2.3	226	DSC	(1)	95B1
2.461 M	7.00	56.41	1.9	169	DSC	(1)	95B1
3.121 M	7.00	48.39	2.2	149	DSC	(1)	95B1
0.953 M	10.55	73.54	2.3	251	DSC	(1)	95B1
1.432 M	10.55	69.10	1.2	205	DSC	(1)	95B1
1.939 M	10.55	61.39	3.7	192	DSC	(1)	95B1

Eglin c, proteinase inhibitor, in the presence of varying GuHCl concentration (continued)

GuHCl Concentration	pH	T_{trs}	ΔCp	ΔH	Approach	Remarks	Ref
2.336 M	10.55	56.00	2.6	169	DSC	(1)	95B1
2.739 M	10.55	48.44	3.4	134	DSC	(1)	95B1
3.311 M	10.55	41.31	2.6	101	DSC	(1)	95B1
	7.0, 10.55		2.50±0.24		DSC	(2)	95B1
	7.0, 10.55		6.49		DSC	(3,4)	95B1

Remarks:
(1) ΔCp from the individual calorimetric scans
(2) ΔCp is the mean value obtained from the individual calorimetric scans
(3) ΔCp from ΔH versus T_{trs}
(4) general expression: $\Delta H = 97.08 + 6.49 \times T_{trs}$

Endothiapepsin

Endothiapepsin complex with inhibitor pepstatin A

pH	T_{trs}	ΔCp	ΔH	Approach	Ref
4.6	75	15.5	983	DSC	95G6

5-Enolpyruvoyl Shikimate-3-Phosphate Synthase

5-Enolpyruvoyl shikimate-3-phosphate synthase

Form	pH	T_{trs}	ΔCp	ΔH	Approach	Remarks	Ref
free enzyme	7.0	55.7±0.3	8.4±0.4	1046±38	DSC		93M12
complex 1	7.0	59.6±0.4	8.8±1.7	1125±33	DSC	(1)	93M12
complex 2	7.0	65.1±0.3	8.8±1.7	1418±38	DSC	(2)	93M12

Remarks:
(1) complex with shikimate-3-phosphate
(2) complex with shikimate-3-phosphate glyphosate

Factor XIII

Factor XIII, recombinant, low (LT) and high (HT) temperature peak, total heat

Transition	pH	T_{trs}	ΔH	Approach	Remarks	Ref
LT	8.6	68.9	1573	DSC	(1)	95K14
LT	9.0	65.4	1356	DSC	(1)	95K14
LT	10.0	55.8	1029	DSC	(2)	95K14
LT	10.4	37.4	715	DSC	(3)	95K14
LT	10.5	35.5	674	DSC	(3)	95K14
LT	8.6	32.0		DSC	(4)	95K14
HT	8.6	91.0	381	DSC	(5)	95K14

Remarks:
(1) high temperature peak was not observed
(2) aggregation after 90°C
(3) aggregation at 80 to 110°C
(4) measured in the presence of 2 M GuHCl the enthalpy of the low temperature transition could not be determined
(5) measured in the presence of 2 M GuHCl, the sample was prepared at room temperature, the low temperature transition was not be observed

Factor XIIIa, recombinant activated factor XIII, low (LT) and high (HT) temperature peaks, resolved by deconvolution

Transition	pH	T_{trs}	ΔH	Approach	Remarks	Ref
LT trans. (1)	3.5	43.1	360	DSC	(1)	95K14
LT trans. (2)	3.5	49.2	360	DSC	(1)	95K14
LT trans. (3)	3.5	66	230	DSC	(1)	95K14
LT trans. (1)	3.3	37.6	272	DSC		95K14
LT trans. (2)	3.3	46.8	318	DSC		95K14
LT trans. (3)	3.3	65.6	247	DSC		95K14
HT trans. (1)	3.3	114.9	406	DSC		95K14
HT trans. (2)	3.3	123.6	385	DSC		95K14
LT trans. (1)	3.1	34.2	255	DSC		95K14
LT trans. (2)	3.1	46.3	280	DSC		95K14
LT trans. (3)	3.1	64.5	243	DSC		95K14
HT trans. (1)	3.1	111.9	410	DSC		95K14
HT trans. (2)	3.1	120.1	397	DSC		95K14
LT trans. (1)	2.9	27	155	DSC		95K14
LT trans. (2)	2.9	41.3	226	DSC		95K14
LT trans. (3)	2.9	60.4	205	DSC		95K14
HT trans. (1)	2.9	107.7	397	DSC		95K14
HT trans. (2)	2.9	115.7	393	DSC		95K14
HT trans. (1)	8.6	81.8	188	DSC	(2)	95K14
HT trans. (2)	8.6	98.6	205	DSC	(2)	95K14
HT trans. (1)	8.6	82.8	184	DSC	(3)	95K14
HT trans. (2)	8.6	99.9	192	DSC	(3)	95K14

Remarks:
(1) the high temperature transition at $T_{trs} = 128°C$ was not resolved by deconvolution
(2) measured in the presence of 2 M GuHCl, the sample was prepared at low temperature
(3) measured in the presence of 2 M GuHCl, the sample was prepared at room temperature

Ferredoxin

Ferredoxin from *Pyrococcus furiosus* (PF) and *Thermotoga maritima* (TM)

Form	T_{trs}	ΔCp	ΔH	Approach	Ref
PF, ox., 4Fe	117.5		48.1	DSC	94K5
PF, red., 4Fe	117		47.7	DSC	94K5
PF, ox., 3Fe	105		46.0	DSC	94K5
TM, ox., 3Fe	102.5		46.0	DSC	94K5
TM, red., 3Fe	102.0		46.0	DSC	94K5

High-potential [4Fe-4S] iron-sulfur protein from *Ectothiorhodospira halophila*

Mutant	pH	T_{trs}	ΔCp	ΔH	Approach	Remarks	Ref
wild-type	7	67.7±0.8	4.5±0.5	−91±8	DSC	(1,2)	95I2
wild-type	9	66.4±0.5		−96±7	DSC	(2)	95I2
wild-type	10	66.5±0.6		−96±8	DSC	(2)	95I2
wild-type	10.8	56.7±0.4		−140±7	DSC	(2)	95I2
wild-type	11	50.3±0.6		−169±6	DSC	(2)	95I2
Tyr12→His	7	42.2±0.5		−139±8	DSC	(2)	95I2
Tyr12→His	9	39.0±0.6		−223±10	DSC	(2)	95I2
Tyr12→Phe	7	54.8±0.5		−89±3	DSC	(2)	95I2
Tyr12→Phe	9	54.4±0.6		−87±10	DSC	(2)	95I2
Tyr12→Phe	10	54.2±0.4		−101±6	DSC	(2)	95I2
Tyr12→Phe	10.8	37.1±0.4		−122±7	DSC	(2)	95I2
Tyr12→Trp	7	37.4±0.5		−99±3	DSC	(2)	95I2

Remarks:
(1) ΔCp from ΔH versus T_{trs} of wild-type
(2) the exothermic transition appears to be associated with the disruption of the prosthetic cluster

Fibrinogen

Fibrinogen and its fragments, results of deconvolution

Transition	pH	T_{trs}	ΔCp	ΔH	Approach	Remarks	Ref
Fibrinogen:							
LT1	3.5	45.8		2259	DSC	(1)	82P4
LT1	3.5	45.5		2364±105	DSC	(1)	92P4
HT1	3.5	90.2		1230	DSC	(2)	82P4
HT2	3.5	100.5		1054	DSC	(3)	82P4
HT1 + HT2	3.5			2238±105	DSC	(6)	92P4
LT1	8.5	56.0		2259	DSC	(1)	82P4
HT1	8.5	95.0		1301	DSC	(2)	82P4
HT2	8.5	>100			DSC	(4)	82P4
X fragment:							
LT1	3.5	45.0		2238	DSC	(1)	82P4
HT1	3.5	~85.0			DSC	(5)	82P4
HT2	3.5	~95.0			DSC	(5)	82P4
D_H fragment:							
LT1	3.5	45.2		1100	DSC	(1)	82P4
HT2	3.5	89.2		498	DSC	(3)	82P4
LT1	8.5	56.0		1464	DSC	(1)	82P4
HT2	8.5	>100			DSC	(4)	82P4

Fibrinogen and its fragments, results of deconvolution (continued)

Transition	pH	T_{trs}	ΔCp	ΔH	Approach	Remarks	Ref
D_L fragment:							
LT1	3.5	45.8		900	DSC	(1)	82P4
HT2	3.5	88.0		444	DSC	(3)	82P4
TSD fragment:							
HT2	3.5	74.0	7.1	347	DSC	(3)	82P4
E fragment:							
HT1	3.5	75.2		849	DSC	(2)	82P4
HT1	8.5	92.2		1109	DSC	(2)	82P4

Remarks:
(1) low temperature transition 1
(2) high temperature transition 1
(3) high temperature transition 2
(4) transition not observed due to aggregation
(5) ΔH of HT1 and HT2 amounts to 2134 kJ/mol
(6) HT1 at 90.5°C, HT2 as shoulder

Fibrinogen, D-fragment

pH	T_{trs}	ΔCp	ΔH	Approach	Ref
3.5	74	7.1	350	DSC	82M1

Fibrinogen, bovine, proteolytic fragments derived from the C-terminal regions, low (LT) and high (HT) temperature transitions

Fragment/Transition		pH	T_{trs}	ΔH	Approach	Remarks	Ref
D_H	LT	3.5	45.0	1074	DSC	(1,2)	95L7
D_H	HT	3.5	89.0	481	DSC	(1,2)	95L7
D_{LA}	LT	3.5	40.0	957	DSC	(1,2)	95L7
D_{LA}	HT	3.5	89.0	451	DSC	(1,2)	95L7
D_L	LT	3.5	45.0	840	DSC	(1,2)	95L7
D_L	HT	3.5	88.0	410	DSC	(1,2)	95L7
D_X	LT	3.5	40.0	681	DSC	(1,2)	95L7
D_X	HT	3.5	88.0	364	DSC	(1,2)	95L7
D_Y	LT	3.5	41.3	460	DSC	(1,2)	95L7
D_Y	HT	3.5	83.5	334	DSC	(1,2)	95L7
D_Z	LT	3.5	33.0	372	DSC	(1,2)	95L7
D_Z	HT	3.5	81.0	297	DSC	(1,2)	95L7
TSD_1	HT	3.5	74.0	293	DSC	(1,3)	95L7
TSD_2	HT	3.5	70.0	255	DSC	(1,3)	95L7
TSD_3	HT	3.5	67.5	251	DSC	(1,3)	95L7

Remarks:
(1) given is the peak maximum temperature and the calorimetric heat of the transition
(2) D domain specification
(3) TSD thermostable region of the D fragment

Fibronectin

Fibronectin, 29 kD N-terminal domain

Ligand	pH	T_{trs}	ΔH	Approach	Remarks	Ref
without	7.2	69.3	117.2	heat	(1)	90K3
heparin	7.2	68.1	311.3	heat	(1)	90K3

Remark:
(1) van't Hoff treatment, T_{trs} was taken from Fig. 7 of Ref. 90K3

Fibronectin, domains

Domain	pH	T_{trs}	ΔCp	ΔH	Approach	Remarks	Ref
domain 1	4.0	74.8		245	DSC	(1,2)	90T1
domain 2	4.0	73.1		195	DSC	(1,2)	90T1
domain 3	7.4	64		175	DSC	(1,3)	90T1
domains 3+4	4.0	60.7		350	DSC	(1,4)	90T1
domains 3+4+5	8.0	63		550	DSC	(1,5)	90T1
domain 6	8.0	86		300	DSC	(1,6)	90T1
domain 7	8.0	62		290	DSC	(1,7)	90T1
domain 8	8.0	66		260	DSC	(1,7)	90T1
domain 9	4.0	70		200	DSC	(1,7)	90T1
domain 10	8.0	56		310	DSC	(1,8)	90T1
domain 11	8.0	60		370	DSC	(1,8)	90T1
domain 12	8.0	90		385	DSC	(1,9)	90T1

Remarks:
(1) data with 10 % accuracy
(2) determined from the 29 kD fragment
(3) determined from the 13 kD fragment
(4) determined from the 30 kD fragment
(5) determined from the 50,70 kD fragment
(6) determined from the 70 kD fragment
(7) determined from the 60 kD fragment
(8) determined from the 60/65 kD fragment
(9) determined from the 31 kD fragment

Fibronectin, 42 kDa gelatin binding fragment

pH	T_{trs}	ΔCp	ΔH	Approach	Remark	Ref
2.5	37.3		259	DSC	(1)	91L8
2.5	37.5		289	DSC		91L8
3.4	51.8		343	DSC		91L8
4.1	56.0		372	DSC		91L8
7.4	64.4		531	DSC		91L8
8.0	64.1		544	DSC		91L8
9.5	60.9		469	DSC		91L8

Remark:
(1) for ΔCp, see table containing specific values

Fibronectin, 42 kDa gelatin binding fragment and subfragments

Fragment	pH	T_{trs}	ΔCp	ΔH	Approach	Remarks	Ref
42 kDa	7.4	64.4		531	DSC	(1–3)	91L8
30 kDa	7.4	65.1		444	DSC	(4)	91L8
21 kDa	7.4	64.2		159	DSC		91L8
16 kDa	7.4	48.6		218	DSC		91L8
13 kDa	7.4	64.7		209	DSC		91L8

Remarks:
(1) additional recordings made in 6 M GuHCl could be resolved into three individual modules (subtransitions): I6 with T_{trs} = 74.3°C and ΔH = 209 kJ/mol, I7 with T_{trs} = 60.1°C and ΔH = 163 kJ/mol, I9 with T_{trs} = 46.4°C and ΔH = 121 kJ/mol
(2) the modules show in the absence of GuHCl thermal transitions at T_{trs} = 122±12°C, 108±12°C, and 94±12°C
(3) ΔCp see table containing specific values
(4) additional recordings made in 6 M GuHCl could be resolved into two individual modules (subtransitions): I6 with T_{trs} = 73.6°C and ΔH = 192 kJ/mol I7, with T_{trs} = 58.6°C and ΔH = 167 kJ/mol

Flagellar Protein, see also hook protein

Flagellin

Flagellin, monomeric, normal type

pH	T_{trs}	ΔCp	ΔH	Approach	Ref
7	47.5–48	28±2	1340±40	DSC	84F1

Flagellin and its proteolytic fragments, resolved into subtransitions

Transition	pH	T_{trs}	ΔCp	ΔH	Approach	Ref
Flagellin						
trans. (1)	7	42.7±0.3	21.4±2.1	306±13	DSC	90U3
trans. (2)	7	46.1±0.1	34.7±8.5	943±25	DSC	90U3
trans. (3)	7	46.3±0.06	22.0±0.55	1326±7	DSC	90U3

Table 2. Enthalpy and Heat Capacity Changes – Molar Values

Flagellin and its proteolytic fragments, resolved into subtransitions (continued)

Transition	pH	T_{trs}	ΔCp	ΔH	Approach	Ref
Fragment F40K						
trans. (1)	7	42.2±0.4	14.2±2.1	261±14	DSC	90U3
trans. (2)	7	45.2±0.2	26.1±6.8	744±28	DSC	90U3
trans. (3)	7	46.0±0.07	22.2±1.2	1260±6	DSC	90U3
Fragment F27K						
trans. (1)	7	48.05±0.03	26.4±1.8	578±6	DSC	90U3
trans. (2)	7	48.5±0.03	1.9±0.5	1007±3	DSC	90U3

Flavodoxin, see Apoflavodoxin

Four-Helix Bundle Proteins, see also de novo and ROP

β-Galactosidase

β-Galactosidase

Form/Ligand	pH	T_{trs}	ΔCp	ΔH	Approach	Remarks	Ref
1 mM MgCl$_2$	7.0	58.0	21.8±2.0	568±43	heat	(1)	93H3
1 mM EDTA	7.0	47.8	14.1±3.0	300±22	heat	(1)	93H3

Remark:
(1) the data were determined by a thermodynamic treatment of temperature-activity data

Gene 32 Protein

Gene 32 protein of phage T4, native Zn^{2+}, and reconstituted forms

Form	pH	T_{trs}	ΔCp	ΔH	Approach	Remarks	Ref
Zn^{2+}	8	55.4±0.2	−3.22±0.08	582±21	DSC	(1)	88K1
apo	8	49.3±0.4	−4.6±1.2	351±29	DSC		88K1
Cd^{2+}	8	53.3±0.2	−12.0±0.4	556±42	DSC		88K1
Co^{2+}	8	56.4±0.2	−15.7±0.4	577±25	DSC		88K1
Zn^{2+}-reco	8	54.9	0	669	DSC	(2)	88K1

Remarks:
(1) native Zn^{2+} form
(2) reconstituted Zn^{2+} form

Gene 32 protein of phage T4, complexed with polynucleotide, native Zn^{2+}, and reconstituted forms

Form	pH	T_{trs}	ΔCp	ΔH	Approach	Remarks	Ref
Zn^{2+}	8	59.9±0.2	−12.1±0.8	636±63	DSC	(1)	88K1
apo	8	49.8±0.2	−1.9±9.2	406±92	DSC	(1)	88K1
Cd^{2+}	8	57.3±0.1	−5.7±13.8	674±8	DSC	(1)	88K1
Co^{2+}	8	60.9±0.6	−9.0±2.6	678±13	DSC	(1)	88K1

Remark:
(1) complexed with stoichiometric amount of poly(dT)

Glucanase

(1,3–1,4)-β-Glucanase, hybrid forms

Mutant	pH	T_{trs}	ΔCp	ΔH	Approach	Remarks	Ref
Measurements in the presence of $CaCl_2$:							
H(A12-M)	6.0	50.5		720	DSC	(1)	94W3
H(A12-M)ΔY13	6.0	52.0/54.8		874	DSC	(1)	94W3
H(A16-M)	6.0	55.9/57.3		828	DSC	(1)	94W3
Measurements in the presence of EDTA:							
H(A12-M)	6.0	49.3		577	DSC	(2)	94W3
H(A12-M)ΔY13	6.0	46.6		611	DSC	(2)	94W3
H(A16-M)	6.0	51.2		669	DSC	(2)	94W3

Remarks:
(1) buffer: 2 mM cacodylate, 1 mM $CaCl_2$, 1.5 M GuHCl
(2) buffer: 2 mM cacodylate, 1 mM EDTA, 1.5 M GuHCl

(1,3–1,4)-β-Glucanase, hybrid forms

Mutant	pH	T_{trs}	ΔCp	ΔH	Approach	Remarks	Ref
AMY:							
trans. (1)	6.0	54.3		310	DSC	(1,2)	96W2
trans. (2)	6.0	58.9		519	DSC	(1,2)	96W2
AMY:							
trans. (1)	6.0	44.2		180	DSC	(1,3)	96W2
trans. (2)	6.0	48.6		481	DSC	(1,3)	96W2
MAC	6.0	50.8		582	DSC	(1,2)	96W2
MAC	6.0	44.5		498	DSC	(1,3)	96W2
hybrid 1:							
trans. (1)	6.0	55.9	9.2	327	DSC	(1,2)	95W4
trans. (2)	6.0	57.3		595	DSC	(1,2)	95W4
hybrid 1	6.0	51.2	8.4	674	DSC	(1,3)	95W4
hybrid 1	6.0	51.2		645	heat	(1,3)	95W4
hybrid 2	6.0	50.5	7.5	653	DSC	(1,2)	95W4
hybrid 2	6.0	49.1		678	heat	(1,2)	95W4
hybrid 2	6.0	49.3	6.7	611	DSC	(1,3)	95W4
hybrid 2	6.0	47.7		607	heat	(1,3)	95W4
hybrid 3	6.0	46.6	7.1	599	DSC	(1,3)	95W4
hybrid 3	6.0	49.7		565	heat	(1,3)	95W4

(1,3–1,4)-β-Glucanase, hybrid forms (continued)

Mutant	pH	T_{trs}	ΔCp	ΔH	Approach	Remarks	Ref
Hybrid 3:							
trans. (1)	6.0	52.0	10.5	364	DSC	(1,2)	95W4
trans. (2)	6.0	54.8		548	DSC	(1,2)	95W4
hybrid 4	6.0	54.7		594	DSC	(1,2)	96W2
hybrïd 4	6.0	48.9		552	DSC	(1,3)	96W2
Hybrid 5:							
trans. (1)	6.0	57.2		314	DSC	(1,2)	96W2
trans. (2)	6.0	58.4		561	DSC	(1,2)	96W2
hybrid 5	6.0	52.1		623	DSC	(1,3)	96W2

Remarks:
(1) the hybrids are composed of N-terminal segments of glucanase from *Bacillus amyloliquefaciens* (AMY) and of C-proximal residues of glucanase from *Bacillus macerans* (MAC)
  hybrid 1: (1–16)AMY.MAC(17–214)
  hybrid 2: (1–12)AMY.MAC(13–214)
  hybrid 3: (1–12)AMY.des-Tyr13MAC(14–214)
  hybrid 4: H(A12-M) ΔY13FMA
  hybrid 5: H(A12-M) ΔF14
(2) buffer: 2 mM sodium cacodylate, 1.5 M GuHCl, 1 mM CaCl$_2$, pH 6.0
(3) buffer: 2 mM sodium cacodylate, 1.5 M GuHCl, 1 mM EDTA, pH 6.0

Glucosamine-6-Phosphate Deaminase

Glucosamine-6-phosphate deaminase from *E. coli*

pH	T_{trs}	ΔCp	ΔH	Approach	Remarks	Ref
7.1	65		5270±490	DSC	(1,2)	93H5

Remarks:
(1) the irreverible transition was analyzed by scan rate dependence
(2) the calorimetric transition can be resolved into six subtransitions

Glucoamylase

Glucoamylase from *Rhizopus*

Transition	pH	T_{trs}	ΔCp	ΔH	Approach	Remarks	Ref
Ligand-free form:							
overall	7.0	54–57	6.5	971	DSC	(1)	91T4
trans. (1)	7.0	54.15		397	DSC	(2)	91T4
trans. (2)	7.0	56.89		563	DSC	(2)	91T4
SGI concentration = 0.149 mM						(3)	
overall	7.0	56–59	18.6	1046	DSC	(4)	91T4
trans. (1)	7.0	56.64		452	DSC	(2)	91T4
trans. (2)	7.0	59.36		623	DSC	(2)	91T4
SGI concentration = 1.49 mM						(3)	
overall	7.0	59–62	27.7	1151	DSC	(4)	91T4
trans. (1)	7.0	59.04		485	DSC	(2)	91T4
trans. (2)	7.0	61.72		703	DSC	(2)	91T4

Glucoamylase from *Rhizopus* (continued)

Transition	pH	T_{trs}	ΔCp	ΔH	Approach	Remarks	Ref
SGI concentration = 14.0 mM						(3)	
overall	7.0	61–64	30.8	1251	DSC	(4)	91T4
trans. (1)	7.0	61.06		493	DSC	(2)	91T4
trans. (2)	7.0	63.84		707	DSC	(2)	91T4

Remarks:
(1) overall transition heat and heat capacity change (average from measurements of protein conc. from 50 to 209 μM, experimental value of ΔH), the peak can be resolved into two subtransitions
(2) obtained by deconvolution
(3) SGI = 5-amino-1,5-dideoxy D-glycopyranose
(4) see remark (1), protein conc. = 50.5 μM

Glucoamylase 1 from *Aspergillus niger*, dependence of thermodynamic quantities on protein conc. (mg/ml), the heat capacity change is given in J/K/g

Concentration	pH	T_{trs}	Δcp	ΔH	Approach	Remark	Ref
12.0	7.8	67.8	0.33	1730	DSC		92W3
10.0	7.8	66.4	0.39	1850	DSC		92W3
8.33	7.8	65.5	0.44	2150	DSC		92W3
4.0	7.8	63.9	0.45	1680	DSC		92W3
	7.8		0.40±0.05	1850±210	DSC	(1)	92W3

Remark:
(1) mean values (±SD)

Glucoamylase from *Aspergillus niger*, fragments

Form	pH	T_{trs}	ΔCp	ΔH	Approach	Remarks	Ref
G2	7.8	65.7	30	1650	DSC	(1,2)	92W3
G1C	7.8	57.4	11	360	DSC	(2,3)	92W3
G1C499	7.8	57.3	10	310	DSC	(2,4)	92W3
G1C509	7.8	54.7	7	310	DSC	(2,5)	92W3

Remarks:
(1) G2 = glucoamylase 2
(2) protein conc. was 0.122 mM
(3) G1C = glucoamylase 1 fragment (471–616)
(4) G1C499 = glucoamylase 1 fragment (499–616)
(5) G1C509 = glucoamylase 1 fragment (509–616)

Glucoamylase from *Aspergillus niger* – effect of β-cyclodextrin

Concentration	pH	T_{trs}	ΔCp	ΔH	Approach	Remarks	Ref
G1C = glucoamylase 1 fragment-(471–616)							
0	7.8	57.4	11	360	DSC	(1,2)	92W3
0.50	7.8	63.3	13	470	DSC	(1,2)	92W3
2.00	7.8	67.7	14	560	DSC	(1,2)	92W3
5.00	7.8	70.3	16	650	DSC	(1,2)	92W3
Glucoamylase 1							
0	7.8	66.4	32	1850	DSC	(1,2)	92W3
2.00	7.8	67.6	38	1880	DSC	(1,2)	92W3
5.00	7.8	69.8	41	1840	DSC	(1,2)	92W3

Remark:
(1) protein conc. was 0.122 mM
(2) specific heat capacity, see table 3

Glucose Oxidase

Glucose oxidase from *Aspergillus niger*

	pH	T_{trs}	ΔCp	ΔH	Remarks	Ref
	3.2	25		3610±560	(1,2)	93H8

Remarks:
(1) measured by isothermal calorimetry
(2) from interaction of glucose oxidase with n-alkylsulfates, extrapolated value

β-Glucosidase

β-Glucosidase from almond

Transition	pH	T_{trs}	ΔCp	ΔH	Approach	Remarks	Ref
overall	4	64–72	14.0	1762	DSC	(1)	91T3
trans. (1)	4	64.6		330	DSC	(2)	91T3
trans. (2)	4	70.0		426	DSC	(2)	91T3
trans. (3)	4	72.3		78.4	DSC	(2)	91T3
overall	5	67–78	19.5	1036	DSC	(1)	91T3
trans. (1)	5	67.6		294	DSC	(1)	91T3
trans. (1)	5	74.3		448	DSC	(2)	91T3
trans. (3)	5	78.2		333	DSC	(2)	91T3
overall	6	69–84	16.9	1171	DSC	(1)	91T3
trans. (1)	6	69.2		311	DSC	(2)	91T3
trans. (2)	6	75.0		391	DSC	(2)	91T3
trans. (3)	6	83.7		527	DSC	(2)	91T3
overall	7	66–84	25.8±14.5	1114±67	DSC	(1)	91T3
trans. (1)	7	66.8±0.4		319±22	DSC	(2)	91T3
trans. (2)	7	75.6±0.2		236±36	DSC	(2)	91T3
trans. (3)	7	83.9±0.2		569±31	DSC	(2)	91T3

β-Glucosidase from almond (continued)

Transition	pH	T_{trs}	ΔC_p	ΔH	Approach	Remarks	Ref
overall	8	64–83	21.5	1059	DSC	(1)	91T3
trans. (1)	8	64.0		300	DSC	(2)	91T3
trans. (2)	8	75.3		235	DSC	(2)	91T3
trans. (3)	8	83.0		619	DSC	(2)	91T3

Remarks:
(1) overall transitional heat and heat capacity change (ΔH is the experimental value), the peak can be resolved into three subtransitions
(2) obtained by deconvolution

Glutamine Synthetase

Glutamine synthetase from *E. coli*

pH	T_{trs}	ΔC_p	ΔH	Approach	Remark	Ref
7.3	51.1		351	heat, v.H.	(1)	89S10

Remark:
(1) see also Ref. 89S10 for the enthalpy change of the enzyme in the presence of various ligands

Glutamine synthetase from *E. coli*

Transition	pH	T_{trs}	ΔC_p	ΔH	Approach	Remarks	Ref
Glutamine synthetase with 1.0 mM Mn^{2+}:							
overall	7.0	51.6±0.1	44.4±3.8	883±17	DSC	(1)	91G2
trans. (1)	7.0	50.4±0.8		289±50	DSC	(1,2)	91G2
trans. (2)	7.0	51.7±0.2		611±38	DSC	(1,2)	91G2
Glutamine synthetase with 1.0 mM Mn^{2+} and 150 mM Gln:							
overall	7.0	58.6±0.1	47.7±2.5	950±21	DSC	(3)	91G2
trans. (1)	7.0	56.9		272	DSC	(2,3)	91G2
trans. (2)	7.0	58.7		761	DSC	(2,3)	91G2
Glutamine synthetase with 10.0 mM Mn^{2+}:							
overall	7.0	43.9±0.2	16.3±7.1	887±8	DSC	(4)	91G2
trans. (1)	7.0	43.2		318	DSC	(2,4)	91G2
trans. (2)	7.0	44.4		590	DSC	(2,4)	91G2

Remarks:
(1) buffer: 50 mM HEPES/KOH, 100 mM KCl, 1.0 mM $MnCl_2$, pH 7.0 at 50°C
(2) resolved by deconvolution
(3) buffer: 50 mM HEPES/KOH, 100 mM KCl, 1.0 mM $MnCl_2$, 150 mM Gln, pH 7.0 at 50°C
(4) buffer: 50 mM HEPES/KOH, 100 mM KCl, 10.0 mM $MnCl_2$, pH 7.0 at 50°C

Glutamine synthetase from *E. coli* in the presence of various ligands and buffers (BU)

Ligand/(BU)	pH	T_{trs}	ΔCp	ΔH	Approach	Remarks	Ref
(HEPES)	7.0	51.6	44.4±3.8	883	DSC	(1–3)	92Z5
MetSox	7.0	63.0		1933	DSC	(1–4)	92Z5
Gln	7.0	58.7		954	DSC	(1–3,5)	92Z5
ADP	7.0	48.5		967	DSC	(1–3,6)	92Z5
(PIPES)	7.1	52.8	36.8±9.2	941	DSC	(1,2,7)	92Z5
ADP	7.1	51.1		937	DSC	(1,2,7)	92Z5
(TES)	7.0	50.9	27.2±4.6	1055	DSC	(1,2,8)	92Z5
(Tris)	7.0		26.8±3.8	1150	DSC	(1,2,9)	92Z5
	7.0	51.6		736±50	DSC	(11)	92Z5

Remarks:
(1) the data refer to unfolding of the glutamine synthetase dodecamer
(2) pH of the buffer measured at 50°C
(3) buffer: 50 mM HEPES, 100 mM KCl, 1.0 mM $MnCl_2$
(4) MetSox = L-methionine(SR)-sulfoximine, degree of saturation = 0.99 at 30°C
(5) Gln = L-glutamine, degree of saturation = 0.96 at 30°C
(6) degree of ADP saturation = 0.99 at 30°C
(7) buffer: 50 mM PIPES, 100 mM KCl, 1.0 mM $MnCl_2$
(8) buffer: 50 mM TES, 100 mM KCl, 1.0 mM $MnCl_2$ (remark 10)
(9) buffer: 50 mM Tris, 100 mM KCl, 1.0 mM $MnCl_2$ (remark 10)
(10) ΔH was taken from Fig. 5 in Ref. 92Z5 (mean value), the enthalpy change was extrapolated to 51.6°C
(11) the enthalpy change was corrected to $\Delta H = 0$ for the protonation enthalpy of the buffers, ΔH of the buffers was determined at 50°C to
     –15.9 kJ/mol for PIPES
     –25.5 kJ/mol for HEPES
     –36.4 kJ/mol for TES
     –51.5 kJ/mol for Tris

Glutamate Dehydrogenase

Glutamate dehydrogenase from *Pyrococcus furiosus*

pH	T_{trs}	ΔCp	ΔH	Approach	Remarks	Ref
7.15	113		1732	DSC	(1–3)	92K9

Remarks:
(1) buffer: imidazole-HCl, 10 mM DTT
(2) the van't Hoff heat amounts to $\Delta H = 1004$ kJ/mol, calorimetric and spectroscopic value
(3) activation of the enzyme takes place at 57°C

Glutamate dehydrogenase from hyperthermophile *Archaeon*, ES4, recombinant

pH	T_{trs}	ΔCp	ΔH	Approach	Remarks	Ref
8.25	113		1732	DSC	(1,2)	93D5

Remarks:
(1) the data refer to MW = 270,000 Da for the hexameric complex
(2) the van't Hoff heat amounts to $\Delta H = 920$ kJ/mol

Glutaredoxin

Glutaredoxin from *E. coli*, oxidized and reduced form

Form	pH	T_{trs}	ΔCp	ΔH	Approach	Remarks	Ref
oxidized	7.0	55	4.3	280	heat	(1)	91S3
reduced	7.0	57	4.3	314	heat	(1)	91S3

Remark:
(1) ΔCp is an estimated value based on the amino acid content of the protein

Glycinin, see plant seed proteins

Glycoprotein

α1-Acid glycoprotein

pH	T_{trs}	ΔCp	ΔH	Approach	Ref
5.25	46.6±0.05	8.8±4.2	252±25	heat, v.H.	90R3

Growth Factor

Growth factor, epidermal

pH	T_{trs}	ΔCp	ΔH	Approach	Ref
7.5	40	0±2	108	GuHCl, v.H.	76H

Acidic fibroblast growth factor, ΔH and ΔCp measured in the presence of a 3-fold weight excess of heparin

pH	T_{trs}	ΔCp	ΔH	Approach	Remarks	Ref
7.2	4	11.7±4.6	−109±84	urea	(1–3)	93V2

Remarks:
(1) from the temperature dependence of ΔG obtained by urea denaturation using linear extrapolation
(2) thermal transition in the absence of heparin at $T_{trs} = 45°C$ (Ref. 93V2)
(3) thermal transition in the presence of a 3-fold excess of heparin at $T_{trs} = 64°C$ (Ref. 93V2)

HBsu

HBsu = DNA-binding histone-like protein from *Bacillus subtilis*

pH	T_{trs}	ΔCp	ΔH	Approach	Remarks	Ref
7.5	61		421	DSC	(1–3)	92W2

Remarks:
(1) measured in the presence of 0.5 M potassium fluoride
(2) data refer to the transition N→2D
(3) T_{trs} in the absence of potassium fluoride is 45°C

Hemopexin

Hemopexin (plasma glycoprotein)

Form	pH	T_{trs}	ΔH	Approach	Ref
native	7.4	53.9±0.3	774±29	DSC	93W5
native + heme	7.4	66.4±0.7	1222±21	DSC	93W5
domain I	7.4	51.9±0.3	393±21	DSC	93W5
domain I + heme	7.4	77.6±0.6	1540±176	DSC	93W5
domain II	7.4	49.3±0.5	586±38	DSC	93W5
domain I + domain II	7.4	54.2/62.4	577/439	DSC	93W5
domain I + domain II + heme	7.4	55.8/74.7	975/1331	DSC	93W5

Hevein

Hevein from latex of rubber

pH	T_{trs}	ΔCp	ΔH	Approach	Remarks	Ref
1.82	73.0	1.54	121.2	DSC		95H5
2.10	74.2	1.86	119.8	DSC		95H5
2.42	74.1	1.71	126.6	DSC		95H5
2.92	77.8	2.09	129.6	DSC		95H5
3.20	83.6	2.60	142.3	DSC		95H5
3.33	82.6	1.76	138.8	DSC		95H5
3.54	84.5	1.84	152.6	DSC		95H5
3.67	90.1	1.88	158.8	DSC		95H5
1.8–3.7	73–90	1.91±0.31		DSC	(1)	95H5
1.8–3.7	73–90	2.34±0.23		DSC	(2)	95H5

Remarks:
(1) ΔCp is the average value from the individual scanning calorimetric recordings
(2) ΔCp is the slope from ΔH versus T_{trs}

Hexokinase

Hexokinase, from rat brain, with and without ligand

Transition	pH	T_{trs}	ΔCp	ΔH	Approach	Remarks	Ref
trans. (1)	8.5	47.9±1.0		456±109	DSC	(1)	90W3
trans. (2)	8.5	50.9±1.4		469±21	DSC	(1)	90W3
ligand glucose:							
trans. (1)	8.5	55.3±0.6		611±29	DSC	(1)	90W3
trans. (2)	8.5	56.7±0.4		640±50	DSC	(1)	90W3
ligand glucose-6-phosphate:							
trans. (1)	8.5	51.1±0.7		519±121	DSC	(1)	90W3
trans. (2)	8.5	61.6±0.7		552±38	DSC	(1)	90W3
ligand N-acetylglucosamine:							
trans. (1)	8.5	48.6±0.5		423±58	DSC	(1)	90W3
trans. (2)	8.5	52.6±0.9		515±79	DSC	(1)	90W3

Remark:
(1) the thermal transition was resolved into the two subtransitions by deconvolution

Hexokinase, from yeast, in the presence of D-glucose

Transition	pH	T_{trs}	ΔCp	ΔH	Approach	Remarks	Ref
trans. (1)	8.5	40.1		300	DSC	(1)	95B2b
trans. (2)	8.5	48.2		380	DSC	(1)	95B2b

Remarks:
(1) the data are deduced from a thermodynamic model using binding constant $K_b = 15000$ M^{-1} and
    $\Delta H_b = 0$ kJ/mol

Hirudin

Hirudin

pH	T_{trs}	ΔCp	ΔH	Approach	Ref
7.0	65		159	heat	91O

Histocompatibility Complex

Class I major histocompatibility complex, empty and filled heterodimer (HD)

Form	pH	T_{trs}	ΔCp	ΔH	Approach	Ref
empty HD	7	45±1	1.3	201	heat	92F1
filled HD	7	56±1	3.8	448	heat	92F1

Histone

Histone GH1 and GH5 fragments

Fragment	pH	T_{trs}	ΔCp	ΔH	Approach	Ref
GH1	7	59.5		182	DSC	82T1
GH5	7	63		213	DSC	82T1

Histone H2A-H2B dimer stability as a function of ionic strength

NaCl Concentration	pH	T_{trs}	ΔCp	ΔH	Approach	Remarks	Ref
0 mM	7.5	42.8	4.2	223	DSC	(1)	95K4
25 mM	7.5	48.4	5.9	259	DSC	(1)	95K4
50 mM	7.5	49.8	5.4	264	DSC	(1)	95K4
100 mM	7.5	50.0	5.9	265	DSC	(1)	95K4
140 mM	7.5	51.7	6.3	280	DSC	(1)	95K4
50 mM	7.5	43.6		273	heat	(1,2)	95K4
100 mM	7.5	48.5		287	heat	(1,2)	95K4
200 mM	7.5	53.7		303	heat	(1,2)	95K4
400 mM	7.5	58.6		328	heat	(1,2)	95K4
1000 mM	7.5	67.4		363	heat	(1,2)	95K4

Remarks:
(1) all thermodynamic data are expressed on the basis of a 14 kDa monomer unit
(2) van't Hoff treatment, the transition was monitored by CD at 222 nm

Table 2. Enthalpy and Heat Capacity Changes – Molar Values

Histone H2A-H2B dimer stability as a function of pH

pH	T_{trs}	ΔCp	ΔH	Approach	Remarks	Ref
4.5	36.4			DSC	(1)	95K4
5.5	43.6	3.8	219	DSC	(1,2)	95K4
6.5	47.4	5.9	256	DSC	(1,2)	95K4
7.5	47.2	6.3	260	DSC	(1,2)	95K4
8.5	49.4	5.9	263	DSC	(1,2)	95K4
9.5	48.2	5.4	256	DSC	(1,2)	95K4
10.5	39.6			DSC	(1,2)	95K4
11.5	37.5			DSC	(1,2)	95K4
4.0	32.0			heat	(2,3)	95K4
4.5	34.3			heat	(2,3)	95K4
5.5	42.4		268	heat	(2,3)	95K4
6.5	45.0		279	heat	(2,3)	95K4
7.5	46.1		282	heat	(2,3)	95K4
8.5	47.3		287	heat	(2,3)	95K4
9.5	46.2		285	heat	(2,3)	95K4
10.5	37.6			heat	(2,3)	95K4
11.5	32.0			heat	(2,3)	95K4

Remarks:
(1) all thermodynamic data are expressed on the basis of a 14 kDa monomer unit
(2) ΔCp from ΔH versus T_{trs} (from pH 5.5 to 9.5) amounts to 5.44 kJ/mol/K
(3) van't Hoff treatment, the transition was monitored by CD at 222 nm

Histone H2A-H2B dimer stability as a function protein concentration

Concentration	pH	T_{trs}	ΔCp	ΔH	Approach	Remarks	Ref
2.0 µM	7.5	43.0		246	heat	(1,2)	95K4
5.0 µM	7.5	45.6		254	heat	(1,2)	95K4
21.6 µM	7.5	48.2		262	heat	(1,2)	95K4
43.3 µM	7.5	49.5		264	heat	(1,2)	95K4
98.6 µM	7.5	50.1		266	heat	(1,2)	95K4
110.0 µM	7.5	51.3	6.3	279	DSC	(1)	95K4

Remarks:
(1) all thermodynamic data are expressed on the basis of a 14 kDa monomer unit
(2) van't Hoff treatment, the transition was monitored by CD at 222 nm

Hook Protein

Flagellar hook protein from *Salmonella typhimurium*

Form/Trans.	pH	T_{trs}	ΔCp	ΔH	Approach	Remarks	Ref
Hook monomer:							
total	7.0	50.9	27	1102	DSC		95V9
trans. (1)	7.0	41.2		184	DSC	(1)	95V9
trans. (2)	7.0	47.6		655	DSC	(1)	95V9
trans. (3)	7.0	50.2		1066	DSC	(1)	95V9
Polyhook:							
total	7.0	69.0	31	2046	DSC	(2)	95V9

Remarks:
(1) result of deconvolution
(2) data per mole of monomer

HPr

HPr, histidine-containing phosphocarrier protein from *Bacillus subtilis*

pH	T_{trs}	ΔCp	ΔH	Remarks	Ref
7.0	73.4±0.2	4.85±0.21	243±7	(1)	95S3
7.0	73.8±0.4	4.85±0.21	248±10	(2)	95S3
7.0	0–40	4.90±0.33		(3)	95S3
7.0	0–40	4.60±0.42		(4)	95S3
7.0		4.98±0.21		(5)	95S3

Remarks:
(1) ΔCp was determined by a combined approach that makes use of ΔG versus T (from urea denaturation) and thermal unfolding in the absence of urea
(2) ΔCp was determined by a global fit of transitions monitored by both urea denaturation at various temperatures and thermal denaturation profiles in the presence and absence of urea
(3) ΔCp was determined from ΔH versus T as a function of urea concentration
(4) ΔCp was determined as in (3) after correction of ΔH for urea binding
(5) ΔCp was determined from the thermal denaturation profile in the presence of 3 M urea

HPr, histidine-containing phosphocarrier protein from *E. coli*

Urea Con-centration	pH	T_{trs}	ΔCp	ΔH	Approach	Remarks	Ref
0.0 M	7.0	63.4	6.23	317	heat	(1)	96N2
1.0 M	7.0	58.5	5.94	271	heat	(1)	96N2
2.0 M	7.0	53.5	5.86	228	heat	(1)	96N2
3.0 M	7.0	47.4	5.48	172	heat	(1,2)	96N2
4.0 M	7.0	37.0	5.31	117	heat	(1,3)	96N2
0 to 4 M	7.0		6.23±0.38		heat	(4)	96N2
0 to 4 M	7.0		6.19±0.25		heat	(5)	96N2
0.0 M	7.0	63.8±0.5	6.07±0.33	296±14	heat	(6)	96N2
0.0 M	7.0	63.1±0.4	5.94±0.38	292±14	heat	(7)	96N2

Remarks:
(1) thermal unfolding in the presence of urea, transition monitored by CD at 222 nm
(2) cold denaturation observed at T_{trs} = –11.4°C with ΔH = –151 kJ/mol
(3) cold denaturation observed at T_{trs} = +1.3°C with ΔH = –108 kJ/mol
(4) from ΔH versus T_{trs}
(5) from ΔCp versus urea conc.
(6) from global data analysis assuming m = ($\delta\Delta G/\delta$[urea]) temperature invariant
(7) from global data analysis assuming m = ($\delta\Delta G/\delta$[urea]) temperature dependent

Immunoglobulin

Bence-Jones protein

pH	T_{trs}	ΔCp	ΔH	Approach	Remark	Ref
7.4	56		293	DSC	(1)	77Z

Remark:
(1) data are calculated per monomer

Human immunoglobulin G

pH	T_{trs}	ΔCp	ΔH	Approach	Remark	Ref
7.4	69.3±0.3		251±14	heat	(1)	95S15

Remark:
(1) global analysis of FTIR spectra based on the intramolecular secondary structure band

Immunoglobulin, light chain, CL fragments

Fragment	pH	T_{trs}	ΔCp	ΔH	Approach	Ref
105–214	7.5	60±0.4		282±8	heat, v.H.	87G2
109–214	7.5	60±0.1		280±2	heat, v.H.	87G2
113–214	7.5	57.5±0.1		258±9	heat, v.H.	87G2

Immunoglobulin, CL fragment, native and modified forms

Form	pH	T_{trs}	ΔCp	ΔH	Approach	Remarks	Ref
native	7.5	61.0±0.5	7.5	278±4	heat		90O1
NFK-Trp187	7.5	46.1±0.2	8.8	230±2	heat	(1)	90O1
Kyn-Trp187	7.5	48.8±0.3	8.8	233±2	heat	(2)	90O1

Remarks:
(1) modified by ozone oxidation of Trp187 to the N-formylkynurenine derivative
(2) modified by ozone oxidation of Trp187 to the kynurenine derivative

Immunoglobulin IgG and a stable conformer obtained by renaturation

Form	pH	T_{trs}	ΔCp	ΔH	Approach	Remarks	Ref
native	7	60–90		3350±210	DSC	(1)	95M6
native'	2	35–60		1830±110	DSC	(2)	95M6
renatured	7	60–90		~2900*	DSC	(3)	95M6

Remarks:
(1) resolved by deconvolution into five subtransitions
(2) native' at pH 2, transition resolved by deconvolution into four subtransitions
(3) renatured IgG from the pH 2 state yields a non-native conformation, the melting peak was resolved by deconvolution into five subtransitions

Immunoglobulin G, cooperative structures, results of deconvolution

Transition	pH	T_{trs}	ΔCp	ΔH	Approach	Ref
Fab fragment:						
trans. (1)	2.5	59		223	DSC	82T2
trans. (2)	2.5	65		470	DSC	82T2
trans. (3)	2.5	71		559	DSC	82T2
trans. (1)	3.0	63		227	DSC	82T2
trans. (2)	3.0	68		474	DSC	82T2
trans. (3)	3.0	75		575	DSC	82T2
trans. (1)	3.51	66		248	DSC	82T2
trans. (2)	3.51	71		483	DSC	82T2
trans. (3)	3.51	77		588	DSC	82T2
pFc' fragment:						
trans. (1)	3.0	51		248	DSC	82T2
trans. (2)	3.0	62		357	DSC	82T2
trans. (1)	3.25	60		260	DSC	82T2
trans. (2)	3.25	67		386	DSC	82T2
trans. (1)	3.51	65		252	DSC	82T2
trans. (2)	3.51	70		491	DSC	82T2
Fc fragment:						
trans. (1)	3.51	54		353	DSC	82T2
trans. (2)	3.51	68		273	DSC	82T2
trans. (3)	3.51	71		273	DSC	82T2
trans. (4)	3.51	73		474	DSC	82T2

Immunoglobulin, deconvolution, MAK33, murine antibody of subtype κ/IgG1, deconvolution

Transition	pH	T_{trs}	ΔCp	ΔH	Approach	Ref
trans. (1)	2.0	66.5		355	DSC	91B6
trans. (2)	2.0	72.1		347	DSC	91B6
trans. (1)	2.7	65.6		346	DSC	91B6
trans. (2)	2.7	73.4		378	DSC	91B6
trans. (1)	7.0	63.9		684	DSC	91B6
trans. (2)	7.0	64.9		883	DSC	91B6
trans. (3)	7.0	69.2		773	DSC	91B6
trans. (4)	7.0	72.5		806	DSC	91B6
trans. (5)	7.0	80.4		497	DSC	91B6

Immunoglobulin Binding Protein, see also Protein G

IgG binding domain of peptostreptococcal protein L, selected by a phage display system

Mutant	pH	T_{trs}	ΔH	Approach	Ref
protein L	6.5	70	285	heat, v.H.	95G12
mutant 17	6.5	40	167	heat, v.H.	95G12
mutant 20	6.5	44	161	heat, v.H.	95G12
mutant 30	6.5	38	182	heat, v.H.	95G12
mutant 37	6.5	54	301	heat, v.H.	95G12
mutant 43	6.5	39	154	heat, v.H.	95G12

Insulin

Insulin

pH	T_{trs}	ΔCp	ΔH	Remark	Ref
9.6	25	2.1±1.2	90.7	(1)	82F2

Remark:
(1) measured by insulin reduction, flow calorimetry

Interleukin

Interleukin-1β

pH	T_{trs}	ΔCp	ΔH	Approach	Remark	Ref
2.0	37	7.9	227	DSC	(1)	94M3
2.2	41		266	DSC		94M3
2.38	44		247	DSC		94M3
2.5	49		330	DSC		94M3
3.0	53		351	DSC		94M3

Remark:
(1) ΔCp from ΔH versus T_{trs}

Interleukin-1β. Temperature depenence of the heat capacity change at protein unfolding in aqueous solution, and related thermodynamic quantities

T	$\Delta H(exp)$	$\Delta S(exp)$	$\Delta Cp(exp)$	Approach	Remarks	Ref
5	7	−99	6.9	DSC	(1–3)	95M3
25	151	401	7.5	DSC	(1–3)	95M3
50	330	1006	7.5	DSC	(1–3)	95M3
75	501	1516	6.2	DSC	(1–3)	95M3
100	640	1903	4.9	DSC	(1–3)	95M3
125	736	2155	2.8	DSC	(1–3)	95M3

Remarks:
(1) the data were taken from Ref. 95M3 with reference to Ref. 94M3
(2) ΔCp was obtained as described in Ref. 90P3
(3) ΔCp in kJ/mol/K, ΔH in kJ/mol and ΔS in J/mol/K

Lac-Repressor

Lac-repressor, headpiece

	pH	T_{trs}	ΔCp	ΔH	Approach	Ref
	8	65	1.255	118±13	DSC	81H2

Laccase

Laccase from the latex of japanese laquer tree, holo and apo protein

Transition	pH	T_{trs}	ΔCp	ΔH	Approach	Remarks	Ref
Holo laccase:							
total	5.6	53.5	20–35	996	DSC	(1,2)	95A2
trans. (1)	5.6	50.0		490	DSC	(3)	95A2
trans. (2)	5.6	54.7		502	DSC	(3)	95A2
total	6.0	51.9±0.5		999±46	DSC	(1)	95A2
trans. (1)	6.0	48.5±1.0		435±46	DSC	(3)	95A2
trans. (2)	6.0	53.1±1.3		494±96	DSC	(3)	95A2
total	6.0	44.6		343	DSC	(1,4)	95A2
trans. (1)	6.0	43.5		364	DSC	(4,5)	95A2
total	7.0	50.7±0.7		1021±38	DSC	(1)	95A2
trans. (1)	7.0	47.9±1.6		435±42	DSC	(3)	95A2
trans. (2)	7.0	51.3±0.7		594±42	DSC	(3)	95A2
Type-2-Cu-depleted laccase:							
total	5.6	55.3		841	DSC	(1)	95A2
trans. (1)	5.6	50.9		331	DSC	(3)	95A2
trans. (2)	5.6	54.8		565	DSC	(3)	95A2
total	6.0	53.9±1.0		879±67	DSC	(1)	95A2
trans. (1)	6.0	50.3±1.4		385±59	DSC	(3)	95A2
trans. (2)	6.0	55.1±1.1		490±67	DSC	(3)	95A2
total	6.0	50.3		866	DSC	(1,6)	95A2
trans. (1)	6.0	49.1		456	DSC	(3,6)	95A2
trans. (2)	6.0	52.9		406	DSC	(3,6)	95A2
total	7.0	51.5±0.7		837±71	DSC	(1)	95A2

Laccase from the latex of japanese laquer tree, holo and apo protein (continued)

Transition	pH	T_{trs}	ΔCp	ΔH	Approach	Remarks	Ref
trans. (1)	7.0	47.6±1.7		335±71	DSC	(3)	95A2
trans. (2)	7.0	51.6±1.0		561±25	DSC	(3)	95A2
Apo laccase:							
total	5.6	43.8		770	DSC	(1)	95A2
trans. (1)	5.6	37.8		301	DSC	(3)	95A2
trans. (2)	5.6	44.7		410	DSC	(3)	95A2
total	6.0	45.5±1.1		602±146	DSC	(1)	95A2
trans. (1)	6.0	44.3±1.9		276±79	DSC	(3)	95A2
trans. (2)	6.0	46.3±0.7		351±54	DSC	(3)	95A2
total	7.0	46.6±2.1		556±151	DSC	(1)	95A2
trans. (1)	7.0	44.6±3.2		268±75	DSC	(3)	95A2
trans. (2)	7.0	48.0±2.1		326±126	DSC	(3)	95A2

Remarks:
(1) total heat, experimental value
(2) ΔCp could be determined from some of the calorimetric recordings
(3) transition resolved into two subtransitions by deconvolution
(4) measured in the presence of 2 M GuHCl
(5) the second subtransition was not detectable
(6) reoxidized by H_2O_2

Laccase from the lignin-degrading basidomycete PM1

Transition	pH	T_{trs}	ΔCp	ΔH	Approach	Remarks	Ref
trans. (1)	6.5	67.0±0.2		385±4	DSC	(1,2)	94C8
trans. (2)	6.5	76.0±0.2		573±8	DSC	(1)	94C8
trans. (3)	6.5	82.9±0.2		351±4	DSC	(1)	94C8

Remarks:
(1) for details of the procedure see Ref. 94C8
(2) the protein shows a marked pretransition with maxima at $T_{trs} = 32.4°C$ and $45.8°C$

α-Lactalbumin

α-Lactalbumin from various species

Species	pH	T_{trs}	ΔCp	ΔH	Approach	Remarks	Ref
bovine	6.3	62±0.2	4±0.8	276±9	DSC		81P1
bovine	7.0	66	4.6	343	heat	(1)	89H1
goat	7.0	68	4.6	381	heat	(1)	89H1
human	7.0	64.5	4.6	343	heat	(1)	89H1
guinea pig	7.0	71	4.6	318	heat	(1)	89H1

Remark:
(1) enthalpy change and transition temperature were taken from Fig. 1 of Ref. 89H1.

α-Lactalbumin from various species, calcium free form

Species	pH	T_{trs}	ΔCp	ΔH	Approach	Remarks	Ref
bovine	7		4.6	159	heat	(1)	86K3
bovine	7.0	40.5	4.6	230	DSC	(2)	89H1
bovine	8		4.95±0.14	187±18	heat, v.H.		84H3
bovine	8		5.2±0.6	175±15	DSC		85P3
goat	7.0	43	4.6	259	heat	(2)	89H1
human	7.0	36	4.6	209	heat	(2)	89H1
guinea pig	7.0	39.5	4.6	151	heat	(2)	89H1

Remarks:
(1) ΔH at 25°C
(2) ΔH and T_{trs} were taken from Fig. 1 of Ref. 89H1

α-Lactalbumin, bovine, various forms and transitions

Form/Trans.	pH	T_{trs}	ΔCp	ΔH	Approach	Remarks	Ref
	8.1	58±2		209	heat	(1)	72B
Ca^{2+}	8.1	62±1		238±21	heat	(2)	85P2
apo	8.1	31±1		155±21	heat, v.H.		85P2
acid	2.5	32±2		130	heat, v.H.		85P2
apo	8.0	43.26	7.9	276±14	DSC	(3)	91X
holo		71.3	6.7	335	DSC	(4)	91X
N→A	8.0	25	1.364±8	32.2±0.4	DSC	(3,5)	91X
N→U	8.0	25	7.619±4	133.1±0.4	DSC	(3,5)	91X

Remarks:
(1) buffer: 0.1 M Tris
(2) transition monitored by fluorescence, 10-fold Ca^{2+} excess concentration
(3) 10 mM borate buffer, 0.2 M NaCl, 1 mM EDTA
(4) in the presence of 40 mM calcium chloride
(5) result of multidimensional analysis of the heat capacity surface

α-Lactalbumin, in the presence of calcium

pH	T_{trs}	ΔCp	ΔH	Approach	Remarks	Ref
7.0	67	6.67		DSC	(1–3)	94H8

Remarks:
(1) ΔCp is the experimental value
(2) Ref. 94H8 contains corrections of ΔCp for volume effects
(3) T_{trs} was taken from the figures in Ref. 94H8

α-Lactalbumin, bovine, in the presence of calcium

Ca²⁺ Concentration	pH	T_{trs}	ΔCp	ΔH	Approach	Remarks	Ref
260 μM	7.5	64.2		301.7	DSC	(1)	96V1
425 μM	7.5	65.5		314.7	DSC	(1)	96V1
550 μM	7.5	66.0		328.9	DSC	(1)	96V1
700 μM	7.5	66.8		290.2	DSC	(1)	96V1
1400 μM	7.5	68.2		316.6	DSC	(1)	96V1
	7.5		3.96		DSC/ITC	(2)	96V1

Remarks:
(1) buffer: 10 mM Tris
(2) ΔCp for unfolding of calcium containing α-lactalbumin without loss of calcium

α-Lactalbumin, native protein and disulfide derivatives

Buffer	pH	T_{trs}	ΔCp	ΔH	Approach	Remarks	Ref
native α-lactalbumin							
Tris, Ca²⁺	8.0	68.9	4.3	318	DSC	(1,2)	96H2
Tris	8.0	64.1	4.3	271	DSC	(1,3)	96H2
Tris, EDTA	8.0	28.5	4.3	180	DSC	(1,4)	96H2
3SS α-lactalbumin						(5)	
Tris, Ca²⁺	8.0	60.5	4.3	246	DSC	(1,2)	96H2
Tris	8.0	55.2	4.3	230	DSC	(1,3)	96H2
Tris, EDTA	8.0	17.4	4.3	41	DSC	(1,4)	96H2
2SS α-lactalbumin						(6)	
Tris, Ca²⁺	8.0	52.6	4.3	81	DSC	(1,2)	96H2
Tris	8.0	38.8	4.3	73	DSC	(1,3)	96H2

Remarks:
(1) ΔCp from ΔH versus T_{trs} for native α-lactalbumin and all disulfide derivatives
(2) 10 mM Tris, 2 mM CaCl₂, pH 8.0
(3) 10 mM Tris, pH 8.0
(4) 10 mM Tris, 1 mM EDTA, pH 8.0
(5) Cys6 and Cys120 reduced and blocked by carboxymethylation
(6) Cys6, Cys120, Cys28 and Cys111 reduced and blocked by carboxymethylation

Transitions of α-lactalbumin, analyzed in terms of a hierarchical cooperative model

pH	T_{trs}	ΔCp	ΔH	Ref
Transition N→I₀				
3.0	25	4.00	122.608	94G3
3.2	25	4.28	109.658	94G3
3.5	25	4.31	103.789	94G3
4.2	25	4.37	98.209	94G3
Transition I₀→U				
3.0	25	1.45	−15.603	94G3
3.2	25	1.99	−19.976	94G3
3.5	25	2.39	−48.893	94G3
4.2	25	2.73	−86.821	94G3
Transition N→U				
3.0	25	5.45	107.003	94G3
3.2	25	6.27	89.677	94G3
3.5	25	6.70	54.895	94G3
4.2	25	7.10	11.386	94G3
5.2	25	7.50	−2.499	94G3

α-Lactalbumin, bovine, various forms

	pH	T_{trs}	ΔCp	ΔH	Approach	Remarks	Ref
holo	7.8	60.9		258	DSC	(1)	92Y2
holo	6.5	67.3–68.4		283–383	DSC	(2)	92E4
apo	6.5	45		109	DSC	(2)	92E4

Remarks:
(1) buffer: 10 mM borate
(2) buffer: 50 mM sodium phosphate, 0.02% azide, scan rate 10 K/min

Bovine α-lactalbumin and apo-α-lactalbumin in different buffer solutions – experimental results

Buffer	pH	T_{trs}	ΔCp	ΔH	Approach	Remarks	Ref
10 mM acetate	3.2	29.6	3.84±0.4	151±15	DSC		94G3
10 mM acetate	3.5	40.8	4.97±0.5	200±10	DSC		94G3
10 mM acetate	4.2	56.8	6.75±0.7	271±14	DSC		94G3
10 mM acetate	5.2	66.8	7.46±0.7	340±17	DSC	(1)	94G3
5 mM Tris-HCl	8.0	64.1	7.1±0.7	295±15	DSC	(1)	94G3
500 mM Tris-HCl	8.0	65.6	7.9±0.8	315±16	DSC	(1)	94G3
10 mM Tris-HCl + 1 mM Ca^{2+}	8.0	67.1	6.8±0.7	314±16	DSC		94G3
Apo-α-lactalbumin							
5 mM Tris-HCl	8.0	28	6.18±0.6	147±26	DSC		94G3
300 mM Tris-HCl	8.0	32.5	5.34±0.5	172±18	DSC		94G3
500 mM Tris-HCl	8.0	38.8	3.91±0.4	200±10	DSC		94G3

Remark:
(1) average ΔCp = 7.48 kJ/mol/K at pH greater than 5

Goat α-lactalbumin, apo form

pH	T_{trs}	ΔCp	ΔH	Approach	Remarks	Ref
7.5	25	3.1±0.3	127±3	heat	(1)	91D7
7.5	25	2.7±0.3	129±12	calorimetry	(1,2)	91D7

Remarks:
(1) apo form in the absence of metal ions
(2) calorimetry, special approach to resolve conformational and Mn^{2+} binding process

α-Lactalbumin, human, various forms

Form	pH	T_{trs}	ΔCp	ΔH	Approach	Remarks	Ref
Ca^{2+}	7.5	68	1.85	180	DSC	(1,2)	91P2
apo	7.5	44	1.85	140	DSC	(2,3)	91P2
	8.1	57±2		230	heat	(4)	72B
	7.4	63.5±0.4		208±15	heat	(5)	95S15

Remarks:
(1) measured in the presence of 2 mM CaCl$_2$
(2) T_{trs} was taken from Fig. 3 in Ref. 91P2
(3) measured in the presence of 3 mM EGTA
(4) buffer: 0.1 M Tris
(5) global analysis of FTIR spectra based on the intramolecular secondary structure band

α-Lactalbumin from various species, in the presence of polyamins

Ligand	pH	T_{trs}	ΔH	Approach	Ref
Bovine calcium-free α-lactalbumin:					
without ligand	7	20	169±11	heat	93M16
9 mM putrescine	7	28	192±5	heat	93M16
9 mM spermidine	7	32	197±6	heat	93M16
9 mM spermine	7	37	204±11	heat	93M16
Bovine calcium-containing α-lactalbumin:					
without polyamines	7	61	297±15	heat	93M16
Human calcium-free α-lactalbumin:					
without ligand	7	23	137±8	heat	93M16
9 mM spermine	7	17	122±6	heat	93M16
Equine calcium-free α-lactalbumin:					
without ligand	7	29	191±46	heat	93M16
9 mM spermine	7	19	181±11	heat	93M16

β-Lactamase

β-Lactamase I from *Bacillus cereus*

pH	T_{trs}	ΔCp	ΔH	Approach	Remarks	Ref
7.15	57.5±0.1		464±25	DSC	(1)	83A3
7.0	51.03		646	DSC	(2)	92A4

Remarks:
(1) strain 569/H, ΔH = 15.5±0.8 J/g
(2) T_{trs} and ΔH were found to be scan rate dependent

β-Lactamases from various bacterial species

	pH	T_{trs}	ΔCp	ΔH	Approach	Remarks	Ref
S. aureus	7	41.8±0.3		418±25	heat, v.H.	(1,2)	95V1
TEM-1	7	50.1±0.2		469±29	heat, v.H.	(1,2)	95V1
Bac. lich.	7	63.2±0.6		686±38	heat, v.H.	(1,2)	95V1

Remarks:
(1) the enzymes are TEM-1 β-lactamase and lactamases produced by *Staphylococcus aureus* and *Bacillus licheniformis*
(2) Ref. 95V1 contains additional data on activation free energy obtained by GuHCl unfolding kinetics

β-Lactamase from *Staphylococcus aureus* PC1

pH	T_{trs}	ΔCp	ΔH	Approach	Remark	Ref
7.5	41.6±1.0		464±63	heat, v.H.	(1)	94R0

Remark:
(1) Ref. 94R0 contains additional data for inhibited β-lactamase

TEM β-lactamase, wild-type and mutants

Mutant	pH	T_{trs}	ΔCp	ΔH	Approach	Remarks	Ref
wild-type	7.0	50.6±0.6	8.8±1.7	456	heat, v.H.	(1,2)	95R2
Glu104→Lys	7.0	52.3±0.4		519	heat, v.H.	(1)	95R2
Gly238→Ser	7.0	46.2±0.3		360	heat, v.H.	(1)	95R2
double mutant (Glu104→Lys and Gly238→Ser)							
	7.0	48±1.3		377	heat, v.H.	(1)	95R2
Arg164→His	7.0	50±0.5		377	heat, v.H.	(1)	95R2
Arg164→Ser	7.0	48.8±0.6		452	heat, v.H.	(1)	95R2
double mutant (Arg164→Ser and Glu240→Lys)							
	7.0	52.5±1		531	heat, v.H.	(1)	95R2
Ser235→Ala	7.0	52.2±0.4		502	heat, v.H.	(1,2)	95R2

Remarks:
(1) ΔCp is an estimated value which is based on the amino acid composition
(2) preliminary T_{trs} values, see Ref. 94D5

β-Lactoglobulin

β-Lactoglobulin

	pH	T_{trs}	ΔCp	ΔH	Approach	Remarks	Ref
			10.8±0.36		urea, calorimetry	(1,2)	78D

Remarks:
(1) for the dependence of transition temperature on the protein concentration, see Ref. 95Q1
(2) flow calorimetry

β-Lactoglobulin from bovine milk, heat denaturation in the presence of urea

Urea Concentration	pH	T_{trs}	ΔCp	ΔH	Approach	Remarks	Ref
0.0 M	2.0	78.0	5.58±0.7	312±15	DSC	(1)	92G5
2.0 M	2.0	70.8	5.82±0.7	243±12	DSC	(1)	92G5
4.4 M	2.0	56.0	7.89±0.8	120±20	DSC	(1)	92G5
0.0 M	2.0	77.8	5.58	312.2	DSC		94G5
2.0 M	2.0	70.6	5.82	243.0	DSC		94G5
4.0 M	2.0	55.8	6.9	120.8	DSC		94G5
0.0 M	2.0	85.0	5.99±0.7	340±15	DSC	(2)	92G5
2.0 M	2.0	81.3	6.14±0.7	296±14	DSC	(2)	92G5
4.4 M	2.0	68.6	6.86±0.7	204±20	DSC	(2)	92G5
0.0 M	2.0	91.0	5.66±0.7	371±14	DSC	(3)	92G5
2.0 M	2.0	84.0	6.29±0.7	313±14	DSC	(3)	92G5
4.4 M	2.0	76.0	6.64±0.7	248±15	DSC	(3)	92G5
	2.0		5.70±0.7		DSC	(4)	92G5
	2.0		4.90±0.4		DSC	(5)	92G5

Remarks:
(1) buffer: 0.1 M KCl/HCl pH 2.0
(2) buffer: 0.1 M sodium phosphate pH 2.0
(3) buffer: 0.2 M sodium phosphate pH 2.0
(4) ΔCp is indicated as the mean value in Ref. 92G5
(5) ΔCp from ΔH versus T_{trs} at different concentrations of phosphate

β-Lactoglobulin from bovine milk, cold denaturation in the presence of urea

Urea Con-centration	pH	T_{trs}	ΔC_p	ΔH	Approach	Remarks	Ref
3.4 M	2.0	–3.0	11.0±1.2	–271±30	DSC	(1)	92G5
4.0 M	2.0	4.0	12.0±1.2	–260±26	DSC	(1)	92G5
4.4 M	2.0	6.3	11.9±1.2	–220±24	DSC	(1)	92G5
3.4 M	2.0	9.0	11.6±1.5	212±31	DSC	(2)	92G5
4.0 M	2.0	13.0	11.9±1.5	208±31	DSC	(2)	92G5
4.4 M	2.0	16.0	13.1±1.5	180±30	DSC	(2)	92G5

Remarks:
(1) direct measurement of cold denaturation
(2) renaturation after cold denaturation

β-Lactoglobulin in various solvents

Solvent	pH	T_{trs}	ΔC_p	ΔH	Approach	Ref
	2.0	81.4	5.4±1.6	262	heat	93P11
	2.0	82.3		414	DSC	93P11
	6.5	80.1	6.0±1.8	232	heat	93P11
	6.5	80.0		412	DSC	93P11
	7.4	65.9	6.2±1.9	167	heat	93P11
2 M urea		69.4	5.7±1.7	181	heat	93P11
2 M urea		72.0		260	DSC	93P11
3 M urea		60.0	5.5±1.7	129	heat	93P11
4 M urea		55.9		105	heat	93P11
2 M methylurea		66.5	5.4±1.6	179	heat	93P11
3 M methylurea		56.1	4.5±1.3	154	heat	93P11
4 M methylurea		45.9	3.7±1.1	132	heat	93P11
2 M N,N'-dimethylurea		58.8	4.8±1.5	178	heat	93P11
2 M N,N'-dimethylurea		64.0		308	DSC	93P11
3 M N,N'-dimethylurea		47.7	4.2±1.3	152	heat	93P11
4 M N,N'-dimethylurea		35.4	4.0±1.2	133	heat	93P11
5 M N,N'-dimethylurea		24.3		115	heat	93P11
2 M ethylurea		54.8	4.1±1.2	161	heat	93P11
2 M ethylurea		66.5		372	DSC	93P11
3 M ethylurea		41.1	3.9±1.2	138	heat	93P11
4 M ethylurea		30.0	3.8±1.1	118	heat	93P11

Lactoperoxidase

Lactoperoxidase

pH	T_{trs}	ΔC_p	ΔH	Approach	Ref
6.05	69.3	36.9±2.3	2168	DSC	86P3

λ-**Repressor**

λ-Repressor, N-terminal domain, wild-type and mutants

	pH	T_{trs}	ΔCp	ΔH	Approach	Ref
wild-type	8	53.4±0.1		274±41	DSC	86H1
wild-type	8	51.5		272	DSC	84H2
Lys4→Gln	8	53.5		293	DSC	84H2
Tyr22→His	8	28.8		126	DSC	84H2
Gln44→Tyr	8	51.4		251	DSC	84H2
Gly46→Ala	8	56.5±0.1		294±44	DSC	86H1
Gly48→Ala	8	58.1±0.1		258±38	DSC	86H1
Ala49→Val	8	38.5		121	DSC	84H2
Ala66→Thr	8	29		167	DSC	84H2
Ile84→Ser	8	37.2		205	DSC	84H2
double mutant (Gly46→Ala and Gly48→Ala)						
	8	59.6±0.1		246±37	DSC	86H1

λ-Repressor, C-terminal domain

	pH	T_{trs}	ΔCp	ΔH	Approach	Ref
	8	70.4±0.08		551±14	DSC	84H2

λ-Repressor, N-terminal domain (res. 1–102), wild-type and mutants

Mutant	pH	T_{trs}	ΔCp	ΔH	Approach	Ref
wild-type	7.0	55.7		259	heat	92L4
Val36→Ile	7.0	59.1		218	heat	92L4
Met40→Ala	7.0	47.1		192	heat	92L4
double mutant (Met40→Val and Val47→Leu)						
	7.0	51.3		172	heat	92L4
double mutant (Val36→Phe and Met40→Leu)						
	7.0	51.6		180	heat	92L4
double mutant (Leu18→Ala and Met40→Ala)						
	7.0	23.2		113	heat	92L4
triple mutant (Val36→Leu, Met40→Val and Val47→Ile)						
	7.0	53.6		184	heat	92L4
triple mutant (Val36→Ile, Met40→Val and Val47→Leu)						
	7.0	53.4		184	heat	92L4
triple mutant (Val36→Ile, Met40→Val and Val47→Ile)						
	7.0	53.7		197	heat	92L4
triple mutant (Val36→Leu, Met40→Leu and Val47→Ile)						
	7.0	59.6		213	heat	92L4
triple mutant (Val36→Phe, Met40→Ala and Val47→Ile)						
	7.0	47.2		159	heat	92L4
triple mutant (Val36→Phe, Met40→Phe and Val47→Phe)						
	7.0	45.4		142	heat	92L4
multiple mutant (Val36→Phe, Met40→Phe, Val47→Ile and Leu65→Phe)						
	7.0	49.1		159	heat	92L4

Table 2. Enthalpy and Heat Capacity Changes – Molar Values

λ Repressor, monomeric, residues 6–85 of the N-terminal domain, heat and cold denaturation

Transition	pH	T_{trs}	ΔCp	ΔH	Approach	Remarks	Ref
heat	8.0	57.2±0.1	6.02±0.13	285±4	heat, urea	(1,2)	96H7
cold	8.0	−16.2±1.5	6.02±0.13	149±13	heat, urea	(1,2)	96H7

Remarks:
(1) from a three-dimensional thermal-urea denaturation profile monitored by CD and NMR
(2) buffer: 20 mM KD_2PO_4, 100 mM NaCl in 99% D_2O

λ Cro Repressor

λ Cro repressor, wild-type

pH	T_{trs}	ΔCp	ΔH	Approach	Ref
5.18	35.5	6.31	146	DSC	92G6
5.48	38.0	6.74	163	DSC	92G6
6.00	44.0	6.92	201	DSC	92G6
7.00	46.2	5.70	221	DSC	92G6

λ Cro repressor, mutant Cys55 = Val55→Cys

pH	T_{trs}	ΔCp	ΔH	Approach	Remarks	Ref
3.0		6.51	199	DSC	(1)	92G6
4.0		6.72	349	DSC	(1)	92G6
4.5		6.23	383	DSC	(1)	92G6
5.0		6.35	408	DSC	(1)	92G6
7.0			458	DSC	(1)	92G6

Remark:
(1) the transition can be resolved into two subtransitions, see following table

λ Cro repressor, mutant Cys55 = Val55→Cys

Transition	pH	T_{trs}	ΔCp	ΔH	Approach	Remarks	Ref
trans. (1)	3.0	37	2.9±0.3	162	DSC	(1)	92G6
trans. (2)	3.0	33	3.1±0.3	50	DSC	(1)	92G6
total	3.0		6.4±0.7	212	DSC	(2)	92G6
trans. (1)	4.0	53	2.9±0.3	205	DSC	(1)	92G6
trans. (2)	4.0	66	3.1±0.3	170	DSC	(1)	92G6
total	4.0		6.4±0.7	375	DSC	(2)	92G6
trans. (1)	4.5	54	2.9±0.3	204	DSC	(1)	92G6
trans. (2)	4.5	77	3.1±0.3	199	DSC	(1)	92G6
total	4.5		6.4±0.7	403	DSC	(2)	92G6
trans. (1)	5.0	54	2.9±0.3	211	DSC	(1)	92G6
trans. (2)	5.0	82	3.1±0.3	205	DSC	(1)	92G6
total	5.0		6.4±0.7	416	DSC	(2)	92G6
trans. (1)	7.0	55	2.9±0.3	219	DSC	(1)	92G6
trans. (2)	7.0	101	3.1±0.3	253	DSC	(1)	92G6
total	7.0		6.4±0.7	473	DSC	(2)	92G6

Remarks:
(1) ΔCp was determined from T_{trs} versus ΔH of the individual recordings
(2) ΔCp is the mean from the individual calorimetric recordings

λ Cro repressor, CNBr-fragment, residues (13–66)

pH	T_{trs}	ΔCp	ΔH	Approach	Ref
3.5	56.8	1.98	129	DSC	92G6
4.5	75.4	2.21	178	DSC	92G6
5.5	82.7	2.31	220	DSC	92G6

λ Cro repressor, tryptic fragment (1–55) of Cys55-Cro

pH	T_{trs}	ΔCp	ΔH	Approach	Ref
3.5	67.4	2.62	148	DSC	92G6
4.0	73.3	3.11	162	DSC	92G6
4.5	81.1	2.80	195	DSC	92G6

λ cI repressor, wild-type, in the presence and absence of DNA

Domain	pH	T_{trs}	ΔH	Approach	Ref
In the absence of DNA:					
N-terminal	7.0	50.7±0.1	213±2	DSC	95M14
C-terminal	7.0	73.3±0.2	345±16	DSC	95M14
In the presence of DNA:					
N-terminal	7.0	59.4±0.2	346±22	DSC	95M14
C-terminal	7.0	73.0±0.2	336±13	DSC	95M14

λ Cro repressor, wild-type and covalently linked dimer Val55→Cys

Transition	pH	T_{trs}	ΔCp	ΔH	Approach	Remarks	Ref
Wild-type							
trans. (1)	7.0	30.5	3.7	180	DSC	(1,3)	96F2
trans. (2)	7.0	138	0	225	DSC	(1,3)	96F2
Mutant Val55→Cys							
trans. (1)	4.5	32.1	2.0	160.0	DSC	(2,3)	96F2
trans. (2)	4.5	99.2	1.5	215.7	DSC	(2,3)	96F2
Mutant Val55→Cys							
trans. (1)	7.0	35.2		166.4	DSC	(2,3)	96F2
trans. (2)	7.0	123.2		225.5	DSC	(2,3)	96F2

Remarks:
(1) the transition was treated by the following three-state model: $N_2 \rightarrow 1/2(I_4) \rightarrow 2U$
(2) the transition was treated by the following three-state model: $N_2 \rightarrow 1/2(I_2)_2 \rightarrow 2U$
(3) for further data and alternative models, see Ref. 96F2

Table 2. Enthalpy and Heat Capacity Changes – Molar Values

Laminin

Laminin, C-terminal α-helical coiled-coil region, recombinant, isolated reduced fragments and disulfide-linked fragments

Fragment, Concentration		pH	T_{trs}	ΔH	Approach	Remarks	Ref
Reduced and alkylated fragments:							
β fragment	54 μM	7.4	46.6±0.2	198±20	heat, v.H.	(1,2)	95K3
γ fragment	65 μM	7.4	−9±3	52±10	heat, v.H.	(1,2)	95K3
β + γ fragment	25 μM	7.4	42±0.2	304±30	heat, v.H.	(1,2)	95K3
β + γ fragment	25 μM	7.4	42±0.2	252±25	DSC	(1,2,5)	95K3
β + γD10	15 μM	7.4	32.3±0.2	225±23	heat, v.H.	(1,2)	95K3
β fragment	12 μM	7.4	38.8±0.2	154±15	heat, v.H.	(1,3)	95K3
β + γ fragment	22 μM	7.4	40.1±0.2	260±26	heat, v.H.	(1,3)	95K3
β + γ fragment	76 μM	7.4	31.9±0.2	306±27	heat, v.H.	(1,4)	95K3
Disulfide-linked fragments:							
γ-γ		7.4	4±3	36±10	heat, v.H.	(1,2)	95A5
β-γ		7.4	60.4±0.2	309±30	heat, v.H.	(1,2)	95A5
β-γ		7.4	60.4±0.2	282±28	DSC	(1,2)	95A5
β-γ		7.4	58.4±0.2	322±26	heat, v.H.	(1,3)	95A5
β-γ		7.4	52.3±0.2	306±27	heat, v.H.	(1,4)	95A5

Remarks:
(1) fragments:
  α-residues 2044 to 2146
  β-residues 1700 to 1786
  γ-residues 1506 to 1607
  γD10-residues 1588 to 1598 of γ deleted
(2) buffer: 5 mM phosphate, pH 7.4
(3) buffer: 5 mM phosphate, 0.1 M NaCl, pH 7.4
(4) buffer: 5 mM phosphate, 1 M urea, pH 7.4
(5) at protein conc. 37, 25, 7.4, and 3.7 μM is T_{trs} 42.6, 42, 38, and 36.4°C, respectively; from the concentration dependence it follows $\Delta H° = 287$ kJ/mol

Lectin

Erythrina corallodendron lectin, in the absence and in the presence of monosaccharides and disaccharides

Ligand	pH	T_{trs}	ΔH	Approach	Remarks	Ref
without ligand	7.4	60.7–62.3	1338±152	DSC	(1–3)	96S12
without ligand	7.4	62.5	1465±161	DSC	(3,4,6)	96S12
without ligand	7.4	63.6	1507±70	DSC	(3,5,6)	96S12
LacNAc 1–15 mM	7.4	61.7–66.9	1555±207	DSC	(1,3,7)	96S12
Lac 3–24 mM	7.4	60.8–65.0	1475±60	DSC	(1,3,7)	96S12
Me-α-Gal 4–24 mM	7.4	62.4–65.6	1048±178	DSC	(1,3,7)	96S12
Me-β-Gal 4–24 mM	7.4	62.2–64.7	1123±36	DSC	(1,3,7)	96S12

Remarks:
(1) scan rate 20 K/h
(2) protein concentration 0.017 to 0.2 mM
(3) the data refer to a A→2B transition, the cooperative ratio varies 2.1 to 2.45
(4) scan rate 45 K/h
(5) scan rate 90 K/h
(6) protein concentration 0.033 mM
(7) abbreviations:
    LacNAc – N-acetyllactosamine
    Lac – lactose
    Me-α-Gal – methyl α-galactoside
    Me-β-Gal – methyl β-galactoside

Leucine Zipper, see also de novo synthesized protein

Leucine zipper from GCN4, in the presence of salt

Salt Concentration	pH	T_{trs}	ΔCp	ΔH	Approach	Remarks	Ref
Reference value without salt:							
	7.0	67.2	1.05±0.08	94.1	DSC	(1–3)	95K6
In the presence of KCl:							
0.04 M	7.0	65.6		94.1	DSC		95K6
0.12 M	7.0	64.3		90.4	DSC		95K6
0.25 M	7.0	62.6		93.3	DSC		95K6
0.50 M	7.0	62.0		93.3	DSC		95K6
0.75 M	7.0	64.0		90.4	DSC		95K6
1.00 M	7.0	64.7		93.3	DSC		95K6
In the presence of NaCl:							
0.04 M	7.0	65.4		92.5	DSC		95K6
0.12 M	7.0	64.6		96.2	DSC		95K6
0.25 M	7.0	63.5		93.7	DSC		95K6
0.50 M	7.0	62.1		93.3	DSC		95K6
0.75 M	7.0	62.7		93.3	DSC		95K6
1.00 M	7.0	64.3		95.8	DSC		95K6
In the presence of choline:							
0.04 M	7.0	65.5		92.0	DSC		95K6
0.12 M	7.0	64.8		91.6	DSC		95K6
0.25 M	7.0	62.7		94.1	DSC		95K6
0.50 M	7.0	61.2		90.8	DSC		95K6
0.75 M	7.0	62.0		94.1	DSC		95K6
1.00 M	7.0	62.6		91.6	DSC		95K6

Leucine zipper from GCN4, in the presence of salt (continued)

Salt Concentration	pH	T_{trs}	ΔCp	ΔH	Approach	Remarks	Ref
In the presence of LiCl:							
0.04 M	7.0	65.2		92.5	DSC		95K6
0.12 M	7.0	63.7		92.5	DSC		95K6
0.25 M	7.0	62.0		90.0	DSC		95K6
0.50 M	7.0	60.3		90.0	DSC		95K6
0.75 M	7.0	59.4		88.3	DSC		95K6
1.00 M	7.0	60.2		90.0	DSC		95K6
In the presence of KF:							
0.04 M	7.0	67.1		93.3	DSC		95K6
0.12 M	7.0	66.8		95.8	DSC		95K6
0.25 M	7.0	66.7		95.8	DSC		95K6
0.50 M	7.0	70.2		99.2	DSC		95K6
0.75 M	7.0	73.8		102.1	DSC		95K6
1.00 M	7.0	77.4		111.7	DSC		95K6
In the presence of KBr:							
0.04 M	7.0	65.2		91.6	DSC		95K6
0.12 M	7.0	62.9		95.0	DSC		95K6
0.25 M	7.0	61.7		92.5	DSC		95K6
0.50 M	7.0	60.4		90.0	DSC		95K6
0.75 M	7.0	59.4		88.3	DSC		95K6
1.00 M	7.0	59.9		89.5	DSC		95K6
Mean value:							
	7.0	60		89.5±3.8	DSC	(4)	95K6

Remarks:
(1) the data refer to two-state dimeric model, i.e., a $N_2 \to 2U$ unfolding transition
(2) reference buffer is 10 mM ACES, 0.25 mM EDTA, pH 7.0; the sequence of the peptide (GCN4–33) is:
   NH$_2$-Arg-Met-Lys-Gln-Leu-Glu-Asp-Lys-Val-Glu-Glu-Leu-Leu-Ser-Lys-Asn-Tyr-His-Leu-Glu-Asn-Glu-Val-Ala-Arg-Leu-Lys-Lys-Leu-Val-Gly-Glu-Arg-COOH
(3) ΔCp from ΔH versus T_{trs} over all data listed in the table
(4) average value of ΔH at 60°C independently of the salt

GCN4–33 coiled-coil mutants

Mutant	pH	T_{trs}	ΔCp	ΔH	Approach	Remarks	Ref
wild-type (Ala24/Ala24)							
	7.0	70.0		100	DSC	(1,2)	96D1
Ser14→Ala/Ser14→Ala							
	7.0	52.8		81	DSC	(1,2)	96D1
Ser14→Gly/Ser14→Gly							
	7.0	64.5		94	DSC	(1,2)	96D1
Ala24→Gly/Ala24/Gly							
	7.0	53.5		82	DSC	(1,2)	96D1
all four mutants							
	7.0	53–70	1.07±0.05		DSC	(1–3)	96D1
	7.0	53–70	1.12±0.03		DSC	(1,2,4)	96D1

Remarks:
(1) the 33 amino acid peptide corresponds to the leucine zipper region of GCN4
(2) measured in 10 mM ACES, 0.25 mM EDTA, pH 7.0
(3) mean from individual DSC scans
(4) from ΔH versus T_{trs}, general expression: $\Delta H(T) = (94.709\pm1.674) + (1.117\pm0.029)\times(T-65)$ in kJ/mol with T in °C

Lipoprotein

Lipoprotein, see also apolipoprotein

Lipoprotein, in dimyristollecithin

	pH	T_{trs}	ΔCp	ΔH	Approach	Ref
		10.5		134	DSC	77T

Lysozyme, CPL1

CPL1 lysozyme and its domains

Transition	pH	T_{trs}	ΔCp	ΔH	Approach	Remarks	Ref
CPL1 lysozyme, molar mass 39.1 kDa:							
overall	7.0	43.5		894	DSC	(1)	93S3
trans. (1)	7.0	43.5		564	DSC	(1)	93S3
trans. (2)	7.0	51.4		309	DSC	(1)	93S3
C-terminal domain C-CPL1, molar mass 18.7 kDa:							
	7.0	42.9		455	DSC		93S3
N-terminal domain F1, molar mass 24 kDa:							
	7.0	52.0		725	DSC		93S3

Remarks:
(1) the overall heat effect can be resolved by deconvolution into the two subsequent transitions trans. (1) and trans. (2)

CPL1 lysozyme in the presence of choline

Choline Concentration	pH	T_{trs}	ΔCp	ΔH	Approach	Remark	Ref
CPL1 lysozyme, molar mass 39.1 kDa:							
0 mM	7.0	43.5		894	DSC	(1)	93S3
10 mM	7.0	51.5		994	DSC		93S3
22 mM	7.0	53.3		1209	DSC		93S3
C-terminal domain C-CPL1, molar mass 18.7 kDa:							
10 mM	7.0	50.9		614	DSC		93S3

Remark:
(1) overall heat effect

Lysozyme, Equine

Equine lysozyme

Transition	pH	T_{trs}	ΔCp	ΔH	Approach	Remark	Ref
N→I	6.0	70	5.8	430	heat	(1)	93N2
I→U		35–68	0.0	160	heat		93N2

Remark:
(1) measured in the presence of 2.5 μM free calcium ΔH was taken from Fig. 5 in Ref. 93N2

Equine lysozyme, in the presence of calcium

Transition	pH	T_{trs}	ΔCp	ΔH	Approach	Ref
0.0 mM calcium:						
trans. (1)	7.5	39.7		103.1	DSC	93V1
trans. (2a)	7.5	55.5		29.7	DSC	93V1
trans. (2b)	7.5	68.6		41.8	DSC	93V1
0.1 mM calcium:						
trans. (1)	7.5	45.2		87.4	DSC	93V1
trans. (2a)	7.5	55.5		52.6	DSC	93V1
trans. (2b)	7.5	69.3		46.3	DSC	93V1
2.0 mM calcium:						
trans. (2a)	7.5	59.0		134.0	DSC	93V1
trans. (2b)	7.5	64.5		148.4	DSC	93V1
10 mM calcium:						
trans. (2a)	7.5	60.0		121.8	DSC	93V1
trans. (2b)	7.5	64.5		167.8	DSC	93V1

Equine lysozyme, holo and apo protein

Transition	pH	T_{trs}	ΔCp	ΔH	Approach	Ref
Holo equine lysozyme in the presence of 1.5 mM Ca^{2+}:						
trans. (1)	4.5	54.73	4.90	204.7	DSC	95G10
trans. (2)	4.5	66.2	2.50	133.3	DSC	95G10
Apo equine lysozyme:						
trans. (1)	4.5	41.5	5.04	153.7	DSC	95G10
trans. (2)	4.5	66.44	2.56	123.8	DSC	95G10
Total heat effect:						
		25	7.2±0.7		DSC	95G10
		50	7.6±0.7		DSC	95G10
		80	6.07	513	DSC	95G10

Lysozyme, HEW

HEN Egg White lysozyme

pH	T_{trs}	ΔCp	ΔH	Approach	Remarks	Ref
1.9	25	6.3±0.8	226±17	GuHCl	(2)	76P1
2	52.03±0.52	6.49±0.5	382±8	DSC		79V
2.5	64.1		481	DSC		95P9
3.8	74.8±0.5		392±30	heat	(4)	92R1
4	77	6.7±0.4	560	DSC		74P1
4	77	6.7±0.4	560	DSC		73K
4.5	78.5	6.57±0.29	585±6	DSC		76P2
4.7	74±1		502	heat	(5)	72B
5	75.5	5.6	509	DSC		82F1

HEN Egg White lysozyme (continued)

pH	T_{trs}	ΔC_p	ΔH	Approach	Remarks	Ref
5.37	76.5		577±29	heat, v.H.		69D
7.0	74		524	heat	(6)	65H
2–6		10.0±2.1		heat	(3)	87L
		6.7±0.8		DSC	(1)	76P1
		5.75		GuHCl, v.H.		70T1

Remarks:
(1) DSC in the presence of GuHCl
(2) isothermal calorimetric titration
(3) measured in the presence of poly(ethyleneglycols)
(4) measured by ^1H NMR
(5) in the presence of 0.1 M KCl
(6) in the presence of ethylene glycol (10% v/v)

HEN Egg White lysozyme

pH	T_{trs}	ΔC_p	ΔH	Approach	Remarks	Ref
2.7	63.4±0.2		464±4	DSC		91G2
		6.39±0.60		DSC	(1–3)	89S5
		5.94±3.10		DSC	(4)	89S5

Remarks:
(1) ΔC_p is from $d(\Delta H)/dT$ over 68 single scans
(2) ΔH can be described by the expression $\Delta H = (434.7\pm4.1)+(6.39\pm0.60)\times(T_{trs}-64)$
(3) the pH dependence of T_{trs} can be expressed by $T_{trs} = (25.1\pm1.4)+(13.0\pm0.5)\times pH$ from pH 2 to 4
(4) ΔC_p from the change in the transition baseline at T_{trs} (mean value over about 60 scans)

Lysozyme HEW, comparison of instruments by the example of lysozyme

pH	T_{trs}	ΔC_p	ΔH	Approach	Remarks	Ref
2.5	60.7		389.5	DSC	(1,2)	91U2
2.5	60.2		390.4	DSC	(1,3)	91U2
2.5	60.8		398.9	DSC	(2,4)	91U2
2.5	59.9		406.8	DSC	(3,4)	91U2

Remarks:
(1) measured on an adiabatic microcalorimeter
(2) experimental values without corrections
(3) with corrections for the time constant of the microcalorimeter
(4) measured on a conductive type microcalorimeter

Lysozyme

pH	T_{trs}	ΔC_p	ΔH	Approach	Remarks	Ref
1.9	56	6.15		DSC	(1–3)	94H8

Remarks:
(1) ΔC_p is the experimental value
(2) Ref. 94H8 contains corrections of ΔC_p for volume effects
(3) T_{trs} was taken from the figures in Ref. 94H8

Table 2. Enthalpy and Heat Capacity Changes – Molar Values

HEN Egg White lysozyme, pH dependence of T_{trs}, ΔCp and ΔH

pH	T_{trs}	ΔCp	ΔH	Approach	Ref
1.5	48	5.5	381	DSC	74P1
1.8	53		404	DSC	74P1
2.0	56	5.4	444	DSC	74P1
2.5	66	6.0	498	DSC	74P1
2.6	69	6.6	523	DSC	74P1
3.0	74.5	6.0	565	DSC	74P1
4.0	77	6.3	561	DSC	74P1
4.5	78.5	6.6	590	DSC	74P1

HEN Egg White lysozyme, temperature dependence of heat capacity change at protein unfolding

T	$\Delta Cp_{(calc)}$	$\Delta Cp_{(exp)}$	Approach	Remarks	Ref
5	8.5	7.0	DSC	(1,2)	90P3
25	9.1	7.5	DSC	(1,2)	90P3
50	8.9	7.3	DSC	(1,2)	90P3
75	7.4	6.7	DSC	(1,2)	90P3
100	5.7	4.7	DSC	(1,2)	90P3
125	3.6	2.6	DSC	(1,2)	90P3

Remarks:
(1) the calculated value is based on heat capacity of the constituent amino acids minus experimental value for the native protein
(2) the experimental value is based on heat capacity of the heat denatured protein minus experimental value for the native protein

Lysozyme HEW, temperature dependence of the heat capacity change at protein unfolding in aqueous solution, and related thermodynamic quantities

T	$\Delta H(exp)$	$\Delta S(exp)$	$\Delta Cp(exp)$	Approach	Remarks	Ref
5	111	164	8.5	DSC	(1–3)	95M3
25	242	618	9.1	DSC	(1–3)	95M3
50	408	1153	8.9	DSC	(1–3)	95M3
75	562	1615	7.4	DSC	(1–3)	95M3
100	683	1954	5.7	DSC	(1–3)	95M3
125	362	2138	3.6	DSC	(1–3)	95M3

Remarks:
(1) the data were taken from Ref. 95M3 with reference to Refs. 93M5 and 93P12
(2) ΔCp was obtained as described in 90P3
(3) ΔCp in kJ/mol/K, ΔH in kJ/mol and ΔS in J/mol/K

Lysozyme HEW, deuterated (D) protein in D_2O and undeuterated (H) protein in H_2O

	pH	T_{trs}	ΔCp	ΔH	Approach	Remarks	Ref
D-lysozyme	1.5–5	50–80	6.7±0.3	310–510	DSC in D_2O	(1,2)	95M2
H-lysozyme	2–5.5	55–76	6.7±0.2	390–590	DSC in H_2O	(1)	95M2

Remarks:
(1) T_{trs} and ΔH were taken from Fig. 1 in Ref. 95M2
(2) $\Delta(\Delta H)$ between D-lysozyme and H-lysozyme denaturation amounts to $-(72\pm15)$ kJ/mol

HEN Egg White lysozyme, wild-type and mutants

Mutant	pH	T_{trs}	ΔH	Approach	Ref
wild-type	3.0	72.2	531	heat, v.H.	92K2
Asn103→Asp	3.0	72.2	540	heat, v.H.	92K2
Asn106→Asp	3.0	72.7	536	heat, v.H.	92K2

Lysozyme HEW, in the presence of various added solutes

Added Solute	pH	n	ΔCp	$\Delta H(0°C)$	Approach	Remarks	Ref
none	2.8–4.0	13	7.15±1.13	19.5	DSC	(1,2)	96L3
0.5 M sucrose	2.6–4.0	12	11.42±1.42	–321.6	DSC	(1,2)	96L3
1.0 M GuHCl	2.4–6.0	22	7.66±0.38	–53.1	DSC	(1,2)	96L3
2.0 M GuHCl	3.2–6.0	15	7.07±0.88	–44.6	DSC	(1,2)	96L3
10% glycerol	2.5–5.0	20	9.79±0.79	–155.0	DSC	(1,2)	96L3

Remarks:
(1) n = number of experiments
(2) ΔH at 0°C
(3) 10% (v/v) glycerol

HEN Egg White lysozyme, in the presence of 1-propanol

Concentration (M)	pH	T_{trs}	ΔCp	ΔH	Approach	Ref
0.67	3.7	70.2	6.5	541	DSC	89S6
1.34	3.7	63.3	6.5	506	DSC	89S6
2.0	3.7	56.8	6.5	473	DSC	89S6
2.67	3.7	49.5	6.5	435	DSC	89S6
4.0	3.7	40	6.5	393	DSC	89S6

HEN Egg White lysozyme, crosslinked between Glu35 and Trp108, in the presence of 1-propanol

Concentration (M)	pH	T_{trs}	ΔCp	ΔH	Approach	Ref
1.34	3.7	80.5	6.5	608	DSC	89S6
2.0	3.7	73.9	6.5	591	DSC	89S6
2.67	3.7	68.2	6.5	532	DSC	89S6
4.0	3.7	59.7	6.5	506	DSC	89S6

HEN Egg White lysozyme, native and modified forms

Form	pH	T_{trs}	ΔCp	ΔH	Approach	Remarks	Ref
native	7.5	57.9±0.1	6.3	308±3	heat	(1)	9OO1
NFK-Trp62	7.5	54.8±0.4	6.3	324±5	heat	(1,2)	9OO1
Kyn-Trp62	7.5	57.6±0.6	6.3	314±8	heat	(1,3)	9OO1

Remarks:
(1) the measurements were made in the presence of 1.5 M GuHCl, ΔCp was taken from Ref. 79P4
(2) modified by ozone oxidation of Trp59 to the N-formylkynurenine derivative
(3) modified by ozone oxidation of Trp59 to the kynurenine derivative

HEN Egg White lysozyme, native and modified protein

Form	pH	T_{trs}	ΔCp	ΔH	Approach	Remarks	Ref
native	3.8	77.5±1.0		460±25	heat	(1)	84D2
modified	3.8	53±2		327±50	heat	(2,3)	91R1

Remarks:
(1) monitored by NMR, for ΔH see also Ref. 91R1
(2) three-disulfide derivative, Cys6-Cys127 disulfide bond reduceed and Cys carboxymethylated
(3) monitored by NMR, measured in 9:1 (v./v.) H_2O and D_2O

HEN Egg White lysozyme, modified, alkylated

Form	pH	T_{trs}	ΔCp	ΔH	Approach	Remark	Ref
native	3.0	72.5	6.21	490±9	DSC		92F5
ethylated	3.0	70.9	6.14	436±8	DSC		92F5
butylated	3.0	69.4	5.96	424±11	DSC		92F5
hexylated	3.0	63.5	6.23	358±13	DSC		92F5
2,2-DMP	3.0	67.7	5.98	409±10	DSC	(1)	92F5
benzylated	3.0	70.2	5.89	425±15	DSC		92F5

Remark:
(1) 2,2-DMP: 2,2-dimethylpropylated lysozyme

HEN Egg White lysozyme, modified, three-disulfide derivative

Form	pH	T_{trs}	ΔCp	ΔH	Approach	Remarks	Ref
unmodified		25.0	6.30±0.21	210±8	DSC		92C4
unmodified	3.8	77.5	6.30±0.21	540	DSC	(1)	92C4
modified		25.0	5.20±0.21	209±6	DSC	(2)	92C4
modified	3.8	52	5.20±0.21	350	DSC	(1,2)	92C4

Remarks:
(1) T_{trs} and ΔH were taken from Figs. 1–3 in Ref. 92C4
(2) Cys6 and Cys127 were specifically reduced and carboxymethylated

Chicken lysozyme, wild-type and hyperthermostable (hs) mutant

Mutant	pH	T_{trs}	ΔH	Approach	Remark	Ref
wild-type	5.0	76.2±0.4	544	DSC		95S7
hs mutant	5.0	86.4±0.5	628	DSC	(1)	95S7

Remark:
(1) hs mutant contains the following six mutations: His15→Leu, Ala31→Val, Ile55→Leu, Ser91→Thr, Asp101→Ser and Arg114→His

Lysozyme, Human

Human lysozyme, wild-type

pH	T_{trs}	ΔCp	ΔH	Approach	Remarks	Ref
1.49	50.0	5.23	363	DSC	(1,2)	92K12
1.80	52.2	8.12	390	DSC	(1,2)	92K12
1.99	55.5	6.53	410	DSC	(1,2)	92K12
2.00	55.6	8.12	420	DSC	(1,2)	92K12
2.10	57.1	5.98	427	DSC	(1,2)	92K12
2.34	59.9	6.02	440	DSC	(1,2)	92K12
2.50	63.0	5.94	459	DSC	(1,2)	92K12
2.79	68.8	5.15	503	DSC	(1,2)	92K12
3.01	71.4	7.66	524	DSC	(1,2)	92K12
3.01	71.4	7.28	509	DSC	(1,2)	92K12
3.03	71.4	7.49	515	DSC	(1,2)	92K12
3.10	73.5	6.44	537	DSC	(1,2)	92K12
3.50	75.1	5.82	523	DSC	(1,2)	92K12
4.00	79.5	6.40	561	DSC	(1,2)	92K12
4.01	79.5	5.77	557	DSC	(1,2)	92K12
4.49	80.3	7.66	575	DSC	(1,2)	92K12
4.49	80.4	6.32	580	DSC	(1,2)	92K12
4.50	80.2	7.24	567	DSC	(1,2)	92K12
4.50	80.3	7.15	579	DSC	(1,2)	92K12

Remarks:
(1) measurement in the absence of Ca^{2+} (for comparison with a mutant protein that exhibits a calcium binding site)
(2) mean of the ΔCp values = 6.65±0.88 kJ/mol/K

Human lysozyme

pH	T_{trs}	ΔCp	ΔH	Approach	Remarks	Ref
4.7	81±2		523	heat	(1)	72B
7.4	70±1		542±25	heat	(2,3)	95S15

Remarks:
(1) in the presence of 0.1 M KCl
(2) global analysis of FTIR spectra based on the intramolecular secondary structure band
(3) a second transition was observed at T_{trs} = 49.7°C

Table 2. Enthalpy and Heat Capacity Changes – Molar Values

Human lysozyme and the amyloidogenic mutant (Ile56→Thr)

Mutant	pH	T_{trs}	ΔCp	ΔH	Approach	Remarks	Ref
wild-type	2.7	64.9±0.5	6.6±0.5	477±4	DSC	(1)	96F6
Ile56→Thr	2.7	52.4	5.2±0.5	425±5	DSC	(1)	96F6
Ile56→Thr	2.67	51.78	4.2	367	DSC	(2)	96F6
Ile56→Thr	2.98	57.96	5.0	390	DSC	(2)	96F6
Ile56→Thr	3.11	60.52	4.8	408	DSC	(2)	96F6

Remarks:
(1) ΔCp from ΔH versus T_{trs}
(2) ΔCp from each calorimetric curve

Human lysozyme, mutant (Gln86→Asp and Ala92→Asp), apo form

pH	T_{trs}	ΔCp	ΔH	Approach	Remarks	Ref
2.00	53.2	7.91	407	DSC	(1–3)	92K12
2.30	57.1	5.65	419	DSC	(1–3)	92K12
2.50	60.8	5.19	435	DSC	(1–3)	92K12
2.75	65.7	6.82	482	DSC	(1–3)	92K12
2.80	66.4	6.40	480	DSC	(1–3)	92K12
3.00	69.5	6.57	495	DSC	(1–3)	92K12
3.10	70.8	7.82	506	DSC	(1–3)	92K12
4.50	76.5	4.69	546	DSC	(1–3)	92K12

Remarks:
(1) the mutant protein has a calcium binding site
(2) measurement in the absence of Ca^{2+}
(3) mean of ΔCp = 6.40±1.09 kJ/mol/K

Human lysozyme, wild-type, in the presence of calcium

Concentration (mM)	pH	T_{trs}	ΔCp	ΔH	Approach	Remarks	Ref
0.0	2.61	80.3	7.15	580	DSC	(1,2)	92K12
1.0	2.36	80.3	6.82	573	DSC	(1,2)	92K12
2.0	2.25	80.2	5.77	577	DSC	(1,2)	92K12
5.0	2.65	80.1	5.02	568	DSC	(1,2)	92K12
10.0	2.24	80.0	4.27	569	DSC	(1,2)	92K12
50.0	2.54	78.9	3.97	562	DSC	(1,2)	92K12

Remarks:
(1) measurement in the presence of Ca^{2+} (for comparison with a mutant protein that exhibits a calcium binding site)
(2) mean of the ΔCp values = 5.48±1.30 kJ/K/mol

Human lysozyme, mutant (Gln86→Asp and Ala92→Asp) measured in the presence of calcium

Concentration (mM)	pH	T_{trs}	ΔCp	ΔH	Approach	Remarks	Ref
0.0	2.28	76.5	4.69	546	DSC	(1,2)	92K12
1.0	1.33	85.5	4.10	566	DSC	(1,2)	92K12
2.0	1.37	86.8	2.38	590	DSC	(1,2)	92K12
5.0	1.30	88.2	2.97	578	DSC	(1,2)	92K12
10.0	1.31	89.2	4.64	590	DSC	(1,2)	92K12
50.0	1.26	90.2	5.44	589	DSC	(1,2)	92K12
1.0	2.24	85.7	3.64	578	DSC	(1,2)	92K12
10.0	2.19	89.2	3.35	595	DSC	(1,2)	92K12

Remarks:
(1) the mutant protein has a calcium binding site
(2) mean of the ΔCp values = 3.89±0.96 kJ/K/mol

Human lysozyme, wild-type and mutants

Mutant	pH	T_{trs}	ΔCp	ΔH	Approach	Remarks	Ref
wild-type	2.80	68.8	6.59	502	DSC		92H3
Pro71→Gly	2.80	64.1	6.63		DSC		92H3
Pro71→Gly		68.8	6.63	505	DSC	(1)	92H3
Pro103→Gly	2.80	68.7	6.61		DSC		92H3
Pro103→Gly		68.8	6.61	501	DSC	(1)	92H3
Asp91→Pro	2.80	67.7	6.57		DSC		92H3
Asp91→Pro		68.8	6.57	483	DSC	(1)	92H3
Ala47→Pro	2.80	69.1	6.53		DSC		92H3
Ala47→Pro		68.8	6.53	500	DSC	(1)	92H3
Val110→Pro	2.80	70.0	6.58		DSC		92H3
Val110→Pro		68.8	6.58	503	DSC	(1)	92H3
double mutant (Pro71→Gly and Pro103-Gly)							
	2.80	64.3	6.60		DSC		92H3
		68.8	6.60	504	DSC	(1)	92H3

Remark:
(1) T_{trs} and ΔH refer to the transition temperature of the wild-type protein

Human lysozyme, wild-type and mutants

Mutant	pH	T_{trs}	ΔCp	ΔH	Approach	Remarks	Ref
wild-type	3.0	71.4	6.49	421	DSC	(1)	92K11
Cys77→Ala	3.0	57.1	7.28	397	DSC	(1)	92K11
double mutant (Cys77→Ala and Cys95→Ala)							
	3.0	56.9	5.86	380	DSC	(1)	92K11

Remark:
(1) ΔCp was taken from the temperature dependence of ΔH

Human lysozyme, wild-type and mutants

Mutant	pH	T_{trs}	ΔCp	ΔH	Approach	Remarks	Ref
wild-type	3	69.89±0.06		506±11	v.H.	(1)	90F
Val110→Pro	3	70.95±0.06		525±12	v.H.	(1)	90F
double mutant (Cys77→Ala and Cys95→Ala)							
	3	55.56±0.05		401±7	v.H.	(1)	90F

Remark:
(1) determined by multidimensional spectroscopy

Human lysozyme, mutant (Cys77→Ala)

Mutant	pH	T_{trs}	ΔCp	ΔH	Approach	Ref
Cys77→Ala	2.30	42.3	7.95	287	DSC	92K11
Cys77→Ala	2.50	45.4	6.74	311	DSC	92K11
Cys77→Ala	2.60	46.4	6.19	322	DSC	92K11
Cys77→Ala	2.75	50.5	6.82	353	DSC	92K11
Cys77→Ala	2.93	55.0	4.73	379	DSC	92K11
Cys77→Ala	3.01	56.8	6.19	398	DSC	92K11
Cys77→Ala	3.06	57.1	6.78	394	DSC	92K11

Remark:
(1) mean value: ΔCp = 6.49±0.92 kJ/mol/K

Human lysozyme, double mutant (Cys77→Ala and Cys95→Ala)

Mutant	pH	T_{trs}	ΔCp	ΔH	Approach	Ref
C77/95→Ala	2.32	42.7	7.03	292	DSC	92K11
C77/95→Ala	2.51	45.6	7.70	310	DSC	92K11
C77/95→Ala	2.60	46.4	6.28	320	DSC	92K11
C77/95→Ala	2.80	52.8	6.19	362	DSC	92K11
C77/95→Ala	3.01	56.7	4.98	385	DSC	92K11
C77/95→Ala	3.06	56.9	5.65	382	DSC	92K11
C77/95→Ala	3.49	63.6	5.82	418	DSC	92K11
C77/95→Ala	4.50	65.6	6.61	426	DSC	92K11

Remark:
(1) mean value: ΔCp = 6.28±0.79 kJ/mol/K

Human lysozyme, wild-type and Ile→Val mutants

Mutant	pH	T_{trs}	ΔCp	ΔH	Approach	Remarks	Ref
wild-type	3.02	70.1	7.4	515	DSC	(1)	95T1
wild-type	2.88	67.6	7.7	490	DSC	(1)	95T1
wild-type	2.72	65.6	4.9	481	DSC	(1)	95T1
wild-type	2.71	65.4	6.7	485	DSC	(1)	95T1
wild-type	2.71	65.3	6.7	481	DSC	(1)	95T1
wild-type	2.53	61.3	5.4	452	DSC	(1)	95T1
wild-type	2.46	61.4	6.7	456	DSC	(1)	95T1
wild-type			6.5±1.1		DSC	(2)	95T1
wild-type			6.6±0.5		DSC	(3)	95T1
Ile23→Val	2.90	67.5	6.9	519	DSC	(1)	95T1
Ile23→Val	2.70	63.7	3.2	502	DSC	(1)	95T1
Ile23→Val	2.53	60.3	6.3	481	DSC	(1)	95T1

Human lysozyme, wild-type and Ile→Val mutants (continued)

Mutant	pH	T_{trs}	ΔCp	ΔH	Approach	Remarks	Ref
Ile23→Val	2.46	59.8	4.7	464	DSC	(1)	95T1
Ile23→Val			5.3±2.0		DSC	(2)	95T1
Ile23→Val			5.8±1.8		DSC	(3)	95T1
Ile56→Val	3.03	67.5	4.8	498	DSC	(1)	95T1
Ile56→Val	2.88	64.4	4.0	460	DSC	(1)	95T1
Ile56→Val	2.70	61.7	4.4	469	DSC	(1)	95T1
Ile56→Val	2.69	61.5	6.6	444	DSC	(1)	95T1
Ile56→Val	2.66	60.6	6.6	444	DSC	(1)	95T1
Ile56→Val	2.53	57.9	7.4	444	DSC	(1)	95T1
Ile56→Val			5.6±1.6		DSC	(2)	95T1
Ile56→Val			5.6±1.7		DSC	(3)	95T1
Ile59→Val	3.03	67.4	4.6	469	DSC	(1)	95T1
Ile59→Val	2.88	64.4	4.0	460	DSC	(1)	95T1
Ile59→Val	2.71	61.3	5.6	435	DSC	(1)	95T1
Ile59→Val	2.70	61.9	4.6	456	DSC	(1)	95T1
Ile59→Val	2.66	60.6	4.8	435	DSC	(1)	95T1
Ile59→Val	2.46	57.7	3.7	423	DSC	(1)	95T1
Ile59→Val			4.6±0.8		DSC	(2)	95T1
Ile59→Val			5.0±1.0		DSC	(3)	95T1
Ile89→Val	3.14	72.0	6.0	540	DSC	(1)	95T1
Ile89→Val	3.14	71.7	6.7	536	DSC	(1)	95T1
Ile89→Val	2.92	67.9	6.8	510	DSC	(1)	95T1
Ile89→Val	2.71	62.9	8.4	464	DSC	(1)	95T1
Ile89→Val	2.46	59.4	5.1	439	DSC	(1)	95T1
Ile89→Val			6.7±1.4		DSC	(2)	95T1
Ile89→Val			8.0±0.3		DSC	(3)	95T1
Ile106→Val	3.14	71.3	5.6	510	DSC	(1)	95T1
Ile106→Val	3.14	71.1	5.9	494	DSC	(1)	95T1
Ile106→Val	3.14	71.0	4.8	494	DSC	(1)	95T1
Ile106→Val	3.10	69.8	6.0	477	DSC	(1)	95T1
Ile106→Val	3.04	68.7	6.5	469	DSC	(1)	95T1
Ile106→Val	2.92	67.0	5.5	464	DSC	(1)	95T1
Ile106→Val	2.85	65.5	6.0	469	DSC	(1)	95T1
Ile106→Val	2.70	63.2	4.8	452	DSC	(1)	95T1
Ile106→Val	2.61	60.6	4.4	444	DSC	(1)	95T1
Ile106→Val	2.46	58.3	3.4	410	DSC	(1)	95T1
Ile106→Val			5.3±1.0		DSC	(2)	95T1
Ile106→Val			5.9±0.7		DSC	(3)	95T1

Remarks:
(1) ΔCp from single calorimetric scans
(2) mean value
(3) ΔCp from ΔH versus T_{trs}

Lysozyme, Pigeon

Pigeon lysozyme

Transition	pH	T_{trs}	ΔCp	ΔH	Approach	Remarks	Ref
N→U	6.0	70	6.0	335	heat	(1,2)	93N2

Remarks:
(1) measured in the presence of 50 µM free calcium ΔH was taken from Fig. 5 in Ref. 93N2
(2) two-state transition in contrast to equine lysozyme

Lysozyme Phage T4

The data are arranged as follows:

a) pH dependence, general expressions,
b) pH dependence, single data,
c) single amino acid replacements,
d) single and multiple amino acid replacements,
e) modified protein, newly engineered disulfide bonds, unnatural amino acids,
f) effect of benzene on protein stability,
g) insertion mutants, permuted protein,
h) mutants that unfold in a three-state process.

a) pH dependence, general expressions

Lysozyme phage T4, wild-type, pH dependence [remark (1)]

pH	T_{trs}	ΔCp	ΔH	Approach	Ref
1.6	32.92	9.75	328	DSC	89K4
1.8	35.6	9.75	363	DSC	89K4
2	38.75	9.75	406	DSC	89K4
2.2	40.55	9.75	426	DSC	89K4
2.34	42.94	9.75	466	DSC	89K4
2.4	43.8	9.75	478	DSC	89K4
2.5	45.03	9.75	479	DSC	89K4
2.7	48.7	9.75	479	DSC	89K4
2.84	51.68	9.75	517	DSC	89K4

Remark:
(1) the data represent averaged experimental values, the general expressions are $T_{trs} = 9.63 + 14.41 \times pH$ (± 0.58) and $\Delta H = 24.98 + 9.75 \times T$ (± 17.6); Refs. 87S2, 89K4

Lysozyme phage T4, temperature dependence of the heat capacity change at protein unfolding in aqueous solution, and related thermodynamic quantities

T	ΔH(exp)	ΔS(exp)	ΔCp(exp)	Approach	Remarks	Ref
5	20	−190	11.0	DSC	(1–3)	95M3
25	240	576	11.0	DSC	(1–3)	95M3
50	499	1413	9.7	DSC	(1–3)	95M3
75	671	1928	6.7	DSC	(1–3)	95M3
100	805	2302	4.0	DSC	(1–3)	95M3
125	856	2439	0.2	DSC	(1–3)	95M3

Remarks:
(1) the data were taken from Ref. 95M3 with reference to Ref. 89K4
(2) ΔCp was obtained as described in Ref. 90P3
(3) ΔCp in kJ/mol/K, ΔH in kJ/mol and ΔS in J/mol/K

Lysozyme phage T4, wild-type and mutants, pH dependence

Mutant	T_{trs}(A)	ΔH(B)	ΔCp(C)	ΔCp(D)	Approach	Ref
wild-type	[1]	−43.97±3.47	10.75±0.08	10.08±0.33	DSC	92L1
Ile3→Glu	[2]	21.84±4.77	10.50±0.13	10.92±1.05	DSC	92L1
Ile3→Leu	[4]	39.71±5.23	9.54±0.13	9.87±0.75	DSC	92L1
Ile3→Phe	[3]	90.12±5.31	8.79±0.13	11.05±1.09	DSC	92L1
Ile3→Pro	[5]	−94.73±10.79	12.38±0.29	15.48±1.46	DSC	92L1
Ile3→Thr	[6]	−35.06±4.77	11.67±0.13	12.47±1.00	DSC	92L1

Explanation:
(A) T_{trs} can be obtained from the following expressions representing the pH dependence of T_{trs} [1] – [6]:
[1] $T_{trs} = A + B \times pH = 9.13 + 14.81 \times pH$ (for pH 1.60 to 2.84)
[2] $T_{trs} = A + B \times pH = (−0.62±0.13)+(−16.84±0.05) \times pH$ (for pH 1.98 to 3.05)
[3] $T_{trs} = A + B \times pH = (−2.29±0.23)+(−18.18±0.10) \times pH$ (for pH 1.96 to 3.01)
[4] $T_{trs} = A + B \times pH = (11.43±0.28)+(14.99±0.12) \times pH$ (for pH 2.0 to 3.19)
[5] $T_{trs} = A + B \times pH = (−14.02±0.24)+(20.39±0.10) \times pH$ (for pH 2.0 to 3.01)
[6] $T_{trs} = A + B \times pH = (−7.67±0.16)+(18.97±0.06) \times pH$ (for pH 2.02 to 2.97)
(B) and (C) General expression for the temperature dependence of $\Delta H(T) = \Delta H(0°C) + \Delta Cp \times T_{trs}$
    (with T_{trs} in °C)
(D) ΔCp obtained from single recordings (mean value)

Lysozyme phage T4, wild-type and mutants, pH dependence

Mutant	pH	T_{trs}	ΔCp	ΔH	Approach	Remarks	Ref
wild-type	1.6–2.84	0	10.75	−43.97	DSC	(1,2)	92H7
Ala82→Pro	1.77–2.98	0	10.29	−1.13	DSC	(1,3)	92H7
Ala93→Pro	1.77–2.98	0	8.54	84.09	DSC	(1,4)	92H7
Gly113→Ala	1.77–2.98	0	12.05	−81.50	DSC	(1,5)	92H7
(Cys54→Thr and Cys97→Ala)	1.76–4.54	0	8.12	68.53	DSC	(1,6)	92H7

Remarks:
(1) T_{trs} = 0°C to facilitate the calculation of $\Delta H(T) = \Delta H(0°C)+ \Delta Cp \times T$ (T in °C)
(2) the pH dependence of T_{trs} = 9.13 + 14.81 × pH
(3) the pH dependence of T_{trs} = 2.26 + 18.13 × pH
(4) the pH dependence of T_{trs} = 3.90 + 17.47 × pH
(5) the pH dependence of T_{trs} = 11.54 + 14.44 × pH
(6) the pH dependence of T_{trs} = 12.59 + 12.49 × pH

Table 2. Enthalpy and Heat Capacity Changes – Molar Values

Lysozyme phage T4, mutant Arg96→His, pH dependence [remark (1)]

pH	T_{trs}	ΔCp	ΔH	Approach	Ref
2	23.15	11.13	223	DSC	89K4
2.1	24.31	11.13	234	DSC	89K4
2.2	27.11	11.13	296	DSC	89K4
2.34	29.64	11.13	276	DSC	89K4
2.4	30.93	11.13	308	DSC	89K4
2.5	33.44	11.13	319	DSC	89K4
2.7	37.46	11.13	389	DSC	89K4
2.84	41.18	11.13	412	DSC	89K4

Remark:
(1) the data represent averaged experimental values, the general expressions are $T_{trs} = -19.84 + 21.31 \times pH$ (± 0.51) and $\Delta H = -35.9 + 11.13 \times T$ (± 18.7); Refs. 87S2, 89K4

Lysozyme phage T4, wild-type and mutants in pos. 157, pH dependence

Mutant	pH	T_{trs}	ΔCp	$\Delta H(0°C)$	Approach	Remarks	Ref
wild-type	1.6–2.84 [2a]		10.75±0.08	−43.97±3.47	DSC	(2)	91C7
wild type	1.6–2.84 [2a]		10.08±0.33		DSC	(1)	91C7
Thr157→Ala	1.6–2.84 [2b]		11.46±0.13	−62.63±5.02	DSC	(2)	91C7
Thr157→Ala	1.6–2.84 [2b]		10.17±0.50		DSC	(1)	91C7
Thr157→Glu	2.02–2.98 [2c]		9.50±0.21	−35.90±9.00	DSC	(2)	91C7
Thr157→Glu	2.02–2.98 [2c]		9.62±1.13		DSC	(1)	91C7
Thr157→Ile	1.8–2.84 [2d]		10.13±0.17	−10.46±1.52	DSC	(2)	91C7
Thr157→Ile	1.8–2.84 [2d]		13.01±0.75		DSC	(1)	91C7
Thr157→Leu	1.71–3.27 [2e]		9.66±0.25	−32.13±10.13	DSC	(2)	91C7
Thr157→Leu	1.71–3.27 [2e]		7.28±1.88		DSC	(1)	91C7
Thr157→Asn	1.7–2.82 [2f]		11.92±0.29	−73.93±11.8	DSC	(2)	91C7
Thr157→Asn	1.7–2.82 [2f]		14.10±0.92		DSC	(1)	91C7
Thr157→Arg	1.8–2.84 [2g]		9.79±0.25	7.87±10.38	DSC	(2)	91C7
Thr157→Arg	1.8–2.84 [2g]		11.09±0.75		DSC	(1)	91C7
Thr157→Val	1.6–2.84 [2h]		11.00±0.08	−59.20±4.18	DSC	(2)	91C7
Thr157→Val	1.6–2.84 [2h]		10.25±0.67		DSC	(1)	91C7

Remarks:
(1) ΔCp is the mean value from single calorimetric recordings
(2) the pH dependence of T_{trs} is given by the following expressions
[2a] wild-type: $T_{trs} = A + B \times pH$, with A= 9.13 and B=14.81
[2b] Thr157→Ala: $T_{trs} = A + B \times pH$, with A=−0.18 and B=17.49
[2c] Thr157→Glu: $T_{trs} = A + B \times pH$, with A=−5.87 and B=19.06
[2d] Thr157→Ile: $T_{trs} = A + B \times pH$, with A=−6.26 and B=18.64
[2e] Thr157→Leu: $T_{trs} = A + B \times pH$, with A=−1.29 and B=16.61
[2f] Thr157→Asn: $T_{trs} = A + B \times pH$, with A=+4.65 and B=15.31
[2g] Thr157→Arg: $T_{trs} = A + B \times pH$, with A=−4.03 and B=19.41
[2h] Thr157→Val: $T_{trs} = A + B \times pH$, with A=−3.80 and B=18.04

b) pH dependence, single data

Lysozyme phage T4, wild-type and mutants

Mutant	pH	T_{trs}	ΔC_p	ΔH	Approach	Remarks	Ref
wild-type	1.4	32.8	8.8	360	heat, v.H.		77E
wild-type	2	36.7	8.8	423	heat, v.H.		77E
wild-type	2	41.9±0.4	8.4±0.8	372±21	heat, v.H.		87M
wild-type	2.03	41.47±0.1		347	heat	(1)	90D1
wild-type	2.2	42.4	8.8	448	heat, v.H.		77E
wild-type	2.5	46	8.8	510	heat, v.H.		77E
wild-type	2.83	51.1±0.15		452	heat	(1)	90D1
wild-type	3	54.5	8.8	544	heat, v.H.		77E
wild-type	3	56.6	9.46	326	heat, v.H.		84H1
wild-type	6.5	64.7±0.5	8.4±0.8	540±38	heat, v.H.		87M
wild-type			8.56		DSC		79P4
Ala41→Val	2.03	43.07±0.1		349	heat	(1)	90D1
Ala41→Val	2.83	52.43±0.1		459	heat	(1)	90D1
Gly77→Ala	2	40.5±0.7	8.4±0.8	356±17	heat, v.H.		87M
Ala82→Pro	2	42.7±0.1	8.4±0.8	377±21	heat, v.H.		87M
Gly77→Ala	6.5	65.6±0.2	8.4±0.8	523±38	heat, v.H.		87M
Ala82→Pro	6.5	66.8±0.7	8.4±0.7	572±38	heat, v.H.		87M
Arg96→His	3	43	10.0	299	heat, v.H.		84H1
Met102→Val	3	43.9	9.75	273	heat, v.H.		84H1
Glu128→Lys	3	51.3	9.41	328	heat, v.H.		84H1
Val131→Ala	2.03	42.03±0.2		355	heat	(1)	90D1
Val131→Ala	2.83	52.0±0.2		445	heat	(1)	90D1
Val131→Thr	2.03	41.06±0.1		345	heat	(1)	90D1
Val131→Thr	2.83	50.95±0.1		445	heat	(1)	90D1
Trp138→Tyr	2.2	36.1	8.8	351	heat, v.H.		77E
Ala146→Thr	3	47.2	10.5	238	heat, v.H.		84H1
double mutant (Ala41→Val and Val131→Ala)							
	2.03	43.67±0.1		436	heat	(1)	90D1
	2.83	52.91±0.1		449	heat	(1)	90D1
triple mutant (Trp126,138,158→Tyr)							
	2.2	35.8	8.8	357	heat, v.H.		77E

Remark:
(1) measured in the presence of 0.2 M KCl, monitored by CD 223 nm

Lysozyme phage T4, pseudo wild-type (Cys54→Thr and Cys97→Ala) and mutants

Mutant	pH	T_{trs}	ΔC_p	ΔH	Approach	Remarks	Ref
wt*	3.0	51.6	7.5	490	heat	(1,3)	91D3
wt*	4.4	64.6	7.5	577	heat	(1,3)	91D3
wt*	5.3	65.3	7.5	565	heat	(1,3)	91D3
wt*	6.5	61.9	7.5	506	heat	(1,3)	91D3
wt*	10.4	47.9	7.5	333	heat	(1,3,4)	91D3
Met102→Lys	3.0	16.6	7.5	52	heat	(1–3)	91D3
Met102→Lys	5.3	45.0	7.5	222	heat	(1–3)	91D3
Met102→Lys	10.4	37.8	7.5	324	heat	(1–4)	91D3

Lysozyme phage T4, pseudo wild-type (Cys54→Thr and Cys97→Ala) and mutants (continued)

Mutant	pH	T_{trs}	ΔCp	ΔH	Approach	Remarks	Ref
Leu133→Asp	3.0	36.7	7.5	200	heat	(1–3)	91D3
Leu133→Asp	4.4	49.6	7.5	247	heat	(1–3)	91D3
Leu133→Asp	6.5	44.0	7.5	213	heat	(1–3)	91D3
Leu133→Asp	10.4	28.9	7.5	197	heat	(1–4)	91D3

Remarks:
(1) measurements in 20 mM potassium phosphate with 25 mM KCl
(2) the mutant protein contains additionally the replacements (Cys54→Thr and Cys97→Ala) of the template (pseudo wild-type, wt*)
(3) ΔCp is based on assumptions supported by independent findings (see Ref. 91D3)
(4) buffer (1) with 0.1 mM EDTA

c) single amino acid replacements

Lysozyme phage T4, site 44 mutants

Mutant	pH	T_{trs}	ΔCp	ΔH	Approach	Remarks	Ref
wt*	3.01	51.68	10.04	477	heat	(1–3)	94B6
Ser44→Ala	3.01	52.89		490	heat	(1–3)	94B6
Ser44→Arg	3.01	52.36		485	heat	(1–3)	94B6
Ser44→Asn	3.01	51.28		460	heat	(1–3)	94B6
Ser44→Asp	3.01	51.36		460	heat	(1–3)	94B6
Ser44→Cys	3.01	51.33		423	heat	(1–3)	94B6
Ser44→Gln	3.01	52.43		485	heat	(1–3)	94B6
Ser44→Glu	3.01	51.68		464	heat	(1–3)	94B6
Ser44→Gly	3.01	50.13		460	heat	(1–3)	94B6
Ser44→His	3.01	51.80		490	heat	(1–3)	94B6
Ser44→Ile	3.01	52.59		469	heat	(1–3)	94B6
Ser44→Leu	3.01	52.77		490	heat	(1–3)	94B6
Ser44→Lys	3.01	52.25		485	heat	(1–3)	94B6
Ser44→Met	3.01	52.60		481	heat	(1–3)	94B6
Ser44→Phe	3.01	51.86		485	heat	(1–3)	94B6
Ser44→Pro	3.01	41.36		310	heat	(1–3)	94B6
Ser44→Thr	3.01	51.71		473	heat	(1–3)	94B6
Ser44→Trp	3.01	51.83		469	heat	(1–3)	94B6
Ser44→Tyr	3.01	52.22		481	heat	(1–3)	94B6
Ser44→Val	3.01	51.97		473	heat	(1–3)	94B6

Remarks:
(1) wt* = pseudo wild-type, containing (Cys54→Thr and Cys97→Ala)
(2) estimated error: T_{trs} ±0.1°C, ΔH ±(4–6)%, ΔCp = 10.04 kJ/mol/K was taken from Ref. 92H10 assuming the same value for all mutants
(3) buffer: 25 mM KCl, 3 mM H_3PO_4, 17 mM KH_2PO_4

Lysozyme phage T4, pseudo wild-type (wt*) and mutants, alanine replacements within α-helix 39–50

Mutant	pH	ΔT	ΔCp	ΔH	Approach	Remarks	Ref
wt*	3.0	0	10.5	494	heat	(1–3)	92H2
wt*	6.7	0	10.5	531	heat	(1–3)	92H2
Asn40→Ala	3.0	1.2		490	heat	(1–3)	92H2
Asn40→Ala	6.7	0.88		523	heat	(1–3)	92H2
Lys43→Ala	3.0	−2.95		423	heat	(1–3)	92H2
Lys43→Ala	6.7	−4.0		452	heat	(1–3)	92H2
Ser44→Ala	3.0	1.2		510	heat	(1–3)	92H2
Ser44→Ala	6.7	0.98		498	heat	(1–3)	92H2
Glu45→Ala	3.0	1.47		523	heat	(1–3)	92H2
Glu45→Ala	6.7	0.04		519	heat	(1–3)	92H2
Leu46→Ala	3.0	−8.39		372	heat	(1–3)	92H2
Leu46→Ala	6.7	−6.4		368	heat	(1–3)	92H2
Asp47→Ala	3.0	−0.81		477	heat	(1–3)	92H2
Asp47→Ala	6.7	−2.75		473	heat	(1–3)	92H2
Lys48→Ala	3.0	−0.93		456	heat	(1–3)	92H2
Lys48→Ala	6.7	−1.68		460	heat	(1–3)	92H2
double mutant (Glu45→Ala and Lys48→Ala)							
	3.0	1.04		506	heat	(1–3)	92H2
	6.7	0.03		531	heat	(1–3)	92H2
multiple Ala replacements in positions 40–49							
	3.0	−8.47		293	heat	(1–3)	92H2
	6.7	−10.7		268	heat	(1–3)	92H2
multiple Ala replacements in positions 40–42, 44–45, 47–49							
	3.0	3.09		490	heat	(1–3)	92H2
	6.7	−1.56		490	heat	(1–3)	92H2

Remarks:
(1) ΔT refers to the pseudo wild-type Cys54→Thr and Cys97→Ala (wt*) having T_{trs} = 51.65°C at pH 3.0 and 62.2°C at pH 6.7
(2) error in T_{trs} ±0.15°C and in ΔH ±17 kJ/mol
(3) measured in 20 mM phosphate with 25 mM KCl at pH 3 and 10 mM phosphate with 0.15 M KCl at pH 6.7

Lysozyme phage T4, site 105 mutants

Mutant	pH	T_{trs}	ΔCp	ΔH	Approach	Remarks	Ref
wild-type	2.1	42.0		364	heat	(1)	93P10
Gln105→Ala	2.1	38.5		322	heat	(1)	93P10
Gln105→Glu	2.1	39.9		343	heat	(1)	93P10
Gln105→Gly	2.1	34.8		301	heat	(1)	93P10
wild-type	5.8	66.3		602	heat	(1)	93P10
Gln105→Ala	5.8	64.7		531	heat	(1)	93P10
Gln105→Glu	5.8	63.3		519	heat	(1)	93P10
Gln105→Gly	5.8	62.4		552	heat	(1)	93P10

Remark:
(1) the estimated error in T_{trs} is ±0.4°C, in ΔH is ±95 kJ/mol

Lysozyme phage T4, alanine mutations in helix 115–123

Mutant	pH	T_{trs}	ΔH	Approach	Remarks	Ref
pseudo wild-type	3.01	51.65	477	heat, v.H.	(1)	95B4
Thr115→Ala	3.01	51.22	456	heat, v.H.	(1)	95B4
Asn116→Ala	3.01	52.14	456	heat, v.H.	(1)	95B4
Ser117→Ala	3.01	55.29	481	heat, v.H.	(1)	95B4
Leu118→Ala	3.01	39.6	314	heat, v.H.	(1)	92E3
Arg119→Ala	3.01	51.12	448	heat, v.H.	(1)	95B4
Met120→Ala	3.01	51.07	469	heat, v.H.	(1)	95B4
Leu121→Ala	3.01	42.5	339	heat, v.H.	(1)	92E3
Gln122→Ala	3.01	50.94	452	heat, v.H.	(1)	95B4
Gln123→Ala	3.01	51.01	464	heat, v.H.	(1)	95B4
double mutant (Thr115→Ala and Ser117→Ala)	3.01	54.49	456	heat, v.H.	(1)	95B4
double mutant (Ser117→Ala and Arg119→Ala)	3.01	55.02	473	heat, v.H.	(1)	95B4
double mutant (Met120→Ala and Gln122→Ala)	3.01	50.90	452	heat, v.H.	(1)	95B4
double mutant (Thr115→Ala and Arg119→Ala)	3.01	51.13	444	heat, v.H.	(1)	95B4
double mutant (Asn116→Ala and Met120→Ala)	3.01	52.23	485	heat, v.H.	(1)	95B4
double mutant (Arg119→Ala and Gln123→Ala)	3.01	51.14	460	heat, v.H.	(1)	95B4
multiple mutant (residues 115–123→Ala except Leu118 and Leu121)	3.01	54.84	427	heat, v.H.	(1)	95B4

Remark:
(1) buffer: 0.0029 M H_3PO_4, 0.017 M KH_2PO_4, 0.025 M KCl, pH 3.01

Lysozyme phage T4, site 117 mutants

Mutant	pH	T_{trs}	ΔCp	ΔH	Approach	Ref
wild-type	3.0	51.70±0.14		490±13	heat	93A2
Ser117→Phe	3.0	56.48±0.16		435±13	heat	93A2
wild-type	5.4	65.09±0.22		561±17	heat	93A2
Ser117→Phe	5.4	67.93±0.15		552±17	heat	93A2

Lysozyme phage T4, site 129 mutants

Mutant	pH	T_{trs}	ΔH	Approach	Remarks	Ref
wild-type	1.99	41.4±0.3	364±42	heat, v.H.	(1)	89K1
Ala129→Val	1.99	38.5±0.5	327±42	heat, v.H.	(1)	89K1

Remark:
(1) measured in 0.2 M KCl

Lysozyme phage T4, site 131 mutants

Mutant	pH	T_{trs}	ΔCp	ΔH	Approach	Remarks	Ref
wild-type	3.01	53.08	10.04	531	heat	(1–4)	94B6
Val131→Ala	3.01	53.74		536	heat	(1–4)	94B6
Val131→Asp	3.01	53.30		515	heat	(1–4)	94B6
Val131→Glu	3.01	53.60		523	heat	(1–4)	94B6
Val131→Gly	3.01	51.28		515	heat	(1–4)	94B6
Val131→Ile	3.01	53.49		531	heat	(1–4)	94B6
Val131→Leu	3.01	53.31		506	heat	(1–4)	94B6
Val131→Met	3.01	53.40		527	heat	(1–4)	94B6
Val131→Ser	3.01	52.96		519	heat	(1–4)	94B6
Val131→Thr	3.01	52.75		510	heat	(1–4)	94B6

Remarks:
(1) wild-type (Val131) is the reference protein, T_{trs} = 53.08°C
(2) estimated error: T_{trs} ±0.1°C, ΔCp ±(4–6)%
(3) buffer: 25 mM KCl, 3 mM H_3PO_4, 17 mM KH_2PO_4
(4) ΔCp = 10.04 kJ/mol/K was taken from Ref. 92H10 and assumed to have the same value for all mutants

Lysozyme phage T4, site 133 mutants

Mutant	pH	T_{trs}	ΔH	Approach	Remarks	Ref
wild-type	1.99	41.4±0.3	364±42	heat, v.H.	(1)	89K1
Leu133→Phe	1.99	40.7±0.5	347±42	heat, v.H.	(1)	89K1

Remark:
(1) measured in 0.2 M KCl

Lysozyme phage T4, wild-type and mutants (X→Pro) and (Gly→Ala)

Mutant	pH	T_{trs}	ΔCp	ΔH	Approach	Remarks	Ref
wild-type	2.0	41.9±0.4		372±21	heat	(1)	92N2
Lys60→Pro	2.0	42.2±0.5			heat	(1)	92N2
Gly77→Ala	2.0	40.5±0.7		356±17	heat	(1)	92N2
Ala82→Pro	2.0	42.7±0.1		377±21	heat	(1)	92N2
Ala93→Pro	2.0	42.3±0.2		389±21	heat	(1)	92N2
Gly113→Ala	2.0	42.3±0.4		414±29	heat	(1)	92N2
Ile3→Pro	3.01	46.2±0.4		427	heat		92D3
wild-type	6.5	64.7±0.5		540±38	heat	(2)	92N2
Lys60→Pro	6.5	64.6±0.5			heat	(2)	92N2
Gly77→Ala	6.5	65.6±0.2		523±38	heat	(2)	92N2
Ala82→Pro	6.5	66.8±0.2		527±38	heat	(2)	92N2
Ala93→Pro	6.5	64.9±0.4		506±42	heat	(2)	92N2
Gly113→Ala	6.5	65.3±0.4		502±42	heat	(2)	92N2

Remarks:
(1) buffer: 0.2 M KCl, 10 mM HCl, and 10 mM H_3PO_4
(2) buffer: 0.15 M KCl, 20 mM potassium phosphate

Lysozyme phage T4, pseudo wild-type (wt*) and mutants X→Ser or Thr

Mutant	pH	T_{trs}	ΔCp	ΔH	Approach	Remarks	Ref
wt*	3.0	51.55±0.15	10.46	473	heat	(1–3)	93B4
Ala41→Ser	3.0	49.78		444	heat	(2,3)	93B4
Ala42→Ser	3.0	44.06		372	heat	(2,3)	93B4
Ala49→Ser	3.0	50.02		460	heat	(2,3)	93B4
Ala73→Ser	3.0	50.28		464	heat	(2,3)	93B4
Ala82→Ser	3.0	50.56		469	heat	(2,3)	93B4
Ala93→Ser	3.0	51.03		469	heat	(2,3)	93B4
Ala98→Ser	3.0	44.08		406	heat	(2,3)	93B4
Ala130→Ser	3.0	48.66		448	heat	(2,3)	93B4
Ala134→Ser	3.0	51.11		464	heat	(2,3)	93B4
Val75→Thr	3.0	47.85		473	heat	(2,3)	93B4
Val87→Thr	3.0	47.00		439	heat	(2,3)	93B4
Val149→Thr	3.0	41.47		322	heat	(2,3)	93B4

Remarks:
(1) wt* = (Cys54→Thr and Cys97→Ala) is the reference protein according to Matsumura and Matthews, Science 243 (1989) 792
(2) ΔCp was taken from Ref. 91Z2 and assumed to be identically for the mutants, the estimated uncertainty in ΔH amounts to ±21 kJ/mol
(3) buffer: 25 mM KCl, 20 mM potassium phosphate, pH 3.0

Lysozyme phage T4, variants and derivatives, X→Cys mutants

Mutant	pH	T_{trs}	ΔT	Approach	Remarks	Ref
Native form:						
parent protein (Cys54→Ser and Cys97→Ser)						
	2.4	36.4		heat	(1,3)	92L6
Asn55→Cys	2.4	32.8	−3.6	heat	(3)	92L6
Leu79→Cys	2.4	31.3	−5.1	heat	(3)	92L6
Arg96→Cys	2.4	27.7	−8.7	heat	(3)	92L6
Arg96→Cys	5.3	58.4		heat	(3)	92L6
Arg119→Cys	2.4	33.4	−3.0	heat	(3)	92L6
Lys135→Cys	2.4	33.8	−2.6	heat	(3)	92L6
Thr142→Cys	2.4	35.9	−0.5	heat	(3)	92L6
Ala146→Cys	2.4	32.3	−4.1	heat	(3)	92L6
Ala146→Cys	5.3	58.3		heat	(3)	92L6
Asp159→Cys	2.4	37.1	+0.7	heat	(3)	92L6
Mixed disulfide:					(2,3)	
Asn55→Cys	2.4	31.4	−1.4	heat	(3)	92L6
Leu79→Cys	2.4	30.6	−0.7	heat	(3)	92L6
Arg96→Cys	2.4	25.6	−1.1	heat	(3)	92L6
Arg96→Cys	5.3	58.1	−0.3	heat	(3)	92L6
Arg119→Cys	2.4	33.9	+0.5	heat	(3)	92L6
Thr142→Cys	2.4	31.7	−4.2	heat	(3)	92L6
Lys135→Cys	2.4	34.1	−0.7	heat	(3)	92L6
Ala146→Cys	2.4	24.5	−7.8	heat	(3)	92L6
Ala146→Cys	5.3	54.9	−3.4	heat	(3)	92L6
Asp159→Cys	2.4	33.9	−3.2	heat	(3)	92L6

Remarks:
(1) reference protein for the cysteine-containing proteins
(2) ΔT is the difference in T_{trs} between the modified protein and the unmodified protein
(3) reproducibility of $T_{trs} = \pm0.2°C$

d) single and multiple amino acid replacements

Lysozyme phage T4, wild-type and mutants, alanine replacements within α-helix 126–134

Mutant	pH	T_{trs}	ΔCp	ΔH	Approach	Ref
wild-type	2.0	40.77±0.2		360	heat, v.H.	91Z2
Glu128→Ala	2.0	41.4		356	heat, v.H.	91Z2
Val131→Ala	2.0	41.8		371	heat, v.H.	91Z2
Leu133→Ala	2.02	40.75	10.0	222	heat, v.H.	92Z2
Double mutants:						
(Asp127→Ala and Glu128→Ala)						
	2.02	40.75	10.0	359	heat, v.H.	92Z2
(Glu128→Ala and Val131→Ala)						
	2.0	42.3		389	heat, v.H.	91Z2
(Val131→Ala and Asn132→Ala)						
	2.0	43.1		343	heat, v.H.	91Z2
Triple mutant						
(Glu128→Ala, Val121→Ala, and Asn132→Ala)						
	2.0	44.2		369	heat, v.H.	91Z2
Multiple mutants:						
(Asp127→Ala, Glu128→Ala, Val131→Ala and Asn132→Ala)						
	2.02	40.75	10.0	361	heat, v.H.	92Z2
(Glu128→Ala, Val131→Ala, Asn132→Ala and Leu133→Ala)						
	2.02	40.75	10.0	267	heat, v.H.	92Z2
(Asp127→Ala, Glu128→Ala, Val131→Ala, Asn132→Ala and Leu133→Ala)						
	2.02	40.75	10.0	260	heat, v.H.	92Z2

Lysozyme phage T4, core repacking variants

Mutant	pH	T_{trs}	ΔCp	ΔH	Approach	Remarks	Ref
wt*	5.4	65.3	10.46	552	heat	(1,2)	96B1
Leu121→Ala	5.4	59.2		460	heat	(1,2)	96B1
Ala129→Leu	5.4	62.1		469	heat	(1,2)	96B1
Ala129→Met	5.4	60.1		460	heat	(1,2)	96B1
Phe153→Ala	5.4	55.6		490	heat	(1,3)	93E2
double mutant (Leu121→Ala and Ala129→Leu)							
	5.4	62.5		494	heat	(1,2)	96B1
double mutant (Leu121→Ala and Ala129→Met)							
	5.4	62.6		596	heat	(1,2)	96B1
double mutant (Leu129→Met and Phe153→Ala)							
	5.4	52.6		393	heat	(1,2)	96B1

Remarks:
(1) wt* is the cysteine-free pseudo wild-type, reference protein
(2) buffer: 0.10 M sodium chloride, 1.4 mM acetic acid, 8.6 mM sodium acetate
(3) data from Ref. 93E2, measured in 0.2 M KCl at pH 5.7, T_{trs} of the wt* amounts to 64.9°C in the buffer and ΔH is 556 kJ/mol

Lysozyme phage T4, methionine substitutions

Mutant	pH	ΔT	ΔH	Approach	Remarks	Ref
pseudo wild-type	5.42	0.0	544	heat	(1–3)	96G2
Ile78→Met	5.42	–3.7	490	heat	(1–3)	96G2
Leu84→Met	5.42	–4.9	460	heat	(1–3)	96G2
Leu91→Met	5.42	–2.0	523	heat	(1–3)	96G2
Leu99→Met	5.42	–1.3	561	heat	(1–4)	96G2
Ile100→Met	5.42	–4.5	523	heat	(1–3)	96G2
Val103→Met	5.42	–3.1	490	heat	(1–3)	96G2
Leu118→Met	5.42	–1.8	544	heat	(1–3)	96G2
Leu121→Met	5.42	–2.1	540	heat	(1–3)	96G2
Leu133→Met	5.42	–1.0	536	heat	(1–3)	96G2
Phe153→Met	5.42	–1.6	536	heat	(1–4)	96G2
multiple mutant including 7 Met replacements						
	5.42	–14.5	402	heat	(1–3,5)	96G2
multiple mutant including 10 Met replacements						
	5.42	–25	176	heat	(1–3,6)	96G2

Remarks:

(1) reference protein is the cysteine-free pseudo wild-type (Cys54→Thr and Cys97→Ala) having
 T_{trs} = 65.3°C
(2) buffer: 1.4 mM acetic acid, 8.6 mM sodium acetate, 0.1 M sodium chloride
(3) estimated error in T_{trs} ±0.2° for single mutants and ±0.5° for multiple mutant, and in ΔH ±20 kJ/mol
(4) see also Ref. 93E2
(5) Leu84→Met, Leu91→Met, Leu99→Met, Leu118→Met, Leu121→Met, Leu133→Met, and Phe153→Met
(6) Ile78→Met, Leu84→Met, Leu91→Met, Leu99→Met, Ile100→Met, Val103→Met, Leu118→Met,
 Leu121→Met, Leu133→Met, and Phe153→Met

Lysozyme phage T4, wild-type and mutants

Mutant	pH	T_{trs}	ΔCp	ΔH	Approach	Ref
wild-type	2.0	40.4		414	heat, v.H.	91D1
wild-type	3.0	52.6		515	heat, v.H.	91D1
wild-type	4.0	63.6		598	heat, v.H.	91D1
wild-type	5.5	66.7		594	heat, v.H.	91D1
wild-type	6.5	64.6		523	heat, v.H.	91D1
Lys60→His	2.0	35.1		314	heat, v.H.	91D1
Lys60→His	6.5	62.6		481	heat, v.H.	91D1
Lys83→His	2.0	36.3		335	heat, v.H.	91D1
Lys83→His	6.5	62.0		510	heat, v.H.	91D1
Ser90→His	2.0	34.4		301	heat, v.H.	91D1
Ser90→His	5.5	62.7		536	heat, v.H.	91D1
Ser90→His	6.5	60.1		490	heat, v.H.	91D1
Thr115→Glu	2.0	38.7		351	heat, v.H.	91D1
Thr115→Glu	3.0	52.0		490	heat, v.H.	91D1
Thr115→Glu	4.0	63.3		548	heat, v.H.	91D1
Thr115→Glu	5.5	66.8		527	heat, v.H.	91D1
Thr115→Glu	6.5	65.3		523	heat, v.H.	91D1
Gln123→Glu	2.0	41.4		343	heat, v.H.	91D1
Gln123→Glu	5.5	66.9		473	heat, v.H.	91D1
Gln123→Glu	6.5	65.8		552	heat, v.H.	91D1
Asn144→Glu	2.0	39.4		347	heat, v.H.	91D1
Asn144→Glu	5.5	66.8		569	heat, v.H.	91D1
Asn144→Glu	6.5	64.5		577	heat, v.H.	91D1

Lysozyme phage T4, wild-type and mutants (continued)

Mutant	pH	T_{trs}	ΔCp	ΔH	Approach	Ref
double mutant (Lys60→His and Leu13→Asp)						
	2.0	32.4		297	heat, v.H.	91D1
	5.5	58.3		439	heat, v.H.	91D1
	6.5	55.9		343	heat, v.H.	91D1
double mutant (Lys83→His and Ala112→Asp)						
	2.0	33.7		305	heat, v.H.	91D1
	5.5	62.0		510	heat, v.H.	91D1
	6.5	59.1		498	heat, v.H.	91D1
double mutant (Ser90→His and Gln122→Asp)						
	2.0	32.7		280	heat, v.H.	91D1
	5.5	61.3		536	heat, v.H.	91D1
	6.5	57.3		485	heat, v.H.	91D1
double mutant (Thr115→Glu and Lys83→Met)						
	2.0	38.9		360	heat, v.H.	91D1
	6.5	64.7		519	heat, v.H.	91D1
double mutant (Asn144→Glu and Lys147→Met)						
	2.0	39.6		360	heat, v.H.	91D1
	5.5	66.6		582	heat, v.H.	91D1
	6.5	64.2		561	heat, v.H.	91D1
double mutant (Cys54→Thr and Cys97→Ala), called wt* in Ref. 91D1						
	2.0	28.5		322	heat, v.H.	91D1
	5.5	65.4		552	heat, v.H.	91D1
	6.5	63.0		548	heat, v.H.	91D1

Lysozyme phage T4, wild-type and mutants

Mutant	pH	T_{trs}	ΔCp	ΔH	Approach	Remarks	Ref
wild-type	3.0	54.1	10.1	547	heat	(1,2)	91D4
Ala98→Val	3.0	39.3		107	heat	(2)	91D4
Val149→Cys	3.0	49.0		459	heat	(2)	91D4
Thr152→Ser	3.0	47.5		446	heat	(2)	91D4
double mutant (Ala98→Val and Thr152→Ser)							
	3.0	40.2		163	heat	(2)	91D4
triple mutant (Ala98→Val, Val149→Cys and Thr152→Ser)							
	3.0	42.0		311	heat	(2)	91D4
triple mutant (Ala98→Val, Val149→Ile and Thr152→Ser)							
	3.0	42.1		290	heat	(2)	91D4

Remarks:
(1) ΔCp was taken from Ref. 89K4
(2) buffer: 20 mM potassium phosphate + 25 mM KCl

Lysozyme phage T4, wild-type, pseudo wild-type (wt*) and mutants

Mutant	pH	T_{trs}	ΔCp	ΔH	Approach	Remarks	Ref
wt*	3.01	51.8	10.5	497	heat	(1,2)	92E3
wild-type	3.01	53.5	10.5	544	heat	(2)	92E3
Leu46→Ala	3.01	43.2	10.5	376	heat	(2,3)	92E3
Leu99→Ala	3.01	36.1	10.5	330	heat	(2,3)	92E3
Leu118→Ala	3.01	39.6	10.5	316	heat	(2,3)	92E3
Leu121→Ala	3.01	42.5	10.5	339	heat	(2,3)	92E3
Leu133→Ala	3.01	42.9	10.5	383	heat	(2,4)	92E3
Phe153→Ala	3.01	39.5	10.5	313	heat	(2,3)	92E3
(Leu99→Ala and Phe153→Ala) of wt*:							
	3.01	10.0	10.5	25	heat	(2,3)	92E3

Remarks:
(1) wt* = (Cys54→Thr and Cys97→Ala)
(2) ΔCp is an estimated value
(3) the mutant protein was constructed using the gene for the pseudo wild-type lysozyme T4 (wt*).
(4) the mutant protein was constructed using the gene of the wild-type lysozyme

Lysozyme phage T4, pseudo wild-type (wt*) and mutants

Mutant	pH	T_{trs}	ΔCp	ΔH	Approach	Remarks	Ref
wt*	3.01	51.76	10.5	498	heat	(1)	93E2
Leu99→Ala	3.01	36.1	10.5	335	heat	(1)	93E2
Leu99→Ile	3.01	47.8	10.5	448	heat	(1)	93E2
Leu99→Met	3.01	49.8	10.5	485	heat	(1)	93E2
Leu99→Phe	3.01	50.7	10.5	456	heat	(1)	93E2
Leu99→Val	3.01	45.2	10.5	423	heat	(1)	93E2
Phe153→Ala	3.01	39.5	10.5	314	heat	(1)	93E2
Phe153→Ile	3.01	50.3	10.5	448	heat	(1)	93E2
Phe153→Leu	3.01	52.4	10.5	481	heat	(1)	93E2
Phe153→Met	3.01	49.4	10.5	448	heat	(1)	93E2
Phe153→Val	3.01	45.7	10.5	356	heat	(1)	93E2
double mutant (Leu99→Ala and Phe153→Ala)							
	3.01	10.0	10.5	25	heat	(1)	93E2
wt*	5.71	64.88	14.6	556	heat	(1)	93E2
Leu99→Ala	5.7	53.5	14.6	460	heat	(1)	93E2
Leu99→Ile	5.7	61.2	14.6	531	heat	(1)	93E2
Leu99→Met	5.7	63.4	14.6	586	heat	(1)	93E2
Leu99→Phe	5.7	64.2	14.6	527	heat	(1)	93E2
Leu99→Val	5.7	59.7	14.6	506	heat	(1)	93E2
Phe153→Ala	5.7	55.6	14.6	490	heat	(1)	93E2
Phe153→Ile	5.7	64.4	14.6	536	heat	(1)	93E2
Phe153→Leu	5.7	65.7	14.6	565	heat	(1)	93E2
Phe153→Met	5.7	63.3	14.6	552	heat	(1)	93E2
Phe153→Val	5.7	60.4	14.6	515	heat	(1)	93E2
double mutant (Leu99→Ala and Phe153→Ala)							
	5.7	42.1	14.6	234	heat	(1)	93E2

Remark:
(1) reference protein is the cysteine-free pseudo wild-type (Cys54→Thr and Cys97→Ala)

Lysozyme phage T4, pseudo wild-type (wt*) and mutants

Mutant	pH	T_{trs}	ΔCp	ΔH	Approach	Remarks	Ref
wt*	5.4	65.12±0.2	14.6	561	heat	(1)	93P9
Arg14→Lys	5.4	65.04±0.3		565±21	heat	(1)	93P9
Glu22→Lys	5.4	66.49±0.3		590±21	heat	(1)	93P9
Thr26→Ser	5.4	66.47±0.3		598±21	heat	(1)	93P9
Asn40→Asp	5.4	66.26±0.3		552±21	heat	(1)	93P9
Ala41→Asp	5.4	65.83±0.3		577±21	heat	(1)	93P9
Arg80→Lys	5.4	64.69±0.3		569±21	heat	(1)	93P9
double mutant (Arg80→Lys and Arg119→His):							
	5.4	63.93±0.3		556±21	heat	(1)	93P9
Ala93→Thr	5.4	65.25±0.3		577±21	heat	(1)	93P9
Gly113→Glu	5.4	65.91±0.3		527±21	heat	(1)	93P9
Arg119→His	5.4	64.38±0.3		552±21	heat	(1)	93P9
Thr151→Ser	5.4	66.05±0.3		590±21	heat	(1)	93P9
Phe153→Leu	5.4	66.00±0.3		573±21	heat	(1)	93P9
Asn163→Asp	5.4	64.62±0.3		590±21	heat	(1)	93P9

Remark:
(1) buffer: 10 mM sodium acetate, 0.1 M NaCl

Lysozyme phage T4, combination of stability enhancing point mutations

Mutant	pH	T_{trs}	ΔCp	ΔH	Approach	Remarks	Ref
wild-type	3.0	53.42	7.53	531	heat	(1,2)	95Z1
	5.4	66.51	10.46	586	heat	(1,2)	95Z1
Ile3→Leu	3.0	55.53		552	heat	(1,2)	95Z1
	5.4	68.20		615	heat	(1,2)	95Z1
Ser38→Asp	3.0	53.88		515	heat	(1,2)	95Z1
	5.4	67.87		586	heat	(1,2)	95Z1
Ala41→Val	3.0	54.52		552	heat	(1,2)	95Z1
	5.4	67.21		590	heat	(1,2)	95Z1
Ala82→Pro	3.0	54.50		523	heat	(1,2)	95Z1
	5.4	67.85		594	heat	(1,2)	95Z1
Asn116→Asp	3.0	54.32		548	heat	(1,2)	95Z1
	5.4	67.87		582	heat	(1,2)	95Z1
Val131→Ala	3.0	54.14		527	heat	(1,2)	95Z1
	5.4	67.38		586	heat	(1,2)	95Z1
Asn144→Asp	3.0	54.36		523	heat	(1,2)	95Z1
	5.4	67.50		594	heat	(1,2)	95Z1
double mutant (Val131→Ala and Ala41→Val)							
	3.0	55.25		515	heat	(1,2)	95Z1
	5.4	68.02		586	heat	(1,2)	95Z1
double mutant (Ser38→Asp and Asn144→Asp)							
	3.0	54.77		498	heat	(1,2)	95Z1
	5.4	68.87		569	heat	(1,2)	95Z1
triple mutant (Ser38→Asp, Ala82→Pro and Asn144→Asp)							
	3.0	56.00		515	heat	(1,2)	95Z1
	5.4	70.26		577	heat	(1,2)	95Z1
multiple mutant (Ser38→Asp, Ala82→Pro, Asn144→Asp and Ile3→Leu)							
	3.0	58.29		577	heat	(1,2)	95Z1
	5.4	71.89		644	heat	(1,2)	95Z1
multiple mutant (Ser38→Asp, Ala82→Pro, Asn144→Asp, Ile3→Leu and Val131→Ala)							
	3.0	58.98		586	heat	(1,2)	95Z1
	5.4	72.55		628	heat	(1,2)	95Z1

Lysozyme phage T4, combination of stability enhancing point mutations (continued)

Mutant	pH	T_{trs}	ΔCp	ΔH	Approach	Remarks	Ref
multiple mutant (Ser38→Asp, Ala82→Pro, Asn144→Asp, Ile3→Leu, Val131→Ala and Ala41→Val)							
	3.0	59.92		582	heat	(1,2)	95Z1
	5.4	73.42		607	heat	(1,2)	95Z1
multiple mutant (Ser38→Asp, Ala82→Pro, Asn144→Asp, Ile3→Leu, Val131→Ala, Ala41→Val and Asn116→Asp)							
	3.0	60.68		582	heat	(1,2)	95Z1
	5.4	74.83		653	heat	(1,2)	95Z1

Remarks:
(1) buffer: 25 mM KCl, 3 mM H_3PO_4, 17 mM KH_2PO_4, pH 3.01, and 100 mM NaCl, 1.4 mM acetic acid, 8.6 mM sodium acetate, pH 5.42
(2) estimated errors in ΔT ±25°C and in ΔH ±21 kJ/mol

Lysozyme phage T4, substitutions in the substrate binding site

Mutant	pH	ΔT	ΔH	Approach	Remarks	Ref
wild-type	5.42	0.0	586	heat, v.H.	(1)	95S5
pseudo wild-type	5.42	0.0	548	heat, v.H.	(2,3)	95S5
Glu11→Phe	5.42	4.3	577	heat, v.H.	(4)	95S5
Glu11→Met	5.42	4.1	561	heat, v.H.	(4)	95S5
Glu11→Ala	5.42	2.6	573	heat, v.H.	(4)	95S5
Glu11→His	5.42	0.1	523	heat, v.H.	(5)	95S5
Glu11→Asn	5.42	−0.6	498	heat, v.H.	(5)	95S5
Asp20→Asn	5.42	3.1	582	heat, v.H.	(5)	95S5
Asp20→Thr	5.42	2.2	565	heat, v.H.	(5)	95S5
Asp20→Ser	5.42	1.6	540	heat, v.H.	(5)	95S5
Asp20→Ala	5.42	−0.6	531	heat, v.H.	(5)	95S5
Gly30→Ala	5.42	0.1	561	heat, v.H.	(4)	95S5
Gly30→Phe	5.42	−4.9	414	heat, v.H.	(4)	95S5
Ser117→Val	5.42	5.1	556	heat, v.H.	(4)	95S5
Ser117→Ile	5.42	4.2	536	heat, v.H.	(4)	95S5
Ser117→Phe	5.42	2.8	552	heat, v.H.	(4,6)	95S5
Asn132→Met	5.42	3.6	561	heat, v.H.	(4)	95S5
Asn132→Phe	5.42	3.3	561	heat, v.H.	(4)	95S5
Asn132→Ile	5.42	3.0	586	heat, v.H.	(4)	95S5
double mutant (Ser117→Ala and Asn132→Ile)						
	5.42	5.3	506	heat, v.H.	(5)	95S5
double mutant (Ser117→Ala and Asn132→Met)						
	5.42	4.7	502	heat, v.H.	(5)	95S5
double mutant (Ser117→Ile and Asn132→Met)						
	5.42	5.5	515	heat, v.H.	(4)	95S5
double mutant (Ser117→Ile and Asn132→Ile)						
	5.42	3.6	489	heat, v.H.	(4)	95S5

Remarks:
(1) wild-type, T_{trs} = 64.48°C
(2) pseudo wild-type, T_{trs} = 65.10°C
(3) $\Delta G_{(w.t.)} - \Delta G_{(wt*)}$ = −2.2 kJ/mol at 64.48°C, ΔCp was taken as 14.6 kJ/mol/K
(4) reference protein is the wild-type, see (1)
(5) reference protein is the pseudo wild-type, see (2,3)
(6) see also mutations concerning position 117 of lysozyme phage T4 and Ref. 93A2

e) Modified protein, newly engineered disulfide bonds, insertion of unnatural amino acids

Lysozyme phage T4, disulfide mutants of pseudo wild-type (wt*)

Mutant	pH	ΔT	ΔCp	ΔH	Approach	Remarks	Ref
Double mutant (Ile3→Cys and Cys54→Thr):							
reduced	2	−1.9	11.7	283	heat	(2)	89M3
oxidized	2	4.8	11.7	302	heat	(2)	89M3
Double mutant (Ile9→Cys and Leu164→Cys):							
reduced	2	−6.5	11.7	235	heat	(1)	89M3
oxidized	2	6.4	11.7	310	heat	(1)	89M3
Double mutant (Thr21→Cys and Thr142→Cys):							
reduced	2	−2.7	11.7	215	heat	(1)	89M3
oxidized	2	11.0	11.7	280	heat	(1)	89M3
Double mutant (Ser90→Cys and Gln122→Cys):							
reduced	2	−5.8	8.4	259	heat	(1)	89M3
oxidized	2	−0.5	8.4	274	heat	(1)	89M3
Double mutant (Asp127→Cys and Arg154→Cys):							
reduced	2	−5.4	8.4	231	heat	(1)	89M3
oxidized	2	−2.4	8.4	144	heat	(1)	89M3

Remark:
(1) ΔT and Δ(ΔG) refer to the pseudo wild-type Cys54→Thr and Cys97→Ala (wt*) which melts at at the same temperature, T_{trs} = 41.9°C at pH 2.0, as the wild-type
(2) ΔT and Δ(ΔG) refer to wild-type

Lysozyme phage T4 with newly engineered disulfide bonds, heat and cold denaturation

Mutant/Form	pH	T_{trs}	ΔCp	ΔH	Approach	Remarks	Ref
double mutant (Ile3→Cys and Cys54→Thr21) with disulfide linkage Cys3→Cys97							
	5	−2.7	9.12	−137	GuHCl	(1)	89C
	5	28	9.12	145	GuHCl	(2)	89C

Remarks:
(1) cold denaturation in the presence of 3 M GuHCl
(2) heat denaturation in the presence of 3 M GuHCl

Lysozyme phage T4, mutant (Ala46→Cys), modified with cystamine (Ala46-SSCH$_2$CH$_2$NH$_3^+$)

pH	T_{trs}	ΔCp	ΔH	Approach	Remarks	Ref
2.44	24.5		197	heat	(1)	92L6
2.85	33.5		284	heat	(1)	92L6
5.13	55.7		477	heat	(2)	92L6
5.32	54.9		432	heat	(1)	92L6

Remarks:
(1) buffer: 20 mM potassium phosphate, 25 mM KCl
(2) buffer: 100 mM potassium acetate

Lysozyme phage T4, mutants with unnatural amino acids

Mutant	T_{trs}	ΔCp	ΔH	Approach	Remarks	Ref
wild-type	43.5±0.2	10.5	401±17	heat	(1)	92M3
Leu133→[1]	47.8±0.3	10.5	410±8	heat	(1,2)	92M3
Leu133→[2]	45.4±0.2	10.5	431±67	heat	(1,3)	92M3
Leu133→[3]	39.5±0.3	10.5	331±17	heat	(1,4)	92M3
Leu133→[4]	32.9±0.5	10.5	264±54	heat	(1,5)	92M3
Leu133→[5]	~31			heat	(6)	92M3

Remarks:
(1) van't Hoff treatment, ΔCp is an estimated value
(2) mutant [1]: Leu133→S-2-amino-3-cyclopentylpropanoic acid
(3) mutant [2]: Leu133→S,S-2-amino-4-methyl hexanoic acid
(4) mutant [3]: Leu133→norvaline
(5) mutant [4]: Leu133→O-methyl serine
(6) mutant [5]: Leu133→ethylglycine

Lysozyme phage T4, spin-labeled mutants

Mutant	pH	ΔT	ΔH	Approach	Remarks	Ref
Asp22→R1	2.95	0.0	418	heat	(1–3)	96M4
Asn40→R1	2.95	−1.2	418	heat	(1–3)	96M4
Ser44→R1	2.95	0.7	464	heat	(1–3)	96M4
Lys48→R1	2.95	−3.2	393	heat	(1–3)	96M4
Gly51→R1	2.95	−2.5	351	heat	(1–3)	96M4
Asp61→R1	2.95	−1.0	377	heat	(1–3)	96M4
Lys65→R1	2.95	−1.0	481	heat	(1–3)	96M4
Asp72→R1	2.95	0.4	397	heat	(1–3)	96M4
Ala74→R1	2.95	0.0	397	heat	(1–3)	96M4
Val75→R1	2.95	−3.2	402	heat	(1–3)	96M4
Arg80→R1	2.95	−4.5	335	heat	(1–3)	96M4
Asn81→R1	2.95	−2.9	393	heat	(1–3)	96M4
Val87→R1	2.95	−11.0	259	heat	(1–3)	96M4
Leu99→R1	2.95	−13.3	268	heat	(1–3)	96M4
Ala130→R1	2.95	−3.1	351	heat	(1–3)	96M4
Val131→R1	2.95	0.5	418	heat	(1–3)	96M4
Asn132→R1	2.95	1.5	490	heat	(1–3)	96M4
Leu133→R1	2.95	−9.5	305	heat	(1–3)	96M4
Lys135→R1	2.95	−1.8	343	heat	(1–3)	96M4
Ile150→R1	2.95	−4.8	356	heat	(1–3)	96M4
Phe153→R1	2.95	−13.0	238	heat	(1–3)	96M4
double mutant (Lys65→R1 and Asn68→Ala)						
	2.95	−2.0	406	heat	(1–3)	96M4
double mutant (Lys65→R1 and Asn68→Gly)						
	2.95	−4.0	381	heat	(1–3)	96M4
multiple mutant (Asp72→R1, Asn68→Ala, Gln69→Ala and Arg76→Ala)						
	2.95	−2.2	381	heat	(1–3)	96M4

Remarks:
(1) reference protein is the pseudo wild-type (Cys54→Thr and Cys97→Ala)
(2) R1 = the corresponding position of the protein was replaced by cysteine and labeled with a sulfhydryl-specific nitroxide reagent
(3) measured in 20 mM potassium phosphate and 25 mM KCl, the transition was monitored by CD at 223 nm

f) Effect of benzene on protein stability

Lysozyme phage T4, pseudo wild-type (Cys54→Thr AND Cys97→Ala) and mutants, effect of benzene on the stability

Benzene concentration	pH	T_{trs}	ΔCp	ΔH	Approach	Remarks	Ref
Pseudo wild-type (Cys54→Thr and Cys97→Ala):							
0.0 mM	3.01	51.6		498	heat	(1)	92E2
10.4 mM	3.01	51.5		474	heat	(1)	92E2
Mutant (Leu99→Ala):							
0.0 mM	3.01	36.2		345	heat	(1,2)	92E2
2.6 mM	3.01	40.7		407	heat	(1,2)	92E2
5.6 mM	3.01	41.6		421	heat	(1,2)	92E2
7.5 mM	3.01	42.2		413	heat	(1,2)	92E2
Mutant (Phe153→Ala):							
0.0 mM	3.01	39.5		303	heat	(1,2)	92E2
3.0 mM	3.01	39.5		300	heat	(1,2)	92E2
7.2 mM	3.01	39.4		295	heat	(1,2)	92E2
Double mutant (Leu99→Ala and Phe153→Ala):							
0.0 mM	4.46	41.7		190	heat	(1,2)	92E2
10.4 mM	4.46	44.7		266	heat	(1,2)	92E2

Remarks:
(1) measurements in 20 mM potassium phosphate with 25 mM KCl in the presence of varying amounts of benzene, reference state is the mutant in the presence of 0.0 mM benzene
(2) the mutant protein contains additionally the replacements (Cys54→Thr and Cys97→Ala) of the template (pseudo wild-type, wt*)

g) Insertion mutants, permuted protein

Lysozyme phage T4, insertion mutants

Mutant	pH	T_{trs}	ΔCp	ΔH	Approach	Remarks	Ref
(Asn40-Ala-Ala41)							
	5.4	57.3		406	heat	(1–3)	93H4
(Asn40→Leu and Leu40-Ala-Ala41)							
	5.4	61.3		510	heat	(1–3)	93H4
(Asn40-Ala-Ala-Ala41)							
	5.4	62.2		506	heat	(1–3)	93H4
(Asn40-Ala-Ala-Ala-Ala41)							
	5.4	60.4		485	heat	(1–3)	93H4
(Asn40-Ala-Ala-Ala-Ala-Ala41)							
	5.45	58.75		464	heat	(1–3)	94H7
(Asn40-Ser-Leu-Asp-Ala41)							
	5.4	63.2		531	heat	(1–3)	93H4
(Ser44-Ala-Glu45)							
	5.4	53.4		230	heat	(1–4)	93H4
(Ser44-Ala-Ala-Glu45)							
	5.4	55.3		339	heat	(1–3)	93H4
(Ser44→Ala and Ala44-Ala-Ala-Glu45)							
	5.4	60.8		477	heat	(1–3)	93H4
(Ser44-Ala-Ala-Ala-Glu45)							
	5.4	59.2		444	heat	(1–3)	93H4

Lysozyme phage T4, insertion mutants (continued)

Mutant	pH	T_{trs}	ΔCp	ΔH	Approach	Remarks	Ref
(Ser44-Ala-Ala-Ala-Glu45)							
	5.45	57.25		448	heat	(1–3)	94H7
(Lys48-Ala-Ala49)							
	5.4	53.8		272	heat	(1–4)	93H4
(Lys48-Ala-Ala-Ala49)							
	5.4	54.7		314	heat	(1–4)	93H4
(Lys48-Ala-Ala-Ala-Ala49)							
	5.45	51.45		226	heat	(1–3)	93H4
(Lys48-His-Pro-Ala49)							
	5.4	58.1		435	heat	(1–3)	93H4
(Lys48-Ala-Ala-Ala49)							
	5.4	50.4		218	heat	(1–4)	93H4
(Lys48-Ala-Ala-Ala-Ala49)							
	5.45	51.45		226	heat	(1–3)	94H7
(Lys48-Ala-Ala-Ala-Ala-Ala49)							
	5.45	52.65		222	heat	(1–3)	94H7
(Asn40→Leu, Lys43→Ala, and Ser44-Ala-Glu45)							
	5.45	59.75		498	heat	(1–3)	94H7
(Asn40-Glu-Ser-Ala41)							
	5.45	63.55		527	heat	(1–39	94H7
(Asn40-Ser-Leu-Asp-Ala41 and Leu46→Ala)							
	5.45	57.05		402	heat	(1–3)	94H7
(Ser44-Ala-Ala-Ala-Glu45 and Leu46→Ala)							
	5.45	53.95		293	heat	(1–3)	94H7
(Lys48-Leu-Pro-Ala49)							
	5.45	57.85		444	heat	(1–3)	94H7
(Asn40-Ala-Ala-Ala41 and Lys48-Leu-Pro-Ala49)							
	5.45	55.35		360	heat	(1–3)	94H7
(Leu39-Ala-Asn40)							
	5.45	62.65		531	heat	(1–3)	94H7
(Ala42-Lys-Lys43)							
	5.45	54.15		268	heat	(1–3)	94H7

Remarks:
(1) the data refer to the pseudo wild-type having T_{trs} = 65.15 °C and ΔH = 569 kJ/mol
(2) ΔCp of the wild-type amounts to 18.8 kJ/mol/K
(3) buffer: 10 mM sodium acetate with 0.1 M NaCl

Lysozyme phage T4, insertion, deletion and extension mutants

Mutant	pH	T_{trs}	ΔH	Approach	Remarks	Ref
Insertions:						
(Phe4-Ala-Glu5)	5.4	56.8	427	heat	(1–6)	96V4
(Arg8-Ala-Ile9)	5.4	46.1	234	heat	(1–6)	96V4
(Asn40-Ala-Ala41)	5.4	57.3	406	heat	(1–7)	96V4
(Ser44-Ala-Glu45)	5.4	53.4	230	heat	(1–7)	96V4
(Lys48-Ala-Ala49)	5.4	53.8	272	heat	(1–7)	96V4
(Glu64-Ala-Lys65)	5.4	54.7	385	heat	(1–6)	96V4
(Asn68-Ala-Gln69)	5.4	52.1	343	heat	(1–6)	96V4
(Ala73-Ala-Ala74)	5.4	52.8	431	heat	(1–6)	96V4
(Val75-Ala-Arg76)	5.4	57.4	481	heat	(1–6)	96V4
(Tyr88-Ala-Asp89)	5.4	53.9	439	heat	(1–6)	96V4
(Arg96-Ala-Ala97)	5.4	36.4	96	heat	(1–6)	96V4

Lysozyme phage T4, insertion, deletion and extension mutants (continued)

Mutant	pH	T_{trs}	ΔH	Approach	Remarks	Ref
(Glu108-Ala-Thr109)	5.4	58.1	439	heat	(1–6)	96V4
(Thr115-Ala-Asn116)	5.4	60.5	490	heat	(1–6)	96V4
(Arg119-Ala-Met120)	5.4	58.6	406	heat	(1–6)	96V4
(Asp127-Ala-Glu128)	5.4	55.1	406	heat	(1–6)	96V4
(Val131-Ala-Asn132)	5.4	57.3	435	heat	(1–6)	96V4
(Asn140-Ala-Gln141)	5.4	59.6	481	heat	(1–6)	96V4
(Asn144-Ala-Arg145)	5.4	63.2	531	heat	(1–6)	96V4
(Lys147-Ala-Arg148)	5.4	59.3	481	heat	(1–6)	96V4
(Arg148-Ala-Val149)	5.4	51.5	368	heat	(1–6)	96V4
(Ile150-Ala-Thr151)	5.4	40.9	142	heat	(1–6)	96V4
Deletions:						
Arg8 deleted	5.4	45.8	188	heat	(1–6)	96V4
Ser44 deleted	5.4	58.1	389	heat	(1–6)	96V4
Ala73 deleted	5.4	53.8	410	heat	(1–6)	96V4
Arg119 deleted	5.4	55.1	397	heat	(1–6)	96V4
Asp127 deleted	5.4	56.8	418	heat	(1–6)	96V4
Extensions:						
(Leu164-Ala-Ala-Ala-Ala)						
	5.4	65.8	548	heat	(1–6)	96V4

Remarks:
(1) the data refer to the cysteine-free pseudo wild-type (wt*, Cys54→Thr and Cys97→Ala)
(2) the thermal stabilities were determined in 0.10 M NaCl, 1.4 mM acetic acid, 8.6 mM sodium acetate, pH 5.42, and in 0.025 M KCl, 2.95 mM H_3PO_4, 17.0 mM KH_2PO_4, pH 3.01
(3) van't Hoff treatment of transitions monitored by CD at 223 nm
(4) ΔCp was taken as 10.46 kJ/mol/K at 53°C for pH 5.4 buffer, and as 7.53 kJ/mol/K at 42°C for pH 3.0 buffer according to Ref. 95Z1
(5) the error in T_{trs} is ±0.15°C near T_{trs} of wt* and increased to approximately ±0.35°C as the ΔH value decreased
(6) the error in ΔH is ±3% for the most stable mutants and the value increased to approximately ±18% for ΔH of 100 kJ/mol approximately the lowest ΔH that can be considered two-state in these data sets
(7) data from Refs. 93H4 and 94H7

Lysozyme phage T4, insertion mutants following Ala73 and Arg148

Mutant	pH	ΔT	ΔH	Approach	Remarks	Ref
(Ala73-Ala)	3.0	−19.7	238	heat	(1–6)	96V4
(Ala73-Ala)	5.4	−12.5	431	heat	(1–6)	96V4
(Ala73-Ala-Ala)	3.0	−29.2	105	heat	(1–6)	96V4
(Ala73-Ala-Ala)	5.4	−18.0	314	heat	(1–6)	96V4
(Ala73-Ala-Ala-Ala)	3.0	−21.6	132	heat	(1–6)	96V4
(Ala73-Ala-Ala-Ala)	5.4	−15.9	289	heat	(1–6)	96V4
(Ala73-Leu)	3.0	−14.2	310	heat	(1–6)	96V4
(Ala73-Leu)	5.4	−9.0	456	heat	(1–6)	96V4
(Ala73-Arg)	3.0	−17.6	285	heat	(1–6)	96V4
(Ala73-Arg)	5.4	−8.2	460	heat	(1–6)	96V4
(Ala73-Val-Leu)	3.0	−5.8	410	heat	(1–6)	96V4
(Ala73-Val-Leu)	5.4	−5.2	517	heat	(1–6)	96V4
(Arg148-Ala)	3.0	−14.4	226	heat	(1–6)	96V4
(Arg148-Ala)	5.4	−13.7	368	heat	(1–6)	96V4
(Arg148-Ala-Ala)	3.0	−19.5	144	heat	(1–6)	96V4
(Arg148-Ala-Ala)	5.4	−17.5	280	heat	(1–6)	96V4
(Arg148-Ala-Ala-Ala)	3.0	−20.0	130	heat	(1–6)	96V4
(Arg148-Ala-Ala-Ala)	5.4	−15.9	301	heat	(1–6)	96V4

Lysozyme phage T4, insertion mutants following Ala73 and Arg148 (continued)

Mutant	pH	ΔT	ΔH	Approach	Remarks	Ref
(Arg148-Ala-Ala-Ala-Ala)						
	3.0			heat	(7)	96V4
	5.4	–23.2	130	heat	(1–6)	96V4
(Arg148-Asp)	3.0	–9.1	293	heat	(1–6)	96V4
(Arg148-Asp)	5.4	–14.7	322	heat	(1–6)	96V4
(Arg148-Ser)	3.0	–17.0	180	heat	(1–6)	96V4
(Arg148-Ser)	5.4	–15.7	326	heat	(1–6)	96V4
(Arg148-Asp-Ser)	3.0	–20.7	100	heat	(1–6)	96V4
(Arg148-Asp-Ser)	5.4	–19.0	251	heat	(1–6)	96V4
(Arg148-Thr-Thr)	3.0	–19.9	88	heat	(1–6)	96V4
(Arg148-Thr-Thr)	5.4	–20.5	230	heat	(1–6)	96V4
(Arg148-Val-Pro)	3.0	–18.0	77	heat	(1–6)	96V4
(Arg148-Val-Pro)	5.4	–22.0	230	heat	(1–6)	96V4

Remarks:
(1) the data refer to the cysteine-free pseudo wild-type (wt*, Cys54→Thr and Cys97→Ala) having T_{trs} = 65.22°C at pH 5.4 and T_{trs} = 51.66°C at pH 3.0
(2) the thermal stabilities were determined in 0.10 M NaCl, 1.4 mM acetic acid, 8.6 mM sodium acetate, pH 5.42, and in 0.025 M KCl, 2.95 mM H_3PO_4, 17.0 mM KH_2PO_4, pH 3.01
(3) van't Hoff treatment of transitions monitored by CD at 223 nm
(4) ΔCp was taken as 10.46 kJ/mol/K at 53°C for pH 5.4 buffer, and as 7.53 kJ/mol/K at 42°C for pH 3.0 buffer according to Ref. 95Z1
(5) the error in T_{trs} is ±0.15°C near T_{trs} of wt* and increased to approximately ±0.35°C as the ΔH value decreased
(6) the error in ΔH is ±3% for the most stable mutants and the value increased to approximately ±18% for ΔH of 100 kJ/mol approximately the lowest ΔH that can be considered two-state in these data sets
(7) not two-state

Lysozyme phage T4, pseudo wild-type (wt*), extended, and permuted form

Mutant	pH	T_{trs}	ΔCp	ΔH	Approach	Remarks	Ref
wt*	2.5	45.0		475	DSC	(1,2)	93Z3
extended	2.5	44.0±0.6		351±21	DSC	(2,3)	93Z3
permuted	2.5	40.1±0.7		285±17	DSC	(2,4)	93Z3

Remarks:
(1) data from Ref. 89K4
(2) buffer: 20 mM potassium phosphate, 25 mM KCl, 0.1 mM DTT
(3) the protein contains at the C-terminus additionally Ser-4*(Gly)-Ala
(4) the permuted protein begins at residue 37 of the pseudo wild-type and ends at residue 36; the chain termini are joined by the linker peptide Ser-4*(Gly)-Ala; additional amino acid replacements are: Pro37→Met, Ser38→Asp, and Asn2→Asp

Lysozyme phage T4, extended (ext.) and circularly permuted (per.) mutants, oxidized (ox.) and reduced (red.) form, data for pH 2.5

Disulfides	Template	Form	N	T_{trs}	ΔC_p	ΔH	Approach	Remarks	Ref
Cys9-Cys164	ext.	ox.	155	55.7	8.8±0.7	368±10	heat	(1,2)	94Z5
Cys9-Cys164	ext.	red.		41.8	7.5±0.4	276±12	heat	(1,2)	94Z5
Cys9-Cys164	per.	ox.	15	49.0	6.7±0.4	188±2	heat	(1,2)	94Z5
Cys9-Cys164	per.	red.		38.1	7.1±1.0	172±5	heat	(1,2)	94Z5
Cys21-Cys142	ext.	ox.	121	59.7	7.9±0.4	431±4	heat	(1,2)	94Z5
Cys21-Cys142	ext.	red.		47.7	8.8±1.0	280±8	heat	(1,2)	94Z5
Cys21-Cys142	per.	ox.	49	50.0	6.7±0.6	276±5	heat	(1,2)	94Z5
Cys21-Cys142	per.	red.		44.4	6.3±1.0	184±3	heat	(1,2)	94Z5
Cys3-Cys97	ext.	ox.	94	54.0	10.9±0.7	322±5	heat	(1,2)	94Z5
Cys3-Cys97	ext.	red.		45.4	10.5±0.4	326±3	heat	(1,2)	94Z5
Cys3-Cys97	per.	ox.	76	55.3	2.1±1.0	188±5	heat	(1,2)	94Z5
Cys3-Cys97	per.	red.		42.5	11.3±1.0	238±14	heat	(1,2)	94Z5

Remarks:
(1) T_{trs} (error ±0.3°C) is given for pH 2.5, thermal stabilities were measured at five pH values (1.70, 2.00, 2.20, 2.50, and 2.70)
(2) N is the length of the loops made by disulfides in the extended and permuted proteins

h) Mutants that unfold in a three-state process

Lysozyme phage T4, pseudo wild-type and mutants that unfold in a three-state process

Mutant	pH	T_{trs}	ΔC_p	ΔH	Approach	Remarks	Ref
wt*	2.40	41.5	7.1	431	DSC	(1–3)	96C3
wt*	2.60	45.4	7.7	468	DSC	(1–3)	96C3
wt*	2.80	49.6	8.0	503	DSC	(1–3)	96C3
wt*	3.00	53.9	6.1	533	DSC	(1–3)	96C3
wt*	3.30	59.2	5.7	561	DSC	(1–3)	96C3
wt*	3.50	61.9	9.8	571	DSC	(1–3)	96C3
wt*	3.70	64.4	9.7	581	DSC	(1–3)	96C3
wt*	2.4–3.7	41.5–64.4	7.7		DSC	(1,6)	96C3
wt*	2.4–3.7	41.5–64.4	6.5		DSC	(1,7)	96C3
Ser44→Ala							
total	3.00	38.2	6.7	302	DSC	(2–4)	96C3
trans. (1)	3.00	31.9		108	DSC	(5)	96C3
trans. (2)	3.00	38.5		225	DSC	(5)	96C3
Ser44→Ala							
total	3.20	43.9	2.7	373	DSC	(2–4)	96C3
trans. (1)	3.20	41.3		160	DSC	(5)	96C3
trans. (2)	3.20	44.3		222	DSC	(5)	96C3
Ser44→Ala							
total	3.30	44.9	7.1	376	DSC	(2–4)	96C3
trans. (1)	3.30	39.6		129	DSC	(5)	96C3
trans. (2)	3.30	45.3		251	DSC	(5)	96C3
Ser44→Ala							
total	3.60	49.6	7.1	400	DSC	(2–4)	96C3
trans. (1)	3.60	45.3		148	DSC	(5)	96C3
trans. (2)	3.60	50.1		259	DSC	(5)	96C3

Lysozyme phage T4, pseudo wild-type and mutants that unfold in a three-state process (continued)

Mutant	pH	T_{trs}	ΔC_p	ΔH	Approach	Remarks	Ref
Ser44→Ala							
total	3.70	50.7	5.5	378	DSC	(2–4)	96C3
trans. (1)	3.70	45.4		120	DSC	(5)	96C3
trans. (2)	3.70	49.6		271	DSC	(5)	96C3
Ser44→Ala							
total	3.70	50.7	10.5	379	DSC	(2–4)	96C3
trans. (1)	3.70	44.0		118	DSC	(5)	96C3
trans. (2)	3.70	51.0		274	DSC	(5)	96C3
Ser44→Ala							
total	4.00	54.1	7.1	442	DSC	(2–4)	96C3
trans. (1)	4.00	50.4		170	DSC	(5)	96C3
trans. (2)	4.00	54.7		279	DSC	(5)	96C3
Ser44→Ala							
total	3.0–4.0	38–54	7.2		DSC	(6)	96C3
trans. (1)	3.0–4.0	32–50	2.5		DSC	(8)	96C3
trans. (2)	3.0–4.0	38–55	3.9		DSC	(8)	96C3
total	3.0–4.0		6.4		DSC	(9)	96C3
Ala42→Lys							
total	3.00	38.6	6.7	301	DSC	(2–4)	96C3
trans. (1)	3.00	34.4		102	DSC	(5)	96C3
trans. (2)	3.00	38.1		219	DSC	(5)	96C3
Ala42→Lys							
total	3.20	42.9	6.5	331	DSC	(2–4)	96C3
trans. (1)	3.20	35.7		107	DSC	(5)	96C3
trans. (2)	3.20	42.9		255	DSC	(5)	96C3
Ala42→Lys							
total	3.30	44.7	6.7	354	DSC	(2–4)	96C3
trans. (1)	3.30	39.6		120	DSC	(5)	96C3
trans. (2)	3.30	44.7		252	DSC	(5)	96C3
Ala42→Lys							
total	3.50	47.7	8.3	383	DSC	(2–4)	96C3
trans. (1)	3.50	41.2		123	DSC	(5)	96C3
trans. (2)	3.50	47.9		269	DSC	(5)	96C3
Ala42→Lys							
total	3.60	49.9	7.1	372	DSC	(2–4)	96C3
trans. (1)	3.60	44.1		109	DSC	(5)	96C3
trans. (2)	3.60	49.7		271	DSC	(5)	96C3
Ala42→Lys							
total	3.70	51.2	7.0	409	DSC	(2–4)	96C3
trans. (1)	3.70	45.5		123	DSC	(5)	96C3
trans. (2)	3.70	51.1		278	DSC	(5)	96C3
Ala42→Lys							
total	4.00	54.2	7.0	440	DSC	(2–4)	96C3
trans. (1)	4.00	50.4		158	DSC	(5)	96C3
trans. (2)	4.00	54.5		280	DSC	(5)	96C3
Ala42→Lys							
total	3.0–4.0	38–54	7.2		DSC	(6)	96C3
trans. (1)	3.0–4.0	34–50	2.8		DSC	(8)	96C3
trans. (2)	3.0–4.0	38–55	3.6		DSC	(8)	96C3
total	3.0–4.0		6.4		DSC	(9)	96C3

Lysozyme phage T4, pseudo wild-type and mutants that unfold in a three-state process (continued)

Mutant	pH	T_{trs}	ΔCp	ΔH	Approach	Remarks	Ref
Lys48→Ala							
total	3.00	41.4	7.9	390	DSC	(2–4)	96C3
trans. (1)	3.00	44.5		187	DSC	(5)	96C3
trans. (2)	3.00	38.7		216	DSC	(5)	96C3
Lys48→Ala							
total	3.50	49.7	8.2	436	DSC	(2–4)	96C3
trans. (1)	3.50	51.4		170	DSC	(5)	96C3
trans. (2)	3.50	47.9		267	DSC	(5)	96C3
Lys48→Ala							
total	4.00	54.4	7.3	430	DSC	(2–4)	96C3
trans. (1)	4.00	55.0		173	DSC	(5)	96C3
trans. (2)	4.00	53.2		271	DSC	(5)	96C3

Remarks:
(1) pseudo wild-type, wt* (Cys54-Thr and Cys97→Ala)
(2) calorimetric enthalpy change
(3) T_{trs} according to a two-state fit, error is ±0.5°C
(4) ΔH is the total heat, experimental value
(5) result of deconvolution
(6) average ΔCp from the individual scans
(7) ΔCp from ΔH versus T_{trs}
(8) ΔCp from ΔH of transition trans. (1) or trans. (2) versus T_{trs}
(9) sum of ΔCp from transitions trans. (1) and trans. (2)

Malate Dehydrogenase

Malate dehydrogenase, wild-type and mutant

Mutant	pH	T_{trs}	ΔH	Approach	Remark	Ref
wild-type	7.2	59.0	430	DSC		94G2
Arg102→Gln	7.2	60.6	440	DSC	(1)	94G2

Remark:
(1) the increase in T_{trs} remains constant over a pH range from 3.6 to 7

Mannose Transporter

Mannose transporter from *E. coli*, hydrophilic subunit, IIA and IIB domain

Mutant	pH	T_{trs}	ΔH	Approach	Ref
IIA domain	7.4	94	565	DSC	94M5
IIB domain	7.4	56		DSC	94M5

Melittin

Melittin, temperature dependence of the heat capacity change

pH	T_{ref}	ΔCp	ΔH	Approach	Remarks	Ref
4.5	80	3.15	332	heat	(1,2)	94H3
4.5	80	3.14	311	heat	(1,3)	94H3
4.5	80	1.73	266	DSC	(1,4)	94H3
4.5	80	1.63	263	DSC	(1,5)	94H3

Remarks:
(1) the temperature dependence of ΔCp can be calculated by $\Delta Cp(T) = \Delta Cp(T_{ref}) + a \times (T-T_{ref}) + b \times (T-T_{ref})^2$
(2) measured by CD varying salt concentration $a = -6.44 \times 10^{-2}$, $b = -5.98 \times 10^{-4}$
(3) measured by CD varying peptide concentration $a = -6.45 \times 10^{-2}$, $b = -6.00 \times 10^{-4}$
(3) measured by DSC varying salt concentration $a = -0.13$, $b = -7.24 \times 10^{-4}$
(4) measured by DSC varying peptide concentration $a = -0.15$, $b = -9.25 \times 10^{-4}$

Methionine Repressor Protein

Methionine repressor protein (MetJ) from *E. coli*

pH	T_{trs}	ΔCp	ΔH	Approach	Remarks	Ref
7.0	53.2±0.2	8.94±0.66	505±28	DSC	(1,2)	92J1

Remarks:
(1) buffer: 25 mM potassium phosphate, 100 mM KCl
(2) temperature function of $\Delta H = 48+8.94(T'_{trs}-273)$ with T'_{trs} in Kelvin, further data for the dependence on ionic strength in Ref. 92J1

Methionine repressor protein (MetJ) from *E. coli*, temperature dependence of the heat capacity change at protein unfolding in aqueous solution, and related thermodynamic quantities

T	$\Delta H(exp)$	$\Delta S(exp)$	$\Delta Cp(exp)$	Approach	Remarks	Ref
5	92		8.0	DSC	(1–3)	95M3
25	270		9.0	DSC	(1–3)	95M3
50	498		8.9	DSC	(1–3)	95M3
75	692		7.0	DSC	(1–3)	95M3
100	832		5.1	DSC	(1–3)	95M3
125	902		2.1	DSC	(1–3)	95M3

Remarks:
(1) the data were taken from Ref. 95M3 with reference to Ref. 92J1
(2) ΔCp was obtained as described in Ref. 90P3
(3) ΔCp in kJ/mol/K, ΔH in kJ/mol and ΔS in J/mol/K

Methionine repressor (MetJ)

pH	T_{trs}	ΔCp	ΔH	Approach	Remarks	Ref
7	59.4		750	DSC	(1,2)	94C10

Remarks:
(1) bound to DNA 16-mer
(2) buffer: 25 mM phosphate, 0.1 M KCl, 1 mM DTT, pH 7

Myb Oncoprotein

DNA-binding domain of the *c-myb* protooncogene product

Form/Trans.	pH	T_{trs}	ΔCp	ΔH	Approach	Remarks	Ref
myb, free form:							
N→I	7.5	42.7±0.3	3.5±0.4	175±3	DSC		93S4
N→D	7.5	51.2±0.1	0.21±0.04	393±5	DSC		93S4
myb-DNA complex:							
N→I	7.5	52.5±0.9	−4.2±0.8	123±4	DSC	(1)	93S4
N→D	7.5	55.7±0.1	4.3±1.5	553±6	DSC	(1)	93S4

Remark:
(1) after subtraction of the transition curve of the excess DNA heat capacity profile

Myoglobin

Myoglobin from various species, different forms

Species/ Form	pH	T_{trs}	ΔCp	ΔH	Approach	Remarks	Ref
armadillo	8	78.8	6.0±1.3	417±24	DSC		90K2
carp	8	68.1	7.6±0.6	370±19	DSC		90K2
chicken	8	25	9.9±0.7	133±7	heat	(1)	85H
horse	8	81.5	7.8±0.5	548±20	DSC		90K2
opossum	8	80.3	5.6±0.9	435±18	DSC		90K2
raccoon	8	82.1	6.8±0.5	465±16	DSC		90K2
rat	8	83.8	6.2±1.0	473±27	DSC		90K2
sperm whale cyanide complex							
	9.5	82.2	11.7	776	heat	(2)	93P7
	9.6		11.7	702	heat	(2)	93L5
	10.5	78.5	11.6	628	DSC		79P4
	10.5	78.5	11.6	628	DSC		74P1
	11	72	11.3	561	DSC		72A
	11	72	11.3	561	DSC		71P
sperm whale apo form	5	61	6.5	222	DSC		88G4

Remarks:
(1) van't Hoff treatment, measurement in the presence of 1 M GuHCl
(2) recombinant protein, measured in the presence of KCN

480   Table 2. Enthalpy and Heat Capacity Changes – Molar Values

Myoglobin, temperature dependence of heat capacity change at protein unfolding

T	$\Delta Cp_{(calc)}$	$\Delta Cp_{(exp)}$	Approach	Remarks	Ref
5	14.4	10.7	DSC	(1–3)	90P3
25	14.0	11.3	DSC	(1–3)	90P3
50	12.8	10.2	DSC	(1–3)	90P3
75	10.3	9.5	DSC	(1–3)	90P3
100	7.9	6.7	DSC	(1–3)	90P3
125	4.3	3.5	DSC	(1–3)	90P3

Remarks:
(1) the calculated value is based on heat capacity of the constituent amino acids minus experimental value for the native protein
(2) the experimental value is based on heat capacity of the heat denatured protein minus experimental value for the native protein
(3) the partial molar heat capacity of hemine is included into the heat capacity change

Myoglobin, temperature dependence of the heat capacity change at protein unfolding in aqueous solution, and related thermodynamic quantities

T	$\Delta H(exp)$	$\Delta S(exp)$	$\Delta Cp(exp)$	Approach	Remarks	Ref
5	−231	−919	14.4	DSC	(1–3)	95M3
25	6	−116	14.0	DSC	(1–3)	95M3
50	291	805	12.8	DSC	(1–3)	95M3
75	555	1595	10.3	DSC	(1–3)	95M3
100	774	2207	7.7	DSC	(1–3)	95M3
125	920	2588	4.4	DSC	(1–3)	95M3

Remarks:
(1) the data were taken from Ref. 95M3 with reference to Refs. 93M5 and 93P12
(2) ΔCp was obtained as described in Ref. 90P3
(3) ΔCp in kJ/mol/K, ΔH in kJ/mol and ΔS in J/mol/K

Myoglobin from horse, holo and apoprotein, molten globule state

Form	pH	T	ΔCp	ΔH	Approach	Remarks	Ref
holo	4.3	74	8.7		DSC		94N1
apo, native	5.3	30	4.0		DSC		94N1
apo, native	5.9	65	4.12±0.02	159±1	heat	(1)	94N1
apo, m.g.	2.0	20	1.80±0.02	5.4±0.1	heat	(2,3)	94N1
apo, m.g.	2.0		1.50±0.02		heat	(2,4)	94N1
apo, m.g.	2.0	20	2.1	7.7	heat	(5)	95H1
apo, m.g.	2.0	10–40	3.1		ITC	(6)	95H1

Remarks:
(1) measured by CD, buffer: 10 mM sodium acetate
(2) m.g. = molten globule state
(3) measured by CD in the presence of various trichloroacetate concentrations
(4) measured in the presence of NaCl
(5) measured in the presence of Na_2SO_4
(6) acid unfolded apomyoglobin titrated with and Na_2SO_4 and $NaClO_4$ by isothermal titration calorimetry (ITC)

Myoglobin from horse, holo and apoprotein, native and molten globule states

Form	pH	T	ΔC_p	ΔH	Approach	Remarks	Ref
holo	4.3	74	8.7		DSC		94N1
apo, native	5.3	30	4.0		DSC		94N1
apo, native	5.9	65	4.12±0.02	159±1	heat	(1)	94N1
apo, m.g.	2.0	20	1.80±0.02	5.4±0.1	heat	(2,3)	94N1
apo, m.g.	2.0		1.50±0.02		heat	(2,4)	94N1
apo, m.g.	2.0	60	0.82	63.3±1.9	heat	(2,4)	95N3
apo, m.g.	2.0	60	1.72	102.8±0.5	DSC	(2,4)	95N3
apo, m.g.	2.0	60	0.69	72.5±0.7	DSC	(2,5)	95N3
apo, m.g.	2.0	20	2.3		heat	(2,5)	95N3
apo, m.g.	2.0	20	1.7		heat	(2,6)	95N3

Remarks:
(1) measured by CD, buffer: 10 mM sodium acetate
(2) m.g. = molten globule state
(3) measured by CD in the presence of various trichloroacetate concentrations
(4) measured in the presence of NaCl
(5) trichloroacetic acid-induced molten globule
(6) NaCl-induced molten globule

Apomyoglobin from horse

	pH	T_{trs}	ΔC_p	ΔH	Approach	Ref
	5.9	65	5.1	195±17	heat	96B2

Myosin

Myosin and its constitutive fragments

Form	pH	T_{trs}	ΔC_p	ΔH	Approach	Remarks	Ref
myosin thick filament	7	41–60		7175±310	DSC	(1,2a)	89B1
	7			7025±410	DSC	(1,2a)	89B1
myosin rod	7	41–60		4430±250	DSC	(1,2a)	89B1
myosin rod	7–9	46–67		3945±244	DSC	(2a,3a)	89L4
myosin rod	7	47–69		3924	DSC	(2b,3a)	89L4
LMM	7	45–66		3103	DSC	(2a,3b)	89L4
LMM	7	46–61		2886	DSC	(2b,3c)	89L4
S-1	7	46.3		1070±60	DSC	(2a)	89B1
S-1	7.9	45		1255	DSC	(2c)	90S6
S-2	7	46–57		777	DSC	(2a,3d)	89L4
S-2	7	49–57		780	DSC	(2b,3d)	89L4
S-2	7.9	40.3	112	1470	DSC	(2c)	90S6
light chains	7	51.0±10.2		356±63	DSC	(2a)	89B1

Remarks:
(1)   the data refer to a dimer
(2a) in the presence of phosphate buffer
(2b) in the presence of pyrophosphate
(2c) in the presence of Tris buffer
(3a) resolved into 6 subtransitions
(3b) resolved into 5 subtransitions
(3c) resolved into 4 subtransitions
(3d) resolved into 2 subtransitions

Table 2. Enthalpy and Heat Capacity Changes – Molar Values

Myosin of bovine heart muscle, subtransitions resolved by deconvolution

Transition	pH	T_{trs}	ΔCp	ΔH	Approach	Ref
trans. (1)	7.0	18		704.44	DSC	90L3
trans. (2)	7.0	41.3		399.02	DSC	90L3
trans. (3)	7.0	43.3		773.76	DSC	90L3
trans. (4)	7.0	45.5		1089	DSC	90L3
trans. (5)	7.0	48.5		1612.8	DSC	90L3
trans. (6)	7.0	54.3		3304.8	DSC	90L3

Myosin, thick filament, deconvolution (first and second subtransition)

Transition	pH	T_{trs}	ΔCp	ΔH	Approach	Remarks	Ref
trans. (1)	6.3	47.3		3368	DSC	(1)	90B3
trans. (2)	6.3	53.9		3669	DSC	(1)	90B3
trans. (1)	6.5	47.4		3397	DSC	(1)	90B3
trans. (2)	6.5	51.9		3669	DSC	(1)	90B3
trans. (1)	6.7	47.3		4100	DSC	(1)	90B3
trans. (2)	6.7	51.3		3180	DSC	(1)	90B3
trans. (1)	7.0	46.9		4058	DSC	(1)	90B3
trans. (2)	7.0	49.2		3209	DSC	(1)	90B3

Remark:
(1) measurements in the presence of 0.13 M KCl

Myosin rod, dimer

pH	T_{trs}	ΔCp	ΔH	Approach	Ref
			4469±314	DSC	90B4
7.0	40–60	112±21		DSC	90B4

Light meromyosin (= LMM)

pH	T_{trs}	ΔCp	ΔH	Approach	Ref
			2448±205	DSC	90B4
7.0	35–65	59±13		DSC	90B4

Myosin subfragment 2 (= S-2)

pH	T_{trs}	ΔCp	ΔH	Approach	Ref
			1925±288	DSC	90B4
7.0	40–60	46±17		DSC	90B4

Myosin rod, deconvolution [remark (1)]

Transition	pH	T_{trs}	ΔCp	ΔH	Approach	Ref
trans. (1)	7.02	42.4		795	DSC	90B4
trans. (2)	7.02	44.7		582	DSC	90B4
trans. (3)	7.02	46.3		1004	DSC	90B4
trans. (4)	7.02	50.4		510	DSC	90B4
trans. (5)	7.02	54.6		674	DSC	90B4
trans. (6)	7.02	55.1		439	DSC	90B4

Remark:
(1) Ref. 90B4 contains further data obtained at other pH values

Myosin rod, subtransition 1

pH	T_{trs}	ΔCp	ΔH	Approach	Ref
7.02	42.4	18.8	795	DSC	90B4

Light meromyosin ($\doteq$ LMM), deconvolution [remark (1)]

Transition	pH	T_{trs}	ΔCp	ΔH	Approach	Ref
trans. (1)	7.03	42.4	15.9	690	DSC	90B4
trans. (2)	7.03	48.8	5.7	293	DSC	90B4
trans. (3)	7.03	49.0	10.9	469	DSC	90B4
trans. (4)	7.03	53.7	6.3	264	DSC	90B4
trans. (5)	7.03	54.8	19.7	849	DSC	90B4

Remark:
(1) further data for other pH values are contained in Tab 3 of Ref. 90B4

Myosin subfragment 2 (= S-2), deconvolution

Transition	pH	T_{trs}	ΔCp	ΔH	Approach	Remarks	Ref
trans. (1)	7.02	42.1	10.9	435	DSC	(1)	90B4
trans. (2)	7.02	44.7	16.3	661	DSC	(1)	90B4
trans. (3)	7.02	49.8	10.5	427	DSC	(1)	90B4

Remark:
(1) further data for other pH values are contained in Tab. 4 of Ref. 90B4

Heavy meromyosin, low temperature transition

pH	T_{trs}	ΔCp	ΔH	Approach	Remark	Ref
8.0	12	≈ 0	117	DSC	(1)	90S5

Remark:
(1) in the presence of 50 mM Tris, 0.1 M KCl, 1 mM $MgCl_2$, and 1 mM adenyl-5'-imidophosphate

Myosin subfragment 1 (S1) and S1-nucleotide complexes

Form	pH	T_{trs}	ΔCp	ΔH	Approach	Remarks	Ref
S1	7.3	47.2±0.1	22.6±0.2	1130±30	DSC	(1)	92L2
S1-ADP	7.3	47.8±0.1	16.0±0.2	1200±30	DSC	(1)	92L2
pPDM-S1	7.3	45.4±0.1	12.3±0.3	1120±30	DSC	(1,2)	92L2
S1-AdoPP(NH)P							
	7.3	53.2±0.1	10.3±0.2	1460±40	DSC	(1,3)	92L2
S1-ADP-Vi	7.3	56.1±0.1	12.3±0.2	1330±40	DSC	(1,4)	92L2

Remarks:
(1) buffer: 10 mM HEPES, 1 mM $MgCl_2$
(2) pPDM = p-phenylene-N,N'-dimaleimide
(3) AdoPP(NH)P = adenosine 5'-[β,γ-imido]triphosphate
(4) Vi = orthovanadate

Myosin subfragment S1 and nucleoside diphosphate complexes

Ligand	pH	T_{trs}	ΔH	Approach	Ref
In the absence of nucleoside diphosphate:					
	7.3	48.6±0.2	1100±80	DSC	95B5
In the presence of 0.2 mM nucleoside diphosphate:					
ADP	7.3	49.2±0.2	1120±80	DSC	95B5
CDP	7.3	49.2±0.2	1120±80	DSC	95B5
UDP	7.3	49.2±0.2	1120±80	DSC	95B5
IDP	7.3	49.4±0.2	1120±80	DSC	95B5
GDP	7.3	49.4±0.2	1130±80	DSC	95B5
In the presence of 0.3 mM PP:					
	7.3	49.4±0.2	1130±80	DSC	95B5
In the presence of 0.2 mM nucleoside diphosphate and 0.2 mM orthovanadate:					
ADP	7.3	58.0±0.2	1330±90	DSC	95B5
CDP	7.3	56.8±0.2	1300±90	DSC	95B5
UDP	7.3	53.0±0.2	1250±90	DSC	95B5
IDP	7.3	49.8±0.2	1130±80	DSC	95B5
GDP	7.3	49.4±0.2	1130±80	DSC	95B5
In the presence of 0.2 mM nucleoside diphosphate, 5 mM sodium fluoride, and beryllium chloride:					
ADP	7.3	56.4±0.2	1280±90	DSC	95B5
CDP	7.3	55.4±0.2	1290±90	DSC	95B5
UDP	7.3	51.0±0.2	1250±90	DSC	95B5
IDP	7.3	49.8±0.2	1130±80	DSC	95B5
GDP	7.3	49.4±0.2	1130±80	DSC	95B5

Myosin II from *Acanthamoeba castellani* and skeletal muscle myosin

Form	pH	T_{trs}	ΔH	Approach	Remarks	Ref
dephospho M-II	7.5	41.7±0.1	4415±145	DSC	(1–3)	95Z4
phospho M-II	7.5	41.7	3850	DSC	(1–3)	95Z4
truncated M-II	7.5	41.7	4060	DSC	(1–4)	95Z4
M-II head fragment	7.5	41.1±0.2	711±84	DSC	(1–3)	95Z4
dephospho M-II+	7.5	41.8/48.5	~5500	DSC	(1–3,5)	95Z4
skeletal myosin	7.5	46.1/53.7	10630	DSC	(2,3,6)	95Z4

Remarks:
(1) M-II: myosin II from *Acanthamoeba castellani*
(2) buffer: 10 mM imidazole/HCl, 0.6 M KCl, 1 mM DTT, pH 7.5
(3) scan rate 60 K/h, Ref. 95Z4 contains further data for a higher scan rate
(4) chymotryptic cleaved myosin II with 66 residues removed from the C-terminus of each heavy chain
(5) dephospho myosin II + 5 mM Mg.AMPPNP
(6) rabbit skeletal muscle myosin

Myosin and rod from carp acclimated at 10 and 30°C

Transition	pH	T_{trs}	ΔCp	ΔH	Approach	Remarks	Ref
Myosin from carp, acclimated at 10°C:							
trans. (1)	8.0	39.9		1230	DSC	(1,2)	95N1
trans. (2)	8.0	47.4		356	DSC	(1,2)	95N1
trans. (1)	6.5	34.7		510	DSC	(1,3)	95N1
trans. (2)	6.5	37.5		1042	DSC	(1,3)	95N1
trans. (3)	6.5	48.3		485	DSC	(1,3)	95N1

Myosin and rod from carp acclimated at 10 and 30°C (continued)

Transition	pH	T_{trs}	ΔCp	ΔH	Approach	Remarks	Ref
trans. (1)	8.0	32.8		305	DSC	(1,3)	95N1
trans. (2)	8.0	34.9		954	DSC	(1,3)	95N1
trans. (3)	8.0	47.4		335	DSC	(1,3)	95N1
Myosin from carp, acclimated at 30°C:							
trans. (1)	6.5	36.3		1146	DSC	(1,3)	95N1
trans. (2)	6.5	40.4		439	DSC	(1,3)	95N1
trans. (3)	6.5	50.9		218	DSC	(1,3)	95N1
trans. (1)	8.0	35.9		1038	DSC	(1,3)	95N1
trans. (2)	8.0	39.7		883	DSC	(1,3)	95N1
trans. (3)	8.0	49.1		381	DSC	(1,3)	95N1
Myosin rod from carp, acclimated at 10°C:							
trans. (1)	8.0	33.0		1042	DSC	(1,2)	95N1
trans. (2)	8.0	44.0		372	DSC	(1,2)	95N1
trans. (1)	6.5	34.5		414	DSC	(1,3)	95N1
trans. (2)	6.5	37.8		849	DSC	(1,3)	95N1
trans. (3)	6.5	51.1		326	DSC	(1,3)	95N1
trans. (1)	8.0	32.9		360	DSC	(1,3)	95N1
trans. (2)	8.0	33.4		611	DSC	(1,3)	95N1
trans. (3)	8.0	44.1		289	DSC	(1,3)	95N1
Myosin rod from carp, acclimated at 30°C:							
trans. (1)	6.5	36.2		531	DSC	(1,3)	95N1
trans. (2)	6.5	40.7		448	DSC	(1,3)	95N1
trans. (3)	6.5	51.9		326	DSC	(1,3)	95N1
trans. (1)	8.0	34.5		397	DSC	(1,3)	95N1
trans. (2)	8.0	39.7		481	DSC	(1,3)	95N1
trans. (3)	8.0	46.7		665	DSC	(1,3)	95N1

Remarks:
(1) buffer
    pH 8.0: 20 mM Tris, 0.6 M KCl, 5 mM $MgCl_2$, 0.1 mM DTT,
    pH 6.5: 20 mM potassium phosphate, 0.5 M KCl, 1 mM EDTA, 0.1 mM DTT
(2) results of deconvolution, the peak was resolved into two subtransitions
(3) results of deconvolution, the peak was resolved into three subtransitions

Myosin from chicken breast muscle

Transition	pH	T_{trs}	ΔCp	ΔH	Approach	Remarks	Ref
trans. (1)	6.5	44.2±0.4		620±31	DSC	(1,2)	94W1
trans. (2)	6.5	47.1±0.2		1005±20	DSC	(1,2)	94W1
trans. (3)	6.5	49.0±0.2		1247±28	DSC	(1,2)	94W1
trans. (4)	6.5	50.7±0.2		1328±71	DSC	(1,2)	94W1
trans. (5)	6.5	52.9±0.3		1014±65	DSC	(1,2)	94W1
trans. (6)	6.5	56.4±0.2		897±45	DSC	(1,2)	94W1
trans. (7)	6.5	58.7±0.5		902±71	DSC	(1,2)	94W1
trans. (8)	6.5	62.6±0.6		748±44	DSC	(1,2)	94W1
trans. (9)	6.5	66.8±0.4		730±32	DSC	(1,2)	94W1
trans. (10)	6.5	70.8±0.4		574±72	DSC	(1,2)	94W1

Remarks:
(1) results of deconvolution
(2) buffer: 50 mM sodium phosphate, 0.6 M NaCl, pH 6.5

Table 2. Enthalpy and Heat Capacity Changes – Molar Values

Myosin from chicken breast muscle, in the presence of pyrophosphate

Transition	pH	T_{trs}	ΔCp	ΔH	Approach	Remarks	Ref
trans. (1)	6.5	44.1±0.4		421±19	DSC	(1,2)	94W1
trans. (2)	6.5	46.1±0.2		727±20	DSC	(1,2)	94W1
trans. (3)	6.5	48.0±0.2		1008±9	DSC	(1,2)	94W1
trans. (4)	6.5	50.0±0.2		1077±21	DSC	(1,2)	94W1
trans. (5)	6.5	52.4±0.3		685±36	DSC	(1,2)	94W1
trans. (6)	6.5	56.3±0.2		925±32	DSC	(1,2)	94W1
trans. (7)	6.5	57.4±0.2		668±97	DSC	(1,2)	94W1
trans. (8)	6.5	60.3±0.2		686±39	DSC	(1,2)	94W1
trans. (9)	6.5	64.9±0.2		635±24	DSC	(1,2)	94W1
trans. (10)	6.5	69.6±0.6		424±41	DSC	(1,2)	94W1

Remarks:
(1) results of deconvolution
(2) buffer: 50 mM sodium phosphate, 5 mM pyrophosphate, 0.6 M NaCl, pH 6.5

Nuclease from Staphylococcus aureus

The data are arranged as follows:

a) pH dependence, single data and general expressions,
b) pH, salt and concentration dependence, detailed experimental data,
c) data obtained on single and multiple mutants, and on hybrid proteins,
d) data obtained in the presence of methanol.

a) pH dependence, single data and general expressions

Staphylococcal nuclease

Form	pH	T_{trs}	ΔCp	ΔH	Approach	Remarks	Ref
	3.98	35		186	DSC		88G5
	4.5	43.3		268	DSC		88G5
	5	48.3		290	DSC		88G5
	6.8	50		332	DSC		88G5
	7	51.7	8.23	353	DSC		88G5
	7.0	53.3	9.6	357	DSC		91C4
	7.0	53.5±0.2		360.2±3.3	heat	(1-3)	95S14
without ligand							
	7	53.4±0.1	9.4±0.4	299±3	DSC		85C1
in the presence of calcium							
	7	56	8.5±0.3	410	DSC	(4)	85C1

Remarks:
(1) the measurements were made on an automated device
(2) van't Hoff heat, mean of 7 determinations
(3) buffer: 25 mM sodium phosphate, 100 mM NaCl, pH 7.0
(4) in the presence of 320 mM calcium

Staphylococcal nuclease, modified

pH	T_{trs}	ΔCp	ΔH	Approach	Remark	Ref
3.98–7	35–52	8.23	DSC	(1,2)	89G4	

Remark:
(1) determined measuring the partial heat capacity of the native protein at pH 7.0 and the heat capacity of the unfolded protein at pH 3.23
(2) the protein contains an additional heptapeptide at the N-terminus: Met-Asp-Pro-Trp-Val-Tyr-Ser-

Staphylococcal nuclease, wild-type and mutants, pH dependence of T_{trs} expressed by a polynomial
$T_{trs} = A_0 + A_1{\times}pH + A_2{\times}pH^2 + A_3{\times}pH^3$

	A_0	A_1	A_2	A_3	pH Range	Ref
wild–type	−260.6	146.9	−23.24	1.232	pH 3.9–7.0	93T1
Leu25→Ala	−311.5	154.9	−23.13	1.167	pH 4.5–7.0	93T1
Val66→Leu	−266.0	151.6	−23.97	1.266	pH 3.9–7.0	93T1
Gly79→Ser	−346.3	189.8	−30.90	1.679	pH 4.1–7.0	93T1
Gly88→Val	−248.5	143.6	−22.80	1.209	pH 3.9–7.0	93T1
Ala90→Ser	−189.9	97.84	−14.05	0.682	pH 4.5–7.0	93T1
His124→Leu	−316.5	190.7	−32.27	1.806	pH 3.6–7.0	93T1
double mutant (Val66→Leu and Gly88→Val)						
	−867.4	476.7	−80.81	4.512	pH 4.0–7.0	93T1
triple mutant (Val66→Leu, Gly79→Ser and Gly88→Val)						
	−351.5	198.6	−32.06	1.714	pH 3.9–7.0	93T1

Staphylococcal nuclease, wild-type and mutants, general expression representing the temperature dependence of ΔH by $\Delta H(T) = \Delta H° + \Delta Cp{\times}T$ (T in °C)

	pH	T_{trs}	ΔCp	$\Delta H°$	Approach	Ref
wild-type	3.9–7.0	30.1–59.0	9.238	−167	DSC	93T1
Leu25→Ala	3.9–7.0	29.9–40.8	10.49	−185	DSC	93T1
Val66→Leu	3.9–7.0	35.3–57.4	8.686	−195	DSC	93T1
Gly79→Ser	3.9–7.0	26.7–45.6	8.209	−172	DSC	93T1
Gly88→Val	3.9–7.0	34.8–56.3	10.37	−296	DSC	93T1
Ala90→Ser	3.9–7.0	26.2–41.8	10.17	−159	DSC	93T1
His124→Leu	3.9–7.0	36.3–57.4	8.368	−113	DSC	93T1

b) pH and concentration dependence, detailed experimental data

Staphylococcal nuclease, wild-type, pH, and concentration dependence detailed experimental results, protein concentration in µM

pH	Concentration	T_{trs}	$\Delta H^{(cal)}$	$\Delta H^{(v.H.)}$	Approach	Ref
7.0	86.0	54.0	337	417	DSC	93T1
	131.0	52.9	318	409	DSC	93T1
	169.0	53.0	349	412	DSC	93T1
	176.7	53.3	331	429	DSC	93T1
	254	52.2	315	426	DSC	93T1
	287	52.0	346	432	DSC	93T1
	347	52.2	332	453	DSC	93T1
	483	51.5	328	470	DSC	93T1
	674	51.1	327	485	DSC	93T1
	904	50.4	328	510	DSC	93T1

Staphylococcal nuclease, wild-type, pH, and concentration dependence detailed experimental results, protein concentration in μM (continued)

pH	Concentration	T_{trs}	$\Delta H^{(cal)}$	$\Delta H^{(v.H.)}$	Approach	Ref
6.5	85.4	52.8	313	383	DSC	93T1
	150.1	52.2	353	388	DSC	93T1
	291	51.8	330	431	DSC	93T1
	497	51.2	307	454	DSC	93T1
	673	50.7	323	463	DSC	93T1
6.0	163.7	51.9	302	397	DSC	93T1
	187.5	51.9	280	410	DSC	93T1
	314	51.3	288	435	DSC	93T1
	315	51.1	297	412	DSC	93T1
	635	50.2	290	453	DSC	93T1
	643	50.1	297	431	DSC	93T1
5.5	298	49.3	284	384	DSC	93T1
	492	48.7	276	384	DSC	93T1
	665	48.2	287	392	DSC	93T1
	994	47.6	281	404	DSC	93T1
5.0	272	48.1	212	378	DSC	93T1
	419	47.3	223	369	DSC	93T1
	580	46.8	235	371	DSC	93T1
	763	46.1	228	374	DSC	93T1
	970	45.9	241	377	DSC	93T1
4.5	361	43.0	204	297	DSC	93T1
	687	42.1	210	309	DSC	93T1
	895	41.6	200	309	DSC	93T1
	1126	41.1	207	310	DSC	93T1
4.2	278	38.5	192	244	DSC	93T1
	461	38.0	192	251	DSC	93T1
	678	37.7	188	254	DSC	93T1
	919	37.2	183	256	DSC	93T1
	1186	37.2	192	254	DSC	93T1
4.0	277	35.5	150	218	DSC	93T1
	464	34.5	157	200	DSC	93T1
	652	33.3	138	208	DSC	93T1
	892	33.4	134	207	DSC	93T1
	1236	32.1	133	203	DSC	93T1
3.9	284	32.7	138	200	DSC	93T1
	426	31.5	126	187	DSC	93T1
	555	31.0	120	191	DSC	93T1
	707	30.8	127	187	DSC	93T1
	951	30.6	125	183	DSC	93T1
	1279	30.1	126	183	DSC	93T1

Staphylococcal nuclease, wild-type, in the presence of NaCl (mM)

pH	NaCl (mM)	T_{trs}	ΔCp	ΔH	Approach	Remarks	Ref
4.10	0	43.2	5.0	206	DSC	(1)	94C1
4.10	100	44.0	7.1	189	DSC	(1,2)	94C1
4.10	100	42.5	5.5	180	DSC	(1)	94C1
4.10	300	42.1	5.4	175	DSC	(1)	94C1
4.10	800	42.0	5.0	185	DSC	(1)	94C1
4.50	100	45.5	9.1	239	DSC	(1)	94C1
4.80	100	48.7	8.3	258	DSC	(1)	94C1
5.00	100	50.8	9.0	306	DSC	(1)	94C1
5.50	100	52.1	9.0	301	DSC	(1)	94C1
6.00	100	53.4	8.2	328	DSC	(1)	94C1
7.00	100	54.1	9.0	324	DSC	(1)	94C1
7.50	100	54.0	9.6	341	DSC	(1)	94C1
8.00	100	53.5	9.7	350	DSC	(1)	94C1
8.50	100	53.8	11	348	DSC	(1)	94C1
4.1–8.5		42–54	13.4			(3)	94C1

Remarks:
(1) errors are ±0.5 K in T_{trs}, ±10% in ΔH, and ±2 kJ K^{-1} mol^{-1} in ΔCp
(2) at pH 4.10, in the presence of 100 mM NaCl, no significant dependence of T_{trs} and ΔH on protein concentration was observed
(3) from ΔH versus T_{trs}

Staphylococcal nuclease, wild-type, in the presence of NaCl (mM)

pH	NaCl	T_{trs1}	ΔH_{trs1}	T_{trs2}	ΔH_{trs2}	Approach	Remarks	Ref
4.1	0	47.1	252	54.0	33	DSC	(1)	94C3
4.1	100	42.0	181	48.3	78	DSC	(1)	94C3
4.1	500	41.0	202	69.7	92	DSC	(1)	94C3
7.0	100	54.1	324			DSC	(1)	94C3

Remark:
(1) results of deconvolution, error in T_{trs} ±0.5°C and in enthalpies ±10%

Staphylococcal nuclease, mutant (Val66→Ala), in the presence of NaCl

pH	NaCl (mM)	T_{trs}	ΔCp	ΔH	Approach	Remarks	Ref
5.00	100	36.3	6.9	160	DSC	(1)	94C1
6.00	100	41.2	8.8	225	DSC	(1)	94C1
7.00	100	42.2	8.2	248	DSC	(1)	94C1
8.00	100	42.4	8.4	252	DSC	(1)	94C1
5.0–8.0		36–42	14.8			(2,3)	94C1

Remarks:
(1) errors are ±0.5 K in T_{trs}, ±10% in ΔH, and ±2 kJ/K/mol in ΔCp
(2) from ΔH versus T_{trs}
(3) for comparison, thermodynamic data at the thermal transition of the A state at pH 2.1 in the presence of 0.8 M Na_2SO_4 are T_{trs} = 31.3°C, ΔH = 31 kJ/mol, and ΔCp = 1 kJ/K/mol according to Ref. 94C2

Table 2. Enthalpy and Heat Capacity Changes – Molar Values

Staphylococcal nuclease, mutant (Val66→Leu), in the presence of NaCl (mM)

pH	NaCl	T_{trs1}	ΔH_{trs1}	T_{trs2}	ΔH_{trs2}	Approach	Remarks	Ref
3.8	0	45.1	157	45.7	133	DSC	(1)	94C3
3.8	20	41.3	140	45.6	108	DSC	(1)	94C3
3.8	50	38.1	141	49.1	102	DSC	(1)	94C3
3.8	100	40.7	141	53.2	96	DSC	(1)	94C3
3.8	150	37.9	145	58.7	102	DSC	(1)	94C3
3.8	200	37.9	145	62.5	108	DSC	(1)	94C3
3.8	300	37.8	148	65.6	117	DSC	(1)	94C3
3.8	400	38.9	145	70.9	116	DSC	(1)	94C3
3.8	500	37.6	143	70.8	112	DSC	(1,2)	94C3
4.1	100	44.5	174	56.7	130	DSC	(1)	94C3
5.0	100	55.8	232	60.3	99	DSC	(1)	94C3
6.0	100	59.5	265	59.4	120	DSC	(1)	94C3
7.0	100	59.8	300	54.7	52	DSC	(1)	94C3
8.0	100	57.1	342			DSC	(1)	94C3
8.0	500	57.7	315			DSC	(1)	94C3

Remarks:
(1) results of deconvolution, error in T_{trs} ±0.5°C and in enthalpies ±10%
(2) ΔCp amounts to 4±1 kJ/mol/K for the first transition and 1±1 kJ/mol/K for the second transition

Staphylococcal nuclease, mutant (Val66→Trp), in the presence of NaCl (mM)

pH	NaCl	T_{trs1}	ΔH_{trs1}	T_{trs2}	ΔH_{trs2}	Approach	Remarks	Ref
6.0	0	47.8	170	43.0	82	DSC	(1)	94C3
6.0	100	45.3	176	60.0	51	DSC	(1)	94C3
6.0	200	43.7	174	71.6	54	DSC	(1)	94C3
6.0	300	43.4	164	77.9	56	DSC	(1)	94C3
6.0	400	43.5	158	66.3	74	DSC	(1)	94C3
6.0	500	42.8	150	67.5	79	DSC	(1)	94C3
7.0	100	46.3	208	50.8	74	DSC	(1)	94C3

Remark:
(1) results of deconvolution, error in T_{trs} ± 0.5°C and in enthalpies ±10%

Staphylococcal nuclease, mutant (Glu75→Val), in the presence of NaCl (mM)

pH	NaCl	T_{trs1}	ΔH_{trs1}	T_{trs2}	ΔH_{trs2}	Approach	Remarks	Ref
5.0	100	40.1	152	65.49	68	DSC	(1)	94C3
6.0	100	45.1	218	75.2	62	DSC	(1)	94C3
7.0	100	47.5	238	60.8	66	DSC	(1)	94C3

Remark:
(1) results of deconvolution, error in T_{trs} ±0.5°C and in enthalpies ±10%

Staphylococcal nuclease (Asp77→Ala), in the presence of NaCl (mM)

pH	NaCl	T_{trs1}	ΔH_{trs1}	T_{trs2}	ΔH_{trs2}	Approach	Remarks	Ref
5.0	100	40.7	151	60.3	68	DSC	(1)	94C3
7.0	0	44.1	206	71.4	83	DSC	(1)	94C3
7.0	100	43.4	210	72.6	88	DSC	(1,2)	94C3
7.0	500	45.3	217	69.7	88	DSC	(1)	94C3

Remarks:
(1) results of deconvolution, error in T_{trs} ± 0.5°C and in enthalpies ±10%
(2) ΔC_p amounts to 6±1 kJ/mol/K for the first transition and 1±1 kJ/mol/K for the second transition

Staphylococcal nuclease, mutant (Gly88→Val), in the presence of NaCl (mM)

pH	NaCl	T_{trs1}	ΔH_{trs1}	T_{trs2}	ΔH_{trs2}	Approach	Remarks	Ref
4.1	100	45.2	153	58.6	95	DSC	(1)	94C3
5.0	100	55.0	222	62.7	89	DSC	(1)	94C3
6.0	100	57.5	276	65.7	91	DSC	(1)	94C3
7.0	100	58.4	291	64.8	58	DSC	(1)	94C3

Remark:
(1) results of deconvolution, error in T_{trs} ± 0.5°C and in enthalpies ±10%

c) Data obtained on single and multiple mutants, and on hybrid proteins

Staphylococcal nuclease, wild-type and mutants

Mutant	pH	T_{trs}	ΔC_p	ΔH	Approach	Remarks	Ref
wild-type	5.31	47.6		260±30	heat	(1)	90A1
Ile18→Met	5.42	45.5		250±30	heat	(1)	90A1
Val23→Phe	5.35	37.4		120±20	heat	(1)	90A1
His46→Tyr	5.38	47.0		290±40	heat	(1)	90A1
Gly79→Ser	5.28	38.1		230±30	heat	(1)	90A1
Leu89→Phe	5.38	40.1		180±30	heat	(1)	90A1
His124→Leu	5.20	56.0		360±50	heat	(1)	90A1
double mutant (Phe76→Val and His124→Leu)							
	5.21	46.8		220±40	heat	(1)	90A1

Remark:
(1) monitored by NMR

Table 2. Enthalpy and Heat Capacity Changes – Molar Values

Nuclease from *Staphylococcus aureus*, mutant (His124→Leu, H124L), oxidized (ox.) and reduced (red.) forms of its engineered disulfide mutants

Mutant/Cys Form	pH	T_{trs}	ΔH	Approach	Remarks	Ref
H124L	5.5	56.0±0.9	378±9	heat	(1,2)	96H4
H124L, Cys80-Cys116 ox.	5.5	58.0±1.9	231±10	heat	(1,2)	96H4
H124L, Cys80-Cys116 red.	5.5	51.2±0.5	221±24	heat	(1,2)	96H4
double mutant (Gly79→Ser and His124→Leu):						
Cys80-Cys116 ox.	5.5	53.9±3.1	282±19	heat	(1,2)	96H4
Cys80-Cys116 red.	5.5	43.1±2.8	138±9	heat	(1,2)	96H4
H124L, Cys79-Cys118 ox.	5.5	64.4±1.7	338±12	heat	(1,2)	96H4
H124L, Cys79-Cys118 red.	5.5	49.0±1.1	289±7	heat	(1,2)	96H4
H124L, Cys77-Cys118 ox.	5.5	54.0±1.8	194±7	heat	(1,2)	96H4
H124L, Cys77-Cys118 red.	5.5	30.0±4.1	157±11	heat	(1,2)	96H4

Remarks:
(1) van't Hoff treatment
(2) measured in D_2O at pH* 5.5

Staphylococcal nuclease, wild-type and mutants

Mutant	pH	T_{trs}	ΔCp	ΔH	Approach	Remarks	Ref
wild-type	7.0	50.6		346±2	heat	(1)	91E1
wild-type	7.0	52.1		334	heat	(2)	91E1
wild-type	7.0	53.3±0.2	7.5	363±6	heat, v.H.		88S8
wild-type	7.0	52.8	9.2	402±8	DSC		88S8
Val66→Leu	7.0	56.0	5.4	314	heat, v.H.		88S8
Ala69→Thr	7.0	41.2	6.7	273	heat, v.H.		88S8
Gly88→Val	7.0	56.1	5.4	297	heat, v.H.		88S8
Pro117→Gly	7.0	55.6		385	heat	(3)	91E1
Pro117→Thr	7.0	51.2		300	heat	(3)	91E1
His124→Leu	7.0	56.4		385	heat	(3)	91E1
Lys116→Gly	7.0	54.4		364	heat	(3)	91E1
double mutant (Ile18→Met and Ala90→Ser)							
	7.0	41.6	7.9	303	heat, v.H.		88S8
double mutant (Val66→Leu and Gly88→Val)							
	7.0	57.6	3.3	234	heat, v.H.		88S8
triple mutant (Val66→Leu, Gly88→Val, and Gly79→Ser)							
	7.0	53.4	2.1	184	heat, v.H.		88S8

Remarks:
(1) in the absence of phosphate
(2) in the presence of phosphate
(3) measurement made in Tris buffer

Staphylococcal nuclease, wild-type and mutants

Mutant	pH	T_{trs}	ΔCp	ΔH	Approach	Remarks	Ref
wild-type	7.0	53.5±0.2		360.2±3.3	heat	(1–3)	95S11
Met32→Leu	7.0	50.3±0.3		339±8	heat	(1–3)	96S10
Met32→Ile	7.0	51.2±0.3		364±8	heat	(1–3)	96S10

Remarks:
(1) the measurements were made on an automated device
(2) van't Hoff heat, mean of 7 determinations
(3) buffer: 25 mM sodium phosphate, 100 mM NaCl, pH 7.0

Staphylococcal nuclease, wild-type and mutants

Mutant	pH	T_{trs1}	ΔH_{trs1}	T_{trs2}	ΔH_{trs2}	Approach	Remarks	Ref
wild-type	7.0	54.1	356			DSC	(1–4)	95C2
wild-type	5.0	50.8	299			DSC	(1–4)	95C2
Leu7→Ala	7.0	50.9	313			DSC	(1–4)	95C2
Leu7→Ala	5.0	47.4	247			DSC	(1–4)	95C2
Val23→Ala	7.0	41.6	278			DSC	(1–4)	95C2
Val23→Ala	5.0	34.9	189			DSC	(1–4)	95C2
Val23→Phe	5.0	44.0	161	55.9	83	DSC	(1–3,5)	95C2
Lys24→Gly	7.0	47.9	327			DSC	(1–4)	95C2
Lys24→Gly	5.0	42.6	253			DSC	(1–4)	95C2
Ala69→Thr	7.0	40.5	282			DSC	(1–4)	95C2
Ala69→Thr	5.0	34.3	186			DSC	(1–4)	95C2
Glu75→Ala	7.0	48.5	241	63.3	82	DSC	(1–3,5)	95C2
Glu75→Ala	5.0	41.9	148	57.1	99	DSC	(1–3,5)	95C2
Glu75→Gly	7.0	38.9	198	65.8	73	DSC	(1–3,5)	95C2
Gly79→Ser	7.0	49.7	227	58.5	70	DSC	(1–3,5)	95C2
Gly79→Ser	5.0	45.2	184	60.3	83	DSC	(1–3,5)	95C2
Gly88→Trp	7.0	58.7	304	64.4	87	DSC	(1–3,5)	95C2
Gly88→Trp	5.0	55.6	244	66.2	105	DSC	(1–3,5)	95C2
Gly88→Trp	4.1	48.8	149	55.3	80	DSC	(1–3,5)	95C2
Leu137→Ala	7.0	45.3	268			DSC	(1–4)	95C2
Leu137→Ala	5.0	39.6	180			DSC	(1–4)	95C2
triple mutant (Val66→Leu, Gly88→Val and Gly79→Ser)								
	7.0	57.6	154	69.1	165	DSC	(1–3,5)	95C2
	5.0	53.7	146	68.7	189	DSC	(1–3,5)	95C2
	4.1	41.4	129	64.9	161	DSC	(1–3,5)	95C2

Remarks:
(1) buffers
   pH 7.0: 20 mM sodium phosphate, 100 mM NaCl, 1 mM EDTA
   pH 5.0: 20 mM NaOAC, 100 mM NaCl, 1 mM EDTA
   pH 4.1: 20 mM glycine-HCl, 100 mM NaCl, 1 mM EDTA
(2) errors in T_{trs} ±0.5°C and in ΔH ±10%
(3) results of peak fit or deconvolution
(4) ΔC_p was 7 kJ/mol/K for the transition
(5) ΔC_p was 6±1 kJ/mol/K and 1±1 kJ/mol/K for the first and second transition, respectively

Staphylococcal nuclease, mutant (Pro117→Gly)

Salt Concentration	pH	T_{trs}	ΔC_p	ΔH	Approach	Remarks	Ref
0.1 M NaCl	7.0	58.2±0.1	8.37±0.42	389±4	DSC	(1,2)	94X
0.1 M NaCl	4.1	43.6±0.2	8.58±0.56	250±8	DSC	(1,3)	94X
0.1 M NaCl	3.5	33.1±0.2	5.69±0.42	138±4	DSC	(1,4)	94X
0.5 M NaCl, subtransitions:							
trans. (1)	3.5	33.9±0.3	4.4±4.2	161±8	DSC	(1,5)	94X
trans. (2)	3.5	60.4±0.4	2.13±0.21	90±10	DSC	(1,6)	94X

Remarks:
(1) the temperature dependence of ΔC_p follows a polynomial expression: $\Delta C_p(T) = \Delta C_p(T0) + b \times (T-T0) + c \times (T^2-T0^2)$; T0 is the temperature at which $\Delta G = 0$, ΔH and ΔC_p are given in the table at T0
(2) b = 208±4 J/K²/mol and c = –0.448±0.004 J/K³/mol
(3) b = 214±2 J/K²/mol and c = –0.452±0.004 J/K³/mol
(4) b = 88±4 J/K²/mol and c = –0.251±0.004 J/K³/mol
(5) b = 112±8 J/K²/mol and c = –0.255±0.004 J/K³/mol
(6) b = 150±8 J/K²/mol and c = –0.251±0.008 J/K³/mol

Table 2. Enthalpy and Heat Capacity Changes – Molar Values

Staphylococcal nuclease, wild-type and tryptophan mutants

Mutant	pH	T_{trs}	ΔCp	ΔH	Approach	Remarks	Ref
wild-type	7.20	51.3±0.1		329±12	heat	(1,2)	96E2
wild-type	7.20	51.3±0.1	10.7±1.0	333±11	heat	(1)	96E2
Val66→Trp*	7.20	49.7±0.9		123±10	heat	(1–3)	96E2
Val66→Trp*	7.20	47.5±0.7	3.4±0.8	118±9	heat	(1,3)	96E2
Val66→Trp	7.20	43.1±0.5		179±14	heat	(1,2)	96E2
Val66→Trp	7.20	43.8±0.5	−6.2±1.3	182±12	heat	(1)	96E2
Val66→Trp, three-state fit without ΔCp:							
trans. (1)	7.20	44.4±0.3		175±11	heat	(1,2)	96E2
trans. (2)	7.20	50.3±0.5		93±13	heat	(1,2)	96E2
Val66→Trp, three-state fit with ΔCp:							
trans. (1)	7.20	44.7±1.0	6.6±1.2	195±14	heat	(1)	96E2
trans. (2)	7.20	44.8±4.0	3.5±1.2	77±23	heat	(1)	96E2
Val66→Trp, three-state fit with fixed parameters for trans. (2):							
trans. (1)	7.20	43.5	7.1	192	heat	(1)	96E2
trans. (2)	7.20	(47.5)	(3.4)	(118)	heat	(1)	96E2

Remarks:
(1) parameters from a global fit of simultaneously acquired data from fluorescence and CD at 222 and 235 nm
(2) ΔCp was not allowed to float
(3) mutant Val66→Trp* does not contain Trp140

Staphylococcal nuclease, wild-type and fragment (1-136)

Mutant	pH	T_{trs}	ΔH	Approach	Remarks	Ref
full length SNase:						
trans. (1)	3.9	39.3	195±20	DSC	(1,2)	94G4
trans. (2)	3.9	66.1	124±15	DSC	(1,2)	94G4
fragment 1-136	3.9		128±15	heat, v.H.	(1)	94G4
	3.9		110±20	DSC	(1,3)	94G4

Remarks:
(1) buffer: 10 mM Na-acetate, 400 mM NaCl, pH 3.9
(2) results of deconvolution
(3) van't Hoff heat from the calorimetric peak

Staphylococcal nuclease, multiple mutants

Mutant	pH	T_{trs}	ΔCp	ΔH	Approach	Remarks	Ref
double mutant (Val66→Leu and Gly88→Val)							
	7.0	56.4–59.2	5.10±0.75	226–262	DSC	(1)	93T1
triple mutant (Val66→Leu, Gly79-Ser and Gly88→Val)							
	7.0	55.6–57.7	9.29±0.84	226–262	DSC	(2)	93T1

Remarks:
(1) at protein concentration 180–1200 µM
(2) at protein concentration 255–355 µM

Staphylococcal nuclease, mutants NC [nuclease-conA, see also remark (1)]

Mutant	pH	T_{trs}	ΔCp	ΔH	Approach	Remarks	Ref
NC	7.0	32.8		204±3	heat	(1)	91E1
NC	7.0	32.2	9.75	203	heat	(1)	91E1
NC-Ser28→Gly	5.3	26.0		180	heat	(1)	91E1
NC-Ser28→Gly	7.0	30.5		192±3	heat	(1)	91E1
NC-Ser28→Gly	7.0	30.1	11.21	192	heat	(1)	91E1

Remark:
(1) mutant nuclease-conA (NC) is a mutant protein containing a six amino acid residue β-turn substitute from concanavalin A, i.e., Ser-Ser-Asn-Gly-Ser-Pro instead of the residues 27–31 Tyr-Lys-Gly-Gln-Pro in the wild-type Staphylococcal nuclease

Staphylococcal nuclease, wild-type and mutants

Mutant	pH	T_{trs}	ΔCp	ΔH	Approach	Remarks	Ref
wild-type	7.0	52.5±0.5	6.8±1.2	309±27	heat	(1,2)	91A2
hybrid and mutant Ser28→Gly	7.0	30.0±0.5	8.6±2.0	174±13	heat	(1–3)	91A2

Remarks:
(1) measured in the absence of calcium and pdTp (3',5'deoxy-thymidine biphosphate) in 10 mM cacodylate buffer
(2) ΔCp was calculated using T_{trs} and ΔH from thermal unfolding, and ΔG from GuHSCN denaturation at 0°C
(3) hybrid protein in which the res. 27–30 of Staphylococcal nuclease (Tyr-Lys-Gly-Gln) are replaced by the residues 160–164 of concanavalin (Ser-Ser-Asn-Gly-Ser)

Staphylococcal nuclease, mutants and cross-linked dimers

Mutant	pH	T_{trs}	ΔH	Approach	Remarks	Ref
A. untreated cysteine standard						
wild-type	7.0	53.0	360	heat	(1–3)	95B11
Gly29→Cys	7.0	48.4	339	heat	(1–3)	95B11
Gly50→Cys	7.0	50.4	322	heat	(1–3)	95B11
Glu57→Cys	7.0	52.1	322	heat	(1–3)	95B11
Ala60→Cys	7.0	49.4	310	heat	(1–3)	95B11
Lys70→Cys	7.0	50.2	314	heat	(1–3)	95B11
Lys78→Cys	7.0	53.4	347	heat	(1–3)	95B11
Arg105→Cys	7.0	41.9	243	heat	(1–3)	95B11
Ala112→Cys	7.0	50.2	322	heat	(1–3)	95B11
Lys134→Cys	7.0	50.1	322	heat	(1–3)	95B11
B. phenylalanine substitution mutants						
Gly29→Phe	7.0	47.3	343	heat	(1–3)	95B11
Gly50→Phe	7.0	50.5	301	heat	(1–3)	95B11
Glu57→Phe	7.0	52.4	335	heat	(1–3)	95B11
Lys70→Phe	7.0	53.0	360	heat	(1–3)	95B11
Lys78→Phe	7.0	52.5	326	heat	(1–3)	95B11
Arg105→Phe	7.0	37.9	218	heat	(1–3)	95B11
Ala112→Phe	7.0	46.4	272	heat	(1–3)	95B11
Lys134→Phe	7.0	51.3	322	heat	(1–3)	95B11

Table 2. Enthalpy and Heat Capacity Changes – Molar Values

Staphylococcal nuclease, mutants and cross-linked dimers (continued)

Mutant	pH	T_{trs}	ΔH	Approach	Remarks	Ref
C. MCA-treated protein					(4)	
wild-type	7.0	52.9	347	heat	(1–3)	95B11
Gly29→Cys	7.0	49.1	310	heat	(1–3)	95B11
Gly50→Cys	7.0	49.1	280	heat	(1–3)	95B11
Glu57→Cys	7.0	52.5	331	heat	(1–3)	95B11
Ala60→Cys	7.0	49.5	314	heat	(1–3)	95B11
Lys70→Cys	7.0	50.3	314	heat	(1–3)	95B11
Lys78→Cys	7.0	52.6	310	heat	(1–3)	95B11
Arg105→Cys	7.0	39.0	218	heat	(1–3)	95B11
Ala112→Cys	7.0	52.9	326	heat	(1–3)	95B11
Lys134→Cys	7.0	53.1	331	heat	(1–3)	95B11
G. BMH dimer					(5)	
Gly29→Cys	7.0	43.4	310	heat	(1–3)	95B11
Gly50→Cys	7.0	47.7	393	heat	(1–3)	95B11
Glu57→Cys	7.0	48.6	464	heat	(1–3)	95B11
Ala60→Cys	7.0	44.6	368	heat	(1–3)	95B11
Lys70→Cys	7.0	41.7	381	heat	(1–3)	95B11
Lys78→Cys	7.0	44.7	205	heat	(1–3)	95B11
Arg105→Cys	7.0	36.7	142	heat	(1–3)	95B11
Ala112→Cys	7.0	47.0	389	heat	(1–3)	95B11
Lys134→Cys	7.0	46.6	368	heat	(1–3)	95B11
H. mustard dimer					(6)	
Lys70→Cys	7.0	42.9	502	heat	(1–3)	95B11

Remarks:
(1) van't Hoff treatment, transition was monitored by fluorescence
(2) buffer: 25 mM sodium phosphate, 100 mM NaCl, 0.5 mM tricarboxyethyl phosphine (TCEP), pH 7.0
(3) error in T_{trs} is estimated to be ±0.3°C and in ΔH ±8 kJ/mol
(4) MCA = ε-maleimidocaproic acid
(5) BMH = 1,6-bismaleimidohexane
(6) mustard = bis(2-chloroethyl)sulfide

d) Data obtained in the presence of methanol

Staphylococcal nuclease, in the presence of methanol, methanol concentration %(v/v), at pH 7.0

	Methanol Concentration	T_{trs}	ΔH	Approach	Remarks	Ref
	0 %	53	340	heat	(1,2)	90N1
	15 %	44	360	heat	(1,2)	90N1
	35 %	32	355	heat	(1,2)	90N1
	50 %	19	305	heat	(1,2)	90N1
	60 %	4	255	heat	(1,2)	90N1
	70 %	–4	175	heat	(1,2)	90N1

Remarks:
(1) van't Hoff treatment
(2) the data were taken from Tab. 1 and Figs. 3–4 of Ref. 90N1

Nucleoside Diphosphate Kinase

Nucleoside diphosphate kinase from *E. coli* and *Dictyostelium discoideum*

Species/ Transition	pH	T_{trs}	ΔH	Approach	Remarks	Ref
Dictyostelium, wild-type						
	7.5	61.8	3284	DSC	(1,2)	96G5
Dictyostelium, mutant Pro105→Gly:						
trans. (1)	7.5	37.8	1540	DSC	(1,3)	96G5
trans. (2)	7.5	46.7	962	DSC	(1,3)	96G5
E. coli						
trans. (1)	7.5	37.5	406	DSC	(1,4)	96G5
trans. (2)	7.5	55.7	1335	DSC	(1,4)	96G5

Remarks:
(1) error in T_{trs} ±0.2°C and in ΔH ±10%
(2) the hexameric enzyme from *Dictyostelium discoideum* displays a single irreversible transition
(3) the hexameric mutant protein dissociates into folded monomers at 38°C before irreversible denaturation occurs
(4) the tetrameric kinase from *E. coli* first dissociates reversibly and then unfolds

Oncoprotein, see also myb oncoprotein

Opsin

Opsin from various species

Species	pH	T_{trs}	ΔCp	ΔH	Approach	Ref
bovine	6.8	59	4.3±0.3	490±10	DSC	88S7
frog	6.8	46	3.5±0.3	158±3	DSC	88S7
rat	6.8	47	5.8±0.4	269±7	DSC	88S7

Orosomucoid

Human orosomucoid

pH	T_{trs}	ΔCp	ΔH	Approach	Ref
7.0	58	18.8	350	heat, v.H.	95K10

Ovalbumin

Ovalbumin

pH	T_{trs}	ΔCp	ΔH	Approach	Ref
7	25	11.3±1.7	216	GuHCl	76A
7	78.9			DSC	90K7

Table 2. Enthalpy and Heat Capacity Changes – Molar Values

Ovalbumin modified, with intact and reduced intrachain disulfide bond

Form	pH	T_{trs}	ΔCp	ΔH	Approach	Ref
intact	7.0	78.9			DSC	91T1
reduced	7.0	72.1			DSC	91T1

Ovalbumin and S-ovalbumin

	pH	T_{trs}	ΔCp	ΔH	Approach	Remark	Ref
native	7.0	77.9		820±38	DSC		95H8
S-ovalbumin	7.0	85.7		812±25	DSC	(1)	95H8

Remark:
(1) S-ovalbumin is more stable than ovalbumin, $\Delta(\Delta G)$ amounts to 12.6 kJ/mol at 25°C

Ovomucoid

Turkey ovomucoid third domain

	pH	T_{trs}	ΔCp	ΔH	Approach	Remarks	Ref
	2.0	58.6±0.4	2.47±0.50	171±1	heat	(1)	93S12
	2.0	58.6±0.4	2.68±0.46	171±1	heat	(2)	93S12

Remarks:
(1) ΔCp was determined from T_{trs} versus pH and denaturant concentration
(2) ΔCp was determined by a global fit to an equation that includes the temperature dependence of ΔH

Ovomucoid third domain, effect of pH

Solvent	pH	T_{trs}	ΔCp	ΔH	Approach	Remarks	Ref
H_2O	2.56	65.7	3.1±0.6	187±5	DSC	(1,2)	95S18
H_2O	3.48	78.7		209±5	DSC	(1)	95S18
H_2O	4.51	85.2		240±2	DSC	(1)	95S18
D_2O	1.50	60.5		171±1	heat	(1,3)	95S18
D_2O	2.10	62.7		179±3	heat	(1,3)	95S18
D_2O	2.50	66.2		183±3	heat	(1,3)	95S18
D_2O	1.99	63.1	2.4±0.5	185±2	DSC	(1,2)	95S18
D_2O	2.52	68.5		200±5	DSC	(1,2)	95S18
D_2O	3.17	75.6		194±6	DSC	(1,2)	95S18
D_2O	3.46	78.7		201±3	DSC	(1,2)	95S18
D_2O	3.51	80.8		230±3	DSC	(1,2)	95S18
D_2O	3.94	85.7		231±3	DSC	(1,2)	95S18
D_2O	4.50	87.4		254±4	DSC	(1,2)	95S18
D_2O	4.96	87.8		261	DSC	(1,2)	95S18

Remarks:
(1) the pH in D_2O is given as the apparent pH
(2) ΔCp from ΔH versus T_{trs}, see Fig. 3 in Ref. 95S18 the average value from 7 DSC measurements amounts to $\Delta Cp = 2.59\pm0.83$ kJ/mol/K
(3) van't Hoff treatment, transition monitored by CD

Ovomucoid third domain, effect of ionic strength, measured in H_2O

Ionic Str.	pH	T_{trs}	ΔCp	ΔH	Approach	Ref
0.010	2.0	58.6±0.4		171±1	heat, v.H.	95S18
0.010	2.0	59.2±0.5		176±11	heat, v.H.	95S18
0.013	2.0	60.3±0.1		166±2	heat, v.H.	95S18
0.014	2.0	64.1±0.2		174±1	heat, v.H.	95S18
0.020	2.0	59.5±0.1		169±2	heat, v.H.	95S18
0.030	2.0	59.8±0.4		172±2	heat, v.H.	95S18
0.030	2.0	61.3±1.6		166±5	DSC/heat, v.H.	95S18
0.110	2.0	63.2±0.4		179±8	heat, v.H.	95S18
0.110	2.0	64.0		180±2	DSC	95S18
0.230	2.0	65.7±1.3		182±4	DSC	95S18
0.710	2.0	68.8±0.4		147±7.5	heat, v.H.	95S18

Ovomucoid third domain, effect of pH

Salt Concentration	pH	T_{trs}	ΔH	Approach	Ref
200 mM KCl	1.50	63.8	177±3	DSC	95S18
200 mM KCl	2.28	68.0	189±7	DSC	95S18
200 mM KCl	2.94	73.5	200±4	DSC	95S18
200 mM KCl	3.24	78.0		DSC	95S18
200 mM KCl	3.61	83.8		DSC	95S18
200 mM KCl	4.13	86.0	238±5	DSC	95S18
200 mM KCl	4.51	88.2	245±3	DSC	95S18

Ovomucoid third domain, results of fits

Solvent	pH	T_{trs}	ΔCp	ΔH	Approach	Ref
D_2O			2.42±0.50			95S18
H_2O			3.10±0.59			95S18
200 mM KCl			2.72±0.13			95S18

Pancreatic Polypeptide

Pancreatic polypeptide from chicken

	T_{trs}	ΔCp	ΔH	Approach	Ref
	25	0.79±0.17	112.5±8.4	heat, v.H.	86K1

Papain

Papain

pH	T_{trs}	ΔCp	ΔH	Approach	Ref
1.9	50.05		458.6	DSC	93S11
2.4	53.80		556.1	DSC	93S11
2.9	69.50		642.7	DSC	93S11
3.5	81.15		864.0	DSC	93S11
3.8	83.8	13.7*	889*	DSC	78T
3.9	85.20		957.7	DSC	93S11

Papain

pH	T_{trs}	ΔCp	ΔH	Approach	Remarks	Ref
2.55	51.0	12.6±1.7	451	DSC	(1–3)	93S8
2.90	60.3	12.6±1.7	556	DSC	(1–3)	93S8
3.18	74.6	12.6±1.7	778	DSC	(1–3)	93S8
3.80	78.5	12.6±1.7	826	DSC	(1–3)	93S8
4.10	81.8	12.6±1.7	882	DSC	(1–3)	93S8

Remarks:
(1) ΔCp is the mean from the calorimetric recordings, from ΔH versus T_{trs} it follows $\Delta Cp = 13.8$ kJ/mol/K
(2) ΔH, i.e., ΔH^{cal} is different from $\Delta H^{v.H.}$
(3) the transition can be resolved into subtransitions:
$T_{trs,1} = 75.5\pm0.3°C$ and $\Delta H_1 = 406\pm29$ kJ/mol,
$T_{trs,2} = 78.9\pm0.2°C$ and $\Delta H_2 = 527\pm38$ kJ/mol,
with the interaction term $\Delta G = 0.02\pm0.17$ kJ/mol (see for comparison the chymopapain example)

Papain, temperature dependence of the heat capacity change at protein unfolding in aqueous solution, and related thermodynamic quantities

T	$\Delta H(exp)$	$\Delta S(exp)$	$\Delta Cp(exp)$	Approach	Remarks	Ref
5	−166	−911	16.9	DSC	(1–3)	95M3
25	164	236	16.0	DSC	(1–3)	95M3
50	535	1438	13.7	DSC	(1–3)	95M3
75	826	2312	9.6	DSC	(1–3)	95M3
100	1015	2840	5.6	DSC	(1–3)	95M3
125	1091	3042	0.5	DSC	(1–3)	95M3

Remarks:
(1) the data were taken from Ref. 95M3 with reference to Ref. 88P3
(2) ΔCp was obtained as described in Ref. 90P3
(3) ΔCp in kJ/mol/K, ΔH in kJ/mol and ΔS in J/mol/K

Papaya proteinase 3

pH	T_{trs}	ΔCp	ΔH	Approach	Ref
1.9	51.66		496.2	DSC	93S11
2.4	57.72		599.1	DSC	93S11
2.9	70.40		679.1	DSC	93S11
3.5	83.65		943.9	DSC	93S11
3.9	86.20		970.3	DSC	93S11

Papaya proteinase 4

	pH	T_{trs}	ΔCp	ΔH	Approach	Ref
	1.9	42.2		413.0	DSC	93S11
	2.4	43.6		465.7	DSC	93S11
	2.9	45.5		556.1	DSC	93S11
	3.5	77.8		790.4	DSC	93S11
	3.9	80.3		826.8	DSC	93S11

Chymopapain

	pH	T_{trs}	ΔCp	ΔH	Approach	Ref
	1.9	69.30		613.0	DSC	93S11
	2.4	71.60		643.9	DSC	93S11
	2.9	78.30		732.6	DSC	93S11
	3.5	85.75		971.5	DSC	93S11
	3.9	89.20		1016.3	DSC	93S11

Chymopapapin

	pH	T_{trs}	ΔCp	ΔH	Approach	Remarks	Ref
	1.20	43.0	8.8±2.1	438	DSC	(1–3)	93S8
	1.91	56.9	8.8±2.1	578	DSC	(1–3)	93S8
	2.20	62.0	8.8±2.1	602	DSC	(1–3)	93S8
	2.35	67.9	8.8±2.1	674	DSC	(1–3)	93S8
	2.56	75.2	8.8±2.1	728	DSC	(1–3)	93S8
	2.90	78.2	8.8±2.1	779	DSC	(1–3)	93S8
	3.18	81.1	8.8±2.1	809	DSC	(1–3)	93S8
	3.45	83.7	8.8±2.1	831	DSC	(1–3)	93S8

Remarks:
(1) ΔCp is the mean from the calorimetric recordings, from ΔH versus T_{trs} it follows $\Delta Cp = 9.6$ kJ/mol/K
(2) ΔH, i.e., ΔH^{cal} is different from $\Delta H^{v.H.}$
(3) the transition can be resolved into subtransitions:
$T_{trs,1} = 71.1±0.3°C$ and $\Delta H_1 = 280±25$ kJ/mol,
$T_{trs,2} = 77.8±0.1°C$ and $\Delta H_2 = 560±33$ kJ/mol,
with interaction term $\Delta G = 4.1±0.2$ kJ/mol (see for comparison the papain example)

Parvalbumin

Parvalbumin in the presence (holo form) and in the absence (apo form) of calcium

Form	pH	T_{trs}	ΔCp	ΔH	Approach	Ref
holo	7	90	4.6±0.5	500±30	DSC	78F
apo	7	35	5.6±0.5	168±12	DSC	78F

Table 2. Enthalpy and Heat Capacity Changes – Molar Values

Pepsin

Pepsin in the presence of NaCl

pH	T_{trs}	ΔCp	ΔH	Approach	Ref
7.1	60.5		406	DSC	81P2

Pepsinogen

Pepsinogen

pH	T_{trs}	ΔCp	ΔH	Approach	Ref
6	66		1134	DSC	81P2
6–8	25	21.8	132	urea	78A
		25.5±1.3			81M1

Pepsinogen, temperature dependence of the heat capacity change at protein unfolding in aqueous solution and related thermodynamic quantities

T	$\Delta H(exp)$	$\Delta S(exp)$	$\Delta Cp(exp)$	Approach	Remarks	Ref
5	−577	−2279	34.2	DSC	(1–3)	95M3
25	72	−19	30.6	DSC	(1–3)	95M3
50	770	2242	25.2	DSC	(1–3)	95M3
75	1301	3739	17.3	DSC	(1–3)	95M3
100	1639	4687	9.8	DSC	(1–3)	95M3
125	1767	5030	0.5	DSC	(1–3)	95M3

Remarks:
(1) the data were taken from Ref. 95M3 with reference to Ref. 88P3
(2) ΔCp was obtained as described in Ref. 90P3
(3) ΔCp in kJ/mol/K, ΔH in kJ/mol and ΔS in J/mol/K

Pepsinogen from swine

Transition	pH	T_{trs}	ΔCp	ΔH	Approach	Remarks	Ref
trans. (1)	6.0	64		623±268	heat	(1)	92M2
trans. (2)	6.0	64		117±8	heat	(1)	92M2
trans. (1)	6.5	61		448±88	heat	(1)	92M2
trans. (2)	6.5	60		138±4	heat	(1)	92M2
trans. (1)	7.0	55		460±67	heat	(1)	92M2
trans. (2)	7.0	60		117±4	heat	(1)	92M2
trans. (1)	7.5	47		276±67	heat	(1)	92M2
trans. (2)	7.5	60		138±42	heat	(1)	92M2
trans. (1)	8.0	44		410±8	heat	(1)	92M2
trans. (2)	8.0	60		113±8	heat	(1)	92M2
trans. (1)	8.5	38		272±29	heat	(1)	92M2
trans. (2)	8.5	57		134±8	heat	(1)	92M2
trans. (1)	9.0	33		126±21	heat	(1)	92M2
trans. (2)	9.0	57		134±8	heat	(1)	92M2

Remark:
(1) van't Hoff analysis, the transition was resolved into two transitions using the parallel model outlined in Ref. 92M2

Pepsinogen, in the presence and in the absence of ethanol

	pH	T_{trs}	ΔCp	ΔH	Approach	Remarks	Ref
	6.4	62.8	24.3±1.7	1038	DSC	(1)	94M1
	6.4	52.1	17.6±1.7	1038	DSC	(2)	94M1

Remarks:
(1) Ref. 94M1 contains further data for other pH values
(2) measured in the presence of 20% ethanol

Pepsinogen, in the presence and in the absence of ethanol

Ethanol	pH	T_{trs}	ΔCp	ΔH	Approach	Remarks	Ref
0 %	6.0	66.2	24.3±1.7	1063	DSC	(1)	95M1
0 %	6.4	62.8		1038	DSC		95M1
0 %	7.2	56.1		816	DSC		95M1
0 %	7.7	55.0		724	DSC		95M1
20 %	5.9	55.8	17.6±1.7	1025	DSC	(1)	95M1
20 %	6.4	52.1		1038	DSC		95M1
20 %	6.8	47.2		925	DSC		95M1
20 %	7.3	39.8		803	DSC		95M1
20 %	8.0	35.3		707	DSC		95M1
20 %	8.2	33.8		711	DSC		95M1

Remark:
(1) ΔCp values obtained from single calorimetric recordings coincide with ΔCp from ΔH versus T_{trs}

Phage P22 Coat Protein

Phage P22 coat protein, wild-type and temperature sensitive mutants

Mutant	pH	T_{trs}	ΔH	Approach	Remark	Ref
wild-type	7.6	40.0±0.1	594±45	DSC	(1)	95G1
Trp48→Gln	7.6	41.2±0.1	318±63	DSC		95G1
Ala108→Val	7.6	37.3±0.1	324±31	DSC		95G1
Thr294→Ile	7.6	43.5±0.1	278±22	DSC		95G1
Phe353→Leu	7.6	40.7±0.1	310±23	DSC		95G1

Remark:
(1) the data refer to wild-type coat protein subunits; coat protein denaturation within the shell lattice proceeds at $T_{trs} = 87°C$ with $\Delta H = 1700$ kJ/mol, see 93G1

Phaseolin

Phaseolin, 7S globulin from french beans

Transition	pH	T_{trs}	ΔCp	ΔH	Approach	Remarks	Ref
trans. (1)	7.0	86±1	12.4±0.6	610±30	DSC	(1)	92B9
trans. (2)	7.0	91±1	12.4±0.6	890±40	DSC	(1)	92B9

Remark:
(1) ΔCp is based on the temperature dependence of ΔH registered from pH 2–11, the value coincides with the mean value obtained from single calorimetric recordings $\Delta Cp = 22.6±5.6$ kJ/mol/K

Phosphoglycerate Kinase

D-phosphoglycerate kinase from yeast

pH	T_{trs}	ΔCp	ΔH	Approach	Ref
5.4	53.5	7.07	690	DSC	87H
5.88	53.9	6.49	716	DSC	87H
6.5	55.8	7.74	993	DSC	87H
7	55.4*	6.78±0.59	854.4±10.9	DSC	87H
7.04	55.7	5.31	898	DSC	87H
7.46	55.2	5.52	894	DSC	87H
7.96	54.7	3.89	880	DSC	87H
8.5	54.5	3.22	835	DSC	87H
9	54.4	3.97	731	DSC	87H

Phosphoglycerate kinase from yeast

pH	T_{trs}	ΔCp	ΔH	Approach	Remark	Ref
7.0		20±3	855±36	DSC	(1)	91G1

Remark:
(1) T_{trs} was found to be scan rate dependent, for the data treatment, see Ref. 91G1

Phosphoglycerate kinase from yeast

pH	T_{trs}	ΔCp	ΔH	Approach	Remarks	Ref
6.5	40	31	586	DSC	(1,2)	92F3

Remarks:
(1) measured in the presence of 0.7 M GuHCl
(2) ΔH was taken from Ref. 89G5

Phosphoglycerate kinase from yeast, heat and cold denaturation

pH	T_{trs}	ΔCp	ΔH	Approach	Remarks	Ref
6.5	40.8±0.3	36±9	570±25	DSC	(1,2)	93D1
6.5	6.3±0.3			DSC	(1,3,4)	93D1
6.5	16.5±0.5			DSC	(1,3,5)	93D1

Remarks:
(1) buffer: 29 mM sodium phosphate, 10 mM EDTA, 1mM DTT, and 0.7 M GuHCl
(2) heat denaturation
(3) cold denaturation, the transition temperature was extrapolated to zero scan rate
(4) first cold denaturation peak
(5) second cold denaturation peak

Phosphoglycerate kinase from yeast, intact protein and domains, heat and cold denaturation

GuHCl Concentration	pH	T_{trs}	ΔC_p	ΔH	Approach	Remarks	Ref
The intact protein:							
0 M	6.5	57.5±0.2		>800	DSC	(1,2)	95G4
0.7 M	6.5	40.8±0.3	36±9	570±25	DSC	(1,2)	93D1
0.7 M	6.5	6.3±0.3			DSC	(1,3,4)	93D1
0.7 M	6.5	16.5±0.5			DSC	(1,3,5)	93D1
Isolated N-terminal domain:							
0 M	6.5	42±1	4.5±2.0	167±23	DSC	(1,2)	95G4
0.6	6.5	6.2±0.5			DSC	(1,6)	95G4
Isolated C-terminal domain:							
0 M	6.5	59.1±0.2		>430±30	DSC	(1,2)	95G4
0.5	6.5	11.9±0.5			DSC	(1,6)	95G4
0.5	6.5	~43			DSC	(1,2)	95G4

Remarks:
(1) buffer: 29 mM sodium phosphate, 10 mM EDTA, 1mM DTT, and the GuHCl concentration indicated
(2) heat denaturation
(3) cold denaturation, the transition temperature was extrapolated to zero scan rate
(4) first cold denaturation peak
(5) second cold denaturation peak
(6) cold denaturation, scan rate 20 K/h

D-Phosphoglycerate kinase from yeast in the presence of ligands

Ligand	pH	T_{trs}	ΔC_p	ΔH	Approach	Remarks	Ref
3-PG	7	54.8–57.0	7.9±0.2	936±10	DSC	(1)	87H
MgADP	7	56.3–58.2	7.7±0.2	953±7	DSC		87H
MgATP	7	54.9–56.5	5.1±0.2	887±5	DSC		87H
sulfate	7	56.1–62.7	9.0±0.2	1000±5	DSC	(2)	89H2
sulfate + 3-PG							
	7	60.3–61.6	9.2±0.5	1025±6	DSC	(3)	89H2
	7	59.3–61.4	9.5±0.1	1047±6	DSC	(4)	89H2
	7	60.6–61.9	8.2±0.3	1041±3	DSC	(5)	89H2
sulfate + MgATP							
	7	60.2–62.3	9.2±0.2	1000±5	DSC	(6)	89H2
	7	58.3–61.1	9.9±0.3	1025±8	DSC	(7)	89H2
	7	60.6–61.4	8.8±0.3	1015±7	DSC	(8)	89H2

Remarks:
(1) abbreviation 3-PG = 3-phosphoglycerate
(2) sulfate concentration 1–120 mM
(3) sulfate concentration 2–80 mM, 10 mM 3-PG
(4) sulfate concentration 10 mM, 2–24 mM 3-PG
(5) sulfate concentration 60 mM, 2–24 mM 3-PG
(6) sulfate concentration 2–80 mM, 10 mM MgATP
(7) sulfate concentration 10 mM, 1–18 mM MgATP
(8) sulfate concentration 60 mM, 1–18 mM MgATP

Phosphoglycerate kinase, wild-type and mutant

Transition	pH	T_{trs}	ΔCp	ΔH	Approach	Remarks	Ref
Wild-type:							
trans. (1)	7.5	51.8		385	DSC	(1)	90B0
trans. (2)	7.5	54.1		546	DSC	(1)	90B0
Mutant (Ala183→Pro):							
trans. (1)	7.5	44.5		357	DSC	(1,2)	90B0
trans. (2)	7.5	54.2		529	DSC	(1,2)	90B0

Remarks:
(1) the transitions were resolved by deconvolution using the sequential model
(2) $\Delta(\Delta G)$ for the mutant protein amounts to –8.4 kJ/mol based on the change in T_{trs} compared with the wild-type

Phosphoglycerate kinase from yeast, wild-type and deletion mutants

Mutant	T_{trs1}	ΔH_{trs1}	T_{trs2}	ΔH_{trs2}	Approach	Remarks	Ref
wild-type	52.3	376	54.1	592	DSC	(1)	95M7
wild-type			62.0	1204	DSC	(1,2)	95M7
Δ(413–415)	45.0	332	54.1	514	DSC	(1,3)	95M7
Δ(413–415)	56.9	329	58.8	753	DSC	(1–3)	95M7

Remarks:
(1) buffer: 20 mM triethanolamine-acetate, pH 7.5
(2) measured in the presence of 5.4 mM ADP, 5.4 mM 3-PG, and 6.5 mM Mg^{2+}
(3) Δ(413–415) = deletion mutant des–(413–415)

Phycocyanin

Phycocyanins from mesophile and thermophile strains

Species	pH	T_{trs}	ΔCp	ΔH	Approach	Remarks	Ref
thermophilic	6	74		753	DSC	(1)	94C5
mesophilic	6	63		414	DSC	(2)	94C5

Remarks:
(1) thermophilic – from *Synechococcus lividus*
(2) mesophilic – from *Phormidium luridum*

Plant Seed Proteins

Glycinin from soybean (see also Phaseolin)

Protein	pH	T_{trs}	ΔH	Approach	Remarks	Ref
11S protein	7.6	99.5	770	DSC	(1,2)	95B6
7S protein	7.6	82	685	DSC	(1,2)	95B6

Remarks:
(1) the paper contains data for seeds at various stages of germination
(2) the data were taken from Figs. 3b and 4 in Ref. 95B6

Plasminogen

Lys-plasminogen and its constitutive fragments, subtransitions

Transition	pH	T_{trs}	ΔH	Approach	Remarks	Ref
Lys-plasminogen:						
trans. (1)	3.4	31.8	205	DSC	(1,2)	84N
trans. (2)	3.4	38.8	170	DSC	(1,2)	84N
trans. (3)	3.4	41.8	200	DSC	(1,2)	84N
trans. (4)	3.4	43.4	215	DSC	(1,2)	84N
trans. (5)	3.4	59.5	135	DSC	(1,2)	84N
trans. (6)	3.4	60	260	DSC	(1,2)	84N
trans. (7)	3.4	71.8	275	DSC	(1,2)	84N
trans. (1)	3.4	33.4	184	DSC	(1,2)	94M17
trans. (2)	3.4	39.6	178	DSC	(1,2)	94M17
trans. (3)	3.4	44.3	228	DSC	(1,2)	94M17
trans. (4)	3.4	46.2	259	DSC	(1,2)	94M17
trans. (5)	3.4	62.2	342	DSC	(1,2)	94M17
trans. (6)	3.4	65.2	126	DSC	(1,2)	94M17
trans. (7)	3.4	74.8	279	DSC	(1,2)	94M17
Glu-plasminogen:						
trans. (1)	3.4	31.9	250	DSC	(1,2)	94M17
trans. (2)	3.4	39.9	199	DSC	(1,2)	94M17
trans. (3)	3.4	43.3	236	DSC	(1,2)	94M17
trans. (4)	3.4	46.9	266	DSC	(1,2)	94M17
trans. (5)	3.4	61.2	349	DSC	(1,2)	94M17
trans. (6)	3.4	65.4	216	DSC	(1,2)	94M17
trans. (7)	3.4	74.5	293	DSC	(1,2)	94M17
Fragment K 1–3:						
trans. (1)	3.4	31.6	220	DSC	(1,3)	84N
trans. (2)	3.4	39.1	185	DSC	(1,3)	84N
trans. (3)	3.4	43.3	200	DSC	(1,3)	84N
trans. (1)	4.0	43.6	245	DSC	(1,3)	84N
trans. (2)	4.0	49.8	240	DSC	(1,3)	84N
trans. (3)	4.0	55.4	230	DSC	(1,3)	84N
trans. (1)	5.4	54.7	275	DSC	(1,3)	84N
trans. (2)	5.4	58	280	DSC	(1,3)	84N
trans. (3)	5.4	63	270	DSC	(1,3)	84N
trans. (1)	7.4	61.9	300	DSC	(1,3)	84N
trans. (2)	7.4	66.5	335	DSC	(1,3)	84N
trans. (3)	7.4	67.7	315	DSC	(1,3)	84N
Fragment K 1:						
trans. (1)	3.4	34.6	215	DSC	(1,4)	84N
Fragment K 4:						
trans. (1)	3.4	42.9	215	DSC	(1,4)	84N
trans. (1)	4.0	50.3	255	DSC	(1,4)	84N
trans. (1)	5.4	54.3	270	DSC	(1,4)	84N
trans. (1)	7.4	61.8	315	DSC	(1,4)	84N
Plasmin heavy chain:						
trans. (1)	3.4	32.3	185	DSC	(1,5)	84N
trans. (2)	3.4	39.9	180	DSC	(1,5)	84N
trans. (3)	3.4	41.7	165	DSC	(1,5)	84N
trans. (4)	3.4	44	235	DSC	(1,5)	84N
trans. (5)	3.4	59.7	280	DSC	(1,5)	84N

Lys-plasminogen and its constitutive fragments, subtransitions (continued)

Transition	pH	T_{trs}	ΔH	Approach	Remarks	Ref
Plasmin light chain:						
trans. (1)	3.4	40.6	125	DSC	(1,6)	84N
trans. (2)	3.4	61.4	240	DSC	(1,6)	84N
Miniplasminogen:						
trans. (1)	3.4	59.3	135	DSC	(1,3)	84N
trans. (2)	3.4	59.8	255	DSC	(1,3)	84N
trans. (3)	3.4	71.7	285	DSC	(1,3)	84N
trans. (1)	3.4	60.8	319	DSC	(1,3)	94M17
trans. (2)	3.4	64.8	131	DSC	(1,3)	94M17
trans. (3)	3.4	73.1	268	DSC	(1,3)	94M17
Microplasminogen:						
trans. (1)	3.4	73.1	175	DSC	(1,6)	94M17
trans. (2)	3.4	73.6	217	DSC	(1,6)	94M17
Miniplasmin:						
trans. (1)	3.4	51.5	150	DSC	(1,3)	84N
trans. (2)	3.4	60.1	260	DSC	(1,3)	84N
trans. (3)	3.4	72.6	260	DSC	(1,3)	84N
C-terminal fragment of miniplasminogen:						
trans. (1)	3.4	58.2	145	DSC	(1,6)	84N
trans. (2)	3.4	58.8	280	DSC	(1,6)	84N
N-terminal fragment of miniplasminogen:						
trans. (1)	3.4	60.9	270	DSC	(1,4)	84N
Fragment K 5:						
trans. (1)	3.4	59.1	265	DSC	(1,4)	84N

Remarks:
(1) data obtained by deconvolution
(2) heat capacity function resolved into 7 subtransitions
(3) heat capacity function resolved into 3 subtransitions
(4) single transition
(5) heat capacity function resolved into 5 subtransitions
(6) heat capacity function resolved into 2 subtransitions

Plasminogen recombinant kringle-1, influence of ε-aminocapronic acid

pH	T_{trs}	ΔCp	ΔH	Approach	Remarks	Ref
7.4	67.6		243	DSC	(1)	91M1
7.4	85.9		460	DSC	(2)	91M1

Remarks:
(1) in the absence of ε-aminocapronic acid
(2) in the presence of ε-aminocapronic acid

Plasminogen kringle-5

Transition	pH	T_{trs}	ΔCp	ΔH	Remark	Ref
		57			(1)	91M1

Remark:
(1) unpublished results of L.C. Sehl and F.J. Castellino, cited in Ref. 91M1

Plasminogen fragments

Mutant	pH	T_{trs}	ΔCp	ΔH	Approach	Remarks	Ref
Lys-plasminogen (Lys77-Asn790)							
trans. (1)	7.4	53.5±1.0		2008±167	DSC	(1,2)	81C1
trans. (2)	7.4	61.2±1.0			DSC	(2)	81C1
fragment (Tyr79-Val353 = kringles 1–3)							
	7.4	62.0±1.0		1087±105	DSC	(2)	81C1
fragment (Val354-Ala439 = kringle–4)							
	7.4	58.0±1.0		364±63	DSC	(2)	81C1
	7.4	71.0±1.5		339±63	DSC	(3)	81C1
Val-plasminogen (Val442-Asn790)							
trans. (1)	7.4	51.0±1.0		594±126	DSC	(1,2)	81C1
trans. (2)	7.4	63.0±1.0			DSC	(2)	81C1

Remarks:
(1) ΔH is the total heat of the first and second transition
(2) measurement in the absence of 6-amino hexanoic acid
(3) measurement in the presence of 6-amino hexanoic acid

Plasminogen Activator

Tissue plasminogen activator

	pH	T_{trs}	ΔCp	ΔH	Approach	Remark	Ref
	4	59.4±0.7		493±40	DSC		88R1
	5	69.5±0.9		920±100	DSC		88R1
	6	72.5±0.7		815±90	DSC		88R1
	7.4	72.8±0.8		1200±120	DSC	(1)	88R1

Remark:
(1) the transition consists of two two-state transitions with T_{trs} = 70.2°C and ΔH = 425 kJ/mol and
T_{trs} = 72.6°C and ΔH = 850 kJ/mol.

Tissue plasminogen activator kringle-2 domain, wild-type and mutants

Mutant	pH	T_{trs}	ΔCp	ΔH	Approach	Ref
wild-type	4.5	64.3±0.8	5±3.8	339±21	DSC	89K2
Val65→Met	4.5	54.4	4.6	284	DSC	89K2

Tissue plasminogen activator, recombinant, single-chain

Mutant	pH	T_{trs}	ΔCp	ΔH	Approach	Remarks	Ref
overall	3.4	64.0		1431	DSC	(1,5)	91N2
trans. (1)	3.4	45.9		192	DSC	(1,5)	91N2
trans. (2)	3.4	53.7		268	DSC	(1,5)	91N2
trans. (3)	3.4	60.2		301	DSC	(1,5)	91N2
trans. (4)	3.4	65.8		364	DSC	(1,5)	91N2
trans. (5)	3.4	76.7		305	DSC	(1,5)	91N2
overall	4.0	67.0		1494	DSC	(1,4)	91N2
trans. (1)	4.0	48.3		176	DSC	(1,4)	91N2
trans. (2)	4.0	58.9		301	DSC	(1,4)	91N2
trans. (3)	4.0	61.4		331	DSC	(1,4)	91N2
trans. (4)	4.0	67.8		393	DSC	(1,4)	91N2

Table 2. Enthalpy and Heat Capacity Changes – Molar Values

Tissue plasminogen activator, recombinant, single-chain (continued)

Mutant	pH	T_{trs}	ΔCp	ΔH	Approach	Remarks	Ref
trans. (5)	4.0	76.0		293	DSC	(1,4)	91N2
overall	7.4	71.1		1728	DSC	(2)	91N2
trans. (2)	7.4	59.9		238	DSC	(1,2)	91N2
trans. (3)	7.4	66.8		418	DSC	(1,2)	91N2
trans. (4)	7.4	70.9		720	DSC	(1,2)	91N2
trans. (5)	7.4	74.8		351	DSC	(1,2)	91N2
overall	7.4	69.2		1824	DSC	(1,3)	91N2
trans. (2)	7.4	57.0		243	DSC	(1,3)	91N2
trans. (3)	7.4	64.9		439	DSC	(1,3)	91N2
trans. (4)	7.4	69.3		741	DSC	(1,3)	91N2
trans. (5)	7.4	74.3		402	DSC	(1,3)	91N2

Remarks:
(1) the transitions correspond to the following domains:
    trans. (1) – serine protease N-terminal part
    trans. (2) – kringle-2
    trans. (3) – kringle-1
    trans. (4) – serine protease C-terminal part
    trans. (5) – epidermal growth factor domain
(2) buffer: 300 mM HEPES, pH 7.4
(3) buffer: 50 mM MOPS with 250 mM NaCl, pH 7.4
(4) buffer: 50 mM glycine, pH 4.0
(5) buffer: 50 mM glycine, pH 3.4

Tissue plasminogen activator, recombinant, two-chain

Mutant	pH	T_{trs}	ΔCp	ΔH	Approach	Remarks	Ref
overall	3.4	64.0		1456	DSC	(1,2)	91N2
trans. (1)	3.4	41.3		188	DSC	(1,2)	91N2
trans. (2)	3.4	51.6		259	DSC	(1,2)	91N2
trans. (3)	3.4	59.8		339	DSC	(1,2)	91N2
trans. (4)	3.4	66.4		372	DSC	(1,2)	91N2
trans. (5)	3.4	78.3		297	DSC	(1,2)	91N2

Remarks:
(1) the transitions correspond to the following domains:
    trans. (1) – serine protease N-terminal part
    trans. (2) – kringle-2
    trans. (3) – kringle-1
    trans. (4) – serine protease C-terminal part
    trans. (5) – epidermal growth factor domain
(2) buffer: 50 mM glycine, pH 3.4

Tissue plasminogen activator, recombinant, and its fragments

Transition	pH	T_{trs}	ΔCp	ΔH	Approach	Remarks	Ref
Recombinant tissue plasminogen activator, single-chain:							
overall	3.4	64.0		1431	DSC	(1,2)	91N2
trans. (1)	3.4	45.9		192	DSC	(1,2)	91N2
trans. (2)	3.4	53.7		268	DSC	(1,2)	91N2
trans. (3)	3.4	60.2		301	DSC	(1,2)	91N2
trans. (4)	3.4	65.8		364	DSC	(1,2)	91N2
trans. (5)	3.4	76.7		305	DSC	(1,2)	91N2

Tissue plasminogen activator, recombinant, and its fragments (continued)

Transition	pH	T_{trs}	ΔCp	ΔH	Approach	Remarks	Ref
50 kDa Subtilisin fragment:							
overall	3.4	57.3		1042	DSC	(1,2)	91N2
trans. (1)	3.4	46.8		201	DSC	(1,2)	91N2
trans. (2)	3.4	54.9		285	DSC	(1,2)	91N2
trans. (3)	3.4	62.7		293	DSC	(1,2)	91N2
trans. (5)	3.4	71.7		264	DSC	(1,2)	91N2
32/35 kDa Subtilisin fragment:							
overall	3.4	57.3		778	DSC	(1,2)	91N2
trans. (2)	3.4	51.3		259	DSC	(1,2)	91N2
trans. (3)	3.4	60.1		289	DSC	(1,2)	91N2
trans. (5)	3.4	72.4		230	DSC	(1,2)	91N2
31 kDa Pepsin fragment:							
overall	3.4	62.0		703	DSC	(1,2)	91N2
trans. (2)	3.4	48.0		155	DSC	(1,2)	91N2
trans. (3)	3.4	61.3		310	DSC	(1,2)	91N2
trans. (5)	3.4	76.1		238	DSC	(1,2)	91N2
35 kDa Pepsin fragment:							
overall	3.4	61.0		661	DSC	(1,2)	91N2
trans. (2)	3.4	38.6		130	DSC	(1,2)	91N2
trans. (3)	3.4	60.9		305	DSC	(1,2)	91N2
trans. (5)	3.4	75.3		226	DSC	(1,2)	91N2
19/23 kDa Pepsin fragment:							
overall	3.4	60.0		410	DSC	(1,2)	91N2
trans. (2)	3.4	47.7		121	DSC	(1,2)	91N2
trans. (3)	3.4	60.8		289	DSC	(1,2)	91N2
Fragment, lacking the finger and EGF domains:							
overall	3.4	52.5		954	DSC	(1,2)	91N2
trans. (1)	3.4	41.1		167	DSC	(1,2)	91N2
trans. (2)	3.4	50.9		251	DSC	(1,2)	91N2
trans. (3)	3.4	57.2		251	DSC	(1,2)	91N2
trans. (4)	3.4	52.2		285	DSC	(1,2)	91N2
Fragment, lacking the finger and EGF domains:							
overall	7.4	68.1		1318	DSC	(1,3)	91N2
trans. (1)	7.4	49.3		230	DSC	(1,3)	91N2
trans. (2)	7.4	59.0		280	DSC	(1,3)	91N2
trans. (3)	7.4	66.3		406	DSC	(1,3)	91N2
trans. (4)	7.4	72.8		402	DSC	(1,3)	91N2
Tissue plasminogen activator, recombinant, single-chain, melting in 6 M GuHCl:							
trans. (6)	3.4	58.3		117	DSC	(1,4)	91N2
32/35 kDa pepsin fragment, melting in 6 M GuHCl:							
trans. (6)	3.4	63.7		109	DSC	(1,4)	91N2

Remarks:
(1) the transitions correspond to the following domains:
    trans. (1) – serine protease N-terminal part
    trans. (2) – kringle-2
    trans. (3) – kringle-1
    trans. (4) – serine protease C-terminal part
    trans. (5) – epidermal growth factor domain
    trans. (6) - finger
(2) buffer: 50 mM glycine, pH 3.4
(3) buffer: 300 mM HEPES , pH 7.4
(4) buffer (2) with 6 M GuHCl, ΔH is not corrected for GuHCl binding

Tissue plasminogen activator, kringle-2, wild-type and mutants in the presence and absence of EACA [remark (1)]

Mutant	pH	T_{trs}(-EACA)	T_{trs}(+EACA)	Approach	Ref
wild-type	8.0	75.4±0.3	86.4±0.4	DSC	92D2
Trp74→Leu	8.0	71.7±0.3	73.9±0.4	DSC	92D2
Trp74→Phe	8.0	71.9±0.4	81.2±0.4	DSC	92D2
Trp74→Ser	8.0	73.7±0.3	76.5±0.3	DSC	92D2
Trp74→Tyr	8.0	73.0±0.3	81.6±0.4	DSC	92D2

Remark:
(1) EACA = ε-aminocaproic acid

Kringle 2 domain of tissue-type plasminogen activator, recombinant, wild-type and Trp mutants

Mutant	pH	T_{trs}	ΔH	Approach	Remarks	Ref
wild-type	8.0	75.6		DSC	(1,2)	94D4
Trp25→Phe	8.0	50.8		DSC	(1,2)	94D4
Trp25→Tyr	8.0	58.0		DSC	(1,2)	94D4

Remarks:
(1) kringle 2 region of tissue-type plasminogen activator, residues 180 to 261
(2) buffer: Tris-acetate, pH 8.0

Plasminogen activator, urokinase-type, protease domain

Transition	pH	T_{trs}	ΔCp	ΔH	Approach	Remarks	Ref
trans. (1)	4.5	76.0±0.1		610±40	DSC	(1,2)	94N3
trans. (2)	4.5	92.3±0.3		290±40	DSC	(1,2)	94N3

Remarks:
(1) the transitions were resolved by deconvolutiom
(2) the protein was inactivated with (L-Glu-L-Gly-L-Arg-chloromethyl ketone)

Protein C

Anticoagulant protein C, natural and recombinant protein

Mutant	pH	T_{trs}	ΔH	Approach	Remarks	Ref
recombinant protein C	8.5	61.5	460	DSC	(1)	95M12
protein C	8.5	62.2	456	DSC	(1)	95M12
dG reco. protein C	8.5	60.5	490	DSC	(1,2)	95M12
dG protein C	8.5	61.9	502	DSC	(1)	95M12
activated protein C	8.5	65.3	477	DSC	(1)	95M12
mod. act. protein C	8.5	63.2	548	DSC	(3)	95M12
recombinant protein C	3.8	43.3	285	DSC	(4)	95M12
recombinant protein C	3.4	91.6	343	DSC		95M12
recombinant protein C	3.2	90.1	364	DSC	(5)	95M12
recombinant protein C	3.0	80.1	310	DSC	(5)	95M12
protein C	3.2	111.0	368	DSC	(5)	95M12
protein C	3.0	105.0	368	DSC	(5)	95M12

Remarks:
(1) the high temperature peak was not observed, measured in the presence of 2 mM EDTA
(2) dG – Gla-domainless protein
(3) modified – D-Phe-Pro-Arg chloromethyl ketone, the protein was measured in the presence of 2 mM EDTA
(4) high temperature transition at 112°C
(5) the low temperature transition was not observed

Anticoagulant protein C, natural and recombinant protein, results of deconvolution

Mutant	T_{trs1}	ΔH_{trs1}	T_{trs2}	ΔH_{trs2}	Approach	Remarks	Ref
rPC	59.0	226	63.6	238	DSC	(1)	95M12
PC	58.8	222	64.5	238	DSC	(1)	95M12
dG-rPC	59.3	276	62.5	238	DSC	(1)	95M12
dG-PC	58.6	259	64.8	238	DSC	(1)	95M12
APC	61.3	247	69.8	243	DSC	(1)	95M12
mod. APC	63.2	527			DSC	(1)	95M12
rPC, pH 3.8	37.8	138	47.8	159	DSC	(2)	95M12
rPC, pH 3.4	84.2	159	91.8	213	DSC	(3)	95M12
rPC, pH 3.2	83.6	167	90.2	218	DSC	(3)	95M12
rPC, pH 3.0	71.6	142	81.6	192	DSC	(3)	95M12
PC, pH 3.2	104.9	180	113.6	226	DSC	(3)	95M12
PC, pH 3.0	99.1	167	108.5	243	DSC	(3)	95M12
PC	53.9	406	70.5	251	DSC	(4)	95M12
PC	52.1	410	69.7	251	DSC	(4)	95M12
rPC	52.1	381	70.0	259	DSC	(5)	95M12
dG-rPC	57.3	255	67.4	238	DSC	(5)	95M12
dG-PC	56.8	259	68.3	247	DSC	(5)	95M12

Abbreviations:
rPC   –   recombinant protein C
PC   –   protein C
dG   –   Gla-domainless protein
APC –   activated protein C
mod.–   modified by D-Phe-Pro-Arg chloromethyl ketone

Remarks:
(1) measured at pH 8.5 in the presence of 2 mM EDTA
(2) deconvolution of the high temperature transition was not performed
(3) the low temperature transition was not observed
(4) measured at pH 8.5 in the presence of 2 mM Ca^{2+}
(5) measured at pH 8.5 in the presence of 1 mM Ca^{2+}

Protein G, Immunoglobulin Binding Protein

Streeptococcal protein G

	pH	T_{trs}	ΔCp	ΔH	Approach	Ref
IgG-binding domain B1:						
	2.30	55.7±0.3	2.6±0.3	178±6	DSC	92A2
	3.51	79.8±0.4	2.6±0.3	237±4	DSC	92A2
	5.40	87.5±0.1	2.6±0.3	258±3	DSC	92A2
IgG-binding domain B2:						
	2.69	58.4±0.3	2.9±0.2	169±5	DSC	92A2
	2.88	64.0±0.1	2.9±0.2	195±1	DSC	92A2
	3.10	65.8±0.7	2.9±0.2	189±8	DSC	92A2
	4.00	76.1±0.4	2.9±0.2	205±11	DSC	92A2
	5.40	79.4±0.2	2.9±0.2	238±5	DSC	92A2

Table 2. Enthalpy and Heat Capacity Changes – Molar Values

Streptococcal protein G, IgG-binding domain, temperature dependence of the heat capacity change at protein unfolding in aqueous solution and related thermodynamic quantities

T	ΔH(exp)	ΔS(exp)	ΔCp(exp)	Approach	Remarks	Ref
5	−4	−103	3.6	DSC	(1–3)	95M3
25	67	145	3.6	DSC	(1–3)	95M3
50	153	422	3.3	DSC	(1–3)	95M3
75	227	643	2.6	DSC	(1–3)	95M3
100	283	801	2.0	DSC	(1–3)	95M3
125	320	897	1.0	DSC	(1–3)	95M3

Remarks:
(1) the data were taken from Ref. 95M3 with reference to Ref. 92A2
(2) ΔCp was obtained as described in Ref. 90P3
(3) ΔCp in kJ/mol/K, ΔH in kJ/mol and ΔS in J/mol/K

Streptococcal protein G, IgG-binding domain

Mutant	pH	T_{trs}	ΔCp	ΔH	Approach	Remarks	Ref
GB1	11.2	25	2.9	83.6	heat	(1–3)	92A3
GB2	11.2	25	2.9	93.9	heat	(1–3)	92A3

Remarks:
(1) the transition was monitored by CD
(2) ΔCp was taken from Ref. 92A2
(3) the mutant proteins differ in six positions: Ile6→Val, Leu7→Ile, Glu19→Lys, Ala24→Glu, Val29→Ala, and Glu42→Val in GB1

B1 immunoglobulin-binding domain from streptococcal protein G, wild-type and mutants

Mutant	pH	T_{trs}	ΔCp	ΔH	Approach	Remarks	Ref
wild-type	5.5	77	2.9	240	heat	(1,2)	95O2
Asp31→Gly	5.5	63		215	heat	(1,2)	95O2
Ala35→Pro	5.5	45		137	heat	(1,2)	95O2
Thr62 deletd 5.5	43			130	heat	(1,2)	95O2
double mutant (Thr60→Ser and Asp55→Gly)							
	5.5	37		105	heat	(1,2)	95O2

Remarks:
(1) ΔCp was taken from Ref. 92A2
(2) buffer: 10 mM MES, 150 mM NaCl, pH 5.5

Immunoglobulin-binding domain B1 from protein G
Mutant AASS (Ile6→Ala, Thr44→Ala, Thr51→Ser AND Thr55→Ser) with a guest site in position 53

Mutant	pH	T_{trs}	ΔCp	ΔH	Approach	Remarks	Ref
AASS-Ala53	5.4	43.8	2.61±0.26	137	heat	(1–3)	94M15
AASS-Thr53	5.4	53.7	2.61±0.26	167	heat	(1–3)	94M15

Remarks:
(1) buffer: 50 mM sodium acetate, 150 mM NaCl, pH 5.4
(2) ΔCp was determined from ΔH versus T_{trs} on several mutants
(3) the data concern the β-sheet propensity of amino acids, see ΔG values

Immunoglobulin-binding domain B1 from protein G (GB1), recombinant and "Chameleon" sequences that adopt context-dependent α or β structure

Mutant	pH	T_{trs}	ΔH	Approach	Remarks	Ref
Ala26→Tyr	5.4	<0		heat	(1–3)	96M8
Ala26→Ile	5.4	59.6	178	heat	(1–3)	96M8
Ala26→Val	5.4	66.5	213	heat	(1–3)	96M8
Tyr45→Ala	5.4	53.4	169	heat	(1–3)	96M8
Tyr45→Val	5.4	61.1	199	heat	(1–3)	96M8
Tyr45→Ile	5.4	61.8	186	heat	(1–3)	96M8
"chameleon-α"	5.4	61.4	174	heat	(1–4)	96M8
"chameleon-α"	5.4		185	DSC	(1–4)	96M8
"chameleon-β"	5.4	39.2	116	heat	(1–3,5)	96M8

Remarks:
(1) van't Hoff treatment
(2) GB1* with residue Lys57 added to the 56-residue construct
(3) measured in 50 mM sodium acetate, 150 mM NaCl, pH 5.4
(4) GB1* containing in position 22–33 the α-helix residues Ala-Trp-Thr-Val-Glu-Lys-Ala-Phe-Lys-Thr-Phe
(5) GB1* containing in position 42–50 the β-sheet residues Ala-Trp-Thr-Val-Glu-Lys-Ala-Phe-Lys-Thr-Phe

Proteinase

Proteinase from *Aspergillus saitoi*

pH	T_{trs}	ΔCp	ΔH	Approach	Remarks	Ref
5.0	65		770	DSC		95T8
5.0			310	heat	(1)	95T8
5.0			176	heat, v.H.	(2)	95T8

Remarks:
(1) van't Hoff heat, from extrapolation to zero heating rate
(2) van't Hoff heat from equilibrium treatment

Proteinase A from *Aspergillus niger*

pH	T_{trs}	ΔCp	ΔH	Approach	Remarks	Ref
5.6	53	13.0±1.3	600	DSC	(1,2)	95F5
		12.8±0.9		DSC	(3)	95F5

Remarks:
(1) ΔCp from ΔH versus T_{trs}
(2) ΔH and T_{trs} were taken from Figs. 3 and 4 in Ref. 95F5
(3) ΔCp from single scanning calorimetric recordings performed at pH 2 to 9

Prothrombin

Prothrombin and fragments in the presence and absence of EDTA, Ca^{2+}, and procoagulant membranes (PSPC)

Protein	total ΔH	T_{trs1}	T_{trs2}	T_{trs3}	T_{trs4}	Approach	Remarks	Ref
Pro + EDTA	1284		58.4	63.9		DSC	(1–3)	94L3
Pro + Ca^{2+}	1176		58.1	64.6	78.4	DSC	(1–3)	94L3
Pro + Ca^{2+} + PSPC (10%)	1075		56.0	70.2	73.6	DSC	(1–3)	94L3
Pro + Ca^{2+} + PSPC (25%)	1017		54.7	70.8	76.8	DSC	(1–3)	94L3
Pre2 + F1.2	887		53.3	59.8		DSC	(1–3)	94L3
Pre2 + F1.2 + Ca^{2+}	1008	47.1	52.4	68.7		DSC	(1–3)	94L3
Pre2 + F1.2 + Ca^{2+} +PSPC (25%)	891		49.6	70.7	75.4	DSC	(1–3)	94L3
Pre1 + EDTA	824	54.0	58.6			DSC	(1–3)	94L3
Pre1 + Ca^{2+}	958	53.4	57.8			DSC	(1–3)	94L3
Pre1 + Ca^{2+} + PSPC (25%)	519	34.2	49.4		56.6	DSC	(1–3)	94L3
Pre2 + F2	803	46.6	51.7			DSC	(1–3)	94L3
Pre2 + F2 + Ca^{2+}	879	(52.5)	51.2		71.8	DSC	(1–3)	94L3
Pre2 + F2 + Ca^{2+} + PSPC (25%)	879	30.4	47.4		69.0	DSC	(1–3)	94L3
Pre2 + F1	619		46.4	62.0		DSC	(1–3)	94L3
Pre2 + F1 + Ca^{2+}	757		47.8	69.4		DSC	(1–3)	94L3
Pre2 + F1 + Ca^{2+} + PSPC (25%)	711	42.3	47.4	67.5	70.9	DSC	(1–3)	94L3
F1.2 + EDTA	439	45.5	59.4			DSC	(1–3)	94L3
F1 + F2 + EDTA	485	45.3	59.7			DSC	(1–3)	94L3
F1.2 + Ca^{2+}	582	45.5	68.9			DSC	(1–3)	94L3
F1.2 + PSPC(25%)	368	46.6	59.6			DSC	(1–3)	94L3
F1.2 + Ca^{2+} + PSPC (25%)	598	46.9	71.8	81.8		DSC	(1–3)	94L3
F1.2 + Ca^{2+} + PSPC (15%)	452	46.7	71.2	75.0		DSC	(1–3)	94L3
F1.2 + Ca^{2+} + PSPC (10%)	351	46.2	71.0	(53.6)		DSC	(1–3)	94L3
F1 + EDTA	230		59.0			DSC	(1–3)	94L3
F1 + Ca^{2+}	314		67.8			DSC	(1–3)	94L3
F1 + Ca^{2+} + PSPC (25%)	351		68.9			DSC	(1–3)	94L3
F1 + Ca^{2+} + PSPC (10%)	218		71.0			DSC	(1–3)	94L3
F2 + EDTA	100	45.8				DSC	(1–3)	94L3
F2 + Ca^{2+}	88	45.5				DSC	(1–3)	94L3
F2 + PSPC(25%)	402	41.2	(33.1)			DSC	(1–3)	94L3
F2 + Ca^{2+} + PSPC (25%)	326	43.1	(39.8)			DSC	(1–3)	94L3

Abbreviations:
Pro   – bovine prothrombin
Pre1   – prethrombin 1
Pre2   – prethrombin 2, catalytic domain
F1   – fragment 1, membrane binding domain
F2   – fragment 2, linker domain
F1.2   – fragment 1.2, covalently linked F1-F2
PSPC – procoagulant membranes

Remarks:
(1) buffer: 50 mM MOPS, 150 mM NaCl, pH 7.4
(2) experimental error is given as ±0.1°C in T_{trs} and ±63 kJ/mol in ΔH
(3) detailed heat values of the individual transitions are given in Ref. 94L3

Rhodopsin

Rhodopsin from various species

Species	pH	T_{trs}	ΔCp	ΔH	Approach	Ref
bovine	6.8	74	14.2±1.1	630±12	DSC	88S7
frog	6.8	67	6.2±0.4	410±10	DSC	88S7
rat	6.8	67	4.3±0.4	416±10	DSC	88S7

Rhodopsin, bovine bleached and unbleached rod outer disk membranes

Form	pH	T_{trs}	ΔCp	ΔH	Approach	Remarks	Ref
unbleached	7.0	71.9±0.4		700±17	DSC	(1)	91K4
bleached	7.0	55.9±0.3		520±17	DSC	(2)	91K4

Remarks:
(1) ΔH per mol rhodopsin
(2) ΔH per mol opsin

Ribulose 1,5-Biphosphate Carboxylase/Oxygenase

Ribulose 1,5-biphosphate carboxylase/oxygenase from lucerne = *Lucerne rubisco*

pH	T_{trs}	ΔCp	ΔH	Approach	Remarks	Ref
7.5	67.1±0.2		656±25	DSC	(1,2)	93B3
10.1	66.4±0.3		631±19	DSC	(1,2)	93B3
5.9–10	62–67	32.5±5.6		DSC	(1,2)	93B3

Remarks:
(1) ΔH and ΔCp are calculated per average subunit (molecular weight = 550,000/16)
(2) Ref. 93B3 contains data for further pH values

Ribonuclease A

Ribonuclease A, bovine pancreatic

pH	T_{trs}	ΔCp	ΔH	Approach	Remarks	Ref
2.0		7.1±1.0		pressure denat.		71H1
2.5	30	8.42	283	DSC		67B
2.8	42.6	9.2	397	heat, v.H.		89P2
3	53.5±0.1	5.02	370±9	DSC		84F3
3.4	54	4.7	340	DSC	(1)	85C2
3.4	54.2	4.7	340	DSC	(4)	85S
3.6	54	5.9	410	DSC		76T
4.1	63.9	4.2±1.1	382	DSC		80J
5	61	4.57	453	DSC	(2)	84F4
5.5	64	5.12	486	DSC		79P4
5.5	64	5.12	486	DSC		74P1
5.8	61.5±0.1	5.02	428±17	DSC		84F3
7	61.3	8.66±0.29	703	DSC		70T2
2–6		5.9±1.3		v.H.	(3)	87L

Ribonuclease A, bovine pancreatic (continued)

pH	T_{trs}	ΔCp	ΔH	Approach	Remarks	Ref
		2.8±0.3		DSC		65B
		9.6±1.7		DSC		67D
		6.3		DSC		69R
		4.2		DSC		73P1
		9.2±2.1				70S

Remarks:
(1) extrapolated to zero protein concentration
(2) dependence on protein concentration
(3) measured in the presence of poly(ethyleneglycol)
(4) from measurements in concentrated protein solution, extrapolated to zero concentration using Figs. 9 and 10 of Ref. 85S

Ribonuclease A, bovine pancreatic

pH	T_{trs}	ΔCp	ΔH	Approach	Remarks	Ref
2.0	38.7±0.5	4.34±0.40	308.2±6.4	DSC	(1–4)	88S5
4.0	62.2±0.6	4.34±0.40	408.9±6.6	DSC	(1–4)	88S5

Remarks:
(1) ΔCp is from d(ΔH)/dT over 69 single scans
(2) ΔCp from the change in the transition baseline at T_{trs} amounts to $\Delta Cp = 3.4\pm0.5$ kJ/mol/K (mean value for 69 single scans)
(3) the pH dependence of T_{trs} (°C) can be expressed by $T_{trs} = 60.1+11.8\times(pH-3.81)$ from pH 2 to 4
(4) ΔH can be described by the expression $\Delta H = (399.9\pm2.5)+(4.3\pm0.2)\times(T_{trs}-60)$

Ribonuclease A

pH	T_{trs}	ΔCp	ΔH	Approach	Remarks	Ref
2.5	45	4.52		DSC	(1–3)	94H8

Remarks:
(1) ΔCp is the experimental value
(2) Ref. 94H8 contains corrections of ΔCp for volume effects
(3) T_{trs} was taken from the figures in Ref. 94H8

Ribonuclease A, bovine pancreatic

pH	T_{trs}	ΔCp	ΔH	Approach	Remarks	Ref
2.0	36.7		275	heat	(1,2)	96L1
2.0	36.7		294	heat	(1,3)	96L1

Remarks:
(1) thermal denaturation monitored by fourth derivative spectroscopy
(2) $\lambda = 283.1$ nm
(3) $\lambda = 285.5$ nm

Ribonuclease A, bovine pancreatic, pH dependence of T_{trs}, ΔCp, and ΔH

pH	T_{trs}	ΔCp	ΔH	Approach	Ref
2.75	42	4.8	385	DSC	74P1
3.00	46	4.4	378	DSC	74P1
3.30	49	4.6	408	DSC	74P1
3.70	55	4.7	432	DSC	74P1
4.00	57	4.3	456	DSC	74P1
5.40	63	4.3	475	DSC	74P1
5.47	64	4.0	487	DSC	74P1

Ribonuclease A, bovine pancreatic, temperature dependence of heat capacity change at protein unfolding

T	$\Delta Cp_{(calc)}$	$\Delta Cp_{(exp)}$	Approach	Remarks	Ref
5	4.6	3.6	DSC	(1,2)	90P3
25	5.2	5.2	DSC	(1,2)	90P3
50	5.3	5.2	DSC	(1,2)	90P3
75	4.3	5.0	DSC	(1,2)	90P3
100	3.5	4.0	DSC	(1,2)	90P3
125	2.0	2.6	DSC	(1,2)	90P3

Remarks:
(1) the calculated value is based on heat capacity of the constituent amino acids minus experimental value for the native protein
(2) the experimental value is based on heat capacity of the heat denatured protein minus experimental value for the native protein

Ribonuclease A, temperature dependence of the heat capacity change at protein unfolding in aqueous solution, and related thermodynamic quantities

T	$\Delta H_{(exp)}$	$\Delta S_{(exp)}$	$\Delta Cp_{(exp)}$	Approach	Remarks	Ref
5	220	641	4.6	DSC	(1–3)	95M3
25	294	896	5.2	DSC	(1–3)	95M3
50	405	1254	5.3	DSC	(1–3)	95M3
75	512	1574	4.3	DSC	(1–3)	95M3
100	603	1826	3.5	DSC	(1–3)	95M3
125	664	1989	2.0	DSC	(1–3)	95M3

Remarks:
(1) the data were taken from Ref. 95M3 with reference to Refs. 93M5 and 93P12
(2) ΔCp was obtained as described in Ref. 90P3
(3) ΔCp in kJ/mol/K, ΔH in kJ/mol and ΔS in J/mol/K

Ribonuclease A, bovine, deuterated (D) protein in D_2O and undeuterated (H) protein in H_2O

	pH	T_{trs}	ΔCp	ΔH	Approach	Re-marks	Ref
D-RNase	2–5.5	40–65	4.8±0.4	275–410	DSC in D_2O	(1,2)	95M2
H-RNase	2.5–6	43–63	5.2±0.3	350–460	DSC in H_2O	(1)	95M2

Remarks:
(1) T_{trs} and ΔH were taken from Fig. 1 in Ref. 95M2
(2) $\Delta(\Delta H)$ between D-RNase and H-RNase denaturation amounts to $-(63\pm13)$ kJ/mol

Ribonuclease A, pressure dependence of ΔCp

Pressure	pH	T_{trs}	ΔCp	ΔH	Approach	Remarks	Ref
1 atm	1.0	42.2	7.49	181	pressure	(1)	95Y2
500 atm	1.0	39.8	6.74	157	pressure	(1)	95Y2
1000 atm	1.0	36.9	5.98	135	pressure	(1)	95Y2
1500 atm	1.0	33.5	5.23	114	pressure	(1)	95Y2
2000 atm	1.0	29.5	4.52	95	pressure	(1)	95Y2

Remark:
(1) buffer: 0.15 M KCl in D_2O, pH was not corrected for isotope effect

Ribonuclease A, phosphate-free

pH	T_{trs}	ΔCp	ΔH	Approach	Ref
7.0	59.5	5.0	393	heat	93A1

Ribonuclease A, bovine, in the presence of 2' CMP

Concentration (mM)	pH	T_{trs}	ΔCp	ΔH	Approach	Ref
0.00	5.5	60.85	4.18	415	DSC	90B6
0.06	5.5	62.41	3.47	450	DSC	90B6
0.15	5.5	63.81	4.39	485	DSC	90B6
0.46	5.5	65.64	5.15	530	DSC	90B6
1.52	5.5	67.65	3.64	547	DSC	90B6
6.09	5.5	69.71	4.10	548	DSC	90B6
15.10	5.5	71.04	4.31	564	DSC	90B6
36.50	5.5	72.11	4.18	567	DSC	90B6

Ribonuclease A, bovine, in the presence of 2' CMP and 200 mM KCl

Concentration (mM)	pH	T_{trs}	ΔCp	ΔH	Approach	Ref
0.00	5.5	61.44	4.31	440	DSC	90B6
0.06	5.5	62.44	2.93	469	DSC	90B6
0.16	5.5	63.30	3.64	479	DSC	90B6
0.62	5.5	65.28	3.89	526	DSC	90B6
2.32	5.5	67.32	3.89	537	DSC	90B6
9.30	5.5	69.51	3.97	541	DSC	90B6
31.00	5.5	71.31	3.51	552	DSC	90B6

Ribonuclease A

pH	T_{trs}	ΔCp	ΔH	Approach	Remarks	Ref
1.80	32.56	3.00	240.2	DSC	(1,2)	96L3
2.00	34.92	5.56	259.7	DSC	(1,2)	96L3
2.20	38.14	1.74	259.7	DSC	(1,2)	96L3
2.40	41.42	2.23	282.6	DSC	(1,2)	96L3
2.50	45.82	2.95	345.1	DSC	(1,2)	96L3
2.60	44.97	1.74	307.8	DSC	(1,2)	96L3
2.75	48.51	0.16	350.6	DSC	(1,2)	96L3

Ribonuclease A (continued)

pH	T_{trs}	ΔC_p	ΔH	Approach	Remarks	Ref
2.80	48.03	3.93	319.3	DSC	(1,2)	96L3
3.00	50.52	4.03	363.4	DSC	(1,2)	96L3
3.50	56.52	4.51	396.6	DSC	(1,2)	96L3
3.70	58.31	6.88	410.4	DSC	(1,2)	96L3
4.00	56.75	7.15	388.7	DSC	(1,2)	96L3
4.50	58.67	6.61	404.1	DSC	(1,2)	96L3
5.00	60.14	3.82	424.2	DSC	(1,2)	96L3
5.50	65.32	5.26	456.3	DSC	(1,2)	96L3
6.00	64.67	8.02	475.8	DSC	(1,2)	96L3
6.00	64.67	7.87	475.8	DSC	(1,2)	96L3
6.50	64.16	5.58	482.6	DSC	(1,2)	96L3
1.8–6.5		4.52±0.59		DSC	(3)	96L3
1.8–6.5	33–64	7.28		DSC	(4)	96L3

Remarks:
(1) measured in 50 mM phosphate buffer
(2) ΔC_p at T_{trs} from individual scans
(3) mean value of ΔC_p from individual scans
(4) from ΔH versus T_{trs}

Ribonuclease A, wild-type and mutations in position 97

Mutant	pH	T_{trs}	ΔH	Approach	Remarks	Ref
wild-type	5.0	63.3	494	heat, v.H.	(1)	96E1
Tyr97→Ala	5.0	29.0	265	heat, v.H.	(1)	96E1
Tyr97→Gly	5.0	30.0	246	heat, v.H.	(1)	96E1
Tyr97→Phe	5.0	53.2	409	heat, v.H.	(1)	96E1

Remarks:
(1) measured in 0.10 M sodium acetate containing 0.10 M NaCl

Ribonuclease A, in the presence of guanidine hydrochloride

GuHCl (M)	pH	T_{trs}	ΔC_p	ΔH	Approach	Ref
0.0	5.0	60.9	6.3	460	DSC	94B3
0.5	5.0	58.0	5.8	445	DSC	94B3
1.0	5.0	52.9	5.0	361	DSC	94B3
2.0	5.0	47.7	5.6	300	DSC	94B3
0.0	6.0	62.2	5.6	498	DSC	94B3
0.5	6.0	53.1	7.4	462	DSC	94B3
1.0	6.0	53.1	7.4	407	DSC	94B3
1.5	6.0	48.3	6.5	373	DSC	94B3
2.0	6.0	42.3	6.3	269	DSC	94B3
2.5	6.0	35.9	4.6	211	DSC	94B3
0.0	7.0	62.8	8.8	520	DSC	94B3
0.5	7.0	59.3	5.6	466	DSC	94B3
1.0	7.0	53.8	5.2	406	DSC	94B3

Ribonuclease A, in the presence of urea

urea (M)	pH	T_{trs}	ΔCp	ΔH	Approach	Ref
0.0	5.0	60.9	6.3	460	DSC	94B3
1.0	5.0	56.7	5.8	382	DSC	94B3
2.0	5.0	51.7	5.0	320	DSC	94B3
3.0	5.0	46.9	5.6	292	DSC	94B3
0.0	6.0	62.2	5.6	498	DSC	94B3
1.0	6.0	58.4	7.5	415	DSC	94B3
2.0	6.0	54.0	5.7	353	DSC	94B3

Ribonuclease A, in the presence of various added solutes

Added Solute	pH	n	ΔCp	$\Delta H(0°C)$	Approach	Remarks	Ref
none	1.8–6.5	18	7.28±0.08	–4.27	DSC	(1,2)	96L3
0.5 M sucrose	2.2–6.5	14	11.3±0.1	–278.9	DSC	(1,2)	96L3
1.0 M sucrose	3.0–6.5	12	13.6±0.1	–501.7	DSC	(1,2)	96L3
1.0 M glycine	3.6–7.0	34	8.54±0.08	–131.8	DSC	(1,2)	96L3
0.5 M GuHCl	3.2–6.0	23	8.08±0.08	–74.8	DSC	(1,2)	96L3
1.0 M GuHCl	3.5–6.0	8	8.79±0.75	–181.3	DSC	(1,2)	96L3
10% glycerol	2.5–6.0	15	7.95±0.67	–132.0	DSC	(1,2)	96L3
0.5 M NaCl	2.8–6.0	11	7.70±0.63	–96.8	DSC	(1–3)	96L3

Remarks:
(1) n = number of experiments
(2) ΔH at 0°C
(3) 10% (v/v) glycerol

Ribonuclease A, bovine pancreatic, effect of osmolytes (sarcosine)

Molal Concentration	pH	T_{trs}	ΔCp	ΔH	Approach	Remarks	Ref
0.0	4.5	57	5.0±0.8	443	DSC	(1)	95P4
2.79	4.5		6.0±0.8		DSC	(1)	95P4
4.06	4.5		6.7±0.9		DSC	(1)	95P4
5.27	4.5		8.8±0.5		DSC	(1)	95P4

Remark:
(1) general expression: $\Delta Cp = 5.30 + (1.39 \cdot 10^{-2}) \times (m_3)^3$ with m_3 in mol/kg

Ribonuclease A, C-peptide analogue, in the presence of sodium dodecylsulfate (SDS)

SDS Concentration	T_{trs}	ΔCp	ΔH	Approach	Ref
1 mM	60	0	40	heat	90W5
5 mM	73	0	37	heat	90W5
10 mM	89	0	29	heat	90W5

Ribonuclease A, in the presence of 50% methanol

pH	T_{trs}	ΔCp	ΔH	Approach	Remarks	Ref
2.2	8.3		234	heat	(1,2)	94R1
2.2	10.2		113	heat	(1,3)	94R1
2.2	20.2		84	heat	(1,4)	94R1
2.2	8.3		205	heat	(1,5)	94R1
2.2	10.2		209	heat	(1,6)	94R1
2.2	9.3		100	heat	(1,7)	94R1

Remarks:
(1) treatment by a two-state model, for a four-state model see Ref. 94R1
(2) transition monitored by CD at 280 nm
(3) transition monitored by CD at 235 nm
(4) transition monitored by CD at 222 nm
(5) transition monitored by absorption at 222 nm
(6) transition monitored by fluorescence at 320 nm
(7) transition monitored by fluorescence at 280 nm

Ribonuclease A in water-free nonane

pH	T_{trs}	Water Content	Approach	Remarks	Ref
8.0	124	6	DSC	(1)	91V2
8.0	111	11	DSC	(1)	91V2
8.0	106	13	DSC	(1)	91V2
8.0	99	16	DSC	(1)	91V2
8.0	92	20	DSC	(1)	91V2
8.0	61	100	DSC	(2)	91V2

Remarks:
(1) given is the residual water content of the ribonuclease powder
(2) reference value, water without nonane

Ribonuclease A from bovine pancreas, free and immobilized enzyme

Form	pH	T_{trs}	ΔCp	ΔH	Approach	Remarks	Ref
free	7.0	65.6	8.9±1.2	463	DSC	(1)	91B1
immobilized enzyme on silica beads:							
trans. (1)	7.0	62.8		292.3	DSC	(2)	91B1
trans. (2)	7.0	66.3		170.6	DSC	(2)	91B1

Remarks:
(1) ΔCp was calculated from ΔH versus T_{trs} given in Ref. 91B1
(2) the main transition could be resolved into the two subtransitions

Ribonuclease A, bovine pancreatic

pH	T_{trs}	ΔCp	ΔH	Approach	Remarks	Ref
5	58.9±0.3		375±6	DSC	(1)	94R3
7	65.6±0.3		460±8	DSC	(1)	94R3
5	62.9±0.3		439±6	DSC	(1,2)	94R3

Remarks:
(1) reference value for immobilized RNase A
(2) in the presence of 3'-CMP

Table 2. Enthalpy and Heat Capacity Changes – Molar Values

Ribonuclease A, bovine pancreatic, immobilized on CPC-silica

pH	T_{trs1}	ΔH_{trs1}	T_{trs2}	ΔH_{trs2}	Approach	Ref
7.0	60.4	232	70.1	195	DSC	94R3
7.5	58.2	268	67.1	193	DSC	94R3
8.0	58.7	273	68.3	171	DSC	94R3
8.2	58.5	268	67.1	163	DSC	94R3

Ribonuclease A in the presence of 2' CMP (cytidine-2'-monophosphate)

Ligand Concentration	pH	T_{trs}	ΔCp	ΔH	Approach	Remark	Ref
0.0 mM	5.5	62.03	see below	455	DSC	(1)	92S14
0.161 mM	5.5	64.97		542	DSC		92S14
0.322 mM	5.5	66.18		549	DSC		92S14
0.962 mM	5.5	68.24		542	DSC		92S14
4.83 mM	5.5	70.82		566	DSC		92S14

Remarks:
(1) the following results were obtained by two-dimensional DSC:

ΔH = 457.3±0.5 kJ/mol
T_{trs} = 61.877±0.013 °C
ΔCp = 4.81±0.25 kJ/mol/K
$\Delta H_{binding}$ = −75.3±2.5 kJ/mol (binding of 2'CMP)
log K = 6.19±0.05 (binding of 2'CMP)
n = 0.9130±0.0026 (stochiometry of ligand binding)

Ribonuclease A, in the presence of subsaturating amounts of CMP

Molar Ratio	pH	T_{trs}	ΔCp	ΔH	Approach	Remarks	Ref
0.0	5.0	61.3	5.9	465	DSC	(1–3)	94B2
0.50 3' CMP	5.0	63.8	7.0	515	DSC	(1,3)	94B2
0.66 2' CMP	5.0	68.0	6.7	536	DSC	(1,3)	94B2
0.0	5.5	61.9	5.5	475	DSC	(1–3)	94B2
0.68 2' CMP	5.5	69.0	6.8	550	DSC	(1,3)	94B2
0.60 2' CMP	5.0	68.7	6.4	565	DSC	(1,4)	94B2

Remarks:
(1) for the results and the thermodynamic parameters of ligand binding see also Ref. 95B2b
(2) RNase A measured without ligand
(3) measured in 0.1 M acetate buffer
(4) measured in 0.1 M acetate buffer with 0.1 M KCl

Ribonuclease A, modified by reductive alkylation

Form	pH	T_{trs}	ΔCp	ΔH	Approach	Ref
native protein	3.0	50.5	4.97	407	DSC	91F2
methylated (94%)	3.0	44.2	5.12	362	DSC	91F2
ethylated (54%)	3.0	46.5	4.85	400	DSC	91F2
n-butylated (40%)	3.0	44.3	5.48	373	DSC	91F2
n-hexylated (37%)	3.0	42.8	5.03	344	DSC	91F2

Remark:
(1) Ref. 91F2 contains further data for varying extent of modification and other pH values

Ribonuclease A, native and modified protein

Mutant	pH	T_{trs}	ΔH	Approach	Remark	Ref
native RNase	4.6	56.2±0.4	446±36	heat, v.H.		94T2
des-[65-72] Rnase	4.6	38.4±0.4	336±33	heat, v.H.	(1)	94T2

Remark:
(1) three-disulfide intermediate lacking the 65-72 disulfide bond

Ribonuclease S and S-Protein

Ribonuclease S and S-Protein

Protein	pH	T_{trs}	ΔCp	ΔH	Approach	Remark	Ref
RNase S	7.0	48.0±0.3	9.2±0.4	531±21	DSC		71H2
RNase S	7.0	46.2±0.3		303±9	heat, v.H.		71H2
RNase S	6.8	45.5		326±29	heat, v.H.		81L
RNase S	7.4	52.6±0.3		268±19	heat	(1)	95S15
S-protein	7	37.6	5.94	230	DSC		70T2
S-protein	7	47.7	8.7	448	DSC		70T2
S-protein	7.0	36.4±0.4	5.9±0.4	226±13	DSC		71H2
S-protein	7.0	36.1±1.2		169±11	heat, v.H.		71H2

Remark:
(1) global analysis of FTIR spectra based on the intramolecular secondary structure band

Ribonuclease A and S

Protein	pH	T_{trs}	ΔCp	ΔH	Approach	Remarks	Ref
RNase A	7.0	62.8	5.5	500	DSC	(1,2)	96C6
RNase A	7.0	64.8	5.6	525	DSC	(1,3)	96C6
RNase A	7.0	66.1	5.5	550	DSC	(1,4)	96C6
RNase S	7.0	50.1	5.1	430	DSC	(1,2)	96C6
RNase S	7.0	51.7	5.0	455	DSC	(1,2)	96C6
RNase S	7.0	52.9	5.1	485	DSC	(1,2)	96C6
RNase S		43–53	5.0		DSC	(1,5)	96C6

Remarks:
(1) protein conc. 0.140 mM
(2) buffer: 10 mM MOPS, 200 mM NaCl
(3) buffer: 30 mM phosphate
(4) buffer: 100 mM phosphate
(5) average from concentration dependent measurements at pH 7.0 and pH 7.5 (see below)

Table 2. Enthalpy and Heat Capacity Changes – Molar Values

Ribonuclease S, dependence on protein concentration at pH 7.0

Concentration (M)	pH	T_{trs}	ΔC_p	ΔH	Approach	Remarks	Ref
5.0×10^{-6}	7.0	43.2		395	heat	(1,2)	96C6
1.0×10^{-5}	7.0	45.0		400	heat	(1,2)	96C6
5.0×10^{-5}	7.0	47.5		415	heat	(1,2)	96C6
7.0×10^{-5}	7.0	48.4	4.8	420	DSC	(1,3)	96C6
1.0×10^{-4}	7.0	49.3	4.9	425	DSC	(1,3)	96C6
1.4×10^{-4}	7.0	50.1	5.1	430	DSC	(1,3)	96C6
2.6×10^{-4}	7.0	51.0	5.2	435	DSC	(1,3)	96C6
3.0×10^{-4}	7.0	51.5	5.1	435	DSC	(1,3)	96C6
5.9×10^{-4}	7.0	52.0	4.9	440	DSC	(1,3)	96C6
7.5×10^{-4}	7.0	52.9	5.0	445	DSC	(1,3)	96C6
8.8×10^{-4}	7.0	53.3	5.4	450	DSC	(1,3)	96C6
	7.0	43–53		435±15		(4)	96C6
	7.0	43–53		449±15		(5)	96C6

Remarks:
(1) buffer: 10 mM MOPS, 200 mM NaCl
(2) van't Hoff treatment, transition monitored by CD
(3) calorimetric enthalpy
(4) averaged value over all measurements
(5) van't Hoff heat from concentration dependence, i.e., ln C versus $1/T_{trs}$

Ribonuclease S, dependence on protein concentration at pH 7.5

Concentration (M)	pH	T_{trs}	ΔC_p	ΔH	Approach	Remarks	Ref
1.0×10^{-5}	7.5	46.0		415	heat	(1,2)	96C6
6.0×10^{-5}	7.5	48.2		420	heat	(1,2)	96C6
1.0×10^{-4}	7.5	49.8	4.9	430	DSC	(1,3)	96C6
1.4×10^{-4}	7.5	50.6	5.0	435	DSC	(1,3)	96C6
1.7×10^{-4}	7.5	51.0	5.2	435	DSC	(1,3)	96C6
2.4×10^{-4}	7.5	51.9	4.9	440	DSC	(1,3)	96C6
4.1×10^{-4}	7.5	52.5	5.1	445	DSC	(1,3)	96C6
6.6×10^{-4}	7.5	53.2	5.0	450	DSC	(1,3)	96C6
	7.5	46–53		440±20		(4)	96C6
	7.5	46–53		465±30		(5)	96C6

Remarks:
(1) buffer: 10 mM MOPS, 200 mM NaCl
(2) van't Hoff treatment, transition monitored by CD
(3) calorimetric enthalpy
(4) averaged value over all measurements
(5) van't Hoff heat from concentration dependence, i.e., ln C versus $1/T_{trs}$

Ribonuclease S, pH dependence

pH	T_{trs}	ΔC_p	ΔH	Approach	Remarks	Ref
4.0	44.6	5.2	395	DSC	(1,2)	96C6
5.0	45.5	5.0	400	DSC	(1,2)	96C6
6.0	46.8	4.9	410	DSC	(1,2)	96C6
6.5	47.2	5.1	415	DSC	(1,2)	96C6

Ribonuclease S, pH dependence (continued)

pH	T_{trs}	ΔC_p	ΔH	Approach	Remarks	Ref
7.0	50.1	5.1	430	DSC	(1,2)	96C6
7.5	50.6	5.0	435	DSC	(1,2)	96C6
8.0	51.0	4.6	415	DSC	(1,2)	96C6
8.5	49.8	4.7	400	DSC	(1,2)	96C6

Remarks:
(1) buffer: 10 mM MOPS, 200 mM NaCl
(2) protein conc. 0.140 mM

Ribonuclease S-protein, dependence on protein concentration

Concentration (mM)	pH	T_{trs}	ΔC_p	ΔH	Approach	Remarks	Ref
0.115	7.0	39.0	2.6	165	DSC	(1)	96G8
0.180	7.0	38.9	2.2	175	DSC	(1)	96G8
0.260	7.0	38.6	2.0	180	DSC	(1)	96G8
0.320	7.0	40.0	2.5	185	DSC	(1)	96G8

Remarks:
(1) buffer: 10 mM MOPS, 200 mM NaCl

Ribonuclease S-peptide/S-protein complexes at different molar ratios (R)

Ratio (R)	pH	T_{trs}	ΔC_p	ΔH	Approach	Remarks	Ref
0.00	7.0	38.6	2.0	180	DSC	(1,2)	96G8
0.25	7.0	49.9	2.5	245	DSC	(1–3)	96G8
0.45	7.0	50.2	3.1	295	DSC	(1–3)	96G8
0.67	7.0	50.8	3.9	340	DSC	(1,2)	96G8
1.10	7.0	51.0	5.1	430	DSC	(1,2)	96G8

Remarks:
(1) buffer: 10 mM MOPS, 200 mM NaCl
(2) S-protein conc. was fixed at 0.260 mM
(3) for the results and the thermodynamic parameters of S-peptide/S-protein complex formation see also Ref. 95B2b

Ribonuclease HI

Ribonuclease HI

pH	T_{trs}	ΔC_p	ΔH	Approach	Remarks	Ref
3.0	50.2	5.87±0.42	392	heat	(1–3)	95Y3

Remarks:
(1) van't Hoff treatment, transition monitored by CD
(2) the paper contains an extended data set from pH 0.5 to pH 4.3 and T_{trs} from 23 to 68°C along with data quantifying reversibility
(3) $\Delta H = a \times T_{trs}(°C) + b$, with $a = 5.87 \pm 0.42$ and $b = 97.36 \pm 19.25$

Ribonuclease HI from *E. coli*, wild-type and mutants

Mutant	pH	T_{trs}	ΔCp	ΔH	Approach	Ref
wild-type	3.0	49.8		410	heat	92K6
Ser68→Ala	3.0	48.3		372	heat	92K6
Ser68→Gly	3.0	41.8		354	heat	92K6
Ser68→Leu	3.0	48.2		351	heat	92K6
Ser68→Thr	3.0	49.1		444	heat	92K6
Ser68→Val	3.0	51.7		413	heat	92K6

Ribonuclease HI from *E. coli*, wild-type and mutants

Mutant	pH	T_{trs}	ΔCp	ΔH	Approach	Remarks	Ref
wild-type	3.0	49.8		410	heat	(1)	92K7
Lys95→Ala	3.0	50.1		478	heat	(1)	92K7
Lys91→Arg	3.0	50.3		498	heat	(1)	92K7
Lys95→Asn	3.0	52.7		455	heat	(1)	92K7
Asp94→Glu	3.0	48.6		401	heat	(1)	92K7
Lys95→Gly	3.0	55.5		428	heat	(1)	92K7
double mutant (Lys95→Ala and Asp94→Glu)							
	3.0	47.8		392	heat	(1)	92K7
double mutant (Lys91→Arg and Lys95→Gly)							
	3.0	55.0		477	heat	(1)	92K7
double mutant (Asp94→Glu and Lys95→Gly)							
	3.0	53.4		530	heat	(1)	92K7
mutant R6	3.0	53.8		438	heat	(1,3)	92K7
wild-type	5.5	52.0		372	heat	(2)	92K7
Lys95→Ala	5.5	52.4		403	heat	(2)	92K7
Lys91→Arg	5.5	52.1		403	heat	(2)	92K7
Lys95→Asn	5.5	55.2		377	heat	(2)	92K7
Asp94→Glu	5.5	50.4		389	heat	(2)	92K7
Lys95→Gly	5.5	58.8		378	heat	(2)	92K7
double mutant (Lys95→Ala and Asp94→Glu)							
	5.5	50.2		366	heat	(2)	92K7
double mutant (Lys91→Arg and Lys95→Gly)							
	5.5	57.8		385	heat	(2)	92K7
double mutant (Asp94→Glu and Lys95→Gly)							
	5.5	57.5		428	heat	(2)	92K7
mutant R6	5.5	57.6		401	heat	(2,3)	92K7

Remarks:
(1) buffer: 10 mM glycine-HCl, 1 mM DTT
(2) buffer: 20 mM sodium acetate, 1 M GuHCl, 1 mM DTT
(3) residues 91–95 of ribonuclease HI from *E. coli* (Lys-Thr-Ala-Asp-Lys) were replaced by (Arg-Thr-Ala-Glu-Gly) from *Thermus thermophilus*

Ribonuclease HI from *E. coli*, wild-type and Asp134-mutants

Mutant	pH	T_{trs}	ΔCp	ΔH	Approach	Remarks	Ref
wild-type	3.0	50.0		449	heat	(1)	94H6
Asp134→Ala	3.0	52.9		438	heat	(1)	94H6
Asp134→Asn	3.0	49.7		462	heat	(1)	94H6
Asp134→Gln	3.0	51.6		475	heat	(1)	94H6
Asp134→Glu	3.0	52.4		450	heat	(1)	94H6
Asp134→His	3.0	51.7		445	heat	(1)	94H6

Ribonuclease HI from *E. coli*, wild-type and Asp134-mutants (continued)

Mutant	pH	T_{trs}	ΔCp	ΔH	Approach	Remarks	Ref
Asp134→Ile	3.0	52.3		438	heat	(1)	94H6
Asp134→Leu	3.0	53.8		494	heat	(1)	94H6
Asp134→Ser	3.0	50.9		444	heat	(1)	94H6
Asp134→Thr	3.0	50.4		448	heat	(1)	94H6
Asp134→Val	3.0	51.1		446	heat	(1)	94H6
wild-type	5.5	53.1	5.18	348	heat	(2,3)	94H6
Asp134→Ala	5.5	58.6		491	heat	(3)	94H6
Asp134→Asn	5.5	56.3		405	heat	(3)	94H6
Asp134→Gln	5.5	57.9		405	heat	(3)	94H6
Asp134→Glu	5.5	56.2		428	heat	(3)	94H6
Asp134→His	5.5	60.1		466	heat	(3)	94H6
Asp134→Ile	5.5	57.7		424	heat	(3)	94H6
Asp134→Leu	5.5	58.6		403	heat	(3)	94H6
Asp134→Ser	5.5	57.0		440	heat	(3)	94H6
Asp134→Thr	5.5	57.0		377	heat	(3)	94H6
Asp134→Val	5.5	57.2		414	heat	(3)	94H6

Remarks:
(1) buffer pH 3.0: 10 mM glycine-HCl
(2) ΔCp = 5.18 kJ/mol/K refers to M. Oobatake, unpubl. results, cited in Ref. 94H6
(3) buffer pH 5.5: 20 mM sodium acetate, 1 M GuHCl, 0.1 M NaCl

Ribonuclease HI, wild-type and mutants

Mutant	pH	T_{trs}	ΔH	Approach	Remarks	Ref
wild-type	3.0	50.0	398	heat	(1,2)	96K1
Asp10→Ala	3.0	58.1	441	heat	(1,2)	96K1
Asp10→Asn	3.0	47.4	379	heat	(1,2)	96K1
Asp10→Glu	3.0	53.8	409	heat	(1,2)	96K1
Asp10→His	3.0	51.5	479	heat	(1,2)	96K1
Asp10→Ser	3.0	52.4	505	heat	(1,2)	96K1
Glu48→Ala	3.0	49.8	383	heat	(1,2)	96K1
Glu48→Asp	3.0	50.8	431	heat	(1,2)	96K1
Glu48→Gln	3.0	49.8	415	heat	(1,2)	96K1
Asp70→Ala	3.0	49.7	358	heat	(1,2)	96K1
Asp70→Asn	3.0	48.8	461	heat	(1,2)	96K1
Asp70→Glu	3.0	51.8	378	heat	(1,2)	96K1
Asp134→Ala	3.0	52.9	438	heat	(1,2)	96K1
Asp134→Asn	3.0	49.7	462	heat	(1,2)	96K1

Remarks:
(1) van't Hoff treatment
(2) buffer: 10 mM glycine/HCl

Ribonuclease HI, wild-type and mutants

Mutant	pH	T_{trs}	ΔCp	ΔH	Approach	Remarks	Ref
wild-type	3.0	49.8		410	heat	(1)	92K8
R1	3.0	47.8		391	heat	(1)	92K8
R2	3.0	47.5		387	heat	(1)	92K8
R4	3.0	53.2		458	heat	(1)	92K8
R5	3.0	52.5		454	heat	(1)	92K8

Ribonuclease HI, wild-type and mutants (continued)

Mutant	pH	T_{trs}	ΔCp	ΔH	Approach	Remarks	Ref
80-Gly-81	3.0	51.0		414	heat	(1)	92K8
R6	3.0	53.8		438	heat	(1)	92K8
R7	3.0	44.1		396	heat	(1)	92K8
Gln113→Pro	3.0	49.2		420	heat	(2)	92K8
R7-E119	3.0	48.7		324	heat	(1)	92K8
R8	3.0	50.2		403	heat	(1)	92K8
R9	3.0	30.9		212	heat	(1)	92K8
R1/R2	3.0	45.3		413	heat	(1)	92K8
R1/R4	3.0	50.2		441	heat	(1)	92K8
R1/R2/R4	3.0	47.7		412	heat	(1)	92K8
R4/R6	3.0	56.5		446	heat	(1)	92K8
R4/R7	3.0	47.8		446	heat	(1)	92K8
R5/R6	3.0	55.8		471	heat	(1)	92K8
R5/R7	3.0	47.9		395	heat	(1)	92K8
R4/R5/R7	3.0	51.8		461	heat	(1)	92K8
R5/R6/R7	3.0	51.7		428	heat	(1)	92K8
R4/R5/R6/R7	3.0	56.0		474	heat	(2)	92K8
wild-type	5.5	52.0		372	heat	(2)	92K8
R1	5.5	50.3		329	heat	(2)	92K8
R2	5.5	48.8		337	heat	(2)	92K8
R4	5.5	56.1		427	heat	(2)	92K8
R5	5.5	56.5		448	heat	(2)	92K8
80-Gly-81	5.5	52.8		408	heat	(2)	92K8
R6	5.5	57.6		363	heat	(2)	92K8
R7	5.5	54.4		377	heat	(2)	92K8
Gln113→Pro	5.5	49.9		339	heat	(2)	92K8
R7-E119	5.5	53.9		431	heat	(2)	92K8
R8	5.5	52.0		372	heat	(2)	92K8
R9	5.5	38.2		190	heat	(2)	92K8
R1/R2	5.5	47.1		294	heat	(2)	92K8
R1/R4	5.5	53.9		371	heat	(2)	92K8
R1/R2/R4	5.5	51.2		440	heat	(2)	92K8
R4/R6	5.5	60.8		511	heat	(2)	92K8
R4/R7	5.5	58.9		448	heat	(2)	92K8
R5/R6	5.5	61.2		454	heat	(2)	92K8
R5/R7	5.5	60.2		452	heat	(2)	92K8
R5/R6/R7	5.5	64.4		448	heat	(2)	92K8
R4/R5/R7	5.5	54.5		480	heat	(2)	92K8
R4/R5/R6/R7	5.5	68.7		481	heat	(2)	92K8

Mutants – Replacements of certain parts of the sequence of E. coli by corresponding parts of Thermus thermophilus
R1: Met(1)-Leu(2) of E. coli replaced by Met(-4)-Asn-Pro-Ser-Pro-Arg(2)
R2: Tyr(28)-Arg-Gly-Arg(31) of E. coli replaced by Phe(28)-His-Ala-His(31)
R4: His(62) of E. coli replaced by Pro(62)
R5: Val(74)-Arg-Gln-Gly-Ile-Thr-Gln(80) of E. coli replaced by Leu(74)-Lys-Lys-Ala-Phe-DEL.-Glu-(80)-Gly(80b) Gly(80b): insertion 80-Gly-81
R6: Lys(91)-Thr-Ala-Asp-Lys(95) of E.coli replaced by Arg(91)-DEL.-DEL.-Glu-Gly(95)
R7: Leu(111)-Gly-Gln-His-Gln-Ile-Lys-Trp-Glu-Trp(120) of E. coli replaced by Met(111)-Ala-Pro-DEL.-Arg-Val-Arg-Phe-His-Phe(120)
    R7-E119: Leu(111)-Gly-Gln-His-Gln-Ile-Lys-Trp-Glu-Trp(120) of E. coli replaced by Met(111)-Ala-Pro-DEL.-Arg-Val-Arg-Phe-His-Phe(120) with Glu(119) instead of His(119)
R8: Ala125→Thr
R9: Thr(149)-Gly-Tyr-Gln-Val-Glu-Val(155) of E. coli replaced by Cys(149)-Pro-Pro-Arg-Ala-Pro-Thr-Leu-Phe-His-Glu-Glu-Ala(161)

Remarks:
(1) buffer: 10 mM glycine, 1 mM DTT
(2) buffer: 20 mM acetate, 1 M GuHCl

Ribonuclease HI, wild-type and mutants

Mutant	pH	ΔT	ΔH	Approach	Remarks	Ref
wild-type	3.0	0.0	443	heat	(1,2)	95A3
Gly23→Ala	3.0	2.3	415	heat	(1,2)	95A3
His62→Pro	3.0	3.4	458	heat	(1,2)	95A3
Val74→Leu	3.0	3.7	496	heat	(1,2)	95A3
Lys95→Gly	3.0	5.7	428	heat	(1,2)	95A3
Asp134→Asn	3.0	−0.3	462	heat	(1,2)	95A3
Asp134→His	3.0	1.7	446	heat	(1,2)	95A3
5N mutant	3.0	12.5	548	heat	(1–3)	95A3
5H mutant	3.0	14.2	455	heat	(1,2,4)	95A3
wild-type	5.5	0.0	342	heat	(1,2)	95A3
Gly23→Ala	5.5	1.8	384	heat	(1,2)	95A3
His62→Pro	5.5	4.1	427	heat	(1,2)	95A3
Val74→Leu	5.5	3.3	382	heat	(1,2)	95A3
Lys95→Gly	5.5	6.8	378	heat	(1,2)	95A3
Asp134→Asn	5.5	3.2	406	heat	(1,2)	95A3
Asp134→His	5.5	7.0	446	heat	(1,2)	95A3
5N mutant	5.5	17.6	542	heat	(1–3)	95A3
5H mutant	5.5	20.2	534	heat	(1,2,4)	95A3

Remarks:
(1) buffers: 10 mM glycine-HCl (pH 3.0) or sodium acetate (pH 5.5) containing 0.1 M NaCl and 1 M GuHCl
(2) van't Hoff treatment, the estimated error in ΔT amounts to ±0.3°C, and in ΔH ±50 kJ/mol
(3) the 5N mutant is a multiple mutant with (Gly23→Ala, His62→Pro, Val74→Leu, Lys95→Gly, and Asp134→Asn)
(4) the 5N mutant is a multiple mutant with (Gly23→Ala, His62→Pro, Val74→Leu, Lys95→Gly, and Asp134→His)

Ribonuclease T1

General Remark. Different extinction coefficients have been reported for ribonuclease T1 (see C.N. Pace et al., Protein Sci. 4 (1995) 2411). Results of calorimetric studies of ribonuclease T1 may be, therefore, influenced by systematic errors in concentration determination (see Ref. 94Y2).

Ribonuclease T1 from *Aspergillus oryzae* and recombinant protein
1) Gln25 isoenzyme

Ribonuclease T1 (Gln25 isoenzyme)

pH	T_{trs}	ΔCp	ΔH	Approach	Remarks	Ref
5.95	52.8	6.23	436	heat, v.H.		89P2
6.0	52.8		436	heat, v.H.		90W2
7	48.25	7.18	398	heat, v.H.		89P2
7	48.3	5.23	404	heat	(1)	89P1
7	48.1±0.3	6.9	406.3±2.5	heat, v.H.		89T2
7	48.3	6.9	402	heat, v.H.		89S8
7	26.95	7.03	194	heat	(2)	89P2
7.9	43.3	6.9	364	heat, v.H.		89P2

Remarks:
(1) value used in data treatment
(2) measurement in the presence of 4.31 M urea

Ribonuclease T1 (Gln25 isoenzyme), calorimetric data

pH	T_{trs}	ΔCp	ΔH	Approach	Remarks	Ref
2.2	45.43±0.15	6.4±3.3	375±3	DSC	(1–4)	92H8
2.8	49.05±0.14	6.9±2.0	399±3	DSC	(1–4)	92H8
4.0	56.71±0.11	10.3±2.8	428±2	DSC	(1–4)	92H8
5.0	56.54±0.01	7.3±0.6	444±2	DSC	(1–4)	92H8
6.0	54.03±0.07	2.9±0.2	423±15	DSC	(1–4)	92H8
7.0	48.91±0.11	5.6±0.6	400±4	DSC	(1–4)	92H8
7.5	46.74±0.07	7.2±2.6	390±5	DSC	(1–4)	92H8
8.0	45.80±0.09	3.7±0.3	383±8	DSC	(1–4)	92H8
9.0	40.73±0.12	1.8±1.0	320±2	DSC	(1–4)	92H8
10.0	39.38±0.19	−1.7±2.3	322±3	DSC	(1–4)	92H8

Remarks:
(1) ΔCp is the value from the single calorimetric recordings, the mean value is $\Delta Cp = 5.02 \pm 1.21$ kJ/mol/K
(2) for all 32 calorimetric experiments it follows: $\Delta H = 68.12 + 6.657 \times T_{trs}$
(3) pH dependence of T_{trs}:
    from pH 5–10: $T_{trs} = 75.03 - 3.675 \times pH$
    from pH 2.2–4.0: $T_{trs} = 31.55 + 6.283 \times pH$
(4) buffers used:
    30 mM glycine (pH 2.2, 2.8, 9.0, 10.0),
    30 mM acetate (pH 4.0, 5.0),
    30 mM PIPES (pH 6.0–8.0)

Ribonuclease T1 (Gln25 isoenzyme)

pH	T_{trs}	ΔCp	ΔH	Approach	Remarks	Ref
5.0	58.8	3.71	466	DSC	(1–3)	94Y2
6.0	54.1	5.35	465	DSC	(1–3)	94Y2
6.5	50.9	5.13	444	DSC	(1–3)	94Y2
7.0	48.8	5.05	453	DSC	(1–3)	94Y2
7.5	47.4	4.84	410	DSC	(1–3)	94Y2
7.9	45.8	4.95	402	DSC	(1–3)	94Y2
8.6	43.4	5.02	396	DSC	(1–3)	94Y2
9.1	42.6	5.09	362	DSC	(1–3)	94Y2
9.7	40.7	5.08	333	DSC	(1–3)	94Y2
9.85	40.6	5.16	333	DSC	(1–3)	94Y2
10.0	39.7	5.31	373	DSC	(1–3)	94Y2

Remarks:
(1) average value $\Delta Cp = 4.97\pm0.42$ kJ/mol/K
(2) from ΔH versus T_{trs}: $\Delta Cp = 5.63$ kJ/mol/K
(3) 10 mM buffer solutions:
    pH 5.0–5.3, sodium acetate
    pH 5.5–7.0, sodium cacodylate-HCl
    pH 7.5–7.9, PIPES-HCl
    pH 8.0–9.0, sodium borate
    pH 9.1–9.9, glycine-NaOH
    pH 10.0, sodium borate-NaOH

Ribonuclease T1, both isoforms, temperature dependence of the heat capacity change at protein unfolding in aqueous solution, and related thermodynamic quantities

T	ΔH(exp)	ΔS(exp)	ΔCp(exp)	Approach	Remarks	Ref
5	173	444	5.5	DSC	(1–3)	95M3
25	281	817	5.3	DSC	(1–3)	95M3
50	410	1233	5.0	DSC	(1–3)	95M3
75	528	1584	4.3	DSC	(1–3)	95M3
100	621	1845	3.0	DSC	(1–3)	95M3
125	672	1976	0.8	DSC	(1–3)	95M3

Remarks:
(1) the data were taken from Ref. 95M3 with reference to Ref. 94Y2
(2) ΔCp was obtained as described in Ref. 90P3
(3) ΔCp in kJ/mol/K, ΔH in kJ/mol and ΔS in J/mol/K

Ribonuclease T1 from *Aspergillus oryzae* and recombinant protein, wild-type and mutants (Gln25 isoenzyme)

Mutant	pH	T_{trs}	ΔCp	ΔH	Approach	Remarks	Ref
wild-type (clone)							
	7	49.3	6.9	389	heat, v.H.		89S8
Gln25→Lys	7	51.7	6.9	448	heat, v.H.		89S8
Glu58→Ala	7	46	6.9	360	heat, v.H.		89S8
double mutant (Gln25→Lys and Glu58→Ala)							
	7	48.8	6.9	389	heat, v.H.		89S8
wild-type	7.5	48				(1)	90N2
double mutant (Tyr24→Cys and Asn84→Cys)							
	7.5	56				(1,2)	90N2

Remarks:
(1) heat and urea gradient electrophoresis
(2) mutant having three disulfide bonds

Ribonuclease T1, native and modified forms

Form	pH	T_{trs}	ΔCp	ΔH	Approach	Remarks	Ref
native	7.5	48.7±0.7	6.9	368±6	heat	(1)	90O1
NFK-Trp59	7.5	25.3±0.1	6.9	344±9	heat	(1,2)	90O1
Kyn-Trp59	7.5	33.9±0.6	6.9	345±6	heat	(1,3)	90O1

Remarks:
(1) ΔCp was taken from Ref. 89P2
(2) modified by ozone oxidation of Trp59 to the N-formylkynurenine derivative
(3) modified by ozone oxidation of Trp59 to the kynurenine derivative

Ribonuclease T1, modified

Form	pH	T_{trs}	ΔH	Approach	Remark	Ref
intact form	4.4	52.3±0.3	530±40	heat, NMR		94K6
CM form	4.4	60.8±0.3	780±50	heat, NMR	(1)	94K6

Remark:
(1) γ-carboxymethylated at Glu58

Ribonuclease T1 (Gln25 isoenzyme), in the presence of NaCl

NaCl	pH	T_{trs}	ΔCp	ΔH	Approach	Remarks	Ref
0.00 M	7.0	48.91	5.6	400	DSC	(1–3)	92H8
0.05 M	7.0	51.05	3.9	416	DSC	(1–3)	92H8
0.10 M	7.0	52.06	6.7	424	DSC	(1–3)	92H8
0.20 M	7.0	53.88	5.5	440	DSC	(1–3)	92H8
0.40 M	7.0	56.49	5.9	449	DSC	(1–3)	92H8
0.60 M	7.0	58.49	5.5	463	DSC	(1–3)	92H8
0.80 M	7.0	60.22	8.8	464	DSC	(1–3)	92H8
1.20 M	7.0	63.19	5.9	470	DSC	(1–3)	92H8
1.60 M	7.0	65.59	5.7	480	DSC	(1–3)	92H8

Remarks:
(1) ΔCp is the value from single calorimetric recordings, the mean value is $\Delta Cp = 5.94 \pm 0.50$ kJ/mol/K
(2) $\Delta Cp = 4.62 \pm 0.71$ kJ/mol/K from ΔH versus T_{trs}
(3) buffer: 30 mM PIPES and NaCl

Ribonuclease T1 (Gln25 isoenzyme), in the presence of $MgCl_2$

$MgCl_2$	pH	T_{trs}	ΔCp	ΔH	Approach	Remarks	Ref
0.00 M	7.0	48.91	5.6	400	DSC	(1–3)	92H8
0.01 M	7.0	52.44	5.1	411	DSC	(1–3)	92H8
0.02 M	7.0	53.52	3.6	414	DSC	(1–3)	92H8
0.03 M	7.0	54.00	3.9	415	DSC	(1–3)	92H8
0.06 M	7.0	55.39	4.3	434	DSC	(1–3)	92H8
0.10 M	7.0	56.56	4.6	433	DSC	(1–3)	92H8
0.20 M	7.0	58.07	3.0	437	DSC	(1–3)	92H8
0.40 M	7.0	60.13	5.1	451	DSC	(1–3)	92H8
0.60 M	7.0	61.86	4.5	458	DSC	(1–3)	92H8
0.80 M	7.0	62.97	4.8	463	DSC	(1–3)	92H8

Remarks:
(1) ΔCp is the value from single calorimetric recordings, the mean value is $\Delta Cp = 4.44 \pm 0.29$ kJ/mol/K
(2) $\Delta Cp = 4.77 \pm 0.48$ kJ/mol/K from ΔH versus T_{trs}
(3) buffer: 30 mM PIPES and NaCl

Ribonuclease T1 (Gln25 isoenzyme), in the presence of nucleotides

pH	T_{trs}	ΔCp	ΔH	Approach	Remarks	Ref
5.0	56.7		490	heat	(4)	79O
5.0	63.1		700	heat	(5)	79O
5.5	60.5	8.2	533	DSC	(1–3)	92H9

Remarks:
(1) measured in the presence of 100 mM 3'-GMP in 50 mM sodium acetate buffer
(2) Ref. 92H9 contains further data on the dependence of ΔH on the concentration of mononucleotides
(3) general expression for the dependence of the thermodynamic quantities T_{trs} and ΔH on the ligand concentration according to the equations $[L]_0 = \exp(A+1000B/T_{1/2})$ and $(\Delta H)_L = (\Delta H_0)_L+(\Delta Cp)_L \times T_{1/2}$ (kJ/mol and kJ/K/mol)

Ligand	Range $[L]_0$	A	B	$(\Delta H_0)_L$	$(\Delta Cp)_L$
2'-GMP	0.4–7.2	180.2	−60.75	246	4.94
3'-GMP	0.4–8.0	188.4	−62.86	340	3.12
5'-GMP	0.6–10.0	185.2	−61.32	25.4	8.43
2'-AMP	1.0–8.0	229.9	−75.79	−29.4	8.69
2'-CMP	1.0–8.0	309.9	−101.94	178	4.78

(4) in the absence of 2'-GMP
(5) in the presence of 143.3 µM 2'-GMP

Ribonuclease T1, (Gln25 isoenzyme) in the presence of ionic ligands

Ligand	pH	T_{trs}	ΔT	ΔH	Approach	Remarks	Ref
Reference value, in the absence of ligand:							
	6.0	52.8	0	436	heat	(1)	90W2
20 mM spermine tetrahydrochloride:							
	6.0	59.9	7.1	427	heat	(1,2)	90W2
20 mM magnesium chloride:							
	6.0	54.7	2.9	404	heat	(1,2)	90W2
20 mM sodium phosphate:							
	6.0	54.0	2.2	423	heat	(1,2)	90W2
20 mM spermidine tetrahydrochloride:							
	6.0	57.1	5.3	460	heat	(1,2)	90W2
20 mM putrescine dihydrochloride:							
	6.0	54.4	2.6	418	heat	(1,2)	90W2
500 mM sodium chloride:							
	6.0	60.2	8.4	473	heat	(1,2)	90W2

Remarks:
(1) van't Hoff treatment, buffer: 50 mM MES
(2) selected examples from concentration dependencies presented in Ref. 90W2

Ribonuclease T1 from *Aspergillus oryzae* and recombinant protein
2) Lys25 isoenzyme

Ribonuclease T1 (Lys25 isoenzyme)

pH	T_{trs}	ΔCp	ΔH	Approach	Remarks	Ref
2	48.5	6.9±1	448±20	heat, v.H.		90K5
2	49.6		459±16	DSC		90K5
2	51.2		452±15	DSC		90K5
2–7	49–62	5.1±0.5		DSC		90K5
3	55.4		481±14	DSC		90K5
4	58.7		489±5	DSC		90K5
4	59.0		503±6	DSC		90K5
5	61.2	4.3±1	531±20	heat, v.H.		90K5
5	62.0		516±20	DSC		90K5
5	61.4		521±8	DSC		90K5
5.3	60.7	6.7±0.8	387	DSC	(1)	92P3
5.3	25	4.94	249	DSC	(2)	92P3
6	58.3		513±10	DSC		90K5
7	51.9		480±9	DSC		90K5
7	53.2		473±16	DSC		90K5
7.0	53.3	6.7±0.8	354	DSC	(2)	92P3

Remarks:
(1) measured in 30 mM MES buffer
(2) from multidimensional analysis varying scan rate and GuHCl concentration. The data refer to 25°C and
    zero denaturant concentration
(3) measured in 30 mM MOPS buffer

Ribonuclease T1 (Lys25 isoenzyme)

pH	T_{trs}	ΔC_p	ΔH	Approach	Remarks	Ref
5.0	61.2	4.78	508	DSC	(1–3)	94Y2
5.3	60.3	4.89	484	DSC	(1–3)	94Y2
5.3	60.2	3.90	503	DSC	(1–3)	94Y2
5.6	58.7	4.07	473	DSC	(1–3)	94Y2
5.8	57.7	3.03	480	DSC	(1–3)	94Y2
6.0	56.9	4.12	470	DSC	(1–3)	94Y2
6.5	54.0	4.98	457	DSC	(1–3)	94Y2
6.5	54.2	5.36	462	DSC	(1–3)	94Y2
7.0	52.1	5.97	445	DSC	(1–3)	94Y2
8.0	48.4	4.90	423	DSC	(1–3)	94Y2
8.6	46.6	4.82	408	DSC	(1–3)	94Y2
9.7	44.0	5.78	393	DSC	(1–3)	94Y2

Remarks:
(1) average value ΔC_p = 4.87±0.64 kJ/mol/K
(2) from ΔH versus T_{trs}: ΔC_p = 5.18 kJ/mol/K
(3) 10 mM buffer solutions
    pH 5.0–5.3, sodium acetate
    pH 5.5–7.0, sodium cacodylate-HCl
    pH 7.5–7.9, PIPES-HCl
    pH 8.0–9.0, sodium borate
    pH 9.1–10.0, sodium borate-NaOH

Ribonuclease T1, temperature dependence of ΔC_p for both isoforms

T	ΔC_p	Approach	Remarks	Ref
5	5.5	DSC	(1)	94Y2
25	5.3	DSC	(1)	94Y2
50	5.0	DSC	(1)	94Y2
75	4.3	DSC	(1)	94Y2
100	3.0	DSC	(1)	94Y2
125	0.8	DSC	(1)	94Y2

Remark:
(1) ΔC_p was determined from heat capacity measurements of the folded and unfolded protein

Ribonuclease T1 (Lys25), upscan and downscan, monitored by fourier transform infrared spectroscopy (FTIR)

Scan	pH	T_{trs}	ΔC_p	ΔH	Approach	Remarks	Ref
upscan	7.0	58.8		351	heat	(1)	93F1
downscan	7.0	57.0		334	heat	(1)	93F1
upscan	7.0	59.0		359	heat	(2)	93F1
downscan	7.0	57.1		333	heat	(2)	93F1
upscan	7.0	59.4		377	heat	(3)	93F1
downscan	7.0	57.9		342	heat	(3)	93F1
upscan	7.0	58.7		355	heat	(4)	93F1
downscan	7.0	57.2		319	heat	(4)	93F1

Remarks:
(1) monitored by FTIR at 1644 cm-1 (irregular structures)
(2) monitored by FTIR at 1624 cm-1 (β-band)
(3) monitored by FTIR at 1585 cm-1 (Asp side chain)
(4) monitored by FTIR at 1515 cm-1 (Tyr band)

Ribonuclease T1 (Lys25 isoenzyme), wild-type and mutants

Mutant	pH	T_{trs}	ΔCp	ΔH	Remarks	Ref
wild-type	7	50.9		460	(1)	92S11
Asn9→Ala	7	48.8		423	(1)	92S11
Tyr11→Phe	7	44.9		423	(1)	92S11
Ser12→Ala	7	47.7		414	(1)	92S11
Ser17→Ala	7	52.6		456	(1)	92S11
Asn36→Ala	7	50.9		464	(1)	92S11
Tyr42→Phe	7	54.3		444	(1)	92S11
Asn44→Ala	7	45.4		381	(1)	92S11
Tyr56→Phe	7	48.8		414	(1)	92S11
Tyr57→Phe	7	49.6		448	(1)	92S11
Ser64→Ala	7	46.3		435	(1)	92S11
Tyr68→Phe	7	46.9		372	(1)	92S11
Asn81→Ala	7	42.3		381	(1)	92S11

Remark:
(1) measurements in 30 mM MOPS buffer

Ribonuclease T1 (Lys25 isoenzyme), wild-type and mutants

Mutant	pH	T_{trs}	ΔCp	ΔH	Approach	Remark	Ref
wild-type	6.0	57.2	4.1±0.9	404	DSC	(1)	94S3
Trp59→Tyr	6.0	54.3		384	DSC		94S3
Tyr24→Trp	6.0	58.8		444	DSC		94S3
double mutant (Trp59→Tyr and Tyr24→Trp)							
	6.0	56.6		421	DSC		94S3
Tyr42→Trp	6.0	56.6		402	DSC		94S3
double mutant (Tyr42→Trp and Trp59→Tyr)							
	6.0	53.7		384	DSC		94S3
Tyr45→Trp	6.0	55.9		447	DSC		94S3
double mutant (Tyr45→Trp and Trp59→Tyr)							
	6.0	52.8		398	DSC		94S3
His40→Thr	6.0	56.7		419	DSC		94S3
double mutant (His40→Thr and Trp59→Tyr)							
	6.0	53.9		387	DSC		94S3
His92→Ala	6.0	55.9		387	DSC		94S3
double mutant (His92→Ala and Trp59→Tyr)							
	6.0	53.2		364	DSC		94S3

Remark:
(1) ΔCp was found to be identical for the wild-type and the mutants

Ribonuclease T1 (Lys25 isoenzyme), wild-type and mutants

Mutant	pH	T_{trs}	ΔH	Approach	Remarks	Ref
wild-type	7.0	57.6	416	heat, v.H.	(1,2)	94F1
wild-type	7.0	57.6	416	heat, v.H.	(1,3)	94F1
Tyr45→Trp	7.0	56.0	373	heat, v.H.	(1,2)	94F1
Tyr45→Trp	7.0	55.7	397	heat, v.H.	(1,3)	94F1
Trp59→Tyr	7.0	53.4	370	heat, v.H.	(1,2)	94F1
Trp59→Tyr	7.0	53.1	385	heat, v.H.	(1,3)	94F1
double mutant (Tyr45→Trp and Trp59→Tyr)						
	7.0	53.1	374	heat, v.H.	(1,2)	94F1
	7.0	52.9	391	heat, v.H.	(1,2)	94F1

Remarks:
(1) given is the pD value instead of pH; uncertainty in T_{trs} ±0.2°C and in ΔH ± 20 kJ/mol
(2) from FTIR spectroscopy, β-band
(3) from FTIR spectroscopy, Tyr band

Ribonuclease T1 (Lys25 isoenzyme), in the presence of NaCl (molar conc.)

Concentration	pH	T_{trs}	ΔCp	ΔH	Approach	Ref
0	5	58.6		529±8	DSC	90K5
0	5	59.5		545±36	DSC	90K5
0.02	5	60.8		522±21	DSC	90K5
0.2	5	63.3		510±21	DSC	90K5
0.2	5	62.8		558±5	DSC	90K5
0.5	5	66.1		530±30	DSC	90K5
1	5	69.5		557±13	DSC	90K5
1	5	70.0		542±11	DSC	90K5
2	5	75.7		516±9	DSC	90K5

Ribonuclease T1, wild-type and disulfide-bond cleaved ribonuclease T1

Form	pH	T_{trs}	ΔCp	ΔH	Approach	Remarks	Ref
wild-type	4	72.4±0.2	6.65	392.0±4.8	DSC	(1)	95H3
wild-type	5	73.2±0.3		390.0±2.5	DSC	(1)	95H3
wild-type	6	70.4±0.3		377.4±2.5	DSC	(1)	95H3
wild-type	7	66.8±0.2		299.9±1.8	DSC	(1)	95H3
wild-type	8	65.4±0.2		266.9±3.4	DSC	(1)	95H3
red-cam	4	38.3±0.2		232.4±5.7	DSC	(1,2)	95H3
red-cam	5	39.8±0.4		257.8±1.8	DSC	(1,2)	95H3
red-cam	6	39.3±0.3		322.4±2.1	DSC	(1,2)	95H3
red-cam	7	36.0±0.1		279.1±1.6	DSC	(1,2)	95H3
red-cam	8	35.8±0.1		243.5±3.1	DSC	(1,2)	95H3
Complexes with 2' GMP:							
wild-type	5	74.4±0.2		454.5±1.7	DSC	(1)	95H3
red-cam	5	43.1±0.2		344.8±2.3	DSC	(1,2)	95H3

Remarks:
(1) measured in the presence of 2 M NaCl, buffer: 100 mM 3,3'-dimethylglutaric acid, 2 M NaCl, pH 4–8
(2) red-cam, reduced carboxyamidomethyl ribonuclease T1

Ribonuclease T1 in buffer and in reverse micelles [AOT/IO, remark (1)]

W_0	pH	T_{trs}	ΔCp	ΔH	Approach	Remarks	Ref
buffer only	7	52.7±0.5	5.0±0.8	409±21	heat	(2,4)	96S6
4.94	7	53.3±0.5	5.0±0.8	408±21	heat	(3)	96S6
6.17	7	51.3±0.5	5.0±0.8	387±21	heat	(3)	96S6
7.40	7	50.0±0.5	5.0±0.8	363±21	heat	(3)	96S6
12.0	7	49.5±0.5	5.0±0.8	345±21	heat	(3)	96S6

Remark:
(1) AOT = bis(2-ethylhexyl)sodium sulfosuccinate IO = isooctane W_0 = moles of entrapped water per mole surfactant
(2) 50 mM cacodylate buffer, pH 7
(3) 150 mM AOT/IO at different W_0 values
(4) ΔCp from ΔH versus T_{trs} at different W_0

ROP

ROP, four-α-helix-bundle protein, temperature dependence of the heat capacity change at protein unfolding in aqueous solution, and related thermodynamic quantities

T	ΔH(exp)	ΔS(exp)	ΔCp(exp)	Approach	Remarks	Ref
5	107		7.9	DSC	(1–3)	95M3
25	265		7.9	DSC	(1–3)	05M3
50	451		7.0	DSC	(1–3)	95M3
75	580		5.0	DSC	(1–3)	95M3
100	601		2.9	DSC	(1–3)	95M3
125	700		0.4	DSC	(1–3)	95M3

Remarks:
(1) the data were taken from Ref. 95M3 with reference to Ref. 93S10
(2) ΔCp was obtained as described in Ref. 90P3
(3) ΔCp in kJ/mol/K, ΔH in kJ/mol and ΔS in J/mol/K

ROP, 4-helix bundle, wild-type

pH	T_{trs}	ΔCp	ΔH	Approach	Remark	Ref
7.0	68.7	4.2		heat	(1)	96P2

Remark:
(1) from ΔH versus T_{trs}

ROP, dimeric four-α-helix-bundle protein

GuHCl (M)	pH	T_{trs}	ΔCp	ΔH	Approach	Remarks	Ref
0	6.0	71.0±0.5	10.3±1.3	580±20	DSC	(1–3)	93S10
0	2–9		8±2.5		DSC	(1,2,4)	93R1
2.5	6.0	51.6±0.3	12.7±0.6	260±20	DSC	(1,5)	93S10

Remarks:
(1) data per mol of dimer
(2) buffer: 10 M sodium phosphate, 10 mM Na_2SO_4, 1 mM EDTA
(3) ΔCp from the individual transitions, this value is the more accurate one according to Ref. 93S10
(4) ΔCp from ΔH versus T_{trs}
(5) the buffer contains additionally 2.5 M GuHCl

ROP, four-helical protein, wild-type and mutants

Mutant	pH	T_{trs}	ΔCp	ΔH	Approach	Remarks	Ref
wild-type	6.0	71.0		580	DSC		93H7
wild-type	6.0	71.0±0.5	10.3±1.3	580±20	DSC	(1,2)	95S13
wild-type	6.0	71.0±0.5	9.2		DSC	(3)	95S13
Leu41→Ala	6.0	52.9±0.5	6.9±1.5	335±20	DSC	(1,2)	95S13
Leu41→Ala	6.0	52.9±0.5	7.3		DSC	(3)	95S13
Leu41→Val	6.0	65.3±0.5	8.3±1.5	461±20	DSC	(1,2)	95S13
Leu41→Val	6.0	65.3±0.5	8.9		DSC	(3)	95S13
Leu48→Ala	6.0	43.4		238	DSC		93H7

Remarks:
(1) data per mol of dimer
(2) ΔCp from single calorimetric scans
(3) ΔCp from heat capacity measurements of native and denatured protein

ROP, wild-type and redesigned four-helix bundle

Mutant	pH	T_{trs}	ΔCp	ΔH	Approach	Remarks	Ref
ROP11	7	74.6±0.7	0.2±2.2	476±32	DSC	(1–3)	94M23
ROP13	7	87.2±0.5	3.9±1.2	348±23	DSC	(2–4)	94M23
ROP21	7	95.4±0.5	3.2±1.5	395±13	DSC	(2–4)	94M23

Remarks:
(1) ROP11 corresponds to the wild-type
(2) data analysis was performed in terms of the model $N_2 \rightarrow 2D$
(3) buffer: 100 mM Na phosphate, 200 mM NaCl, 1 mM DTT
(4) ROP13 and ROP21 are redesigned four-helix bundles in which idealized layers are created that contain Ala and Leu

Four-helix-bundle protein, monomeric ROP, influence of varying loop connections

Loop Connections	pH	T_{trs}	ΔH	Approach	Remarks	Ref
$(Gly)_4, (Gly)_4$	7	74.5	280	heat	(1)	95P8
$(Gly)_4, (Gly)_5$	7	74.3	272	heat	(1)	95P8
$(Gly)_5, (Gly)_4$	7	73.8	259	heat	(1)	95P8
$(Gly)_5, (Gly)_5$	7	74.5	289	heat	(1)	95P8

Remark:
(1) measured by CD at 222 nm in 10 mM sodium phosphate, 350 mM NaCl, 1 mM DTT

ROP variants, hydrophobic core design

Mutant	pH	T_{trs}	ΔCp	ΔH	Approach	Remarks	Ref
wild-type	7	73.9±0.8	2.3±1.6	459±27	DSC	(1,2)	96M11
Ala_2Leu_2-8-rev							
	7	94.3±1.4	6.2±3.0	431±18	DSC	(1,2,3)	96M11
Leu_2Ala_2-8	7	114.5±0.6		526±21	DSC	(1,2,3)	96M11
Ala_2Met_2-8	7	49.2±1.7	1.1±1.7	397±43	DSC	(1,2,3)	96M11

Remarks:
(1) buffer: 100 mM sodium phosphate, 200 mM NaCl, 1 mM DDT for the wild type
(2) ratio of van't Hoff enthalpy to calorimetric enthalpy from 0.5 to 0.6 as expected for a dimeric protein
(3) the mutant description refers to position and number of replacements, see also Ref. 96M11

Rubredoxin

Rubredoxin from *Pyrococcus furiosus*

Form	pH	T_{trs}	ΔCp	ΔH	Approach	Ref
oxidized		113		94.1	DSC	94K5
reduced		102		92.0	DSC	94K5
Zn form		123		99.6	DSC	94K5

Sac7 Protein

Sac7 DNA binding protein, native and recombinant

Form	pH	T_{trs}	ΔH	Approach	Remarks	Ref
native	4.0	86.8	208	DSC	(1)	95M10
native	6.0	99.4		DSC	(2)	95M10
recombinant	4.0	80.3	222	DSC	(1)	95M10
recombinant	6.0	92.7		DSC	(2)	95M10

Remarks:
(1) buffer: 0.05 M potassium acetate, 0.3 M KCl
(2) buffer: 0.01 M potassium phosphate, 0.1 M KCl, 1 mM EDTA

Serum Albumin

Serum Albumin, bovine (BSA)

pH	T_{trs}	ΔCp	ΔH	Approach	Ref
7.15	62	25±5	750	DSC	75L

Serum albumin, human (HSA)

pH	T_{trs}	ΔCp	ΔH	Approach	Remarks	Ref
7.4	25		550		(1)	93K11
7.4	63.24±0.43	16.7±1.7	372±5	DSC	(2)	95P2
7.4	63.4±0.3		438±39	heat	(3)	95S15

Remarks:
(1) from transfer of the protein to urea
(2) the transition temperature amounts to T_{trs} = 62.15±0.17 °C as monitored by fluorescence and to T_{trs} = 63.24±1.1°C as monitored by hydrolase activity
(3) global analysis of FTIR spectra based on the intramolecular secondary structure band

Serum albumin, human, undefatted

	pH	T_{trs}	ΔC_p	ΔH	Approach	Remarks	Ref
	7	64.6		1152	DSC	(1)	88S6
	7	72	35.8	1435	DSC	(2)	88S6
	7	64–66		1163±17	DSC	(3)	88R3
	7	78.22±0.03		1435±20	DSC	(4)	88R3

Remarks:
(1) in the absence of ligand
(2) in the presence of N-Ac-L-tryptophanate and caprylate
(3) concentration dependence, T_{trs} varies from 63.8 to 66.4°C
(4) in the presence of long chain fatty acid

Serum albumin, bovine and human, defatted, overall transition heat

Species	pH	T_{trs}	ΔC_p	ΔH	Approach	Remarks	Ref
Bovine							
	7.0	64.3		839	DSC	(1)	90Y1
	7.0	59.2–78.9		987	DSC	(2,3)	85T1
Human							
	7.0	59.4–80.1		1007	DSC	(2,4)	85T1

Remarks:
(1) measurement made in 2% protein containing solution in the presence of 0.1 M NaCl
(2) measurement made in 10 mM phosphate buffer
(3) Residues 1–581, the transition can be resolved into three subtransitions (see below)
(4) Residues 1–585, the transition can be resolved into three subtransitions (see below)

Serum albumin, bovine and human, defatted, fragments overall transition heat

Species/Res.	pH	T_{trs}	ΔC_p	ΔH	Approach	Remarks	Ref
bovine, residues 198–581							
	7.0	56.4–71.7		548	DSC	(1,2)	85T1
human, residues 198–585							
	7.0	65.6–79.2		606	DSC	(1,2)	85T1

Remark:
(1) the transition can be resolved into two subtransitions (see below)
(2) measurement made in 10 mM phosphate buffer

Serum albumin, bovine and human, defatted, subtransitions

Species/Res.	pH	T_{trs}	ΔC_p	ΔH	Approach	Remarks	Ref
Bovine, residues 1–581:							
trans. (1)	7.0	59.2		315	DSC	(1)	85T1
trans. (2)	7.0	67.9		311	DSC	(1)	85T1
trans. (3)	7.0	78.9		361	DSC	(1)	85T1
Bovine, fragment, residues 198–581:							
trans. (2)	7.0	56.4	4.8	248	DSC	(1,2)	85T1
trans. (3)	7.0	71.7	3.0	300	DSC	(1,2)	85T1

Serum albumin, bovine and human, defatted, subtransitions (continued)

Species/Res.	pH	T_{trs}	ΔCp	ΔH	Approach	Remarks	Ref
Human, residues 1–585:							
trans. (1)	7.0	59.4		377	DSC	(1)	85T1
trans. (2)	7.0	67.4		303	DSC	(1)	85T1
trans. (3)	7.0	80.1		327	DSC	(1)	85T1
Human, fragment, residues 198–585:							
trans. (2)	7.0	65.6		270	DSC	(1)	85T1
trans. (3)	7.0	79.2		336	DSC	(1)	85T1

Remarks:
(1) measurement made in 10 mM phosphate buffer, data were obtained by deconvolution
(2) ΔCp from the difference in ΔH and T_{trs} of the fragment and the correponding peaks of the intact protein; the procedure is supported by identical helix content of protein and fragments

Bovine serum albumin, untreated and SH-protected protein

Transition	pH	T_{trs}	ΔCp	ΔH	Approach	Remarks	Ref
authentic BSA:							
trans. (1)	5.0	61.4		210	DSC	(1,2)	95B2a
trans. (2)	5.0	69.0		535	DSC	(1,2)	95B2a
trans. (1)	5.5	63.8		297	DSC	(1,2)	95B2a
trans. (2)	5.5	69.8		623	DSC	(1,2)	95B2a
trans. (1)	6.0	65.4		370	DSC	(1,2)	95B2a
trans. (2)	6.0	68.2		730	DSC	(1,2)	95B2a
trans. (1)	6.5	66.4		396	DSC	(1,2)	95B2a
trans. (2)	6.5	69.5		780	DSC	(1,2)	95B2a
trans. (1)	7.0	67.7		340	DSC	(1,2)	95B2a
trans. (2)	7.0	71.5		810	DSC	(1,2)	95B2a
trans. (1)	7.5	67.4		385	DSC	(1,2)	95B2a
trans. (2)	7.5	71.6		820	DSC	(1,2)	95B2a
trans. (1)	8.0	64.0		325	DSC	(1,2)	95B2a
trans. (2)	8.0	68.7		880	DSC	(1,2)	95B2a
SH group protected:							
trans. (1)	7.0	62.9		372	DSC	(1–3)	95B2a
trans. (2)	7.0	67.7		672	DSC	(1–3)	95B2a
trans. (1)	8.0	63.9		320	DSC	(1–3)	95B2a
trans. (2)	8.0	68.3		690	DSC	(1–3)	95B2a

Remarks:
(1) the protein was highly purified by HPLC
(2) results of deconvolution
(3) protein treated with iodacetamide

Bovine serum albumin, in the presence of salts

Salt/Transition	pH	T_{trs}	ΔH	Approach	Remarks	Ref
NaCl/trans. (1)	7.0	56.5	790	DSC	(1–3)	91Y1
NaCl/trans. (2)	7.0	63		DSC	(1)	91Y1
NaSCN	7.0	64	935	DSC	(3,4)	91Y1

Remarks:
(1) in the presence of 0.01 M NaCl
(2) overall transition heat of the two transitions
(3) the data were taken from Figs. 7 and 8 of Ref. 91Y1
(4) in the presence of 0.01 M NaSCN (single transition)

Table 2. Enthalpy and Heat Capacity Changes – Molar Values

Serum albumin, bovine (BSA), in the presence of sodium dodecyl sulfate (SDS) – selected values

Ratio (A)	pH	T_{trs}	ΔCp	ΔH	Approach	Remarks	Ref
0	7.0	55+63		720	DSC	(1)	92Y1
10	7.0	87		1320	DSC	(1)	92Y1
30	7.0	83		600	DSC	(1)	92Y1

Remarks:
(A) molar ratio SDS/BSA
(1) the data were taken from Figs. 4 and 5 in Ref. 92Y1

Serum Retinol Binding Protein

Serum retinol binding protein

pH	T_{trs}	ΔCp	ΔH	Approach	Remark	Ref
7.4	78.0±0.4	10.6±1.7	837±42	DSC	(1)	92M6

Remark:
(1) treatment of the unfolding equilibrium by a thermodynamic model that includes ligand dissociation: NR→U + R (R = retinol)

Signal Transduction Protein

SEM-5, signal transduction protein from *Caenorhabditis elegans*, SH3 domain

pH	T_{trs}	ΔCp	ΔH	Approach	Ref
7.3	73.4		270	heat, v.H.	94L5

Spectrin

Spectrin, SH3 domain

pH	T_{trs}	ΔCp	ΔH	Approach	Remarks	Ref
2.0	34	3.7	93	DSC	(1–3)	94V2
2.25	40	3.6	114	DSC	(1–3)	94V2
2.5	47	3.4	139	DSC	(1–3)	94V2
2.75	54	3.2	162	DSC	(1–3)	94V2
3.0	58	3.0	174	DSC	(1–3)	94V2
3.5	63	2.9	188	DSC	(1–3)	94V2
3.5	63±0.8		188±15	DSC		94V2
3.5	60±6.0		196±3	heat	(4)	94V2
4.0	66	2.8	197	DSC	(1–3)	94V2

Remarks:
(1) low ionic strength buffer, 10 mM glycine or 10 mM acetate
(2) ΔCp may be approximated by: $\Delta Cp(T) = -5.95+0.087\times T-(1.795\times 10^{-4})\times T^2$ with T in K
(3) from linear regression it follows $\Delta Cp = 3.4\pm 0.1$ kJ/K/mol
(4) transition monitored by fluorescence

Spectrin, SH3 domain, temperature dependence of the heat capacity change at protein unfolding in aqueous solution, and related thermodynamic quantities

T	ΔH(exp)	ΔS(exp)	ΔCp(exp)	Approach	Remarks	Ref
5	−32	−166	5.0	DSC	(1–3)	95M3
25	52	126	4.8	DSC	(1–3)	95M3
50	146	428	4.3	DSC	(1–3)	95M3
75	222	653	3.6	DSC	(1–3)	95M3
100	274	797	2.7	DSC	(1–3)	95M3
125	296	856	1.6	DSC	(1–3)	95M3

Remarks:
(1) the data were taken from Ref. 95M3 with reference to Ref. 94V2
(2) ΔCp was obtained as described in Ref. 90P3
(3) ΔCp in kJ/mol/K, ΔH in kJ/mol and ΔS in J/mol/K

Spermadhesin

Spermadhesin

Mutant	pH	T_{trs}	ΔH	Approach	Remarks	Ref
aSFP	7.0	78.6	660	DSC	(1)	95M13
PSP	7.0	60.5	439	DSC	(2)	95M13

Remarks:
(1) aSFP = bovine acidic seminal fluid protein
(2) PSP corresponds to the PSP-I/PSP-II heterodimer of porcine seminal plasma proteins I and II

Stefin

Stefin A and B, low molecular weight cysteine protease inhibitors

Protein	pH	T_{trs}	ΔCp	ΔH	Approach	Remarks	Ref
stefin A	5.0	90.8	6.28	473	DSC	(1)	92Z1
stefin A	6.5	94.5	6.28	485.5	DSC	(1)	92Z1
stefin A	8.1	94.5	6.28	515	DSC	(1)	92Z1
stefin B	5.0	50.2	6.70±0.8	293	DSC		92Z1
stefin B	6.5	64.9	6.28	326.5	DSC		92Z1
stefin B	8.1	65.9	6.70±0.8	339	DSC		92Z1

Remark:
(1) ΔCp is an average value

Stefin A and B, recombinant human protein

	pH	T_{trs}	ΔCp	ΔH	Approach	Ref
stefin A	8.1	94.5	6.3	510	DSC	94J3
stefin B	5.3	50.2	6.7	293	DSC	94J3

Stellacyanin

Apostellacyanin from the latex of japanese laquer treee

	pH	T_{trs}	ΔCp	ΔH	Approach	Ref
	7.2	52.7		481	DSC	95A2

Streeptococcal Protein G, see protein G

Streptokinase

Streptokinase

	pH	T_{trs}	ΔCp	ΔH	Approach	Ref
	7.4	46.1±0.9		410±46	DSC	89R

Streptokinase from *Streptococcus equisimilis*, domains

Transition	pH	T_{trs}	ΔCp	ΔH	Approach	Remarks	Ref
trans. (1)	4.0	42.3		407	DSC	(1)	92W1
trans. (1)	7.5	45.9±0.4		431±18	DSC	(2)	92W1
trans. (2)	7.5	60.1±1.3		306±16	DSC	(3)	92W1

Remarks:
(1) single thermal transition
(2) first of two subsequent thermal transitions
(3) second of two subsequent thermal transitions

Streptokinase, proteolytic fragments

Fragment/ TRS	pH	T_{trs}	ΔCp	ΔH	Approach	Remarks	Ref
Fragment SK1:							
overall	7.5	46.1		668	DSC	(1,2)	92M5
trans. (1)	7.5	46.1		450±10	DSC/deconv.		92M5
trans. (2)	7.5	47.3		219±10	DSC/deconv.		92M5
Fragments Tr27 and Th26:							
	7.5	45.4		448±75	DSC	(1,3–5)	92M5
Fragments Tr17 and Th16:							
	7.5	42.8		326±33	DSC	(1,6,7)	92M5

Remarks:
(1) for ΔCp, see specific values in TABLE 3
(2) SK1:residues 1–59/60–293
(3) Tr27: residues 60–293
(4) Th26: residues 63–291
(5) van't Hoff heat $\Delta H^{v.H.} = 356 \pm 54$ kJ/mol
(6) Tr17: residues 148–293
(7) Th16: residues 151–291

Streptokinase, results of deconvolution

Transition	pH	T_{trs}	ΔCp	ΔH	Approach	Remarks	Ref
total	3.4			–	DSC		96M6
trans. (1)	3.4	18.2		138	DSC	(1)	96M6
trans. (2)	3.4	33.7		163	DSC	(1)	96M6
trans. (3)	3.4	84.2		276	DSC		96M6
trans. (4)	3.4	93		870	DSC	(1)	96M6
total	3.6			982	DSC		96M6
trans. (1)	3.6	26.0		201	DSC		96M6
trans. (2)	3.6	37.1		184	DSC		96M6
trans. (3)	3.6	90.1		288	DSC		96M6
trans. (4)	3.6	99.1		309	DSC		96M6
total	3.8			1003	DSC		96M6
trans. (1)	3.8	30.2		234	DSC		96M6
trans. (2)	3.8	39.6		213	DSC		96M6
trans. (3)	3.8	97.8		263	DSC	(1)	96M6
trans. (4)	3.8	105.8		293	DSC	(1)	96M6
total	4.2			–	DSC		96M6
trans. (1)	4.2	38.1		247	DSC		96M6
trans. (2)	4.2	40.6		234	DSC		96M6
trans. (3)	4.2	n.o.			DSC	(2)	96M6
trans. (4)	4.2	n.o.			DSC	(2)	96M6
total	6.5			1103	DSC		96M6
trans. (1)	6.5	48.6		368	DSC		96M6
trans. (2)	6.5	57.8		242	DSC		96M6
trans. (3)	6.5	92.8		230	DSC		96M6
trans. (4)	6.5	102.5		263	DSC		96M6
total	6.7			1116	DSC		96M6
trans. (1)	6.7	49.1		364	DSC		96M6
trans. (2)	6.7	60.6		238	DSC		96M6
trans. (3)	6.7	88.9		242	DSC		96M6
trans. (4)	6.7	99.9		272	DSC		96M6
total	7.0			1079	DSC		96M6
trans. (1)	7.0	49.1		385	DSC		96M6
trans. (2)	7.0	61.8		247	DSC		96M6
trans. (3)	7.0	68.6		188	DSC		96M6
trans. (4)	7.0	87.2		259	DSC		96M6
total	8.5			1112	DSC		96M6
trans. (1)	8.5	47.3		359	DSC		96M6
trans. (2)	8.5	58.7		251	DSC		96M6
trans. (3)	8.5	68.4		272	DSC		96M6
trans. (4)	8.5	82.0		230	DSC		96M6

Remarks:
(1) the data are characterized in Ref. 96M6 as less accurate values
(2) n.o. = not observed due to aggregation

Table 2. Enthalpy and Heat Capacity Changes – Molar Values

Streptokinase, 37-kDa fragment, results of deconvolution

Transition	pH	T_{trs}	ΔCp	ΔH	Approach	Ref
total	7.0			444	DSC	96M6
trans. (1)	7.0	48.4		264	DSC	96M6
trans. (2)	7.0	48.2		180	DSC	96M6
total	8.2			473	DSC	96M6
trans. (1)	8.2	46.5		268	DSC	96M6
trans. (2)	8.2	46.6		205	DSC	96M6
total	8.5			681	DSC	96M6
trans. (1)	8.5	46.0		305	DSC	96M6
trans. (2)	8.5	45.6		163	DSC	96M6
trans. (3)	8.5	27.0		213	DSC	96M6

Streptokinase, 17-kDa fragment, results of deconvolution

Transition	pH	T_{trs}	ΔCp	ΔH	Approach	Ref
total	7.0			263	DSC	96M6
trans. (1)	7.0	47.7		263	DSC	96M6
total	8.2			305	DSC	96M6
trans. (1)	8.2	47.6		305	DSC	96M6
total	8.5			510	DSC	96M6
trans. (1)	8.5	47.3		309	DSC	96M6
trans. (3)	8.5	26.7		201	DSC	96M6

Streptokinase and its fragments

Fragment	pH	T_{trs}	ΔCp	ΔH	Approach	Remarks	Ref
Intact streptokinase, residues Ile1-Lys414:							
trans. (1)	7.0	44.8±0.3		367±31	heat	(1–3)	96C17
trans. (2)	7.0	61.4±0.2		315±18	heat	(1–3)	96C17
Fragment A2-B-C, residues Ala64-Tyr380:							
trans. (1)	7.0	45.0±0.3		209±12	heat	(1,2)	96C17
trans. (2)	7.0	67.5±0.4		333±37	heat	(1,2)	96C17
Fragment B-C, residues Lys147-Tyr380:							
trans. (1)	7.0	42.2±0.2		320±16	heat	(1,2)	96C17
trans. (2)	7.0	67.0±0.3		194±11	heat	(1,2)	96C17
Fragment A2-B, residues Ala64-Leu292:							
trans. (1)	7.0	45.3±0.3		275±17	heat	(1,2)	96C17
Fragment B, residues Lys147-Leu292:							
trans. (1)	7.0	46.0±0.2		365±18	heat	(1,2)	96C17
Fragment C, residues Asp288-Tyr380:							
trans. (2)	7.0	72.3±0.5		195±8	heat	(1,2)	96C17

Remark:
(1) thermal unfolding monitored by far-UV CD and ^{1}H-NMR
(2) buffer: 20 mM phosphate pH 7.0
(3) the results are interpreted in terms of three domains with domain B that unfolds at about 46°C and domains A and C responsible for the high temperature transition

Subtilisin BPN'

Subtilisin BPN', wild-type and mutants

Mutant	pH	T_{trs}	ΔC_p	ΔH	Approach	Remark	Ref
wild-type	8.0	58.5±0.1	20.1±1.7	369±21	DSC		89P4
combined	8.0	72.8±0.2	16.3±2.5	439±1	DSC	(1)	89P4

Remark:
(1) combined variant = Met50→Phe, Asn76→Asp, Gly169→Ala Gln206→Cys (ox.), Tyr217→Lys, and Asn218→Ser

Subtilisin BPN', wild-type and mutants

Mutant	pH	T_{trs}	ΔC_p	ΔH	Approach	Remarks	Ref
wild-type	8.0	75.2			DSC	(1)	90E2
Gln19→Glu	8.0	76.8			DSC	(1)	90E2
Val26→Arg	8.0	74.7			DSC	(1)	90E2
Ser89→Glu	8.0	75.6			DSC	(1)	90E2
Ala98→Lys	8.0	75.4			DSC	(1)	90E2
Thr164→Arg	8.0	59.8			DSC	(1)	90E2
Leu235→Arg	8.0	74.9			DSC	(1)	90E2
Gln271→Glu	8.0	73.7			DSC	(1)	90E2
double mutant (Gln19→Glu and Gln271→Glu)							
	8.0	74.0			DSC	(1)	90E2

Remark:
(1) measured in the presence of 2 mM $CaCl_2$ and 2 mM N-dansyl-3-amino-benzeneboronic acid (inhibitor)

Subtilisin BPN', mutants

Mutant	pH	T_{trs}	ΔC_p	ΔH	Approach	Remarks	Ref
S12	9.63	63.5		585	DSC	(1,2)	92B7
S15	9.63	63.0		585	DSC	(3)	92B7

Remarks:
(1) multiple mutations: (Met50→Phe, Tyr217→Lys, Asn218→Ser, and Ser221→Cys)
(2) apo form of the protein
(3) residues 75–83 deleted, the calcium binding ability of the protein is abolished

Subtilisin Inhibitor

Subtilisin inhibitor

pH	T_{trs}	ΔC_p	ΔH	Approach	Ref
7	80.1*		511±11	DSC	81T

Table 2. Enthalpy and Heat Capacity Changes – Molar Values

Subtilisin inhibitor from *Streptomyces albogriseolus*

	pH	T_{trs}	ΔCp	ΔH	Approach	Remarks	Ref
Heat denaturation:							
	2.45	21.6	7.6±0.5	126	heat	(1–4)	91T2
	2.72	31.6	7.6±0.5	106	heat	(1–4)	91T2
	2.87	35.7	7.6±0.5	218	heat	(1–4)	91T2
	3.03	38.9	7.6±0.5	250	heat	(1–4)	91T2
	3.07	50.2	8.45±0.63	313±23	DSC	(5)	91T2
	3.21	45.8	7.6±0.5	313	heat	(1–4)	91T2
Cold denaturation:							
	2.45	13.8		−190	heat	(6)	91T2
	2.72	4.1		−145	heat	(6)	91T2
	2.87	−1.8		−114	heat	(6)	91T2
	3.03	−3.6		−141	heat	(6)	91T2
	3.21	−10.7		−203	heat	(6)	91T2

Remarks:
(1) average of data obtained by CD measurement at 222, 217, 212, 202, and 197 nm treatment by the dimer-monomer equilibrium $N_2 \rightarrow 2U$
(2) experiments were made at 0.368 mg/ml protein conc.
(3) the temperature dependence of ΔH can be expressed by $\Delta H = -44.4 + (7.61 \pm 0.46) \times T_{trs}$
(4) the pH dependence of the transition temperature can be expressed by $T_{trs} = -(53.0 \pm 5.1) + (30.7 \pm 1.8) \times pH$ for pH 2.45–3.21
(5) experiment was made at protein conc. 3.90 mg/ml treatment by the dimer-monomer equilibrium $N_2 \rightarrow 2U$
(6) the pH dependence of the transition temperature can be expressed by $T_{trs} = (89.7 \pm 7.0) - (31.3 \pm 2.4) \times pH$ for pH 3.21–2.45

Subtilisin inhibitor from *Streptomyces*, effect of intersubunit disulfide bond

Mutant	pH	T_{trs}	ΔH	Approach	Remarks	Ref
wild-type	3.0	46.58	361.6	DSC	(1,2)	94T3
wild-type	3.2	53.36	444.3	DSC	(1,2)	94T3
wild-type	7.0	82.21	796.2	DSC	(1,2)	94T3
wild-type	9.5	80.70	777.8	DSC	(1,2)	94T3
Asp83→Cys	3.0	62.98	303.9	DSC	(1,2)	94T3
Asp83→Cys	3.2	67.34	369.5	DSC	(1,2)	94T3
Asp83→Cys	7.0	95.30	648.9	DSC	(1,2)	94T3
Asp83→Cys	9.5	94.97	634.3	DSC	(1,2)	94T3
Asp83→Asn	3.0	45.64	377.7	DSC	(1,2)	94T3
Asp83→Asn	3.2	51.28	460.7	DSC	(1,2)	94T3
Asp83→Asn	7.0	83.90	813.4	DSC	(1,2)	94T3
Asp83→Asn	9.5	82.58	795.0	DSC	(1,2)	94T3

Remarks:
(1) for ΔCp, see specific values, table 3
(2) the pH dependence of T_{trs} in the acidic region may be represented by the following expressions wild type: $T_{trs} = -55.06(\pm 8.42) + 33.88(\pm 2.70) \times pH$, Asp83→Cys: $T_{trs} = -2.36(\pm 1.65) + 21.78(\pm 0.58) \times pH$, Asp83→Asn: $T_{trs} = -39.02(\pm 7.64) + 28.22(\pm 2.48) \times pH$

Superoxide Dismutase

Superoxide dismutase, wild-type and mutants

Mutant/Transition		pH	T_{trs}	ΔH	Approach	Remarks	Ref
wild-type	trans. (1)	7.8	74.9	268.4	DSC	(1)	90L1
wild-type	trans. (2)	7.8	83.6	665.3	DSC	(1)	90L1
Cys111→Ser	trans. (1)	7.8	77.7	409.3	DSC	(1,2)	90L1
Cys111→Ser	trans. (2)	7.8	84.5	687.8	DSC	(1,2)	90L1
Cys6→Ala	trans. (1)	7.8	75.7	257.4	DSC	(1,3)	90L1
Cys6→Ala	trans. (2)	7.8	84.1	881.2	DSC	(1,3)	90L1
double mutant (Cys6→Ala and Cys111→Ser)							
	trans. (1)	7.8	77.3	330.6	DSC	(1)	90L1
	trans. (2)	7.8	82.8	444.8	DSC	(1)	90L1

Remarks:
(1) two subsequent transitions resolved by deconvolution
(2) Cys6 unchanged

Superoxide dismutase, wild-type and Ile58 mutant

Transition	pH	T_{trs}	ΔCp	ΔH	Approach	Remarks	Ref
wild-type:							
trans. (2)	7.8	70.0±0.7		594±50	DSC	(1)	96B9
trans. (3)	7.8	88.9±0.5		379±26	DSC	(1)	96B9
Ile58→Thr:							
trans. (1)	7.8	34.3±2.4		272±105	DSC	(1)	96B9
trans. (2)	7.8	56.4±0.5		270±8	DSC	(1)	96B9
trans. (3)	7.8	72.4±1.1		799±42	DSC	(1)	96B9

Remark:
(1) ΔH is given per mole of tetramer

Tailspike Protein

Tailspike protein of P22 bacteriophage, wild-type and mutants

Mutant	pH	T_{trs}	ΔH	Approach	Remarks	Ref
wild-type	7.4	88.4±0.3	6320±600	DSC	(1)	89S11
Gly177→Arg	7.4	87.9±0.3	6440±600	DSC	(1)	89S11
Thr235→Ile	7.4	88±0.3	5860±600	DSC	(1)	89S11
Gly244→Arg	7.4	87.5±0.3	6780±600	DSC	(1)	89S11
Glu309→Val	7.4	87.8±0.3	5690±600	DSC	(1)	89S11
Arg285→Lys	7.4	85.9±0.3	6190±600	DSC	(1)	89S11
Gly323→Asp	7.4	88.5±0.3	6780±600	DSC	(1)	89S11
Arg382→Ser	7.4	83.2±0.3	4980±600	DSC	(1)	89S11

Remark:
(1) the data refer to the trimer

Taka-Amylase

Taka-amylase A

pH	T_{trs}	ΔCp	ΔH	Approach	Remark	Ref
7	62	36.4±4.1	2251±40	DSC	(1)	87F

Remark:
(1) measurement in the absence of calcium

Tendamistat

Tendamistat, α-amylase inhibitor

pH	T_{trs}	ΔCp	ΔH	Approach	Remarks	Ref
2.0	68.3	2.89±0.81	236	DSC	(1)	92R2
5.0	93	2.89±0.81	307	DSC	(1)	92R2
7.0	81.6	2.89±0.81	274	DSC	(1)	92R2

Remark:
(1) ΔCp was found to be temperature dependent based on an analysis of heat capacity functions

Tendamistat, α-amylase inhibitor, temperature dependence of the heat capacity change at protein unfolding in aqueous solution, and related thermodynamic quantities

T	$\Delta H(exp)$	$\Delta S(exp)$	$\Delta Cp(exp)$	Approach	Remarks	Ref
5	−22	−213	3.5	DSC	(1–3)	95M3
25	70	109	3.6	DSC	(1–3)	95M3
50	176	452	3.6	DSC	(1–3)	95M3
75	262	711	2.9	DSC	(1–3)	95M3
100	321	877	2.3	DSC	(1–3)	95M3
125	351	955	1.4	DSC	(1–3)	95M3

Remarks:
(1) the data were taken from Ref. 95M3 with reference to Ref. 92R2
(2) ΔCp was obtained as described in Ref. 90P3
(3) ΔCp in kJ/mol/K, ΔH in kJ/mol and ΔS in J/mol/K

Tendamistat

Mutant	pH	T_{trs}	ΔCp	ΔH	Approach	Ref
wild-type	7.0	81.6	2.9	274	DSC	93H7
double mutant (Cys45→Ala and Cys73→Ala)						
	7.0	59.0	2.8	213	DSC	93H7

Tendamistat, α-amylase inhibitor, wild-type and disulfide mutants

Mutant	pH	T_{trs}	ΔCp	ΔH	Approach	Remarks	Ref
wild-type	2.0	68.3	2.89±0.81	236	DSC	(1)	92R2
wild-type	5.0	93	2.89±0.81	307	DSC	(1)	92R2
wild-type	7.0	81.6	2.9	274	DSC		93H7
wild-type	7.0	81.6	2.89±0.81	274	DSC	(1)	92R2
wild-type	7.0	81.6	2.9	196	DSC	(2,3)	95V8
(Cys11→Ala and Cys27→Ala)							
	7.0	42.7	2.2	135	DSC	(2,3)	95V8
(Cys11→Ala and Cys27→Leu)							
	6.0	55.9		201	DSC		95V8
	6.0	55.5		199	DSC		95V8
	7.0	48.6		186	DSC		95V8
	7.0	49.7		192	DSC		95V8
	7.0	50.0		187	DSC		95V8
	7.0	49.7		185	DSC		95V8
	7.0	49.5	2.0	187	DSC	(2,3)	95V8
(Cys11→Ala and Cys27→Ser)							
	7.0	50.4	2.2	186	DSC	(2,3)	95V8
(Cys11→Ala and Cys27→Thr)							
	6.0	54.8		223	DSC		95V8
	6.0	55.6		224	DSC		95V8
	7.0	52.6		212	DSC		95V8
	7.0	52.5		214	DSC		95V8
	7.0	52.5	2.3	213	DSC	(2,3)	95V8
	8.0	41.5		194	DSC		95V8
	8.0	43.1		189	DSC		95V8
(Cys45→Ala and Cys73→Ala)							
	5.0	67.7		239	DSC		95V8
	6.0	65.3		234	DSC		95V8
	7.0	59.0	2.8	213	DSC		93H7
	7.0	59.1		228	DSC		95V8
	7.0	59.2		229	DSC		95V8
	7.0	57.6		219	DSC		95V8
	7.0	58.6	2.3	225	DSC	(2,3)	95V8
	8.0	50.3		201	DSC		95V8
	8.0	49.8		199	DSC		95V8

Remarks:
(1) ΔCp was found to be temperature dependent based on an analysis of heat capacity functions
(2) error in T_{trs} ±0.5°C, in ΔH ±5%
(3) ΔCp is the mean value derived from individual scans

Thermolysin

Thermolysin and thermolysin fragment

Form	pH	T_{trs}	ΔCp	ΔH	Approach	Ref
native	7.5	90		1336±50	DSC	88S2
fragment 206–316						
	7.5	66	7.9	272	heat, v.H.	82V

Thermolysin fragment 255–316
thermodynamic parameters derived from concentration dependent measurements

Protein Concentration	T_{trs}	ΔH_{cal}	$\Delta H_{v.H.}$	ΔCp	Approach	Remarks	Ref
0.19 mg/ml	67.1	177	307		DSC	(1,2,4)	94C9
0.29 mg/ml	68.0	200	306		DSC	(1,2,4)	94C9
0.72 mg/ml	70.2	176	352		DSC	(1,2,4)	94C9
0.94 mg/ml	70.7	204	320	1.8	DSC	(1,2,4)	94C9
1.08 mg/ml	70.9	178	318	3.2	DSC	(1,2,4)	94C9
1.37 mg/ml	71.4	170	349	2.0	DSC	(1,3,4)	94C9
2.21 mg/ml	72.4	199	346	2.4	DSC	(1,2,4)	94C9
4.55 mg/ml	73.7	199	363	2.4	DSC	(1,2,4)	94C9
average		192±13	347±22	2.4±0.5		(5,6)	94C9

Remarks:
(1) measured in 20 mM phosphate, 0.1 M NaCl, pH 7.5
(2) scan rate 2 K/min
(3) scan rate 0.5 K/min
(4) unfolding equilibrium according to: folded dimer→2 unfolded monomers
(5) averages weighted by sample concentration
(6) global analysis of the calorimetric heat capacity curves carried out by multi-dimensional fitting to the model $N_2 \rightarrow 2N \rightarrow 2U$ is given in Ref. 95A9

Thermolysin, fragment 205–316

	pH	T_{trs}	ΔCp	ΔH	Approach	Remarks	Ref
	7.5	68±0.8		276±13	DSC	(1,2)	96C16
	7.5	67.8		282	DSC	(3)	96C16

Remarks:
(1) thermal unfolding according to a $N_2 \rightarrow 2N \rightarrow 2U$ model
(2) average from concentration dependent measurements
(3) from multiple fit including the dissociation equilibrium

Thioredoxin

Thioredoxin from *E. coli*, oxidized and reduced form

Form	pH	T_{trs}	ΔCp	ΔH	Approach	Ref
oxidized	7.0	86		435	DSC	91B3
reduced	7.0	75		351	DSC	91B3

Thioredoxin from *E. coli*, oxidized form, in the presence of NaCl

NaCl Concentration	pH	T_{trs}	ΔCp	ΔH	Approach	Ref
0.1 M	7.0	87.0	6.9±0.2	447±5	DSC	92S2
0.5 M	7.0	87.0	6.9±0.2	433±5	DSC	92S2
1.0 M	7.0	87.1	6.9±0.2	410±1	DSC	92S2
1.5 M	7.0	88.0	6.9±0.2	418±3	DSC	92S2

Thioredoxin from *E. coli*, oxidized, dependence on pH and protein concentration

Concentration (μM)	pH	T_{trs}	ΔCp	ΔH	Approach	Remarks	Ref
297	6.00	86.85	6.4	320	DSC		93L1
197	6.00	87.82	3.0	340	DSC		93L1
98	6.00	88.44	4.1	349	DSC		93L1
758	6.50	84.12	6.4	409	DSC		93L1
293	6.50	85.56	7.6	420	DSC		93L1
211	6.50	85.94	6.9	438	DSC		93L1
160	6.50	86.14	7.8	429	DSC		93L1
66	6.50	86.43	7.4	444	DSC		93L1
413	6.75	84.63	6.0	419	DSC		93L1
257	6.75	85.24	5.3	433	DSC		93L1
212	6.75	85.54	5.0	429	DSC		93L1
189	6.75	85.33	6.1	396	DSC		93L1
577	7.00	84.75	5.1	333	DSC		93L1
576	7.00	84.78	5.5	342	DSC		93L1
318	7.00	85.41	6.1	366	DSC		93L1
290	7.00	85.46	7.1	353	DSC		93L1
227	7.00	85.42	4.4	399	DSC		93L1
227	7.00	85.18	5.4	412	DSC		93L1
193	7.00	85.80	3.4	383	DSC		93L1
189	7.00	85.82	4.9	356	DSC		93L1
98	7.00	86.01	5.4	400	DSC		93L1
1043	7.25	81.95	7.2	385	DSC		93L1
738	7.25	82.50	7.4	379	DSC		93L1
456	7.25	83.40	4.9	403	DSC		93L1
343	7.25	83.74	4.9	379	DSC		93L1
161	7.25	84.66	4.2	398	DSC		93L1
103	7.25	85.05	4.0	405	DSC		93L1
498	7.50	82.11	7.6	357	DSC		93L1
369	7.50	82.73	10.8	388	DSC		93L1
297	7.50	83.40	7.5	405	DSC		93L1
217	7.50	83.48	7.9	407	DSC		93L1
1146	7.75	80.49	10.0	327	DSC		93L1
414	7.75	82.17	10.8	390	DSC		93L1
348	7.75	82.08	11.0	354	DSC		93L1
183	7.75	83.16	14.4	335	DSC		93L1
175	7.75	83.37	6.0	360	DSC		93L1
129	7.75	83.97	6.5	373	DSC		93L1
121	7.75	84.51	7.4	349	DSC		93L1
84	7.75	83.71	13.5	367	DSC		93L1
80	7.75	83.90	9.5	382	DSC		93L1
501	8.00	81.17	11.4	347	DSC		93L1
312	8.00	82.00	11.3	343	DSC		93L1
57	8.00	83.30	16.1	359	DSC		93L1
57–1146	6.5–8.0	81.2–86.4	11.3±2.8	327–420	DSC	(1,2)	93L1
57–1146	6.5–8.0	81.2–86.4	7.9±2.6	327–420	DSC	(1,3)	93L1

Remarks:
(1) ΔCp is based on 40 calorimetric determinations listed above (except pH 6.0)
(2) ΔCp = 8.08±2.80 kJ/mol/K, obtained from various forms, is considered in Ref. 94L1 as the best value. See remark (3) for the thioredoxin Cys double mutant below.
(3) mean of the individual scans

Thioredoxin from *E. coli*, reduced form, dependence on pH and protein concentration

Concentration (µM)	pH	T_{trs}	ΔCp	ΔH	Approach	Remarks	Ref
409	6.0	72.87	9.8	244	DSC	(1)	94L1
291	6.0	74.16	7.2	279	DSC	(1)	94L1
196	6.0	74.99	7.2	328	DSC	(1)	94L1
428	7.0	72.71	10.1	317	DSC	(1)	94L1
324	7.0	73.36	6.4	283	DSC	(1)	94L1
308	7.0	73.09	10.4	318	DSC	(1)	94L1
239	7.0	73.69	8.5	326	DSC	(1)	94L1
164	7.0	74.17	7.8	349	DSC	(1)	94L1

Remark:
(1) for ΔCp, see the thioredoxin Cys double mutant below

Thioredoxin cysteine mutant (Cys32→Ser and Cys35→Ser), dependence on pH and protein concentration

Concentration (µM)	pH	T_{trs}	ΔCp	ΔH	Approach	Remarks	Ref
565	6.0	76.48	4.6	198	DSC	(1–3)	94L1
440	6.0	76.84	5.7	217	DSC	(1–3)	94L1
392	6.0	77.25	3.8	212	DSC	(1–3)	94L1
313	6.0	77.67	3.8	216	DSC	(1–3)	94L1
1013	7.0	73.15	9.5	331	DSC	(1–3)	94L1
706	7.0	73.75	9.8	382	DSC	(1–3)	94L1
423	7.0	74.44	9.5	392	DSC	(1–3)	94L1
382	7.0	74.71	9.0	375	DSC	(1–3)	94L1
285	7.0	75.28	5.0	350	DSC	(1–3)	94L1
255	7.0	74.65	13.6	401	DSC	(1–3)	94L1
161	7.0	75.48	10.6	382	DSC	(1–3)	94L1
500	8.0	71.71	10.4	351	DSC	(1–3)	94L1
370	8.0	72.10	11.3	367	DSC	(1–3)	94L1
327	8.0	72.38	9.0	339	DSC	(1–3)	94L1
241	8.0	72.53	10.2	355	DSC	(1–3)	94L1
188	8.0	72.71	10.0	363	DSC	(1–3)	94L1

Remarks:
(1) ΔCp for both reduced wild-type and Cys double mutant amounts to 9.46±1.84 based on the individual scans, and 11.3±2.9 based on ΔH versus T_{trs}
(2) data obtained at pH 6 were not included into the determination of ΔCp
(3) if structural integrity between various forms of thioredoxin is assumed, and the results from Ref. 93L1 for the oxidized protein are included, it follows ΔCp = 8.08±2.80 based on 57 individual scans. This value is regarded as the best value in Ref. 94L1

Thioredoxin from *E. coli*, mutant (Leu78→Lys)

Protein Concentration	pH	T_{trs}	ΔCp	ΔH	Approach	Remarks	Ref
387 mM	6.50	71.78	2.9	362	DSC	(1,2)	95L2
239 mM	6.50	72.28	4.2	357	DSC	(1,2)	95L2
229 mM	6.50	72.22	4.5	367	DSC	(1,2)	95L2
86 mM	6.50	72.58	8.2	374	DSC	(1,2)	95L2
76 mM	6.50	73.09	2.3	323	DSC	(1,2)	95L2
817 mM	7.00	70.32	2.9	374	DSC	(1,2)	95L2
481 mM	7.00	71.18	2.7	379	DSC	(1,2)	95L2
302 mM	7.00	72.11	3.1	403	DSC	(1,2)	95L2
235 mM	7.00	72.00	2.7	379	DSC	(1,2)	95L2
152 mM	7.00	72.49	5.8	386	DSC	(1,2)	95L2
109 mM	7.00	72.77	4.7	410	DSC	(1,2)	95L2
78 mM	7.00	72.89	4.6	345	DSC	(1,2)	95L2
463 mM	7.50	71.79	2.7	387	DSC	(1,2)	95L2
325 mM	7.50	71.98	2.3	390	DSC	(1,2)	95L2
240 mM	7.50	72.33	2.7	412	DSC	(1,2)	95L2
74 mM	7.50	73.70	6.2	361	DSC	(1,2)	95L2

Remarks:
(1) based on the potential of thioredoxin to dimerize in both native and denatured forms, calorimetric scans were fit to a two-state model with incomplete dissociation/association
(2) ΔCp from ΔH versus T_{trs} amounts to 3.93±1.63 kJ/mol/K

Thioredoxin from *E. coli*, mutant (Leu78→Arg)

Protein Concentration	pH	T_{trs}	ΔCp	ΔH	Approach	Remarks	Ref
455 mM	6.50	68.72	4.2	254	DSC	(1,2)	95L2
266 mM	6.50	69.48	5.7	253	DSC	(1,2)	95L2
172 mM	6.50	69.85	5.8	261	DSC	(1,2)	95L2
113 mM	6.50	70.15	6.9	254	DSC	(1,2)	95L2
91 mM	6.50	70.13	5.5	267	DSC	(1,2)	95L2
531 mM	7.00	68.60	3.6	310	DSC	(1,2)	95L2
365 mM	7.00	68.94	3.9	294	DSC	(1,2)	95L2
253 mM	7.00	69.31	3.7	320	DSC	(1,2)	95L2
229 mM	7.00	69.58	3.6	336	DSC	(1,2)	95L2
188 mM	7.00	69.86	4.4	316	DSC	(1,2)	95L2
103 mM	7.00	70.43	2.5	326	DSC	(1,2)	95L2
455 mM	7.50	68.71	3.9	260	DSC	(1,2)	95L2
266 mM	7.50	69.37	4.2	267	DSC	(1,2)	95L2
178 mM	7.50	69.54	6.7	254	DSC	(1,2)	95L2
87 mM	7.50	70.33	4.1	296	DSC	(1,2)	95L2

Remarks:
(1) based on the potential of thioredoxin to dimerize in both native and denatured forms, calorimetric scans were fit to a two-state model with incomplete dissociation/association
(2) ΔCp from ΔH versus T_{trs} amounts to 3.93±1.63 kJ/mol/K

Thymidylate Synthase

Thymidylate synthase, in the presence of ligands

Ligand	pH	T_{trs}	ΔH	Approach	Remarks	Ref
without ligand	7.4	44.9±0.3	653±17	DSC	(1)	96C8
dUMP	7.4	50.5±0.5	661±25	DSC	(1,2)	96C8
FdUMP	7.4	49.8±1.0	644±17	DSC	(1,3)	96C8
dGMP	7.4	45.1±0.6	656±17	DSC	(1,4)	96C8
1843U	7.4	47.4±0.9	665±21	DSC	(1,5,6)	96C8

Remarks:
(1) buffer: 30 mM HEPES, 5 mM MgCl$_2$, 1 mM DTT, 10% ethylene glycol
(2) dUMP = 2'-deoxyuridine 5'-monophosphate
(3) FdUMP = 5-fluoro-2'-deoxyuridine 5'-monophosphate
(4) dGMP = 2'-deoxyguanosine 5'-monophosphate
(5) 1843U = (S)-2-(5-(((1,2-dihydro-3-methyl-1-oxobenzo[f]quinazolin-9-yl]methyl)amino)-1-oxo-2-isoindo-
    linyl)glutamic acid, folate analogue inhibitor
(6) Ref. 96C8 contains additional thermodynamic data concerning the ternary complex

Thyroid Transcription Factor

Thyroid transcription factor 1 (TTF-1) homeodomain

pH	T_{ref}	ΔC_p	ΔH	Approach	Remark	Ref
7.5	25	0.33	106	heat, v.H.	(1)	94D2

Remark:
(1) buffer: 10 mM phosphate, 100 mM NaF, pH 7.5

Tobacco Mosaic Virus

Tobacco mosaic virus

pH	T_{trs}	ΔC_p	ΔH	Approach	Remarks	Ref
5.6	55–80		750	DSC	(1)	92M8
7.0	70–85		750	DSC	(1)	92M8
8.0	80–90		750	DSC	(1)	92M8

Remark:
(1) ΔH was resolved by deconvolution into three transitions with ΔH_1 = 190 kJ/mol, $\Delta H_2 = \Delta H_3$ = 290 kJ/mol

Toxins

Cholera toxin, A and B subunit

Subunit	pH	T_{trs}	ΔC_p	ΔH	Approach	Remark	Ref
A subunit	7.5	51		377	DSC		88G2
B subunit	7.5	74		2175	DSC	(1)	88G2

Remark:
(1) data per pentamer of the protein

Cholera toxin, B-subunit

	pH	T_{trs}	ΔCp	ΔH	Approach	Remarks	Ref
	7.5	76.5±0.3		1372±31	DSC	(1,2)	91B2
	6.0	74.9±0.2		1146±33	DSC	(1,2)	91B2
	5.5	73.4		920	DSC	(1,2)	91B2
	5.0	69.0±0.1		561±42	DSC	(1,2)	91B2
	4.0	53.2±0.1		439±17	DSC	(1,2)	91B2

Remarks:
(1) data per pentamer of the protein
(1) the cooperative ratio increases from 0.29 at pH 7.5
   to 0.86 at pH 4.0

Cytotoxin α-sarcin, in the presence of lipid

Lipid	pH	T_{trs}	ΔCp	ΔH	Approach	Remarks	Ref
without	7.0	52.6		569	DSC	(1)	95G3
DMPG	7.0	49.1		322	DSC	(2)	95G3
DOPG	7.0	48.9		439	DSC	(3)	95G3
egg-PG	7.0	47.7		351	DSC	(4)	95G3

Remarks:
(1) buffer: 50 mM MOPS, 0.1 M NaCl, 1 mM EDTA, pH 7.0
(2) dimyristoylphosphatidylglycerol (220:1 lipid/protein molar ratio)
(3) dioleoylphosphatidylglycerol (220:1 lipid/protein molar ratio)
(4) egg phosphatidylglycerol (220:1 lipid/protein molar ratio)

Diphtheria toxin (D.T.) and its domains

Domain	pH	T_{trs}	ΔCp	ΔH	Approach	Remarks	Ref
intact D.T.	8.0	54.9		1234	DSC		90R1
A-domain	8.0	44.2	14.6±8.4	460±63	DSC	(1)	90R1
B-domain	8.0	57.9	18.0±5.0	795±84	DSC	(1)	90R1

Remark:
(1) ΔH refers to T_{trs} = 54.9°C of the intact diphtheria toxin.

8 kDa cytotoxin from sea anemone

	pH	T_{trs}	ΔCp	ΔH	Approach	Remarks	Ref
	7.4	70–79		210±8	DSC	(1)	94Z4

Remark:
(1) thermal denaturation is scan rate dependent; ΔH refers to a two-state kinetic model

Transcarbamoylase

Ornithine transcarbamoylase of *E. coli*, wild-type and mutants

Mutant	pH	T_{trs}	ΔCp	ΔH	Approach	Remarks	Ref
wild-type	8.3	62.5		2510	DSC	(1)	96M12
Ser55→His	8.3	62.3		2800	DSC	(1)	96M12
Lys86→Gln	8.3	69.7		2890	DSC	(1)	96M12
Arg319→Ala	8.3	69.6		2970	DSC	(1)	96M12
Ala325→Gly	8.3	63.0		2760	DSC	(1)	96M12
Leu331 del.	8.3	58.3		2010	DSC	(1)	96M12

Remark:
(1) non-coincidence of calorimetric and van't Hoff heat

Transcription Factor

Transcription factor LFB1, dimerization domain (B1-DIM)

	pH	T_{trs}	ΔCp	ΔH	Approach	Remarks	Ref
	7.5	65	3.1	209	heat	(1–3)	91D6

Remarks:
(1) the data refer to the unfolding of the dimer, i.e., $N_2 \rightarrow 2U$
(2) the measurements were made at different protein concentrations
(3) measurement in the presence of fluoride

Transcription factor GCN4, 56 res. fragment

Concentration (μM)	pH	T_{trs}	ΔCp	ΔH	Approach	Remarks	Ref
1	7.06	41.4		142	heat	(1–3)	93T4
2	7.06	47.8		153	heat	(1–3)	93T4
5	7.06	51.2		150	heat	(1–3)	93T4
10	7.06	53.5		144	heat	(1–3)	93T4
20	7.06	54.8		143	heat	(1–3)	93T4
156	7.06	66.2	0.66±0.63	144	DSC	(1,3)	93T4
335	7.06	68.6	0.46±0.63	145	DSC	(1,3)	93T4
504	7.06	70.2	0.56±0.63	146	DSC	(1,3)	93T4

Remarks:
(1) transition $D_2 \rightarrow 2U$, data per mole of monomer
(2) monitored by CD
(3) buffer: 12.5 mM potassium phosphate, 25 mM KCl, and 0.25 mM EDTA

Transferrin

Transferrin N-terminal half-molecule, apo-form

Mutant	pH	T_{trs}	ΔCp	ΔH	Approach	Remarks	Ref
wild-type	7.5	66.4	20	927	DSC	(1)	93L7
Asp63→Cys	7.5	66.8	28	715	DSC	(1)	93L7
Asp63→Ser	7.5	65.6	16	690	DSC	(1)	93L7
Gly65→Arg	7.5	65.3	16	720	DSC	(1)	93L7
Lys206→Gln	7.5	65.5	28	720	DSC	(1)	93L7
His207→Glu	7.5	65.6	19	750	DSC	(1)	93L7
N-lobe	7.5	68.35	30	983	DSC	(1)	93L7

Remark:
(1) estimated uncertainties: for $\Delta H = \pm 50$, $T_{trs} = \pm 0.5$, and $\Delta Cp = \pm 5$

Transferrin in N-terminal half-molecule, holo-form

Mutant	pH	T_{trs}	ΔCp	ΔH	Approach	Remarks	Ref
wild-type	7.5	86.0	17	1477	DSC	(1)	93L7
Asp63→Cys	7.5	80.8		1146	DSC	(1)	93L7
Asp63→Ser	7.5	74.3	18	1038	DSC	(1)	93L7
Gly65→Arg	7.5	76.7	13	1100	DSC	(1)	93L7
Lys206→Gln	7.5	90.9	13	1205	DSC	(1)	93L7
His207→Glu	7.5	87.7	25	1305	DSC	(1)	93L7

Remark:
(1) estimated uncertainties: for $\Delta H = \pm 50$, $T_{trs} = \pm 0.5$, and $\Delta Cp = \pm 5$

Transferrin; apo human serum transferrin

Transition	pH	T_{trs}	ΔCp	ΔH	Approach	Remarks	Ref
trans. (1)	7.5	57.62	30.1	636	DSC	(1,2)	94L7
trans. (2)	7.5	68.35	30.1	983	DSC	(1,2)	94L7

Remarks:
(1) buffer: 500 mM HEPES, 25 mM $NaHCO_3$
(2) resolved by deconvolution

Transferrin; apo ovotransferrin

	pH	T_{trs}	ΔCp	ΔH	Approach	Ref
	7.5	60.24	55.2	1314	DSC	94L7

Remark:
(1) buffer: 500 mM HEPES, 25 mM $NaHCO_3$

Transferrin Receptor

Transferrin receptor, human, extracellular fragment

	pH	T_{trs}	ΔCp	ΔH	Approach	Remarks	Ref
	5.6	55±0.5		125±8	heat	(1,2)	94H1
	5.6	55±1		117±8	heat	(1,3)	94H1
	5.6	55±2		96±17	heat	(1,4)	94H1
	7.4	70±0.5		456±25	heat	(1,2)	94H1
	7.4	71±0.5		402±25	heat	(1,3)	94H1
	7.4	71±3		167±29	heat	(1,4)	94H1

Remarks:
(1) monitored by FTIR. Given is the van't Hoff heat. The transition is irreversible
(2) transition assigned to aggregated structures
(3) transition assigned to α-helix/β-sheet
(4) transition assigned to amide II/amide II'

Tropomyosin

αα-Tropomyosin, various modified forms

Form	pH	T_{trs}	ΔCp	ΔH	Approach	Remarks	Ref
reduced	7.4	49.64	10.9±0.8	1440±38	DSC	(1)	91S7
cross-linked	7.4	52.15	20.5±5.0	1218±46	DSC	(1)	91S7
reduced-CM	7.4	46.26	42.3	1452	DSC	(1,2)	91S7
reduced-CAM	7.4	47.39	13.4	1427	DSC	(1,3)	91S7

Remarks:
(1) deconvolution see specific values in table 3
(2) SH groups carboxymethylated
(3) SH groups carboxyamidomethylated

ββ-Tropomyosin, various modified forms

Form	pH	T_{trs}	ΔCp	ΔH	Approach	Remarks	Ref
reduced	7.4	42.99	6.3	1188	DSC	(1)	91S7
cross-linked	7.4	44.76	31.8	1289	DSC	(1)	91S7

Remark:
(1) deconvolution see specific values in table 3

Tropomyosin, homodimers and heterodimer

Dimer	pH	T_{trs}	ΔCp	ΔH	Approach	Remarks	Ref
αα	7.0	40.0		678	heat	(1)	91L4
αβ	7.0	43		941	heat	(1)	91L4
ββ	7.0	42.6		410	heat	(1)	91L4

Remark:
(1) in the presence of 0.5 M NaCl, 10 mM sodium phosphate, 1 mM EDTA, and 0.5 mM DTT

Tropomyosin

Transition	pH	T_{trs}	ΔCp	ΔH	Approach	Remarks	Ref
N-(1-Pyrenyl)-iodacetamide-tropomyosin, high salt conditions:							
trans. (1)	7.5	38.5		125	heat	(1,2)	94I
trans. (2)	7.5	45.5		375	heat	(1,2)	94I
Reduced tropomyosin, high salt conditions:							
trans. (1)	7.5	44.5		105	heat	(1,2)	94I
trans. (2)	7.5	46		480	heat	(1,2)	94I
Disulfide cross-linked tropomyosin, high salt conditions:							
trans. (1)	7.5	33.5		145	heat	(1,3)	94I
trans. (2)	7.5	53		315	heat	(1,3)	94I
Disulfide cross-linked tropomyosin, low salt conditions:							
trans. (1)	7.5	30		230	heat	(1,4)	94I
trans. (2)	7.5	44.5		295	heat	(1,4)	94I
Disulfide cross-linked tropomyosin, methanol conditions:							
trans. (1)	7.5	47		210	heat	(1,5)	94I
trans. (2)	7.5	58.5		295	heat	(1,5)	94I

Remarks:
(1) trans. (1) = local transition trans. (2) = global transition
(2) high salt: 0.5 M NaCl, 1 mM EDTA, 20 mM sodium phosphate 1 mM DTE
(3) high salt: 0.5 M NaCl, 1 mM EDTA, 20 mM sodium phosphate
(4) low salt: 2 mM sodium phosphate, 1 mM EDTA
(5) methanol: 20% methanol, 0.5 M NaCl, 1 mM EDTA, 20 mM sodium phosphate

α-Tropomyosin variants

Transition	pH	T_{trs}	ΔCp	ΔH	Approach	Remarks	Ref
Mutant 2a6b9d:							
trans. (1)	7.5	25±1		88±8	heat	(1–3)	95G9
trans. (3)	7.5	32±1		675±75	heat	(1–3)	95G9
Mutant 2b6b9a:							
trans. (1)	7.5	30±3		79±4	heat	(1–3)	95G9
trans. (3)	7.5	44		582±67	heat	(1–3)	95G9
Mutant 2b6b9d:							
trans. (1)	7.5	30		100	heat	(1–3)	95G9
trans. (2)	7.5	39		435	heat	(1–3)	95G9
trans. (3)	7.5	48		787	heat	(1–3)	95G9
Mutant 2a6a9d:							
trans. (1)	7.5	30±10		79±13	heat	(1–3)	95G9
trans. (2)	7.5	32±1		397±201	heat	(1–3)	95G9
trans. (3)	7.5	40±3		448±113	heat	(1–3)	95G9
Mutant 2a6a9a:							
trans. (1)	7.5	28		88	heat	(1–3)	95G9
trans. (2)	7.5	33		506	heat	(1–3)	95G9
trans. (3)	7.5	43		573	heat	(1–3)	95G9
Mutant 2a6zip9d:							
trans. (1)	7.5	24±4		75±13	heat	(1–3)	95G9
trans. (2)	7.5	32±1		360±146	heat	(1–3)	95G9
trans. (3)	7.5	45±1		460±113	heat	(1–3)	95G9

α-Tropomyosin variants (continued)

Transition	pH	T_{trs}	ΔCp	ΔH	Approach	Remarks	Ref
Mutant 2a6zip9a:							
trans. (1)	7.5	21		96	heat	(1–3)	95G9
trans. (2)	7.5	33		297	heat	(1–3)	95G9
trans. (3)	7.5	50±1		423±17	heat	(1–3)	95G9
Mutant 2b6b9da:							
trans. (1)	7.5	30±2.5		105	heat	(1–3)	95G9
trans. (2)	7.5	39		466	heat	(1–3)	95G9
trans. (3)	7.5	47		745±38	heat	(1–3)	95G9
Mutant 2b6b9ad:							
trans. (1)	7.5	31		96	heat	(1–3)	95G9
trans. (3)	7.5	45		653	heat	(1–3)	95G9
Mutant C2b6b9a:							
trans. (1)	7.5	32±3		79±4	heat	(1–3)	95G9
trans. (3)	7.5	44		661±25	heat	(1–3)	95G9
Mutant C2zip6b9a:							
trans. (1)	7.5	23±1		79±13	heat	(1–3)	95G9
trans. (2)	7.5	36±8		803±573	heat	(1–3)	95G9
trans. (3)	7.5	45±1		820±301	heat	(1–3)	95G9
Mutant C2rc6b9a:							
trans. (1)	7.5	19±1		159±63	heat	(1–3)	95G9
trans. (2)	7.5	25±7		142±71	heat	(1–3)	95G9
trans. (3)	7.5	41		799±29	heat	(1–3)	95G9

Remarks:
(1) for the detailed mutant description, see Ref. 95G9
(2) buffer: 10 mM sodium phosphate, 500 mM NaCl 1 mM EDTA, 1mM DTT, pH 7.5
(3) transition monitored by CD at 222 nm, analysis of the transition curve assuming three independent helix-coil transitions

Homodimeric chicken gizzard tropomyosins (CG-TM)

Transition	pH	T_{trs}	ΔCp	ΔH	Approach	Remarks	Ref
Reduced ββ-CG-TM:							
trans. (1)	7.4	42.66±0.32			DSC	(1–3)	96O1
trans. (2)	7.4	49.62±0.17			DSC	(1–3)	96O1
total values			2.82±1.66	299±18	DSC	(1–3)	96O1
Oxidized ββ-CG-TM:							
total values	7.4	50.23±0.55	2.40±1.52	253±9	DSC	(1–4)	96O1
Reduced γγ-GC-TM:							
total values	7.4	42.80±0.67	3.74±0.93	320±19	DSC	(1–4)	96O1
Oxidized γγ-GC-TM:							
total values	7.4	46.45±0.22	2.60±0.90	300±17	DSC	(1–3,5)	96O1

Remarks:
(1) for further details, see specific values in table 3
(2) measured in 50 mM sodium phosphate, 500 mM NaCl, 10 mM EDTA, pH 7.4
(3) T_{trs} is the weighted mean temperature
(4) resolved into three subtransitions, see remark (1)
(5) resolved into four subtransitions, see remark (1)

Troponin C

Troponin C from rabitt skeletal muscle at varying solvent conditions, transitions resolved by deconvolution

Component Added	pH	T_{trs}	ΔH	Approach	Remarks	Ref
2.0 mM EDTA	7.25	57	235±10	DSC	(1)	85T2
2.0 mM EDTA	7.0	55		heat	(5,6)	83B3
10.0 mM EDTA, 7.0 mM CaCl$_2$						
total	7.25		360±20	DSC	(1,2)	85T2
trans. (1)	7.25	57	240±10	DSC	(1,2)	85T2
trans. (2)	7.25	32	115±10	DSC	(1,2)	85T2
10.0 mM EDTA, 9.0 mM CaCl$_2$						
total	7.25		510±40	DSC	(1,3)	85T2
trans. (1)	7.25	73	230±10	DSC	(1,3)	85T2
trans. (2)	7.25	57	126±15	DSC	(1,3)	85T2
trans. (3)	7.25	45	113±15	DSC	(1,3)	85T2
10.0 mM EDTA, 9.3 mM CaCl$_2$						
total	7.25		510±40	DSC	(1,4)	85T2
trans. (1)	7.25	105	240±10	DSC	(1,4)	85T2
trans. (2)	7.25	62	105±15	DSC	(1,4)	85T2
trans. (3)	7.25	57	115±15	DSC	(1,4)	85T2
0.1 mM CaCl$_2$						
total	7.25		250±30	DSC	(1)	85T2
trans. (1)	7.25	>120	n.d.	DSC	(1)	85T2
trans. (2)	7.25	82	120±15	DSC	(1)	85T2
trans. (3)	7.25	79	120±15	DSC	(1)	85T2
0.5 mM CaCl$_2$						
total	7.25		240±30	DSC	(1)	85T2
trans. (1)	7.25	>120	n.d.	DSC	(1)	85T2
trans. (2)	7.25	91	245±15	DSC	(1)	85T2
2.0 mM EGTA, 0.3 mM MgCl$_2$						
total	7.25		500±40	DSC	(1)	85T2
trans. (1)	7.25	75	180±15	DSC	(1)	85T2
trans. (2)	7.25	66	240±15	DSC	(1)	85T2
2.0 mM EGTA, 2.0 mM MgCl$_2$						
total	7.25		510±50	DSC	(1)	85T2
trans. (1)	7.25	88	190±15	DSC	(1)	85T2
trans. (2)	7.25	70	250±15	DSC	(1)	85T2
2.0 mM EDTA, 100 mM NaCl						
total	7.25		360±20	DSC	(1)	85T2
trans. (1)	7.25	70	250±10	DSC	(1)	85T2
trans. (2)	7.25	41	120±15	DSC	(1)	85T2
2.0 mM EDTA, 200 mM NaCl						
total	7.25		365±20	DSC	(1)	85T2
trans. (1)	7.25	75	250±10	DSC	(1)	85T2
trans. (2)	7.25	50	115±15	DSC	(1)	85T2
4.0 mM EDTA, 500 mM NaCl						
total	7.25		360±20	DSC	(1)	85T2
trans. (1)	7.25	81	250±15	DSC	(1)	85T2
trans. (2)	7.25	60	120±15	DSC	(1)	85T2

Troponin C from rabitt skeletal muscle at varying solvent conditions, transitions resolved by deconvolution (continued)

Component Added	pH	T_{trs}	ΔH	Approach	Remarks	Ref
2.0 mM EDTA, 200 mM KCl						
total	7.25		350±20	DSC	(1)	85T2
trans. (1)	7.25	66	255±15	DSC	(1)	85T2
trans. (2)	7.25	37	115±15	DSC	(1)	85T2
2.0 mM EDTA, 400 mM KCl						
total	7.25		360±20	DSC	(1)	85T2
trans. (1)	7.25	70	245±15	DSC	(1)	85T2
trans. (2)	7.25	41	120±15	DSC	(1)	85T2

Remarks:
(1) buffer: 10 mM cacodylate, pH 7.25, with the indicated substances
(2) free calcium conc. ~$0.45 \cdot 10^{-7}$ M
(3) free calcium conc. ~$1.8 \cdot 10^{-7}$ M
(4) free calcium conc. ~$2.7 \cdot 10^{-7}$ M
(5) buffer: 20 mM HEPES, pH 7.0, with the indicated substances
(6) transition monitored by CD at 222 nm

Troponin C from rabitt skeletal muscle, fragment TR-1 (residues 9 to 84) at varying solvent conditions

Component Added	pH	T_{trs}	ΔH^{cal}	$\Delta H^{v.H.}$	Approach	Remarks	Ref
2.0 mM EDTA	7.25	54	224	210	DSC	(1)	85T2
2.0 mM EDTA	7.0	60			heat	(4,5)	83B3
2.0 mM EDTA, 100 mM NaCl							
	7.25	66	206	206	DSC	(1)	85T2
10.0 mM EDTA, 9.0 mM CaCl$_2$							
	7.25	62	226	218	DSC	(1,2)	85T2
10.0 mM EDTA, 9.3 mM CaCl$_2$							
	7.25	77	206	210	DSC	(1,3)	85T2
2.0 mM EGTA, 2.0 mM MgCl$_2$							
	7.25	70	214	206	DSC	(1)	85T2

Remarks:
(1) buffer: 10 mM cacodylate, pH 7.25, with the indicated substances
(2) free calcium conc. ~$1.8 \cdot 10^{-7}$ M
(3) free calcium conc. ~$2.7 \cdot 10^{-7}$ M
(4) buffer: 20 mM HEPES, pH 7.0, with the indicated substances
(5) transition monitored by CD at 222 nm

Troponin C from rabitt skeletal muscle, fragment TR-2 (residues 89 to 159) at varying solvent conditions

Components Added	pH	T_{trs}	ΔH^{cal}	$\Delta H^{v.H.}$	Approach	Remarks	Ref
2.0 mM EDTA	7.25	33	84	96	DSC	(1)	85T2
2.0 mM EDTA	7.0	~20			heat	(5,6)	83B3
2.0 mM EDTA, 100 mM NaCl							
	7.25	45	122	110	DSC	(1)	85T2
10.0 mM EDTA, 8.5 mM CaCl$_2$							
	7.25	43	189	105	DSC	(1,2)	85T2
10.0 mM EDTA, 9.0 mM CaCl$_2$							
	7.25	75	230	151	DSC	(1,3)	85T2
10.0 mM EDTA, 9.3 mM CaCl$_2$							
	7.25	100	227	218	DSC	(1,4)	85T2
2.0 mM EGTA, 2.0 mM MgCl$_2$							
	7.25	73	176	164	DSC	(1)	85T2

Remarks:
(1) buffer: 10 mM cacodylate, pH 7.25, with the indicated substances
(2) free calcium conc. $\sim 1.2 \cdot 10^{-7}$ M
(3) free calcium conc. $\sim 1.8 \cdot 10^{-7}$ M
(4) free calcium conc. $\sim 2.7 \cdot 10^{-7}$ M
(5) buffer: 20 mM HEPES, pH 7.0, with the indicated substances
(6) transition monitored by CD at 222 nm

Synthetic peptides corresponding to calcium-binding sites III (SCIII)
and IV (SCIV) of troponin C, heterodimer

pH	T_{trs}	ΔCp	ΔH	Approach	Ref
7.2	60	183		heat, v.H.	94S4

Trypsin

β-Trypsin

pH	T_{trs}	ΔCp	ΔH	Approach	Remark	Ref
4.2	65.0±0.2	11.1±1.3	688±13	heat, v.H.		95B8
5	70	11.9	255	DSC	(1)	79P4

Remark:
(1) transition temperature from Fig. 9 of Ref. 79P4, other data from Tab. 6 in Ref. 79P4

Trypsinogen

Bovine trypsinogen

pH	T_{trs}	ΔCp	ΔH	Approach	Ref
4.2	66.0±0.2	12.9±1.3	625±13	heat, v.H.	95B8

Trypsinogen, free and complexed with calcium, with the dipeptide Ile-Val, and with pancreatic trypsin inhibitor (Kunitz)

Form	pH	T_{trs}	ΔCp	ΔH	Approach	Remarks	Ref
free	5.8	65.7±0.2	12.4±0.8	607±15	heat	(1)	94B9
Ca^{2+}	5.8	68.3±0.3	12.2±0.3	687±17	heat	(1)	94B9
Ca^{2+} + Ile-Val							
	5.8	68.2±0.2		728±18	heat, v.H.		94B9
BPTI	5.8	66.8±0.3		706±17	heat, v.H.		94B9
Ca^{2+} + BPTI	5.8	68.4±0.3		745±19	heat, v.H.		94B9
Ca^{2+} + BPTI + Ile-Val							
	5.8	67.9±0.3		746±20	heat, v.H.		94B9

Remark:
(1) ΔCp was determined by an approach that combines thermal and denaturant-induced unfolding according Ref. 89P2

Trypsin Inhibitor

Trypsin inhibitor, bovine pancreatic

pH	T_{trs}	ΔCp	ΔH	Approach	Remark	Ref
7	102.9	0.096	293	DSC		83M2
4	100	3	430	DSC	(1)	79P4

Remark:
(1) transition temperature from Fig. 9 of Ref. 79P4, other data from Tab. 6 in Ref. 79P4

Trypsin inhibitor, bovine pancreatic, temperature dependence of the heat capacity change at protein unfolding in aqueous solution, and related thermodynamic quantities

T	$\Delta H(exp)$	$\Delta S(exp)$	$\Delta Cp(exp)$	Approach	Remarks	Ref
5	72	87	2.8	DSC	(1–3)	95M3
25	130	288	3.0	DSC	(1–3)	95M3
50	200	514	2.6	DSC	(1–3)	95M3
75	259	690	2.1	DSC	(1–3)	95M3
100	303	809	1.3	DSC	(1–3)	95M3
125	323	864	0.3	DSC	(1–3)	95M3

Remarks:
(1) the data were taken from Ref. 95M3 with reference to Ref. 93M4
(2) ΔCp was obtained as described in Ref. 90P3
(3) ΔCp in kJ/mol/K, ΔH in kJ/mol and ΔS in J/mol/K

Bovine pancreatic trypsin inhibitor (BPTI), wild-type and mutant

Mutant	pH	T_{trs}	ΔCp	ΔH	Approach	Remark	Ref
wild-type	1.0	86.6	1.42	293	DSC	(1)	93H7
wild-type	3.0	94.7	1.2	309	DSC		93H7
double mutant (Cys14→Ala and Cys38→Ala)							
	3.0	67.9	2.5	211	DSC		93H7

Remark:
(1) temperature function of ΔCp can be obtained from Cp,N and Cp,D, the heat capacity functions of native and denatured protein, respectively:
Cp,N = a + b×T with a = 0.521 kJ/mol/K and b = 0.034 kJ/mol/K²
Cp,D = c + d×(T–T_{trs}) with c = 14.205 kJ/mol/K and d = –0.046 kJ/mol/K²

Trypsin inhibitor, bovine pancreatic

pH	T_{trs}	ΔCp	ΔH	Approach	Remarks	Ref
2.0	86.0		281	DSC		93M4
2.5	90.0		290	DSC		93M4
3.0	100.5		300	DSC		93M4
3.5	101.5		323	DSC		93M4
4.0	104.0		321	DSC		93M4
4.9	104.5		312	DSC		93M4
	5	2.8		DSC	(1)	93M4
	25	3.0		DSC	(1)	93M4
	50	2.6		DSC	(1)	93M4
	75	2.1		DSC	(1)	93M4
	100	1.3		DSC	(1)	93M4
	125	0.3		DSC	(1)	93M4

Remark:
(1) ΔCp is based on heat capacity measurements on native, unfolded, and reduced trypsin inhibitor

Bovine pancreatic trypsin inhibitor, wild-type and mutants

Mutant	pH	T_{trs}	ΔCp	ΔH	Approach	Remarks	Ref
wild-type	2.0	87	1.7	293	DSC	(1,3)	93K6
Phe22→Ala	2.0	78		285	DSC	(3)	93K6
Tyr23→Ala	2.0	54		188	DSC	(3)	93K6
Tyr35→Gly	2.0	69		159	DSC	(3)	93K6
Asn43→Gly	2.0	55	0.8	192	DSC	(2,3)	93K6
Phe45→Ala	2.0	49		151	DSC	(3)	93K6

Remarks:
(1) ΔCp is an unpublished value from Ref. 93M3, cited in Ref. 93K6, ΔCp is temperature dependent: ΔCp = 2.9 at 25°C, ΔCp = 1.7 at 87°C, ΔCp = 0.3 at 125°C
(2) ΔCp is an unpublished value from Ref. 93M3, cited in Ref. 93K6
(3) the uncertainties are T_{trs} = ±0.5, ΔH = ±4 kJ/mol, and ΔCp = ±0.42 kJ/K/mol

Trypsin inhibitor, bovine pancreatic, wild-type and mutants

Mutant	pH	T_{trs}	ΔCp	ΔH	Approach	Ref
wild-type	2.0	87		293	DSC	93K5
Asn43→Gly	2.0	55		192	DSC	93K5
Asn44→Gly	2.0	66		205	DSC	93K5
Gly37→Ala	2.0	66		197	DSC	93K5
Tyr35→Gly	2.0	69		159	DSC	93K5

Bovine pancreatic trypsin inhibitor, alanine mutants

Mutant	pH	$\Delta(T_{trs})$	$\Delta(\Delta H)$	Approach	Remarks	Ref
wild-type	3.0	0.0	0	heat	(1,2)	96C5
Thr11→Ala	3.0	−0.3	−18	heat	(2)	96C5
Gly12→Ala	3.0	−16.1	−148	heat	(2)	96C5
Lys15→Ala	3.0	−4.2	−2	heat	(2)	96C5
Arg17→Ala	3.0	−4.7	9	heat	(2)	96C5
Ile19→Ala	3.0	−4.8	−5	heat	(2)	96C5
Phe33→Ala	3.0	−16.5	−269	heat	(2)	96C5
Tyr35→Ala	3.0	−16.2	−133	heat	(2)	96C5
Gly36→Ala	3.0	−11.7	−61	heat	(2)	96C5
Gly37→Ala	3.0	−4.2	−276	heat	(2)	96C5
Arg39→Ala	3.0	0	2	heat	(2)	96C5

Remarks:
(1) wild-type, reference values: T_{trs} = 94.2°C and ΔH = 353 kJ/mol
(2) thermal unfolding monitored by CD, van't Hoff treatment

Basic pancreatic trypsin inhibitor, intact and cleaved forms

Form	pH	T_{trs}	ΔCp	ΔH	Approach	Remarks	Ref
intact		94.5	1.21	309	DSC	(1)	96K9
cleaved Lys15,Ala	5.0	92±0.2	−0.26±.42	238±13	heat	(2,3)	96K9
cleaved Met52,Arg	5.0	27±0.2	−2.05±.42	169±8	heat	(2,4)	96K9

Remarks:
(1) reference value from 93H7
(2) measured in 50 mM acetate buffer pH 5.0
(3) cleaved between Lys15-Ala16 by hydrolysis
(4) cleaved between Met52-Arg53 by CNBr

Trypsin inhibitor (Kunitz) from soybean

	pH	T_{trs}	ΔCp	ΔH	Approach	Remarks	Ref
	3.0	62	7.5±2.5	340±30	DSC	(1,2)	90B9
	7.0	59	11.0±0.5	429±6	DSC	(3,4)	95F4
	2–11	38–50	7.1±0.5		DSC	(5)	89V
	2–11	38–50	6.6±0.2		DSC	(6)	89V

Remarks:
(1) the data are dependent on the heating rate
(2) the data were taken from Fig. 2 of Ref. 90B9
(3) from scan rate dependence
(4) ΔH is the net enthalpy corrected for the heat of ionization
(5) average value determined from calorimetric recordings
(6) value determined from the temperature dependence of ΔH

Soybean trypsin inhibitor, intact and cleaved forms

Form	pH	T_{trs}	ΔC_p	ΔH	Remarks	Ref
intact	6.4	65±0.2	10.3±1.3	134±8	(1,2)	96K9
cleaved Arg63,Ile	6.4	58±0.2	5.4±1.3	197±13	(1–3)	96K9
cleaved Met84,Leu	6.4	66±0.2	4.6±1.3	293±13	(1,2,4)	96K9

Remarks:
(1) data from thermal denaturation and GuHCl-induced unfolding
(2) measured in 50 mM MES, 0.3 M KCl, pH 6.4
(3) cleaved between Arg63-Ile64 by hydrolysis
(4) cleaved between Met84-Leu85 by CNBr

Tryptophan Repressor

Tryptophan repressor from *E. coli*

	pH	T_{trs}	ΔC_p	ΔH	Approach	Ref
	7.5	90.3±0.35		383±22	DSC	88B1

Tryptophan Synthase

Tryptophan synthase, α-subunit, wild-type and mutants

Mutant	pH	T_{trs}	ΔC_p	ΔH	Approach	Remarks	Ref
wild-type	7	62.4		97±5	DSC	(1)	82Y2
wild-type	7	61.2	10.3	82.4	heat	(1)	84O
wild-type	7.8	57.8±0.5	8.8±1.7	392±25	DSC		80M2
wild-type	9.3	50.4	10.3	82.4	heat	(1)	84O
Glu49→Gln	7	59.4		97±5	DSC	(1)	82Y2
Glu49→Gln	7.2	59.6	11.2	59.8	heat	(1)	84O
Glu49→Gln	9.3	62.4		97±5	DSC	(1)	82Y2
Glu49→Gln	9.3	59.6	11.2	59.8	heat	(1)	84O
Glu49→Ser	7	60.5		97±5	DSC	(1)	82Y2
Glu49→Ser	9.3	58.6		97±5	DSC	(1)	82Y2
Gly211→Arg	7.8	58±0.2	11.7±3.8	464±17	DSC		80M2
Gly211→Glu	7.8	59.6±0.6	11.3±2.9	456±25	DSC		80M2

Remark:
(1) ΔH at 25°C

Tryptophan synthase from *E. coli*, α-subunit, wild-type and mutants

Mutant	pH	T_{trs}	ΔCp	ΔH	Approach	Remarks	Ref
wild-type	9	54.1	19.2	504	DSC		91Y2
Pro57→Ala	9	54.0	19.7	507	DSC	(1)	91Y2
Pro62→Ala	9	52.7	20.5	517	DSC	(1)	91Y2
Pro96→Ala	9	47.8	15.1	474	DSC	(1)	91Y2
Pro132→Ala	9	51.5	20.1	507	DSC	(1)	91Y2
Pro132→Gly	9	51.8	21.3	488	DSC	(1)	91Y2
Pro207→Ala	9	47.4	20.5	398	DSC	(1)	91Y2

Remark:
(1) ΔH refers to $T_{trs} = 54.1°C$ of the wild-type protein

Tryptophan synthase from *E. coli*, α-subunit

pH	T_{trs}	ΔCp	ΔH	Approach	Remarks	Ref
8.7	54.3	17.2	500	DSC	(1,2)	93O1

Remarks:
(1) ΔCp was determined in the presence of urea, at urea concentrations below 0.8 M, other data in the absence of urea
(2) ΔH was taken from Fig. 3 in Ref. 93O1

Tryptophan synthase from *E. coli*, α-subunit, wild-type and mutants

Mutant	pH	T_{trs}	ΔCp	ΔH	Approach	Ref
wild-type	7.2	59.5	10.9	515	DSC	92L3
Ser6→Pro	7.2	54.3			DSC	92L3
Pro21→Ser	7.2	55.2			DSC	92L3
Phe22→Ser	7.2	60.3			DSC	92L3
Thr24→Met	7.2	61.8			DSC	92L3
Pro28→Leu	7.2	54.9±0.3			DSC	92L3
Pro28→Ser	7.2	53.1±0.0			DSC	92L3
Ser33→Leu	7.2	51.9			DSC	92L3
Glu49→Gly	7.2	57.7			DSC	92L3
Gly51→Asp	7.2	55.0			DSC	92L3
Pro53→His	7.2	56.2			DSC	92L3
Pro53→Thr	7.2	55.2			DSC	92L3
Asp56→Gly	7.2	57.5			DSC	92L3
Asp60→Gly	7.2	56.9±0.1			DSC	92L3
Pro62→Gln	7.2	58.9			DSC	92L3
Thr63→Lys	7.2	56.2±0.2			DSC	92L3
Gln65→Arg	7.2	58.9			DSC	92L3
Ala67→Val	7.2	56.8			DSC	92L3
Pro78→Leu	7.2	61.4			DSC	92L3
Pro78→Ser	7.2	57.9			DSC	92L3
Met101→Ile	7.2	63.2			DSC	92L3
Met101→Thr	7.2	67.1±0.2			DSC	92L3
Met101→Val	7.2	67.1			DSC	92L3
Tyr102→Cys	7.2	60.9			DSC	92L3

Tryptophan synthase from *E. coli*, α-subunit, wild-type and mutants (continued)

Mutant	pH	T_{trs}	ΔCp	ΔH	Approach	Ref
Tyr102→His	7.2	60.3±0.3			DSC	92L3
Asp112→Asn	7.2	56.4±0.1			DSC	92L3
Asp112→Gly	7.2	54.4			DSC	92L3
Phe114→Leu	7.2	56.2			DSC	92L3
Tyr115→Cys	7.2	57.2			DSC	92L3

Tryptophan synthase, from *E. coli*, α subunit, wild-type and cysteine mutants

Mutant	pH	T_{trs}	ΔCp	ΔH	Approach	Remarks	Ref
wild-type	9.0	53.7±0.94	20.1±1.7	394±29	DSC	(1,2)	96H6
Cys81→Ala	9.0	51.8±0.16	17.2±3.4	349±12	DSC	(1,2)	96H6
Cys81→Gly	9.0	48.7±0.38	28.5±8.4	436±29	DSC	(1,2)	96H6
Cys81→Ser	9.0	49.1±0.39	21.0±6.7	388±10	DSC	(1,2)	96H6
Cys81→Val	9.0	49.1±0.42	18.0±5.0	240±25	DSC	(1,2)	96H6
Cys118→Ala	9.0	49.2±0.65	13.8±1.7	344±18	DSC	(1,2)	96H6
Cys118→Ser	9.0	45.0±0.23	9.6±5.4	316±18	DSC	(1,2)	96H6
Cys118→Val	9.0	49.3±0.22	10.1±10.5	305±30	DSC	(1,2)	96H6
Cys154→Ala	9.0	50.6±0.03	17.2±1.7	305±7	DSC	(1,2)	96H6
Cys154→Ser	9.0	45.3±0.93	14.7±4.6	194±15	DSC	(1,2)	96H6
Cys154→Val	9.0	50.2±0.26	10.9±1.7	299±9	DSC	(1,2)	96H6
In the presence of IPP:							
wild-type	9.0	55.2±0.82	15.9±1.7	386±28	DSC	(1–3)	96H6
Cys154→Ser	9.0	46.6±0.71	17.6±1.3	312±5	DSC	(1–3)	96H6

Remarks:
(1) T_{trs} at pH 9.0 determined from the pH dependence of T_{trs} given in Ref. 96H6
(2) ΔH and ΔCp were obtained from a linear least square fit of the experimental data for each mutant
(3) protein measured in the presence of indole 3-propanol phosphate (IPP) at a 20-fold molar excess

Tryptophan synthase α-subunit, from various species

Species	pH	T_{trs}	ΔCp	ΔH	Approach	Remarks	Ref
E. coli	7	62*	9.8	84	heat	(1–3)	84Y2
Salmonella	7	55*	9.8	167	heat	(2–4)	84Y2
hybrid	7	55*	9.8	155	heat	(2,3,5)	84Y2

Remarks:
(1) species = *E. coli*
(2) ΔH at 25°C
(3) T_{trs} was taken from the pH dependence in Fig. 3 in Ref. 84Y2
(4) species = *Salmonella typhimurium*
(5) hybrid = hybrid *E. coli* and *Salmonella typhimurium*

Tryptophan synthase, α subunit, from *E. coli* and *Salmonella typhimurium*

Species	pH	T_{trs}	ΔCp	ΔH	Approach	Remarks	Ref
E. coli	9	54.0±0.58	19.3±0.8	499±6.3	DSC	(1)	96H5
Lys109→Asn	9	52.5±0.16	13.4±4.9	406±16.8	DSC	(2)	96H5
Salmonella	9	47.6	19.3	499	DSC	(3)	96H5

Remarks:
(1) *E. coli*, wild-type, see also Ref. 90S8
(2) mutant Lys109→Asn from *E. coli, wild-type*
(3) *Salmonella typhimurium*, wild-type, see also Ref. 91Y2

Table 2. Enthalpy and Heat Capacity Changes – Molar Values

Tryptophan synthase from *Salmonella typhimurium*, α-subunit wild-type and mutants

Mutant	pH	T_{trs}	ΔCp	ΔH	Approach	Remarks	Ref
wild-type	7.0	56	10.4±1.2	523±13	heat	(1)	91K2
Glu49→Phe	7.0	60		347±8	heat	(1)	91K2
Glu49→Gln	7.0	58		628±17	heat	(1)	91K2
Glu49→Asp	7.0	50		485±17	heat	(1)	91K2
Asp60→Ala	7.0	53		527±13	heat	(1)	91K2
Asp60→Glu	7.0	55		690±4	heat	(1)	91K2
Asp60→Asn	7.0	53		619±13	heat	(1)	91K2
Asp60→Tyr	7.0	48.6		1163±247	heat	(2,3)	91K2
Asp60→Tyr	7.0	51.9		1100±619	heat	(2,4)	91K2

Remarks:
(1) two-state fit of the transition
(2) three-state fit of the transition
(3) first of two subsequent transitions (N→I)
(4) second of two subsequent transitions (I→U)

Tubulin

GDP-tubulin and GTP-tubulin, in the presence of ligands

Magnesium	Ligands	pH	T_{trs}	ΔH	Approach	Remarks	Ref
GDP-tubulin:							
0.5 mM		6.7	54.8	724±30	DSC	(1,2)	93D4
1.5 mM		6.7	55.6	777±54	DSC	(1,2)	93D4
4.0 mM		6.7	56.2	849±60	DSC	(1,2)	93D4
2.0 mM	Taxotere	6.7	55.1	765±53	DSC	(1,2)	93D4
3.0 mM	Taxotere	6.7	59.5	950±66	DSC	(1–3)	93D4
3.5 mM	Taxotere	6.7	59.5	1000±80	DSC	(1,2,4)	93D4
4.0 mM	Taxotere	6.7	66.0	1309±75	DSC	(1,2)	93D4
0.5 mM	Taxol	6.7	54.0	828±58	DSC	(1,2)	93D4
4.0 mM	Taxol	6.7	64.4	1305±95	DSC	(1,2)	93D4
3.0 mM	Taxol	6.7	62.8	1440±97	DSC	(2,5)	93D4
4.0 mM	Taxol	6.7	65.5	1441±95	DSC	(2,5)	93D4
6.0 mM	Taxol	6.7	60.1	1146±80	DSC	(2,5)	93D4
GTP-tubulin:							
6.0 mM		6.7	61.3	1096±66	DSC	(2,5)	93D4

Remarks:
(1) buffer: 10 mM phosphate, 1 mM EDTA
(2) ΔCp varies from 44±4 kJ/K/mol, in the pesence of glycerol or at 3 mM Mg^{2+}, to 20±2 kJ/K/mol at 4 mM Mg^{2+} in the absence of glycerol
(3) bimodal denaturation profile with additional T_{trs} = 63.3°C
(4) bimodal denaturation profile with additional T_{trs} = 64.4°C
(5) buffer: 10 mM phosphate, 1 mM EDTA, 3.4 M glycerol

Tumor Suppressor

I) Tumor suppressor p53, tetrameric fragment containing the oligomerization domain (p53tet), thermal transition temperature of p53tet unfolding as a function of pH, salt, and tetramer concentration (table part I)

No	NaCl Concentration	Buffer	pH	p53tet Concentration	T_{trs}	Approach	Remarks	Ref
1	250 mM	glycine	3.0	67.0 µM	54.5	DSC	(1,2)	95J2
2	250 mM	acetate	4.0	73.5 µM	73.7	DSC	(1,2)	95J2
3	250 mM	phosphate	6.0	30.3 µM	82.9	DSC	(1,2)	95J2
4	250 mM	phosphate	7.0	57.0 µM	85.3	DSC	(1,2)	95J2
5	0 mM	glycine	3.0	63.8 µM	34.3	DSC	(1,2)	95J2
6	0 mM	acetate	4.0	70.5 µM	69.5	DSC	(1,2)	95J2
7	0 mM	acetate	4.0	93.5 µM	71.8	DSC	(1,2)	95J2
8	0 mM	acetate	4.0	145.8 µM	75.3	DSC	(1,2)	95J2
9	0 mM	phosphate	6.0	41.0 µM	81.8	DSC	(1,2)	95J2
10	0 mM	phosphate	7.0	39.0 µM	84.4	DSC	(1,2)	95J2

Remarks:
(1) p53tet is the tetramer concentration
(2) see also table part II (below)

II) Tumor suppressor p53, tetrameric fragment containing the oligomerization domain (p53tet), thermal transition temperature of p53tet unfolding as a function of pH, salt, and tetramer concentration (table part II)

No	T_{trs}	$T°$	$\Delta H(T°)$	$\Delta Cp(T°)$	Approach	Remarks	Ref
1	54.5	95.5	178	1.67	DSC	(1–5)	95J2
2	73.7	112.6	207	1.78	DSC	(1–5)	95J2
3	82.9	123.6	218	1.67	DSC	(1–5)	95J2
4	85.3	124.4	219	1.75	DSC	(1–5)	95J2
5	34.3	82.1	156	1.92	DSC	(1–5)	95J2
6	69.5	110.8	193	1.67	DSC	(1–5)	95J2
7	71.8	110.0	196	1.69	DSC	(1–5)	95J2
8	75.3	110.0	197	1.62	DSC	(1–5)	95J2
9	81.8	122.7	218	1.64	DSC	(1–5)	95J2
10	84.8	124.7	220	1.76	DSC	(1–5)	95J2

Remarks:
(1) the No refers to table part I (above)
(2) thermal unfolding of p53tet is a two-state process of the tetramer to unfolded monomers ($N_4 \rightarrow 4U$)
(3) ΔH and ΔCp represent data per monomer
(4) $T°$ is the reference temperature at which $\Delta G°$ for unfolding and dissociation is zero
(5) $T°$ does not correspond to the transition temperature T_{trs}

Tyrosine Kinase

SH3 domain of Bruton's tyrosine kinase in the presence of various buffers and salts

Salt	pH	T_{trs}	ΔH	Approach	Remarks	Ref
20 mM NaCl	3.0	50±2	109±50	heat, v.H.	(1)	96C11
20 mM NaCl	4.0	74±2	146±50	heat, v.H.	(1)	96C11
20 mM NaCl	5.0	81±2	197±50	heat, v.H.	(2)	96C11
20 mM NaCl	6.0	82±2	218±50	heat, v.H.	(3)	96C11
20 mM NaCl	7.0	62±2	79±50	heat, v.H.	(3)	96C11
500 mM Na$_2$SO$_4$	7.0	82±2	234±50	heat, v.H.	(3)	96C11
500 mM NaCl	7.0	79±2	163±50	heat, v.H.	(3)	96C11

Remarks:
(1) buffer: 20 mM citrate
(2) buffer: 20 mM acetate
(3) buffer: 20 mM sodium phosphate

Ubiquitin

Ubiquitin, bovine

pH	T_{trs}	ΔCp	ΔH	Approach	Remarks	Ref
1.2–3.2		3.9±1.3		DSC	(1,2)	93W4
1.2–3.2		4.8±0.5		DSC	(1,3)	93W4
	25		10	DSC	(1,4)	93W4
1.2–5.3		0		DSC	(5,6)	93W4
	25		198±30	DSC	(4,5)	93W4

Remarks:
(1) measured in aqueous solution
(2) ΔCp from single calorimetric recordings
(3) ΔCp from ΔH versus T_{trs}, T_{trs} varies from 54 to 73°C
(4) ΔH extrapolated value
(5) measured in 40% methanol
(6) T_{trs} varies from 21 to 42°C

Ubiquitin, from bovine red blood cells

pH	T_{trs}	ΔCp	ΔH	Approach	Remarks	Ref
2.0	57.0		196	DSC	(1)	94W5
2.2	58.1		185	DSC	(1)	94W5
2.5	61.6		219	DSC	(1)	94W5
2.75	66.3		243	DSC	(1)	94W5
3.0	74.1		265	DSC	(1)	94W5
3.25	79.0		273	DSC	(1)	94W5
3.5	85.1		288	DSC	(1)	94W5
4.0	90.0		302	DSC	(1)	94W5
	5	5.8		DSC	(2)	94W5
	25	5.7		DSC	(2)	94W5
	50	5.1		DSC	(2)	94W5
	75	3.8		DSC	(2)	94W5
	100	2.5		DSC	(2)	94W5
	125	0.8		DSC	(2)	94W5

Remarks:
(1) ΔCp is temperature dependent, see data below
(2) ΔCp was determined from heat capacity measurements of the folded and unfolded protein

Ubiquitin, from bovine red blood cells, temperature dependence of the heat capacity change at protein unfolding in aqueous solution, and related thermodynamic quantities

T	ΔH(exp)	ΔS(exp)	ΔCp(exp)	Approach	Remarks	Ref
5	-88	-444	5.8	DSC	(1–3)	95M3
25	27	-44	5.7	DSC	(1–3)	95M3
50	162	393	5.1	DSC	(1–3)	95M3
75	273	727	3.8	DSC	(1–3)	95M3
100	351	959	2.5	DSC	(1–3)	95M3
125	393	1068	0.8	DSC	(1–3)	95M3

Remarks:
(1) the data were taken from Ref. 95M3 with reference to Ref. 94W5
(2) ΔCp was obtained as described in Ref. 90P3
(3) ΔCp in kJ/mol/K, ΔH in kJ/mol and ΔS in J/mol/K

Urate Oxidase

Urate oxidase from *Aspergillus flavus*, recombinant

pH	T_{trs1}	ΔH_{trs1}	T_{trs2}	ΔH_{trs2}	Approach	Ref
9.50	49.7	344	60.6	247	DSC	95B3
9.05	49.7	321	68.9	112	DSC	95B3
8.80	49.9	335	69.3	107	DSC	95B3
8.50	49.7	330	69.3	124	DSC	95B3
7.80	50.1	346	68.1	273	DSC	95B3
7.45	48.1	310	68.4	242	DSC	95B3
7.25	47.8	301	68.7	223	DSC	95B3
7.20	47.7	317	68.9	201	DSC	95B3
6.95	44.7	214	65.0	68	DSC	95B3
6.10	40.0	18	66.1	264	DSC	95B3
5.95	32.6	29	65.4	281	DSC	95B3
5.80			66.2	262	DSC	95B3

Xylanase

Xylanase from *Streptomyces halstedii*, different enzymes

Transition	pH	T_{trs}	ΔCp	ΔH	Approach	Remarks	Ref
Enzyme Xys1L:							
trans. (1)	3.6	38.5±0.2	9.2	276±17	DSC	(1–3)	94R5
trans. (2)	3.6	43.5±0.4	17.2	351±17	DSC	(1–3)	94R5
trans. (3)	3.6	47.6±0.1	28.0	452±21	DSC	(1–3)	94R5
trans. (1)	6.3	51.7±0.2		418±8	DSC	(1,2)	94R5
trans. (2)	6.3	59.3±0.2		628±21	DSC	(1,2)	94R5
trans. (3)	6.3	64.3±0.1		933±25	DSC	(1,2)	94R5
trans. (1)	7.5	52.4±0.3		389±29	DSC	(1,2)	94R5
trans. (2)	7.5	55.4±0.4		573±33	DSC	(1,2)	94R5
trans. (3)	7.5	58.4±0.1		782±33	DSC	(1,2)	94R5
trans. (1)	8.9	44.5±0.1		322±17	DSC	(1,2)	94R5
trans. (2)	8.9	51.5±0.1		494±21	DSC	(1,2)	94R5
trans. (3)	8.9	54.6±0.1		657±25	DSC	(1,2)	94R5

578   Table 2. Enthalpy and Heat Capacity Changes – Molar Values

Xylanase from *Streptomyces halstedii*, different enzymes (continued)

Transition	pH	T_{trs}	ΔCp	ΔH	Approach	Remarks	Ref
Enzyme Xys1S:							
trans. (1)	4.1	43.2±0.3		238±13	DSC	(1,4)	94R5
trans. (2)	4.1	50.3±0.2		418±17	DSC	(1,4)	94R5
trans. (1)	7.5	60.5±0.2		414±29	DSC	(1,4)	94R5
trans. (2)	7.5	64.0±0.1		837±17	DSC	(1,4)	94R5
trans. (1)	8.9	46.6±0.2		301±13	DSC	(1,4)	94R5
trans. (2)	8.9	53.8±0.1		569±8	DSC	(1,4)	94R5

Remarks:
(1) Xys1L is the xylanase with a molecular mass of 45 kDa, Xys1S is the xylanase with a molecular mass of 35 kDa
(2) the thermal transition of Xys1L was resolved by deconvolution into three subtransitions
(3) ΔCp is from ΔH versus T_{trs}
(4) the thermal transition of Xys1S was resolved by deconvolution into two subtransitions

Table 3.
Enthalpy and Heat Capacity Changes – Specific Values

Acyl Carrier Protein

Acyl carrier protein

Salt	pH	T_{trs}	Δcp	Δh	Approach	Remarks	Ref
low salt	6.1	52.7		18.2	DSC	(1,2)	94M12
low salt Ca^{2+}	6.1	64.3		30.0	DSC	(1,3)	94M12
high salt	6.1	63.2		17.7	DSC	(1,4)	94M12

Remarks:
(1) Δcp, see molar values in table 2
(2) buffer: 50 mM sodium acetate, 0.5 mM DTT
(3) buffer: 50 mM sodium acetate, 0.5 mM DTT, 8.4 mM $CaCl_2$
(2) buffer: 240 mM sodium acetate, 0.5 mM DTT

Adenosine Triphosphate Synthase (F_1F_0) complex, see cytochrome c oxidase

Albumin, see also serum albumin

Alkaline Phosphatase

Alkaline phosphatase

pH	T_{trs}	Δc_p	Δh	Approach	Ref
7.5	52	0.59	17.6	DSC	79C

α-Amylase, see also taka-amylase

α-Amylase from hyperthermophilic archaebacterium *Pyrococcus furiosus*

pH	T_{trs}	Δcp	Δh	Approach	Remark	Ref
8.0	112	0.35		DSC	(1)	93L2

Remark:
(1) the transition can be resolved by deconvolution into three subtransitions at 102, 109, and 112°C, respectively; the temperature values were taken from Figs. 4 and 5 of Ref. 93L2

Apolipoprotein B

Apolipoprotein B-100, in the presence and in the absence of lipid, resolved into subtransitions

Transition	pH	T_{trs}	Δh	Approach	Remarks	Ref
Apolipoprotein, in lipid:						
trans. (1)	10	31.1	3.14	DSC	(1)	90W1
trans. (2)	10	57.1	0.84	DSC	(1)	90W1
trans. (3)	10	73.5	4.14	DSC	(1)	90W1
Apolipoprotein, solubilized:						
trans. (1)	10	49.7	4.73	DSC	(2)	90W1
trans. (2)	10	56.4	10.67	DSC	(2)	90W1
trans. (3)	10	66.6	2.26	DSC	(2)	90W1

Remarks:
(1) apolipoprotein in its native lipid environment, low density lipoprotein
(2) apolipoprotein in a soluble, lipid-free complex with sodium deoxycholate

Arabinose-Binding Protein

L-Arabinose-binding protein (*E. coli*)

pH	T_{trs}	Δc_p	Δh	Approach	Ref
7.4	53.5	0.4±0.01	19.2±0.2	DSC	83F2

Aspartate Aminotransferase

Aspartate aminotransferase, multiple subform α

Form	pH	T_{trs}	Δc_p	Δh	Approach	Ref
apoenzyme	8.2	67	0.49	22.8	DSC	81R
pyridoxal	6	78.8	0.3	25.3	DSC	81R
pyridoxal phosphate	8.2	78.5	0.36	26.9	DSC	81R
pyridoxamine	8.2	74	0.44	24.1	DSC	81R

Aspartate Transcarbamoylase

Aspartate transcarbamoylase (dodecamer)

pH	T_{trs}	Δc_p	Δh	Approach	Ref
7	64–76		20.2±0.7	DSC	88E

Bacteriorhodopsin

Bacteriorhodopsin, white and purple membranes

Form	pH	T_{trs}	Δc_p	Δh	Approach	Remarks	Ref
bacterioopsin in white membrane							
	7.0	79±0.5	0.54	8.0±0.4	DSC		90M1
bacteriorhodopsin in purple membrane							
	7.0	74±0.5		2.7±0.2	DSC	(1)	90M1
	7.0	97±0.5	0.35	13.8±0.8	DSC	(2)	90M1
bacterioopsin reconstituted with all-trans-retinal							
	7.0	101±0.5	0.35	16.7±1.3	DSC		90M1

Remarks:
(1) pretransition
(2) main transition

Barnase

Barnase

pH	T_{trs}	Δc_p	Δh	Approach	Remark	Ref
5.5	55.3	0.29±0.04	39.2	DSC	(1)	93M2

Remark:
(1) Δc_p is independent of pH in the range from 3.8–5.5, Δh was calculated from the molar values given in Ref. 93M2 using M = 12,382 Da

Binase

Ribonuclease from *Bacillus intermedius 7P* (binase)

pH	T_{trs}	Δc_p	Δh	Approach	Ref
7	57.6	1.2±0.3		DSC	87P4

Carbonic Anhydrase

Carbonic anhydrase B

pH	T_{trs}	Δc_p	Δh	Approach	Remark	Ref
–	67	0.58		DSC		79P4
9.5	68	0.55±0.02	30	DSC	(1)	86T1

Remark:
(1) the transition temperature was taken from Fig. 2 and Δh from Fig. 3 of Ref. 86T1.

Carboxypeptidase

Carboxypeptidase A, bovine pancreatic enzyme and procarboxypeptidase (propeptidase)

Form	pH	T_{trs}	Δc_p	Δh	Approach	Ref
enzyme	7.5	71±2	0.48±0.09	20±2	DSC	88S3
propeptidase	7.5	71±1	0.25±0.09	18±1	DSC	88S3

Carboxypeptidase B from porcine pancreas

pH	T_{trs}	Δc_p	Δh	Approach	Remarks	Ref
7.5	68.1		21	DSC	(1,2)	91C6
9.0	65.6		21	DSC	(1,3)	91C6

Remarks:
(1) scan rate 1 K/min. Ref. 91C6 contains further data for different scan rates
(2) buffer: 1 mM phosphate
(3) buffer: 1 mM pyrophposphate

Procarboxypeptidase B from porcine pancreas

pH	T_{trs}	Δc_p	Δh	Approach	Remarks	Ref
7.5	68.9		19	DSC	(1,2)	91C6
9.0	69.0		19	DSC	(1,3)	91C6

Remarks:
(1) scan rate 1 K/min. Ref. 91C6 contains further data for different scan rates
(2) buffer: 1 mM phosphate
(3) buffer: 1 mM pyrophposphate

Chymopapain, see papain

Chymotrypsin

α-Chymotrypsin

pH	T_{trs}	Δc_p	Δh	Approach	Ref
3.8	57	0.5±0.08		DSC	74P1
3.8	57	0.5±0.08		DSC	79P4
3.8	57	0.5±0.08		DSC	74T3

Chymotrypsin in the presence of poly(ethylene glycol)

Concentration	pH	T_{trs}	Δc_p	Δh	Approach	Ref
0 (%, w/w)		54		18.4	DSC	9402
10 (%, w/w)		60		22.6	DSC	9402

Clathrin

Clathrin, coated vesicles and reformed baskets

Form	pH	T_{trs}	Δc_p	Δh	Approach	Remarks	Ref
coated ves.	6.5	55.9±0.1		11.5±1	DSC		89S4
reformed b.	6.5	53.1±0.1		9.1±0.3	DSC	(1)	89S4
reformed b.	6.5	56.3±0.1			DSC	(1)	89S4

Remark:
(1) reformed baskets show two transitions. The given enthalpy change represents the overall heat effect

Cold-Shock Protein

CspB, cold-shock protein from *Bacillus subtilis*

Buffer	pH	T_{trs}	Δcp	Δh	Approach	Remarks	Ref
50 mM HE-PES	7.5	52.8	0.5±0.1	14.9	DSC	(1)	94M4
50 mM ImH	7.5	51.0	0.5±0.1	14.2	DSC	(2)	94M4
50 mM NaP	7.5	53.4	0.5±0.1	20.8	DSC	(3)	94M4

Remarks:
(1) unfolding in HEPES buffer follows a dimer→monomer transition (N_2→2U)
(2) unfolding in imidazole (ImH) buffer follows a dimer→monomer transition (N_2→2U)
(3) unfolding in Na phosphate (NaP) buffer follows a two-state transition (N→U)

Collagen

Collagen from rat skin

pH	T_{trs}	Δc_p	Δh	Approach	Remarks	Ref
	59.15±0.5		5.96±0.57	DSC	(1,2)	84F2

Remarks:
(1) the data refer to the main peak.
(2) the transition temperature is the mean value from Tab. 1 in Ref. 84F2

Collagen from rat tail tendon

Solvent	T_{trs}	Δc_p	Δh	Approach	Remarks	Ref
water	61.63		54.52±0.92	DSC	(1,2)	95M15
acetic acid, 0.5 M	37.02		51.55±0.70	DSC	(1,2)	95M15

Remarks:
(1) Δh refers to dry tendon, data were adjusted to zero scan rate
(2) position and shape of the denaturation endotherm are governed by the kinetics of an irreversible process

Collagen from polar cod skin

pH	T_{trs}	Δc_p	Δh	Approach	Remarks	Ref
3.8	20.5		77	DSC	(1,2)	92S12

Remarks:
(1) Δh is the specific enthalpy in J/g, denaturation enthalpy per one mole amino acid residues amounts to $\Delta H = 7$ kJ/mol
(2) the transition was resolved into three transitions at $T_{trs1} = 18.6°C$, $T_{trs2} = 20.3°C$, and $T_{trs3} = 20.7°C$ which were attributed to domains of 33 kDa, 230 kDa, and 97 kDa

Calf skin collagen type I, salt dependence (T_{trs1} = main transition, T_{trs2} = pretransition, scan rate 0.5°C/min)

Salt Concentration (mM)	T_{trs1}	Δh_1	T_{trs2}	Δh_2	Approach	Remarks	Ref
without LiCl	40.8	45.56	35.2	5.06	DSC	(1)	96K6
5	39.4	42.43	34.1	7.66	DSC		96K6
10	38.6	41.59	34.0	7.70	DSC		96K6
20	38.4	39.29	34.0	9.00	DSC		96K6
50	37.6	41.67	33.9	6.69	DSC		96K6
200	36.1	31.92	33.2	15.65	DSC		96K6
400	35.1	40.17	32.3	5.86	DSC		96K6
600	34.7	27.91	32.1	17.74	DSC		96K6
800	39.7	34.23			DSC		96K6
NaCl							
5	39.4	45.34	34.2	3.81	DSC		96K6
10	38.7	41.63	33.8	7.32	DSC		96K6
20	38.1	40.88	33.2	7.95	DSC		96K6
50	37.0	38.33	33.0	8.91	DSC		96K6
200	34.7	32.59	32.2	13.64	DSC		96K6

Calf skin collagen type I, salt dependence (T_{trs1} = main transition, T_{trs2} = pretransition, scan rate 0.5°C/min) (continued)

Salt Concentration (mM)	T_{trs1}	Δh_1	T_{trs2}	Δh_2	Approach	Remarks	Ref
400	33.1	36.94	30.8	8.62	DSC		96K6
450	33.7	32.34	31.0	11.34	DSC		96K6
500	40.3	32.76	32.9	8.28	DSC		96K6
1000	46.0	20.42		7.74	DSC		96K6
KCl							
5	39.4	42.30	34.1	7.78	DSC		96K6
50	35.8	34.94	32.2	11.63	DSC		96K6
NH$_4$Cl							
10	38.5	41.38	33.8	7.99	DSC		96K6
20	36.8	37.87	32.1	9.54	DSC		96K6
50	35.9	31.46	32.7	15.06	DSC		96K6
200	33.8	36.28	31.5	10.00	DSC		96K6
400	32.0	34.06	30.1	10.88	DSC		96K6
600	32.2	33.26	30.1	10.79	DSC		96K6
750	40.3	25.61	36.4	7.87	DSC		96K6
CaCl$_2$							
5	39.2	41.42	34.1	7.53	DSC		96K6
10	38.0	40.54	33.5	7.82	DSC		96K6
20	35.6	38.95	31.8	8.03	DSC		96K6
50	34.9	33.97	31.5	11.88	DSC		96K6
200	32.7	34.02	29.8	11.42	DSC		96K6
300	31.7	32.72	28.9	11.97	DSC		96K6
400	37.3	27.82	35.3	8.08	DSC		96K6
NaSCN							
5	39.4	41.55	34.1	7.66	DSC		96K6
10	38.6	41.92	34.0	7.03	DSC		96K6
20	36.0	37.45	32.1	10.00	DSC		96K6
50	34.2	36.65	31.3	9.46	DSC		96K6
100	31.4	27.74	29.3	15.65	DSC		96K6
150	30.3	35.82	28.0	7.24	DSC		96K6
200	34.8	33.76			DSC		96K6
NaH$_2$PO$_4$							
10	38.6	41.84	33.8	7.49	DSC		96K6
20	37.8	39.25	33.5	9.46	DSC		96K6
50	37.6	35.98	33.4	12.01	DSC		96K6
200	37.3	34.35	34.4	13.43	DSC		96K6
500	37.8	21.51	36.1	16.02	DSC		96K6
750	45.1	20.42	42.9	6.15	DSC		96K6
Na$_2$HPO$_4$							
10	39.6	41.88	34.2	8.20	DSC		96K6
20	40.2	33.60	37.0	13.39	DSC		96K6
50	40.3	36.53	36.7	12.26	DSC		96K6
100	40.7	34.31	37.8	15.40	DSC		96K6
300	42.0	31.34	38.3	10.38	DSC		96K6
500	49.7	39.92	38.3	1.88	DSC		96K6
Na$_2$SO$_4$							
5	39.1	41.38	34.0	7.70	DSC		96K6
10	38.0	40.54	34.3	7.91	DSC		96K6

Calf skin collagen type I, salt dependence (T_{trs1} = main transition, T_{trs2} = pretransition, scan rate 0.5°C/min) (continued)

Salt Concentration (mM)	T_{trs1}	Δh_1	T_{trs2}	Δh_2	Approach	Remarks	Ref
20	36.1	32.93	33.3	14.27	DSC		96K6
50	35.5	34.48	32.9	12.38	DSC		96K6
75	35.5	33.51	33.1	13.05	DSC		96K6
100	35.4/42.2		32.9		DSC	(2)	96K6
150	32.3/45.4		29.7		DSC	(2)	96K6
200	46.0	30.38			DSC		96K6
250	48.4	33.58			DSC		96K6
Li_2SO_4							
5	39.0	41.34	34.4	6.78	DSC		96K6
10	37.7	40.42	34.1	6.61	DSC		96K6
20	36.6	40.25	33.9	6.57	DSC		96K6
50	35.9	38.58	34.1	7.45	DSC		96K6
100	35.6	38.33	34.1	7.53	DSC		96K6
200	35.8	46.36	34.6		DSC	(2,3)	96K6
300	36.2/43.7	47.28	35.0		DSC	(2,3)	96K6
400	37.5/43.9	47.86			DSC	(2,3)	96K6
500	40.0/44.4	50.00			DSC	(2,3)	96K6
600	42.5/44.7	45.69			DSC	(2,3)	96K6
750	43.7/44.9	50.33			DSC	(2,3)	96K6
900	47.2	48.37			DSC	(2,3)	96K6
NaN_3							
5	39.5	40.88	34.5	8.91	DSC		96K6

Remarks:
(1) Δcp was determined from Δh versus T_{trs}, $\Delta cp = 0.473$ J/g/K over all data points in the range from 32 to 42°C
(2) double peak
(3) total heat instead of Δh_1

Cytochrome b$_5$

Cytochrome b$_5$ from rabbit, fragment 1–90

pH	T_{trs}	Δc_p	Δh	Approach	Ref
7.4	70.7±0.3	0.55±0.08		DSC	83B1
7.4	70.7±0.3	0.55±0.08		DSC	80P

Cytochrome c

Bovine cytochrome c, ferric form

pH	T_{trs}	Δc_p	Δh	Approach	Ref
–	–	0.58		DSC	79P4
–	–	0.58		DSC	74P1
4.8	78±0.3	0.61±0.02		DSC	86P2
3–6	–	0.61		DSC	89P8

Cytochrome c

	pH	T_{trs}	Δcp	Δh	Approach	Remarks	Ref
	3.2	60	0.11		DSC	(1–3)	94H8

Remarks:
(1) Δcp is the experimental value
(2) Ref. 94H8 contains corrections of Δcp for volume effects
(3) T_{trs} was taken from the figures in Ref. 94H8

Cytochrome c from horse heart, chemically modified by maleylation, modification in percent labelled lysines

Modification	pH	T	Δh	Approach	Ref
0.0 %	7.5	82.0	10.59±1.09	DSC	87I
9.0±0.5 %	7.5	78.8±1.3	9.12±1.05	DSC	87I
17.2 %	7.5	75.6±1.6	9.10±1.38	DSC	87I
21.5±0.9 %	7.5	71.7±0.4	9.20±0.84	DSC	87I
27.7±1.7 %	7.5	70.7±0.4	9.62±1.55	DSC	87I
51.7±1.1 %	7.5	67.9±1.1	8.58±0.33	DSC	87I
77.5±3.1 %	7.5	66.4±1.5	4.56±2.43	DSC	87I
96.4±2.5 %	7.5	66.7±0.6	4.94±1.76	DSC	87I

Cytochrome c Oxidase

Cytochrome c oxidase and adenosine triphosphate synthase (F_1F_0) complex in detergent dispersed form

Form	pH	T_{trs}	Δc_p	Δh	Approach	Remarks	Ref
CcO	7.4	57		10.6	DSC	(1)	92Q
F_1F_0	7.4	43.5/59.5		5.4	DSC	(2)	92Q
CcO + F_1F_0	7.4	57		8.7	DSC	(3)	92Q

Remarks:
(1) CcO = cytochrome c oxidase
(2) adenosine triphosphate synthase (F_1F_0) complex
(3) equimolar mixture of CcO and F_1F_0

Dihydrofolate Reductase

Dihydrofolate reductase (DHFR), bovine and *E. coli*

	pH	T_{trs}	Δc_p	Δh	Approach	Remark	Ref
	6.8	42–78	0.66		DSC	(1)	95S1

Remark:
(1) the specific heat capacity change was determined from Δh versus T_{trs} of *E. coli* and bovine DHFR and their complexes with NADPH, trimethoprim and methotrexate

"Dry" Proteins

Protein	T_{trs}	Δc_p	Δh	Approach	Remarks	Ref
bovine albumin	66.2		106.3	DSC	(1)	95G7
γ-globulin	73.7		130.6	DSC	(1)	95G7
ovalbumin	68.5		116.1	DSC	(1)	95G7

Remark:
(1) Ref. 95G7 contains further data for different pretreatments of the samples

Erythrocyte Band 3 Protein

Erythrocyte band 3 protein, reconstituted with lipids

pH	T_{trs}	Δc_p	Δh	Approach	Ref
8	47–66	0.1–0.5	7–33	DSC	88M1

Fibrinogen

Fibrinogen and its fragments

Transition	pH	T_{trs}	Δcp	Δh	Approach	Remarks	Ref
D fragment:							
LT1		35–55	0.13		DSC	(1,2)	82P4
HT2		73–90	0.04		DSC	(1,3)	82P4
E fragment:							
HT1		55–92	0.06		DSC	(1,4)	82P4

Remarks:
(1) data were taken from from Fig. 12 in Ref. 82T2
(2) low temperature transition 1
(3) high temperature transition 2
(4) high temperature transition 1

Fibrinogen, bovine, proteolytic fragment derived from the C-terminal region

Fragment	pH	T_{trs}	Δc_p	Δh	Approach	Ref
D_H	3.5	45.0	0.25		DSC	95L7

Fibronectin

Fibronectin, 42kDa gelatin binding fragment

pH	T_{trs}	Δc_p	Δh	Approach	Remarks	Ref
2.5	37.3		5.9	DSC		91L8
2.5	37.5		6.7	DSC		91L8
3.4	51.8		7.9	DSC		91L8
4.1	56.0	0.21±0.04	8.4	DSC	(1)	91L8
7.4	64.4		12.1	DSC		91L8
8.0	64.1		12.1	DSC		91L8
9.5	60.9		10.5	DSC		91L8
2.5–9.5	37.3–60.9	0.201		DSC	(2)	91L8

Remarks:
(1) from calorimetric recordings, mean value from several runs
(2) from the temperature dependence of Δh

Galactosidase

β-Galactosidase, wild-type and mutants

Mutant	pH	T_{trs}	Δc_p	Δh	Approach	Remarks	Ref
wild-type	7.5	49.8		17.7	DSC	(1)	90E1
Tyr503→Phe	7.5	49.5		18.0	DSC	(1)	90E1
Glu461→Gln	7.5	53.2		18.8	DSC	(1)	90E1
wild-type	7.5	57.1		23.4	DSC	(2)	90E1
Tyr503→Phe	7.5	54.4		20.6	DSC	(2)	90E1
Glu461→Gln	7.5	54.1		18.8	DSC	(2)	90E1
wild-type	7.5	53.1		20.7	DSC	(3)	90E1
Tyr503→Phe	7.5	54.3		19.2	DSC	(3)	90E1
Glu461→Gln	7.5	55.6			DSC	(3)	90E1
wild-type	7.5	60.5		>21.8	DSC	(4)	90E1
Tyr503→Phe	7.5	60.6		>20.9	DSC	(4)	90E1
Glu461→Gln	7.5	58.2			DSC	(4)	90E1

Remarks:
(1) in the presence of 1 mM EDTA
(2) in the presence of 1 mM magnesium
(3) in the presence of 1 mM phenylethyl thio-β-D galactopyranoside
(4) in the presence of 1 mM magnesium and 1 mM phenylethyl thio-β-D-galactopyranoside

Galactoside-Binding Protein

Galactoside-binding protein in the presence of Ca^{2+}

pH	T_{trs}	Δc_p	Δh	Approach	Remark	Ref
7.8	61	0.5	11.2	DSC	(1)	81S2

Remark:
(1) data are calculated per cooperative unit AB_2

Globulins

Plant proteins from *Amaranth* and *Quinoa*

Form	pH	T_{trs}	Δc_p	Δh	Approach	Remarks	Ref
amaranth, salt soluble protein:							
	7.2	59.0±0.8		7.00±1.0	DSC	(1)	96G6
amaranth, salt soluble protein + urea:							
	7.2	47.5±0.7		2.00±0.8	DSC	(1)	96G6
amaranth globulins:							
	7.2	101.0±1.0		4.25±0.9	DSC	(1)	96G6
amaranth globulins + 3 M urea:							
	7.2	84.0±1.3		2.08±0.9	DSC	(1)	96G6
bovine γ-globulins:							
	7.2	112.0±2.5		4.61±1.3	DSC	(1)	96G6
quinoa, salt soluble protein:							
	7.2	58.0±1.5		10.0±1.6	DSC	(1)	96G6
quinoa, salt soluble protein + urea:							
	7.2	46.3±1.7		4.30±1.1	DSC	(1)	96G6

Remark:
(1) buffer: 10 mM phosphate and 0.4 M NaCl

Glucoamylase

Glucoamylase 1 from *Aspergillus niger*
dependence of thermodynamic quantities on protein concentration (mg/ml)

Concen-tration	pH	T_{trs}	Δc_p	Δh	Approach	Remarks	Ref
12.0	7.8	67.8	0.33		DSC	(1)	92W3
10.0	7.8	66.4	0.39		DSC	(1)	92W3
8.33	7.8	65.5	0.44		DSC	(1)	92W3
4.0	7.8	63.9	0.45		DSC	(1)	92W3
	7.8		0.40±0.05		DSC	(2)	92W3

Remarks:
(1) for enthalpy, see molar values
(2) mean value (± standard deviation)

Glucoamylase from *Aspergillus niger*, fragments

Form	pH	T_{trs}	Δc_p	Δh	Approach	Remarks	Ref
G2	7.8	65.7	0.44	23.9	DSC	(1,2)	92W3
G1C	7.8	57.4	0.45	14.5	DSC	(2,3)	92W3
G1C499	7.8	57.3	0.59	18.8	DSC	(2,4)	92W3
G1C509	7.8	54.7	0.57	24.8	DSC	(2,5)	92W3

Remarks:
(1) G2 = glucoamylase 2
(2) protein conc. was 0.122 mM
(3) G1C = glucoamylase 1 fragment (471–616)
(4) G1C499 = glucoamylase 1 fragment (499–616)
(5) G1C509 = glucoamylase 1 fragment (509–616)

Glucoamylase from *Aspergillus niger* and glucoamylase 1 fragment (471–616), effect of β-cyclodextrin (β-CD)

β-CD Concentration	pH	T_{trs}	Δc_p	Δh	Approach	Remarks	Ref
Glucoamylase 1 fragment (471–616):							
0 mM	7.8	57.4	0.45		DSC	(1)	92W3
0.50 mM	7.8	63.3	0.52		DSC	(1)	92W3
2.00 mM	7.8	67.7	0.56		DSC	(1)	92W3
5.18 mM	7.8	70.3	0.62		DSC	(1)	92W3
Glucoamylase 1:							
0 mM	7.8	66.4	0.40		DSC	(1)	92W3
2.00 mM	7.8	67.6	0.47		DSC	(1)	92W3
5.00 mM	7.8	69.8	0.51		DSC	(1)	92W3

Remark:
(1) protein conc. was 0.122 mM, for molar thermodynamic quantities, see Table 2

Glucoamylase from *Aspergillus niger*, isoenzymes G1 and G2

Transition	pH	T_{trs}	Δh	Approach	Remarks	Ref
Isoenzyme G1:					(1)	
trans. (1)	7.0	51.9±1.7	3.17±0.60	DSC		95T6
trans. (2)	7.0	56.2±0.2	5.48±0.53	DSC		95T6
trans. (3)	7.0	58.7±0.4	6.95±0.26	DSC		95T6
trans. (4)	7.0	61.9±0.3	7.62±0.16	DSC		95T6
trans. (5)	7.0	65.0±0.2	7.55±0.12	DSC		95T6
Isoenzyme G2:					(2,3)	
trans. (1)	7.0	52.4±0.6	4.44±0.25	DSC		95T6
trans. (3)	7.0	58.2±0.1	6.48±0.44	DSC		95T6
trans. (4)	7.0	62.3±0.2	8.07±0.33	DSC		95T6
trans. (5)	7.0	65.6±0.4	8.44±0.31	DSC		95T6
Isoenzyme G1 (27.4 μM) in the presence of β-cyclodextrin:					(4)	
trans. (1)	7.0	52.0±1.1	3.82±0.26	DSC		95T6
trans. (2)	7.0	60.3 to 69.6	7.05±0.85	DSC		95T6
trans. (3)	7.0	57.7±0.4	6.60±0.29	DSC		95T6
trans. (4)	7.0	61.4±0.4	7.86±0.38	DSC		95T6
trans. (5)	7.0	64.6±0.4	7.91±0.22	DSC		95T6
Isoenzyme G2 (13.5 μM) in the presence of 1.25 mM β-cyclodextrin:					(3,5)	
trans. (1)	7.0	51.8	5.22	DSC		95T6
trans. (3)	7.0	58.0	7.42	DSC		95T6
trans. (4)	7.0	62.6	9.07	DSC		95T6
trans. (5)	7.0	66.1	8.99	DSC		95T6

Glucoamylase from *Aspergillus niger*, isoenzymes G1 and G2 (continued)

Transition	pH	T_{trs}	Δh	Approach	Remarks	Ref
Isoenzyme G2 (13.5 µM) in the presence of 5.01 mM β-cyclodextrin:					(3,5)	
trans. (1)	7.0	51.9	4.62	DSC		95T6
trans. (3)	7.0	58.1	6.62	DSC		95T6
trans. (4)	7.0	62.4	8.47	DSC		95T6
trans. (5)	7.0	65.8	8.88	DSC		95T6

Remarks:
(1) 5 independent measurements at protein concentration from 11.5 to 94.3 µM, the melting peak was resolved by deconvolution into 5 subtransitions
(2) 3 independent measurements at protein concentration from 13.5 to 27.5 µM, the melting peak was resolved by deconvolution into 4 subtransitions
(3) G2 is the shorter form of the enzyme which lacks the C-terminal 100 amino acids (starch-binding domain)
(4) 7 independent measurements at ligand concentration from 0.157 to 7.52 mM, the melting peak was resolved by deconvolution into 5 subtransitions
(5) resolved into 4 subtransitions

Glucoamylase from *Aspergillus niger*, isoenzymes G1 and G2, in the presence of 1-deoxynojirimycin

Transition	pH	T_{trs}	Δh	Approach	Remarks	Ref
Isoenzyme G1 (25.6 µM), ligand conc. 0.096 mM:						
trans. (1)	7.0	53.9	3.72	DSC		95T6
trans. (2)	7.0	57.9	6.10	DSC		95T6
trans. (3)	7.0	62.1	7.23	DSC		95T6
trans. (4)	7.0	65.4	8.21	DSC		95T6
trans. (5)	7.0	68.2	8.06	DSC		95T6
Isoenzyme G1 (25.6 µM), ligand conc. 0.478 mM:						
trans. (1)	7.0	56.4	4.11	DSC		95T6
trans. (2)	7.0	59.0	5.12	DSC		95T6
trans. (3)	7.0	65.0	7.19	DSC		95T6
trans. (4)	7.0	68.4	8.74	DSC		95T6
trans. (5)	7.0	71.3	8.69	DSC		95T6
Isoenzyme G1 (25.6 µM), ligand conc. 2.39 mM:						
trans. (1)	7.0	65.5	2.60	DSC		95T6
trans. (2)	7.0	59.1	5.07	DSC		95T6
trans. (3)	7.0	68.9	6.71	DSC		95T6
trans. (4)	7.0	72.1	8.35	DSC		95T6
trans. (5)	7.0	74.9	8.81	DSC		95T6
Isoenzyme G1 (25.6 µM), ligand conc. 6.10 mM:						
trans. (1)	7.0	70.1	5.20	DSC		95T6
trans. (2)	7.0	61.4	5.34	DSC		95T6
trans. (3)	7.0	73.2	7.56	DSC		95T6
trans. (4)	7.0	76.2	8.62	DSC		95T6
trans. (5)	7.0	78.6	7.69	DSC		95T6
Isoenzyme G2 (30.4 µM), ligand conc. 0.112 mM:					(1,2)	
trans. (1)	7.0	54.6	5.30	DSC		95T6
trans. (3)	7.0	60.9	8.80	DSC		95T6
trans. (4)	7.0	64.7	10.0	DSC		95T6
trans. (5)	7.0	67.6	9.30	DSC		95T6

Glucoamylase from *Aspergillus niger*, isoenzymes G1 and G2, in the presence of 1-deoxynojirimycin (continued)

Transition	pH	T_{trs}	Δh	Approach	Remarks	Ref
Isoenzyme G2 (30.4 µM), ligand conc. 0.488 mM:					(1,2)	
trans. (1)	7.0	58.6	4.82	DSC		95T6
trans. (3)	7.0	64.0	8.37	DSC		95T6
trans. (4)	7.0	67.4	10.1	DSC		95T6
trans. (5)	7.0	70.3	10.1	DSC		95T6
Isoenzyme G2 (30.4 µM), ligand conc. 1.79 mM:					(1,2)	
trans. (1)	7.0	59.7	4.93	DSC		95T6
trans. (3)	7.0	66.4	8.62	DSC		95T6
trans. (4)	7.0	70.0	11.0	DSC		95T6
trans. (5)	7.0	73.2	11.0	DSC		95T6
Isoenzyme G2 (30.4 µM), ligand conc. 3.10 mM:					(1,2)	
trans. (1)	7.0	65.5	5.44	DSC		95T6
trans. (3)	7.0	69.4	8.86	DSC		95T6
trans. (4)	7.0	72.3	10.6	DSC		95T6
trans. (5)	7.0	74.5	10.7	DSC		95T6

Remarks:
(1) G2 is the shorter form of the enzyme which lacks the C-terminal 100 amino acids (starch-binding domain)
(2) the melting peak was resolved by deconvolution into 4 subtransitions

Glucose Oxidase

Glucose oxidase from *Aspergillus niger*, free and immobilized enzyme

Form	pH	T_{trs}	Δc_p	Δh	Approach	Ref
free enzyme	7.9	57.9		38.6	DSC	89I1
immobilized	7.9	60.5		51.8	DSC	89I1

Glucose oxidase from *Aspergillus niger*

pH	T_{trs}	Δcp	Δh	Approach	Remarks	Ref
3.2	25		23±4	ITC	(1,2)	93H8

Remarks:
(1) measured by isothermal calorimetry
(2) from interaction of glucose oxidase with n-alkylsulfates, extrapolated value

Glutamine Synthetase

Manganese glutamine synthetase from *E. coli*

pH	T	Δc_p	Δh	Approach	Remarks	Ref
7.3	37		59±17	ITC	(1,2)	95Z3

Remarks:
(1) enthalpy of urea-induced dissociation and unfolding measured by isothermal titration calorimetry
(2) the enthalpy of binding of urea to the protein was corrected for

Glycinin, see plant seed protein

Heat Shock Protein

Heat shock proteins Hsp25 and Hsp27

Transition	pH	T_{trs}	Δc_p	Δh	Approach	Ref
Hsp25						
trans. (1)	7.2	31.5±1.0	0.07	0.71	DSC	95D6
trans. (2)	7.2	69.6		6.1	DSC	95D6
Hsp27						
trans. (1)	7.2	38.2±0.8	0.03	0.46	DSC	95D6
trans. (2)	7.2	71.1		4.5	DSC	95D6

Hemolysin

Hemolysin of *Vibrio parahemolyticus*

pH	T_{trs}	Δc_p	Δh	Approach	Ref
8.15	51.4	0.13	23.4	DSC	83U

Histone

Histone GH1 and GH5 fragment

Fragment	pH	T_{trs}	Δc_p	Δh	Approach	Ref
GH1	7	59.5	0.54		DSC	82T1
GH5	7	63	0.5		DSC	82T1

Immunoglobulin

Bence-Jones protein

pH	T_{trs}	Δc_p	Δh	Approach	Remark	Ref
7.4	56	0.46		DSC	(1)	77Z

Remark:
(1) calculated per monomer

Immunoglobulin, deconvolution MAK33, murine antibody of subtype κ/IgG1

pH	T_{trs}	Δc_p	Δh	Approach	Remarks	Ref
2.0	66–72		4.7	DSC	(1)	91B6
2.7	65–73		4.7	DSC	(1)	91B6
7.0	64–80		24.2	DSC	(2)	91B6

Remarks:
(1) resolved into two transitions by deconvolution
(2) resolved into five transitions by deconvolution

Insulin

Insulin

pH	T_{trs}	Δc_p	Δh	Remark	Ref
9.6	25	0.35±0.21	15.9	(1)	82F2

Remark:
(1) insulin reduction, flow calorimetry

Insulin, des-pentapeptide (B26-B30) insulin

pH	T_{trs}	Δc_p	Δh	Approach	Remark	Ref
	60		10.5	DSC	(1)	93H9

Remark:
(1) measured in 20% acetic acid

Lac-Repressor

Lac-repressor, headpiece

pH	T_{trs}	Δc_p	Δh	Approach	Ref
8	65	0.21		DSC	81H2

Laccase

Laccase from the latex of japanese laquer tree, holo and apo protein

	pH	T_{trs}	Δc_p	Δh	Approach	Ref
holo laccase	7.0	50.7±0.7		9.2	DSC	95A2
type-2-Cu-depleted laccase						
	7.0	51.5±0.7		7.5	DSC	95A2
apo laccase	7.0	46.6±2.1		5.0	DSC	95A2

α-Lactalbumin

α-Lactalbumin, bovine

pH	T_{trs}	Δc_p	Δh	Approach	Ref
6.3	62±0.2	0.28±0.06	19.7±0.6	DSC	81P1
7.0	59		17.6	DSC	85P1

α-Lactalbumin, in the presence of calcium

pH	T_{trs}	Δcp	Δh	Approach	Remarks	Ref
7.0	67	0.47		DSC	(1–3)	94H8

Remarks:
(1) Δcp is the experimental value
(2) Ref. 94H8 contains corrections of Δcp for volume effects
(3) T_{trs} was taken from the figures in Ref. 94H8

β-Lactoglobulin

β-Lactoglobulin

pH	T_{trs}	Δc_p	Δh	Approach	Ref
2	83.2	0.288±0.003		DSC	89L2
5.5	80	0.288±0.003		DSC	89L2
4.0	82		13.4	DSC	85P1
6.0	75		10.0	DSC	85P1

β-Lactoglobulin B

pH	T_{trs}	Δc_p	Δh	Approach	Ref
1.5–3.0	77–89	0.65±0.05		DSC	92A5

β-Lactoglobulin from bovine milk, heat denaturation

Urea Concentration	pH	T_{trs}	Δc_p	Δh	Approach	Remarks	Ref
	2.0		0.31±0.04		DSC	(1)	92G5
	2.0		0.27±0.02		DSC	(2)	92G5
4.4 M	2.0		0.39±0.04		DSC	(3)	92G5
4.4 M	2.0		0.36±0.04		DSC	(4)	92G5
4.4 M	2.0		0.67±0.07		DSC	(5)	92G5

Remarks:
(1) Δc_p is given as mean value in Ref. 92G5 (heat denaturation)
(2) Δc_p from Δh versus T_{trs} at different concentrations of phosphate (heat denaturation)
(3) heat denaturation, mean from single recordings
(4) heat denaturation, from Δh versus T_{trs}
(5) cold denaturation, mean from single recordings

Lactoperoxidase

Lactoperoxidase

pH	T_{trs}	Δc_p	Δh	Approach	Ref
6.05	69.3	0.47±0.03	27.6	DSC	86P3

Lipoprotein

Human plasma lipoprotein

Transition	pH	T_{trs}	Δh	Approach	Remarks	Ref
Low-density lipoprotein, donor #1:						
trans. (1)	7.2	30.7	2.6	DSC	(1,2)	95P6
trans. (3)	7.2	79.6	2.6	DSC	(1,4)	95P6
Lipoprotein(a), donor #1:						
trans. (1)	7.2	24.7	4.6	DSC	(1,2)	95P6
trans. (2)	7.2	55.5	9.2	DSC	(1,3)	95P6
trans. (3)	7.2	79.5	2.7	DSC	(1,4)	95P6
Low-density lipoprotein, donor #2:						
trans. (1)	7.2	29.3	2.7	DSC	(1,2)	95P6
trans. (3)	7.2	80.3	2.5	DSC	(1,4)	95P6
Lipoprotein(a), donor #2:						
trans. (1)	7.2	18.7	6.1	DSC	(1,2)	95P6
trans. (2)	7.2	56.0	7.1	DSC	(1,3)	95P6
trans. (3)	7.2	81.3	2.7	DSC	(1,4)	95P6
Low-density lipoprotein, treated with DTT, donor #2:						
trans. (1)	7.2	29.3	3.7	DSC	(1,2)	95P6
trans. (3)	7.2	79.3	1.5	DSC	(1,4)	95P6
Remnant lipoprotein particle, donor #2, remark (5):						
trans. (1)	7.2	18.6	2.8	DSC	(1,2)	95P6
trans. (3)	7.2	79.3	1.6	DSC	(1,4)	95P6
Apolipoprotein(a):						
trans. (2)	7.2	54.4	8.8	DSC	(1,3)	95P6

Remarks:
(1) buffer: 10 mM phosphate, 0.9% (w/v) NaCl, 0.01% (w/v) EDTA, 50 µg/ml gentamicin base, pH 7.2
(2) the first peak arises from core-located apolar lipid, Δh is given in J/g of cholesteryl ester
(3) the second peak was attributed to apolipoprotein(a), Δh is given in J/g of apolipoprotein(a)
(4) the third transition corresponds to apolipoprotein-B100 unfolding, Δh is given in J/g of apolipoprotein-B100
(5) remnant lipoprotein particle obtained upon reduction and subsequent removal of apolipoprotein(a) from lipoprotein(a)

Lysozyme HEW

HEN Egg White lysozyme

pH	T_{trs}	Δc_p	Δh	Approach	Ref
4	77	0.46±0.04		DSC	74P1
4	77	0.46±0.04		DSC	73K
4.5	78.5	0.46±0.02		DSC	76P2
2–4		0.48		DSC	89P8

Lysozyme, HEN Egg White, temperature dependence of Δc_p [remark (1)]

T	Δc_p	Approach	Ref
0	0.42	DSC	89P8
25	0.48	DSC	89P8
50	0.45	DSC	89P8
75	0.36	DSC	89P8

Remark:
(1) the data were taken from Fig. 11 of Ref. 89P8; the curve can be approximated by the following polynomial expression: $Cp(t) = 4.184 \times (a0 + a1 \times t + a2 \times t^2)$, with temperature t in °C, a0 = 0.4265 ($\pm$0.0065), a1 = 3.042×10^{-3} ($\pm 0.301 \times 10^{-3}$), and a2 = -5.233×10^{-5} ($\pm 0.289 \times 10^{-5}$)

Lysozyme

pH	T_{trs}	Δcp	Δh	Approach	Remarks	Ref
1.9	56	0.57		DSC	(1–3)	94H8

Remarks:
(1) Δcp is the experimental value
(2) Ref. 94H8 contains corrections of Δcp for volume effects
(3) T_{trs} was taken from the figures in Ref. 94H8

Lysozyme Phage T4

Lysozyme, phage T4, wild-type and mutants

Mutant	pH	T_{trs}	Δc_p	Δh	Approach	Ref
wild-type			0.46		DSC	79P4
wild-type	1.4–3.0	32.8–54.5	0.46		heat, v.H.	77E
Trp138→Tyr	2.2	36.1	0.46		heat, v.H.	77E
triple mutant (Trp126,138,158→Tyr)						
	2.2	35.8			heat, v.H.	77E

Muscle Protein, see also myosin

Muscle protein from the aductor muscle of *Molina*

pH	T_{trs}	Δc_p	Δh	Approach	Ref
	50.5/72.5		10.5	DSC	94P1

Myelin

Bovine-brain myelin membrane

pH	T_{trs}	Δc_p	Δh	Approach	Remark	Ref
7.0	80.3±0.2		4.7±0.6	DSC	(1)	92R4

Remark:
(1) the transition was assigned to thermal denaturation of myelin proteolipid and DM–20 protein

Myoglobin

Myoglobin, sperm whale, cyanomet form

pH	T_{trs}	Δc_p	Δh	Approach	Ref
10.5	78.5	0.65		DSC	79P4
10.5	78.5	0.65		DSC	74P1
11	72	0.63		DSC	72A
11	72	0.63		DSC	71P

Myoglobin, sperm whale, temperature dependence of Δc_p [remark (1)]

T	Δc_p	Approach	Ref
0	0.6	DSC	89P8
25	0.66	DSC	89P8
50	0.63	DSC	89P8
75	0.51	DSC	89P8

Remark:
(1) the data were taken from Fig. 11 of Ref. 89P8; the curve can be approximated by the following polynomial expression: $Cp(t) = 4.184 \times (a0 + a1 \times t + a2 \times t^2)$, with temperature t in °C, a0 = 0.5999 (±0.0045),
a1 = 3.911×10^{-3} (±0.208×10^{-3}), and a2 = $-6.702 + 10^{-5}$ (±0.200×10^{-5})

Myosin

Myosin from rabbit muscle

Transition	pH	T_{trs}	Δcp	Δh	Approach	Remarks	Ref
overall	7.0	55		13.9±1.2	DSC		77W
major trs.	6.2	52	0.25±0.13	13.4±2.5	DSC	(1)	80S2
major trs.			0.71±0.21		DSC	(2)	80S2
small trs.	6.2			2.5±1.3	DSC		80S2
major trs.	7.4	46		9.2±1.7	DSC		80S2

Remarks:
(1) Δcp from individual scanning calorimetric recordings at various pH values
(2) Δcp from Δh versus T_{trs}

Myosin fragments

Fragment	pH	T_{trs}	Δc_p	Δh	Approach	Remarks	Ref
LMM	6.5	43–63	0.25±0.05		DSC	(1)	79P2
LMM	6.5	43–63	0.25±0.05		DSC	(1)	79P3
LMM		50	0.25±0.05	22.5	DSC	(1,2)	82P2
LF-3	6.5	51–56	0.25±0.05		DSC		79P2
LF-3	6.5	51–56	0.25±0.05		DSC	(3)	79P3
LF-3		53	0.25±0.05	24.3	DSC	(2)	82P2

Remarks:
(1) LMM = light meromyosin
(2) the given enthalpy change is an extrapolated value, the specific melting enthalpy calculated for 100%
helicity
(3) the enthalpy change for the first of the two subsequent transitions at 51°C amounts to 670 kJ/mol, and for
the second one at 56°C 640 kJ/mol

Myosin, thick filament

pH	T_{trs}	Δc_p	Δh	Approach	Remark	Ref
6.3–7.5	46–55	0.46±0.13	14.4±0.6	DSC	(1)	90B3

Remark:
(1) in the presence of 0.17 M KCl

Myosin rod

pH	T_{trs}	Δc_p	Δh	Approach	Ref
6.1–8.0	35–65	0.46±0.13	17.7±1.3	DSC	90B4

Light meromyosin (= LMM)

pH	T_{trs}	Δc_p	Δh	Approach	Ref
6.1–8.0	35–65		17.5±1.5	DSC	90B4

Myosin subfragment 2 (= S–2)

pH	T_{trs}	Δc_p	Δh	Approach	Ref
6.2	40–60		16.8±2.5	DSC	90B4

Nuclease from Staphylococcus Aureus

Staphylococcal nuclease

pH	T_{trs}	Δc_p	Δh	Approach	Remarks	Ref
7.0	20	0.45		DSC	(1,2)	89G4

Remarks:
(1) determined measuring the partial heat capacity of the native protein at pH 7.0, and the heat capacity of the unfolded protein at pH 3.23
(2) the protein contains an additional heptapeptide at the N-terminus: Met-Asp-Pro-Trp-Val-Tyr-Ser

Papain

Papain

pH	T_{trs}	Δc_p	Δh	Approach	Remark	Ref
3.80	78.5	0.59		DSC	(1)	93S8

Remark:
(1) for more detailed data, see molar values

Chymopapain

	pH	T_{trs}	Δc_p	Δh	Approach	Remark	Ref
	2.90	78.2	0.40		DSC	(1)	93S8

Remark:
(1) for more detailed data, see molar values

Paramyosin

Paramyosin and paramyosin fragments

Fragment	pH	T_{trs}	Δc_p	Δh	Approach	Remarks	Ref
paramyosin			≤0.53	24	DSC		79P2
TRC-1			0.29	27	DSC		79P2
TRC-1		77	0.29	26.3	DSC	(1)	82P2
TRC-2	3.0	83	0.42	27	DSC		79P2
TRC-2		83	0.42	27	DSC	(1)	82P2

Remark:
(1) Δh is an extrapolated value, the specific melting enthalpy calculated for 100% helicity

Parvalbumin

Parvalbumin in the presence (holo form) and in the absence (apo form) of calcium

Form	pH	T_{trs}	Δc_p	Δh	Approach	Ref
holo	7	90	0.40±0.04		DSC	78F
apo	7	35	0.49±0.04		DSC	78F

Pepsin

Pepsin in the presence of NaCl

	pH	T_{trs}	Δc_p	Δh	Approach	Ref
	7.1	60.5	0.544±0.042		DSC	78A

Pepsinogen

Pepsinogen

	pH	T_{trs}	Δc_p	Δh	Approach	Ref
	6	66	0.607±0.042		DSC	81P2

Plant Protein, see also Globulins

Plant Seed Protein

Pland seed protein from *Vicia faba* and *Soybean*

Species	pH	T_{trs}	Δc_p	Δh	Approach	Remark	Ref
vicia faba	7.6	78.5	0.30±0.03	17±1	DSC	(1)	85D2
soybean	7.6	80	0.18±0.02	21	DSC		85D1

Remark:
(1) data per protomer, in the absence of NaCl

Glycinin from soybean

pH	T_{trs}	Δc_p	Δh	Approach	Remarks	Ref
7.0	97.8±0.3		7.4±0.4	DSC	(1,2)	89M2

Remarks:
(1) at protein conc. 1.0 %
(2) an additional exothermic peak was observed in the presence of 2-mercaptoethanol

Plasminogen

Plasminogen, human, various fragments

pH	T_{trs}	Δc_p	Δh	Approach	Ref
		0.35–0.45		DSC	84N

Protein G, Immunoglobulin Binding Protein

Streptococcal protein B, B1 domain, mutant in position 53

Mutant	pH	T_{trs}	Δc_p	Δh	Approach	Ref
Thr53	5.2	58.2±0.9	0.38±0.13	23.4±1.7	DSC	96S7
Thr53→Ala	5.2	68.4±0.4	0.50±0.08	29.7±2.5	DSC	96S7

Ribonuclease A

Ribonuclease A, bovine pancreatic

pH	T_{trs}	Δc_p	Δh	Approach	Ref
4		0.38		DSC	89P8
5.5	64	0.38		DSC	79P4
5.5	64	0.38		DSC	74P1
		0.33		DSC	73P1
		0.38		DSC	74T2

Table 3. Enthalpy and Heat Capacity Changes – Specific Values

Ribonuclease A, bovine pancreatic, temperature dependence of Δc_p [remark (1)]

T	Δc_p	Approach	Ref
0	0.25	DSC	89P8
25	0.34	DSC	89P8
50	0.35	DSC	89P8
75	0.29	DSC	89P8

Remark:
(1) the data were taken from Fig. 11 of Ref. 89P8; the curve can be approximated by the following polynomial expression: $Cp(t) = 4.184 \times (a0 + a1 \times t + a2 \times t^2)$, with temperature t in °C, a0 = 0.2604 (±0.0059),
a1 = 4.142×10^{-3} (±0.274×10^{-3}), and a2 = -4.942×10^{-5} (±0.264×10^{-5}).

Ribonuclease A

pH	T_{trs}	Δcp	Δh	Approach	Remarks	Ref
2.5	45	0.33		DSC	(1–3)	94H8

Remarks:
(1) Δcp is the experimental value
(2) Ref. 94H8 contains corrections of Δcp for volume effects
(3) T_{trs} was taken from the figures in Ref. 94H8

Ribonuclease T1

Lys25-ribonuclease T1

pH	T_{trs}	Δc_p	Δh	Approach	Ref
2–7	49–62	0.46		DSC	90K5

Ribonuclease *Bacillus Intermedius 7P*, see binase

Serum Albumin

Serum albumin, bovine

pH	T_{trs}	Δc_p	Δh	Approach	Ref
7.0	60		11.3	DSC	85P1
7.15	62	0.38±0.08		DSC	75L

Serum albumin, human, undefatted

pH	T_{trs}	Δc_p	Δh	Approach	Remarks	Ref
7	64.6		17.3	DSC	(1)	88S6
7	72	0.54	21.6	DSC	(2)	88S6
7	63.8–66.4		17.49±0.25	DSC	(3)	88R3
7	78.22±0.03		21.6±0.3	DSC	(4)	88R3

Remarks:
(1) in the absence of ligands
(2) in the presence of N-Ac-L-tryptophanate and caprylate
(3) data from the concentration dependence
(4) in the presence of long chain fatty acid

Serum albumin, human, undefatted, in the absence and presence of stabilizers

Stabilizer/ Concentration	pH	$T_{trs(1)}$	$T_{trs(2)}$	Δh	Approach	Remarks	Ref
none	6.4	67.7	78.4	23.4	DSC	(1–4)	84R
none	7.0	66.8	78.5	20.7	DSC	(1–4)	84S2
none	7.4	65.3	78.6	19.7	DSC	(1–4)	84R
N-acetyl-D,L-tryptophanate:							
4.0 mM	7.0	69.6	78.9	22.0	DSC	(1–4)	84S2
8.0 mM	7.0	71.0	78.8	23.0	DSC	(1–4)	84S2
16.0 mM	7.0	72.7	79.0	22.8	DSC	(1–4)	84S2
30.0 mM	7.0	74.2	~79.2	23.7	DSC	(1–4)	84S2
N-acetyl-L-tryptophanate:							
4.0 mM	7.0	69.2	78.5	22.4	DSC	(1–4)	84R
30.0 mM	7.0	73.7	~78.9	23.5	DSC	(1–4)	84R
N-acetyl-D-tryptophanate:							
4.0 mM	7.0	69.5	79.0	21.8	DSC	(1–4)	84R
30.0 mM	7.0	74.3	~79.4	24.4	DSC	(1–4)	84R
L-tryptophanate:							
26.0 mM	7.0	67.8	77.9	20.8	DSC	(1–4)	84R
D-tryptophanate:							
27.0 mM	7.0	66.6	77.8	20.0	DSC	(1–4)	84R
caprylate:							
1.0 mM	7.0	69.7	78.9	23.1	DSC	(1–4)	84S2
2.0 mM	7.0	72.2	79.3	24.0	DSC	(1–4)	84S2
4.0 mM	7.0	75.3	~80.3	24.5	DSC	(1–4)	84S2
8.0 mM	7.0	~79.5	81.9	25.3	DSC	(1–4)	84S2
16.0 mM	7.0	83.8	83.8	26.8	DSC	(1–4)	84S2
29.0 mM	7.0	85.9	85.9	27.1	DSC	(1–4)	84S2
73.0 mM	7.0	87.3	87.3	27.4	DSC	(1–4)	84S2
N-acetyl-D,L-tryptophanate and caprylate:							
4 mM + 4 mM	7.0	75.9	~80.0	23.9	DSC	(1–4)	84S2
4 mM + 4 mM	7.4	75.3	~80.4	23.8	DSC	(1–4)	84R

Remarks:
(1) previuosly unheated, undefatted albumin monomer in 145 mM Na^+, containing 1.5 mol endogenous fatty acid/mol monomer
(2) DSC heating rate 14.2 K/h
(3) $T_{trs(1)}$ and $T_{trs(2)}$ refer to the peak maxima of the endotherms
(4) standard deviations: ±0.14 in T_{trs}, ±1.2 J/g in Δh

Stellacyanin

Apostellacyanin from the latex of japanese laquer tree

pH	T_{trs}	Δc_p	Δh	Approach	Ref
7.2	52.7		23.8	DSC	95A2

Streptokinase

Streptokinase from *Streptococcus equisimilis*, domains

Transition	pH	T_{trs}	Δc_p	Δh	Approach	Remarks	Ref
trans. (1)	4.0	42.3	0.23±0.06		DSC	(1)	92W1
trans. (1)	7.5	45.9±0.4	0.21		DSC	(2)	92W1
trans. (2)	7.5	60.1±1.3	0.38		DSC	(3)	92W1

Remarks:
(1) single thermal transition
(2) first of two subsequent thermal transitions
(3) second of two subsequent thermal transitions

Streptokinase, proteolytic fragments

Fragment	pH	T_{trs}	Δc_p	Δh	Approach	Remarks	Ref
SK1	7.5	46.1	0.25±0.02	20.3	DSC	(1)	92M4
Tr27, Th26	7.5	45.4	0.25±0.02	16.9±2.8	DSC	(2,3)	92M4
Tr17, Th16	7.5	42.8	0.21±0.04	19.4±2.0	DSC	(4,5)	92M4

Remarks:
(1) SK1:residues 1-59/60–293
(2) Tr27: residues 60–293
(3) Th26: residues 63–291
(4) Tr17: residues 148–293
(5) Th16: residues 151–291

Subtilisin Inhibitor

Subtilisin inhibitor

	pH	T_{trs}	Δc_p	Δh	Approach	Ref
	7	80.1*	0.2±0.01		DSC	81T

Subtilisin inhibitor from *Streptomyces*, effect of intersubunit disulfide bond

Mutant	pH	T_{trs}	Δc_p	Approach	Remarks	Ref
wild-type	3–9.5	46–81	0.531±0.007	DSC	(1)	94T3
Asp83→Cys	3–9.5	63–95	0.421±0.008	DSC	(1)	94T3
Asp83→Asn	3–9.5	45–83	0.532±0.008	DSC	(1)	94T3

Remark:
(1) for T_{trs} and ΔH see molar values, table 2

Streptomyces subtilisin inhibitor, mutants concerning Met103

Mutant	pH	T_{trs}	Δc_p	Δh	Approach	Remarks	Ref
Met103→Ala	3.07	44.58	0.544	15.9	DSC	(1)	95T3
Met103→Ala	3.13	46.34	0.393	17.2	DSC	(1)	95T3
Met103→Ala	3.16	48.09	0.381	17.8	DSC	(1)	95T3
Met103→Ala	3.21	49.56	0.415	18.3	DSC	(1)	95T3
Met103→Ala	3.25	51.20	0.383	19.5	DSC	(1)	95T3
Met103→Ala	7.00	77.98	0.272	33.4	DSC	(1)	95T3
Met103→Ala	9.20	75.75	0.280	32.6	DSC	(1)	95T3
Met103→Ala	9.50	75.52	0.519	32.5	DSC	(1)	95T3
Met103→Ala			0.395±0.092		DSC	(1,2,4)	95T3
Met103→Gly	7.00	68.83	0.325	27.6	DSC	(1)	95T3
Met103→Gly	7.00	68.64	0.377	27.8	DSC	(1)	95T3
Met103→Gly	9.50	66.22	0.527	26.6	DSC	(1)	95T3
Met103→Gly	9.50	66.30	0.448	26.7	DSC	(1)	95T3
Met103→Gly	9.50	66.49	0.339	26.8	DSC	(1)	95T3
Met103→Gly			0.403±0.084		DSC	(1,2,4)	95T3
Met103→Ile	3.07	40.22	0.552	11.4	DSC	(1)	95T3
Met103→Ile	3.13	42.47	0.464	12.5	DSC	(1)	95T3
Met103→Ile	3.16	45.13	0.388	14.1	DSC	(1)	95T3
Met103→Ile	3.21	46.17	0.456	14.2	DSC	(1)	95T3
Met103→Ile	3.25	48.34	0.397	15.2	DSC	(1)	95T3
Met103→Ile	7.00	78.32	0.235	30.7	DSC	(1)	95T3
Met103→Ile	9.20	75.99	0.295	29.1	DSC	(1)	95T3
Met103→Ile	9.50	75.93	0.332	29.6	DSC	(1)	95T3
Met103→Ile			0.390±0.102		DSC	(1,2,4)	95T3
Met103→Leu	3.07	51.67	0.431	17.0	DSC	(3)	95T3
Met103→Leu	3.16	54.40	0.439	19.9	DSC	(3)	95T3
Met103→Leu	3.17	53.86	0.387	19.2	DSC	(3)	95T3
Met103→Leu	3.21	55.68	0.439	20.1	DSC	(3)	95T3
Met103→Leu	3.25	57.60	0.359	21.0	DSC	(3)	95T3
Met103→Leu	7.00	82.51	0.261	33.3	DSC	(1)	95T3
Met103→Leu	9.20	80.27	0.225	32.5	DSC	(1)	95T3
Met103→Leu	9.50	80.22	0.308	31.8	DSC	(1)	95T3
Met103→Leu			0.351±0.089		DSC	(1,2,4)	95T3
Met103→Val	3.07	43.64	0.469	14.2	DSC	(1)	95T3
Met103→Val	3.13	46.38	0.444	16.1	DSC	(1)	95T3
Met103→Val	3.21	49.29	0.431	17.5	DSC	(1)	95T3
Met103→Val	3.25	51.07	0.406	18.1	DSC	(1)	95T3
Met103→Val	7.00	78.65	0.251	32.5	DSC	(1)	95T3
Met103→Val	9.20	76.27	0.322	31.1	DSC	(1)	95T3
Met103→Val	9.50	76.37	0.296	31.1	DSC	(1)	95T3
Met103→Val			0.374±0.083		DSC	(1,2,4)	95T3

Remarks:
(1) the data were analyzed according to the scheme $N_2 \rightarrow 2D$, protein concentration was 1.89 to 2.33 mg/ml
(2) mean value ± standard deviation
(3) the data were analyzed according to the scheme $N_2 \rightarrow I_2 \rightarrow 2D$, with $T_{trs1} = (49.1 \pm 0.1)°C$
   and $\Delta h_1 = 5.10 \pm 0.75$ J/g, $T_{trs2} = (54.4 \pm 0.1)°C$ and $\Delta h_2 = 12.97 \pm 0.71$ J/g, for further details, see Ref. 95T3
(4) linear fit of Δh_{conf} (excluding the heat of protonation of His43) versus $t_{1/2}$ gives
   Met103→Ala: $\Delta h_{conf} = -(7.70 \pm 0.33) + (0.531 \pm 0.004) \times t$
   Met103→Val: $\Delta h_{conf} = -(7.87 \pm 0.29) + (0.510 \pm 0.004) \times t$
   Met103→Ile: $\Delta h_{conf} = -(8.95 \pm 0.29) + (0.506 \pm 0.004) \times t$
   Met103→Leu: $\Delta h_{conf} = -(8.03 \pm 0.84) + (0.502 \pm 0.004) \times t$
   wild-type: $\Delta h_{conf} = -(8.99 \pm 0.21) + (0.531 \pm 0.004) \times t$
   Met103→Gly: $\Delta h_{conf} = -7.61 + 0.515 \times t$ (data for pH ≥7)

Streptomyces subtilisin inhibitor, mutants concerning Val13

Mutant	pH	T_{trs}	Δc_p	Δh	Approach	Remarks	Ref
Val13→Ala	7.00	69.01	0.346	29.4	DSC	(1)	95T2
Val13→Ala	9.50	66.18	0.370	27.3	DSC	(1)	95T2
Val13→Ala	9.50	66.95	0.523	27.4	DSC	(1)	95T2
Val13→Ala	9.50	67.15	0.464	27.2	DSC	(1)	95T2
Val13→Ala	9.50	67.99	0.473	27.8	DSC	(1)	95T2
Val13→Ala			0.435±0.075		DSC	(1–3)	95T2
Val13→Gly	7.00	54.51	0.380	14.7	DSC	(1)	95T2
Val13→Gly	9.50	55.33	0.342	15.6	DSC	(1)	95T2
Val13→Gly	9.50	55.35	0.333	16.1	DSC	(1)	95T2
Val13→Gly			0.352±0.025		DSC	(1–3)	95T2
Val13→Ile	3.07	47.15	0.360	15.1	DSC	(1)	95T2
Val13→Ile	3.16	49.94	0.353	16.2	DSC	(1)	95T2
Val13→Ile	3.21	51.07	0.329	17.2	DSC	(1)	95T2
Val13→Ile	7.00	79.82	0.336	31.3	DSC	(1)	95T2
Val13→Ile	9.50	78.87	0.331	32.0	DSC	(1)	95T2
Val13→Ile			0.342±0.014		DSC	(1–3)	95T2
Val13→Leu	3.09	43.03	0.351	12.3	DSC	(1)	95T2
Val13→Leu	3.16	46.26	0.274	13.8	DSC	(1)	95T2
Val13→Leu	3.21	46.58	0.408	13.8	DSC	(1)	95T2
Val13→Leu	7.00	77.58	0.383	30.1	DSC	(1)	95T2
Val13→Leu	7.00	77.70	0.278	29.7	DSC	(1)	95T2
Val13→Leu	9.50	74.47	0.489	28.1	DSC	(1)	95T2
Val13→Leu	9.50	75.42	0.352	28.1	DSC	(1)	95T2
Val13→Leu	9.50	75.50	0.402	28.3	DSC	(1)	95T2
Val13→Leu	9.50	75.42	0.344	28.2	DSC	(1)	95T2
Val13→Leu	9.50	76.50	0.377	28.4	DSC	(1)	95T2
Val13→Leu			0.366±0.063		DSC	(1–3)	95T2
Val13→Met	7.00	69.63	0.418	26.4	DSC	(1)	95T2
Val13→Met	9.50	69.09	0.317	26.2	DSC	(1)	95T2
Val13→Met	·9.50	69.02	0.376	26.1	DSC	(1)	95T2
Val13→Met			0.371±0.052		DSC	(1–3)	95T2
Val13→Phe	7.00	69.16	0.385	24.5	DSC	(1)	95T2
Val13→Phe	9.50	67.61	0.444	22.9	DSC	(1)	95T2
Val13→Phe	9.50	68.31	0.333	23.2	DSC	(1)	95T2
Val13→Phe	9.50	68.12	0.448	22.7	DSC	(1)	95T2
Val13→Phe	9.50	68.84	0.337	24.0	DSC	(1)	95T2
Val13→Phe			0.390±0.075		DSC	(1–3)	95T2

Remarks:
(1) the data were analyzed according to the scheme $N_2 \rightarrow 2D$, the scheme $N_2 \rightarrow 2N' \rightarrow 2D$ is also considered in Ref. 95T2, protein concentration was 1.18 to 4.66 mg/ml
(2) mean value ± standard deviation
(3) linear fit of Δh_{conf} versus $t_{1/2}$ gives
Val13→Ile: $\Delta h_{conf} = -(9.20\pm0.92) + (0.515\pm0.017)\times t$
Val13→Leu: $\Delta h_{conf} = -(9.37\pm0.54) + (0.502\pm0.008)\times t$
wild-type: $\Delta h_{conf} = -(8.99\pm0.21) + (0.531\pm0.004)\times t$
Val13→Ala: $\Delta h_{conf} = -6.95 + 0.515\times t$
Val13→Phe: $\Delta h_{conf} = -11.80 + 0.515\times t$
Val13→Gly: $\Delta h_{conf} = -12.89 + 0.515\times t$
Val13→Met: $\Delta h_{conf} = -9.46 + 0.515\times t$

Streptomyces subtilisin inhibitor, mutants concerning Met73

Mutant	pH	T_{trs}	Δc_p	Δh	Approach	Remarks	Ref
Met73→Ala	3.10	51.33	0.306	17.5	DSC	(1)	95T4
Met73→Ala	3.13	52.78	0.427	18.1	DSC	(1)	95T4
Met73→Ala	3.18	53.57	0.423	18.2	DSC	(1)	95T4
Met73→Ala	3.20	54.34	0.326	18.9	DSC	(1)	95T4
Met73→Ala	3.25	57.24	0.314	20.0	DSC	(1)	95T4
Met73→Ala	7.00	83.37	0.213	34.0	DSC	(1)	95T4
Met73→Ala	9.20	81.04	0.229	32.8	DSC	(1)	95T4
Met73→Ala	9.50	80.88	0.333	32.4	DSC	(1)	95T4
Met73→Ala			0.321±0.078		DSC	(1–3)	95T4
Met73→Asp	3.03	51.35	0.321	17.3	DSC	(1)	95T4
Met73→Asp	3.07	53.20	0.341	17.9	DSC	(1)	95T4
Met73→Asp	3.10	53.81	0.252	18.5	DSC	(1)	95T4
Met73→Asp	3.13	54.46	0.336	18.8	DSC	(1)	95T4
Met73→Asp	3.19	56.93	0.216	19.7	DSC	(1)	95T4
Met73→Asp	3.21	57.13	0.303	19.6	DSC	(1)	95T4
Met73→Asp	3.22	57.30	0.273	19.6	DSC	(1)	95T4
Met73→Asp	3.25	58.18	0.344	20.3	DSC	(1)	95T4
Met73→Asp	7.00	85.29	0.265	34.8	DSC	(1)	95T4
Met73→Asp	9.20	83.00	0.203	33.3	DSC	(1)	95T4
Met73→Asp	9.50	83.20	0.256	33.8	DSC	(1)	95T4
Met73→Asp	9.50	83.29	0.259	33.5	DSC	(1)	95T4
Met73→Asp			0.281±0.048		DSC	(1–3)	95T4
Met73→Glu	3.09	50.93	0.307	17.7	DSC	(1)	95T4
Met73→Glu	3.18	53.22	0.498	18.6	DSC	(1)	95T4
Met73→Glu	3.21	53.62	0.306	19.0	DSC	(1)	95T4
Met73→Glu	3.23	55.63	0.284	19.9	DSC	(1)	95T4
Met73→Glu	7.00	84.20	0.217	34.8	DSC	(1)	95T4
Met73→Glu	9.50	81.76	0.531	33.1	DSC	(1)	95T4
Met73→Glu			0.357±0.126		DSC	(1–3)	95T4
Met73→Gly	3.10	49.40	0.349	17.3	DSC	(1)	95T4
Met73→Gly	3.13	50.50	0.364	17.9	DSC	(1)	95T4
Met73→Gly	3.20	52.63	0.303	18.7	DSC	(1)	95T4
Met73→Gly	7.00	82.02	0.215	34.7	DSC	(1)	95T4
Met73→Gly	7.00	81.93	0.343	34.4	DSC	(1)	95T4
Met73→Gly	9.20	79.85	0.329	33.3	DSC	(1)	95T4
Met73→Gly	9.50	79.83	0.354	33.3	DSC	(1)	95T4
Met73→Gly			0.323±0.051		DSC	(1–3)	95T4
Met73→Ile	3.07	46.30	0.481	16.4	DSC	(1)	95T4
Met73→Ile	3.10	47.28	0.354	17.1	DSC	(1)	95T4
Met73→Ile	3.13	48.56	0.337	17.6	DSC	(1)	95T4
Met73→Ile	3.18	50.68	0.321	18.3	DSC	(1)	95T4
Met73→Ile	3.19	50.88	0.254	18.5	DSC	(1)	95T4
Met73→Ile	3.22	51.42	0.390	18.7	DSC	(1)	95T4
Met73→Ile	7.00	80.43	0.254	34.4	DSC	(1)	95T4
Met73→Ile	9.20	78.33	0.202	33.1	DSC	(1)	95T4
Met73→Ile	9.50	78.28	0.381	32.9	DSC	(1)	95T4
Met73→Ile			0.331±0.085		DSC	(1–3)	95T4
Met73→Leu	3.10	49.64	0.353	17.8	DSC	(1)	95T4
Met73→Leu	3.13	50.21	0.367	18.1	DSC	(1)	95T4
Met73→Leu	3.19	52.67	0.308	19.0	DSC	(1)	95T4
Met73→Leu	3.22	53.52	0.317	19.8	DSC	(1)	95T4
Met73→Leu	3.25	54.74	0.318	20.3	DSC	(1)	95T4
Met73→Leu	7.00	81.75	0.267	34.9	DSC	(1)	95T4
Met73→Leu	9.20	79.68	0.256	33.8	DSC	(1)	95T4

Streptomyces subtilisin inhibitor, mutants concerning Met73 (continued)

Mutant	pH	T_{trs}	Δc_p	Δh	Approach	Remarks	Ref
Met73→Leu	9.50	79.65	0.329	33.6	DSC	(1)	95T4
Met73→Leu			0.314±0.038		DSC	(1–3)	95T4
Met73→Lys	3.03	51.19	0.469	17.5	DSC	(1)	95T4
Met73→Lys	3.03	51.10	0.494	18.5	DSC	(1)	95T4
Met73→Lys	3.07	53.07	0.370	19.2	DSC	(1)	95T4
Met73→Lys	3.10	53.48	0.371	19.7	DSC	(1)	95T4
Met73→Lys	3.13	54.72	0.321	20.1	DSC	(1)	95T4
Met73→Lys	3.19	56.41	0.295	20.7	DSC	(1)	95T4
Met73→Lys	3.20	57.22	0.394	21.3	DSC	(1)	95T4
Met73→Lys	3.21	57.29	0.365	21.3	DSC	(1)	95T4
Met73→Lys	3.22	57.78	0.331	22.0	DSC	(1)	95T4
Met73→Lys	3.25	58.91	0.332	22.6	DSC	(1)	95T4
Met73→Lys	7.00	82.86	0.304	35.4	DSC	(1)	95T4
Met73→Lys	9.20	80.94	0.192	34.4	DSC	(1)	95T4
Met73→Lys	9.50	80.87	0.431	34.1	DSC	(1)	95T4
Met73→Lys			0.359±0.079		DSC	(1–3)	95T4
Met73→Val	3.07	47.79	0.444	16.9	DSC	(1)	95T4
Met73→Val	3.10	49.01	0.349	17.7	DSC	(1)	95T4
Met73→Val	3.13	49.09	0.325	17.6	DSC	(1)	95T4
Met73→Val	3.19	51.64	0.313	18.6	DSC	(1)	95T4
Met73→Val	3.20	52.22	0.337	19.2	DSC	(1)	95T4
Met73→Val	3.23	52.47	0.350	19.3	DSC	(1)	95T4
Met73→Val	3.25	53.66	0.361	20.0	DSC	(1)	95T4
Met73→Val	7.00	81.04	0.264	34.7	DSC	(1)	95T4
Met73→Val	9.20	78.93	0.239	33.2	DSC	(1)	95T4
Met73→Val	9.50	78.95	0.307	33.3	DSC	(1)	95T4
Met73→Val			0.329±0.056		DSC	(1–3)	95T4

Remarks:
(1) the data were analyzed according to the scheme N_2→2D protein concentration was 1.60 to 2.45 mg/ml
(2) mean value ± standard deviation
(3) linear fit of Δh_{conf} versus $t_{1/2}$ gives

    wild-type: $\Delta h_{conf} = -(8.99 \pm 0.21) + (0.531 \pm 0.004) \times t$
    Met73→Lys: $\Delta h_{conf} = -(9.41 \pm 0.33) + (0.540 \pm 0.004) \times t$
    Met73→Asp: $\Delta h_{conf} = -(10.00 \pm 0.42) + (0.523 \pm 0.004) \times t$
    Met73→Glu: $\Delta h_{conf} = -(8.58 \pm 0.33) + (0.515 \pm 0.004) \times t$
    Met73→Gly: $\Delta h_{conf} = -(8.87 \pm 0.29) + (0.527 \pm 0.004) \times t$
    Met73→Ala: $\Delta h_{conf} = -(9.41 \pm 0.42) + (0.519 \pm 0.008) \times t$
    Met73→Val: $\Delta h_{conf} = -(8.49 \pm 0.25) + (0.531 \pm 0.004) \times t$
    Met73→Ile: $\Delta h_{conf} = -(8.08 \pm 0.33) + (0.527 \pm 0.004) \times t$
    Met73→Leu: $\Delta h_{conf} = -(8.59 \pm 0.25) + (0.536 \pm 0.004) \times t$

Taka-Amylase

Taka-Amylase A

pH	T_{trs}	Δc_p	Δh	Approach	Remark	Ref
7	62	0.687	42.5	DSC	(1)	87F

Remark:
(1) data in the absence of calcium

Thermolysin

Thermolysin

	pH	T_{trs}	Δc_p	Δh	Approach	Remark	Ref
	7.5	90	0.6*	38.6±1.5	DSC	(1)	88S2

Remark:
(1) Δc_p is an estimated value, taken from Fig. 1 of Ref. 88S2.

Thermolysin, fragment 255–316
thermodynamic parameters derived from concentration dependent measurements

Concentration (mg/ml)	pH	T_{trs}	Δc_p	Δh	Approach	Remarks	Ref
0.19–4.55	7.5	67.1–73.7	0.36±0.08	29±2	DSC	(1,2)	94C9

Remarks:
(1) for further details, see molar values
(2) averages weighted by sample concentration

Toxins

Toxins of different origin

Species	pH	T_{trs}	Δc_p	Δh	Approach	Ref
cytotoxin I	4.5	75*	0.26±0.02	32*	DSC	84K2
neurotoxin I	5.7	75*	0.32	29.5*	DSC	84K2
neurotoxin II	5	96*	0.26±0.02	37.5*	DSC	84K2

Tropomyosin

α-Tropomyosin and tropomyosin fragments

Protein/ Fragment	T_{trs}	Δc_p	Δh	Approach	Remarks	Ref
tropomyosin	58	0.42	27.4	DSC		78P
tropomyosin	58	0.42	27.4	DSC		82P1
C-fragment	46	0.38	27.2	DSC		78P
C-fragment	46	0.38	27.2	DSC		82P1
N-fragment	58	0.19	18.7	DSC		78P
N-fragment	58	0.19	18.7	DSC		82P1
res. 11–189	58	0.27	26.3	DSC	(1)	82P2
res. 190–284	46	0.38	28.9	DSC	(1)	82P2
CNIA 11-127	57	0.27	28.4	DSC	(1)	82P2
Cy2 190-284	42	0.30	27.8	DSC	(1)	82P2

Remark:
(1) Δh is an extrapolated value, the specific melting enthalpy calculated for 100% helicity

αα-Tropomyosin, reduced protein (deconvolution)

Transition	pH	T_{trs}	Δc_p	Δh	Approach	Ref
trans. (1)	7.4	33.5±0.7		2.01±0.33	DSC	91S7
trans. (2)	7.4	46.9±0.2		8.74±0.42	DSC	91S7
trans. (3)	7.4	53.9±0.2		5.02±0.84	DSC	91S7
trans. (4)	7.4	55.8±0.3		5.94±0.96	DSC	91S7

αα-Tropomyosin, cross-linked protein (deconvolution)

Transition	pH	T_{trs}	Δc_p	Δh	Approach	Ref
trans. (1)	7.4	31.75±0.62		3.10±0.46	DSC	91S7
trans. (2)	7.4	53.92±0.31		7.45±0.21	DSC	91S7
trans. (3)	7.4	58.16±0.52		7.95±0.25	DSC	91S7

αα-Tropomyosin, reduced protein, SH groups carboxymethylated (deconvolution)

Transition	pH	T_{trs}	Δc_p	Δh	Approach	Ref
trans. (1)	7.4	32.2		3.05	DSC	91S7
trans. (2)	7.4	39.7		7.41	DSC	91S7
trans. (3)	7.4	53.2		6.57	DSC	91S7
trans. (4)	7.4	56.8		4.35	DSC	91S7

αα-Tropomyosin, reduced protein, SH groups carboxyamidomethylated (deconvolution)

Transition	pH	T_{trs}	Δc_p	Δh	Approach	Ref
trans. (1)	7.4	32.7		2.47	DSC	91S7
trans. (2)	7.4	43.2		8.58	DSC	91S7
trans. (3)	7.4	53.5		4.52	DSC	91S7
trans. (4)	7.4	55.0		5.86	DSC	91S7

ββ-Tropomyosin, reduced protein (deconvolution)

Transition	pH	T_{trs}	Δc_p	Δh	Approach	Ref
trans. (1)	7.4	31.9		2.85	DSC	91S7
trans. (2)	7.4	39.7		6.19	DSC	91S7
trans. (3)	7.4	47.0		6.82	DSC	91S7
trans. (4)	7.4	55.0		2.05	DSC	91S7

ββ-Tropomyosin, cross-linked protein (deconvolution)

Transition	pH	T_{trs}	Δc_p	Δh	Approach	Ref
trans. (1)	7.4	32.0		3.14	DSC	91S7
trans. (2)	7.4	36.7		3.72	DSC	91S7
trans. (3)	7.4	44.6		5.73	DSC	91S7
trans. (4)	7.4	55.7		6.49	DSC	91S7

Homodimeric chicken gizzard tropomyosins (CG-TM)

Transition	pH	T_{trs}	Δh	Approach	Remarks	Ref
Reduced ββ-CG-TM:						
trans. (1)	7.4	42.67±0.32	1.55±0.09	DSC	(1)	96O1
trans. (2)	7.4	49.62±0.17	2.98±0.19	DSC	(1)	96O1
Oxidized ββ-CG-TM:						
trans. (1)	7.4	42.63±1.04	0.99±0.11	DSC	(1)	96O1
trans. (2)	7.4	51.58±0.11	1.75±0.03	DSC	(1)	96O1
trans. (3)	7.4	54.19±0.49	1.30±0.31	DSC	(1)	96O1
Reduced γγ-GC-TM:						
trans. (1)	7.4	37.15±0.07	0.76±0.07	DSC	(1)	96O1
trans. (2)	7.4	43.27±0.17	1.78±0.17	DSC	(1)	96O1
trans. (3)	7.4	44.30±0.07	2.31±0.07	DSC	(1)	96O1
Oxidized γγ-GC-TM:						
trans. (1)	7.4	38.15±0.10	0.97±0.10	DSC	(1)	96O1
trans. (2)	7.4	43.67±0.04	1.85±0.04	DSC	(1)	96O1
trans. (3)	7.4	53.45±0.08	0.84±0.08	DSC	(1)	96O1
trans. (4)	7.4	54.59±0.05	0.90±0.05	DSC	(1)	96O1

Remark:
(1) measured in 50 mM sodium phosphate, 500 mM NaCl, 10 mM EDTA, pH 7.4

Troponin C

Troponin C

pH	T_{trs}	Δc_p	Approach	Remark	Ref
7.2	37–84	0.25±0.08	DSC	(1)	80T

Remark:
(1) three transitions resolved by deconvolution at 27, 58, and 84°C with enthalpy change of 146, 246, and 251 kJ/mol, respectively

Trypsin

β-Trypsin

pH	T_{trs}	Δc_p	Δh	Approach	Remark	Ref
5	70	0.5	10.9	DSC	(1)	79P4

Remark:
(1) Δh = 10.9 J/g at 25°C. T_{trs} was taken from Fig. 9 of Ref. 79P4, the other data from Tab. 6 of Ref. 79P4

Trypsin Inhibitor

Trypsin inhibitor, bovine pancreatic

pH	T_{trs}	Δc_p	Δh	Approach	Remark	Ref
4	100	0.46		DSC	(1)	79P4
4.5	100		46.4	DSC		93H9
7	102.9	0.015		DSC		83M2

Remark:
(1) T_{trs} was taken from Fig. 9 of Ref. 79P4, the other data from Tab. 6 of Ref. 79P4

Kunitz-type soybean trypsin inhibitor

pH	T_{trs}	Δcp	Δh	Approach	Remarks	Ref
7.0	59	0.5±0.04	200±0.3	DSC	(1,2)	95F4

Remarks:
(1) from scan rate dependence
(2) Δh is the net enthalpy corrected for the heat of ionization

Tryptophan Repressor

Tryptophan repressor from *E. coli*

pH	T_{trs}	Δc_p	Approach	Ref
7.5	90.3±0.35	0.25±0.11	DSC	88B1

Tryptophan Synthase

Tryptophan synthase from various species, α-subunit

Species	pH	T_{trs}	Δcp	Δh	Approach	Ref
E. coli	8.98	54.2	0.63	17.1	DSC	90S8
S. thyphim.	9.01	46.2	0.63	12.7	DSC	90S8
hybrid	9.0	47.6	0.63	13.0	DSC	90S8

Explanations:
E. coli   : *Escherichia coli*
S. typhim.: *Salmonella typhimurium*
hybrid   : interspecies hybrid with the C-terminal domain from *E. coli* and the N-terminal domain from *S. typhimurium*

Tryptophan synthase, α-subunit, wild-type and mutants

Mutant	pH	T_{trs}	Δc_p	Approach	Ref
wild-type	7	62.4	0.42±0.04	DSC	82Y2
Glu49→Gln	7	59.4	0.42±0.04	DSC	82Y2
Glu49→Gln	9.3	62.4	0.42±0.04	DSC	82Y2
Glu49→Ser	7	60.5	0.42±0.04	DSC	82Y2
Glu49→Ser	9.3	58.6	0.42±0.04	DSC	82Y2

**References
and Index of Proteins**

References (Tables 1 – 3)

64B1   Barnard, E.A.: J. Mol. Biol. 10 (1964) 235.
64B2   Brandts, J.F.: J. Amer. Chem. Soc. 86 (1964) 4291.
64B3   Brandts, J.F.: J. Amer. Chem. Soc.86 (1964) 4302.
64H   Hvidt, A.: C. R. Trav. Lab. Carlsberg 34 (1964) 299.
64S   Sophianopoulos, A.J., Weiss, B.J.: Biochemistry 3 (1964) 1920.
64T   Tanford, C.: J. Amer. Chem. Soc. 86 (1964) 2050.
65B   Beck, K., Gill, S.J., Downing, M.: J. Amer. Chem. Soc. 87 (1965) 901.
65H   Hamaguchi, K., Sakai, H.: J. Biochem. 57 (1965) 721.
66H   Hvidt, A., Nielsen, S.O.: Adv. Protein Chem. 21 (1966) 287.
67B   Brandts, J.F., Hunt, L.: J. Amer. Chem. Soc. 89 (1967) 4826.
67D   Danford, R., Krakauer, H., Sturtevant, J.M.: Rev. Sci. Instruments 38 (1967) 484.
67H   Hermans Jr., J., Acampora, G.: J. Amer. Chem. Soc. 89 (1967) 1547.
68S   Schechter, A.N., Epstein, C.J.: J. Mol. Biol. 35 (1968) 567.
69A   Aune, K.C., Tanford, C.: Biochemistry 8 (1969) 4586.
69B1   Biltonen, R.L., Lumry, R.: J. Amer. Chem. Soc. 91 (1969) 4256.
69B2   Brandts, J.F: in: Structure and Stability of Biological Macromolecules, ed. by S.N.
        Timasheff and G. Fasman, Marcel Dekker Inc., New York, 1969, p. 213.
69D   Delben, F., Crescenzi, V.: Biochim. Biophys. Acta 164 (1969) 619.
69H   Hartley, R.W.: Biochemistry 8 (1969) 2929.
69L   Lumry, R., Biltonen, R.: in: Structure and Stability of Biological Macromolecules,
        ed. by S.N. Timasheff and G. Fasman, Marcel Dekker Inc., New York, 1969, p. 65.
69R   Reeg C.: Dissertation, Boulder Colorado 1969, see also: Biochemistry 9 (1970)
        2666.
70A1   Abrashi, H.: C. R. Trav. Lab. Carlsberg 37 (1970) 129.
70A2   Abrashi, H.: C. R. Trav. Lab. Carlsberg 37 (1970) 107.
70B   Biltonen, R.: personal communication to J.F. Brandts, cited in: Biochemistry 9
        (1970) 2294.
70J   Jackson, W.M., Brandts, J.F.: Biochemistry 9 (1970) 2294.
70S   Salahuddin, A., Tanford, C.: Biochemistry 9 (1970) 1342.
70T1   Tanford, C., Aune, K.C.: Biochemistry 9 (1970) 206.
70T2   Tsong, T.Y., Hearn, R.P., Wrathall, D.P., Sturtevant, J.M.: Biochemistry 9 (1970)
        2666.
71H1   Hawley, S.A.: Biochemistry 100 (1971) 2436.
71H2   Hearn, R.P., Richards, F.M., Sturtevant, J.M., Watt, G.D.: Biochemistry 10 (1971)
        806.
71O   Ottesen, M.: Methods Biochem. Anal. 20 (1971) 135.
71P   Privalov, P.L., Khechinashvili, N.N., Atanasov, B.P.: Biopolymers 10 (1971) 1865.
72A   Atanasov, B.P., Privalov, P.L., Khechinashvili, N.N.: Molekularnaya Biol. 6 (1972)
        33.
72B   Barel, A.O., Priells, J.P., Maes, E., Looze, Y., Léonis, J.: Biochim. Biophys. Acta
        257 (1972) 288.
72N   Nakanishi, M., Tsuboi, M., Ikegami, A.: J. Mol. Biol. 70 (1972) 351.
72P1   Puett, D.: Biochemistry 11 (1972) 1980.
72P2   Puett, D.: Biochemistry 11 (1972) 4304.
73I   Ivanov, V.I., Bocharov, A.L., Volkenstein, M.V., Karpeisky, M.Ya., Mora, S., Oki-
        na, E.I., Yudina, L.V.: Eur. J. Biochem. 40 (1973) 519.
73K   Khechinashvili, N.N., Privalov, P.L., Tiktopulo, E.I.: FEBS Lett. 30 (1973) 57.
73P1   Privalov, P.L., Tiktopulo, E.I., Khechinashvili, N.N.: Int. J. Peptide Protein Res. 5
        (1973) 229.
73P2   Puett, D., Friebele, E., Hammonds Jr., R.G.: Biochim. Biophys. Acta 328 (1973)
        261.
73P3   Puett, D.: J. Biol. Chem. 248 (1973) 4623.
73P4   Puett, D.: J. Biol. Chem. 248 (1973) 3566.
73R   Rowe, E.S., Tanford, C.: Biochemistry 12 (1973) 4822.
73S   Sugai, S., Yashiro, H., Nitta, K.: Biochim. Biophys. Acta 328 (1973) 35.
73T   Tsong, T.Y.: Biochemistry 12 (1973) 2209.
74G1   Greene Jr., R.F., Pace, C.N.: J. Biol. Chem. 249 (1974) 5388.

74G2   Greene, R.F.: personal communication to C.N. Pace, cited in: Biochemistry 13 (1974) 1289.

74K   Knapp, J.A., Pace, C.N.: Biochemistry 13 (1974) 1289.

74H   Holladay, L.A., Hammonds Jr., R.G., Puett, D.: Biochemistry 15 (1976) 1653.

74P1   Privalov, P.L., Khechinashvili, N.N.: J. Mol. Biol. 86 (1974) 665.

74P2   Privalov, P.L., Khechinashvili, N.N.: Biofizika 19 (1974) 14.

74T1   Takase K., Nitta, K., Sugai, S.: Biochim. Biophys. Acta 371 (1974) 352.

74T2   Tiktopulo, E.I., Privalov, P.L.: Biophys. Chem. 1 (1974) 349.

74T3   Tischenko, V.M., Tiktopulo, E.I., Privalov, P.L.: Biofizika 19 (1974) 400.

75B   Brandts, J.F.: personal communication to C.N. Pace, cited in: Crit. Rev. Biochem. 3 (1975) 1.

75L   Leibman, D.Ya., Tiktopulo, E.I., Privalov, P.L.: Biofizika 20 (1975) 376.

75P   Pace, C.N.: Crit. Rev. Biochem. 3 (1975) 1.

76A   Ahmad, F., Salahuddin, A.: Biochemistry 15 (1976) 5168.

76H   Holladay, L.A., Savage Jr., C.R., Cohen S., Puett, D.: Biochemistry 15 (1976) 2624.

76K1   Kita, N., Kuwajima, K., Nitta, K., Sugai, S.: Biochim. Biophys. Acta 427 (1976) 350.

76K2   Kuwajima, K., Nitta, K., Yoneyama, M., Sugai, S.: J. Mol. Biol. 106 (1976) 359.

76P1   Pfeil, W., Privalov, P.L.: Biophys. Chem. 4 (1976) 33.

76P2   Pfeil, W., Privalov, P.L.: Biophys. Chem. 4 (1976) 23.

76P3   Pfeil, W., Privalov, P.L.: Biophys. Chem. 4 (1976) 41.

76R1   Rowe, E.S.: Biochemistry 15 (1976) 905.

76R2   Robson, B., Pain, R.H.: Biochem. J. 155 (1976) 331.

76T   Tall, A.R., Shipley, G.G., Small, D.M.: J. Biol. Chem. 251 (1976) 3749.

77B   Burgess, R.J., Pain, R.H.: Biochem. Soc. Trans. 5 (1977) 692.

77C1   Chen, C.-H., Kao, O.H.W., Berns, D.S.: Biophys. Chem. 7 (1977) 81.

77C2   Chlebowsky, J.F., Mabrey, S.: J. Biol. Chem. 252 (1977) 7042.

77C3   Contaxis, C.C., Bigelow, C.C., Zarkadas, C.G.: Canad. J. Biochem. 55 (1977) 325.

77E   Elwell, M.L., Schellman, J.A.: Biochim. Biophys. Acta 494 (1977) 367.

77K   Kawaguchi, H., Noda, H.: J. Biochem. 81 (1977) 1307.

77M   McLendon, G.: Biochem. Biophys. Res. Commun. 77 81977) 959.

77N   Nojima, H., Ikai, A., Oshima, T., Noda, H.: J. Mol. Biol. 116 (1977) 429.

77T   Tall, A.R., Small, D.M., Deckelbaum, R.J., Shipley, G.G.: J. Biol. Chem. 252 (1977) 4701.

77W   Wright, D.J., Leach, I.B., Wilding, P.: J. Sci. Food Agric. 28 (1977) 557.

77Z   Zavyalov, V.P., Troitsky, G.V., Khechinashvili, N.N., Privalov, P.L.: Biochim. Biophys. Acta 492 (1977) 102.

78A   Ahmad, F., McPhie, P.: Biochemistry 17 (1978) 241.

78D   DiPaola, G., Belleau, B.: Canad. J. Chem. 56 (1978) 848.

78F   Filimonov, V.V., Pfeil, W., Tsalkova, T.N., Privalov, P.L.: Biophys. Chem. 8 (1978) 117.

78J   Jackson, M.B., Sturtevant, J.M.: Biochemistry 17 (1978) 911.

78M1   McLendon, G., Smith, M.: J. Biol. Chem. 253 (1978) 4004.

78M2   McLendon, G., Sandberg, K.: J. Biol. Chem. 253 (1978) 3913.

78N1   Nojima, H., Hon-nami, K., Oshima, T., Noda, H.: J. Mol. Biol. 122 (1978) 33.

78N2   Nozaka, M., Kuwajima, K., Nitta, K., Sugai, S.: Biochemistry 17 (1978) 3753.

78P   Potekhin, S.A., Privalov, P.L.: Biofizika 23 (1978) 219.

78T   Tiktopulo, E.I., Privalov, P.L.: FEBS Lett. 91 (1978) 57.

79C   Chlebowsky, J.F., Mabrey, S., Falk, M.C.: J. Biol. Chem. 254 (1979) 5745.

79G   Goto, Y., Hamaguchi, K.: J. Biochem. 86 (1979) 1433.

79H   Hon-nami, K., Oshima, T.: Biochemistry 18 (1979) 5693.

79O   Oobatake, M., Takahashi, S., Ooi, T.: J. Biochem. 86 (1979) 55.

79P1   Pace, C.N., Vanderburgh, K.E.: Biochemistry 18 (1979) 288.

79P2   Potekhin, S.A., Privalov, P.L.: Molekularnaya Biol. 13 (1979) 666.

79P3   Potekhin, S.S., Trapkov, V.A., Privalov, P.L.: Biofizika 24 (1979) 46.

79P4   Privalov, P.L.: Adv. Protein Chem. 33 (1979) 167.

79P5   Privalov, P.L., Tsalkova, T.N.: Nature 280 (1979) 693.

79P6   Privalov, P.L., Tiktopulo, E.I., Tischenko, V.M.: J. Mol. Biol. 127 (1979) 203.

79R   Richarz, R., Sehr, P., Wagner, G., Wüthrich, K.: J. Mol. Biol. 130 (1979) 19.

79V   Velicelebi, G., Sturtevant, J.M.: Biochemistry 18 (1979) 1180.
79W   Wagner, G., Kalb, A.J., Wüthrich, K.: Eur. J. Biochem. 95 (1979) 249.
79Y   Yutani, K., Ogasahara, K., Suzuki, M., Sugino, Y.: J. Mol. Biol. 85 (1979) 915.
80A   Adams, B., Burgess, R.J., Carrey, E.A., MacIntosh, I.R., Mitchinson, C., Thomas, R.M., Pain, R.H.: in: Protein Folding, ed. by R. Jaenicke, Elsevier, North-Holland Biomed. Press, 1980, p. 447.
80B   Bendzko, P., Pfeil, W.: Acta Biol. Med. Germ. 39 (1980) 47.
80C   Cupo, P., El-Deiry, W., Whithney, P.L., Awad Jr., W.M.: J. Biol. Chem. 255 (1980) 10828.
80E   Edelstein, C., Scanu, A.M.: J. Biol. Chem. 255 (1980) 5747.
80J   Jacobson, A.L., Turner, C.L.: Biochemistry 19 (1980) 4534.
80K   Knox, D.G., Rosenberg, A.: Biopolymers 19 (1980) 1049.
80M1 Mantulin, W.W., Rohde, M.F., Gotto, A.M., Pownall, H.J.: J. Biol. Chem. 255 (1980) 8185.
80M2 Matthews, C.R., Crisanti, M.M., Gepner, G.L., Velicelebi, G., Sturtevant, J.M.: Biochemistry 19 (1980) 1290.
80P   Pfeil, W., Bendzko, P.: Biochim. Biophys. Acta 626 (1980) 73.
80S1 Schellman, J.A., Hawkes, R.B.: in: Protein Folding, ed. by R. Jaenicke, Elsevier, North-Holland Biomed. Press, 1980, p. 331.
80S2 Swenson, C.A., Ritchie, P.A.: Biochemistry 19 (1980) 5371.
80T   Tsalkova, T.N., Privalov, P.L.: Biochim. Biophys. Acta 624 (1980) 196.
81C1 Castellino, F.J., Ploplis, V.A., Powell, J.R., Strickland, D.K.: J. Biol. Chem. 256 (1981) 4778.
81C2 Contaxis, C.C., Bigelow, C.C.: Biochemistry 20 (1981) 1618.
81D   Doster, W., Hess, B.: Biochemistry 20 (1981) 772.
81H1 Harrington, J.P., Biochim. Biophys. Acta 671 (1981) 85.
81H2 Hinz, H.-J., Cossmann, M., Beyreuther, K.: FEBS Lett. 129 (1981) 246.
81L   Labhardt, A.M.: Biopolymers 20 (1981) 1459.
81M1 Mateo, P.L., Privalov, P.L.: FEBS Lett. 123 (1981) 189.
81M2 Matthews, C.R., Crisanti, M.M.: Biochemistry 20 (1981) 784.
81N   Nall, B.T., Landers, T.A.: Biochemistry 20 (1981) 5403.
81P1 Pfeil, W.: Biophys. Chem. 13 (1981) 181.
81P2 Privalov, P.L., Mateo, P.L., Khechinashvili, N.N., Stepanov, V.M., Revina, L.P.: J. Mol. Biol. 152 (1981) 445.
81R   Relimpio, A., Iriarte, A., Chlebowsky, J.F., Martinez-Carrion, M.: J. Biol. Chem. 256 (1981) 4478.
81S1 Schnarr, M., Maurizot, J.-C.: Biochemistry 21 (1981) 6164.
81S2 Strickland, D.K., Andersen, T.L., Hill, R.L., Castellino, F.J.: Biochemistry 20 (1981) 5294.
81T   Takahashi, K., Sturtevant, J.M.: Biochemistry 20 (1981) 6185.
82A   Ahmad, F., Bigelow, C.C.: J. Biol. Chem. 257 (1982) 12935.
82B   Brems, D.N., Cass, R., Stellwagen, E.: Biochemistry 21 (1982) 1488.
82C   Cho, K.C., Poon, H.T., Choy, C.L.: Biochim. Biophys. Acta 701 (1982) 206.
82F1 Fujita, Y., Iwasa, Y., Noda, Y.: Bull. Chem. Soc. Japan 55 (1982) 1896.
82F2 Fukada, H., Takahashi, K.: Biochemistry 21 (1982) 1570.
82H   Hollecker, M., Creighton, T.E.: Biochim. Biophys. Acta 701 (1982) 395.
82M1 Medved, L.V., Privalov, P.L., Ugarova, T.P.: FEBS Lett. 146 (1982) 339.
82M2 Müller, K., Lüdemann, H.-D., Jaenicke, R.: Biophys. Chem. 16 (1982) 1.
82O   Okabe, N., Fujita, E., Tomita, K.-I.: Biochim. Biophys. Acta 700 (1982) 165.
82P1 Potekhin, S.A., Privalov, P.L.: Biofizika 23 (1982) 219.
82P2 Potekhin, S.A., Privalov, P.L.: J. Mol. Biol. 159 (1982) 519.
82P3 Privalov, P.L.: Adv. Protein Chem. 35 (1982) 1.
82P4 Privalov, P.L., Medved, L.V.: J. Mol. Biol. 159 (1982) 665.
82S1 Sugiyama, T., Miki, N., Miura, R., Miyake, Y., Yamano, T.: Biochim. Biophys. Acta 706 (1982) 42.
82S2 Sumi, A., Hamaguchi, K.: J. Biochem. 92 (1982) 823.
82T1 Tiktopulo, E.I., Privalov, P.L., Odintsova, T.I., Ermokhina, T.M., Krasheninnikov, I.A., Aviles, F.X., Cary, P.D., Crane- Robinson, C.: Eur. J. Biochem. 122 (1982) 327.

82T2 Tischenko, V.M., Zav'yalov, V.P., Medggyesi, G.A., Potekhin, S.A., Privalov, P.L.: Eur. J. Biochem. 126 (1982) 517.

82T3 Tombs, M.P., Blake, G.G.: Biochim. Biophys. Acta 700 (1982) 81.

82V   Vita, C., Fontana, A.: Biochemistry 21 (1982) 5196.

82Y1 Yutani, K., Ogasahara, K., Kimura, A., Sugino, Y.: J. Mol. Biol. 160 (1982) 387.

82Y2 Yutani, K., Khechinashvili, N.N., Lapshina, E.A., Privalov, P.L., Sugino, Y.: Int. J. Peptide Protein Res. 20 (1982) 331.

83A1 Ahmad, F., Contaxis, C.C., Bigelow, C.C.: J. Biol. Chem. 258 (1983) 7960.

83A2 Ahmad, F.: J. Biol. Chem. 258 (1983) 11143.

83A3 Arnold, L.D., Viswanatha, T.: Biochim. Biophys. Acta 749 (1983) 192.

83B1 Bendzko, P., Pfeil, W.: Biochim. Biophys. Acta 742 (1983) 669.

83B2 Bismuto, E., Colonna, G., Irace, G.: Biochemistry 22 (1983) 4165.

83B3 Brzeska, H., Venyaminov, S.V., Grabarek, Z., Drabikowski, W.: FEBS Lett. 153 (1983) 169.

83C   Cupo, F.F., Pace, C.N.: Biochemistry 22 (1983) 2654.

83F1 Flanagan, M.A., Garcia-Moreno, B., Friend, S.H., Feldman, R.J., Scouloudi, H., Gurd, F.R.N.: Biochemistry 22 (1983) 6027.

83F2 Fukada, H., Sturtevant, J.M., Quiocho, F.A.: J. Biol. Chem. 258 (1983) 13193.

83M1 Mitchinson, C., Pain, R.H., Vinson, J.R., Walker, T.: Biochim. Biophys. Acta 743 (1983) 31.

83M2 Moses, E., Hinz, H.-J.: J. Mol. Biol. 170 (1983) 765.

83S   Saito, Y., Wada, A.: Biopolymers 22(1983) 2105.

83T   Thompson, R.E., Morrical, S.W., Campbell, D.P., Carper, W.R.: Biochim. Biophys. Acta 745 (1983) 279.

83U   Uedaira, H., Honda, T., Takeda, Y., Miwatani, T., Uedaira, H., Ohsaka, A.: Int. Sympos. on Thermodynamics of Proteins and Biological Membranes, Granada, 1983, Abstr. P-II-20.

83Z   Zuniga, E.H., Nall, B.T.: Biochemistry 22 (1983) 1430.

84A   Ahmad, F.: J. Biol. Chem. 259 (1984) 4183.

84D1 Desmadril, M., Mitraki, A., Betton, J.M., Yon, J.M.: Biochem. Biophys. Res. Commun. 118 (1984) 416.

84D2 Dobson, C.M., Evans, P.A.: Biochemistry 23 (1984) 4267.

84F1 Fedorov, O.V., Khechinashvili, N.N., Kamiya R., Asakura, S.: J. Mol. Biol. 175 (1984) 83.

84F2 Flandin, F., Buffevant, C., Herbage, D.: Biochim. Biophys. Acta 791 (1984) 205.

84F3 Fujita, Y., Noda, Y.: Bull. Chem. Soc. Japan 57 (1984) 1891.

84F4 Fujita, Y., Noda, Y.: Bull. Chem. Soc. Japan 57 (1984) 2177.

84H1 Hawkes, R., Grütter, M.G., Schellman, J.: J. Mol. Biol. 175 (1984) 195.

84H2 Hecht, M.H., Sturtevant, J.M., Sauer, R.T.: Proc. Natl. Acad. Sci. USA 81 (1984) 5685.

84H3 Hiraoka, Y., Sugai, S.: Int. J. Peptide Protein Res. 23 (1984) 535.

84K1 Kelley, R.F., Stellwagen, E.: Biochemistry 23 (1984) 5095.

84K2 Khechinashvili, N.N., Tsetlin, V.I.: Molekularnaya Biol. 18 (1984) 786.

84M   Matthews, C.R., Perry, K.M., Touchette, N.A.: 188th Natl. Meeting ACS 1984, see also: Biochemistry 23 (1984) 3358.

84N   Novokhatny, V.V., Kudinov, S.A., Privalov, P.L.: J. Mol. Biol. 179 (1984) 215.

84O   Ogasahara, K., Yutani, K., Suzuki, M., Sugino, Y.: Int. J. Peptide Protein Res. 24 (1984) 147.

84R   Ross, P.D., Finlayson, J.S., Shrake, A.: Vox. Sanguinis 47 (1984) 19.

84S1 Santoro, M.M., Miller, J.F., Bolen, D.W.: ASBC/AAI Annual Meeting 1984, see: Fed. Proc. 43 (1984) 1838.

84S2 Shrake, A., Finlayson, J.S., Ross, P.D.: Vox. Sanguinis 47 (1984) 7.

84Y1 Yutani, K., Ogasahara, K., Aoki, K., Kakuno, T., Sugino, Y.: J. Biol. Chem. 259 (1984) 14076.

84Y2 Yutani, K., Sato, T., Ogasahara, K., Miles, E.W.: Arch. Biochem. Biophys. 229 (1984) 448.

85A   Ashikara, Y., Arata, Y., Hamaguchi, K.: J. Biochem. 97 (1985) 517.

85B   Bryant, C., Strottman, J.M., Stellwagen, E.: Biochemistry 24 (1985) 3459.

85C1   Calderon, R.O., Stolovich, N.J., Gerlt, J.A., Sturtevant, J.M.: Biochemistry 24 (1985) 6044.
85C2   Craig, S., Hollecker, M., Creighton, T.E., Pain, R.H.: J. Mol. Biol. 185 (1985) 681.
85D1   Danilenko, A.N., Grozav, E.K., Bibkov, T.M., Grinberg V.Ya., Tolstoguzov, V.B.: Int. J. Macromol. 7 (1985) 109.
85D2   Danilenko, A., Rogova, E.I., Bibkov, T.M., Grinberg, Y.Ya., Tolstoguzov, V.B.: Int. J. Peptide Protein Res. 26 (1985) 5.
85F1   Franks, F., Hatley, R.H.M.: Cryo-Letters 6 (1985) 171.
85F2   Fukada, H., Takahashi, K., Sturtevant, J.M.: Biochemistry 24 (1985) 5109.
85H   Holladay, L.A.: Biophys. Chem. 22 (1985) 281.
85M1   Mantulin, W.W., Pownall, H.J.: Biochim. Biophys. Acta 836 (1985) 215.
85M2   Mitchinson, C., Pain, R.H.: J. Mol. Biol. 184 (1985) 331.
85P1   Paulsson, M., Hegg, P.-O., Castberg, H.B.: Thermochim. Acta 95 (1985) 435.
85P2   Permyakov, E.A., Morozova, L.A., Burstein, E.A.: Biophys. Chem. 21 (1985) 21.
85P3   Pfeil, W., Sadowski, M.: Studia Biophys. 109 (1985) 163.
85S   Sochova, I.V., Belopolskaya, T.V., Smirnova, O.I.: Biophys. Chem. 22 (1985) 323.
85T1   Tiktopulo, E.I., Privalov, P.L., Borisenko, S.N., Troitsky, G.V.: Molekularnaya Biol. 19 (1985) 1072.
85T2   Tsalkova, T.N., Privalov, P.L.: J. Mol. Biol. 181 (1985) 533.
85V   Vita, C., Fontana, A., Chaiken, I.M.: Eur. J. Biochem. 151 (1985) 191.
86B   Beasty, A.M., Hurle, M.R., Manz, J.T., Stackhouse T., Onuffer, J.J., Matthews, C.R.: Biochemistry 25 (1986) 2965.
86C1   Chetverin, A.B., Khechinashvili, N.N., Filimonov, V.V.: FEBS Lett. 205 (1986) 185.
86C2   Corbett, R.J.T., Ahmad, F., Roche, R.S.: Biochem. Cell Biol. 64 (1986) 953.
86G   Goto, Y., Hamaguchi, K.: Biochemistry 25 (1986) 2821.
86H1   Hecht, M.H., Sturtevant, J.M., Sauer, R.T.: Proteins: Structure, Function, and Genetics 1 (1986) 43.
86H2   Hurle, M.R., Tweedy, N.B., Matthews, C.R.: Biochemistry 25 (1986) 6356.
86I   Ikeguchi, M., Kuwajima, K., Sugai, S.: J. Biochem. 99 (1986) 1191.
86K1   Kanazawa, I., Hamaguchi, K.: J. Biochem. 100 (1986) 207.
86K2   Kikuchi, H., Goto, Y., Hamaguchi, K.: Biochemistry 25 (1986) 2009.
86K3   Kuwajima, K., Harushima, Y., Sugai, S.: Int. J. Peptide Protein Res. 27 (1986) 18.
86M   Masson, P., Goasdoue, J.-L.: Biochim. Biophys. Acta 869 (1986) 304.
86P1   Pace, C.N.: Methods. Enzymol. 131 (1986) 266.
86P2   Pfeil, W.: in: Thermodynamic Data for Biochemistry and Biotechnology, ed. by H.-J. Hinz, Springer-Verlag, Heidelberg, New York, 1986, p.349
86P3   Pfeil, W., Ohlsson, P.I.: Biochim. Biophys. Acta 872 (1986) 72.
86R   Ramdas, L., Sherman, F., Nall, B.T.: Biochemistry 25 (1986) 6952.
86S1   Saigo, S.: J. Biochem. 100 (1986) 157.
86S2   Sauer, R.T., Hehir, K., Stearman, R.S., Weiss, M.A., Jeitler-Nilsson, A., Suchanek, E.G., Pabo, C.O.: Biochemistry 25 (1986) 5992.
86S3   Shortle, D., Meeker, A.K.: Proteins: Structure, Function, and Genetics 1 (1986) 81.
86S4   Shortle, D.: J. Cell. Biochem. 30 (1986) 281.
86T1   Tatunashvili, L.V., Privalov, P.L.: Biofizika 31 (1986) 578.
86T2   Thomson, J.A., Bigelow, C.C.: Biochem. Cell Biol. 64 (1986) 993.
86T3   Touchette, N.A., Perry, K.M., Matthews, C.R.: Biochemistry 25 (1986) 5445.
86W   Williams, T.C., Corson, D.C., Oikawa, K., McCubbin, W.D., Kay, C.M., Sykes, B.D.: Biochemistry 25 (1986) 1835.
87A   Alber, T., Dao-pin, S., Wilson, K., Wozniak, J.A., Cook,S.P., Matthews, B.W.: Nature 330 (1987) 41.
87B1   Becktel, W.J., Baase, W.A.: Biopolymers 26 (1987) 619.
87B2   Brouillette, C.G., Muccio, D.D., Finney, T.K.: Biochemistry 26 (1987) 7431.
87C   Craig, S., Schmeissner, U., Wingfield, P., Pain, R.H.: Biochemistry 26 (1987) 3570.
87F   Fukada, H., Takahashi, K., Sturtevant, J.M.: Biochemistry 26 (1987) 4063.
87G1   Goto, Y., Tsunenaga, M., Kawata, Y., Hamaguchi, K.: J. Biochem. 101 (1987) 319.
87G2   Goto, Y., Hamaguchi, K.: Biochemistry 26 (1987) 1879.
87G3   Gray, T.M., Matthews, B.W.: J. Biol. Chem. 262 (1987) 16858.

87G4   Grütter, M.G., Gray, T.M., Weaver, L.H., Alber, T., Wilson, K., Matthews, B.W.: J. Mol. Biol. 315.

87H   Hu, C.Q., Sturtevant, J.M.: Biochemistry 26 (1987) 178.

87I   Ismond, M.A.H., Murray, E.D., Arntfield, S.D.: Biochemistry Internat. 15 (1987) 245.

87K1   Kelley, R.F., Shalongo, W., Jagannadham, M.V., Stellwagen, E.: Biochemistry 26 (1987) 1406.

87K2   Kelley, R.F., Richards, F.M.: Biochemistry 26 (1987) 6765.

87K3   Kelly, L., Holladay, L.A.: Biophys. Chem. 27 (1987) 77.

87L   Lee, L.L.-Y., Lee, J.C.: Biochemistry 26 (1987) 7813.

87M   Matthews, B.W., Nicholson, H. Becktel, W.J.: Proc. Natl. Acad. Sci. USA 84 (1987) 6663.

87P1   Pantoliano, M.W., Ladner, R.C., Bryan, P.N., Rollence M.L., Wood, J.F., Poulos, T.L.: Biochemistry 26 (1987) 2077.

87P2   Perry, K.M., Onuffer, J.J., Touchette, N.A., Herndon, C.S., Gittelman, M.S., Matthews, C.R., Chen, J.-T., Mayer, R.J., Taira, K., Benkovic, S.J., Howell, E.E., Kraut, J.: Biochemistry 26 (1987) 2674.

87P3   Perry, K.M., see 87P2.

87P4   Protasevich, I.I., Platonov, A.L., Pavlovsky, A.G., Esipova, N.G.: J. Biomol. Structure and Dynamics 4 (1987) 885.

87R   Radding, J.A.: Biochemistry 26 (1987) 3530.

87S1   Schwarz, H., Hinz, H.-J., Mehlich, A., Tscheche, H., Wenzel, H.R.: Biochemistry 26 (1987) 3544.

87S2   Sturtevant, J.M.: Ann. Rev. Phys. Chem. 38 (1987) 463.

87T   Tsunenaga, M., Goto, Y., Kawata, Y., Hamaguchi, K.: Biochemistry 26 (1987) 6044.

87V1   Van de Ven, F.J.M., Hilbers, C.W.: Biochemistry 26 (1987) 5548.

87V2   Villafranca, J.E., Howell, E.E., Oatley, S.J., Xuong, N., Kraut, J.: Biochemistry 26 (1987) 2182.

87W   White, T.B., Berget, P.B., Nall, B.T.: Biochemistry 26 (1987) 4358.

87Y   Yutani, K., Ogasahara, K., Tsujita, T., Sugino, Y.: Proc. Natl. Acad. Sci. USA 84 (1987) 4441.

88A   Alber, T.A., Bell, J.A., Dao-Pin, S., Nicholson, H., Wozniak, J.A., Cook, S., Matthews, B.W.: Science 239 (1988) 631.

88B1   Bae, S.-J., Chou, W.-Y., Matthews, K., Sturtevant, J.M.: Proc. Natl. Acad. Sci. USA 85 (1988) 6731.

88B2   Bolen, D.W., Santoro, M.M.: Biochemistry 27 (1988) 8069.

88C   Chardot, T., Mitraki, A., Amigues, Y., Desmadril, M., Betton, J.M., Yon, J.M.: FEBS Lett. 228 (1988) 65.

88E   Edge, V., Allewell, N.M., Sturtevant, J.M.: Biochemistry 27 (1988) 8081.

88F   Franks, F.: Proc. Miami Bio/Technol. Winter Sympos. (1988) p. 15.

88G1   Ghosaini, L.R., Brown, A.M., Sturtevant, J.M.: Biochemistry 27 (1988) 5257.

88G2   Goins, B, Freire, E.: Biochemistry 27 (1988) 2046.

88G3   Goto, Y., Ichimura, N., Hamaguchi, K.: Biochemistry 27 (1988) 1670.

88G4   Griko, Yu.V., Privalov, P.L., Venyaminov, S.Yu., Kutyshenko, V.P.: J.Mol. Biol. 202 (1988) 127.

88G5   Griko, Yu.V., Privalov, P.L., Sturtevant, J.M., Venyaminov, S.Yu.: Proc. Natl. Acad. Sci. USA 85 (1988) 3343.

88H   Hickey, D.R., McLendon, G., Sherman, F.: J. Biol. Chem. 176 (1988) 18298.

88K1   Keating, K.M., Ghosaini, L.R., Giedroc, D.P., Williams, K.R., Coleman, J.E., Sturtevant, J.M.: Biochemistry 27 (1988) 5240.

88K2   Kella, N.K.D., Kinsella, J.E.: Int. J. Peptide Protein Res. 32 (1988) 396.

88K3   Kellis Jr., J.T., Nyberg, K. Sali, D., Fersht, A.R.: Nature 333 (1988) 784.

88K4   Kelly, L., Simmons, J.H., Heck, T., Holladay, L.A.: Int. J. Peptide Protein Res. 31 (1988) 281.

88K5   Kuwajima, K., Sakuraoka, A., Fueki, S., Yoneyama, M., Sugai, S.: Biochemistry 27 (1988) 7419.

88L   Long, R.C., Hawkridge, F.M., Chlebowski, J.F., Hartzell, C.R.: J. Electroanalyt. Chem. 256 (1988) 111.

88M1   Maneri, L.R., Low, P.S.: J. Biol. Chem. 263 (1988) 16170.

88M2   Matsumura, M., Becktel, W.J., Matthews, B.W.: Nature 334 (1988) 406.

88M3   Matsumara, M., Yahanda, S., Yasumura, S., Yutani, K., Aiba, S.: Eur. J. Bichem. 171 (1988) 715.

87N   Nicholson, H., Becktel, W.J., Matthews, B.W.: Nature 336 (1988) 651.

88P1   Pace, C.N., Grimsley, G.R.: Biochemistry 27 (1988) 3242.

88P2   Pace, C.N., Grimsley, G.R. Thomson, J.A., Barnett, B.J.: J. Biol. Chem. 263 (1988) 11820.

88P3   Privalov, P.L., Gill, S.J.: Adv. Protein Chem. 39 (1988) 191.

88R1   Radek, J.T., Castellino, F.J.: Arch. Biochem. Biophys. 267 (1988) 776.

88R2   Regan, L., DeGrado, W.F.: Science 241 (1988) 976.

88R3   Ross, P.D., Shrake, A.: J. Biol. Chem. 263 (1988) 11196.

88S1   Sali, D., Bycroft, M., Fersht, A.R.: Nature 335 (1988) 740.

88S2   Sanchez-Ruiz, J.M., Lopez-Lacomba, J.L., Cortijo, M. Mateo, P.L.: Bichemistry 27 (1988) 1648.

88S3   Sanchez-Ruiz, J.M., Lopez-Lacomba, J.L., Mateo, P.L., Villanova, M., Serra, M.A., Aviles, F.X.: Eur. J. Biochem. 176 (1988) 225.

88S4   Santoro, M.M., Bolen, D.W.: Biochemistry 27 (1988) 8063.

88S5   Schwarz, F.P., Kirchhoff, W.H.: Thermochim. Acta 128 (1988) 267-295.

88S6   Shrake, A., Ross, P.D.: J. Biol. Chem. 263 (1988) 15392.

88S7   Shnyrov, V.L., Berman, A.L.: Biomed. Biochim. Acta 47 (1988) 355.

88S8   Shortle, D., Meeker, A.K., Freire, E.: Biochemistry 27 (1988) 4761.

88S9   Stackhouse, T.M., Onuffer, J.J., Matthews, C.R., Ahmed, S.A., Miles, E.W.: Biochemistry 27 (1988) 824.

88S10   Stearman, R.S., Frankel, A.D., Freire, E., Liu, B., Pabo, C.O.: Biochemistry 27 (1988) 7571.

88W1   Wendt, B., Hofman, T., Martin, S.R., Bayley, P., Brodin, P., Grundstroem, T., Thulin, E., Linse, S., Forsèn, S.: Eur. J. Biochem. 175 (1988) 439.

88W2   Wetzel, R., Perry, L.J., Baase, W.A., Becktel, W.J.: Proc. Natl. Acad. Sci. USA 85 (1988) 401.

88W3   Wingfield, P., Graber, P., Craig, S., Pain, R.H.: Eur. J. Biochem. 173 (1988) 65.

88W4   Wood, L.C., White, T.B., Ramdas, L., Nall, B.T.: Biochemistry 27 (1988) 8562.

89B1   Bertazzon, A., Tsong, T.Y.: Biochemistry 28 (1989) 9784.

89B2   Bowie, J.U., Sauer, R.T.: Biochemistry 28 (1989) 7139.

89B3   Brouillette, C.G., McMichens, R.B., Stern, L.J., Khorona, H.G.: Proteins: Structure, Function, and Genetics 5 (1989) 38.

89C   Chen, B., Schellman, J.A.: Biochemistry 28 (1989) 685.

89D   Das, G., Hickey, D.R., McLendon D., McLendon G., Sherman, F.: Proc. Natl. Acad. Sci. USA 86 (1989) 496.

89G1   Garvey, E.P., Matthews, C.R.: Biochemistry 28 (1989) 2083.

89G2   Garvey, E.P., Swank, J., Matthews, C.R.: Proteins: Structure, Function, and Genetics 6 (1989) 259.

89G3   Goldenberg, D.P., Frieden, R.W., Haack, J.A., Morrison, T.B.: Nature 338 (1989) 127.

89G4   Griko, Yu.V., Privalov, P.L., Venyaminov, S.Yu.: Biofizika 34 (1989) 940.

89G5   Griko, Yu.V., Venyaminov, S.Yu., Privalov, P.L.: FEBS Lett. 244 (1989) 276.

89H1   Harushima, Y, Sugai, S.: Biochemistry 28 (1989) 8568.

89H2   Hu, C.Q., Sturtevant, J.M.: Biochemistry 28 (1989) 813.

89H3   Hughson, F.M., Baldwin, R.L.: Biochemistry 28 (1989) 4415.

89H4   Hynes, T.R., Kautz, R.A., Goodman, M.A., Gill, J.F., Fox, R.O.: Nature 339 (1989) 73.

89I1   Ichijo, H., Uedaira, H., Suehiro, T., Nagasawa, J., Yamauchi, A., Aisaka, N.: Agr. Biol. Chem. 53 (1989) 833.

89I2   Isaacs, B.S., Brew, S.A., Ingham, K.C.: Biochemistry 28 (1989) 842.

89K1   Karpusas, M., Baase, W.A., Matsumura, M., Matthews, B.W.: Proc. Natl. Acad. Sci. USA 86 (1989) 8237.

89K2   Kelley, R.F., Cleary, S.: Biochemistry 28 (1989) 4047.

89K3   Kellis Jr., J.T., Nyborg, K., Fersht, A.R.: Biochemistry 28 (1989) 4914.

89K4   Kitamura, S., Sturtevant, J.M.: Biochemistry 28 (1989) 3788.

89L1   Langsetmo, K., Fuchs, J., Woodward, C.: Biochemistry 28 (1989) 3211.
89L2   Lapanje, S., Poklar, N.: Biophys. Chem. 34 (1989) 155.
89L3   Lin, T.-Y., Kim, P.S.: Biochemistry 28 (1989) 5282.
89L4   Lopez-Lacomba, J.L., Guzman, M., Cortijo, M., Mateo, P.L., Aguirre, R., Harvey, S.C., Cheung, H.C.: Biopolymers 28 (1989) 2143.
89L5   Luntz, T.L., Schejter, A., Garber, E.A.E., Margoliash, E.: Proc. Natl. Acad. Sci. USA 86 (1989) 3524.
89M1   Maglova, L., Atanasov, B., Keszthelyi, L.: Biochim. Biophys. Acta 975 (1989) 271.
89M2   Marshall, W.E., Zarins, Z.M.: J. Agr. Food Chem. 37 (1989) 869.
89M3   Matsumura, M., Becktel, W.J., Levitt, M., Matthews, B.W.: Proc. Natl. Acad. Sci. USA 86 (1989) 6562.
89M4   Matouschek, A., Kellis Jr., J.T., Serrano, L., Fersht, A.R.: Nature 340 (1989) 122.
89M5   McPhie, P.: Biochem. Biophys. Res. Commun. 158 (1989) 115.
89M6   Medved, L.V., Busby, T.F., Ingham, K.C.: Biochemistry 28 (1989) 5408.
89N   Nicholson, H., Soederlind, E., Tonrud, D.F., Matthews, B.W.: J. Mol. Biol. 210 (1989) 181.
89P1   Pace, C.N., Shirley, B.A., Thomson, J.A.: in Protein Structure and Function: a practical approach, ed. by T.E. Creighton, IRL Press, Oxford, 1989, p. 311.
89P2   Pace, C.N., Laurents, D.V.: Biochemistry 28 (1989) 2520.
89P3   Pakula, A.A., Sauer, R.T.: Proteins: Structure, Function, and Genetics 5 (1989) 202.
89P4   Pantoliano, M.W., Whitlow, M., Wood, J.F., Dodd, S.W., Hardmann, K.D., Rollence, M.L., Bryan, P.N.: Biochemistry 28 (1989) 7205.
89P5   Parsell, D.A., Sauer, R.T.: J. Biol. Chem. 264 (1989) 7590.
89P6   Perry, K.M, Onuffer, J.J., Gittelman, M.S., Barmat, L., Matthews, C.R.: Biochemistry 28 (1989) 7961.
89P7   Potekhin, S.A., Pfeil, W.: Biophys. Chem. 34 (1989) 55.
89P8   Privalov, P.L., Tiktopulo, E.I., Venyaminov, S.Yu., Griko, Yu.V, Makhatadze, G.I., Khechinashvili, N.N.: J. Mol. Biol. 205 (1989) 737.
89R   Radek, J.T., Castellino, F.J.: J. Biol. Chem. 264 (1989) 9915.
89S1   Sanbongi, Y., Igarashi, Y., Kodama, T.: Biochemistry 28 (1989) 9574.
89S2   Sandberg, W.S., Terwilliger, T.C.: Science 245 (1989) 54.
89S3   Santucchi, R., Giartosio, A., Ascoli, F.: Arch. Biochem. Biophys. 275 (1989) 496.
89S4   Schwarz, F.P., Steer, C.J., Kirchhoff, W.H.: Arch. Biochem. Biophys. 273 (1989) 433.
89S5   Schwarz, F.P.: Thermochim. Acta 147 (189) 71.
89S6   Segawa, S.-I., Sugihara, M., Maeda, T., Mitsuhisa, Y., Kodama, M., Seki, S., Sakiyama, M.: Biopolymers 28 (1989) 1033.
89S7   Serrano, L., Fersht, A.R.: Nature 342 (1989) 296.
89S8   Shirley, B.A., Stanssens, P., Steyaert, J., Pace, C.N.: J. Biol. Chem. 264 (1989) 11621.
89S9   Shortle, D., Meeker, A.K.: Arch. Biochem. Biophys. 272 (1989) 103.
89S10   Shrake, A., Fisher, M.T., McFarland, P.J., Ginsburg, A.: Biochemistry 28 (1989) 6281.
89S11   Sturtevant, J.M., Yu, M., Haase-Pettingell, C., King, J.: J. Biol. Chem. 264 (1989) 19693.
89T1   Tandon, S., Horowitz, P.M.: J. Biol. Chem. 264 (1989) 9859.
89T2   Thomson, J.A., Shirley, B.A., Grimsley, G.R., Pace, C.N.: 264 (1989) 11614.
89V   Varfolomeeva, E.P., Burova, T.V., Grinberg, V.Ya., Tolstoguzov, V.B.: Molekularnaya Biol. 23 (1990) 1263.
90A1   Alexandrescu, A.T., Hinck, A.P., Markley, J.L.: Biochemistry 29 (1990) 4516.
90A2   Akke, M., Forsèn, S.: Proteins: Structure, Function, and Genetics 8 (1990) 23.
90A3   Anderson, D.E., Becktel, W.J., Dahlquist, F.W.: Biochemistry 29 (1990) 2403.
90B0   Bailey, J.M., Lin, L.-N., Brandts, J.F., Mas, M.T.: J. Protein Chem. 9 (1990) 59.
90B1   Ballery, N., Minard, P. Desmadril, M., Betton, J.-M., Perahia, D., Mouawad, L., Hall, L., Yon, J.M.: Protein Engng. 3 (1990) 199.
90B2   Bertazzon, A., Tian, G.H., Lamblin, A., Tsong, T.Y.: Biochemistry 29 (1990) 291.
90B3   Bertazzon, A., Tsong, T.Y.: Biochemistry 29 (1990) 6447.
90B4   Bertazzon, A., Tsong, T.Y.: Biochemistry 29 (1990) 6453.
90B5   Borden, K.L.B., Richards, F.M.: Biochemistry 29 (1990) 3071.

90B6 Brandts, J.F., Lin, L.-N.: Biochemistry 29 (1990) 6927.
90B7 Brems, D.N., Brown, P.L., Heckenlaible, L.A., Frank, B.H.: Biochemistry 29 (1990) 9289.
90B8 Brems, D.N., Brown, P.L., Becker, G.W.: J. Biol. Chem. 265 (1990) 5504.
90B9 Burova, T.V., Varfolomeeva, E.P., Grinberg, V.Ya, Suchkov, V.V., Papkov, V.S., Bauve, H., Tolstoguzov, V.B.: Biofizika 35 (1990) 222.
90B10 Burz, D.S., Allewell, N.M., Ghosaini, L., Hu, C.Q., Sturtevant, J.M.: Biophys. Chem. 37 (1990) 31.
90D1 Dao-pin, S., Baase, W.A., Matthews, B.W.: Proteins: Structure, Function, and Genetics 7 (1990) 198.
90D2 Di Patti, M.C.B., Musci, G., Giartosio, A., D'Alessio, S., Calabrese, L.: J. Biol. Chem. 265 (1990) 21016.
90E1 Edwards, R.A., Jacobson, A.L., Huber, R.E.: Biochemistry 29 (1990) 11001.
90E2 Erwin, C.R., Barnett, B.L., Oliver J.D., Sullivan J.F.: Protein Engng. 4 (1990) 87.
90F Furukawa, K., Kidokoro, S., Nakamura, H., Kikuchi, M., Wada, A.: Int. Conf. Protein Engineering, Kobe (Japan) 1989, Abstr. in: Protein Engng. 3 (1990) 353.
90G1 Gekko, K., Ito, H.: J. Biochem. 107 (1990) 572.
90G2 Gittelman, M.S., Matthews, C.R.: Biochemistry 29 (1990) 7011.
90G3 Goto, Y., Takahashi, N., Fink, A.L.: Biochemistry 29 (1990) 3480.
90H1 Horovitz, A., Serrano, L., Avron, B., Bycroft, M., Fersht, A.R.: J. Mol. Biol. 216 (1990) 1031.
90H2 Hurle, M.R., Marks, C.B., Kosen, P.A., Anderson, S., Kuntz, I.D.: Biochemistry 29 (1990) 4410.
90J Jandu, S.K., Ray, S., Brooks, L., Leatherbarrow, R.J.: Biochemistry 29 (1990) 6264.
90K1 Kelley, R.; cited as a personal communication to J.A. Wells: Biochemistry 29 (1990) 8509.
90K2 Kelly, L., Holladay, L.A.: Biochemistry 29 (1990) 5062.
90K3 Khan, M.Y., Medow, M.S., Newman, S.A.: Biochem. J. 270 (1990) 33.
90K4 Kiefhaber, T., Quaas, R., Hahn, U., Schmid, F.X.: Biochemistry 29 (1990) 3053.
90K5 Kiefhaber, T., Schmid, F.X., Renner, M., Hinz, H.-J., Hahn U., Quaas, R.: Biochemistry 29 (1990) 8250.
90K6 Kono, M., Sen, A.C., Chakrabarti, B.: Biochemistry 29 (1990) 464.
90K7 Koseki, T., Kitabatake, N., Doi, E.: J. Biochem. 107 (1990) 389.
90K8 Kuwajima, K., Ikeguchi, M., Sugawara, T., Hiraoka, Y., Sugai, S.: Biochemistry 29 (1990) 8240.
90L1 Lepock, J.R., Frey, H.E., Hallewell, R.A.: J. Biol. Chem. 265 (1990) 21612.
90L2 Lindsay, C.D., Pain, R.H.: Eur. J. Biochem. 192 (1990) 133.
90L3 Lörinczi D., Hoffmann U., Poto L., Belagyi, J., Laggner, P.: Gen. Physiol. Biophys. 9 (1990) 589.
90L4 Lyu, P.C., Liff, M.I., Marky, L.A., Kallenbach, N.R.: Science 250 (1990) 669.
90M1 Maglova, L., Guleva, D., Chekulaeva, L., Atanasov, B.: Biochim. Biophys. Acta 1017 (1990) 217.
90M2 Malcolm, B.A., Wilson, K., Matthews, B.W., Kirsch, J.F., Wilson, A.C.: Nature 345 (1990) 86.
90M3 Matouschek, A., Kellis Jr., J.T., Serrano, L., Bycroft, M., Fersht, A.R.: Nature 346 (1990) 440.
90M4 McNutt, M., Mullins, L.S., Raushel, F.M., Pace, C.N.: Biochemistry 29 (1990) 7572.
90M5 Missiakas, D., Betton, J.-M., Minard, P., Yon, J.M.: Biochemistry 29 (1990) 8683.
90N1 Nakano, T., Fink, A.L.: J. Biol. Chem. 265 (1990) 12356.
90N2 Nishikawa, S., Adiwinata, J., Morioka, H., Fujimura, T., Tanaka, T., Uesugi, S., Hakoshima T., Tomita, K., Nakagawa, S., Ikehara, M.: Protein Engng. 3 (1990) 443.
90O1 Okajima, T., Kawata, Y., Hamaguchi, K.: Biochemistry 29 (1990) 9168.
90O2 O'Neil, K.T., DeGrado, W.F., Science 250 (1990) 646-651
90O3 Otlewski, J., Laskowski Jr., H.: cited as a personal communication to J.A. Wells: Biochemistry 29 (1990) 8509.
90P1 Pace, C.N., Laurents, D.V., Thomson, J.A.: Biochemistry 29 (1989) 2564.
90P2 Pakula, A.A., Sauer, R.T.: Nature 334 (1990) 363.

90P3   Privalov, P.L., Makhatadze, G.I.: J. Mol. Biol. 213 (1990) 385.
90R1   Ramsay, G., Freire, E.: Biochemistry 29 (1990) 8677.
90R2   Reidhaar-Olson, J.F., Parsell, D.A., Sauer, R.T.: Biochemistry 29 (1990) 7563.
90R3   Rojo-Dominguez, A., Zubillaga-Luna, R., Hernandez-Arana, A.: Biochemistry 29 (1990) 8689.
90R4   Ropson, I.J., Gordon, J.I., Frieden, C.: Biochemistry 29 (1990) 9591.
90S1   Sasaki, T., Kaiser, E.T.: Biopolymers 29 (1990) 79.
90S2   Savini, I., D'Alessio, S., Giartosio, A., Morpurgo, L., Avigliano, L.: Eur. J. Biochem. 190 (1990) 491.
90S3   Serrano, L., Horovitz, A., Avron, B., Bycroft, M., Fersht, A.R.: Biochemistry 29 (1990) 9343.
90S4   Shortle, D., Stites, W.E., Meeker, A.K.: Biochemistry 29 (1990) 8033.
90S5   Shriver, J.W.: Arch. Biochem. Biophys. 283 (1990) 472.
90S6   Shriver, J.W., Kamath, U.: Biochemistry 29 (1990) 2556.
90S7   Sondek, J., Shortle, D.: Proteins: Structure, Function, and Genetics 7 (1990) 299.
90S8   Sugisaki, Y., Ogasahara, K., Miles, E.W., Yutani, K.: Thermochim. Acta 163 (1990) 117.
90S9   Sugrue, R., Marston, F.A.O., Lowe, P.A., Freedman, R.B.: Biochem. J. 271 (1990) 541.
90T1   Tatunashvili, L.V., Filimonov, V.V., Privalov, P.L., Metsis, M.L., Koteliansky, V.E., Ingham, K.C., Medved, L.V.: J. Mol. Biol. 211 (1990) 161.
90T2   Tweedy, N.B., Hurle, M.R., Chrunyk, B.A., Matthews, C.R.: Biochemistry 29 (1990) 1539.
90U1   Uedaira, H., Kidokoro, S., Iwakura, M., Honda, S., Ohashi, S.: Thermochim. Acta 163 (1990) 123.
90U2   Uedaira, H., Kidokoro, S.-I., Iwakura, M., Honda, S., Ohashi, S.: Enzyme Engng. 10, Vol. 613 of the Annals New York Acad. Sci. (1990) 352.
90U3   Uedaira, H., Kidokoro, S., Vonderviszt, F., Namba, K.: Int. Conf. Protein Engineering, Kobe (Japan) 1989, Abstr. in: Protein Engng. 3 (1990) 340.
90W1   Walsh, M.T., Atkinson, D.: J. Lipid Res. 31 (1990) 1051.
90W2   Walz Jr., F.G., Kitareewan, S.: J. Biol. Chem. 265 (1990) 7127.
90W3   White, T.K., Kim, J.Y., Wilson, J.E.: Arch. Biochem. Biophys. 276 (1990) 510.
90W4   Wilson, K.S., Vorgias, C.E., Tanaka, I., White, S.W., Kimura, M.: Protein Engng. 4 (1990) 11.
90W5   Wu, C.-S.C., Yang, J.T.: Biopolymers 30 (1990) 381.
90Y1   Yamasaki, M., Yano, H., Aoki, K.: Int. J. Biol. Macromol. 12 (1990) 263.
90Y1   Yon, J.M., Desmadril, M., Betton, J.M., Minard, P., Ballery, N., Missiakas, D., Gaillard-Miran, S., Perahia, D., Mouawad, L.: Biochimique 72 (1990) 417.
91A1   Ahrweiler, P.M., Frieden, C.: Biochemistry 30 (1991) 7801.
91A2   Antonino, L.C., Kautz, R.A., Nakano, T., Fox, R.O., Fink, A.L.: Proc. Natl. Acad. Sci. USA 88 (1991) 7715.
91B1   Battistel, E., Bianchi, D., Rialdi, G.: Pure Appl. Chem. 63 (1991) 1483.
91B2   Bhakuni V., Xie D., Freire, G.: Biochemistry 30 (1991) 5055.
91B3   Bolen W., Santoro M., cited as a personal communication in: Sandberg V.A., Kren B., Fuchs J.A., Woodward C.: Biochemistry 30 (1991) 5475.
91B4   Brems, D.N., Brown, P.L., Nakagawa, S.H., Tager, H.S.: J. Biol. Chem. 266 (1991) 1611.
91B5   Brown, E.D., Yada, R.Y.: Biochim. Biophys. Acta 1076 (1991) 406.
91B6   Buchner, J., Renner, M., Lilie, H., Hinz, H.-J., Jaenicke, R., Kiefhaber, T., Rudolph, R.: Biochemistry 30 (1991) 6922.
91C1   Caffrey, M.S., Daldal, F., Holden, H.M., Cusanovich, M.A.: Biochemistry 30 (1991) 4119.
91C2   Caffrey, M.S., Cusanovich, M.A.: Biochemistry 30 (1991) 9238.
91C3   Cavazza, B., Brizzolara, G., Lazzarini, G., Patrone, E., Piccardo, M., Barboro, P., Parodi, S., Pasini, A., Balbi, C.: Biochemistry 30 (1991) 9060.
91C4   Chen, H.M., You, J.L., Markin V.S., Tsong, T.Y.: J. Mol. Biol. 220 (1991) 771.
91C5   Conejero-Lara, F., Mateo, P.L., Aviles, F.X., Sanchez-Ruiz, J.M.: Biochemistry 30 (1991) 2067.

91C6   Conejero-Lara, F., Sanchez-Ruiz, J.M., Mateo, P.L., Burgos, F.J., Vendrell, J., Aviles, F.X.: Eur. J. Biochem. 200 (1991) 663.
91C7   Conelly, P., Ghosaini, L., Hu, C.-Q., Kitamura, S., Tanaka, A., Sturtevant, J.M.: Biochemistry 30 (1991) 1887.
91D1   Dao-pin, S., Sauer, U., Nicholson H., Matthews, B.W.: Biochemistry 30 (1991) 7142.
91D2   Dao-pin, S., Söderlind, E., Baase, W.A., Wozniak, J.A., Sauer, U., Matthews, B.W.: J. Mol. Biol. 221 (1991) 873.
91D3   Dao-pin, S., Anderson, D.E., Baase, W.A., Dahlquist, F.W., Matthews, B.W.: Biochemistry 30 (1991) 11521.
91D4   Daopin, S., Alber, T., Baase, W.A., Wozniak, J.A., Matthews, B.W.: J. Mol. Biol. 221 (1991) 647.
91D5   Das, B.K., Agarwal, S.K., Khan, M.Y.: Biochim. Biophys. Acta 1076 (1991) 343.
91D6   De Francesco, R., Pastore, A., Vecchio, G., Cortese, R.: Biochemistry 30 (1991) 143.
91D7   Desmet, J., Tieghem, E., Van Dael, H., Van\Cauwelaert, F.: Eur. J. Biophys. 20 (1991) 263.
91D8   Dirr, H.W., Reinemer, P.: Biochem. Biophys. Res. Commun. 180 (1991) 294.
91E1   Eftink, M.R., Ghiron, C.A., Kautz, R.A., Fox, R.O.: Biochemistry 30 (1991) 1193.
91F1   Feng, Y., Sligar, S.G.: Biochemistry 30 (1991) 10150.
91F2   Fujita, Y., Noda, Y.: Int. J. Peptide Protein Res. 38 (1991) 445.
91G1   Galisteo, M.L., Mateo, P.L., Sanchez-Ruiz, J.M.: Biochemistry 30 (1991) 2061.
91G2   Ginsburg, A., Zolkiewski, M.: Biochemistry 30 (1991) 9421.
91G3   Gitelson, G.I., Griko, Yu.V., Kurochkin, A.V., Rogov, V.V., Kutyshenko, V.P., Kirpichnikov, M.P., Privalov, P.L.: FEBS Lett. 289 (1991) 201.
91G4   Goodenough, P.W., Bhat, K.M., Collins, M.E., Perry, B.N., Pickersgill, R.W., Summer, I.G., Warwicker, J., de Haas, G.H., Verheij, H.M.: Protein Engng. 4 (1991) 929.
91H1   Hickey, D.R., Berghuis, A.M., Lafond, G., Jaeger, J.A., Cardillo, T.S., McLendon, D., Das, G., Sherman, F., Brayer, G.D., McLendon, G.: J. Biol. Chem. 266 (1991) 11686.
91H2   Hirose, M., Yamashita, H.: J. Biol. Chem. 266 (1991) 1463.
91H3   Horovitz A., Serrano L., Fersht, A.R.: J. Mol. Biol. 219 (1991) 5.
91H4   Hughson, F.M., Barrick, D., Baldwin, R.L.: Biochemistry 30 (1991) 4113.
91I   Ishiwata, A., Kawata, Y., Hamaguchi, K.: Biochemistry 30 (1991) 7766.
91J1   Jackson, S.E., Fersht, A.R.: Biochemistry 30 (1991) 10428.
91J2   Jackson, S.E., Fersht, A.R.: Biochemistry 30 (1991) 10436.
91K1   Kanaya S., Katayanagi K., Morikawa K., Inoue H., Ohtsuka E., Ikehara M.: Eur. J. Biochem. 198 (1991) 437.
91K2   Kanzaki, H., McPhie, P., Miles, E.W.: Arch. Biochem. Biophys. 284 (1991) 174.
91K3   Kawata Y., Hamaguchi K.: Biochemistry 30 (1991) 4367.
91K4   Khan, S.M.A., Bolen, W., Hargrave, P.A., Santoro, M.M., McDowell, J.H.: Eur. J. Biochem. 200 (1991) 53.
91K5   Klemm, J.D., Wozniak, J.A., Alber, T., Goldenberg, D.P.: Biochemistry 30 (1991) 589.
91K6   Kuwajima, K., Garvey, E.P., Finn, B.E., Matthews, C.R., Sugai, S.: Biochemistry 30 (1991) 7693.
91L1   Langsetmo, K., Fuchs, J.A., Woodward, C.: Biochemistry 30 (1991) 7603.
91L2   Lavecchia, R., Zugaro, M.: FEBS Lett. 292 (1991) 162.
91L3   Le Bihan, T., Gicquaud C.: Biochem. Biophys. Res. Commun. 181 (1991) 542.
91L4   Lehrer, S.S., Stafford III, W.F.: Biochemistry 30 (1991) 5682.
91L5   Leòn, D.A., Dostmann, W.R.G., Taylor, S.S.: Biochemistry 30 (1991) 3035.
91L6   Liang, H., Terwilliger, T.C.: Biochemistry 30 (1991) 2772.
91L7   Lin, T.-Y., Kim, P.S.: Proc. Natl. Acad. Sci. USA 88 (1991) 10573.
91L8   Litvinovich, S.V., Strickland, D.K., Medved, L.V., Ingham, K.C.: J. Mol. Biol. 217 (1991) 563.
91L9   Lohner, K., Esser, A.F.: Biochemistry 30 (1991) 6620.
91M1   Menhart, N., Sehl, L.C., Kelley, R.F., Castellino, F.J.: Biochemistry 30 (1991) 1948.

91M2 Muga, A., Mantsch, H.H., Surewicz, W.K.: Biochemistry 30 (1991) 7219.
91N1 Nicholson, H., Anderson, D.E., Dao-pin, S., Matthews, B.W.: Biochemistry 30 (1991) 9816.
91N2 Novokhatny, V.V., Ingham, K.C., Medved, L.V.: J. Biol. Chem. 266 (1991) 12994.
91O   Otto, A., Seckler, R.: Eur. J. Biochem. 202 (1991) 67.
91P1 Pantoliano, M.W., Bird, R.E., Johnson, S., Asel, E.D., Dodd, S.W., Wood, J.F., Hardman, K.D.: Biochemistry 30 (1991) 10117.
91P2 Permyakov, E.A., Shnyrov, V.L., Kalinichenko, L.P., Kuchar, A., Reyzer, I.L., Berliner, L.J.: J. Protein Chem. 10 (1991) 577.
91P3 Peterson, C.B., Schachman, H.K.: Proc. Natl. Acad. Sci. USA 88 (1991) 458.
91P4 Poole, L.B., Loveys, D.A., Hale, S.P., Gerlt, J.A.: Biochemistry 30 (1991) 3621.
91R1 Radford, S.E., Woolfson, D.N., Martin, S.R., Lowe, G., Dobson, C.M.: Biochem. J. 273 (1991) 211.
91R2 Reece, L.J., Nichols, R., Ogden R.C., Howell, E.E.: Biochemistry 30 (1991) 10895.
91S1 Sali, D., Bycroft, M., Fersht, A.R.: J. Mol. Biol. 220 (1991) 779.
91S2 Sandberg, W.S., Terwilliger, T.C.: Proc. Natl. Acad. Sci. USA 88 (1991) 1706.
91S3 Sandberg, V.A., Kren, B., Fuchs, J.A., Woodward, C.: Biochemistry 30 (1991) 5475.
91S4 Scholtz, J.M., Marqusee, S., Baldwin, R.L., York, E.J., Stewart, J.M., Santoro, M., Bolen, D.W.: Proc. Natl. Acad. Sci. USA 88 (1991) 2854.
91S5 Serrano, L., Bycroft, M., Fersht, A.R.: J. Mol. Biol. 218 (1991) 465.
91S6 Stites, W.E., Gittis, A.G., Lattman, E.E., Shortle, D.: J. Mol. Biol. 221 (1991) 7.
91S7 Sturtevant, J.M., Holtzer, M.E., Holtzer, A.: Biopolymers 31 (1991) 489.
91S8 Sugawara, T., Kuwajima, K., Sugai, S.: Biochemistry 30 (1991) 2698.
91T1 Takahashi, N., Koseki, T., Doi, E., Hirose, M.: J. Biochem. 109 (1991) 846.
91T2 Tamura, A., Kimura, K., Takahara, H., Akasaka, K.: Biochemistry 30 (1991) 11307.
91T3 Tanaka, A.: Agr. Biol. Chem. 55 (1991) 2773.
91T4 Tanaka, A.: Netsu Sokutei 18 (1991) 77.
91T5 Teschke, C.M., Kim, J., Song, T., Park, S., Park, C., Randall, L.L.: J. Biol. Chem. 266 (1991) 11789.
91U1 Ueda, T., Yamada, H., Sakamoto, N., Abe, Y., Kawano, K., Terada, Y., Imoto, T.: J. Biochem. 110 (1991) 719.
91U2 Uedaira, H., Kidokoro, S.: Thermochim. Acta 183 (1991) 323.
91V1 VanOsdol, W.W., Mayorga, O.L., Freire, E.: Biophys. J. 59 (1991) 48.
91V2 Volkin, D.B., Staubli, A., Langer, R., Klibanov, A.M.: Biotechnol. Bioengng. 37 (1991) 843.
91V3 Vriend, G., Berendsen, H.J.C., van der Zee, J.R., van den Burg, B., Venema, G., Eijsink, G.H.: Protein Engng. 4 (1991) 941.
91W1 Walsh, M.T., Sen, A.C., Chakrabarti B.: J. Biol. Chem. 266 (1991) 20079.
91W2 Wedler, F.C., McLean, M.A.: Biochim. Biophys. Acta: 1076 (1991) 161.
91W3 Windsor, W.T., Syto, R., Le, H.V., Trotta, P.P.: Biochemistry 30 (1990) 1259.
91X   Xie, D., Bhakuni, V., Freire, E.: Biochemistry 30 (1991) 10673.
91Y1 Yamasaki, M., Yano, H., Aoki, K.: Int. J. Biol. Macromol. 13 (1991) 322.
91Y2 Yutani, K., Hayashi, S., Sugisaki, Y., Ogasahara, K.: Proteins: Structure, Function, and Genetics 9 (1991) 90.
91Z1 Zabin, H.B., Terwilliger, T.C.: J. Mol. Biol. 219 (1991) 257.
91Z2 Zhang, X.-J., Baase, W.A., Matthews, B.W.: Biochemistry 30 (1991) 2012.
92A1 Ahmad, F., Yadav, S., Taneja, S.: Biochem. J. 287 (1992) 481.
92A2 Alexander, P., Fahnestock, S., Lee, T., Orban, J., Bryan, P.: Biochemistry 31 (1992) 3597.
92A3 Alexander, P., Orban, J., Bryan, P.: Biochemistry 31 (1992) 7243.
92A4 Arriaga, P., Menendez, M., Villacorta, J.M., Laynez, J.: Biochemistry 31 (1992) 6603.
92A5 Azuaga, A.I., Galisteo, M.L., Mayorga, O.L., Cortijo, M., Mateo, P.L.: FEBS Lett. 309 (1992) 258.
92B1 Baker, J.O., Tatsumoto, K., Grohmann, K., Woodward, J., Wichert, J.M., Shoemaker, S.P., Himmel, M.E.: Appl. Biochem. Biotechnol. 34/35 (1992) 217.
92B2 Banik, U., Saha, R., Mandal, N.C., Bhattacharyya, B., Roy, S.: Eur. J. Biochem. 206 (1992) 15.

92B3   Bell, J.A., Becktel, W.J., Sauer, U., Baase, W.A., Matthews, B.W.: Biochemistry 31 (1992) 3590.

92B4   Betz, S.F., Pielak, G.J.: Biochemistry 31 (1992) 12337.

92B5   Brems, D.N., Brown, P.L., Bryant, C., Chance, R.E., Green, L.K., Long, H.B., Miller, A.A., Millican, R., Shields, J.E., Frank, B.H,: Protein Engng. 5 (1992) 519.

92B6   Briggs, M.S., Roder, H.: Proc. Natl. Acad. Sci. USA 89 (1992) 2017.

92B7   Bryan, P., Alexander, P., Strausberg, S., Schwarz, F., Lan, W., Gilliland, G., Gallagher, D.T.: Biochemistry 31 (1992) 4937.

92B8   Bryant, C., Strohl, M., Green, L.K., Long, H.B., Alter, L.A., Pekar, A.H., Chance, R.E., Brems, D.N.: Biochemistry 31 (1992) 5692.

92B9   Burova, T.V., Grinberg, N.V., Grinberg, V.Ya., Tolstoguzov, V.B., Schlesier, B., Müntz, K.: Int. J. Biol. Macromol. 14 (1992) 2.

92C1   Chen, B.-L., Baase, A., Nicholson, H., Schellman, J.A.: Biochemistry 31 (1992) 1464.

92C2   Chen, X., Rambo, R., Matthews, C.R.: Biochemistry 31 (1992) 2219.

92C3   Cladera, J., Galisteo, M.L., Sabes, M., Mateo, P.L., Padros, E.: Eur. J. Biochem. 207 (1992) 581.

92C4   Cooper, A., Eyles, S.J., Radford, S.E., Dobson, C.M.: J. Mol. Biol. 225 (1992) 939.

92D1   Derst, C., Henseling, J., Röhm, K.H.: Protein Engng. 5 (1992) 785.

92D2   Serrano, V.S.De, Castellino, F.J.: Biochemistry 31 (1992) 3326.

92D3   Dixon, M.M., Nicholson, H., Shewchuk, L., Baase, W.A., Matthews, B.W.: J. Mol. Biol. 227 (1992) 917.

92E1   Eder, J., Wilmanns, M.: Biochemistry 31 (1992) 4437.

92E2   Eriksson, A.E., Baase, W.A., Wozniak, J.A., Matthews, B.W.: Nature 355 (1992) 371.

92E3   Eriksson, A.E., Baase, W.A., Zhang, X.-J., Heinz, D.W., Blaber, M., Baldwin, E.P., Matthews, B.W.: Science 256 (1992) 178.

92E4   Eynard, L., Iametti, S., Relkin, P., Bonomi, F.: J. Agr. Food Chem. 40 (1992) 1731.

92F1   Fahnestock, M.L., Tamir, I., Narhi, L., Bjorkman, P.J.: Science 258 (1992) 1658.

92F2   Fernando, T., Royer, C.A.: Biochemistry 31 (1992) 6683.

92F3   Freire, E., Murphy, K.P., Sanchez-Ruiz, J.M., Galisteo, M.L., Privalov, P.L.: Biochemistry 31 (1992) 250.

92F4   Fu, L., Freire, E.: Proc. Natl. Acad. Sci. USA 89 (1992) 9335.

92F5   Fujita, Y., Noda, Y.: Int. J. Peptide Protein Res. 40 (1992) 103.

92G1   Gloss, L.M., Planas, A., Kirsch, J.F.: Biochemistry 31 (1992) 32.

92G2   Goto, Y., Hagihara, Y.: Biochemistry 31 (1992) 732.

92G3   Grant, S.K., Deckman, I.C., Culp, J.S., Minnich, M.D., Brooks, I.S., Hensley, P., Debouck, C., Meek, T.D.: Biochemistry 31 (1992) 9491.

92G4   Green, S.M., Meeker, A.K., Shortle, D.: Biochemistry 31 (1992) 5717.

92G5   Griko, Y.V., Privalov, P.L.: Biochemistry 31 (1992) 8810.

92G6   Griko, Y.V., Rogov, V.V., Privalov, P.L.: Biochemistry 31 (1992) 12701.

92H1   Hagihara, Y., Kataoka, M., Aimoto, S., Goto, Y.: Biochemistry 31 (1992) 11908.

92H2   Heinz, D.W., Baase, W.A., Matthews, B.W.: Proc. Natl. Acad. Sci. USA 89 (1992) 3751.

92H3   Herning T., Yutani, K., Inaka, K., Kuroki, R., Matsushima, M., Kikuchi, M.: Biochemistry 31 (1992) 7077.

92H4   Hickey, D.R., McLendon, G., Sherman, F.: in: Stability of Protein Pharmaceuticals, Part B, ed. by T.J. Ahern and M.C. Manning, Plenum Press New York, 1992, p. 183.

92H5   Horovitz, A., Fersht, A.R.: J. Mol. Biol. 224 (1992) 733.

92H6   Horovitz, A., Matthews, J.M., Fersht, A.R.: J. Nol. Biol. 227 (1992) 560.

92H7   Hu, C.-Q, Kitamura, S., Tanaka, A., Sturtevant, J.M.: Biochemistry 31 (1992) 1643.

92H8   Hu, C.-Q., Sturtevant, J.M., Thomson, J.A., Erickson, R.E., Pace, C.N.: Biochemistry 31 (1992) 4876.

92H9   Hu, C.-Q., Sturtevant, J.M.: J. Phys. Chem. 96 (1992) 4052.

92H10   Hurley, J.H., Baase, W.A., Matthews, B.W.: J. Mol. Biol. 224 (1992) 1143.

92I1   Ikeguchi, M., Sugai, S., Fujino, M., Sugawara, T., Kuwajima, K.: Biochemistry 31 (1992) 12695.

92I2   Inoue, M., Yamada, H., Hashimoto, Y., Yasukochi, T., Hamaguchi, K., Miki, T., Horiuchi, T., Imoto, T.: Biochemistry 31 (1992) 8816.

92I3   Inoue, M., Yamada, H., Yasukochi, T., Kuroki, R., Miki, T., Horiuchi, T., Imoto, T.: Biochemistry 31 (1992) 5545.

92J1   Johnson, C.M., Cooper, A., Stockley, P.G.: Biochemistry 31 (1992) 9717.

92J2   Jongh, H.H.J.de, Killian, J.A., de Kruiff, B.: Biochemistry 31 (1992) 1636.

92K0   Kahn, T.W., Sturtevant, J.M., Engelmann, D.M.: Biochemistry 31 (1992) 8829.

92K1   Kanaya, E., Kikuchi, M.: J. Biol. Chem. 267 (1992) 15111.

92K2   Kato, A., Tanimoto, S., Muraki, Y., Kobayashi, K., Kumagai, I.: Biosci. Biotech. Biochem. 56 (1992) 1424.

92K3   Kiefhaber, T., Schmid, F.X.: J. Mol. Biol. 224 (1992) 231.

92K4   Kiefhaber, T., Schmid, F.X., Willaert, K., Engelborghs, Y., Chaffotte, A.: Protein Sci. 1 (1992) 1162.

92K5   Kiefhaber, T., Grunert H.-P., Hahn, U., Schmid, F.X.: Proteins: Structure, Function, and Genetics 12 (1992) 171.

92K6   Kimura, S., Oda, Y., Nakai, T., Katayanagi, K., Kitakuni, E., Nakai, C., Nakamura, H., Ikehara, M., Kanaya, S.: Eur. J. Biochem. 206 (1992) 337.

92K7   Kimura, S., Kanaya, S., Nakamura, H.: J. Biol. Chem. 267 (1992) 22014.

92K8   Kimura, S., Nakamura, H., Hashimoto, T., Oobatake M., Kanaya, S.: J. Biol. Chem. 267 (1992) 21535.

92K9   Klump, H., DiRuggiero, J., Kessel, M., Park, J.-B., Adams, M.W.W., Robb, F.T.: J. Biol. Chem. 267 (1992) 22681.

92K10   Kuroda, Y., Kidokoro, S., Wada, A.: J. Mol. Biol. 223 (1992) 1139.

92K11   Kuroki, R., Inaka, K., Taniyama, Y., Kidokoro, S., Matsushima, M., Kikuchi, M., Yutani, K.: Biochemistry 31 (1992) 8323.

92K12   Kuroki, R., Kawakita, S., Nakamura, H., Yutani, K.: Proc. Natl. Acad. Sci. USA 89 (1992) 6803.

92L1   Ladbury, J.E., Hu, C.-Q., Sturtevant, J.M.: Biochemistry 31 (1992) 10699.

92L2   Levitsky, D.I., Shnyrov, V.L., Khvorov, N.V., Bukatina, A.E., Vedenkina, N.S., Permyakov, E.A., Nikolaeva, O.P., Poglazov, B.F.: Eur. J. Biochem. 209 (1992) 829.

92L3   Lim, W.K., Brouillette, C. Hardman, J.K.: Arch. Biochem. Biophys. 292 (1992) 34.

92L4   Lim, W.A., Farruggio, D.C., Sauer, R.T.: Biochemistry 31 (1992) 4324.

92L5   Loewenthal, R., Sancho, J., Fersht, A.R.: J. Mol. Biol. 224 (1992) 759.

92L6   Lu, J., Baase, W.A., Muchmore, D.C., Dahlquist, F.W.: Biochemistry 31 (1992) 7765.

92M1   Matouschek, A., Serrano, L., Fersht, A.R.: J. Mol. Biol. 224 (1992) 819.

92M2   McPhie, P., Shrager, R.I.: Arch. Biochem. Biophys.: 293 (1992) 46.

92M3   Mendel, D., Ellman, J.A., Chang, Z., Veenstra, D.L., Kollman, P.A., Schultz, P.G.: Science 256 (1992) 1798.

92M4   Meiering, E.M., Serrano, L., Fersht, A.R.: J. Mol. Biol. 225 (1992) 585.

92M5   Misselwitz, R., Kraft, R., Kostka, S., Fabian, H., Welfle, K., Pfeil, W., Welfle, H.: Int. J. Biol. Macromol. 14 (1992) 107.

92M6   Muccio, D.D., Waterhous, D.V., Fish, F., Brouillette, C.G.: Biochemistry 31 (1992) 5560.

92M7   Mücke, M., Schmid, F.X.: Biochemistry 31 (1992) 7848.

92M8   Mutombo, K., Michels, B., Ott, H., Cerf, R., Witz, J.: Eur. J. Biophys. 21 (1992) 77.

92N1   Newbold, R.J., Hewson, R., Whitford, D.: FEBS Lett. 314 (1992) 419.

92N2   Nicholson, H., Tronrud, D.E., Becktel, W.J., Matthews, B.W.: Biopolymers 32 (1992) 1431.

92P1   Pace, C.N., Laurents, D.V., Erickson, R.E.: Biochemistry 31 (1992) 2728.

92P2   Perry, K.M., Pookanjanatavip, M., Zhao, J., Santi, D.V., Stroud, R.M.: Protein Sci. 1 (1992) 796.

92P3   Plaza del Pino, I.M., Pace, C.N., Freire, E.: Biochemistry 31 (1992) 11196.

92P4   Procyk, R., Medved, L., Engelke, K.J., Kudryuk, B., Blombäck, B.: Biochemistry 31 (1992) 2273.

92Q   Qiu, Z.-H., Yu, L., Yu, C.-A.: Biochemistry 31 (1992) 3297.

92R1   Radford, S.E., Buck, M., Topping, K.D., Dobson, C.M., Evans, P.A.: Proteins: Structure, Function, and Genetics 14 (1992) 237.

92R2   Renner, M., Hinz, H.-J., Scharf, M., Engels, J.W.: J. Mol. Biol 223 (1992) 769.
92R3   Rudolph, R., Siebendritt R., Kiefhaber T.: Protein Sci. 1 (1992) 654.
92R4   Ruiz-Sanz, J., Ruiz-Cabello, J., Lopez-Mayorga, O., Cortijo, M., Mateo, P.L.: Eur. Biophys. J. 21 (1992) 169.
92S1   Sancho, J., Serrano, L., Fersht, A.R.: Biochemistry 31 (1992) 2253.
92S2   Santoro, M.M., Bolen, D.W.: Biochemistry 31 (1992) 4901.
92S3   Santoro, M.M., Liu Y., Khan, S.M.A., Hou L.-X., Bolen, D.W.: Biochemistry 31 (1992) 5278.
92S4   Sauer, U.H., Dao-pin, S., Matthews, B.W.: J. Biol. Chem. 267 (1992) 2393.
92S5   Schejter, A., Luntz, T.L., Koshy, T.I., Margoliash, E.: Biochemistry 31 (1992) 8336.
92S6   Schultz, D.A., Baldwin, R.L.: Protein Sci. 1 (1992) 910.
92S7   Sen, A.C., Walsh, M.T., Chakrabarti, B.: J. Biol. Chem. 267 (1992) 11898.
92S8   Serrano, L., Kellis Jr., J.T., Cann, P., Matouschek, A., Fersht, A.R.: J. Mol. Biol. 224 (1992) 783.
92S9   Serrano, L., Matouschek, A., Fersht, A.R.: J. Mol. Biol. 224 (1992) 805.
92S10 Serrano, L., Sancho, J., Hirshberg, M., Ferst. A.R.: J. Mol. Biol. 227 (1992) 544.
92S11 Shirley, B.A., Stanssens, P., Hahn, U., Pace, C.N.: Biochemistry 31 (1992) 725.
92S12 Shnyrov, V.L., Lubsandorzhieva, V.C., Zhadan, G.G., Permyakov, E.A.: Biochem. Internat. 26 (1992) 211.
92S13 Sondek, J., Shortle, D.: Proteins: Structure, Function, and Genetics 13 (1992) 132.
92S14 Straume, M., Freire, E.: Analyt. Biochem. 203 (1992) 259.
92T1   Taniyama, Y., Ogasahara, K., Yutani, K., Kikuchi, M.: J. Biol. Chem. 267 (1992) 4619.
92T2   Texter, F.L., Spencer, D.B., Rosenstein, R., Matthews, C.R.: Biochemistry 31 (1992) 5687.
92T3   Timm, D.E., Neet, K.E.: Protein Sci. 1 (1992) 236.
92W1   Welfle, K., Pfeil, W., Misselwitz, R., Gerlach, D., Welfle H.: Int. J. Biol. Macromol. 14 (1992) 19.
92W2   Welfle, H., Misselwitz, R., Welfle, K., Groch, N., Heinemann, U.: Eur. J. Biophys. 204 (1992) 1049.
92W3   Williamson, G., Belshaw, N.J., Noel, T.R., Ring, S.G., Williamson, M.P.: Eur. J. Biochem. 207 (1992) 661.
92Y1   Yamasaki, M., Yano, H., Aoki, K.: Int. J. Biol. Macromol. 14 (1992) 305.
92Y2   Yutani, K., Ogasahara K., Kuwajima, K.: J. Mol. Biol. 228 (1992) 347.
92Z1   Zerovnik, E., Lohner, K., Jerala, R., Laggner, P., Turk, V.: Eur. J. Biochem. 210 (1992) 217.
92Z2   Zhang, X.-J., Baase, W.A., Matthews, B.W.: Protein Sci. 1 (1992) 761.
92Z3   Zhang, J., Liu, Z.-P., Jones, T.A., Gierasch, L.M., Sambrook, J.F.: Proteins: Structure, Function, and Genetics 13 (1992) 87.
92Z4   Zhou, N.E., Kay, C.M., Hodges, R.S.: J. Biol. Chem. 267 (1992) 2664.
92Z5   Zolkiewski, M., Ginsburg, A.: Biochemistry 31 (1992) 11991.
93A1   Ahmad, F., in: Thermostability of Enzymes, ed. by M.N. Gupta, Narosa Publishing House, New Delhi, 1993, p. 96.
93A2   Anderson, D.E., Hurley, J.H., Nicholson, H., Baase, W.A., Matthews, B.W.: Protein Sci. 2 (1993) 1285.
93A3   Auld, D.S., Young, G.B., Saunders, A.J., Doyle, D.F., Betz, S.F., Pielak, G.J.: Protein Sci. 2 (1993) 2187.
93B1   Baldwin, E.P., Hajiseyedjavadi, O., Baase, W.A., Matthews, B.W.: Science 262 (1993) 1715.
93B2   Barrick, D., Baldwin, R.L.: Biochemistry 32 (1993) 3790.
93B3   Beghin, V., Bizot, H., Audebrand, M., Lefebvre, J., Libouga, D.G., Douillard, R.: Int. J. Biol. Macromol. 15 (1993) 195.
93B4   Blaber, M., Lindstrom, J.D., Gassner, N., Xu, J., Heinz, D.W., Matthews, B.W.: Biochemistry 32 (1993) 11363.
93B5   Blaber, M., Zhang, X., Matthews, B.W.: Science 260 (1990) 1637.
93B6   Bowler, B.E., May, K., Zaragoza, T., York, P., Dong, A., Caughey, W.S.: Biochemistry 32 (1993) 183.
93B7   Bryant, C., Spencer, D.B., Miller, A., Bakaysa D.L., McCune, K.S., Maple, S.R., Pekar, A.H., Brems, D.N.: Biochemistry 32 (1993) 8075.

93B8   Burke, C.J., Volkin, D.B., Mach, H., Middaugh, C.R.: Biochemistry 32 (1993) 6419.

93C1   Caffrey, M.S., Cusanovich, M.A.: Arch. Biochem. Biophys. 304 (1993) 205.

93C2   Cheng, X., Gonzalez, M.L., Lee, J.C.: Biochemistry 32 (1993) 8130.

93C3   Chrunyk, B.A., Wetzel, R.: Protein Engng. 6 (1993) 733.

93C4   Chrunyk, B.A., Evans, J., Lillquist, J., Young, P., Wetzel, R.: J. Biol. Chem. 268 (1993) 18053.

93C5   Chun, S.-Y., Strobel, S., Bassford Jr., P., Randall, L.I.: J. Biol. Chem. 268 (1993) 20855.

93C6   Clark, A.C., Sinclair, J.F., Baldwin, T.O.: J. Biol. Chem. 268 (1993) 10773.

93C7   Clarke, J., Fersht, A.R.: Biochemistry 32 (1993) 4322.

93D1   Damschun, G., Damaschun, H., Gast, K., Misselwitz, R., Müller, J.J., Pfeil, W., Zirwer, D.: Biochemistry 32 (1993) 7739.

93D2   DeFelippis, M.R., Alter, L.A., Pekar, A.H., Havel, H.A., Brems, D.N.: Biochemistry 32 (1993) 1555.

93D3   DeKoster, G.T., Robertson, A.D., Stock, A.M., Stock, J.B.: in: Techniques in Protein Chemistry IV, ed. by R.H. Angeletti, Academic Press, San Diego, 1993, p. 533.

93D4   Diaz, J.F., Menéndez, M., Andreu, J.M.: Biochemistry 32 (1993) 10067.

93D5   DiRuggiero, J., Robb, F.T., Jagus, R., Klump, H.H., Borges, K.M., Kessel, M., Mai, X., Adams, M.W.W.: J. Biol. Chem. 268 (1993) 17767.

93E1   Egan, D.A., Logan, T.M., Liang, H., Matayoshi, E., Fesik, S. W., Holzman, T.F.: Biochemistry 32 (1993) 1920.

93E2   Eriksson, A.E., Baase, W.A., Matthews, B.W.: J. Mol. Biol. 229 (1993) 747.

93F1   Fabian, H., Schultz, C., Naumann, D., Landt, O., Hahn, U., Saenger, W.: J. Mol. Biol. 232 (1993) 967.

93F2   Fazili, K.M., Mir, M.M., Qasim, M.A.: Biochem. Mol. Biol. Internat. 31 (1993) 807.

93F3   Filimonov, V.V., Prieto, J., Martinez, J.C., Bruix, M., Mateo, P.L., Serrano, L.: Biochemistry 32 (1993) 12906.

93G1   Galisteo, M.L., King, J.: Biophys. J. 65 (1993) 227.

93G2   Galisteo, M.L., Sanchez-Ruiz, J.M.: Eur. Biophys. J. 22 (1993) 25.

93G3   Gekko, K., Yamagami, K., Kunori, Y., Ichihara, S., Kodama, M., Iwakura, M.: J. Biochem. 113 (1993) 74.

93G4   Ghadiri, M.R., Case, M.A.: Angew. Chem. 105 (1993) 1663.

93G5   Gittis, A.G., Stites, W.E., Lattman, E.E.: J. Mol. Biol. 232 (1993) 718.

93G6   Goldenberg, D.P., Bekeart, L.S., Laheru, D.A., Zhou, J.D.: Biochemistry 32 (1993) 2835.

93G7   Green, S.M., Shortle, D.: Biochemistry 32 (1993) 10131.

93H1   Hagihara, Y., Aimoto, S., Fink, A.L., Goto, Y.: J. Mol. Biol. 231 (1993) 180.

93H2   Handel, T.N., Williams, S.A., DeGrado, W.F.: Science 261 (1993) 879.

93H3   Hei, D.J., Clark, D.S.: Biotechnol. Bioengng. 42 (1993) 1245.

93H4   Heinz, D.W., Baase, W.A., Dahlquist, F.W., Matthews, B.W.: Nature 361 (1993) 561.

93H5   Hernandez-Arana, A., Rojo-Dominguez, A., Altamirano, M.M., Calcagno, M.L.: Biochemistry 32 (1993) 3644.

93H6   Hewson, R., Newbold, R.J., Whitford, D.: Protein Engng. 6 (1993) 953.

93H7   Hinz, H.-J., Steif, C., Vogl, T., Meyer, R., Renner, M., Ledermüller, R.: Pure Appl. Chem. 65 (1993) 947.

93H8   Housaindokht, M.R., Moosavi-Movahedi, A.A., Moghadasi, J., Jones, M.N.: Int. J. Biol. Macromol. 15 (1993) 337.

93H9   Hua, Q.X., Ladbury, J.E., Weiss, M.A.: Biochemistry 32 (1993) 1433.

93I1   Ishikawa, K., Nakamura, H., Morikawa, K., Kimura, S., Kanaya, S.: Biochemistry 32 (1993) 7136.

93I2   Ishikawa, K., Nakamura, H., Morikawa, K., Kanaya, S.: Biochemistry 32 (1993) 6171.

93I3   Iwakura, M., Jones, B.E., Falzone, C.J., Matthews, C.R.: Biochemistry 32 (1993) 13566.

93J1   Jackson, S.E., Moracci, M., elMasry, N., Johnson, C.M., Fersht, A.R.: Biochemistry 32 (1993) 11259.

93J2   Jennings, P.A., Wright, P.E.: Science 262 (1993) 892.
93J3   Jiang, N., Frieden, C.: Biochemistry 32 (1993) 11015.
93J4   Jones, C.M., Henry, E.R., Hu, Y., Chan C.-K., Luck, S.D., Bhuyan, A., Roder, H., Hofrichter, J., Eaton, W.A.: Proc. Natl. Acad. Sci. USA 90 (1993) 11860.
93K1   Kaarsholm, N.C., Norris, K., Jorgensen, R.J., Mikkelsen, J., Ludvigsen, S., Olsen, O.H., Sorensen, A.R., Havelund, S.: Biochemistry 32 (1993) 10773.
93K2   Kamtekar, S., Schiffer, J.M., Xiong, H., Babik, J.M., Hecht, M.H.: Science 262 (1993) 1680.
93K3   Khorasanizadeh, S., Peters, I.D., Butt, T.R., Roder, H.: Biochemistry 32 (1993) 7054.
93K4   Kim, C.A., Berg, J.M: Nature 362 (1993) 267.
93K5   Kim, K.-S., Fuchs, J.A., Woodward, C.K.: Biochemistry 32 (1993) 9600.
93K6   Kim, K.-S., Tao, F., Fuchs, J., Danishefsky, A.T., Housset, D., Wlodawer, A., Woodward, C.: Protein Sci. 2 (1993) 588.
93K7   Kim, K.-S., Woodward, C.: Biochemistry 32 (1993) 9609.
93K8   Kiss, R.S., Ryan, R.O., Hicks, L.D., Oikawa, K., Kay, C.M.: Biochemistry 32 (1993) 7872.
93K9   Koide, S., Dyson, H.J., Wright, P.E.: Biochemistry 32 (1993) 12299.
93K10   Konno, T., Morishima, I.: Biochim. Biophys. Acta 1162 (1993) 93.
93K11   Kranjc, Z., Lapanje, S.: unpublished data, cited in Kranjc, Z., Lapanje, S.: Int. J. Peptide Protein Res. 42 (1993) 320.
93L1   Ladbury, J.E., Wynn, R., Hellinga, H.W., Sturtevant, J.M.: Biochemistry 32 (1993) 7526.
93L2   Laderman, K.A., Davis, B.R., Krutzsch, H.C., Lewis, M.S., Griko, Y.V., Privalov, P.L., Anfinsen, C.B.: J. Biol. Chem. 268 (1993) 24394.
93L3   Lascu, I., Deville-Bonne, D., Glaser, P., Veron, M.: J. Biol. Chem. 268 (1993) 20268.
93L4   LeBihan, T., Gicquaud, C.: Biochem. Biophys. Res. Comun. 194 (1993) 1065.
93L5   Lin, L., Kallenbach, N.R.: Proc. 1993 Miami Bio/Technology Winter Sympos., Abstr. Protein Engng. 6 (1993) 18.
93L6   Lin, L., Pinker, R.J., Kallenbach, N.R.: Biochemistry 32 (1993) 12638.
93L7   Lin, L.-N., Mason, A.B., Woodworth, R.C., Brandts, J.F.: Biochem. J. 293 (1993) 517.
93L8   Loh, S.N., Prehoda, K.E., Wang, J., Markley, J.L: Biochemistry 32 (1993) 11022.
93L9   Luthra, M., Balasubramanian, D.: J. Biol. Chem. 268 (1993) 18119.
93M1   Mach, H., Ryan, J.A., Burke, C.J., Volkin, D.B., Middaugh, C.R.: Biochemistry 32 (1992) 7703.
93M2   Makarov, A.A., Protasevich, I.I., Kuznetsova, N.V., Fedorov, B.B., Korolev, S.V., Struminskaya, N.K., Bazhulina, N.P., Leshchinskaya, I.B., Hartley, R.W., Kirpichnikov, M.P., Yakovlev, G.I., Esipova, N.G.: J. Biomol. Struct. & Dynamics 10 (1993) 1047.
93M3   Makhatadze, G.I.: unpubl. data, cited in Ref. 93K6: Kim, K.-S., Tao, F., Fuchs, J., Danishefsky, A.T., Housset, D., Wlodawer, A., Woodward, C., Protein Sci. 2 (1993) 588.
93M4   Makhatadze, G.I., Kim, K.-S., Woodward, C., Privalov, P.L.: Protein Sci. 2 (1993) 2028.
93M5   Makhatadze, G.I., Privalov, P.L.: J. Mol. Biol. 232 (1993) 639.
93M6   Mann, C.J., Matthews, C.R.: Biochemistry 32 (1993) 5282.
93M7   Mårtensson, L.-G., Johnsson, B.-H.: Biochemistry 32 (1993) 224.
93M8   Matthews, S.J., Jandu, S.K., Leatherbarrow, R.J.: Biochemistry 32 (1993) 657.
93M9   Mayo, S.L., Baldwin, R.L.: Science 262 (1993) 873.
93M10   Mayr, L.M., Landt, O., Hahn, U., Schmid, F.X.: J. Mol. Biol. 231 (1993) 897.
93M11   Mayr, L.M., Schmid, F.X.: Biochemistry 32 (1993) 7994.
93M12   Merabet, E.K., Walker, M.C., Yuen, H.K., Sikorski, J.A.: Biochim. Biophys. Acta 1161 (1993) 272.
93M13   Milla, M.E., Brown, B.M., Sauer, R.T.: Protein Sci. 2 (1993) 2198.
93M14   Montgomery, D., Jordan, R., McMacken, R., Freire, E.: J. Mol. Biol. 232 (1993) 680.

93M15   Morjana, N.A., McKeone, B.J., Gilbert, H.F.: Proc. Natl. Acad. Sci. USA 90 (1993) 2107.

93M16   Morozova, L., Desmet, J., Joniau, M.: Eur. J. Biochem. 218 (1993) 303.

93M17   Muga, A., Gonzalez-Manas, J.M., Lakey, J.H., Pattus, F., Surewicz, W.K.: J. Biol. Chem. 268 (1993) 1553.

93M18   Myer,Y.P.: in: Techniques in Protein Chemistry IV, ed. by R.H. Angeletti, Ed., Academic Press, San Diego, 1993, p. 509.

93N1   Nakanao, T., Antonino, L.C., Fox, R.O., Fink, A.L.: Biochemistry 32 (1993) 2534.

93N2   Nitta, K., Tsuge, H., Iwamoto, H.: Int. J. Peptide Protein. Res. 41 (1993) 118.

93O1   Ogasahara, K., Matsushita, E., Yutani, K.: J. Mol. Biol. 234 (1993) 1197.

93O2   Osmark, P., Sørensen, P., Poulsen, F.M.: Biochemistry 32 (1993) 11007.

93O3   O'Shea, E.K., Lumb, K.J., Kim, P.S.: Current Biol. 3 (1993) 658.

93P1   Pappa, H.S., Cass, A.E.G.: Eur. J. Biochem. 212 (1993) 227.

93P2   Pessi, A., Bianchi, E., Crameri, A., Venturini, S., Tramontano, A., Sollazzo, M.: Nature 362 (1993) 367.

93P3   Pfeil, W.: Protein Sci. 2 (1993) 1497.

93P4   Pfeil, W., Nölting, B.O., Jung, C.: Biochemistry 32 (1993) 8856.

93P5   Pfeil, W., Ristau, O.: unpublished results (1993).

93P6   Pineda, T., Churchich, J.E.: J. Biol. Chem. 268 (1993) 20218.

93P7   Pinker, R.J., Rose, G.D., Kallenbach, N.R.: Proc. 1993 Miami Bio/Technology Winter Sympos., Abstr. Protein Engng. 6 (1993) 18.

93P8   Pinker, R., Lin L., Rose, G.D., Kallenbach, N.R.: Protein Sci. 2 (1993) 1099.

93P9   Pjura, P., Matsumura, M., Baase, W.A., Matthews, B.W.: Protein Sci. 2 (1993) 2217.

93P10   Pjura, P., McIntosh, L.P., Wozniak, J.A., Matthews, B.W.: Proteins: Structure, Function, and Genetics 15 (1993) 401.

93P11   Poklar, N., Vesnaver, G., Lapanje, S.: Biophys. Chem. 47 (1993) 143.

93P12   Privalov, P.L., Makhatadze, G.I.: J. Mol. Biol. 232 (1993) 660.

93R1   Robinson, C.R., Sligar, S.G.: Protein Sci. 2 (1993) 826.

93R2   Ryan, R.O., Oikawa, K., Kay, C.M.: J. Biol. Chem. 268 (1993) 1525.

93S0   Saab-Rincón, G., Froebe, C.L., Matthews, C.R.: Biochemistry 32 (1993) 13981.

93S1   Sandberg, W.S., Terwilliger, T.C.: Proc. Natl. Acad. Sci. USA 90 (1993) 8367.

93S2   Sanz J.M., Fersht, A.R.: Biochemistry 32 (1993) 13584.

93S3   Sanz, J.M., Garcia, J.L., Laynez, J., Usobiaga, P., Menendez, M.: J. Biol. Chem. 268 (1993) 6125.

93S4   Sarai, A., Uedaira, H., Morii, H., Yasukawa, T., Ogata, K., Nishimura, Y., Ishii, S.: Biochemistry 32 (1993) 7759.

93S5   Schreiber, G., Fersht, A.R.: Biochemistry 32 (1993) 11195.

93S6   Serrano, L., Day, A.G., Fersht, A.R., J.: Mol. Biol. 233 (1993) 305.

93S7   Shimizu, A., Ikeguchi, M., Sugai, S.: Biochemistry 32 (1993) 13198.

93S8   Solís-Mendiola, S., Rojo-Domínguez, A., Hernández-Arana, A.: Biochim. Biophys. Acta 1203 (1993) 121.

93S9   Staniforth, R.A., Burston, S.G., Smith, C.J., Jackson, G.S., Badcoe, I.G., Atkinson, T., Holbrook, J., Clarke, A.R.: Biochemistry 32 (1993) 3842.

93S10   Steif, C., Weber, P., Hinz, H.-J., Flossdorf, J., Cesareni, G., Kokkinidis, M.: Biochemistry 32 (1993) 3867.

93S11   Sumner, I.G., Harris, G.W., Taylor, M.A.J., Pickersgill, R.W., Owen, A.J., Goodenough, P.W.: Eur. J. Biochem. 214 (1993) 129.

93S12   Swint, L., Robertson, A.D.: Protein Sci. 2 (1993) 2037.

93T1   Tanaka, A., Flanagan, J.: Sturtevant, J.M., Protein Sci. 2 (1993) 567.

93T2   Tao, F., Fuchs, J.A., Woodward, C.: in: Techniques in Protein Chemistry IV, ed. by R. Angletti, Academic Press, San Diego, 1993, p. 512. data cited in Ref. 93K6: Kim, K.-S., Tao, F., Fuchs, J., Danishefsky, A.T., Housset, D., Wlodawer, A., Woodward, C.: Protein Sci. 2 (1993) 588.

93T3   Tao, F., Fuchs, J., Woodward, C.: in: Techniques in Protein Chemistry IV, ed. by R.H. Angeletti, Academic Press, San Diego, 1993, p. 549.

93T4   Thompson, K.S., Vinson, C.R., Freire, E.: Biochemistry 32 (1993) 5491.

93T5   Tissot, A., Diplomarbeit (1993), cited in: Oliveberg et al., Biochemistry 34 (1995) 9424.

93T6 Tsuji, T., Chrunyk, B.A., Chen, X., Matthews, C.R.: Biochemistry 32 (1993) 5566.

93T7 Tweedy, N.B., Nair, S.K., Paterno, S.A., Fierke, C.A., Christianson, D.W.: Biochemistry 32 (1993) 10944.

93U Ueda, T., Tamura, T., Maeda Y., Hashimoto, Y., Miki, T., Yamada, H., Imoto, T.: Protein Engng. 6 (1993) 183.

93V1 Van Dael, H., Haezebrouck, P., Morozova, L., Arico-Muendel, C., Dobson, C.M.: Biochemistry 32 (1993) 11886.

93V2 Volkin, D.B., Tsai, P.K., Dabora, J.M., Gress, J.O., Burke, C.J., Linhardt, R.J., Middaugh, C.R.: Arch. Biochem. Biophys. 300 (1993) 30.

93V3 Vuilleumier, S., Sancho, J., Loewenthal, R., Fersht, A.R.: Biochemistry 32 (1993) 10303.

93W1 Ward, L.D., Hammacher, A., Zhang, J.-G., Weinstock, J., Yasukawa, K., Morton, C.J., Norton, R.S., Simpson, R.J.: Protein Sci. 2 (1993) 1472.

93W2 Wang, Z.-Y., Freire, E., McCarty, R.E.: J. Biol. Chem. 268 (1993) 20785.

93W3 Welfle, H., Misselwitz, R., Welfle, K., Schindelin, H., Scholtz, A.S., Heinemann, U.: Eur. J. Biochem. 217 (1993) 849.

93W4 Woolfson D.N., Cooper, A., Harding, M.M., Williams, D.H., Evans, P.A.: J. Mol. Biol. 229 (1993) 502.

93W5 Wu, M.-L., Morgan, W.T.: Biochemistry 32 (1993) 7216.

93W6 Wunderlich, M., Jaenicke, R., Glockshuber, R.: J. Mol. Biol. 233 (1993) 559.

93W7 Wynn, R., Richards, F.M.: Protein Sci. 2 (1993) 395.

93W8 Wynn, R., Richards, F.M.: Biochemistry 32 (1993) 12922.

93Y1 Yamada, H., Ueda, T., Imoto, T.: J. Biochem. 114 (1993) 398.

93Y2 Yu, C.-A., Steidl, J.R., Yu, L.: Biochim. Biophys. Acta 736 (1993) 226.

93Z1 Zapun, A., Bardweel, J.C.A., Creighton, T.E.: Biochemistry 32 (1993) 5083.

93Z2 Zhang, J.-G., Reid, G.E., Moritz, R.L., Ward, L.D., Simpson, R.J.: Eur. J. Biochem. 217 (1993) 53.

93Z3 Zhang, T., Bertelsen, E., Benvegnu, D., Alber, T.: Biochemistry 32 (1993) 12311.

93Z4 Zhu, B.-Y., Zhou, N.E., Kay, C.M., Hodges, R.S.: Protein Sci. 2 (1993) 383.

93Z5 Zhuang, P., Yin, M., Holland, J.C., Peterson, C.B., Howell, E.E.: J. Biol. Chem. 268 (1993) 22672.

94A1 Ahmad, F., Taneja, S., Yadav, S., Haque, S.E.: J. Biochem. 115 (1994) 322.

94A2 Ahmad, Z., Ahmad, F.: Biochim. Biophys. Acta 1207 (1994) 223.

94B1 Bagel'ova, J., Antalik, M., Bona, M.: Biochem. J. 297 (1994) 99.

94B2 Barone, G., Del Vecchio, P., Fessas, D., Giancola, C., Graziano, G., Riccio, A: J. Thermal Analysis 41 (1994) 1263.

94B3 Barone, G., Del Vecchio, P., Fessas, D., Giancola, C., Graziano, G., Riccio, A: J. Thermal Analysis 41 (1994) 1357.

94B4 Barrick, D., Hughson, F.M., Baldwin, R.L.: J. Mol. Biol. 237 (1994) 588.

94B5 Beernink, P.T., Tolan, D.R.: Protein Sci. 3 (1994) 1383.

94B6 Blaber, M., Zhang X.-J., Lindstrom J.D., Pepiot, S.D., Baase, W.A., Matthews, B.W.: J. Mol. Biol. 235 (1994) 600.

94B7 Bottomley, S.P., Poopplewell, A.G., Scawen, M. Wan, T., Sutton, B.J., Gore, M.G.: Protein Engng. 7 (1994) 1463.

94B8 Bromberg, S., LiCata, V.J., Mallikarachchi, D., Allewell, N.M.: Protein Sci. 3 (1994) 1236.

94B9 Bulaj, G., Otlewski, J.: Eur. J. Biochem. 223 (1994) 939.

94C1 Carra, J.H., Anderson, E.A., Privalov, P.L.: Protein Sci. 3 (1994) 944.

94C2 Carra, J.H., Anderson, E.A., Privalov, P.L.: Protein Sci. 3 (1994) 952.

94C3 Carra, J.H., Anderson, E.A., Privalov, P.L.: Biochemistry 33 (1994) 10842.

94C4 Chakrabartty, A., Kortemme, T., Baldwin, R.L.: Protein Sci. 3 (1994) 843.

94C5 Chen, C.-H., Roth, L.G., MacColl, R., Berns, D.S.: Biophys. Chem. 50 (1994) 313.

94C6 Chen, J., Matthews, K.S.: Biochemistry 33 (1994) 8728.

94C7 Cohen, D.S., Pielak, G.J.: Protein Sci. 3 (1994) 1253.

94C8 Coll, P.M., Pérez, P., Villar, E., Shnyrov, V.L.: Biochem. Mol. Biol. Internat. 34 (1994) 1091.

94C9 Conejero-Lara, F., De Filippis, V., Fontana, A., Mateo, P.L.: FEBS Lett. 344 (1994) 154

94C10 Cooper, A., McAlpine, A., Stockley, P.G.: FEBS Lett. 348 (1994) 41.

94C11   Cornish, V.W., Kaplan, M.I., Veenstra, D.L., Kollman, P.A., Schultz, P.G.: Biochemistry 33 (1994) 12022.

94C12   Creighton, T.E., Shortle, D.: J. Mol. Biol. 242 (1994) 670.

94D1   Dabora, J.M., Marqusee, S.: Protein Sci. 3 (1994) 1401.

94D2   Damante, G., Tell, G., Leonardi, A., Fogolari, F., Bortolotti, N., Di Lauro, R., Formisano, S.: FEBS Lett. 354 (1994) 293.

94D3   De Prat Gay, G., Ruiz-Sanz, J., Fersht, A.R.: Biochemistry 33 (1994) 7964.

94D4   De Serrano, V.S., Castellino, F.J.: Biochemistry 33 (1994) 1340.

94D5   Dubus, A., Wilkin, J.-M., Raquet, X., Normark, S., Frère, J.-M.: Biochem. J. 301 (1994) 485.

94E1   Eftink, M.R.: Biophys. J. 66 (1994) 482.

94E2   Eftink, M.R., Helton, K.J., Beavers, A., Ramsay, G.D.: Biochemistry 33 (1994) 10220.

94E3   Elöve, G.A., Bhuyan, A.K., Roder, H.: Biochemistry 33 (1994) 6925.

94F1   Fabian, H., Schultz, C., Backmann, J., Hahn, U., Saenger, W., Mantsch, H.H., Naumann, D.: Biochemistry 33 (1994) 10725.

94F2   Ferrari, M.E., Lohman, T.M.: Biochemistry 33 (1994) 12896.

94F3   Frisch, C., Kolmar, H., Fritz, H.-J.: Biol. Chem. Hoppe-Seyler 375 (1994) 353.

94G1   Gay, G. de Prat, Johnson, C.M., Fersht, A.R.: Protein Engng. 7 (1994) 103.

94G2   Goward, C.R., Miller, J., Nicholls, D.J., Irons, L.I., Scawen, M.D., O'Brien, R., Chowdhry, B.Z.: Eur. J. Biochem. 224 (1994) 249.

94G3   Griko, Yu.V., Freire, E., Privalov, P.L.: Biochemistry 33 (1994) 1889.

94G4   Griko, Yu.V., Gittis, A., Lattman, E.E., Privalov, P.L.: J. Mol. Biol. 243 (1994) 93.

94G5   Griko, Yu.V., Kutyshenko, V.P.: Biophys. J. 67 (1994) 356.

94G6   Griko, Yu.V., Makhatadze, G.I., Privalov, P.L., Hartley, R.W.: Protein Sci. 3 (1994) 669.

94G7   Gross, M., Furter-Graves, E.M., Wallimann, T., Eppenberger, H.M., Furter, R.: Protein Sci. 3 (1994) 1058.

94H1   Hadden, J.M., Bloemendal, M., Haris, P.I., van Stokkum, I.H.M., Chapman, D., Srai, S.K.S.: FEBS Lett. 350 (1994) 235.

94H2   Hagihara, Y., Tan, Y., Goto, Y.: J. Mol. Biol. 237 (1994) 336.

94H3   Hagihara, Y., Oobatake, M., Goto, Y.: Protein Sci. 3 (1994) 1418.

94H4   Haltia, T., Semo, N., Arrondo, J.L.R., Goñi, F.M., Freire, E.: Biochemistry 33 (1994) 9731.

94H5   Hamada, D., Kidokoro, S.-I., Fukada, H., Takahashi, K., Goto, Y.: Proc. Natl. Acad. Sci. USA 91 (1994) 10325.

94H6   Haruki, M., Noguchi, E., Nakai, C., Liu, Y.-Y., Oobatake, M., Itaya, M., Kanaya, S.: Eur. J. Biochem. 220 (1994) 623.

94H7   Heinz, D.W., Baase, W.A., Zhang, X.-J., Blaber, M., Dahlquist, F.W., Matthews, B.W.: J. Mol. Biol. 236 (1994) 869.

94H8   Hinz, H.-J., Vogl, T., Meyer, R.: Biophys. Chem. 52 (1994) 275.

94H9   Horvath, L.A., Sturtevant, J.M., Prestegard, J.H.: Protein Sci. 3 (1994) 103.

94H10   Huang, Y., Beeser, S., Guillemette, J.G., Storms, R.K., Kornblatt, J.A.: Eur. J. Biochem. 223 (1994) 155.

94H11   Hynes, T.R., Hodel, A., Fox, R.O.: Biochemistry 33 (1994) 5021.

94I   Ishii, Y.: Eur. J. Biochem. 221 (1994) 705.

94J1   Jackson, S.E., Fersht, A.R.: Biochemistry 33 (1994) 13880.

94J2   Jasanoff, A., Davis, B., Fersht, A.R.: Biochemistry 33 (1994) 6350.

94J3   Jerala, R., Zerovnik, E., Lohner, K., Turk, V.: Protein Engng. 7 (1994) 977.

94J4   Jung, C., Pfeil, W., Köpke, K., Schulze, H., Ristau, O.: in Cytochrome P450, Biochemistry, Biophysics and Molecular Biology, ed. by M.C. Lechner, John Libbey Eurotext, Paris, 1994, p. 543.

94K1   Katakura, Y., Totsuka, M., Ametani, A., Kaminogawa, S.: Biochim. Biophys. Acta 1207 (1994) 58.

94K2   Keitel, T., Meldgaard, M., Heinemann, U.: Eur. J. Biochem. 222 (1994) 203.

94K3   King, L.: Arch. Biochem. Biophys. 311 (1994) 251.

94K4   Kitakuni, E., Kuroda, Y., Oobatake, M., Tanaka, T., Nakamura, H.: Protein Sci. 3 (1994) 831.

94K5   Klump, H.H., Adams, M.W.W., Robb, F.T.: Pure Appl. Chem. 66 (1994) 485.

94K6   Kojima, M., Mizukoshi, T., Miyano, H., Suzuki, E., Tanokura, M., Takahashi, K.: FEBS Lett. 351 (1994) 389.

94K7   Kolmar, H., Frisch, C., Kleemann, G., Götze, K., Stevens, F.J., Fritz, H.-J.: Biol. Chem. Hoppe-Seyler 375 (1994) 61.

94K8   Komar-Panicucci, S., Weis, D., Bakker, G., Qiao, T., Sherman, F., McLendon, G.: Biochemistry 33 (1994) 10556.

94K9   Koshy, T.I., Luntz, T.L., Plotkin, B., Schejter, A., Margoliash, E.: Biochem. J. 299 (1994) 347.

94K10   Kreimer, D.I., Szosenfogel, R., Goldfarb, D., Silman, I., Weiner, L.: Proc. Natl. Acad. Sci USA 91 (1994) 12145.

94K11   Kuszewski, J., Clore, G.M., Gronenborn, A.M.: Protein Sci. 3 (1994) 1945.

94K12   Kwon, K.-S., Kim, J., Shin, H.S., Yu, M.-H.: J. Biol. Chem. 269 (1994) 9627.

94L1   Ladbury, J.E., Kishore, N., Hellinga, H.W., Wynn, R., Sturtevant, J.M.: Biochemistry 33 (1994) 3688.

94L2   Lawrence, D.A., Olson, S.T., Palaniappan, S., Ginsburg, D.: Biochemistry 33 (1994) 3643.

94L3   Lentz, B.R., Zhou, C.-M., Wu, J.R.: Biochemistry 33 (1994) 5460.

94L4   Liggins, J.R., Sherman, F., Mathews, A.J., Nall, B.T.: Biochemistry 33 (1994) 9209.

94L5   Lim, W.A., Fox, R.O., Richards, F.M.: Protein Sci. 3 (1994) 1261.

94L6   Lin, L., Pinker, R.J., Phillips, G.N., Kallenbach, N.R.: Protein Sci. 3 (1994) 1430.

94L7   Lin, L.-N., Mason, A.B., Woodworth, R.C., Brandts, J.F.: Biochemistry 33 (1994) 1881.

94L8   Lin, T.-Y., Timasheff, S.N.: Biochemistry 33 (1994) 12695.

94M1   Makarov, A.A., Protasevich, I.I., Grishina, I.B., Lobachov, V.M., Bazhulina, N.P., Esipova, N.G.: Molekularnaya Biol. 28 (1994) 1346.

94M2   Makarov, A.A., Protasevich, I.I., Lobachov, V.M., Kirpichnikov, G.I., Yakovlev, G.I., Gilli, R.M., Briand, C.M., Hartley, R.W.: FEBS Lett. 354 (1994) 251.

94M3   Makhatadze, G.I., Clore, G.M., Gronenborn, A.M., Privalov, P.L.: Biochemistry 33 (1994) 9327.

94M4   Makhatadze, G.I., Marahiel, M.: Protein Sci. 3 (1994) 2144.

94M5   Markovic-Housley, Z., Cooper, A., Lustig, A., Flükiger, K., Stolz,, B., Erni, B.: Biochemistry 33 (1994) 10977.

94M6   Marqusee, S., Sauer, R.T.: Protein Sci. 3 (1994) 2217.

94M7   Martinez, J.C., Harrous, M.E., Filimonov, V.V., Mateo, P.L., Fersht, A.R.: Biochemistry 33 (1994) 3919.

94M8   elMasry, N.F., Fersht, A.R.: Protein Engng. 7 (1994) 777.

94M9   Matouschek, A., Matthews, J.M., Johnson, C.M., Fersht, A.: Protein Engng. 7 (1994) 1089.

94M10   Mayr, E.-M., Jaenicke, R., Glockshuber, R.: J. Mol. Biol. 235 (1994) 84.

94M11   Mayr, L.M., Willbold, D., Landt, O., Schmid, F.X.: Protein Sci. 3 (1994) 327.

94M12   Merino, J.M., Moller, J.V., Gutiérrez-Merino, C.: FEBS Lett. 343 (1994) 155.

94M13   Milla, M.E., Sauer, R.T.: Biochemistry 33 (1994) 1125.

94M14   Milla, M.E., Brown, B.M., Sauer, R.T.: Structural. Biol. 1 (1994) 518.

94M15   Minor Jr., D.L., Kim, P.S.: Nature 367 (1994) 660.

94M16   Minor Jr., D.L., Kim, P.S.: Nature 371 (1994) 264.

94M17   Misselwitz, R., Welfle, K., Welfle, H.: Int. J. Biol. Macromol. 16 (1994) 187.

94M18   Monera, O.D., Kay, C.M., Hodges, R.S.: Biochemistry 33 (1994) 3862.

94M19   Monera, O.D., Kay, C.M., Hodges, R.S.: Protein Sci. 3 (1994) 1984.

94M20   Mücke, M., Schmid, F.X.: J. Mol. Biol. 239 (1994) 713.

94M21   Mücke, M., Schmid, F.X.: Biochemistry 33 (1994) 14608.

94M22   Muñoz, V., Lopez, E.M., Jager, M., Serrano, L.: Biochemistry 33 (1994) 5858.

94M23   Munson, M., O'Brien, R., Sturtevant, J.M., Regan, L.: Protein Sci. 3 (1994) 2015.

94N1   Nishii, I., Kataoka, M., Tokunaga, F., Goto, Y.: Biochemistry 33 (1994) 4903.

94N2   Nobbs, T.J., Cortes, A., Gelpi, J.L., Holbrook, J.J., Atkinson, T., Scawen, M.D., Nicholls, D.J.: Biochem. J. 300 (1994) 491.

94N3   Nowak, U.K., Cooper, A., Saunders, D., Smith, R.A.G., Dobson, C.M.: Biochemistry 33 (1994) 2951.

94O1   Oliveberg, M., Vuilleumier, S., Fersht, A.R.: Biochemistry 33 (1994) 8826.

94O2   Otamiri, M., Adlercreutz, P., Mattiason, B.: Biotechnol. Bioengng. 44 (1994) 73.

94O3   Otzen, D.E., Itzhaki, L.S., ElMasry, N.F., Jackson, S.E., Fersht, A.R.: Proc. Natl. Acad. Sci. USA 91 (1994) 10422.

94P1   Paredi, M.E., Tomas, M.C., Crupkin, M., Anon, M.C.: J. Agric. Food Chem. 42 (1994) 873.

94P2   Pineda, T., Churchich, J.E.: Biochim. Biophys. Acta 1207 (1994) 173.

94P3   Poklar, N., Vesnaver, G., Lapanje, S.: J. Protein Chem. 13 (1994) 323.

94P4   Potekhin, S.A., Medvedkin, V.N., Kashparov, I.A., Venyaminov, S.Yu.: Protein Engng. 7 (1994) 1097.

94P5   Protasova, N.Yu., Kireeva, M.L., Murzina, N.V., Murzin, A.G., Uversky, V.N., Gryaznova, O.I., Gudkov, A.T.: Protein Engng. 7 (1994) 1373.

94R0   Rahil, J., Pratt, R.F.: Biochemistry 33 (1994) 116.

94R1   Ramsay, G., Eftink, M.R.: Biophys. J. 66 (1994) 516.

94R2   Regan, L., Rockwell, A., Wasserman, Z., DeGrado, W.: Protein Sci. 3 (1994) 2419.

94R3   Rialdi, G., Battistel, E.: Proteins: Structure, Function, and Genetics 19 (1994) 120.

94R4   Rock, F.L., Li, X., Chong, P., Ida, N., Klein, M: Biochemistry 33(1994) 5146.

94R5   Ruiz-Arribas, A., Santamaria, R.I., Zhadan, G.G., Villar, E., Shnyrov, V.L.: Biochemistry 33 (1994) 13787.

94S1   Sanz, J.M., Johnson, C.M., Fersht, A.R.: Biochemistry 33 (1994) 11189.

94S2   Schreiber, G., Buckle, A.M., Fersht, A.R.: Structure 2 (1994) 945.

94S3   Schubert, W.-D., Schluckebier, G., Backmann, J., Granzin, J., Kisker, C., Choe, H.-W., Hahn, U., Pfeil, W., Saenger, W.: Eur. J. Biochem. 220 (1994) 527.

94S4   Shaw, G.S., Hodges, R.S., Kay, C.M., Sykes, B.D.: Protein Sci. 3 (1994) 1010.

94S5   Sirangelo, I., Bismuto, E., Irace, G.: FEBS Lett. 338 (1994) 11.

94S6   Smith, C.K., Withka, J.M., Regan, L.: Biochemistry 33 (1994) 5510.

94S7   Staniforth, R.A., Cortes, A., Burston, S.G., Atkinson. T., Holbrook, J.J., Clarke, A.R.: FEBS Lett. 344 (1994) 129.

94S8   Steipe, B., Schiller, B., Plückthun, A., Steinbacher, S.: J. Mol. Biol. 240 (1994) 188.

94S9   Stites, W.E., Meeker, A.K., Shortle, D.: J. Mol. Biol. 235 (1994) 27.

94S10   Szpikowska, B.K., Beechem, J.M., Sherman, M.A., Mas, M.T.: Biochemistry 33 (1994) 2217.

94T1   Taddei, N., Buck, M., Broadhurst, R.W., Stefani, M., Ramponi, G., Dobson, C.M.: Eur. J. Biochem. 225 (1994) 811.

94T2   Talluri, S., Rothwarf, D.M., Scheraga, H.A.: Biochemistry 33 (1994) 10437.

94T3   Tamura, A., Kojima, S., Miura, K., Sturtevant, J.M.: Biochemistry 33 (1994) 14512.

94T4   Tanaka, T., Kuroda, Y., Kimura, H., Kidokoro, S., Nakamura, H.: Protein Engng. 7 (1994) 969.

94T5   Taneja, S., Ahmad, F.: Biochem. J. 303 (1994) 147.

94T6   Timm, D.E., de Haseth, P.L., Neet, K.E.: Biochemistry 33 (1994) 4667.

94V1   Venugopal, M.G., Ramshaw, J.A.M., Braswell, E., Zhu, D., Brodsky, B.: Biochemistry 33 (1994) 7948.

94V2   Viguera, A.R., Martinez, J.C., Filimonov, V.V., Mateo, P.L., Serrano, L.: Biochemistry 33 (1994) 2142.

94V3   Vuilleumier, S., Fersht, A.R.: Eur. J. Biochem. 221 (1994) 1003.

94W1   Wang, S.F., Smith, D.M.: J. Agric. Food Chem. 42 (1994) 2665.

94W2   Weers, P.M.M., Kay, C.M., Oikawa, K., Wientzek, M., van der Horst, D.J., Ryan, R.O.: Biochemistry 33 (1994) 3617.

94W3   Welfle, K., Misselwitz, R., Welfle, H., Simon, O., Politz, O., Borris, R.: J. Biomol. Struct. & Dynamics 11 (1994) 1417.

94W4   Williamson, R.A., Bartels, H., Murphy, G., Freedman, R.B.: Protein Engng. 7 (1994) 1035.

94W5   Wintrode, P.L., Makhatadze, G.I., Privalov, P.L.: Proteins: Structure, Function, and Genetics 18 (1994) 246.

94X   Xie, D., Fox, R., Freire, E.: Protein Sci. 3 (1994) 2175.

94Y1   Yan, Y., Erickson, B.W.: Protein Sci. 3 (1994) 1069.

94Y2   Yu, Y., Makhatadze, G.I., Pace, C.N., Privalov, P.L.: Biochemistry 33 (1994) 3312.

94Z1   Zahn, R., Axmann, S.E., Rücknagel, K.-P., Jaeger, E., Laminet, A.A., Plückthun, A.: J. Mol. Biol. 242 (1994) 150.

94Z2   Zahn, A., Plückthun, A.: J. Mol. Biol. 242 (1994) 165.

94Z3   Zapun, A., Cooper, L., Creighton, T.E.: Biochemistry 33 (1994) 1907.

94Z4   Zhadan, G.G., Shnyrov, V.L.: Biochem. J. 299 (1994) 731.
94Z5   Zhang, T., Bertelsen, E., Alber, T.: Structural Biol. 1 (1994) 434.
94Z6   Zhou, N.E., Kay, C.M., Hodges, R.S.: Protein Engng. 7 (1994) 1365.
94Z7   Zhuang, P., Eisenstein, E., Howell, E.E.: Biochemistry 33 (1994) 4237.
95A1   Agashe, V.R., Udgaonkar, J.B.: Biochemistry 34 (1995) 3286.
95A2   Agostinelli, E., Cervoni, L., Giartosio, A., Morpurgo, L.: Biochem. J. 306 (1995) 697.
95A3   Akasako, A., Haruki, M., Oobatake, M., Kanaya, S.: Biochemistry 34 (1995) 8115.
95A4   Antalik, M., Bagelova, J.: Gen. Physiol. Biophys. 14 (1995) 19.
95A5   Antonsson, P., Kammerer, R.A., Schulthess, T., Hänisch, G., Engel, J.: J. Mol. Biol. 250 (1995) 74.
95A6   Aronsson, G., Mårtensson, L.-G., Carlsson, U., Johnsson, B.-H.: Biochemistry 34 (1995) 2153.
95A7   Arroyo-Reyna, A., Hernandez-Arana, A.: Biochim. Biophys. Acta 1248 (1995) 123.
95A8   Assemat, K., Alzari, P.M., Clément-Métral, J.: Protein Sci. 4 (1995) 2510.
95A9   Azuaga, A.I., Conejero-Lara, F., Rivas, G., De Filippis, V., Fontana, A., Mateo, P.L.: Biochim. Biophys. Acta 1252 (1995) 95.
95B1   Bae, S.-J., Sturtevant, J.M.: Biophys. Chem. 55 (1995) 247.
95B2a   Barone, G., Capasso, S., Del Vecchio, P., De Sena, C., Fessas, D., Giancola, C., Graziano, G., Tramonti, P.: J. Thermal Analysis 45 (1995) 1255.
95B2b   Barone, G., Catanzano, F., Del Vecchio, P., Giancola, C., Graziano, G.: Pure Appl. Chem. 67 (1995) 1867.
95B3   Bayol, A., Dupin, P., Boe, J.F., Claudy, P., Létoffé, J.M.: Biophys. Chem. 54 (1995) 229.
95B4   Blaber, M., Baase, W.A., Gassner, N., Matthews, B.W.: J. Mol. Biol. 246 (1995) 317.
95B5   Bobkov, A.A., Levitsky, D.I.: Biochemistry 34 (1995) 9708.
95B6   Bogomolov, A.A., Bikbov, T.M.: Biochimia (Russ.) 60 (1995) 626.
95B7   Breton, J., La Fiura, A., Bertolero, F., Orsini, G., Valsasina, B., Ziliotto, R., De Filippis, V., Polverino de Laureto, P., Fontana, A.: Eur. J. Biochem. 227 (1995) 573.
95B8   Bulaj, G., Otlewski, J.: J. Mol. Biol. 247 (1995) 701.
95B9   Burova, T.V., Bernhardt, R., Pfeil, W.: Protein Sci. 4 (1995) 909.
95B10   Byrne, M.P., Manuel, R.L., Lowe, L.G., Stites, W.E.: Biochemistry 34 (1995) 13949.
95B11   Byrne, M.P., Stites, W.E.: Protein Sci. 4 (1995) 2545.
95C1   Cardamone, M., Puri, N.K., Brandon, M.R.: Biochemistry 34 (1995) 5773.
95C2   Carra, J.H., Privalov, P.L.: Biochemistry 34 (1995) 2034.
95C3   Choi, S.-G., O'Donnell, S.E., Sarken, K.D., Hardman, J.K.: J. Biol. Chem. 270 (1995) 17712.
95C4   Choi, S.-G., Hardman, J.K.: J. Biol. Chem. 270 (1995) 28177.
95C5   Clarke, J., Henrick, K., Fersht, A.R.: J. Mol. Biol. 253 (1995) 493.
95C6   Clarke, J., Hounslow, A.M., Fersht, A.R.: J. Mol. Biol. 253 (1995) 505.
95D1   Davidson, A.R., Lumb, K.J., Sauer, R.T.: Nature Sructural Biol. 2 (1995) 856.
95D2   DeFelippis, M.R., Kilcomons, M.A., Lents, M.P., Youngman, K.M., Havel, H.A.: Biochim. Biophys. Acta 1247 (1995) 35.
95D3   DeKoster, G.T., Robertson, A.D.: J. Mol. Biol. 249 (1995) 529.
95D4   Desjarlais, J.R., Handel, T.M.: Protein Sci. 4 (1995) 2006.
95D5   Diamond, D.L., Strobel, S., Chun, S.-Y., Randall, L.L.: Protein Sci. 4 (1995) 1118.
95D6   Dudich, I.V., Zav'yalov, V.P., Pfeil, W., Gaestel, M., Zav'yalova, G.A., Denesyuk, A.I., Korpela, T.: Biochim. Biophys. Acta 1253 (1995) 163.
95F1   Fairman, R., Chao, H.-G., Mueller, L., Lavoie, T.B., Shen, L., Novotny, J., Matsueda, G.R.: Protein Sci. 4 (1995) 1457.
95F2   Florens, L., Bianco, P., Haladjian, J., Bruschi, M., Protasevich, I., Makarov, A.: FEBS Lett. 373 (1995) 280.
95F3   Frank, M.K., Clore, G.M., Gronenborn, A.M.: Protein Sci. 4 (1995) 2605.
95F4   Fukada, H., Kitamura, S., Takahashi, K.: Thermochim. Acta 266 (1965) 365.
95F5   Fukada, H., Takahashi, K., Sorai, M., Kojima, M., Tanokura, M., Takahashi, K.: Thermochim. Acta 267 (1995) 373.
95G1   Galisteo, M.L., Gordon, C.L., King, J.: J. Biol. Chem. 270 (1995) 16595.

95G2   Garcia, P., Desmadril, M., Minard, P., Yon, J.M.: Biochemistry 34 (1995) 397.
95G3   Gasset, M., Mancheño, J.M., Laynez, J., Lacadena, J., Fernández- Ballester, G., del Pozo, A.M., Oñaderra, M., Gavilanes J.G.: Biochim. Biophys. Acta 1252 (1995) 126.
95G4   Gast, K., Damaschun, G., Desmadril, M., Minard, P., Müller- Frohne, M., Pfeil, W., Zirwer, D.: FEBS Lett. 358 (1995) 247.
95G5   Genov, N., Filippi, B., Dolashka, P., Wilson, K.S., Betzel, C.: Int. J. Peptide Protein Res. 45 (1995) 391.
95G6   Gómez, J., Freire, E.: J. Mol. Biol. 252 (1995) 337.
95G7   Gorinstein, S., Zemser, M., Friedman, M., Chang, Sh.-M.: Int. J. Peptide Protein Res. 45 (1995) 248.
95G8   Gorovits, B.M., Horowitz, P.M.: J. Biol. Chem. 270 (1995) 28551.
95G9   Greenfield, N.J., Hitchcook-DeGregori, S.E.: Biochemistry 34 (1995) 16797.
95G10   Griko, Yu.V., Freire, E., Privalov, G., Van Dael, H., Privalov, P.L.: J. Mol. Biol. 252 (1995) 447.
95G11   Gross, M., Lustig, A., Wallimann, T., Furter, R.: Biochemistry 34 (1995) 10350.
95G12   Gu, H., Yi, Q., Bray, S.T., Riddle, D.S., Shiau, A.K., Baker, D.: Protein Sci. 4 (1995) 1108.
95G13   Gulliver, G.A., Rumbley, C.A., Carrero, J., Voss Jr., E.W.: Biochemistry 34 (1995) 5158.
95H1   Hamada, D., Fukada, H., Takahashi, K., Goto, Y.: Thermochim. Acta 266 (1965) 385.
95H2   Hammen, P.K., Scholtz, J.M., Anderson, W., Waygood, E.B., Klevit, R.E.: Protein Sci. 4 (1995) 936.
95H3   Haun, M.F., Wirth, M., Rüterjans, H.: Eur. J. Biochem. 227 (1995) 516.
95H4   Helms, L.R., Wetzel, R.: Protein Sci. 4 (1995) 2073.
95H5   Hernandez-Arana, A., Rojo-Dominguez, A., Soriano-Garcia, M., Rodriguez-Romero, A.: Eur. J. Biochem. 228 (1995) 649.
95H6   Herrmann, L., Bowler, B.E., Dong, A., Caughey, W.S.: Biochemistry 34 (1995) 3040.
95H7   Huang, G.S., Oas, T.G.: Biochemistry 34 (1995) 3884.
95H8   Huntington, J.A., Patston, P.A., Gettins, P.G.W.: Protein Sci. 4 (1995) 613.
95H9   Hurley, J.K., Caffrey, M.S., Markley, J.L., Cheng, H., Xia, B., Chae, Y.K., Holden, H.M., Tollin, G.: Protein Sci. 4 (1995) 58.
95I1   Itzhaki, L.S., Otzen, D.E., Fersht, A.R.: J. Mol. Biol. 254 (1995) 260.
95I2   Iwagami, S.G., Creagh, A.L., Haynes, C.A., Borsari, M., Felli, I.C., Piccioli, M., Eltis, L.D.: Protein Sci. 4 (1995) 2562.
95J1   Johnson, C.M., Fersht, A.R.: Biochemistry 34 (1995) 6795.
95J2   Johnson, C.R., Morin, P.E., Arrowsmith, C.H., Freire, E.: Biochemistry 34 (1995) 5309.
95J3   Jongh, H.H.J. de, Goormaghtigh, E., Ruysschaert, J.-M.: Biochemistry 34 (1995) 172.
95K1   Kahana, E., Gratzer, W.B.: Biochemistry 34 (1995) 8110.
95K2   Kalnin, N.N., Kuwajima, K.: Proteins: Structure, Function, and Genetics 23 (1995) 163.
95K3   Kammerer, R.A., Antonsson, P., Schulthess, T., Fauser, C., Engel, J.: J. Mol. Biol. 250 (1995) 64.
95K4   Karantza, V., Baxevanis, A.D., Freire, E., Moudrianakis, E.N: Biochemistry 34 (1995) 5988.
95K5   Kato, A., Shimizu, T., Saga, S.: FEBS Lett. 371 (1995) 17.
95K6   Kenar, K.T., Garcia-Moreno, B., Freire, E.: Protein Sci. 4 (1995) 1934.
95K7   Khurana, R., Hate, A.T., Nath, U., Udgaonkar, J.B.: Protein Sci. 4 (1995) 1133.
95K8   Kiefhaber, T., Baldwin, R.L.: J. Mol. Biol. 252 (1995) 122.
95K9   Klug, C.S., Su, W., Liu, J., Klebba, P.E., Feix, J.B.: Biochemistry 34 (1995) 14230.
95K10   Kodicek, M., Infanzon, A., Karpenko, V.: Biochim. Biophys. Acta 1246 (1995) 10.
95K11   Kohn, W.D., Kay, C.M., Hodges, R.S.: Protein Sci. 4 (1995) 237.
95K12   Kragelund, B.B., Robinson, C.V., Knudsen, J., Dobson, C.M., Poulsen, F.M.: Biochemistry 34 (1995) 7217.

95K13   Kreimer, D.I., Shnyrov, V.L., Villar, E., Silman, I., Weiner, L.: Protein Sci. 4 (1995) 2349.

95K14   Kurochkin, I.V., Procyk, R., Bishop, P.D., Yee, V.C., Teller, D.C., Ingham, K.C., Medved, L.V.: J. Mol. Biol. 248 (1995) 414.

95L1   La Rosa, C., Milardi, D., Grasso, D., Guzzi, R., Sportelli, L.: J. Phys. Chem. 99 (1995) 14864.

95L2   Ladbury, J.E., Wynn, R., Thomson, J.A., Sturtevant, J.M.: Biochemistry 34 (1995) 2148.

95L3   Leder, L., Berger, C., Bornhauser, S., Wendt, H., Ackermann, F., Jelesarov, I., Bosshardt, H.R.: Biochemistry 34 (1995) 16509.

95L4   Li, Y.-J., Rothwarf, D.M., Scheraga, H.A.: Nature Structural Biol. 2 (1995) 489.

95L5   Linske-O'Connell, L.I., Sherman, F., McLendon, G.: Biochemistry 34 (1995) 7094.

95L6   Linske-O'Connell, L.I., Sherman, F., McLendon, G.: Biochemistry 34 (1995) 7103.

95L7   Litvinovich, S.V., Henschen, A.H., Krieglstein, K.G., Ingham, K.C., Medved, L.V.: Eur. J. Biochem. 229 (1995) 605.

95L8   Loh, S.N., Kay, M.S., Baldwin, R.L.: Proc. Natl. Acad. Sci. USA 92 (1995) 5446.

95L9   Lörinczy, D., Belágyi, J.: Thermochim. Acta 259 (1995) 153.

95L10   López-Hernández, E., Serrano, L.: Proteins: Structure, Function, and Genetics 22 (1995) 340.

95L11   Lowther, W.T., Majer, P., Dunn, B.M.: Protein Sci. 4 (1995) 689.

95L12   Lumb, K.J., Kim, P.S.: Biochemistry 34 (1995) 8642.

95L13   Luo, J., Iwakura, M., Matthews, C.R.: Biochemistry 34 (1995) 10669.

95M1   Makarov, A.A., Protasevich, I.I., Bazhulina, N.P., Esipova, N.G.: FEBS Lett. 357 (1995) 58.

95M2   Makhatadze, G.I., Clore, G.M., Gronenborn, A.M.: Nature Structural Biol. 2 (1995) 852.

95M3   Makhatadze, G.I., Privalov, P.I.: Adv. Protein Chem. 47 (1995) 307.

95M4   Marmorino, J.L., Pielak, G.J.: Biochemistry 34 (1995) 3140.

95M5   Martinez, J.C., Filimonov, V.V., Mateo, P.L., Schreiber, G., Fersht, A.R.: Biochemistry 34 (1995) 5224.

95M6   Martsev, S.P., Kravchuk, Z.I., Vlasov, A.P., Lyakhnovich, G.V.: FEBS Lett. 361 (1995) 173.

95M7   Mas, M.T., Chen, H.-H., Aisaka, K., Lin, L.-N., Brandts, J.F.: Biochemistry 34 (1995) 7931.

95M8   Matthews, B.W.: Adv. Protein Chem. 46 (1995) 249.

95M9   Matthews, J.M., Fersht, A.R.: Biochemistry 34 (1995) 6805.

95M10   McAfee, J.G., Edmondson, S.P., Datta, P.K., Shriver, J.W., Gupta, R.: Biochemistry 34 (1995) 10063.

95M11   McGee, W.A.: unpublished results, cited in Ref. 95V3

95M12   Medved, L.V., Orthner, C.L., Lubon, H., Lee, T.K., Drohan, W.N., Ingham, K.C.: J. Biol. Chem. 270 (1995) 13652.

95M13   Menéndez, M., Gasset, M., Laynez, J., López-Zumel, C., Usobiaga, P., Töpfer-Petersen, E., Calvete, J.J.: Eur. J. Biochem. 234 (1995) 887.

95M14   Merabet, E., Ackers, G.K.: Biochemistry 34 (1995) 8554.

95M15   Miles, C.A., Burjanadze, T.V., Bailey, A.J.: J. Mol. Biol. 245 (1995) 437.

95M16   Misselwitz, R., Hausdorf, G., Welfle, K., Höhne, W.E., Welfle, H.: Biochim. Biophys. Acta 1250 (1995) 9.

95M17   Moosavi-Movahedi, A.A., Nazari, K.: Int. J. Biol. Macromol. 17 (1995) 43.

95M18   Motoshima, H., Ueda, T., Hashimoto, Y., Tsutsumi, M., Imoto, T.: J. Biochem. 118 (1995) 1138.

95N1   Nakaya, M., Watabe, S.: Biochemistry 34 (1995) 3114.

95N2   Nath, U., Udgaonkar, J.B.: Biochemistry 34 (1995) 1702.

95N3   Nishii, I., Kataoka, M., Goto, Y.: J. Mol. Biol. 250 (1995) 223.

95N4   Nölting, B., Golbik, R., Fersht, A.R.: Proc. Natl. Acad. Sci. USA 92 (1995) 10668.

95O1   Oliveberg, M., Arcus, V.L., Fersht, A.R.: Biochemistry 34 (1995) 9424.

95O2   O'Neil, K.T., Hoess, R.H., Raleigh, D.P., DeGrado, W.F.: Proteins: Structure, Function, and Genetics 21 (1995) 11.

95O3   Otzen, D.E., Fersht, A.R.: Biochemistry 34 (1995) 5718.

95P1   Pfeil, W., Burova, T.V., Beckert, V., Uhlmann, H., Bernhardt, R.: Protein Sci. 4
        Suppl. 1 (1995) 68.
95P2   Picó, G.: Biochem. Mol. Biol. Int. 36 (1995) 1017.
95P3   Pielak, G.J., Auld, D.S., Beasley, J.R., Betz, S.F., Cohen, D.S., Doyle, D.F., Finger,
        S.A., Fredericks, Z.L., Hilgen-Willis, S., Saunders, A.J., Trojak, S.K.: Biochemistry
        34 (1995) 3268.
95P4   Plaza del Pino, I.M., Sanchez-Ruiz, J.M.: Biochemistry 34 (1995) 8621.
95P5   Pons, J., Planas, A., Querol, E.: Protein Engng. 8 (1995) 939.
95P6   Prassl, R., Schuster, B., Abuja, P.M., Zechner, M., Kostner, G.M., Laggner, P.:
        Biochemistry 34 (1995) 3795.
95P7   Predki, P.F., Nayak, L.M., Gottlieb, M.B.C., Regan, L.: Cell 80 (1995) 41.
95P8   Predki, P.F., Regan, L.: Biochemistry 34 (1995) 9834.
95P9   Privalov, G., Kavina, V., Freire, E., Privalov, P.L.: Analyt. Biochem. 232 (1995) 79.
95P10   Pullen, K., Rajagopal, P., Branchini, B.R., Huffine, M.E., Reizer, J., Saier Jr., M.H.,
        Scholtz, J.M., Klevitt, R.E.: Protein Sci. 4 (1995) 2478.
95Q1   Qi, X.L., Brownlow, S., Holt, C., Sellers, P.: Biochim. Biophys. Acta 1248 (1995)
        43.
95Q2   Qin, W., Sanishvili, R., Plotkin, B., Schejter, A., Margoliash, E.: Biochim. Biophys.
        Acta 1252 (1995) 87.
95R0   Raleigh, D.P., Betz, S.F., DeGrado, W.F.: J. Amer. Chem. Soc. 117 (1995) 7558.
95R1   Ramsay, G., Ionescu, R., Eftink, M.R.: Biophys. J. 69 (1995) 701.
95R2   Raquet, X., Vanhove, M., Lamotte-Brasseur, J., Goussard, S., Courvalin, P., Frère,
        J.-M.: Proteins: Structure, Function, and Genetics 23 (1995) 63.
95R3   Reedstrom, R.J., Royer, C.A.: J. Mol. Biol. 253 (1995) 266.
95R4   Rialdi, G., Battistel, E.: J. Thermal Analysis 45 (1995) 631.
95R5   Ritco-Vonsovici, M., Minard, P., Desmadril, M., Yon, J.M.: Biochemistry 34
        (1995) 16543.
95R6   Ritco-Vonsovici, M., Mouratou, B., Minard, P., Desmadril, M., Yon, J.M., Andri-
        eux, M., Leroy, E., Guittet, E.: Biochemistry 34 (1995) 833.
95R7   Ruiz-Sanz, J., de Prat Gay, G., Otzen, D.E., Fersht, A.R.: Biochemistry 34 (1995)
        1695.
95S1   Sasso, S., Protasevich, I., Gilli, R., Makarov, A., Briand, C.: J. Biomol. Structure &
        Dynamics 12 (1995) 1023.
95S2   Schindler, T., Herrler, M., Marahiel, M.A., Schmid, F.X.: Nature Structural Biol. 2
        (1995) 663.
95S3   Scholtz, J.M.: Protein Sci. 4 (1995) 35.
95S4   Sherman, M.A., Beechem, J.M., Mas, M.T.: Biochemistry 34 (1995) 13934.
95S5   Shoichet, B.K., Baase, W.A., Kuroki, R., Matthews, B.W.: Proc. Natl. Acad. Sci.
        USA 92 (1995) 452.
95S6   Shih, P., Holland, D.R., Kirsch, J.F.: Protein Sci. 4 (1995) 2050.
95S7   Shih, P., Kirsch, J.F.: Protein Sci. 4 (1995) 2063.
95S8   Shortle, D.: Adv. Protein Chem. 46 (1995) 217.
95S9   Sluis-Cremer, N., Dirr, H.: FEBS Lett. 371 (1995) 94.
95S10   Smith, D.D.S., Pratt, K.A., Sumner, I.G., Henneke, C.M.: Protein Engng. 8 (1995)
        13.
95S11   Spuergin, P., Abele, U., Schulz, G.E.: Eur. J. Biochem. 231 (1995) 405.
95S12   Steer, B.A., Merrill, A.R.: Biochemistry 34 (1995) 7225.
95S13   Steif, C., Hinz, H.-J., Cesareni, G.: Proteins: Structure, Function, and Genetics 23
        (1995) 83.
95S14   Stites, W.E., Byrne, M.P., Aviv, J., Kaplan, M., Curtis, P.M.: Analyt. Biochem. 227
        (1995) 112.
95S15   Stokkum, I.H.M. van, Linsdell, H., Haden, J.M., Haris, P.I., Chapman, D., Bloe-
        mendal, M.: Biochemistry 34 (1995) 10508.
95S16   Stone, M.J., Ruf, W., Miles, D.J., Edgington, T.S., Wrihght, P.E.: Biochem. J. 310
        (1995) 605.
95S17   Surewicz, W.K., Olesen, P.R.: Biochemistry 34 (1995) 9655.
95S18   Swint-Kruse, L., Robertson, A.D.: Biochemistry 34 (1995) 4724.
95T1   Takano, K., Ogasahara, K., Kaneda, H., Yamagata, Y., Fujii, S., Kanaya, E., Kiku-
        chi, M., Oobatake, M., Yutani, K.: J. Mol. Biol. 254 (1995) 62.

95T2   Tamura, A., Kojima, S., Miura, K., Sturtevant, J.M.: J. Mol. Biol. 249 (1995) 636.
95T3   Tamura, A., Sturtevant, J.M.: J. Mol. Biol. 249 (1995) 625.
95T4   Tamura, A., Sturtevant, J.M.: J. Mol. Biol. 249 (1995) 646.
95T5   Tamura, Y., Gekko, K.: Biochemistry 34 (1995) 1878.
95T6   Tanaka, A., Fukada, H., Takahashi, K.: J. Biochem. 117 (1995) 1024.
95T7   Taylor, L.S., York, P., Williams, A.C., Edwards, H.G.M., Mehta, V., Jackson, G.S., Badcoe, I.G., Clarke, A.R.: Biochim. Biophys. Acta 1253 (1995) 39.
95T8   Tello-Solis, S.R., Hernandez-Arana, A.: Biochem. J. 311 (1995) 969.
95T9   Terwilliger, T.C.: Adv. Protein Chem. 46 (1995) 177.
95T10   Tomizawa, H., Yamada, H., Hashimoto, Y., Imoto, T.: Protein Engng. 8 (1995) 1023.
95U   Uchiyama, H., Perez-Prat, E.M., Watanabe, K., Kumagai, I., Kuwajima, K.: Protein Engng. 8 (1995) 1153.
95V1   Vanhove, M., Houba, S., Lamotte-Brasseur, J., Frere, J.-M.: Biochem. J. 308 (1995) 859.
95V2   Vanhove, M., Raquet, X., Frère, J.-M.: Proteins: Structure, Function, and Genetics 22 (1995) 110.
95V3   Veeraraghavan, S., Rodriguez-Ghidarpour, S., MacKinnon, C., McGee, W.A., Pierce, M.M., Nall, B.T.: Biochemistry 34 (1995) 12892.
95V4   Vidugiris, G.J.A., Markley, J.L., Royer, C.A.: Biochemistry 34 (1995) 4909.
95V5   Viguera, A.R., Blanco, F.J., Serrano, L.: J. Mol. Biol. 247 (1995) 670.
95V6   Villegas, V., Azuaga, A., Catasús, L., Reverter, D., Mateo, P.L., Avilés, F.X., Serrano, L.: Biochemistry 34 (1995) 15105.
95V7   Villegas, V., Viguera, A.R., Avilés, F.X., Serrano, L.: Folding & Design 1 (1995) 29.
95V8   Vogl, T., Brengelmann, R., Hinz, H.-J., Scharf, M., Lötzbeyer, M., Engels, J.W.: J. Mol. Biol. 254 (1995) 481.
95V9   Vonderviszt, F., Závodsky, P., Ishimura, M., Uedaira, H., Namba, K.: J. Mol. Biol. 251 (1995) 520.
95W1   Waldburger, C.D., Schildbach, J.F., Sauer, R.T.: Structural Biology 2 (1995) 122.
95W2   Walter, S., Hubner, B., Hahn, U., Schmid, F.X.: J. Mol. Biol. 252 (1995) 133.
95W3   Ward, L.D., Matthews, J.M., Zhang, J.-G., Simpson, R.J.: Biochemistry 34 (1995) 11652.
95W4   Welfle, K., Misselwitz, R., Welfle, H., Politz, O., Borriss, R.: Eur. J. Biochem. 229 (1995) 726.
95W5   Wintrode, P.L., Griko, Y.V., Privalov, P.L.: Protein Sci. 4 (1995) 1528.
95W6   Wu, J., Long, D.G., Weis, R.M.: Biochemistry 34 (1995) 3056.
95W7   Wynn, R., Anderson, C.L., Richards, F.M., Fox, R.O.: Protein Sci. 4 (1995) 1815.
95Y1   Yakovlev, G.I., Moiseyev, G.P., Protasevich, I.I., Ranjbar, B., Bocharov, A.L., Kirpichnikov, M.P., Gilli, R.M., Hartley, R.W., Makarov, A.A.: FEBS Lett. 366 (1995) 156.
95Y2   Yamaguchi, T., Yamada, H., Akasaka, K.: J. Mol. Biol. 250 (1995) 689.
95Y3   Yamasaki, T., Kanaya, S., Oobatake, M.: Thermochim. Acta 267 (1995) 379.
95Y4   Yamasaki, K., Ogasahara, K., Yutani, K., Oobatake, M., Kanaya, S.: Biochemistry 34 (1995) 16552.
95Y5   Yao, M., Bolen, D.W.: Biochemistry 34 (1995) 3771.
95Y6   Youngman, K.M., Spencer, D.B., Brems, D.N., DeFelippis, M.R.: J. Biol. Chem. 270 (1995) 19816.
95Z1   Zhang, X., Baase, W.A., Shoichet, B.K., Wilson, K.P., Matthews, B.W.: Protein Engng. 8 (1995) 1017.
95Z2   Zitzewitz, J.A., Bilsel, O., Luo, J., Jones, B.E., Matthews, C.R.: Biochemistry 34 (1995) 12812.
95Z3   Zolkiewski, M., Nosworthy, N.J., Ginsburg, A.: Protein Sci. 4 (1995) 1544.
95Z4   Zolkiewski, M., Redowicz, M.J., Korn, E.D., Ginsburg, A.: Arch. Biochem. Biophys. 318 (1995) 207.
96A1   Ahmad, Z., Yadav, S., Ahmad, F., Khan, N.Z.: Biochim. Biophys. Acta 1294 (1996) 63.
96A2   Arai, M., Kuwajima, K.: Folding & Design 1 (1996) 275.
96A3   Arunachalam, U., Kellis Jr., J.T.: Biochemistry 35 (1996) 11379.

96A4   Azuaga, A.I., Sepulcre, F., Padrós, E., Mateo, P.L.: Biochemistry 35 (1996) 16328.
96B1   Baldwin, E., Xu, J., Hajiseyedjavadi, O., Baase, W.A., Matthews, B.W.: J. Mol. Biol. 259 (1996) 542.
96B2   Ballew, R.M., Sabelko, J., Gruebele, M.: Proc. Natl. Acad. Sci. USA 93 (1996) 5759.
96B3   Barteri, M., Gaudiano, M.C., Santucci, R.: Biochim. Biophys. Acta 1295 (1996) 51.
96B4   Betton, J.-M., Hofnung, M.: J. Biol. Chem. 271 (1996) 8046.
96B5   Betz, S.F., DeGrado, W.F.: Biochemistry 35 (1996) 6955.
96B6   Betz, S.F., Marmorino, J.L., Saunders, A.J., Doyle, D.F., Young, G.B., Pielak, G.J.: Biochemistry 35 (1996) 7422.
96B7   Boice, J.A., Dieckmann, G.R., DeGrado, W.F., Fairman, R.: Biochemistry 35 (1996) 14480.
96B8   Boice, J.A., Fairman, R.: Protein Sci. 5 (1996) 1776.
96B9   Borgstahl, G.E.O., Parge, H.E., Hickey, M.J., Johnson, M.J., Boissinot, M., Halle-well, R.A., Lepock, J.R., Cabelli, D.E., Tainer, J.A.: Biochemistry 35 (1996) 4287.
96B10   Burton, R.E., Huang, G.S., Daugherty, M.A., Fullbright, P.W., Oas, T.G.: J. Mol. Biol. 263 (1996) 311.
96B11   Burova, T.V., Beckert, V., Uhlmann, H., Ristau, O., Bernhardt, R., Pfeil, W.: Protein Sci. 5 (1996) 1890.
96C1   Cai, K., Schirch, V.: J. Biol. Chem. 271 (1996) 2987.
96C2   Cai, K., Schirch, V.: J. Biol. Chem. 271 (1996) 27311.
96C3   Carra, J.H., Murphy, E.C., Privalov, P.L.: Biophys. J. 71 (1996) 1994.
96C4   Cashikar, A.G., Rao, N.M.: J. Biol. Chem. 271 (1996) 4741.
96C5   Castro, M.J.M., Anderson, S.: Biochemistry 35 (1996) 11435.
96C6   Catanzano, F., Gianola, C., Graziano, G., Barone, G.: Biochemistry 35 (1996) 13378.
96C7   Chamberlain, A.K., Handel, T.M., Marqusee, S.: Nature Structural Biol. 3 (1996) 782.
96C8   Chen, C.-H., Davis, R.A., Maley, F.: Biochemistry 35 (1996) 8786.
96C9   Chen, H., Li, Y., Panda, T., Buehler, F.U., Ford, C., Reilly, P.J.: Protein Engng. 9 (1996) 499.
96C10   Chen, L., Hodgson, K.O., Doniach, S.: J. Mol. Biol. 261 (1996) 658.
96C11   Chen, Y.-J., Lin, S.-C., Tzeng, S.-R., Patel, H.V., Lyu, P.-C., Cheng, J.-W.: Proteins: Structure, Function, and Genetics 26 (1996) 465.
96C12   Clark, N.S., Dodd, I., Mossakowska, D.E., Smith, R.A.G., Gore, M.G.: Protein Engng. 9 (1996) 877.
96C13   Clarke, J., Fersht, A.R.: Folding & Design 1 (1996) 243.
96C14   Colón, W., Elöve, G.A., Wakem, L.P., Sherman, F., Roder, H.: Biochemistry 35 (1996) 5538.
96C15   Colón, W., Roder, H.: Nature Structural Biol. 3 (1996) 1019.
96C16   Conejero-Lara, F., Mateo, P.L.: Biochemistry 35 (1996) 3477.
96C17   Conejero-Lara, F., Parrado, J., Azuaga, A.I., Smith, R.A.G., Ponting, C.P., Dobson, C.M.: Protein Sci. 5 (1996) 2583.
96C18   Cooper, J.B., Saward, S., Erskine, P.T., Badasso, M.O., Wood, S.P., Zhang, Y., Young, D.: FEBS Lett. 387 (1996) 105.
96D1   D'Aquino, J.A., Gómez, J., Hilser, V.J., Lee, K.H., Amzel, L.M., Freire, E.: Proteins: Structure, Function, and Genetics 25 (1996) 143.
96D2   Doyle, D.F., Waldner, J.C., Parikh, S., Alcazar-Roman, L., Pielak, G.J.: Biochemistry 35 (1996) 7403.
96E1   Eberhardt, E.S., Wittmayer, P.K., Templer, B.M., Raines, R.T.: Protein Sci. 5 (1996) 1697.
96E2   Eftink, M.R., Ionescu, R., Ramsay, G.D., Wong, C.-Y., Wu, J.Q., Maki, A.H.: Biochemistry 35 (1996) 8084.
96E3   Erhard, B., Misselwitz, R., Welfle, K., Hausdorf, G., Glaser, R.W., Schneider-Mergener, J., Welfle, H.: Biochemistry 35 (1996) 9097.
96F1   Fairman, R., Chao, H.-G., Lavoie, T.B., Villafranca, J.J., Matsueda, G.R., Novotny, J.: Biochemistry 35 (1996) 2824.
96F2   Filimonov, V.V., Rogov, V.V.: J. Mol. Biol. 255 (1996) 767.

96F3   Fong, S., Hamill, S.J., Proctor, M., Freund, S.M.V., Benian, G.M., Chotia, C., By-croft, M., Clarke, J.: J. Mol. Biol. 264 (1996) 624.

96F4   Frech, C., Wunderlich, M., Glockshuber, R., Schmid, F.X.: Biochemistry 35 (1996) 11386.

96F5   Frisch, C., Kolmar, H., Schmidt, A., Kleemann, G., Reinhardt, A., Pohl, E., Usón, I., Schneider, T.R., Fritz, H.-J.: Folding & Design 1 (1996) 431.

96F6   Funahashi, J., Takano, K., Ogasahara, K., Yamagata, Y., Yutani, K.: J. Biochem. 120 (1996) 1216.

96G1   Garrett, J.B., Mullins, L.S., Raushel, F.M.: Protein Sci. 5 (1996) 204.

96G2   Gassner, N.C., Baase, W.A., Matthews, B.W.: Proc. Natl. Acad. Sci. USA 93 (1996) 12155.

96G3   Genzor, C.G., Beldarrain, A., Gómez-Moreno, C., López-Lacomba, J.L., Cortijo, M., Sancho, J.: Protein Sci. 5 (1996) 1376.

96G4   Gesierich, U., Pfeil, W.: FEBS Lett. 393 (1996) 151.

96G5   Giartosio, A, Erent, M., Cervoni, L., Moréra, Janin, J., Konrad, M., Lascu, I.: J. Biol. Chem. 271 (1996) 17845.

96G6   Gorinstein, S., Zemser, M., Paredes-López, O.: J. Agric. Food Chem. 44 (1996) 100.

96G7   Gray, T.M., Arnoys, E.J., Blankespoor, S., Born, T., Jagar, R., Everman, R., Plowman, D., Stair, A., Zhang, D.: Protein Sci. 5 (1996) 742.

96G8   Graziano, G., Catanzano, F., Gianola, C., Barone, G.: Biochemistry 35 (1996) 13386.

96G9   Gronenborn, A.M., Frank, M.K., Clore, G.M.: FEBS Lett. 398 (1996) 312.

96G10   Gupta, R., Yadav, S., Ahmad, F.: Biochemistry 35 (1996) 11925.

96H1   Hamada, D., Kuroda, Y., Kataoka, M., Aimoto, S., Yoshimura, T., Goto, Y.: J. Mol. Biol. 256 (1996) 172.

96H2   Hendrix, T.M., Griko, Y., Privalov, P.L.: Protein Sci. 5 (1996) 923.

96H3   Hendsch, Z.S., Jonsson, T., Sauer, R.T., Tidor, B.: Biochemistry 35 (1996) 7621.

96H4   Hinck, A.P., Truckses, D.M., Markley, J.L.: Biochemistry 35 (1996) 10328.

96H5   Hiraga, K., Yutani, K.: Eur. J. Biochem. 240 (1996) 63.

96H6   Hiraga, K., Yutani, K.: Protein Engng. 9 (1996) 425.

96H7   Huang, G.S., Oas, T.G.: Biochemistry 35 (1996) 6173.

96I   Iwakura, M., Honda, S.: J. Biochem. 119 (1996) 414.

96J1   Jelesarov, I., Bosshard, H.R.: J. Mol. Biol. 263 (1996) 344.

96J2   Johnson, J.L., Raushel, F.M.: Biochemistry 35 (1996) 10223.

96J3   Jonsson, T., Waldburger, C.D., Sauer, R.T.: Biochemistry 35 (1996) 4795.

96K1   Kanaya, S., Oobatake, M., Liu, Y.: J. Biol. Chem. 271 (1996) 32729.

96K2   Kay, M.S., Baldwin, R.L.: Nature Structural Biol. 3 (1996) 439.

96K3   Khorasanizadeh, S., Peters, I.D., Roder, H.: Nature Struct. Biol. 3 (1996) 193.

96K4   Kim, K., Cistola, D.P., Frieden, C.: Biochemistry 35 (1996) 7553.

96K5   Knapp, S., Karshikoff, A., Berndt, K.D., Christova, P., Atanasov, B., Ladenstein, R.: J. Mol. Biol. 264 (1996) 1132.

96K6   Komsa-Penkova, R., Koynova, R., Kostov, G., Tenchov, B.G.: Biochim. Biophys. Acta 1297 (1996) 171.

96K7   Kragelund, B.B., Højrup, P., Jensen, M.S., Schjerling, C.K., Juul, E., Knudsen, J., Poulsen, F.M.: J. Mol. Biol. 256 (1996) 187.

96K8   Kranz, J.K., Lu, J., Hall, K.B.: Protein Sci. 5 (1996)1567.

96K9   Krokoszynska, I., Otlewski, J.: J. Mol. Biol. 256 (1996) 793.

96K10   Kwon, W.S., Da Silva, N.A., Kellis Jr., J.T.: Protein Engng. 9 (1996) 1197.

96L1   Lange, R., Bee, N., Mozhaev, V.V., Frank, J.: Eur. Biophys. J. 24 (1996) 284.

96L2   Lee, K.N., Park, S.D., Yu, M.-H.: Nature Structural Biol. 3 (1996) 497.

96L3   Liu, Y., Sturtevant, J.M.: Biochemistry 35 (1996) 3059.

96L4   López-Hernández, E., Serrano, L.: Folding & Design 1 (1996) 43.

96M1   Mainfroid, V., Mande, S.C., Hol, W.G.J., Martial, J.A., Goraj, K.: Biochemistry 35 (1996) 4110.

96M2   McCrary, B.S., Edmondson, S.P., Shriver, J.W.: J. Mol. Biol. 264 (1996) 784.

96M3   McGee, W.A., Rosell, F.I., Liggins, J.R., Rodriguez-Ghidarpour, S., Luo, Y., Chen, J., Brayer, G.D., Mauk, A.G., Nall, B.T.: Biochemistry 35 (1996) 1995.

96M4   Mchaourab, H.S., Lietzow, M.A., Hideg, K., Hubbell, W.L.: Biochemistry 35 (1995) 7692.

96M5   McKnight, C.J., Doering, D.S., Matsudaira, P.T., Kim, P.S.: J. Mol. Biol. 260 (1996) 126.

96M6   Medved, L.V., Solovjov, D.A., Ingham, K.C.: Eur. J. Biochem. 239 (1996) 333.

96M7   Mer, G., Hietter, H., Lefèvre, J.-F.: Nature Structural Biol. 3 (1996) 45.

96M8   Minor Jr., D.L., Kim, P.S.: Nature 380 (1996) 730.

96M9   Mok, Y-K., De Prat Gay, G., Butler, P.J., Bycroft, M.: Protein Sci. 5 (1996) 310.

96M10   Muñoz, V., Cronet, P., López-Hernández, E., Serrano, L.: Folding & Design 1 (1996) 167.

96M11   Munson, M., Balasubramanian, S., Fleming, K.G., Nagi, A.D., O'Brien, R., Sturtevant, J.M., Regan, L.: Protein Sci. 5 (1996) 1584.

96M12   Murata, L.B., Schachman, H.K.: Protein Sci. 5 (1996) 709.

96N1   Neira, J.L., Davis, B., Ladurner, A.G., Buckle, A.M., De Prat Gay, G., Fersht, A.R.: Folding & Design 1 (1996) 89.

96N2   Nicholson, E.M., Scholtz, J.M.: Biochemistry 35 (1996) 11369.

96O1   O'Brien, R., Sturtevant, J.M., Wrabl, J., Emerson-Holtzer, M., Holtzer, A: Biophys. J. 70 (1996) 2403.

96O2   Ohmae E., Kurumiya, T., Makino, S., Gekko, K.: J. Biochem. 120 (1996) 946.

96O3   Oliveberg, M., Fersht, A.R.: Biochemistry 35 (1996) 2738.

96O4   Otlewski, J., Sywula, A., Kolasinski, M., Krowarsch, D.: Eur. J. Biochem. 242 (1996) 601.

96P1   Parker, M.J., Spencer, J., Jackson, G.S., Burston, S.G., Hosszu, L.L.P., Craven, C.J., Waltho, J.P., Clarke, A.R.: Biochemistry 35 (1996) 15740.

96P2   Predki, P.F., Agrawal, V., Brünger, A.T., Regan, L.: Nature Structural Biol. 3 (1996) 54.

96P3   Prinsen, C.F.M., Veerkamp, J.H.: Biochem. J. 314 (1996) 253.

96R1   Ramachandran, S., Udgaonkar, J.B.: Biochemistry 35 (1996) 8776.

96R2   Robinson, C.R., Sauer, R.T.: Biochemistry 35 (1996) 13878.

96R3   Rohl, C.A., Chakrabartty, A., Baldwin, R.L.: Protein Sci. 5 (1996) 2623.

96S1   Saab-Rincón, G., Gualfetti, P.J., Matthews, C.R.: Biochemistry 35 (1996) 1988.

96S2   Sarkar, D., DasGupta, C.: Biochim. Biophys. Acta 1296 (1996) 85.

96S3   Schindler, T., Mayr, L.M., Landt, O., Schmid, F.X.: Eur. J. Biochem. 241 (1996) 516.

96S4   Schindler, T., Schmid, F.X.: Biochemistry 35 (1996) 16833.

96S5   Seale, J.W., Gorovits, B.M., Ybarra, J., Horowitz, P.M.: Biochemistry 35 (1996) 4079.

96S6   Shastry, M.C.R., Eftink, M.R.: Biochemistry 35 (1996) 4094.

96S7   Smith, C.K., Bu, Z., Anderson, K.S., Sturtevant, J.M., Engelman, D.M., Regan, L.: Protein Sci. 5 (1996) 2009.

96S8   Sosnick, T.R., Jackson, S., Wilk, R.R., Englander, S.W., DeGrado, W.F.: Proteins: Structure, Function, and Genetics 24 (1996) 427.

96S9   Sosnick, T.R., Mayne, L., Englander, S.W.: Proteins: Structure, Function, and Genetics 24 (1996) 413.

96S10   Spencer, D.S., Stites, W.E.: J. Mol. Biol. 257 (1996) 497.

96S11   Su, Z.-D., Arooz, M.T., Chen, H.M., Gross, C.J., Tsong, T.Y.: Proc. Natl. Acad. Sci. USA 93 (1996) 2539.

96S12   Surolia, A., Sharon, N., Schwarz, F.P.: J. Biol. Chem. 271 (1996) 17697.

96S13   Szeltner, Z., Polgár, L.: J. Biol. Chem. 271 (1996) 5458.

96S14   Szpikowska, B.K., Mas, M.T.: Arch. Biochem. Biophys. 335 (1996) 1273.

96T1   Tan, Y.-J., Oliveberg, M., Fersht, A.R.: J. Mol. Biol. 264 (1996) 377.

96T2   Thapar, R., Nicholson, E.M., Rajagopal, P., Waygood, E.B., Scholtz, J.M., Klevit, R.E.: Biochemistry 35 (1996) 11268.

96T3   Tissot, A.C., Vuilleumier, S., Fersht, A.R.: Biochemistry 35 (1996) 6786.

96T4   Tong, J.C., Zhu, L.Q., Yang, F.Y.: Biochemistry 35 (1996) 9460.

96U1   Ueda, T., Iwashita, H., Hashimoto, Y., Imoto, T.: J. Biochem. 119 (1996) 157.

96U2   Usobiaga, P., Medrano, F.J., Gasset, M., García, J.L., Saiz, J.L., Rivas, G., Laynez, J., Menéndez, M.: J. Biol. Chem. 271 (1996) 6832.

96V1	Vanderheeren, G., Hanssens, I., Meijberg, W., Van Aerschot, A.: Biochemistry 35 (1996) 16753.

96V1   Vanderheeren, G., Hanssens, I., Meijberg, W., Van Aerschot, A.: Biochemistry 35 (1996) 16753.

96V2   Vanhove, M., Raquet, X., Palzkill, T., Pain, R., Frère, J.-M.: Proteins: Structure, Function, and Genetics 25 (1996) 104.

96V3   Veeraraghavan, S., Holzman, T.F., Nall, B.T.: Biochemistry 35 (1996) 10601.

96V4   Vetter, I.R., Baase, W.A., Heinz, D.W., Xiong, J.-P., Snow, S., Matthews, B.W.: Protein Sci. 5 (1996) 2399.

96V5   Vidugiris, G.J.A., Truckses, D.M., Markley, J.L., Royer, C.A.: Biochemistry 35 (1996) 3857.

96W1   Wang, Z., Mottonen, J., Goldsmith E.J.: Biochemistry 35 (1996) 16443.

96W2   Welfle, K., Misselwitz, R., Politz, O., Borriss, R., Welfle, H.: Protein Sci. 5 (1996) 2255.

96Y   Yu, Y., Monera, O.D., Hodges, R.S., Privalov, P.L.: Biophys. Chem. 59 (1996) 299.

96Z   Zahn, R., Perrett, S., Fersht, A.R.: J. Mol. Biol. 261 (1996) 43.

Index of Proteins

Printing and binding: Druckerei Triltsch, Würzburg

R.K. Scopes
Protein Purification
Principles and Practice
3rd ed. 1994. XIX, 380 pp. 165 figs.
(Springer Advanced Texts
in Chemistry)
Hardcover DM 98,-
ISBN 3-540-94072-3

The third edition of this classic
guide to protein purification
updates methods, principles
and references. As in the wide-
ly-acclaimed earlier editions,
Scopes guides both the novice
and the experienced researcher
from theory to application.
Using the book, the reader is
able to integrate methods effec-
tively into optimum protocols
for the task at hand. Reviews of
earlier editions of **Protein
Purification** described it as
"good practical advice that is
presented in a pleasantly read-
able form" (*Analytical Biochemistry*), "well organized and writ-
ten clearly" (*American Scien-
tist*), and "should be on every
laboratory shelf where protein
are being handled or purified...
a feast and a genuine pleasure
to read" (*Nature*).

M. Holtzhauer
Biochemische
Labormethoden
3., korr. Aufl. 1997. XIV, 249 S.
15 Abb., 79 Tab.
(Springer Labormanual)
Brosch. DM 48,-
ISBN 3-540-62435-X

M. Holtzhauer
Methoden in
der Proteinanalytik
1996. XXIV, 467 S. 208 Abb., 65 Tab.
Geb. DM 98,-
ISBN 3-540-60210-0

**Please order from
Springer-Verlag Berlin**
Fax: + 49 / 30 / 8 27 87- 301
e-mail: orders@springer.de
or through your bookseller

Springer-Verlag, P. O. Box 31 13 40, D-10643 Berlin, Germany.

New series

Group VII of the Landolt-Börnstein New Series is devoted to the physical properties of biological systems. It begins with volume VII/1 covering the nucleic acids which are of central importance in all processes involving gene expression. This field has become of practical consequence during the last decade through the development of genetic engineering. Because of the amount of the data available, volume VII/1 had to be divided into several subvolumes.

P.T. Haromy, W.N. Hunter, O. Kennard, W. Saenger, M. Sundaralingam

Crystallographic and Structural Data I / Kristallographische und strukturelle Daten I

1989. IX, 360 pp. 48 figs.
(Landolt-Börnstein: Numerical Data and Functional Relationships in Science and Technology - New Series, GG 7 Vol. 1 PT a)
Hardcover DM 1080
ISBN 3-540-18875-4

C. Altona, J.J. Butzow, G.L. Eichhorn, H. Eisenberg, M.D. Frank-Kamenetskii, S.M. Freier, W. Guschlbauer, C.W. Hilbers, W.C. Johnson, H.H. Klump, W.L. Peticolas, Y.A. Shin, N. Sugimoto, D.H. Turner, J.A.L.I. Walters

Spectroscopic and Kinetic Data. Physical Data I / Spektroskopische und kinetische Daten. Physikalische Daten I

1990. XII, 445 pp. 134 figs.
(Landolt-Börnstein: Numerical Data and Functional Relationships in Science and Technology - New Series, GG 7 Vol. 1 PT c)
Hardcover DM 1360
ISBN 3-540-19428-2

M. Bansal, W.R. Bauer, P.A. Kollman, S.C. Kowalczykowski, R. Lavery, W.K. Olson, D. Porschke, A. Psoda, B. Pullman, D. Riesner, V. Sasisekharan, D. Shugar, A.R. Srinivasan, G. Steger, U. Wähnert, K.L. Wierzchowski, C. Zimmer

Physical Data II/Physikalische Daten II

Theoretical Investigations/Theoretische Untersuchungen
1990. XIII, 486 pp. 408 figs.
(Landolt-Börnstein: Numerical Data and Functional Relationships in Science and Technology - New Series, GG 7 Vol. 1 PT d)
Hardcover DM 1500
ISBN 3-540-52454-1

K. Aoki, S. Arnott, R. Chandrasekaran, G.A. Jeffrey, D. Moras, S. Neidle

Crystallographic and Structural Data II / Kristallographische und strukturelle Daten II

1989. X, 348 pp. 171 figs.
(Landolt-Börnstein: Numerical Data and Functional Relationships in Science and Technology - New Series, GG 7 Vol. 1 PT b)
Hardcover DM 1040
ISBN 3-540-50492-3

Please order from
Springer-Verlag Berlin
Fax: + 49 / 30 / 8 27 87- 301
e-mail: orders@springer.de
or through your bookseller

Errors and omissions excepted.
Prices subject to change without notice.
In EU countries the local VAT is effective.

Springer

Springer-Verlag, P. O. Box 31 13 40, D-10643 Berlin, Germany.